AeroDef Manufacturing Summit and Exposition 2014

Long Beach, California, USA
25 – 27 February 2014

ISBN: 978-1-5108-5900-5

Printed from e-media with permission by:

Curran Associates, Inc.
57 Morehouse Lane
Red Hook, NY 12571

proceedings.com

Some format issues inherent in the e-media version may also appear in this print version.

Printed by Curran Associates, Inc. (2018)

Society of Manufacturing Engineers (SME)
One SME Drive
Dearborn, Michigan 48128
USA

Phone: 800-733-4763 or 313-425-3000
Fax: 313-425-3400

www.sme.org

Additional copies of this publication are available from:

Curran Associates, Inc.
57 Morehouse Lane
Red Hook, NY 12571 USA
Phone: 845-758-0400
Fax: 845-758-2633
Email: curran@proceedings.com
Web: www.proceedings.com

The Key Leaders

The **AeroDef Manufacturing Intelligence Series** gives you exclusive access to the most influential people in aerospace and defense manufacturing today. Hear from leading professionals who are making the investment and policy decisions that will have a profound impact on the industry's future. You'll get insight on important issues, like what's in the R&D pipeline at the OEMs and what funding is available through new federal manufacturing initiatives. Our Exclusive Keynote Presentation and Featured Speaker programs take place on The Deck, a central area of the exposition floor that encourages interaction among speakers, presenters and attendees. It's high-level access and intelligence you'll get only at AeroDef Manufacturing.

The Newest Technologies

The **AeroDef Manufacturing Summit and Exposition** gives you direct access to the most exciting technologies, products and applications in manufacturing today. It's strategically designed to give you a comprehensive view of a technology's application and potential – hear it from a speaker or conference presenter, see it demonstrated on an expo floor arranged by technology zones. Choose from 60 technical sessions across four tracks - Composites, Metals & Advanced Materials, Digital & Additive Manufacturing/3D Printing, Integrated Assembly & Robotics, and Precision Machining & Tooling. If it's a technology that can advance the interests of the aerospace and defense industry, you'll see it at AeroDef Manufacturing.

The Critical Issues

Aerospace and defense manufacturing is in a period of flux, when each decision reverberates across the extended manufacturing enterprise. The **Executive Insights Series** are panel discussions focused on key issues, identified by an advisory committee of leading OEM professionals and suppliers. How do we effectively manage the extended supply chain and new technologies, secure our networks, applications and programs? What can suppliers do to avert the economic consequences of the federal sequester? What technologies are the OEMs investing in now and researching for the future? Industry leaders are coming to AeroDef Manufacturing because they know they'll find an engaged and thoughtful audience.

AERODEF MANUFACTURING INTELLIGENCE SERIES

Join us for our three-day **Manufacturing Intelligence Series** and get exclusive access to the top people in Aerospace and Defense! Get an insider's look at the critical issues of the day during our interactive **Executive Insights Series** panel discussions and **Featured Speaker** presentations, our **Exclusive Keynote Presentation**, and **AeroDef Take Off!** and **AeroDef Afterburner** networking events. All are included with your registration.

EXCLUSIVE KEYNOTE SPEAKER

Wednesday, February 26, 8:00 – 9:00 am, The Deck

Gerould Young
Director, Materials & Manufacturing Technology, Boeing Research & Technology

"The Lure of New Materials and the Corresponding Reality of Manufacturing at Rate and Cost" 1

Mr. Young leads the Materials and Manufacturing Technology Organization for Boeing Research & Technology. This team of 1,000 engineers and scientists is developing next-generation materials and manufacturing processes for existing and future Boeing products. He started at Boeing on the B2 program using 3D CAD to integrate systems routings into the structural design. He moved to the National Aerospace Plane program, where he designed, analyzed and tested a Mach 25 nose tip thermal protection concept. His expertise in computational fluids and non-linear finite element analysis took him to simulating manufacturing processes, impacts, explosions and evaluation of structural designs for commercial and defense products including F22, JSF and 777. He was promoted to management in Boeing's Structures Research and Development Organization and later held a functional management job maintaining the loads and dynamics skill base. He led three Boeing Joint Strike Fighter (JSF) technology programs focused on affordable titanium product forms casting, welding and forming. He led Boeing's enterprise-wide development for Structural Analysis and Loads processes producing such tools as the Common Structures Workstation and the Engineering Materials Database. Prior to his current role, Mr. Young was Director of Materials and Structures Technology, where he led Boeing's long-term research and development for next-generation structural technologies, which included materials, analysis tools and concept optimization. He holds a BS and MS in Mechanical Engineering from the University of Utah and is a graduate of the Advanced Management Program at University of Pennsylvania's Wharton Business School. He also holds four patents and has authored a number of technical papers.

SPONSORED BY SAP

SAP

EXECUTIVE INSIGHTS SERIES: TUESDAY

Tuesday, February 25, 9:00 - 9:45 am, Ballroom

"Sequestration: OEM & Supplier Perspectives." How deep cuts in defense program are driving OEMs and their suppliers to rethink business strategies.

Panelist: Kellie Johnson, President, Ace Clearwater Enterprises

Ms. Johnson graduated from the University of Southern California in 1982 with a BA in International Relations and joined the family business in 1984. At that time, the company had annual revenues of approximately $10 million, primarily in aerospace sheet metal forming and fabrication. She took over operations in 1985 and was promoted to president in 1989. Focusing on speed, quality and innovation, ACE Clearwater Enterprises has become a preferred supplier for most OEMs. Specializing in complex formed and welded assemblies, the company has grown to nearly 200 employees with current (2013) projected revenue at $32 million. Ms. Johnson is on the board of the National Association of Manufacturers (NAM), as well as the NAM Executive Committee. She is a member of the board of trustees for the Manufacturing Institute. In 2006, she was appointed by Secretary of Commerce Carlos Gutierrez to serve on the Manufacturing Council and was reappointed to serve on the council under Secretary Locke with her term expiring in 2012, during which time she chaired the subcommittee on competitiveness.

Panelist: Mary Simmerman, Vice President, F-35 Supply Chain Management, Lockheed Martin Aeronautics

Ms. Simmerman is the vice president of Supply Chain Management for the F-35 Program. She joined Lockheed Martin in April 2011. In this role, she establishes and executes acquisition strategies to support program plans while maintaining procurement system integrity. She also ensures that suppliers execute their production commitments, achieve their affordability and international goals and provide a full range of sustainment services. She came to Lockheed Martin from Bell Helicopter-Textron, where she last served as the vice president of Global Supply Chain Management and Textron International's Mexico operations from 2006 to 2011. In addition to Bell Helicopter-Textron, she has held corporate and executive leadership positions with Northrop Grumman and the Boeing company supporting space, as well as civil and military programs for both fixed wing and rotorcraft platforms.

NOT PICTURED:

Major General Dennis M. Kenneally, USA (retired), Executive Director, Southwest Defense Alliance (SWDA)

David Herbst, Vice Chairman, Southwest Defense Alliance (SWDA)

SPONSORED BY PAR SYSTEMS

PAR SYSTEMS

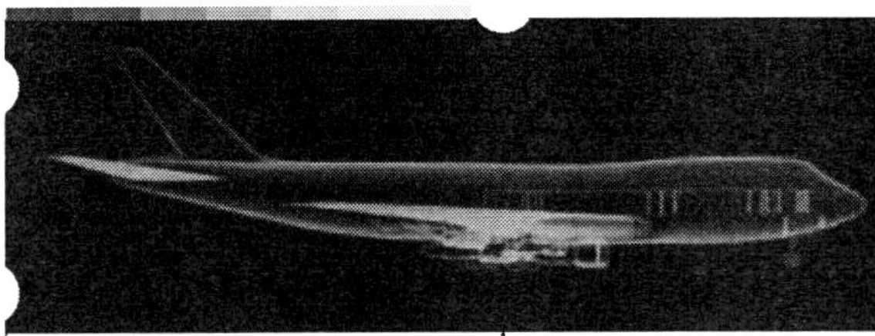

EXECUTIVE INSIGHTS SERIES: WEDNESDAY

Wednesday, February 26, 1:00 - 2:00 pm, The Deck

"Cyber Security and Counterfeit Parts & Materials: An Escalating Global Challenge for Aerospace & Defense Manufacturers." This panel will address issues that will help aerospace and defense manufacturers deal more effectively with outside cyber threats and recognize the increasing number of counterfeit parts in the supply chain.

Moderator: Richard M. (Dickie) George, Senior Advisor for Cyber Security, Johns Hopkins University Applied Physics Lab

Mr. George is the senior advisor for cyber security at the Johns Hopkins University Applied Physics Lab (APL). At the lab, he works on a number of projects sponsored by the U.S. government. He works with senior management at the lab on cyber strategy for protection of critical national systems. He is also the lab's representative to the I3P, a consortium of universities, national labs and non-profit institutions dedicated to strengthening the cyber infrastructure of the United States. Prior to joining APL, he worked at the National Security Agency as a mathematician from 1970 until his retirement in 2011. While at NSA, he wrote more than 125 technical papers on crypto mathematical subjects, and served in a number of positions: analyst, and technical director at the division, office, group, and directorate level. He served as the technical director of the Information Assurance Directorate for eight years until his retirement.

Panelist: Barry Birdsong, Division Chief, Parts and Materials Engineering, Missile Defense Agency

Mr. Birdsong is the Parts, Materials & Processes (PMP) division manager for the Missile Defense Agency. He has served in this position since April 2006. In this role he is responsible for MDA requirements and policy regarding counterfeit parts, parts procurement, radiation effects, part selection, screening, and qualification criteria, failure analysis, DMSMS, COTS strategies, lead-free prohibitions, materials engineering and implementation of the Department of Defense corrosion prevention policy. He is also the manager of the MDA Parts and Materials Advisory Group (PMAG) that includes the MDA Failure Analysis Laboratories, the MDA Advisory Program and the Directorate Lead for Supply Chain Risk Management. In addition, he is program manager for over $25 million in Small Business Innovative Research (SBIR) contracts related to parts and materials technologies.

SPONSORED BY PLEX SYSTEMS

PLEX
THE MANUFACTURING CLOUD

Panelist: Stephen Dill, Chief Architect, Cyber Security, Lockheed Martin Information Systems & Global Services (IS&GS)

Mr. Dill has been working in communications systems design, operations and maintenance, and information security fields for 40 years, and has been with Lockheed Martin since 1991. Mr. Dill is a Lockheed Martin fellow and cyber security architect in the Center for Cyber Security Innovation supporting the IS&GS business area and, occasionally, the rest of Lockheed Martin. He was recently the Cyber Security Solution Family Chief Architect, responsible for the coordination and execution of Cyber Security Research and Development in IS&GS. He has received certifications as information security specialist, program manager and information systems architect.

Panelist: Walter Keller, CEO, Nokomis

Mr. Keller founded Nokomis in 2002 and has cultivated state-of-the-art expertise and game-changing innovation in the area of non-contact automated detection, identification and diagnostics of electronics. Under his leadership, Nokomis has developed unique ultra-sensitive RF and microwave capabilities that facilitate the sensing of unintended electronics emissions with substantially improved sensitivity, acuity and confidence over other available technologies. These technologies have been proven to be exceptionally capable for anti-tamper and counterfeit mitigation applications. Serving as principal investigator on 12 related efforts funded by the Navy, Air Force, Army, Defense Threat Reduction Agency (DTRA), and National Aeronautics and Space Administration (NASA) provides Mr. Keller with a breadth of experience in the detection and exploitation of electronic device electromagnetic emissions for defense applications.

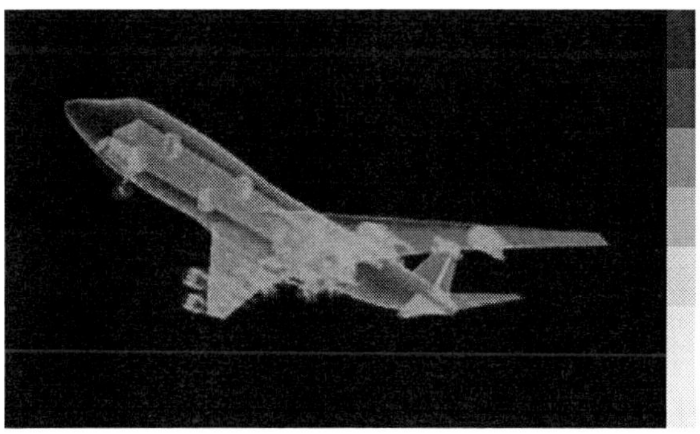

AERODEF MANUFACTURING INTELLIGENCE SERIES
(continued)

EXECUTIVE INSIGHTS SERIES: THURSDAY

Thursday, February 27, 8:30 AM – 9:30 AM, The Deck

"The New Manufacturing Innovation Institutes: How Will They Help Aerospace & Defense Manufacturers?" An update on current and future initiatives of the federal program to promote investment and production in the U.S. Includes participants from the National Institute of Science and Technology Advanced Manufacturing Office, America Makes – the National Additive Manufacturing Innovation Institute (NAMII), Digital Manufacturing & Design Innovation Institute, and Lightweight & Modern Metals Manufacturing Institute.

Moderator: Michael F. Molnar, Director, NIST Advanced Manufacturing Office, Director, Advanced Manufacturing National Program Office, SME 2014 President

Panelist: Ralph Resnick, Founding Director, NAMII

Mr. Resnick joined the National Center for Defense Manufacturing & Machining (NCDMM) in September 2008 as vice president, chief technology officer, and director, Corporate Development. A professional with over 35 years of manufacturing experience, Mr. Resnick is responsible for providing technical leadership within NCDMM, promoting its mission with external stakeholders and adding new insight to the NCDMM business strategy. Before joining NCDMM, he was chief technology officer for both The Ex One Corporation and Extrude Hone. He was a major contributor in establishing Extrude Hone and Ex One as innovative leaders in advanced manufacturing process research and subsequent technology transition to the world's factory floors. He holds a number of patents for designs of capacitance-based sensors and metrology systems and has made many presentations at leading technical conferences. He has been the principal investigator for many projects at the Department of Defense, Department of Energy, National Institute of Standards and Technology's (NIST) Advanced Technology Program, National Science Foundation, National Center for Manufacturing Science, and Small Business Innovation Research (SBIR). He is a fellow of the Society of Manufacturing Engineers (SME) and presently serves on SME's International Awards and Recognition Committee.

Thursday, February 27, 1:00 PM - 2:00 PM, The Deck

"3D Modeling Manufacturing World: Significant Advancements in 3D Manufacturing as it Relates to the Future of Aerospace & Defense." Focus on the current applications and potential of 3D technology to be a transformational agent in aerospace and defense manufacturing.

Moderator: Al Sanders, Ph.D., President & CEO, Design-Vantage Technologies

Dr. Sanders is chairman of the National Defense Industry Association Advanced Manufacturing Engineering Capabilities Committee and also president and CEO of Design-Vantage Technologies, LLC. He founded the company in 2013 after working over 20 years in industry with the mission of helping clients transform their businesses to adapt to the changing manufacturing landscape and gain access to the same modeling and simulation tools and technologies for engineering and manufacturing used by Fortune 500 companies to mitigate risk and streamline their new product development practices. Dr. Sanders brings his clients a wide depth and breadth of industry experience in mechanical and electronic system development and has hands-on experience in disciplines that include technology development, systems engineering, component design and analysis, modeling and simulation, advanced design methods, advanced manufacturing, and six sigma. He as been a technical advisor to the Arizona Commerce Authority, which is in the process of finalizing an award contract with NIST to launch a new MEP center in Arizona that will help small- to medium-sized manufacturing companies in the state transform their businesses into factories of the future.

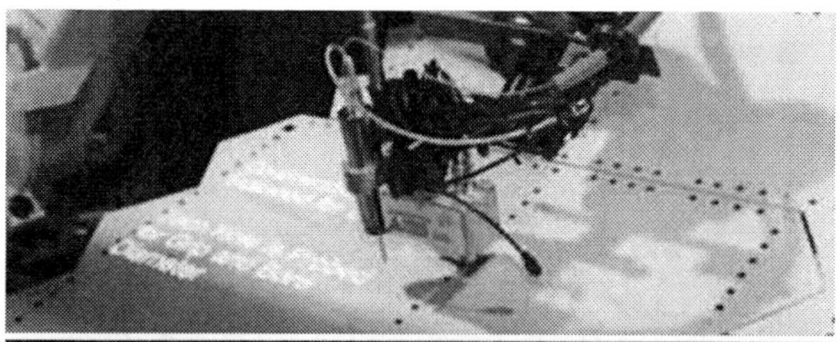

FEATURED SPEAKERS

Tuesday, February 25, 8:00 AM - 8:45 AM, Ballroom

"The Economic Impact of Sequestration to the Industrial Base"
The Southwest Defense Alliance (SWDA) is a public benefit corporation formed in 1998. Headed by Executive Director Maj. Gen. Dennis Kenneally (retired), SWDA works to preserve and enhance critical defense missions and assets in the southwest United States. These significant operational, training and testing assets are linked by unique communications and satellite systems enabling them to work together in many ways. SWDA engages in communication efforts to ensure that local, state and federal elected officials in Arizona, California, Nevada, New Mexico, Texas and Utah have the information needed to be effective advocates.

David Herbst, Vice Chairman, Southwest Defense Alliance (SWDA)

Major General Dennis M. Kenneally, USA (retired)

Major General (ret) Dennis M. Kenneally is Executive Director of the Southwest Defense Alliance, a non-profit corporation of six southwestern states dedicated to protecting America's defense and industrial base. He was appointed by Governor Jerry Brown to the governor's military council and is a former member of the State Threat Assessment Advisory Group of the California Emergency Management Agency. A founding partner of Business Associates International, LLC, he serves on the board of directors of Maytag Aircraft Corporation of the Mercury Air Group, Inc., and Final Mile Logistics. Before his appointments, he worked as a contractor for the U.S. Department of Justice, as a consultant to the Chief Executive Officer of Maxim Systems, Inc., a Department of Defense contractor and as Chairman of the Board of the National Defense Industrial Association of San Diego. General Kenneally served in two Presidential administrations as Associate Administrator of the Veterans Administration (3rd ranking position) and was later appointed as the Deputy Assistant Secretary of the Air Force for Reserve Affairs. A decorated combat veteran, General Kenneally is the former Commanding General of the California Army National Guard, where he commanded the largest, most tasked Army National Guard force in the nation.

Tuesday, February 25, 3:00 PM, The Deck

Yancy D. Mailes, Director, History & Museum Program, Air Force Global Strike Command

Mr. Mailes is responsible for establishing policy and preserving the Command's institutional memory and its material heritage. The Command's six wings control the nation's entire inventory of Minuteman III intercontinental ballistic missiles, B-2 and B-52 bomber aircraft. He ensures that historians concentrate on documenting recent history (both at home station and in the deployed area), while maintaining the Command's unclassified and classified archives, and answering leadership and public inquiries. He also directs the Command's heritage program through museum exhibits, heritage displays, well-maintained airparks, and educational outreach programs.

Thursday, February 27, from 8:00 AM - 8:30 AM, The Deck

Michael F. Molnar, Director, NIST Advanced Manufacturing Office, Director, Advanced Manufacturing National Program Office, SME 2014 President

"The New Manufacturing Innovation Institutes"

Michael F. Molnar, FSME, CMfgE, PE, likes to be introduced simply as "a manufacturing guy from industry" with nearly 30 years of experience in advanced manufacturing. In 2011, he was named the first Chief Manufacturing Officer of the National Institute of Standards and Technology (NIST). Today he leads the NIST Advanced Manufacturing Office for extramural manufacturing programs and also serves as the director of the interagency Advanced Manufacturing National Program Office. As called for by the Advanced Manufacturing Partnership initiative, the AMNPO's mission is to foster industry-led partnerships and to form a "whole of government" approach to strengthen competitiveness and innovation in U.S. manufacturing. His experience includes leadership roles in advanced manufacturing, metrology, manufacturing systems, quality, technology development, sustainability and industrial energy efficiency. His credentials include service as a Federal Fellow in the White House Office of Science and Technology Policy, and election as Fellow of both ASME and the SME. He is a licensed Professional Engineer, a Certified Manufacturing Engineer and a Certified Energy Manager.

NETWORKING OPPORTUNITIES: TUESDAY & WEDNESDAY

Get energized for your AeroDef Manufacturing experience and join us on The Deck for the Take Off! It launches when the summit and expo opens on Tuesday, February 25, at 3:00. Hear an inspiring speaker and get a chance to meet the people making a difference in manufacturing, including and presenters, exhibitors and fellow colleagues. Get another opportunity to meet the presenters, exhibitors and colleagues at the AeroDef Afterburner on Wednesday from 3:30-5:30 pm, on The Deck. Sometimes the most valuable contacts you make takes place after work hours.

TECHNICAL SESSIONS

FANUC Precision Machining & Tooling and Integrated Assembly & Robotics sessions are sponsored by FANUC

DAY ONE: TUESDAY, FEBRUARY 25

Predictive Tools for Optimized Composite Manufacturing
10:00 AM – 11:45 AM

10:00 AM - 10:30 AM
Operations Intelligence to Help Define Composite Processes, Characteristic Measurements and Production Predictive Analytics
Chuck Buckley, Director, Dassault Systemes 38

10:35 AM - 11:05 AM
Influence of ICME in the Design of New Generation Composite Components 52
Adi Sholapurwalla, Vice President of Technology, ESI Group

11:10 AM - 11:40 AM
Virtual Design and Virtual Testing for Predictive Engineering of Composites Structures 68
Rani Richardson, CATIA Composites Consultant, Dassault Systemes

Drilling & Machining for Integrated Assembly
10:00 AM – 11:45 AM

10:00 AM - 10:30 AM
C-130J Automated Drilling Systems 76
R. Barrett Kotnik, Applications Engineer, Lockheed Martin Aeronautics Co.

10:35 AM - 11:05 AM
Understanding One Shot Power Feed Drilling of CFRP/Al Stack for Aircraft Structure 92
Abdelatif Atarsia, Ph.D., Aerospace Manager, MAPAL, Inc.

11:10 AM - 11:40 AM
Integrated CNCs and Robots 104
Roger Hart, Engineering Manager, Siemens Industry, Inc.

Metallic Additive Manufacturing
10:00 AM – 11:45 AM

10:00 AM - 10:30 AM
Advances and New Technologies in Additive Manufacturing of Metals
Ian D. Harris, Ph.D., Technology Leader, EWI 113

10:35 AM - 11:05 AM
Surface Finish Enhancement of Ti-6Al-4V Components Fabricated via Electron-Beam Additive Manufacturing 135
Eric J. Fodran, Ph.D., Metallurgical and Manufacturing Engineer, Northrop Grumman

11:10 AM - 11:40 AM
Simulation and Modeling of the Metal Laser Powder Bed Fusion Process to Accelerate Certification N/A
Wayne King, Ph.D., Director Accelerated Certification of Additively Manufactured Metals, Lawrence Livermore National Laboratory

Precision Machining: Innovative Metalworking Solutions
10:00 AM – 11:45 AM

10:00 AM - 10:30 AM
Enabling Tomorrow's Aerospace Technology Today 159
Mark Lohmeier, Aero Engine Program Manager, FANUC America Corporation

High Accuracy Platforms for Machining of Aerospace Components 177
Randy Von Moll, Director, Fives Cincinnati

11:10 AM - 11:40 AM
Handheld EDM Fastener Removal System 186
David Siedal, LM Fellow, Lockheed Martin/Manufacturing Technology

Machining of Composites
1:00 PM - 2:45 PM

1:00 PM - 1:30 PM
Drilling of Composites
Scott A. Daggett, Aerospace Specialist, OSG Tap & Die, Inc. 201

1:35 PM - 2:05 PM
Tips and Tricks for Drilling and Trimming Carbon Fiber and New Composite Materials 223
Mike MacArthur, Vice President of Engineering, RobbJack Corporation

2:10 PM - 2:40 PM
Cryogenic Machining of Titanium & Supply Chain Insertion 235
Jeff Langevin, Applications Engineer, Lockheed Martin Aeronautics
Jason Huff, Procurement Engineer, Lockheed Martin Aeronautics

Supply Chain Management
1:00 PM - 2:45 PM

1:00 PM - 1:30 PM
Planning and Designing a Lean Manufacturing Facility 245
Chris Williams, Project Manager, Burns & McDonnell

1:35 PM - 2:05 PM
Minimizing Energy Risk in A&D Manufacturing 250
Darren McKnight, Chairman, Board of Advisors, ZF Energy Development

2:10 PM - 2:40 PM
TBD

Quality, Measurement & Inspection
1:00 PM – 2:45 PM

1:00 PM - 1:30 PM
Machine and Production Uptime Through Predictive Maintenance Practices N/A
Thomas Hoenig, President, GTI Spindle Technology, Inc.
Lisa Bailey-Beavers, VP Sales and Marketing, GTI Spindle Technology, Inc.

1:35 PM - 2:05 PM
Automated CCFX Inspection 258
Chad L. Victorino, Manufacturing Technology Engineer, Northrop Grumman

2:10 PM - 2:40 PM
Automated Metrology for Quality Control N/A
Rohit Khanna, President, 3D Infotech

DAY TWO: WEDNESDAY, FEBRUARY 26

2:50 PM - 3:20 PM
Reducing Autoclave Cycle Times Through the Use of Pressurized Water Mold Temperature Controllers 442
Kip Petrykowski, Business Development Manager - Composites, SINGLE Temperature Controls, Inc.

3:25 PM - 3:55 PM
Composite Processing Utilizing Real-Time Visibility of Viscoelastic Properties During Cure 456
Louis C. Dorworth, Division Manager, Abaris Training Resources, Inc.

Robotic Applications
2:15 PM – 4:00 PM

2:15 PM - 2:45 PM
Vision Guided Robotic Assembly System 471
David Siedal, LM Fellow, Lockheed Martin/Manufacturing Technology

2:50 PM - 3:20 PM
Automated Robotic Sanding 497
Jerry Brimer, Manufacturing Engineer, Coatings Automation, Northrop Grumman Aerospace Systems

3:25 PM - 3:55 PM
Advanced Robotic Applications N/A
Chris Kolb, Aerobotix (invited)

Additive Manufacturing - Industry Collaboration
2:15 PM – 4:00 PM

2:15 PM - 2:45 PM
Advanced Additive Manufacturing Materials 501
Luke Rodgers, Senior Materials Engineer, Stratasys

2:50 PM - 3:20 PM
CSI: More Effective Manufacturing Collaboration for the Model-Based Environment 514
John Gray, Solution Manager, ITI TranscenData

3:25 PM - 3:55 PM
America Makes - The National Additive Manufacturing Innovation Institute: Investments in Additive 529
Rob Gorham, Deputy Directory - Technology Development, National Additive Manufacturing Innovation Institute (NAMII)

Precision Machining - Controls & Optimization
2:15 PM – 4:00 PM

2:15 PM - 2:45 PM
Real-Time Adaptive Control for Tool Optimization 550
Robert P. Caron, President, Caron Engineering, Inc.

2:50 PM - 3:20 PM
Innovative Processes and Products for Machining Titanium Aerospace Components 561
Michael Standridge, Industry Specialist Aerospace, Sandvik Coromant
Brian Hoefler, Research & Development Aerospace Team Leader, Sandvik Coromant

3:25 PM - 3:55 PM
Practical Adaptive Machining
Daniel J. Braley, Manufacturing Technology Engineer, Northrop Grumman Aerospace Systems 586

DAY THREE: THURSDAY, FEBRUARY 27

Assembling Composites
9:45 AM – 11:30 AM

9:45 AM - 10:15 AM
Automated Drilling and Countersinking of Composite Titanium Material Stacks One Shot 593
George N. Bullen, CPIM, President & CEO, Smart Blades, Inc.

10:20 AM - 10:50 AM
Composite Tooling Assembly Strategies 602
James Diedesch, Senior Manufacturing Engineer, Janicki Industries

10:55 AM - 11:25 AM
TBD
Dan Day, The Boeing Company

Accuracy in Integration & Assembly
9:45 AM – 11:30 AM

9:45 AM - 10:15 AM
Dimensional Engineering for Determinant Assembly: Applying Dimensional Engineering Tools to Optimize Assembly Processes 615
Don Jasurda, VP of Sales, Dimensional Control Systems, Inc.

10:20 AM - 10:50 AM
Developments in Robot Accuracy for Aerospace Manufacturing N/A
Chris Blanchette, National Account Manager - Distribution, FANUC America Corporation

10:55 AM - 11:25 AM
Computer Aided Inspection for Better Precision Parts, Much Faster, at Lower Cost & with Greater Confidence 624
Bill Greene, Founder/Vice President of Business Development, Level3 Inspection

Additive Manufacturing: Aerospace Case Studies
9:45 AM – 11:30 AM

9:45 AM - 10:15 AM
3D Printing of a Rocket Engine 636
Youping Gao, Discipline Lead, Aerojet Rocketdyne

10:20 AM - 10:50 AM
Additive Manufacturing for High Pressure Hydraulic Valve Manifolds 650
Gerhard Hummel, Dipl.-Ing. MSc., Hydraulics System Engineer / Designated Certification Specialist, Airbus Operations GmbH
Frank Schubert, Dipl.-Ing., Scientific staff member, Technical University Chemnitz
Alexander Altmann, Dipl.-Ing. (FH), R&T Project Manager, Liebherr-Aerospace Lindenberg GmbH

10:55 AM - 11:25 AM
Additive Metal Applications for Gas Turbine Engine Components 659
Greg Morris, Strategy/Business Development, Additive Technologies, GE Aviation

Advanced Technologies
9:45 AM – 11:30 AM

9:45 AM - 10:15 AM
MicroBlasting Technologies: Optimizing Control and Automation
Colin Weightman, Director of Technology, Comco, Inc. 675

10:20 AM - 10:50 AM
Air Pollution Control Equipment for Aerospace Coating and Finishing 710
Scott Gagliostro, Sales Engineer, Munters

10:55 AM – 11:25 AM
Continuous Inspection for Web-Line Production Using Non-Contact Ultrasound 719
Anuj Bhardwaj, President & CEO, The Ultran Group

Additional Papers:

Synthesis, structure, and behaviour of a new CVD- TiB2 coating with extraordinary properties for high performance applications 149

Breaking the Ground in the Execution of the Mission 734
R. Resnick

Air Force Global Strike Command 750
Y. Mailes

Scanning Technology for the Production of Legacy Aircrafts 758
C. Rene-Bazin

Additive Manufacturing at Lockheed Martin Aeronautics 766
B. Rosenberger

Additive Manufacturing/3D Printing Resource Center – Booth 339

Explore the growing number of applications by visiting the Additive Manufacturing/3D Printing Resource Center. On display you will find parts representing major technologies including stereolithography, laser sintering and fused deposition modeling. You will also find information on how this technology can lower production expenses, speed product development, enable new designs, and change how and what you manufacture.

- Morning Session: **Tactical information**—A look at the technology landscape, what's available and what you need to know to get started.
- Afternoon Sessions: **Strategic information**—How to map out your plan; choosing technologies, services and when to get in.

Meet the Additive Manufacturing/3D Printing Knowledge Bar Host

Todd Grimm, president of T. A. Grimm & Associates, was recently named as one of TCT Magazine's 20 most influential in the additive manufacturing (AM) industry. A consultant, writer, author, speaker and advisor, he is editor for ENGINEERING.com and columnist for The TCT Magazine, and the author of User's Guide to Rapid Prototyping. Todd serves on the board of the Additive Manufacturing Users Group (AMUG) as its AM industry advisor and has served as chairman of the Society of Manufacturing Engineers' (SME) community for additive manufacturing.

Additive Manufacturing Certificate Program

Review Course: Monday, February 24th, 9am – 3pm, UC – Irvine

Exam: Tuesday, February 25th, 12:00pm – 3:30pm, Long Beach Convention Center

The review course will include instruction on foundational additive manufacturing principles and will be supported by observation of additive manufacturing applications in action. Course attendees will participate in practice exercises that incorporate concepts and applications from the lecture and lab.

Completing the exam and receiving the Additive Manufacturing Certificate serves as verifiable proof of your foundational knowledge. While the review course is recommended for exam candidates, it is not required to sit the exam. Knowledge and expertise with multiple materials, processes, and equipment are necessary. It is strongly advised that all potential examinees take the pretest on the website to determine readiness. Additional information may be found on the website: www.sme.org/rtam-certificateprogram.

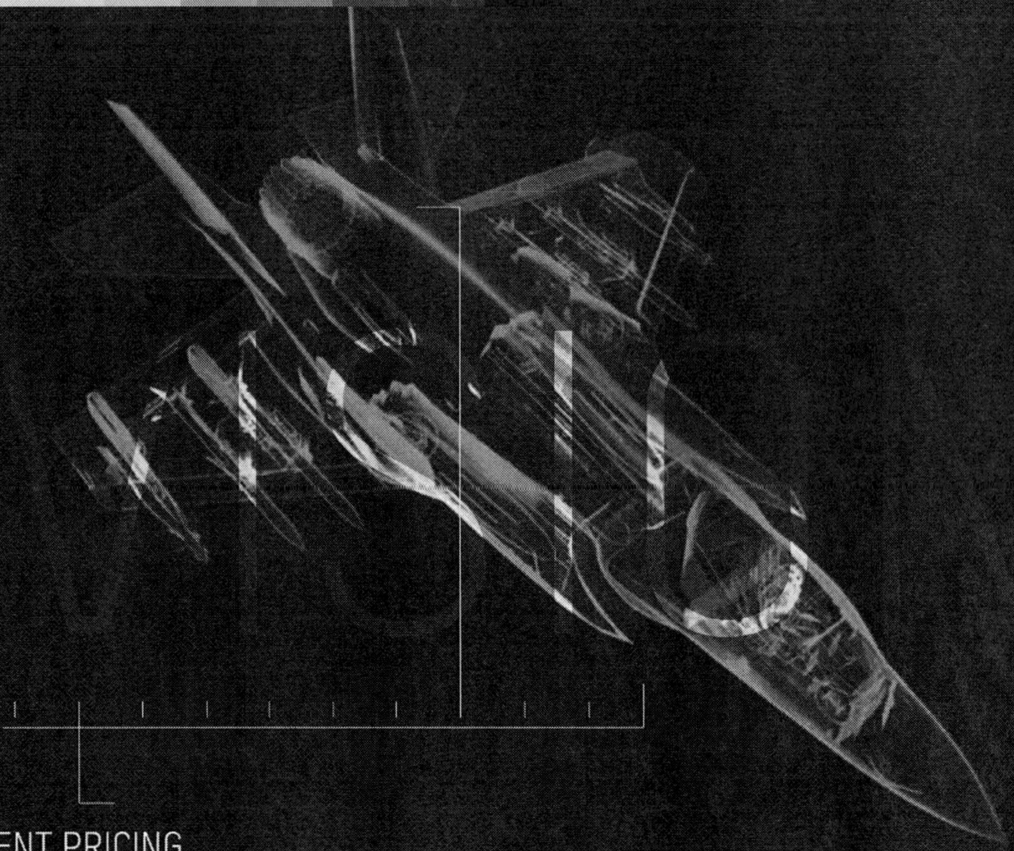

EVENT PRICING

AERODEF MANUFACTURING Event Purchase Options	List Price	Academic/ Government*	Student**
Full Summit	$650	$550	$400
One Day Only	$250	$200	$150
Early Bird Discount On Full Summit Only (Offer Expires: January 24)	$600	N/A	N/A
Exhibits Plus – discounts do not apply Includes: Exhibit hall access; preferred seating at Keynotes & Panel Discussions; lunch voucher; networking receptions, exclusive registration line Note: does not include the technical session program	$99	$99	$99
Exhibits Only (BEFORE February 21)	$0	$0	$0
Exhibits Only (AFTER February 21)	$50	$50	$50
Additive Manufacturing Certificate Review Course	$249	$249	$249
Additive Manufacturing Certificate Exam	$159	$159	$159
Additive Manufacturing Certificate Review Course & Exam Bundle	$379	$379	$379

NOTE: SME Members receive a 10% discount off the list price with the exception of Exhibits Plus.
*Academic & Government Pricing: To qualify for the reduced Military or Government Conference Fees you must submit a letter on appropriate military/government letterhead signed by your Commanding Officer or supervisor to bbonnell@sme.org. Educators may submit either a letter on university letterhead signed by the dean or registrar confirming status or a class schedule inclusive of the institution/year/instructor.
**Full-Time Students: To qualify for student rates, submit a letter on university letterhead signed by the dean or registrar confirming full-time student status or provide a copy of your student I.D. to bbonnell@sme.org.

Tuesday, February 25, 2014 | Show Hours (3:00 PM - 7:00 PM)

Featured Speaker (8:00 AM - 8:45 AM) | Ballroom — SPONSORED BY PAR SYSTEMS
"The Economic Impact of Sequestration to the Industrial Base"
David Herbst, Vice Chairman, Southwest Defense Alliance (SWDA)
Major General Dennis M. Kenneally, USA (retired), Executive Director, Southwest Defense Alliance (SWDA)

Panel Discussion (9:00 AM - 9:45 AM) | Ballroom — SPONSORED BY PAR SYSTEMS
"Sequestration: OEM & Supplier Perspectives"
David Herbst, Vice Chairman, Southwest Defense Alliance (SWDA)
Kellie Johnson, President, Ace Clearwater Enterprises
Major General Dennis M. Kenneally, USA (retired), Executive Director, Southwest Defense Alliance (SWDA)
Mary A. Simmerman, Vice President, F-35 Supply Chain Management, Lockheed Martin Aeronautics

AM Technical Sessions (10:00 AM - 11:45 AM)

Composites, Metals & Advanced Materials	Digital & Additive Manufacturing/3D Printing	Integrated Assembly & Robotics (SPONSORED BY FANUC)	Precision Machining & Tooling (SPONSORED BY FANUC)

LUNCH BREAK (11:45 AM - 1:00 PM)

PM Technical Sessions (1:00 PM - 2:45 PM)

Composites, Metals & Advanced Materials	Digital & Additive Manufacturing/3D Printing	Integrated Assembly & Robotics (SPONSORED BY FANUC)	Precision Machining & Tooling (SPONSORED BY FANUC)

Take Off (3:00 PM) | The Deck
Featured Speaker: Yancy Mailes, Director, History & Museum Program, Air Force Global Strike Command

Wednesday, February 26, 2014 | Show Hours (9:00 AM - 4:00 PM)

Keynote (8:00 AM - 9:00 AM) | The Deck — SPONSORED BY SAP
Gerould Young, Director, Materials & Manufacturing Technology, Boeing Research & Technology

AM Technical Sessions (9:15 AM -11:00 AM)

Composites, Metals & Advanced Materials	Digital & Additive Manufacturing/3D Printing	Integrated Assembly & Robotics (SPONSORED BY FANUC)	Precision Machining & Tooling (SPONSORED BY FANUC)

LUNCH BREAK (11:00 AM - 1:00 PM)

Panel Discussion (1:00 PM - 2:00 PM) | The Deck — SPONSORED BY PLEX
"Cyber Security and Counterfeit Parts & Materials: An Escalating Global Challenge for Aerospace & Defense Manufacturers"
Barry Birdsong, Division Chief, Parts and Materials Engineering, Missile Defense Agency
Stephen Dill, Chief Architect, Cyber Security, Lockheed Martin Information Systems & Global Services
Walter Keller, CEO, Nokomis, Inc.

Break (2:00 PM - 2:15 PM)

PM Technical Sessions (2:15 PM - 4:00 PM)

Composites, Metals & Advanced Materials	Digital & Additive Manufacturing/3D Printing	Integrated Assembly & Robotics (SPONSORED BY FANUC)	Precision Machining & Tooling (SPONSORED BY FANUC)

AeroDef Afterburner (3:30 PM - 5:30 PM) | The Deck

Thursday, February 27, 2014 | Show Hours (9:00 AM - 3:00 PM)

Featured Speaker (8:00 AM - 8:30 AM) | The Deck
"The New Manufacturing Innovation Institutes"
Michael F. Molnar, Director, NIST Advanced Manufacturing Office

Panel Discussion (8:30 AM - 9:30 AM) | The Deck
"The New Manufacturing Innovation Institutes: How Will They Help Aerospace & Defense Manufacturers"
Michael F. Molnar, Director, NIST Advanced Manufacturing Office
Ralph Resnick, Founding Director, NAMII

AM Technical Sessions (9:45 AM - 11:30 AM)

Composites, Metals & Advanced Materials	Digital & Additive Manufacturing/3D Printing	Integrated Assembly & Robotics (SPONSORED BY FANUC)	Precision Machining & Tooling (SPONSORED BY FANUC)

LUNCH BREAK (11:30 AM - 1:00 PM)

Panel Discussion (1:00 PM - 2:00 PM) | The Deck
"3D Modeling Manufacturing World: Significant Advancements in 3D Manufacturing as it Relates to the Future of Aerospace & Defense"
Al Sanders, Ph.D., President & CEO, Design-Vantage Technologies

33.7669° N, 118.1883° W 2.25–2.27.2014

THE DECK

AeroDef Manufacturing's revolutionary show floor is divided into Technology Zones reflecting the entire manufacturing enterprise – from design to manufacture to sustainability. At the center of it all is The Deck, the hub of the expo show floor and location for keynote and featured speakers, daytime panel discussions and afternoon networking receptions. The Deck reflects the kind of commitment to innovation and professional collaboration AeroDef Manufacturing has come to represent.

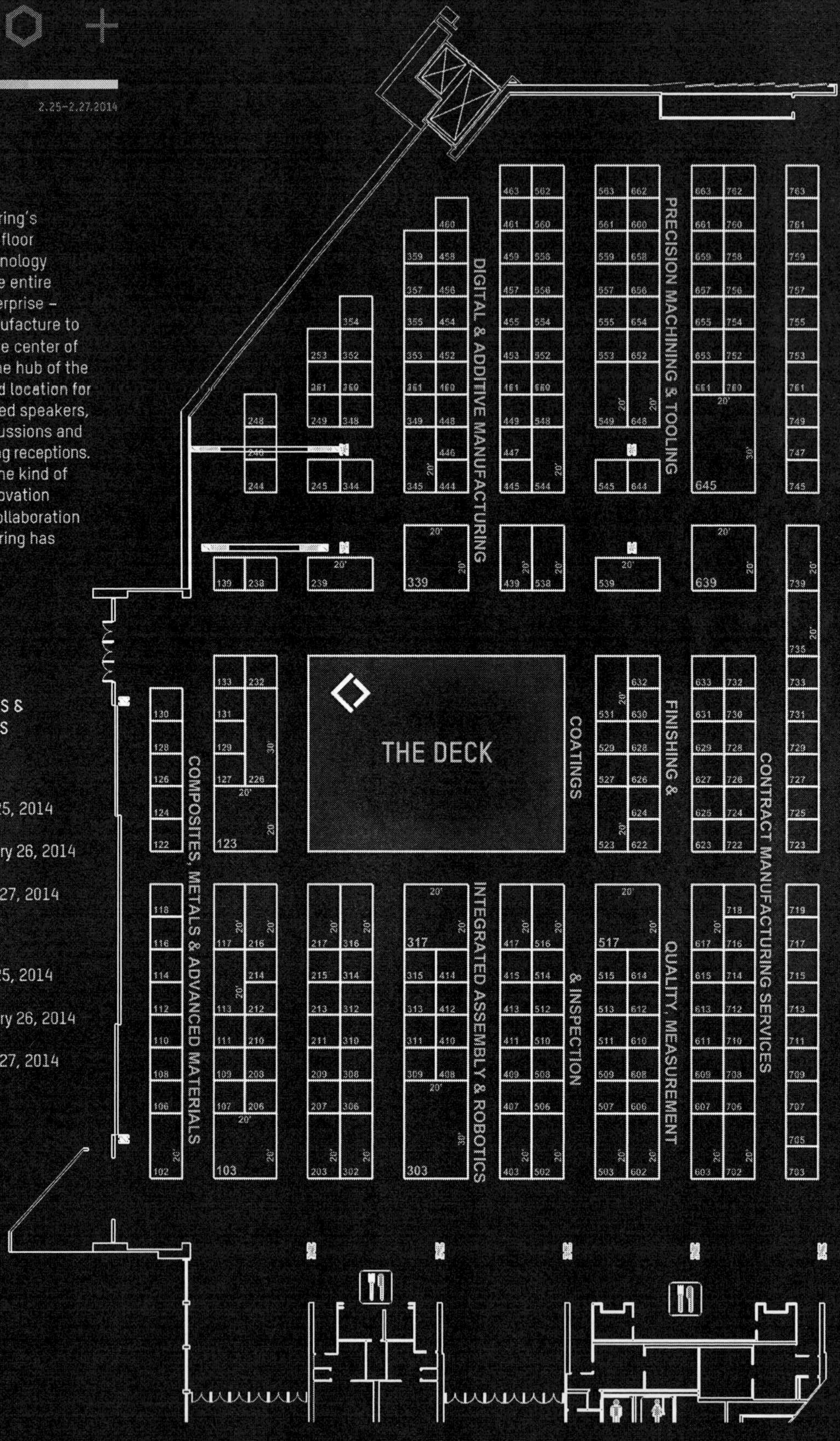

EVENT HOURS

KEYNOTE ADDRESSES & FEATURED SPEAKERS
Daily, 8:00 am

SUMMIT
Tuesday, February 25, 2014
8:00 AM - 3:00 PM
Wednesday, February 26, 2014
8:00 AM - 4:00 PM
Thursday, February 27, 2014
8:00 AM - 2:00 PM

EXPOSITION
Tuesday, February 25, 2014
3:00 PM - 7:00 PM
Wednesday, February 26, 2014
9:00 AM - 4:00 PM
Thursday, February 27, 2014
9:00 AM - 3:00 PM

EXHIBITOR LIST

List as of December 19, 2013

3D Infotech
3-D printing / Additive Manufacturing Resource Center powered by SME
6dms, Inc.
A.J. Dralle
Abaris Training Resources, Inc.
Abrasive Technology
Acme Manufacturing Co.
Aerospace Facilities Group, Inc.
Alvord-Polk Tool
AMAMCO Tool
AMS Corehog
Applied Aerospace Structures Corp.
Assembly Guidance Systems, Inc.
Blum LMT, Inc.
Breton SPA
Bullen Ultrasonics
Callaway Carbon, LLC
Caltec International, LLC
Carr Lane Manufacturing Co.
CERATIZIT USA, Inc.
CMS North America, Inc.
Col-Met Spray Booths, Inc.
Craig Tools, Inc.
Delcam
Delta Sigma Co.
Demag
DE-STA-CO
Dinsmore & Associates, Inc.
DPSS Lasers, Inc.
Duna USA, Inc.
Dustcontrol, Inc.
Duroair Technologies, Inc.
Durr Systems, Inc.
eldec Induction, LLC
Epicor Software Corp.
Epilog Laser Corp.
ErectaStep, LLC
Fagor Automation Corp.
FANUC FA America
FARO Technologies, Inc.
FineLine Prototyping
Fischer Precise USA
Fives Machining Systems, Inc.
FlackTek, Inc.
Flow International Corp.
Foam Molders, Inc.
FOBA Laser Marking + Engraving
Form Grind Corp.
Fowler Precision Instruments
GFMI Aerospace & Defense
Global Finishing Solutions
GPI Prototype & Manufacturing Services, Inc.
Guhring, Inc.
Hangsterfer's Laboratories, Inc.
Harmony Castings, LLC
Harvest Technologies
Hexagon Metrology, Inc.
Hexagon Metrology, Inc. - TESA Group
iBASEt
ICAM Technologies
IKONICS Micro-Machining Solutions
Incodema, Inc.
IntelliFinishing
InterTrade Industries, LLC
IPI Solutions, Ltd.
Janicki Industries, Inc.
Jensen Industries, Inc.
JG&A Metrology Center
Kaeser Compressors, Inc.
Kaman Aerospace Group
Kuka Systems North America
Lancer Systems Lp
LANS Company, Inc.
LAP Laser, LLC
Laser Technology, Inc.
LaserStar Technologies Corp.
Level 3 Inspection, LLC & Smart Inspection Systems
Linear Mold & Engineering
LPI, Inc.
LSP Technologies
M Torres Disenos Industriales
MA Ford Manufacturing Co., Inc.
Machine Tools Supply
MAPAL, Inc.
Marubeni Citizen-Cincom
Mazak Corporation
Melin Tool Company
Metrologic Group Service, Inc.
Microtek Laboratories
MISTRAS Group, Inc.
Mitsubishi Materials USA
Moldex3D
Munters Corp.
Nabertherm
NDT Systems and Dantec Dynamics
Niagara Cutter
Oak Ridge National Laboratory
Omative North America
ONExia, Inc.
OSG Tap & Die
Oxford Performance Materials
Pacific Coast Composites
PaR Systems, Inc.
Parpas America Corp.
Permabond Engineering Adhesives
Plasma Technology, Inc.
Plastics International
Plex Systems
PQ Systems, Inc.
Pratt & Whitney Measurement Systems, Inc.
Pro Ultrasonics
Q-PLUS Labs
Quality Tech Tool Corp.
Rapid Sheet Metal
Rayotek Scientific, Inc.
Renishaw, Inc.
RePliForm, Inc.
Riverside Machine & Engineering
RMB Products, Inc.
RT Instruments, Inc.
SAERTEX USA, LLC
Sandstrom Coating Technologies
SAP, Inc.
Sea-Lap, LLC
Seco Tools, Inc.
Servometer
Siemens Industry, Inc.
SoCal Machine Tools
Solid Concepts, Inc.
Southern Virginia Product Advancement Center
Southwestern Industries, Inc.
Spectrum Plastics Group
Stanley Engineered Fastening
Stratasys, Inc.
Sumitomo Electric Carbide, Inc.
Systems & Materials Research Corporation
TDK-Lambda Americas, Inc.
Techni Waterjet
TenCate Advanced Composites USA
Thermwood Corp.
THK America, Inc.
Tiodize Co., Inc.
US Shop Tools
Wisconsin Oven Corp.
YMW Taps USA
Zimmermann, Inc.

BOEING Engineering, Operations & Technology
BR&T

Research & Technology

The Lure of New Materials and The Reality of Manufacturing at Rate and Cost

Gerould Young
Director Materials & Manufacturing Technology
Boeing Research & Technology
AERODEF Manufacturing
Feb 26 – Long Beach, California

Commercial Transport Performance Improvement
Materials Contribution

Engineering, Operations & Technology | BR&T — Materials & Manufacturing Technology

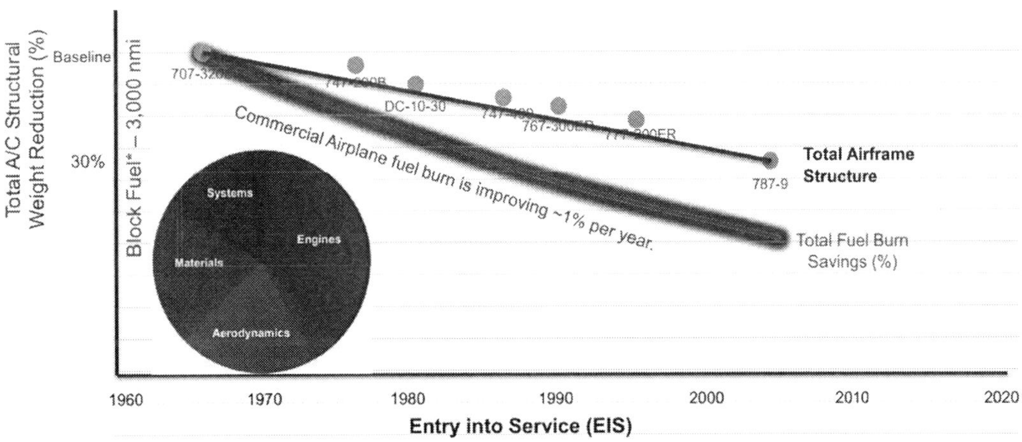

Materials Improvements Pace Airplane Performance Improvements

*Block Fuel = gals/seat over 3,000 miles
E Kaduce, 2012, The Boeing Company, based on publically-available data

Fighter/Attack A/C Performance Improvements
Materials Contribution

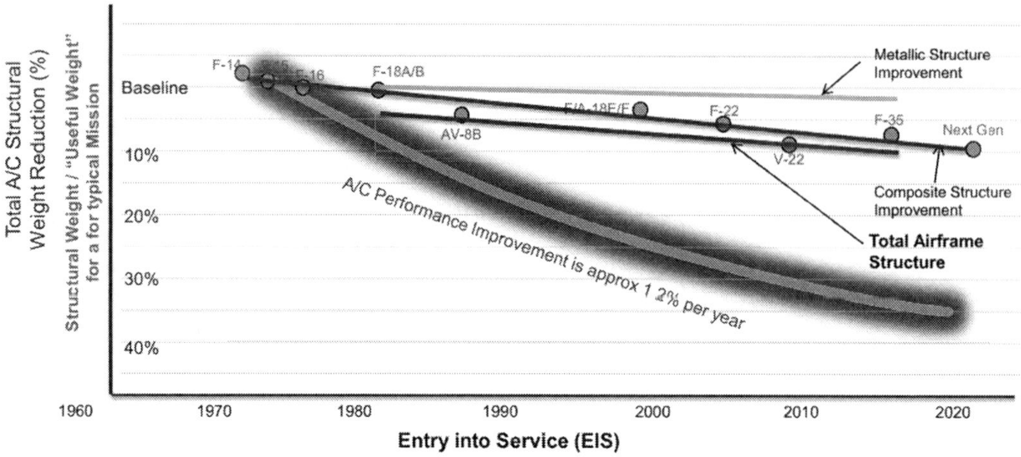

Materials Improvements Pace Airplane Performance Improvements

Effect of Optimization
Broader Distribution and Increased Number

Product	Al%	# Alloys for Al	Ti%	Steel%	Polymer Comp.%	Other%
F-4	66	2	7	22	0	5
F-15	50	4	33	8	2	7
F/A-18 A-D	50	3	13	16	10	11
F/A-18 E/F	31	3	21	14	19	15
C-17	78	5	11	6	5	0
F-22	16	3	35	7	24	18
747	81	6	4	13	1	1
757	78	6	6	12	3	1
767	80	6	2	14	3	1
777	70	7	7	11	11	1
787	20	10	15	10	50	5

Historical Trends of Composites

Engineering, Operations & Technology | BR&T   Materials & Manufacturing Technology

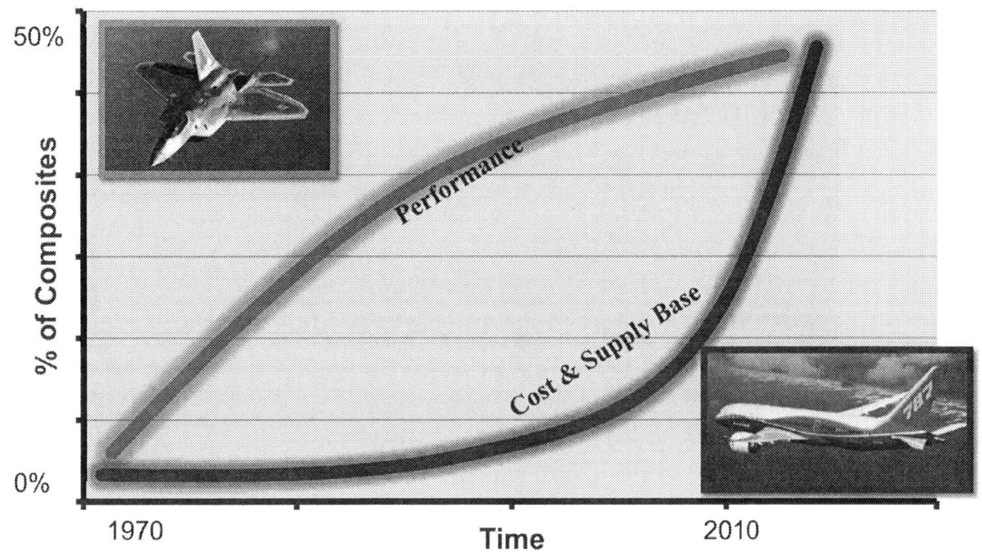

Boeing Experience – Volume on a Large Scale

Engineering, Operations & Technology | BR&T   Materials & Manufacturing Technology

Platform	Percent Composites	Total Wt (lbs)	Approx Composite Wt (lbs)	Approx Delivery Rate	Wt (lbs/Month)	# Delivered	Total Wt Composites Delivered (lbs)
C-17	8%	277,000	22,714	1.5		218	4,951,652
B-2	High					20	
F-18 c/d	10%	24,700	2,470			1,450	3,581,500
777	10%	300,000	30,000	7	210,000	1066	31,980,000
F-22	20%	31,700	6,340	6		339	2,149,260
F-18 e/f	18%	30,500	5,490	4	21,960	500	2,745,000
V-22	43%	33,140	14,250	1	14,250	160	2,280,000
787	50%	250,000	125,000	5	625,000	130	16,250,000
Total					871,210		63,934,360

- **Boeing Has Fielded More than 63 Million Pounds of Composite Structure**
- **Boeing Will Field Nearly 10 Million Additional Pounds Every Year**

Industrialization of Composites

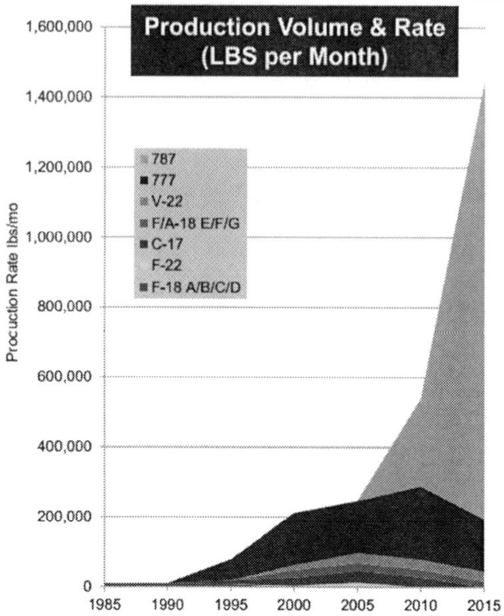

New Materials Drive Higher Costs

Non-Recurring Costs

- Materials Characterization
- Design
- Allowables
- Stress Analysis
- Developmental Testing

Table 4.3

Late 1990s Nonrecurring Engineering Cost Ratios[a]

Material	Average	Minimum/Maximum
Aluminum	1.00	1.00/1.00
Aluminum-lithium	1.00	1.00/1.00
Titanium	1.00	1.00/1.00
Steel	1.05	1.00/1.10
Carbon-epoxy	1.33	1.00/2.00
Carbon-BMI	1.38	1.10/2.00
Carbon-thermoplastic	1.33	1.20/1.40

[a]Late 1980s aluminum = 1.00.

Source Rand Report R-4016AF 1991 & 2001 Update

Recurring Costs

Table 18

MATERIAL COST FACTORS

Time Period/ Material Type	Buy-to-Fly Ratio	Raw Material $/lb (FY90$)	Material[a] $/lb (FY90$)
Late 1980s			
Aluminum	2.5	11	27
Al-lithium	4.2	17	72
Titanium	3.0	26	76
Steel	2.1	8	18
Graphite/epoxy	1.9	69	130
Graphite/bismaleimide	1.9	78	146
Graphite/thermoplastic	1.9	91	173
Mid-1990s			
Aluminum	2.2	10	22
Al-lithium	2.7	9	25
Titanium	3.0	24	72
Steel	2.1	8	18
Graphite/epoxy	1.8	57	102
Graphite/bismaleimide	1.8	61	111
Graphite/thermoplastic	1.8	66	119

[a]No burden added. The average industry burden is 15 percent.

Building Block Approach

Engineering, Operations & Technology | BR&T — Materials & Manufacturing Technology

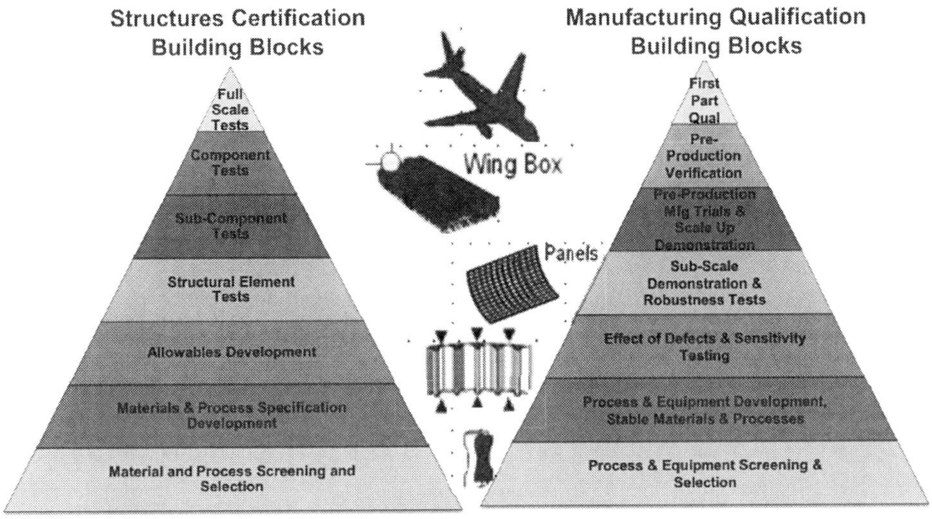

How Can We Reduce the Costs of Structural Certification and Manufacturing Qualification?

Composite Material Development Cost Challenges For Certification

Engineering, Operations & Technology | BR&T — Materials & Manufacturing Technology

- **Design Guides for Composites**
- **Multi-Functional Properties**
- **Failure Predictions**
- **Standardized Test Methods**

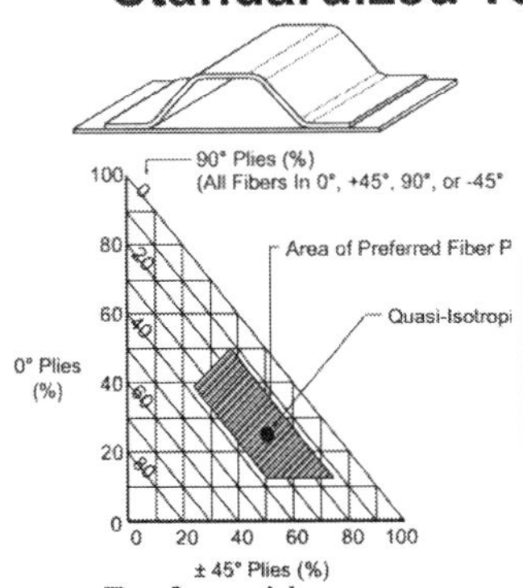

Preferred Lay-ups

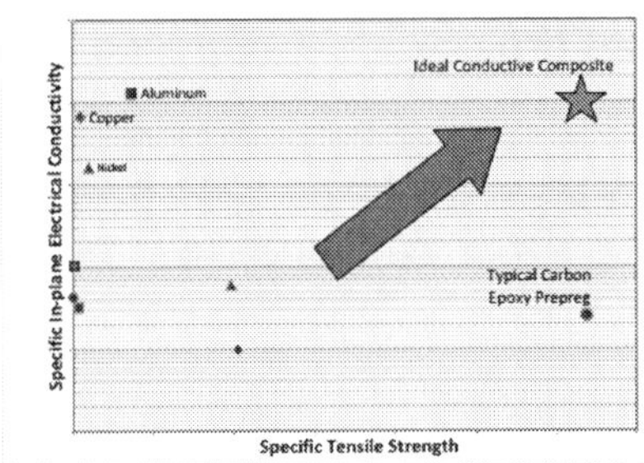

2 Layups – 2 Different Failures

Composite Manufacturing Cost Challenges for Qualification

Engineering, Operations & Technology | BR&T — Materials & Manufacturing Technology

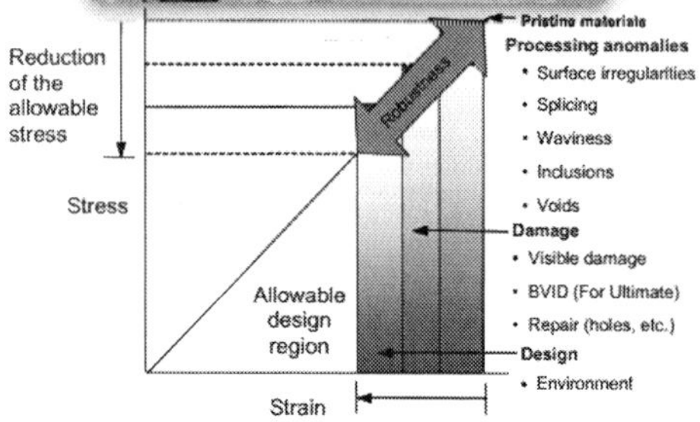

- Reduced process variability
- Materials developed concurrently with understanding of manufacturing tolerances and quality
- Speed & scale at reduced cost & increased quality
- New materials which use existing process, tooling and facilities
- Flexibility to Support Design Space Expansion

Achieving Higher Rates & Low Cost In Aerospace

Engineering, Operations & Technology | BR&T — Materials & Manufacturing Technology

- **High Cost of Composites May Be Reduced As Supply Base and Fabrication Rates Increase**
- **We Are Designing to Reduce Certification and Manufacturing Qualification Costs**
- **We Are Reducing Long Composite Cure Times to Meet Rate**

- **Can We Automate Large Scale Integrated Composite Structure Fabrication?**
- **As We Move From Manual Production To Automated Production Will Rate Increase and Cost Be Reduced?**

Aerospace vs Automobile Automation

Engineering, Operations & Technology | BR&T — Materials & Manufacturing Technology

	Automobile	Aerospace	Why is it more difficult?
Parts/Components	Automotive Hood	Aircraft Wing panel	
Size	6ft X 5ft	12ft X 60ft to 22ft X 100ft	Difficult to maintain accuracy of placement on 60 or 100 ft
Material	Steel	Aluminum, Composite	Is it easier for assembly to have small steel parts vs large composite or al
Weight	30lbs to 60lbs	1,000lbs to 3,000lbs	Takes gantry vs. robot
Tolerances	0.01's	0.001's	Higher tolerances required for Aircraft
Operational speed	70mph	500mph	Smother aerodynamic surfaces required for aerospace components
Life expectancy	5 to 10 years	30 to 50 years in operation	Replacing automotive lines every 5-7 years, aircraft equipment 20-30 years
Assembly System			
Type	Robot & single function EE	C-Frame machine with multi function head	Larger factory floor required and more complex process control
Part/system motion	Robot fixed, part on AGV	Machine on rail, part fixed on pogo system	Larger factory, energy consumption, Lower speed than robot
Process steps	Welding or riveting	Drilling/counters, sealing, riveting, inspecting,	Robots are doing one process, machines multiple processing
Investment	$250,000/robot	$5,000,000/machine	Much larger investment required for aerospace equipment

Historical progression of Automation

Engineering, Operations & Technology | BR&T — Materials & Manufacturing Technology

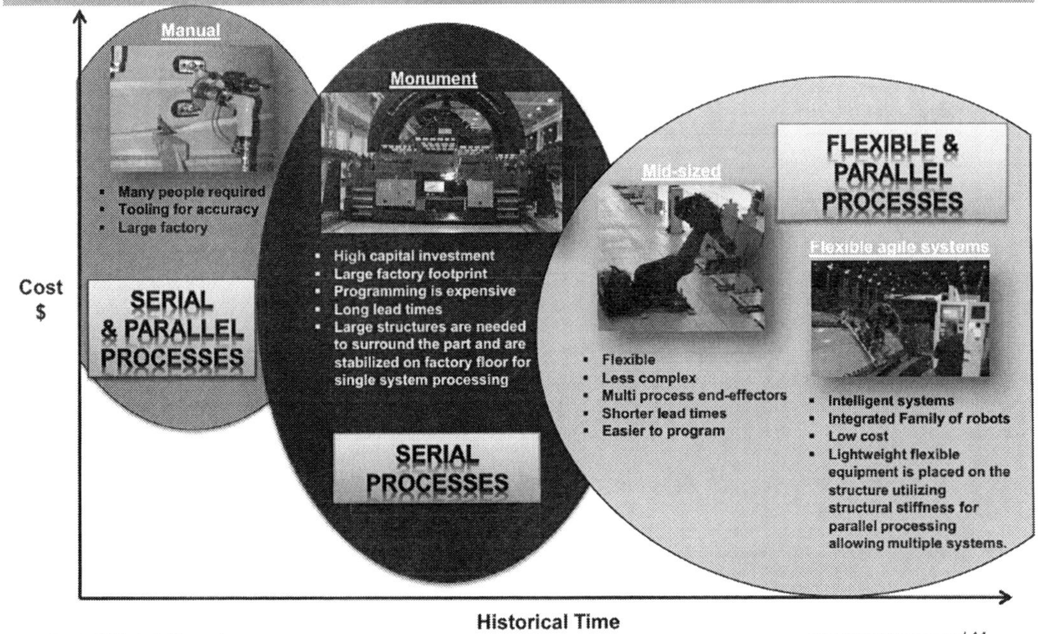

Gerould Young, 3/4/2014, Filename.ppt | 14

Automation of Aerospace Machining

- Single Axis
- Single Process

- Multiple Axis
- Single Process

- Computer Controlled

- Multiple Axis
- Multiple Tool
- Computer Controlled

Today's Machining Capability

- Six Axis
- Computer controlled
- Multiple Processes
- Large integrated structure

Typical Titanium Fabrication Costs

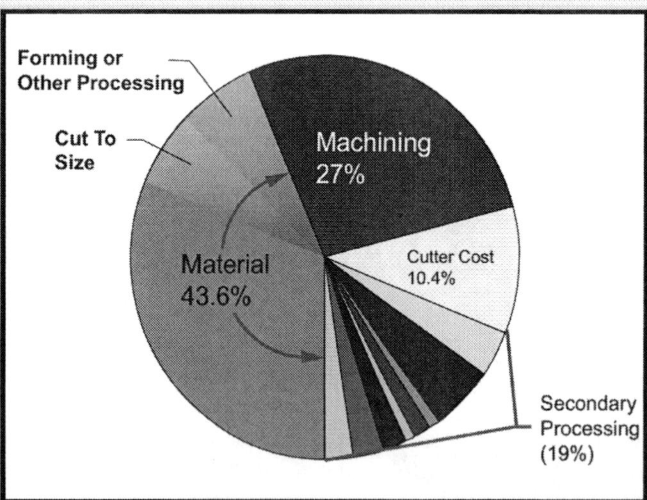

Primary Cost Drivers:

- Material
- Machining
- Cutters
  Tool Life Aluminum (1,000 Hrs)
  Tool Life Titanium (minutes)

Automation Opportunities:

- Deburr
- Dye Penetrant
- Dim. Inspection
- Shot-Peening
- Finish

Automation for Composite Fabrication

Engineering, Operations & Technology | BR&T   Materials & Manufacturing Technology

- Material Placement
  - Complex contours
  - Layup complexity – direction, ply drop-offs, ramps
  - Integrated structures - stringers
  - Tolerances at high laydown rates - laps gaps
  - Hybrid laminates layups – tape and cloth, tape width
- Bagging Processes
  - Vacuum bag on integrated structure
  - Pressure intensifiers and mandrel placement
  - Sensor placement & Leak verification
- Inspection
  - 100% Inspection
  - Wrinkles, FOD, Waviness, Voids
- Sealing and Coating Applications
  - Fay, Fillet and Edge Sealing
  - Protective coatings
- Drilling and Fastener Installation
  - Interfaces & fit up with ply drops, ramps & bag side surface
  - Small tolerance
  - Composite to metal joints

High Rate Production Requires High Quality Components

Automation Delivers Lay Down Cost Reduction

Engineering, Operations & Technology | BR&T   Materials & Manufacturing Technology

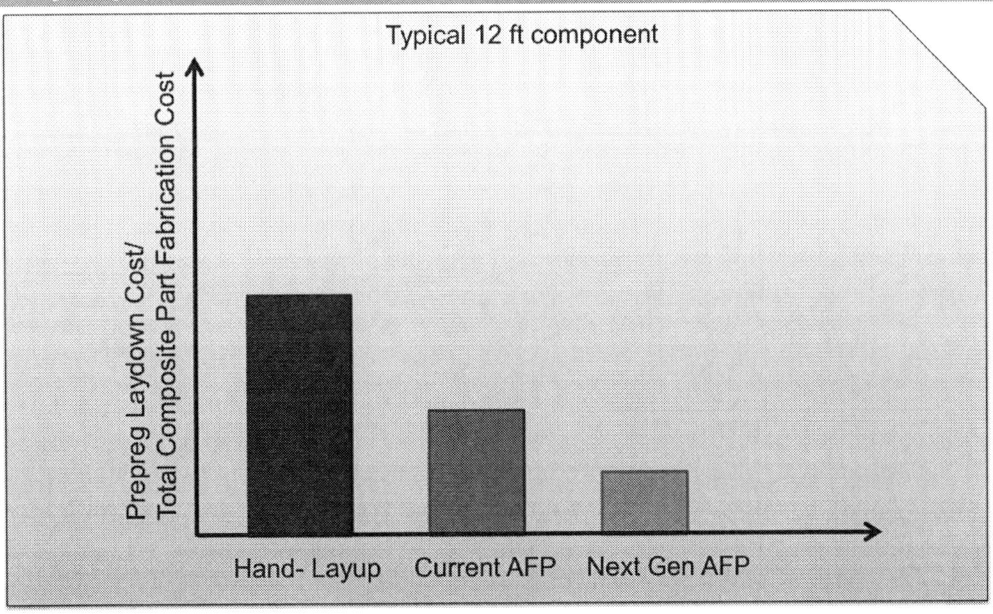

Built-up Assemblies via Rivet Automation

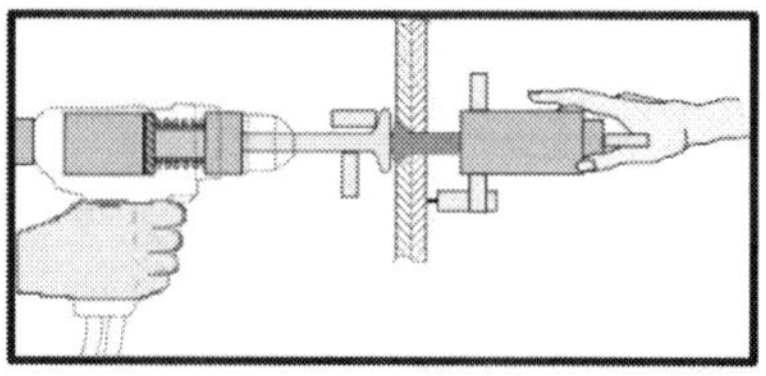

Characteristics of Manual Riveting

- Mechanics use intuition and experience to control force, position and final rivet quality
- Mechanics can verify and correct the rivet quality using their vision in real time

Replacing human riveting personnel with dual robotic system, requires generating detail engineering knowledge of the rivet forming process parameters.

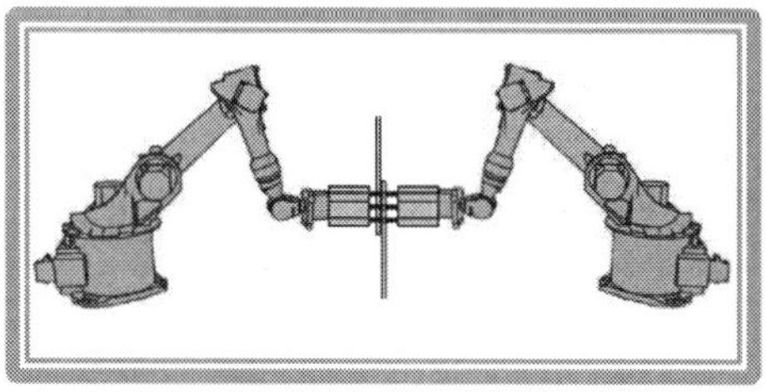

Characteristics of robotic riveting:

- Tremendous amount of knowledge/data is required to characterize the manual process
- Sensors are installed onto hand tools (Gun, Bucking bar) and parameters measured
- Process parameters for robotics are defined entered into programs
- All process parameters have to be entered into robot program
- A variety of sensors are built into Multi Function End Effectors needed to control process parameters
- Robotic riveting quality (tolerance range) must be verified

Trends in Aerospace Assembly

Yesterday

Monuments in the factory to achieve tolerance and serial automation

Today

Significant investment in dedicated assembly tooling requiring large floor space and a significant touch labor content

Future

*Agile Assembly - Networked Enabled/ Reconfigurable/Autonomous Assembly **

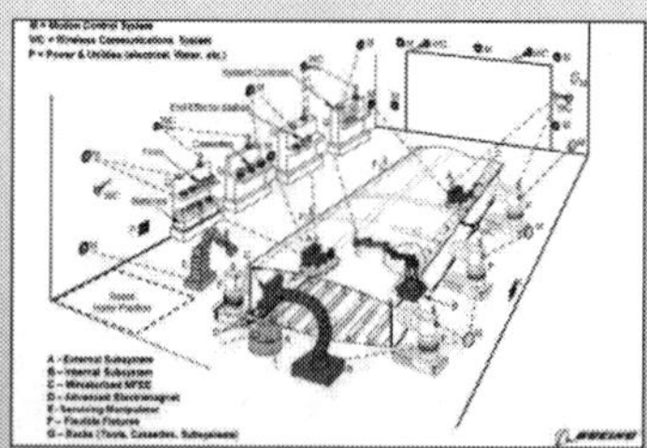

Impact of Automation on Aerospace Costs

Engineering, Operations & Technology | BR&T — Materials & Manufacturing Technology

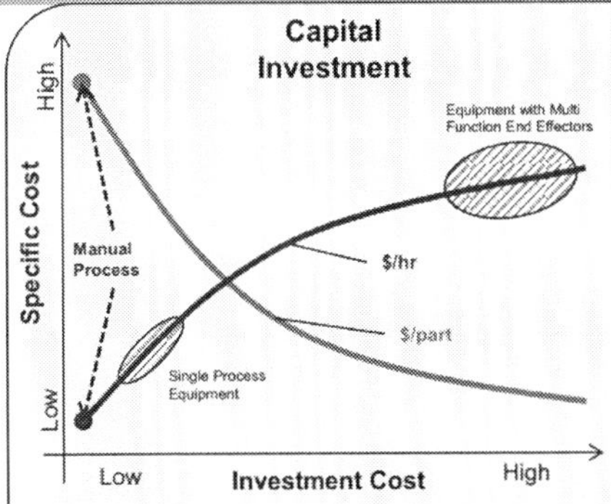

- Automated equipment can reduce cost
- Single process automation has limited impact on part cost
- Multi-process automation can provide part cost benefits

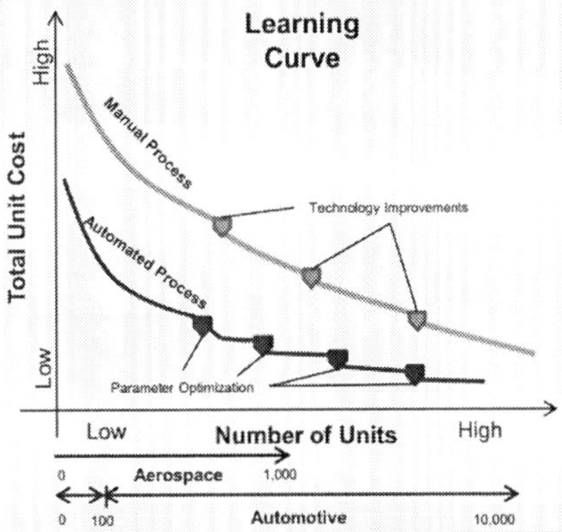

- Manual learning curve for aircraft manufacturing relies on training and technology
- Automation learning curve generates major unit cost reductions quicker but depends on parameter optimization

Capital cost optimization and automation experience required

Challenges for Future Aerospace Factory

Engineering, Operations & Technology | BR&T — Materials & Manufacturing Technology

- The lure of new materials
- Design with materials and manufacturing costs and quality in mind
- Automation may improve cost, rate and quality
- Scale, complexity and structural integration make automation challenging for fabrication processes
- Aerospace automation demands agile flexible intelligent systems with advanced sensors
- Automation optimization for capital cost, experience, rate and number of units has a “industry” learning curve
- New skills in simulation, sensors, data, computers, integration, robotics are required

PerformanceAscent

Sequestration: OEM & Supplier Perspectives

for

AERODEF MANUFACTURING

produced by sme

February 25, 2014

achieve excellence

David J. Blanco

- Founder and President of Performance Ascent, Inc.
- Training and Consultancy company focused on A&D supplier development
- Board of Examiners – Baldrige National Program
- Former Judge – California Baldrige Program
- Board of Directors – California Baldrige Program
- Advisory Council – California Assembly Select Committee on Aerospace
- LAEDC *Jobs Defense Council*
- University of Redlands – Management Studies
- Peter F. Drucker Graduate Management Center

PerformanceAscent

achieve excellence

National Security

"We do not find the sequestration levels to be what we need to defend the country and implement our strategy," Kendall told reporters at the Singapore Airshow.

Frank Kendall, Under Secretary of Defense for Acquisition, Technology and Logistics (Reuters – February 11, 2014)

PerformanceAscent
achieve excellence

Sequestration National Research

George Mason University – **July 2012**

- California Job Loss – 225,464 – Direct & Indirect
- Virginia Job Loss – 207,571 – Direct & Indirect
- Texas Job Loss – 159,473 – Direct & Indirect
- One-Half of the estimated 2 million jobs lost will be small businesses

FAA Budget Impact on the US Economy - **Aug 2012**

- Economic losses from $10 to $20 billion annually
- Up to $1 billion in lost tax revenue annually

PerformanceAscent
achieve excellence

Industrial Base Disruptions

- New wave of supplier consolidations
- Bottlenecks & strains in the supply chain
- Loss critical work skills
- Investment decisions on hold
- Loss of US capacity for commercial production
- Civilian certification employees furloughed
- Longer decision-making
- Shift from best value to lowest price
- RFP release dates slower
- 17% shift from B2G to B2B
- No new hiring

PerformanceAscent
achieve excellence

Direct Supplier Feedback

- Our 100 jobs are at risk
- RFP award on hold
- Contract award cancelled
- Reduced profit to 1%
- Reduced back office support
- RFP award delayed a year+
- Assumed a cash-risk position
- Customer holding back $$$
- Customer terms increased
- Procurement reduced 10%
- Customers planning for worse
- Reduced workforce
- Layoff lost skills
- Work on hold or scrapped
- Stopped investment & hiring
- Large order now only 5 parts
- Planning to exit defense work
- Customer taking back work
- $2 million inventory hold
- Increased off-shore competition

PerformanceAscent
achieve excellence

Final Thoughts

- Customers
- Suppliers
- Collaborations

Performance Ascent
achieve excellence

Final Thoughts

"Winners Never Quit, and Quitters Never Win"

...........Vince Lombardi

Performance Ascent
achieve excellence

Thank You

David J. Blanco

Direct: 626-437-7914

DBlanco@PerformanceAscent.com

PerformanceAscent
achieve excellence

BLUEPRINT FOR ACTION
AeroDef Manufacturing Summit & Exhibition
February 27, 2014
The New Manufacturing
Innovation Institutes
Mike Molnar
Advanced Manufacturing
National Program Office
www.manufacturing.gov
NSF
NASA

Interagency Advanced Manufacturing
National Program Office (AMNPO)
Executive Office of the President
NASA
NSF
SBA
USDA
Advanced Manufacturing National Program Office
(housed at DOC - NIST)

Challenge: US losing leadership in Advanced Products

U.S. Trade Balance for Advanced Technology Products

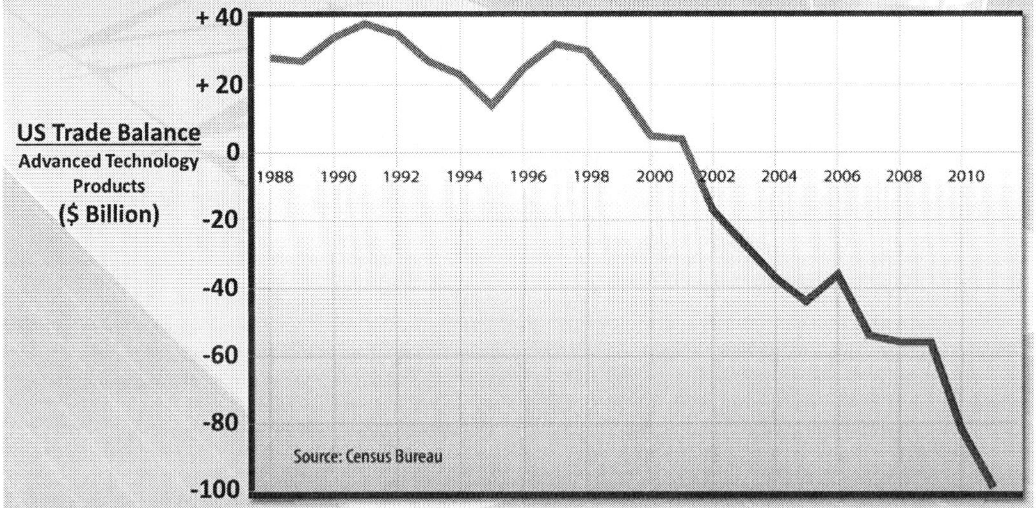

Products invented here, now made elsewhere - not driven by labor cost

Manufacturing Policy Milestones
January 2012
March 2012
August 2012
June 2011
REPORT TO THE PRESIDENT ON ENSURING AMERICAN LEADERSHIP IN ADVANCED MANUFACTURING
February 2012
July 2012
REPORT TO THE PRESIDENT ON CAPTURING DOMESTIC COMPETITIVE ADVANTAGE IN ADVANCED MANUFACTURING
January 2013
NATIONAL NETWORK FOR MANUFACTURING INNOVATION: A PRELIMINARY DESIGN

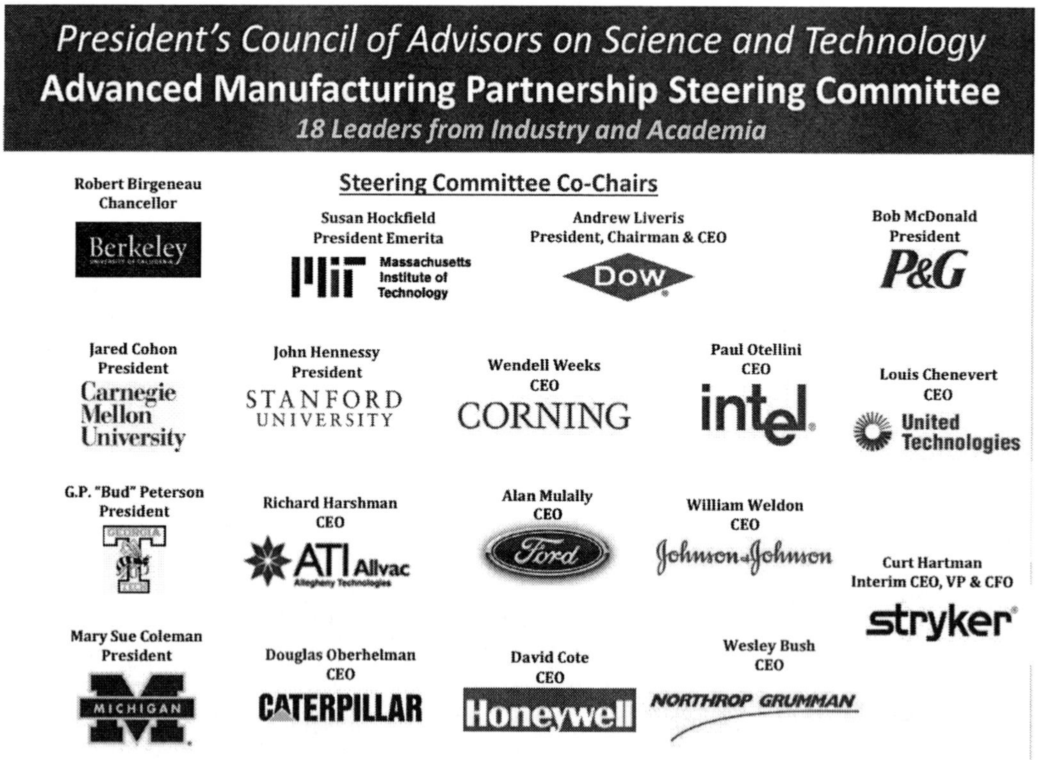
President's Council of Advisors on Science and Technology
Advanced Manufacturing Partnership Steering Committee
18 Leaders from Industry and Academia
Steering Committee Co-Chairs
Susan Hockfield
President Emerita
Massachusetts Institute of Technology
Andrew Liveris
President, Chairman & CEO
DOW
Robert Birgeneau
Chancellor
Berkeley
Bob McDonald
President
P&G
Jared Cohon
President
Carnegie Mellon University
John Hennessy
President
STANFORD UNIVERSITY
Wendell Weeks
CEO
CORNING
Paul Otellini
CEO
intel
Louis Chenevert
CEO
United Technologies
G.P. "Bud" Peterson
President
Richard Harshman
CEO
ATI Allvac
Allegheny Technologies
Alan Mulally
CEO
Ford
William Weldon
CEO
Johnson & Johnson
Curt Hartman
Interim CEO, VP & CFO
stryker
Mary Sue Coleman
President
MICHIGAN
Douglas Oberhelman
CEO
CATERPILLAR
David Cote
CEO
Honeywell
Wesley Bush
CEO
NORTHROP GRUMMAN

AMP Mission and Results

AMP Mission:

- Encourage approaches that sustain and grow U.S. leadership in advanced manufacturing, making the U.S. a magnet for jobs and investment; fostering broad, long-term collaboration among industry, academia, and government partners to drive advances in U.S. innovation and workforce capabilities.

Inaugural AMP achievements:

- Issued 16 recommendations across:
  - Enabling innovation
  - Securing the talent pipeline
  - Improving the business climate
- Spurred critical national initiatives, including the **National Network for Manufacturing Innovation** (NNMI)

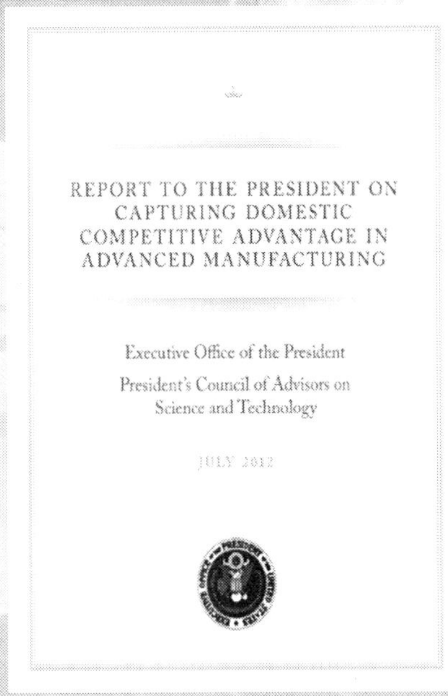

7

The "Scale-up" Gap or Missing Middle

Basic R&D

Common terms
The "valley of death"
The "missing Bell Labs"
The "industrial commons"

Commercialization

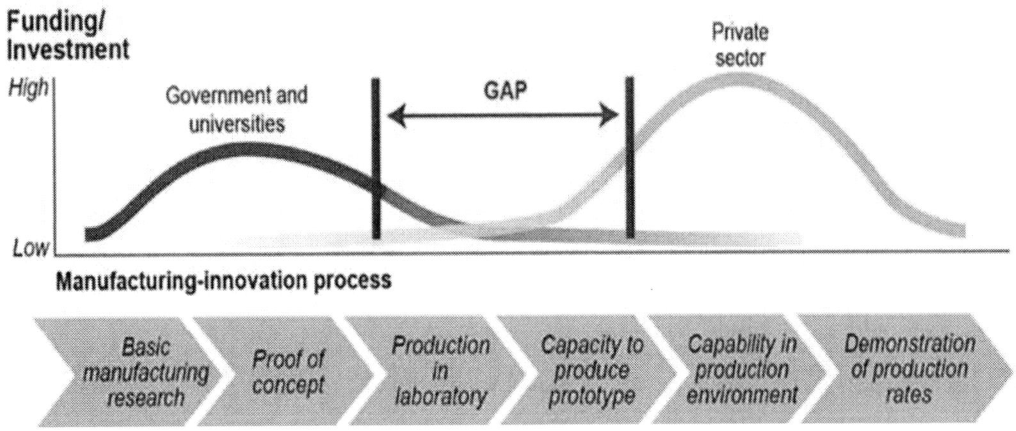

Why a network of Institutes for Manufacturing Innovation?

Goal: strengthen competitiveness and job-creating power of U.S. manufacturing

- Enable better, faster "scale-up" when innovations are presented to U.S. industry

Bridge the gap between fundamental discovery and large volume manufacturing

- create a space where Industry and Academia can work together, thereby ...
- Reducing the complexity, high risk and long time horizons that prevent **individual** companies from investing in Technology Readiness Level (TRL) 4-7 research
- Allow small and medium sized companies and start-ups access to the best minds and equipment to develop new innovative ideas.

These institutes create that space, to "de-risk" new technologies and get them into production here in the U.S., with benefits to everyone from start-ups to large corporations.

Develop a workforce with skills needed for advanced manufacturing

- The United States lags behind competitor nations regarding the skilled workforce needed for advanced manufacturing.
- These Institutes provide a space with state of the art equipment, where work force training can take place, providing workers with the computer and equipment control skills necessary for modern manufacturing.

The NNMI Story Today...

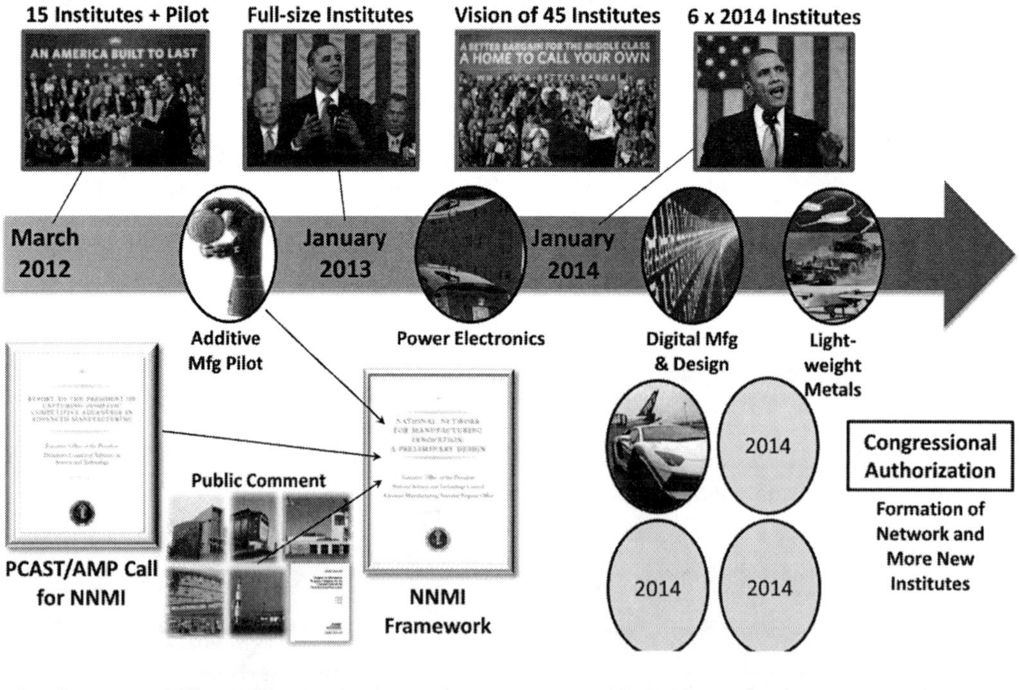

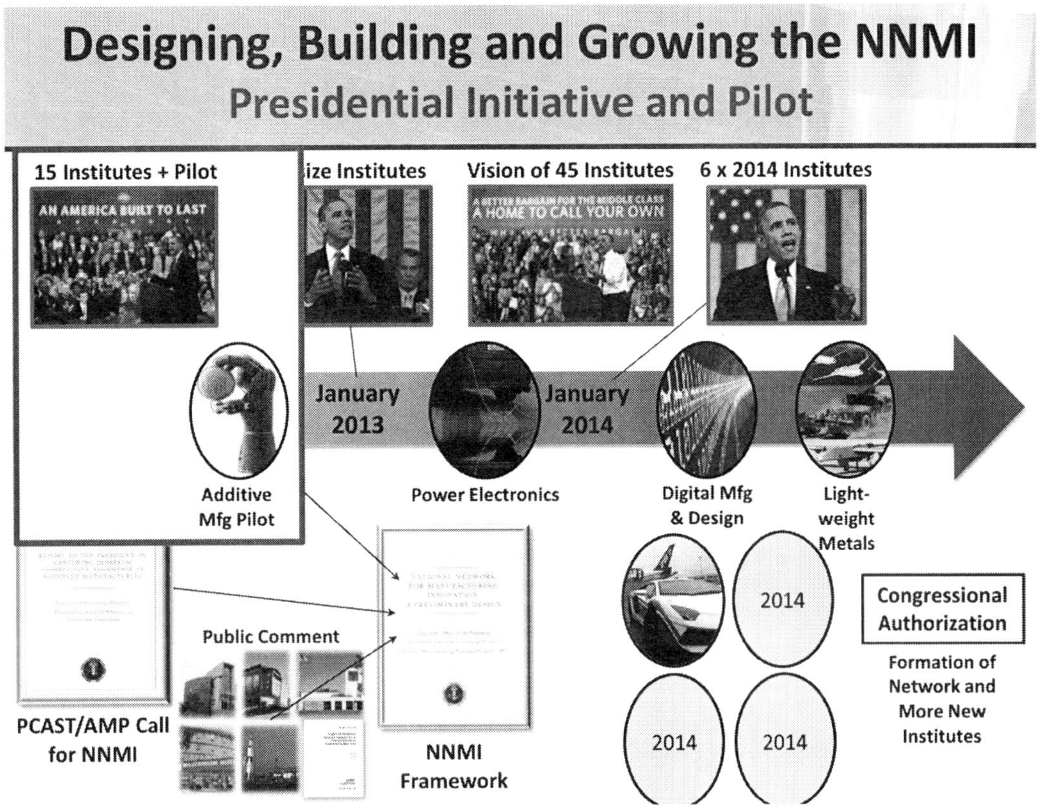

Designing, Building and Growing the NNMI
Presidential Initiative and Pilot
15 Institutes + Pilot
ize Institutes
Vision of 45 Institutes
6 x 2014 Institutes
AN AMERICA BUILT TO LAST
A HOME TO CALL YOUR OWN
January 2013
January 2014
Additive Mfg Pilot
Power Electronics
Digital Mfg & Design
Light-weight Metals
2014
2014
2014
Congressional Authorization
Formation of Network and More New Institutes
Public Comment
PCAST/AMP Call for NNMI
NNMI Framework

National Network for Manufacturing Innovation
AN AMERICA BUILT TO LAST
"Sparking this network of innovation across the country, it will create jobs and will keep America leading in manufacturing..."
President Obama, March 9, 2012
• President asks Congress to authorize initial network of up to 15 Manufacturing Innovation Institutes
• President directs Agencies to work together on Pilot Institute, while designing Institutes with input from Industry and Academia

Additive Manufacturing Innovation Institute Youngstown Ohio

Prime Awardee: National Center for Defense Manufacturing and Machining

- Initial $30M federal investment matched by $40M industry, state/local
- Strong leveraging of equipment, existing resources
- Strong business development
- Ties to many organic facilities
- Tiered membership-based model, low cost to small business and nonprofits

Why Additive Manufacturing? High Potential for Transformative Impact

Projected AM Sales (products and services)

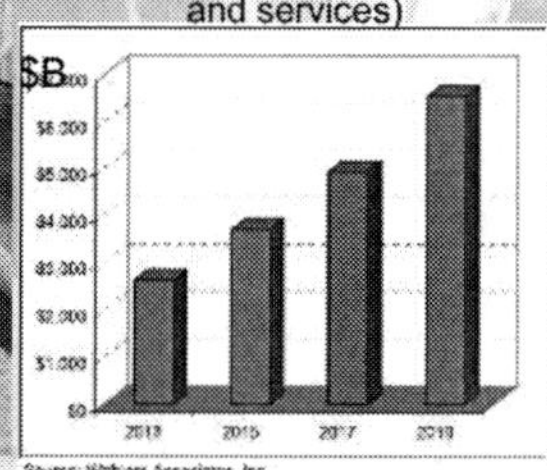

"20% of output of 3D printers is now final products, rather than prototypes. By 2020 it may be 50%." – *The Economist (2011)*

Government agency investments and interest

Consumer Product Market

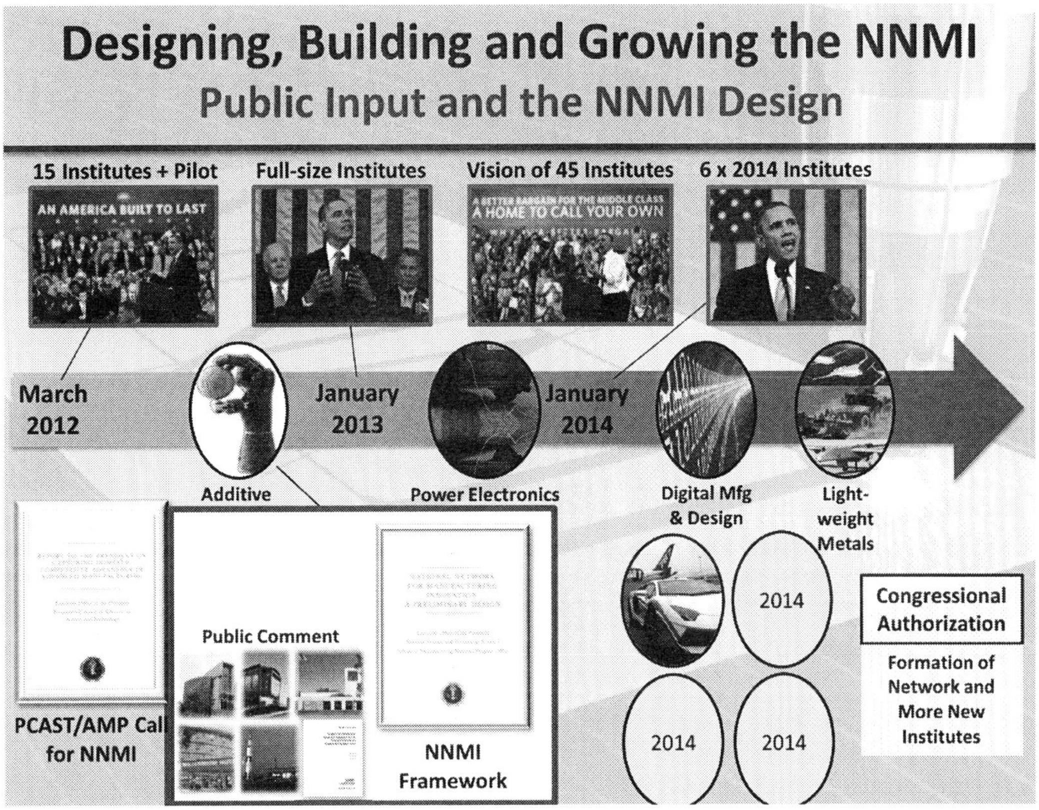
Designing, Building and Growing the NNMI
Public Input and the NNMI Design
15 Institutes + Pilot
Full-size Institutes
Vision of 45 Institutes
6 x 2014 Institutes
AN AMERICA BUILT TO LAST
A HOME TO CALL YOUR OWN
March 2012
January 2013
January 2014
Additive
Power Electronics
Digital Mfg & Design
Light-weight Metals
Public Comment
PCAST/AMP Call for NNMI
NNMI Framework
2014
2014
2014
Congressional Authorization
Formation of Network and More New Institutes

Public Engagement on Design
Workshops & Request for Information
Broad & Diverse Stakeholder Input
1,200 voices on the NNMI Design!
All Other 10%
Economic Development 6%
Research & non-profits 8%
Federal State & Local Gov't 14%
Industry 31%
Academia 31%
Rensselaer Polytechnic Institute
Troy New York
Cuyahoga Community College
Cleveland Ohio
University of Colorado
Boulder, Colorado
National Academies Beckman Center
Irvine California
U.S. Space and Rocket Center
Huntsville, Alabama

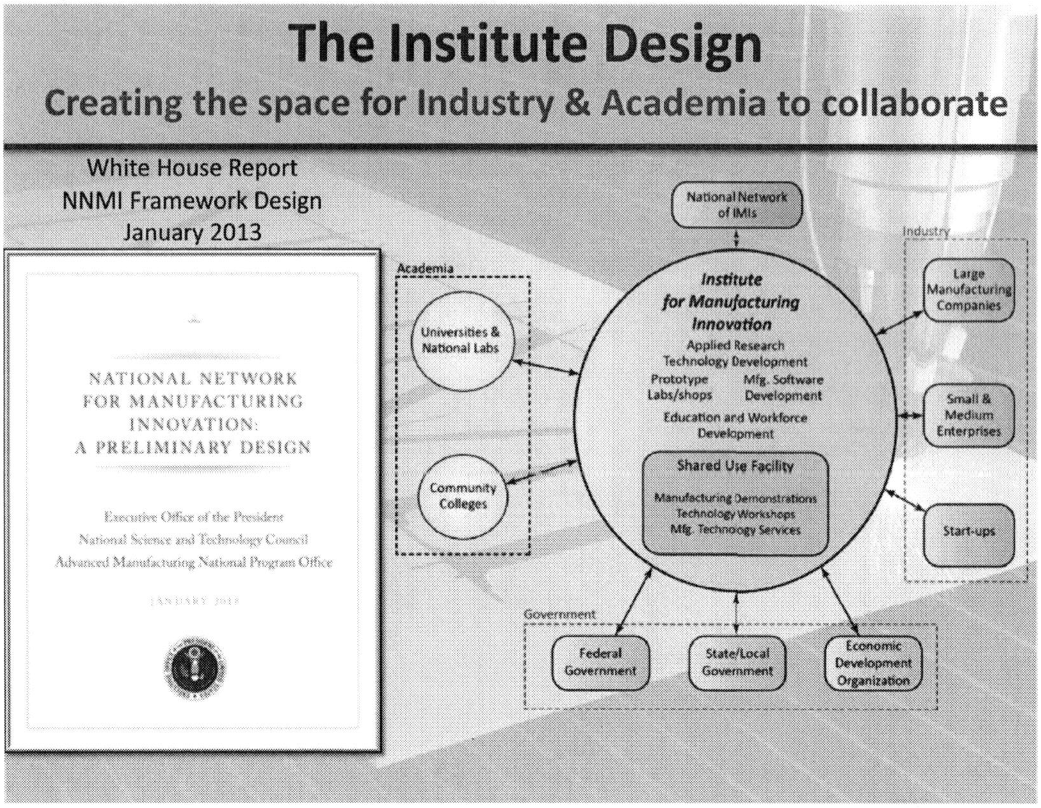
The Institute Design
Creating the space for Industry & Academia to collaborate
White House Report
NNMI Framework Design
January 2013
NATIONAL NETWORK FOR MANUFACTURING INNOVATION: A PRELIMINARY DESIGN
Executive Office of the President
National Science and Technology Council
Advanced Manufacturing National Program Office
National Network of IMIs
Academia
Universities & National Labs
Community Colleges
Institute for Manufacturing Innovation
Applied Research
Technology Development
Prototype Labs/shops
Mfg. Software Development
Education and Workforce Development
Shared Use Facility
Manufacturing Demonstrations
Technology Workshops
Mfg. Technology Services
Industry
Large Manufacturing Companies
Small & Medium Enterprises
Start-ups
Government
Federal Government
State/Local Government
Economic Development Organization

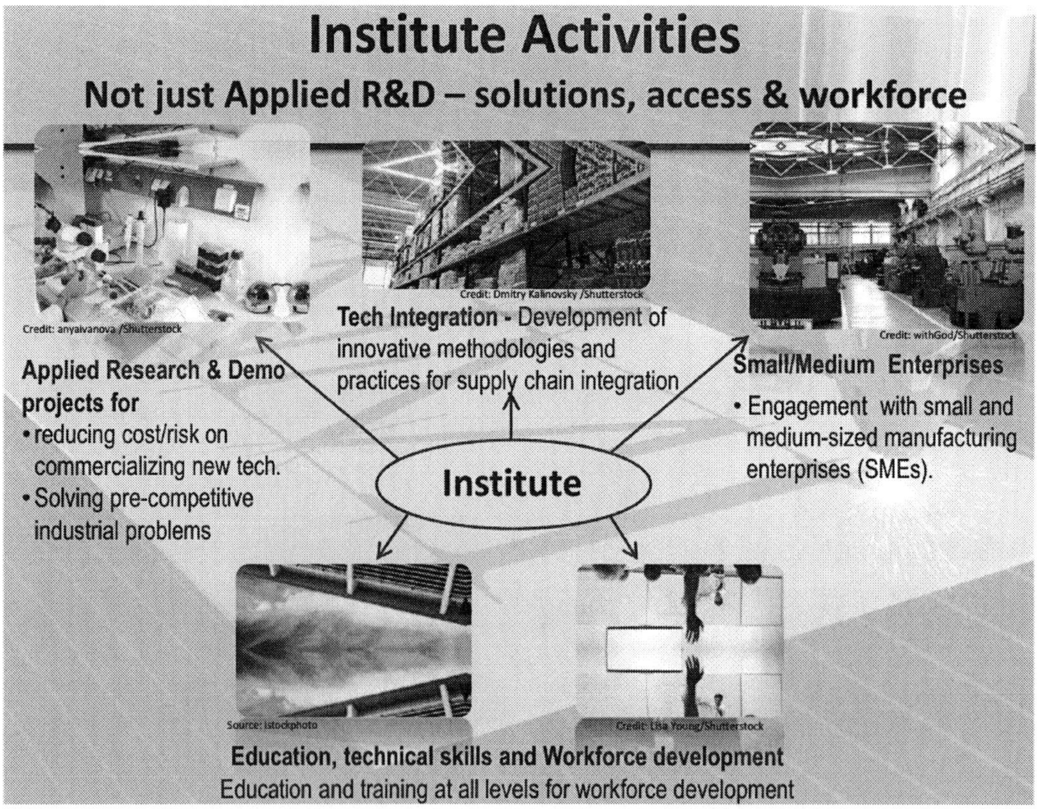
Institute Activities
Not just Applied R&D – solutions, access & workforce
Credit: anyaivanova /Shutterstock
Credit: Dmitry Kalinovsky /Shutterstock
Credit: withGod/Shutterstock
Tech Integration - Development of innovative methodologies and practices for supply chain integration
Applied Research & Demo projects for
• reducing cost/risk on commercializing new tech.
• Solving pre-competitive industrial problems
Small/Medium Enterprises
• Engagement with small and medium-sized manufacturing enterprises (SMEs).
Institute
Source: istockphoto
Credit: Lisa Young/Shutterstock
Education, technical skills and Workforce development
Education and training at all levels for workforce development

So what does a Manufacturing Innovation Institute actually do?

How does an Industry-led Public-Private Partnership plan, develop and "de-risk" new technologies and materials?

How can these help your company develop new products and processes for sustained competitiveness?

America Makes

America Makes
National Additive Manufacturing Innovation Institute

Why America Makes?

America Makes creates mechanisms for collaboration...

Pooling Resources / Pooling Risks

Public/Private Funded Projects
Crowd Funded Projects

Cooperative Development of
- Training
- Assessments
- Case Studies

Solving Problems Collaboratively

Work Shops, Working Groups, Projects

Cooperative Development of
- Material Specs
- Process Specs
- Material Databases
- Design Rules
- Application Guides

Leveraging Community Knowledge

Knowledge Base, Online Collaboration Tools, Databases, Specifications, Application Guides, Curriculum

20

America Makes
National Additive Manufacturing Innovation Institute

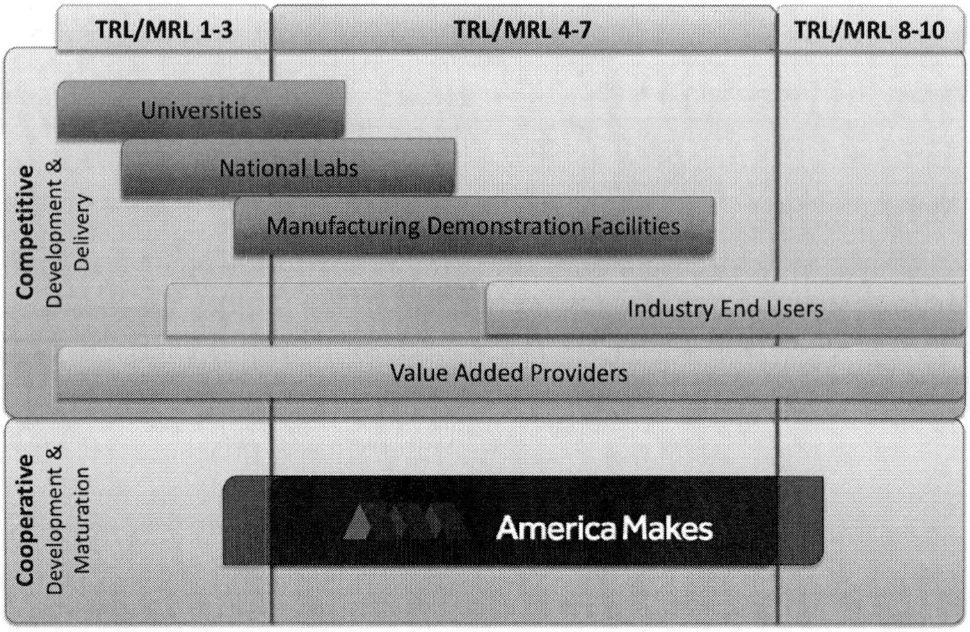

America Makes
National Additive Manufacturing Innovation Institute

Governance Structure – Shared Leadership

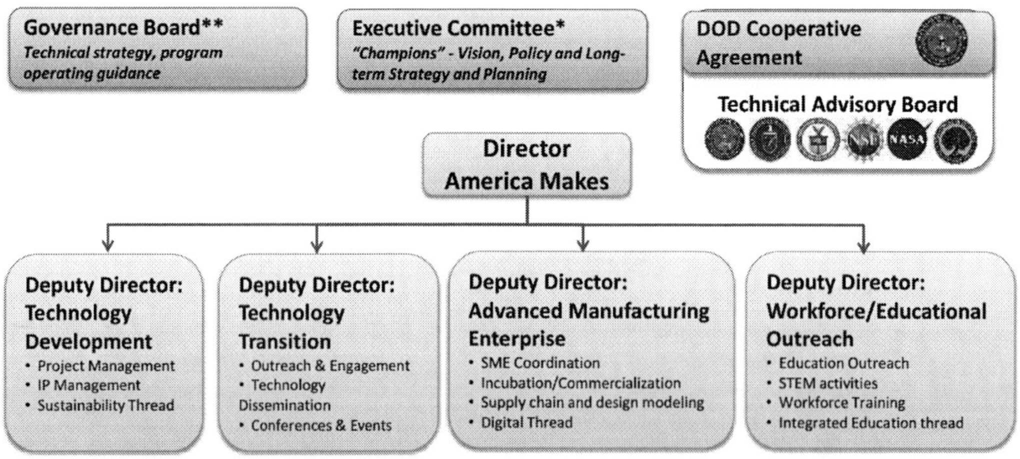

- ***Executive Committee (11)**: Industry, for-profit organization (2), Non-profit association (2), Academic (2), Government (3), At-large (2)
- ****Governance Board (35):** Lead and Full Members, states ex-officio

22

America Makes
National Additive Manufacturing Innovation Institute

Membership Benefits

Community

- Formalized community looking to leverage AM Technologies
- Leverage learning curves
- Cooperatively solve similar issues
- Direct future equipment, material, and software development
- Technical strategy input

Work Force Development

- Pre-K to grey training resources
- Immersive training/learning at America Makes facility

Projects

- Lead America Makes directed project calls
- Team with America Makes & members to win additional government project calls
- Use America Makes resources for targeted company projects

Enterprise Data

- Processing & Application data
- Material properties
- Supply chain
- Analysis

23

America Makes
National Additive Manufacturing Innovation Institute

Technology Roadmap: v1.0

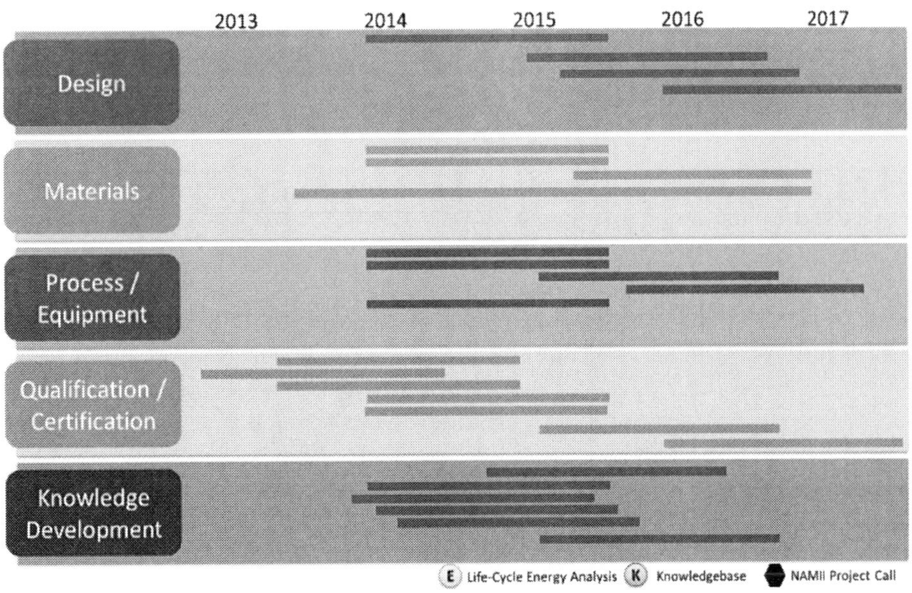

25

Project Call #2

- Eight *Priority Topics* Selected Based on Crowd-Sourced Roadmapping Process

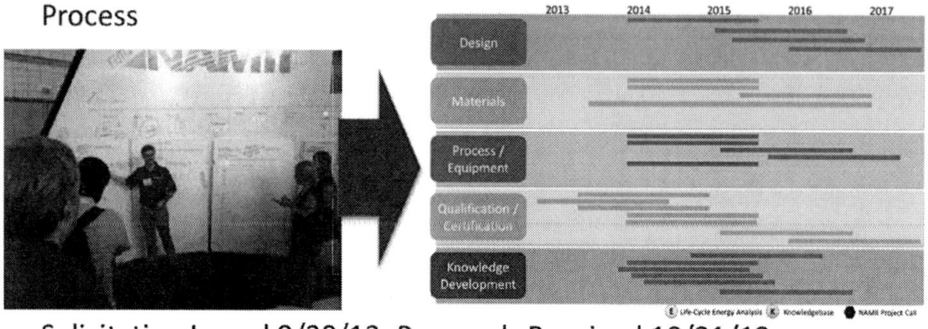

- Solicitation Issued 8/29/13; Proposals Received 10/31/13
- Announced 15 Project Awards 1/21/14
- Awarding $9M of America Makes Funding with $10.3 million matching cost share from awarded project teams = $19.3M total funding

26

Designing, Building and Growing the NNMI
New Manufacturing Innovation Institutes

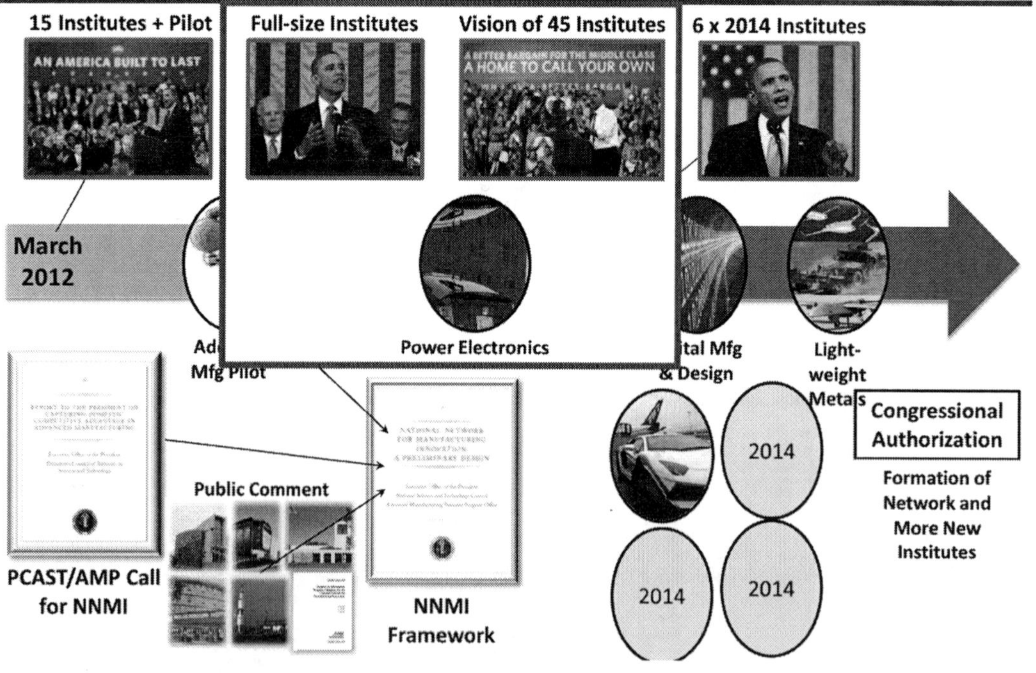

NNMI Vision – 45 institutes

AP Photo/Susan Walsh

"In my State of the Union Address, I also asked Congress to build on a successful pilot program and create 15 manufacturing innovation institutes that connect businesses, universities, and federal agencies to turn communities left behind by global competition into global centers of high-tech jobs.

"Today, I'm asking Congress to build on the bipartisan support for this idea and triple that number to 45 – creating a network of these hubs and guaranteeing that the next revolution in manufacturing is Made in America."

July 30, 2013

With Congressional Legislation

- Open competition on ANY topic proposed by Industry and Academia
- Selection of topics made on merit
  - let best proposals of greatest impact to US industry move ahead
  - Institutes by Administrative Action limited to topics Federal agencies need
- Creates capability for enough institutes to form a value-added network
- Provides stable funding and certainty for consortia – path to sustainability

Next Generation Power Electronics Manufacturing Innovation Institute

Lead: North Carolina State University

ABB, Arkansas Power Electronics International, Avogy, Cree, Deere & Company, Delphi Automotive, Delta Products, DfR Solutions, GridBridge, Hesse Mechatronics,, II-VI, IQE, Monolith Semiconductor, RF Micro Devices, Toshiba International, Transphorm, United Silicon Carbide, Vacon, Arizona State University, Florida State University, University of California-Santa Barbara, Virginia Tech, National Renewable Energy Lab, Naval Research Lab

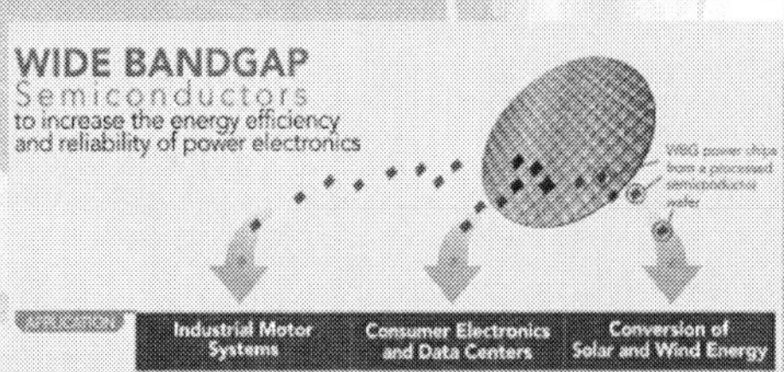

<u>***Mission:***</u> Develop advanced manufacturing processes that will enable large-scale production of wide bandgap semiconductors, which allow power electronics components to be smaller, faster and more efficient than silicon.

Poised to revolutionize the energy efficiency of power control and conversion

President Obama
North Carolina State University, January 15, 2014

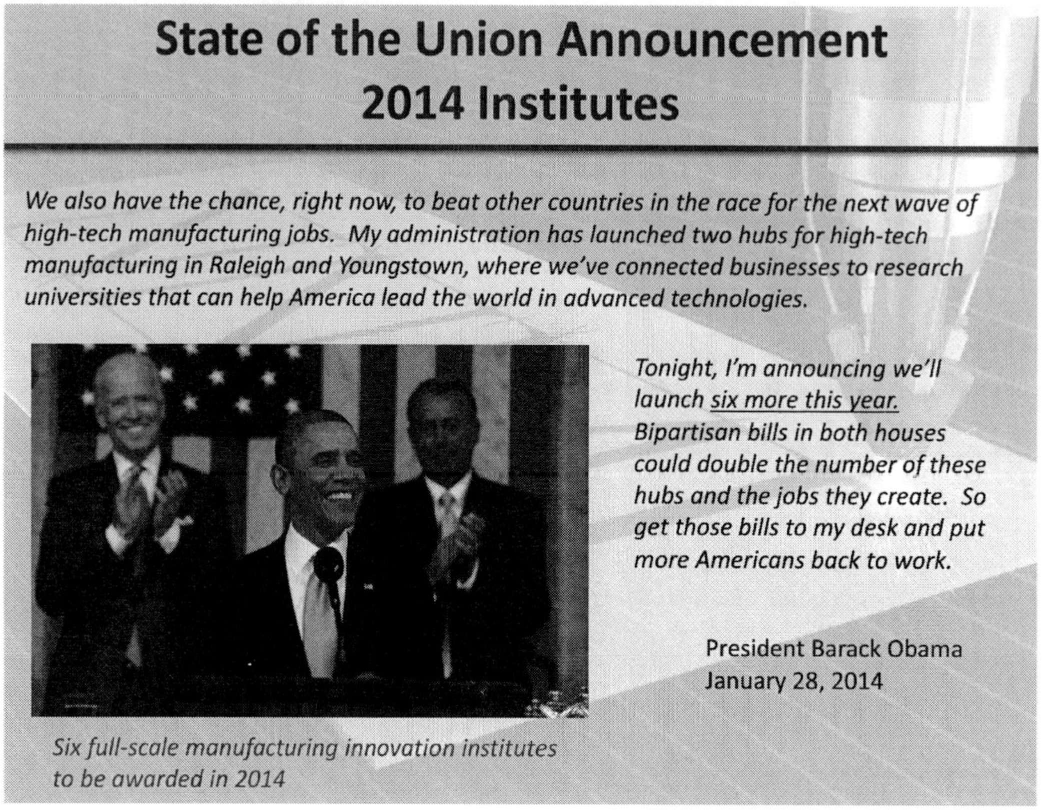
State of the Union Announcement
2014 Institutes
We also have the chance, right now, to beat other countries in the race for the next wave of high-tech manufacturing jobs. My administration has launched two hubs for high-tech manufacturing in Raleigh and Youngstown, where we've connected businesses to research universities that can help America lead the world in advanced technologies.
Tonight, I'm announcing we'll launch six more this year. Bipartisan bills in both houses could double the number of these hubs and the jobs they create. So get those bills to my desk and put more Americans back to work.
President Barack Obama
January 28, 2014
Six full-scale manufacturing innovation institutes to be awarded in 2014

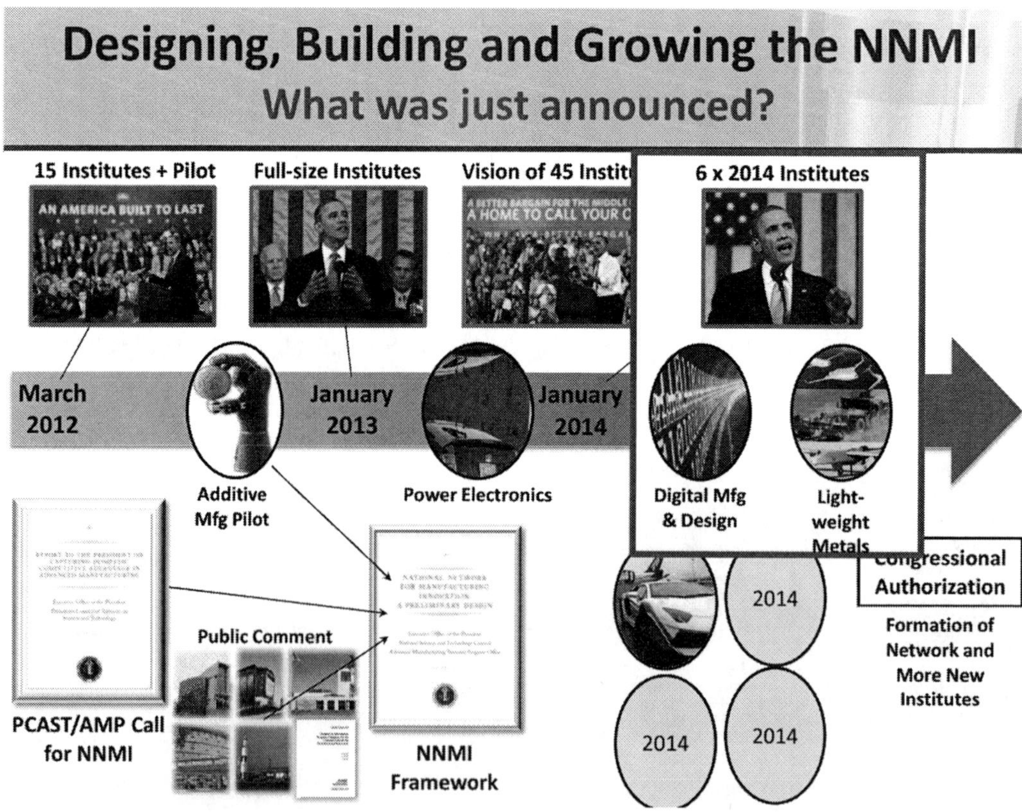
Designing, Building and Growing the NNMI
What was just announced?
15 Institutes + Pilot
AN AMERICA BUILT TO LAST
Full-size Institutes
Vision of 45 Instit
6 x 2014 Institutes
March 2012
January 2013
January 2014
Additive Mfg Pilot
Power Electronics
Digital Mfg & Design
Light-weight Metals
2014
2014
2014
Congressional Authorization
Formation of Network and More New Institutes
Public Comment
PCAST/AMP Call for NNMI
NNMI Framework

New Institutes Announced, February 25, 2014

"That's what these new hubs are all about. They're partnerships – they bring together companies and universities to develop cutting-edge technology, train workers to use that technology, and make sure research is turned into real-world products made by American workers."

President Barack Obama
February 25, 2014

Lightweight and Modern Metals Manufacturing Innovation Institute

Lead: EWI
Hub location: Canton, Michigan
Regional location: I-75 Corridor

- **34 Industry Partners**
- **9 Universities and Labs**
- **17 Other Organizations**

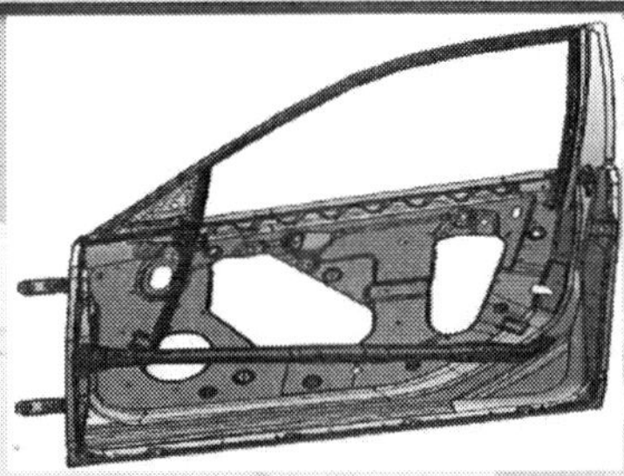

Mission: Provide the National focus on expanding US competitiveness and innovation , and facilitating the transition of these capabilities and new technologies to the industrial base for full-scale application.

Positioned to expand the US Industrial base for new products and technologies for commercial and USG demands that utilize new, lightweight high-performing metals

Digital Manufacturing and Design Innovation Institute

Lead: UI Labs

Hub location: Chicago, Illinois

- 41 Companies
- 23 Universities and Labs
- 9 Other Organizations

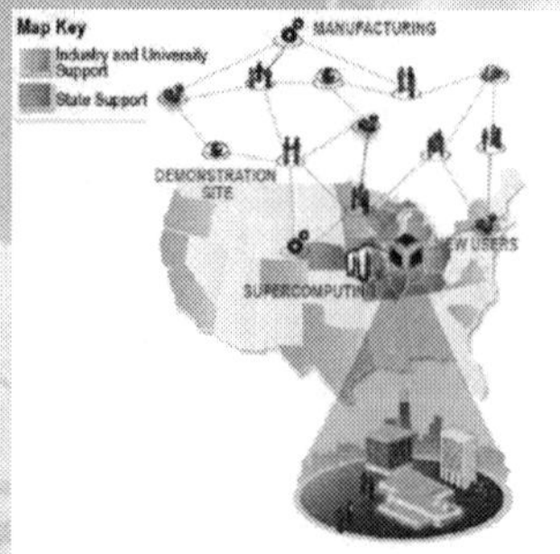

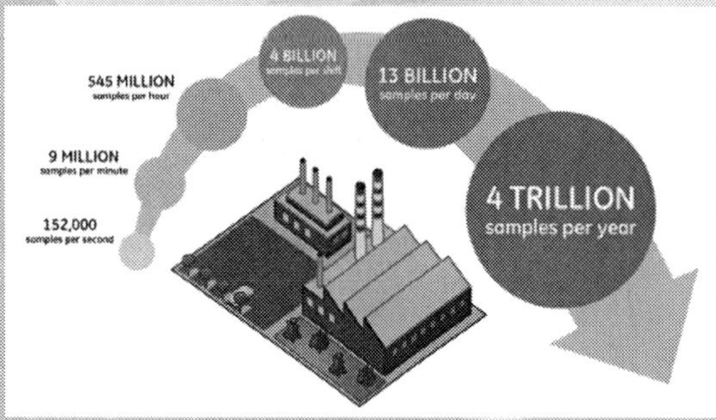

Mission: ***Establish a state-of-the-art proving ground that links IT tools, standards, models, sensors, controls, practices and skills, and transition these tools to the U.S. design & manufacturing base for full-scale application***

Over 3:1 Industry Cost Share

Funding Opportunity Announcement:
Advanced Composites Manufacturing Innovation Institute

Objective

Develop and demonstrate innovative technologies that will, within 10 years, make advanced fiber-reinforced polymer composites at…

50% Lower Cost

Using 75% Less Energy

And reuse or recycle >95% of the material

Application	Estimated Current CFC Cost	Institute CFC Cost Reduction Target (2018)[88]	CFC Ultimate Cost Target (2024)	CFC Tensile Strength	CFC Stiffness	Production Volume Cycle Time
Vehicles (Body Structures)	$26-33/kg	>35%	<$11/kg by 2025 ~60%	0.85GPa (123ksi)	96GPa (14Msi)	100,000 units/yr <3min cycle time (carbon) <5min cycle time (glass)
Wind (Blades)	$26/kg	>25%	$17/kg ~35%	1.903 GPA (276ksi)	134GPa (19.4Msi)	10,000 units/yr (at >60m length blades)
Compressed Gas Storage (700 bar – Type IV)	$20-25/kg	>30%	$10-15/kg ~50%	2.55 GPa (370ksi)	135 GPa (20Msi)	500,000 units/yr (carbon fiber)

https://eere-exchange.energy.gov/Default.aspx#FoaId72dc3a64-6537-4d57-b524-e2778bc65e03

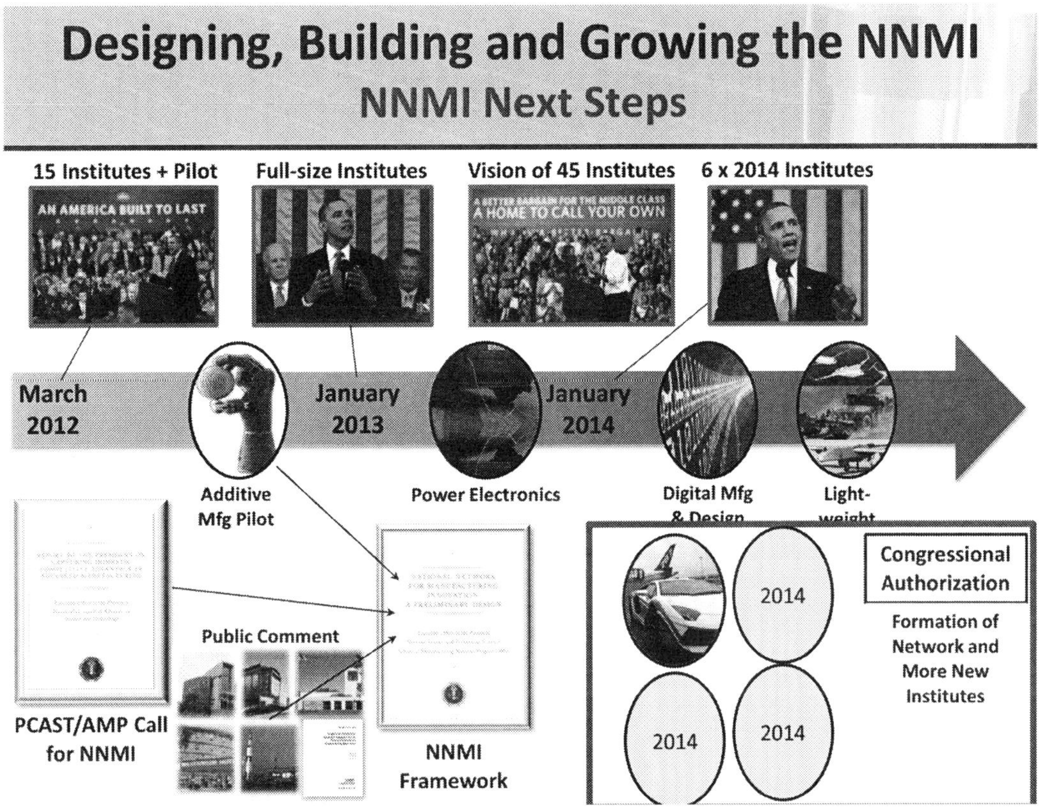
Designing, Building and Growing the NNMI
NNMI Next Steps
15 Institutes + Pilot
Full-size Institutes
Vision of 45 Institutes
6 x 2014 Institutes
AN AMERICA BUILT TO LAST
A HOME TO CALL YOUR OWN
March 2012
January 2013
January 2014
Additive Mfg Pilot
Power Electronics
Digital Mfg & Design
Light-weight
PCAST/AMP Call for NNMI
Public Comment
NNMI Framework
2014
2014
2014
Congressional Authorization
Formation of Network and More New Institutes

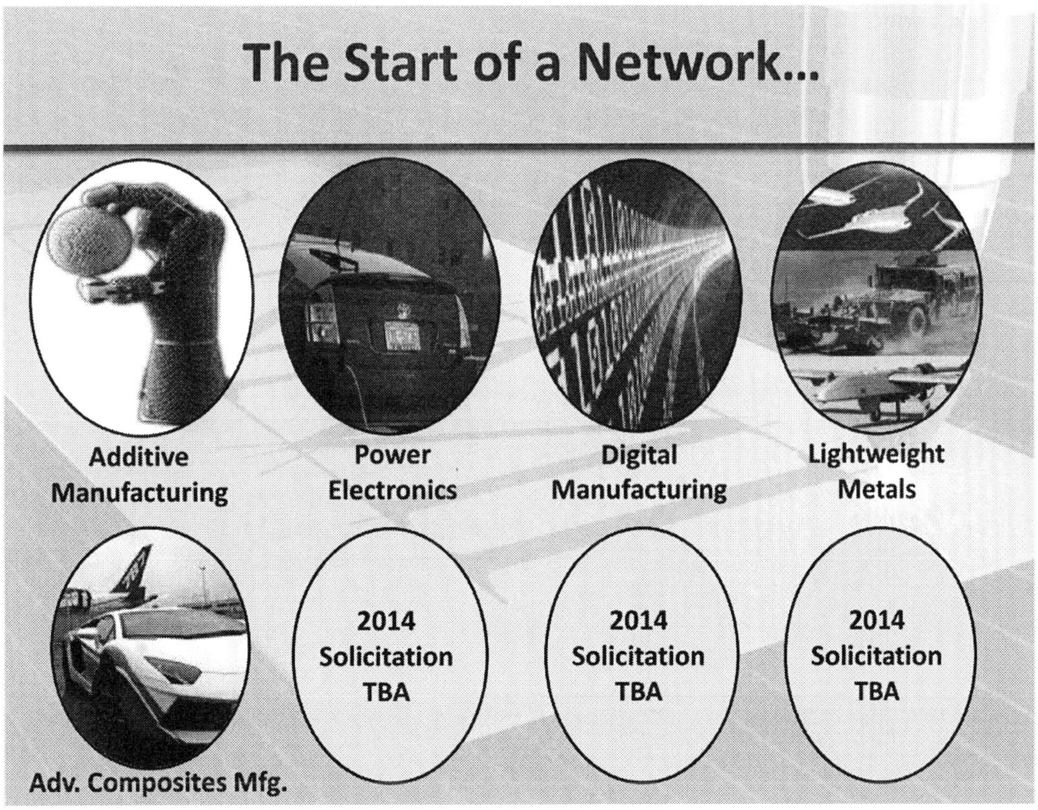
The Start of a Network…
Additive Manufacturing
Power Electronics
Digital Manufacturing
Lightweight Metals
Adv. Composites Mfg.
2014 Solicitation TBA
2014 Solicitation TBA
2014 Solicitation TBA

NNMI Bipartisan/Bicameral Legislation

Revitalize American Manufacturing & Innovation Act of 2013

Lead Sponsors

Sen. Sherrod Brown D Ohio	Sen. Roy Blunt R Missouri	Rep. Tom Reed R NY-23	Rep. Joe Kennedy D MA-4

Senate Commerce Committee Hearing Nov. 13, 2013

House Science Committee, Subcommittee on Research & Technology Hearing Dec. 12 2013

Joint press release: "Their landmark bill would establish a Network for Manufacturing Innovation to position the United States, once again, as the global leader in advanced manufacturing and ensure that the U.S. can out-innovate the rest of the world while creating thousands of high-paying, high-tech manufacturing jobs."

Potential Future NNMI Topics

Public input identified 135 unique topics

With congressional authorization will come open solicitation.

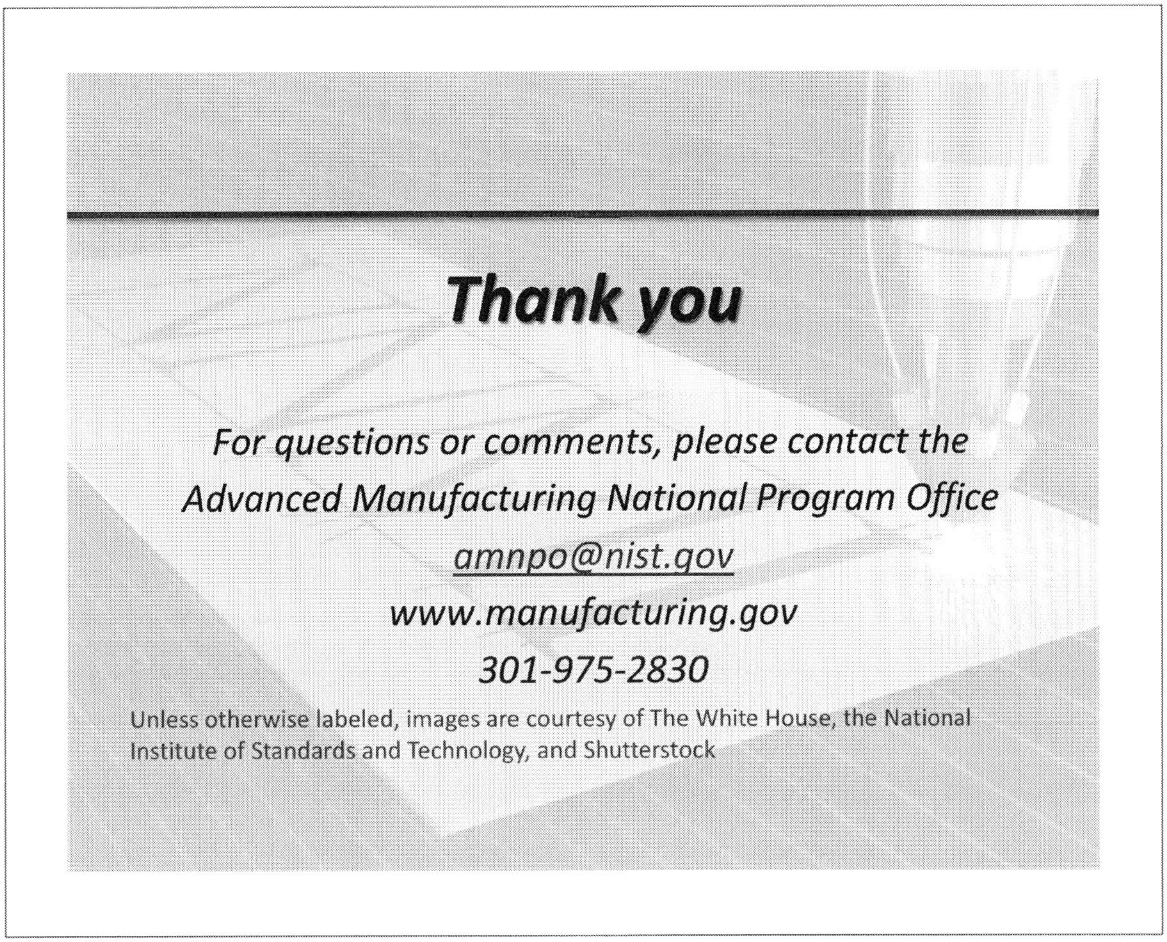
Thank you
For questions or comments, please contact the
Advanced Manufacturing National Program Office
amnpo@nist.gov
www.manufacturing.gov
301-975-2830
Unless otherwise labeled, images are courtesy of The White House, the National Institute of Standards and Technology, and Shutterstock

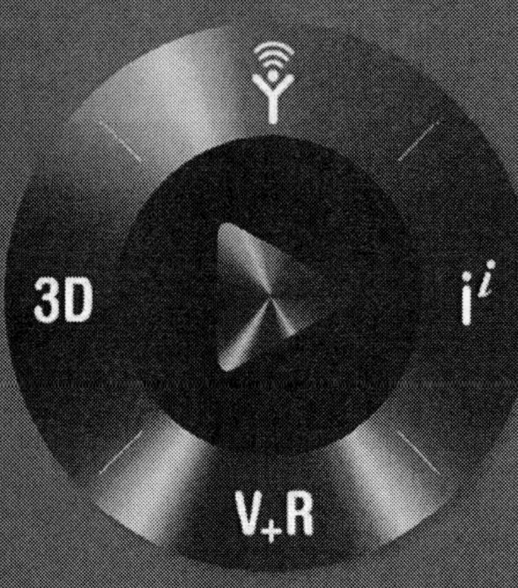

3DEXPERIENCE

Intelligent Manufacturing of Composite Structure

AeroDef Manufacturing 2014
February 25-27

DASSAULT SYSTEMES | IF WE ask the right questions we can change the world

The Magic Kingdom…….

- Challenge for today---

1. Finish this presentation in 30 minutes
2. Introduce a new innovation that is "magic"

2

DASSAULT SYSTEMES | IF WE ask the right questions we can change the world

DELMIA OI AT BOEING AEROSTRUCTURES

BOEING

Composite Landing Gear Door :

▷ 40% reduction of **cost** of repair and rework

▷ **cycle time** reduced from 89 days to 59 days

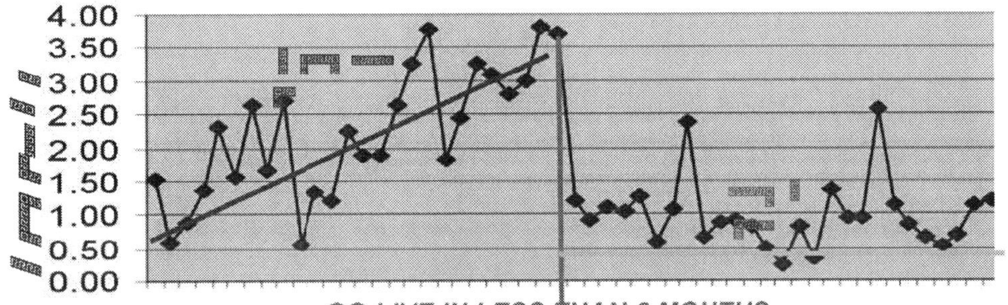

GO LIVE IN LESS THAN 6 MONTHS

DASSAULT SYSTEMES | IF WE ask the right questions we can change the world

3

DELMIA OI AT AIRBUS

Problem

Porosity defects on skins of A320 Vertical Tail Plane (Stade, Germany)

Benefits

Used DELMIA Operations Intelligence to reduce the number of defects by 75%

"*DELMIA Operations Intelligence* **is a really fantastic tool, from the moment that you come up against a problem in a complex manufacturing process that involves large quantities of parameters."**

Bill Black, Chief Quality Officer of the EADS Group

DASSAULT SYSTEMES | IF WE ask the right questions we can change the world

4

From: PIMBLETT Dave
Sent: Thursday, September 12, 2013 10:23 AM
To: BOHN Ted; BUCKLEY Chuck
Subject: OI & XXXX

I just spent 2 weeks in XXXXX's Italian plant, the last week with XXXX (the tier one supplier).
We had a great workshop with XXXX, The results from a week of analysis with PRD was spectacular and they are convinced that they need it to solve the huge problems they have with composite production.
XXXX had 4 chemists/scientists who work in Bills organization at XXXX for 4 weeks; they came up with some suggestions but nothing concrete.
Our 4 hour investigation into the problem of setting up the mandrel for production gave better results than they and a team of statisticians came up with.

When we compared results the XXXX team agreed with everything we found. The Client was very happy as they want to solve this without XXXX placing enormous demands on them with no clear results (like paying for 4 scientists for 4 weeks)

From model to simulated behavior to real behavior

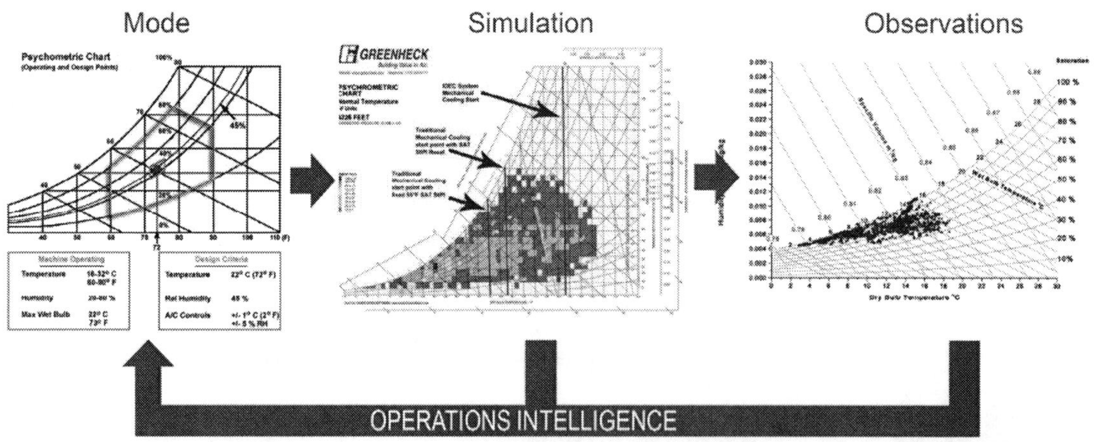

WHEN TO APPLY OPERATIONS INTELLIGENCE:

WHERE COST OF FAILURE IN PRODUCTION (REWORK, SCRAP) OR OPERATIONS IS HIGH (FIELD FAILURE)

1. **High Engineering Content**
2. **Small Batch runs or unique end items**
3. **Highly variable process**
4. **Complex Configuration Management**

DASSAULT SYSTEMES | IF WE ask the right questions we can change the world

7

From simple metal processes to highly sensitive multivariate processes

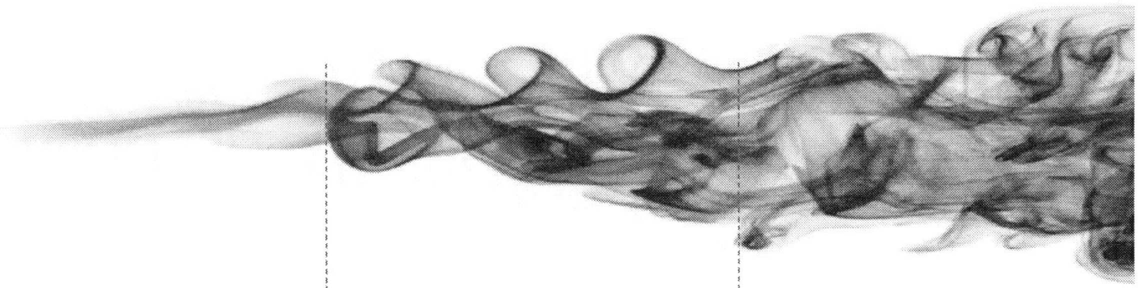

Linear	Complex	Highly Sensitive
Spreadsheet	Specialized Learning Algorithms	Real-time Dynamic Process Control

DASSAULT SYSTEMES | IF WE ask the right questions we can change the world

8

Unpredictability increases with complexity

- Potentially dozens or hundreds of process steps, each with multiple parameters
- High influence of the raw material characteristics (physical properties, formulation, composition, expiration, storage times and conditions)
- High influence of the operating conditions (product or process specifications) and the environment (humidity, temperature, etc.)
- Chemical reactions are highly non-linear, non-reversible phenomea and very difficult to predict
- Because of the highly multivariate and non-linear nature of the processes, Theoretical Models and Statistical Process control are not effective to predict and eliminate defects.

3 WAYS TO IMPROVE PRODUCTION METRICS

PRODUCT/RECIPE ENGINEERING ↘

PROCESS ENGINEERING ↙

↑ Shop Floor Productivity
↑ Equipment Utilization
↑ Product Quality
↓ Time to market
↑ Predictability
↑ Yield

↑

OPERATIONS INTELLIGENCE QUALITY ENGINEERING

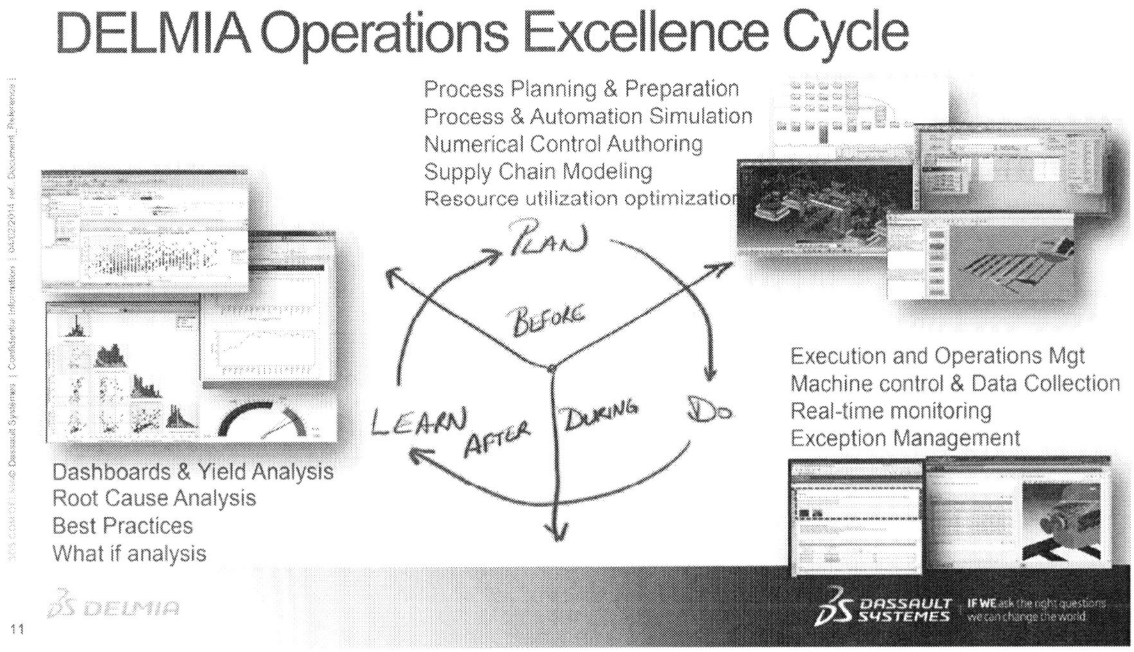
DELMIA Operations Excellence Cycle
Process Planning & Preparation
Process & Automation Simulation
Numerical Control Authoring
Supply Chain Modeling
Resource utilization optimization
PLAN
BEFORE
LEARN
AFTER
DURING
DO
Execution and Operations Mgt
Machine control & Data Collection
Real-time monitoring
Exception Management
Dashboards & Yield Analysis
Root Cause Analysis
Best Practices
What if analysis
DELMIA
DASSAULT SYSTEMES
11

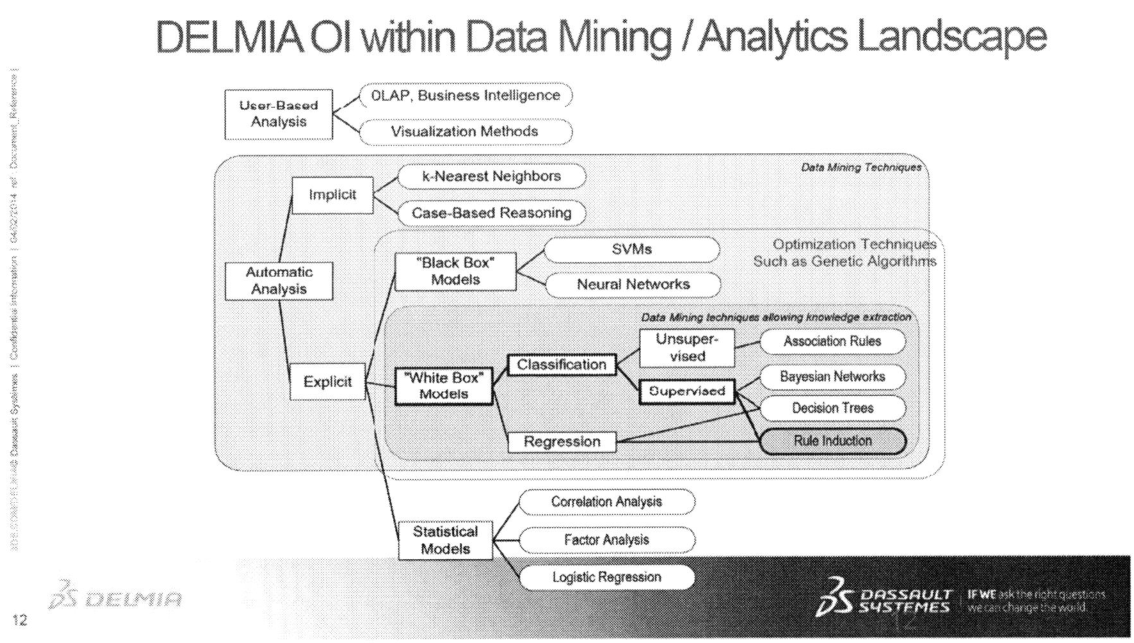
DELMIA OI within Data Mining / Analytics Landscape
User-Based Analysis
OLAP, Business Intelligence
Visualization Methods
Data Mining Techniques
Implicit
k-Nearest Neighbors
Case-Based Reasoning
Automatic Analysis
Optimization Techniques Such as Genetic Algorithms
"Black Box" Models
SVMs
Neural Networks
Data Mining techniques allowing knowledge extraction
Explicit
"White Box" Models
Classification
Unsuper-vised
Supervised
Association Rules
Bayesian Networks
Decision Trees
Rule Induction
Regression
Statistical Models
Correlation Analysis
Factor Analysis
Logistic Regression
DELMIA
DASSAULT SYSTEMES
12

Temporary adjustments to controllable parameters

13

Release constraints when context allows it

14

LEARNING by EXPERIENCE

Resin Rate	Temperature	Vacuum	Autoclave #	*RESULT*
			❶	OK
		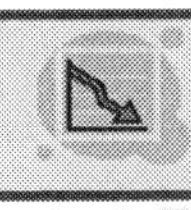	❷	NOT OK
	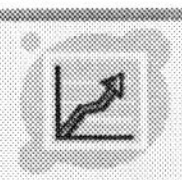		❶	OK
			❶	NOT OK

FINDING PATTERNS

NO STATISTICS, NO EQUATIONS, NO MATHS

JUST LOGIC

DELMIA

DASSAULT SYSTEMES IF WE ask the right questions we can change the world.

15

A complement to Statistics for the most complex situations

OI Learning Engine

ANALYZING THE LINES OF THE TABLE

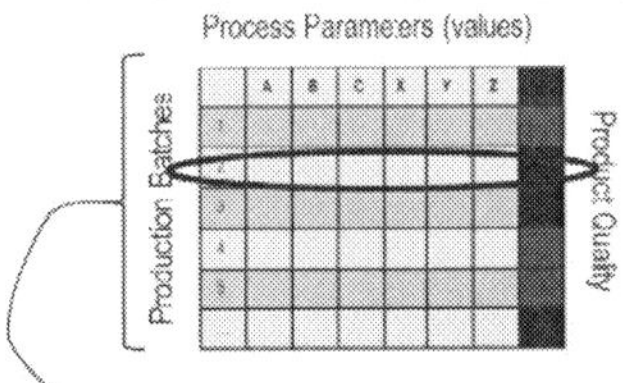

Logical (Non-Statistical) Analysis: Looking for common patterns among batches, independently of number of variables

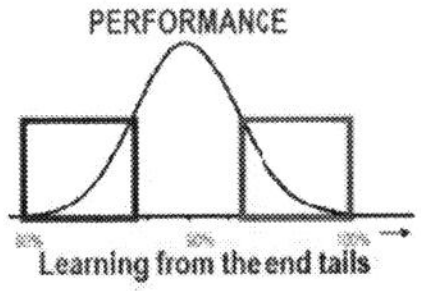

Learning from the end tails

Statistical Tools

ANALYZING THE COLUMNS OF THE TABLE

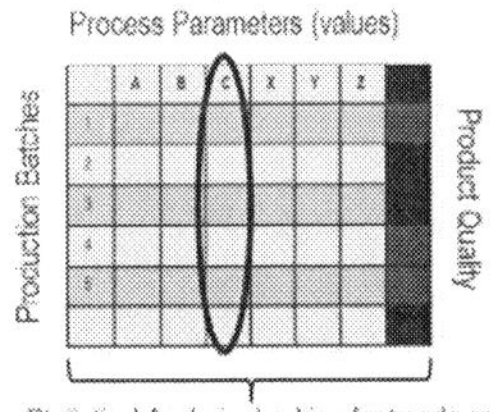

Statistical Analysis: Looking for trends and correlations among parameters, independently of number of samples

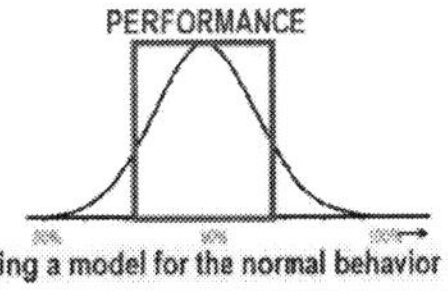

Creating a model for the normal behavior

DELMIA

DASSAULT SYSTEMES IF WE ask the right questions we can change the world.

16

Overview of the Operations Intelligence Algorithms

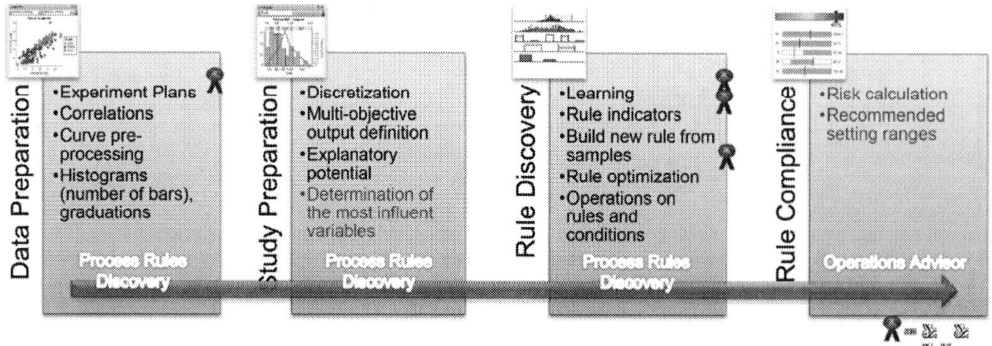

- The above algorithms are either home-made or inspired from state-of-the-art algorithms that appear most likely to meet the objectives of the function
- Example of algorithms used: genetic algorithms, fuzzy logic inspired algorithm, naïve Bayesian algorithm, ANOVA, Partial Least Squares (PLS)…

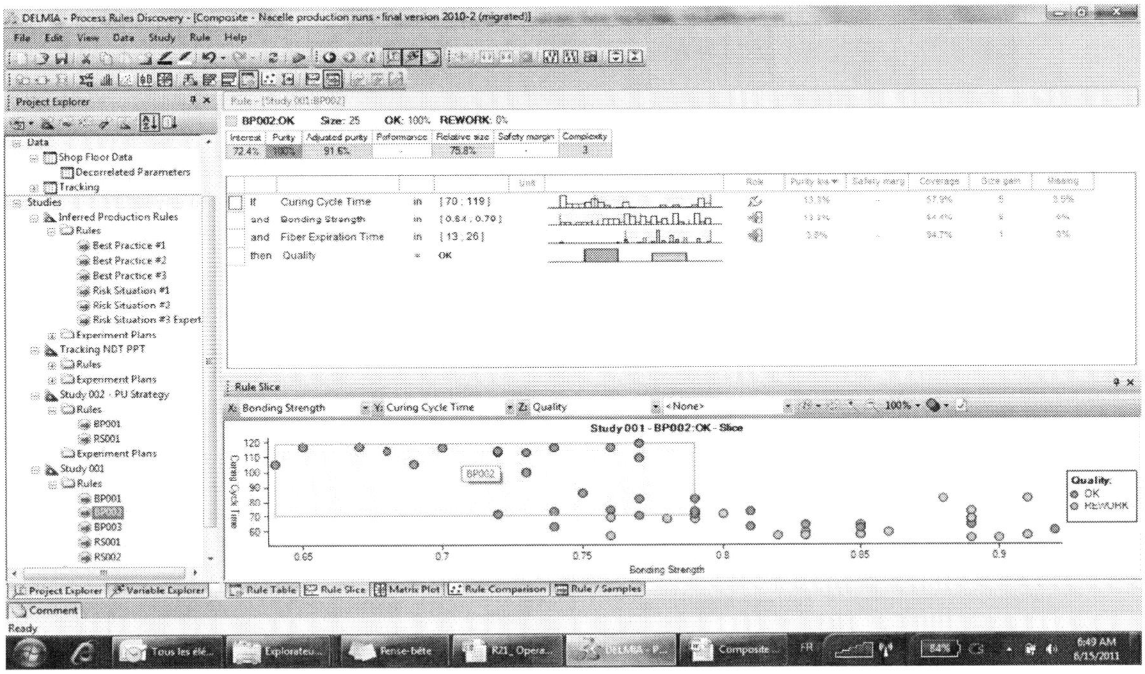

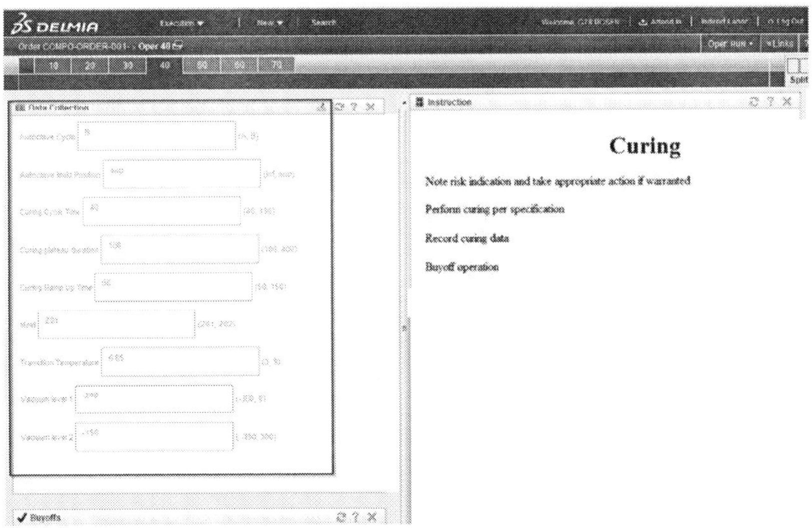

Manufacturing Execution

Data Collection

Features:

- Operational data collection
- Procedurally enforces that data is within control/specification for *individual* collection points based on planning

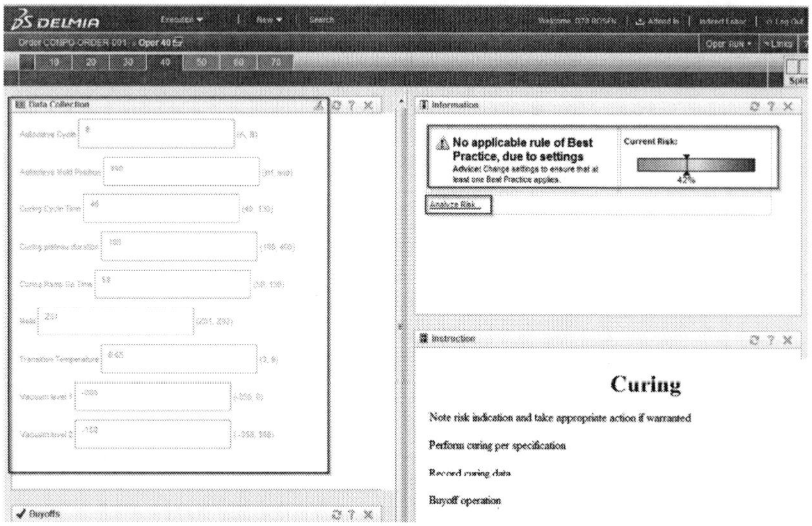

Operations Advisor

Risk Assessment

Features:

- Multi-variable process can still produce unplanned results when combining variables – materials, process, environment, etc.
- Show current risk based on combined data collected at a Check Point or operation
- Analyze Risk allows the operator to find a solution to reduce risk

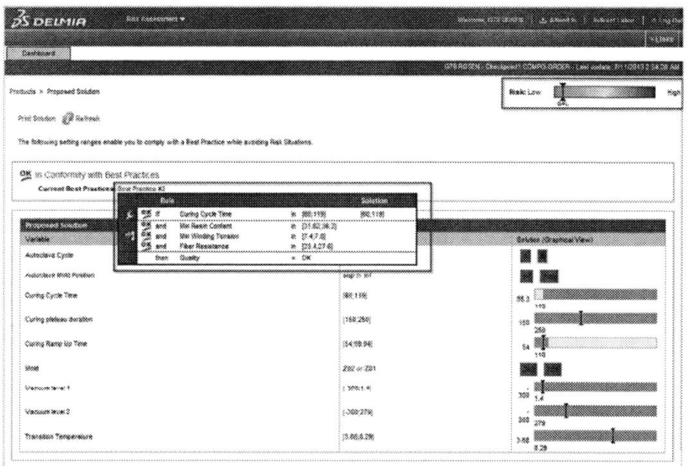

Operations Advisor

Risk Assessment

Features:

- Find Solution based on Best Practices Rules
- Suggest variables that are not in compliance with best practice and how to correct
- Reduce risk of failure, scrap, rework and delivery commitments while increasing yields

Manager Report

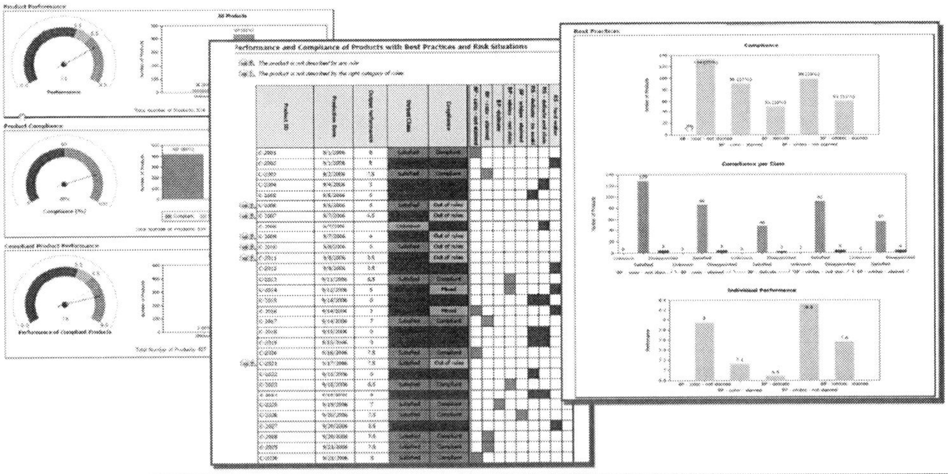

Operations Intelligence *as a closed-loop mechanism*

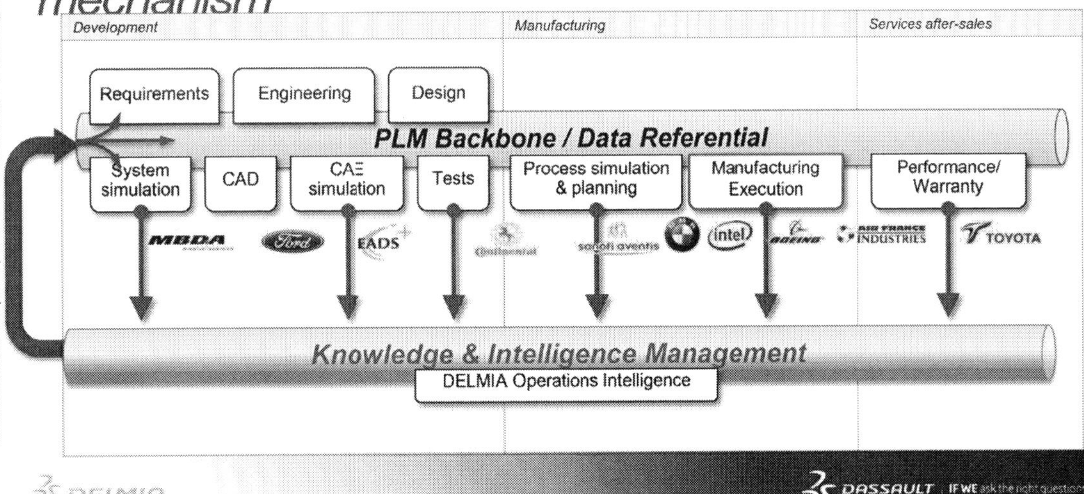

Changing times require new innovations

DELMIA

DASSAULT SYSTEMES | IF WE ask the right questions we can change the world.

25

DELMIA Operations Intelligence

What does this mean to you??

- **Dramatically reduce** the cost of composite scrap and repair
- Accelerate throughput----RATE
- Explanatory Rules managed by Process Owners....not statisticians
- Feedback loop to Design Engineering
- Integrate with Simulation Engines to avoid building ANY bad products
- Drive Supplier Compliance

DELMIA

DASSAULT SYSTEMES | IF WE ask the right questions we can change the world.

26

IF WE
TM
3DS.COM ©Dassault Systèmes | Confidential Information | 04/02/2014 | ref.: Document_Reference |
27
DASSAULT SYSTEMES
IF WE ask the right questions
we can change the world

get it right®

Influence of ICME in the Design of New Generation Composite Components

Ramesh Dwarampudi*, Adi Sholapurwalla, Olivier Morisot

www.esi-group.com

 1

get it right®

Effects of Manufacturing on Product Performance

Fiber Reorientation during Draping

Observations during Test:

- Design and loads are symmetric
- Failure occurs slightly off-center, on left side

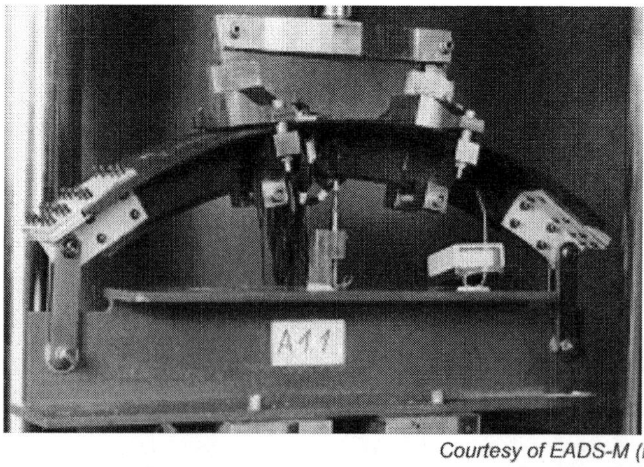

Courtesy of EADS-M (EC FALCOM project)

www.esi-group.com

 2

Effects of Manufacturing on Product Performance

Fiber Reorientation during Draping

get it right®

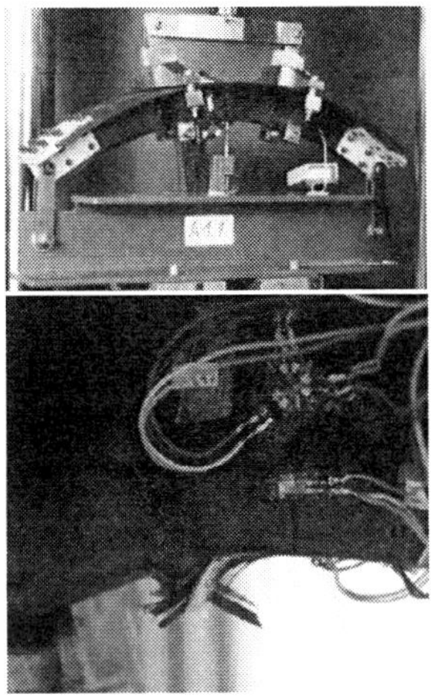

Assuming Ideal Fiber Orientation: Discrepancy between experiment and simulation

Accounting for preforming leads to the correct location of failure

Courtesy of EADS-M (EC FALCOM project)

Effects of Manufacturing on Product Performance

Porosities during Infusion

get it right®

- Potential Voids during Infusion
  - Macro voids Inter-tow
  - Micro voids intra-tow

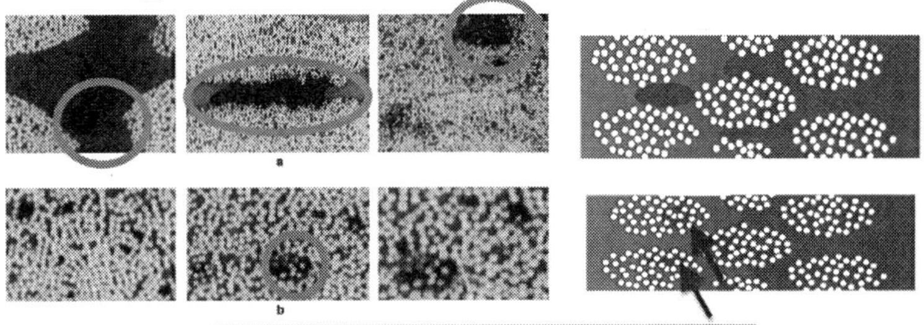

Degradation of Mechanical Performance

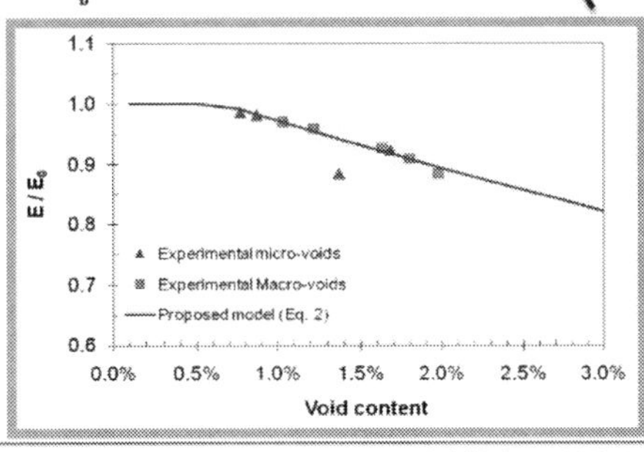

get it right®

The End-to-End Virtual Prototyping

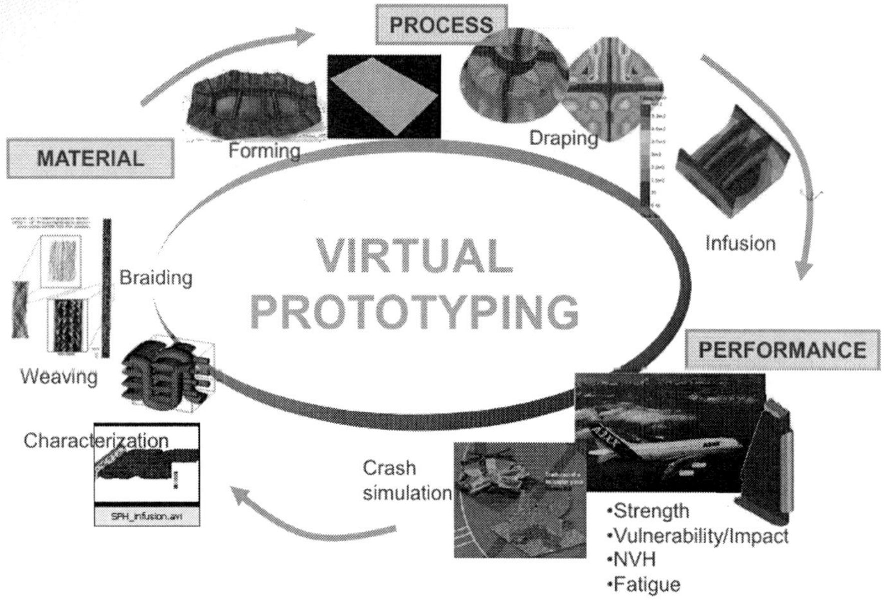

www.esi-group.com

 5

get it right®

Composite Manufacturing Simulation

www.esi-group.com

get it right®

End-to-End ESI Composites Solution

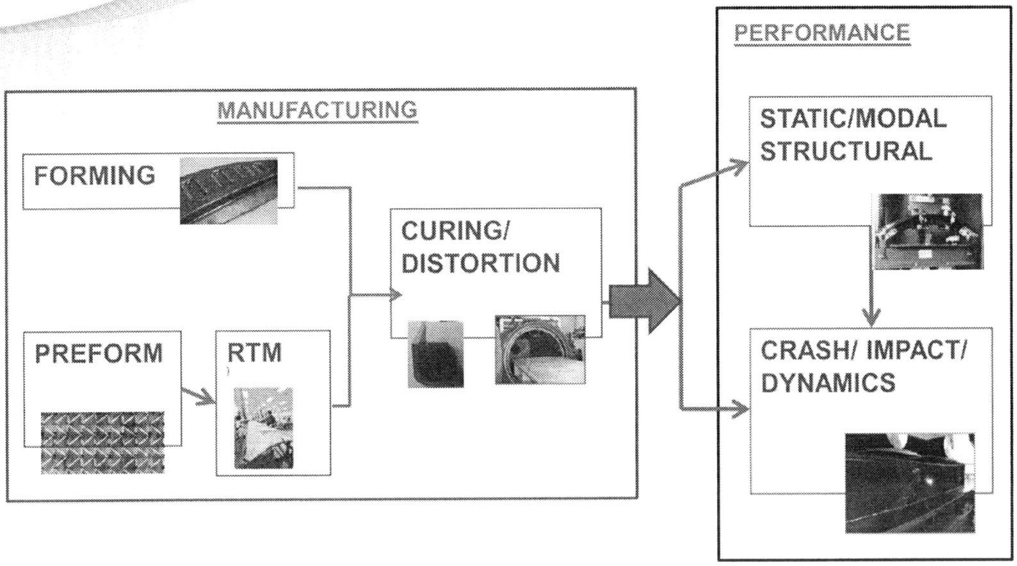

get it right®

The Main Continuous Fibers Manufacturing Options

- **Pre-Pregs**
  - Lay pre-cut individual plies to build the laminate
  - OR forming of complete pre-impregnated laminate
  - Cure at high temperature and pressure in an Autoclave
  - OR curing in forming tools (OOA)
- **LCM** (Liquid Composites Molding)
  - Lay pre-cut dry fabrics to build the laminate
  - OR Forming of dry fabrics
  - Injection/Infusion of dry fabrics
  - Curing in Injection/Infusion tools

Draped preform
Coutesy Renault

Autoclave

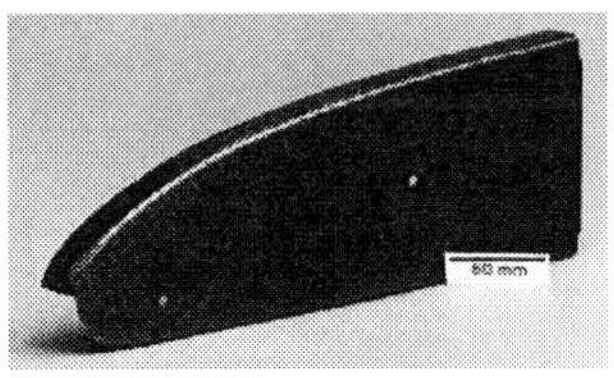
LCM Bracket Courtesy: CRC-ACS 2000+

Pre-Preg Forming Draping Simulation

get it right®

- Evaluate different forming strategies:
  - Stamping, diaphragm (single or double) forming, thermoforming
  - Clamping conditions, process parameters (tool velocity, temperature, pressure…)
  - Different Lay-up strategies

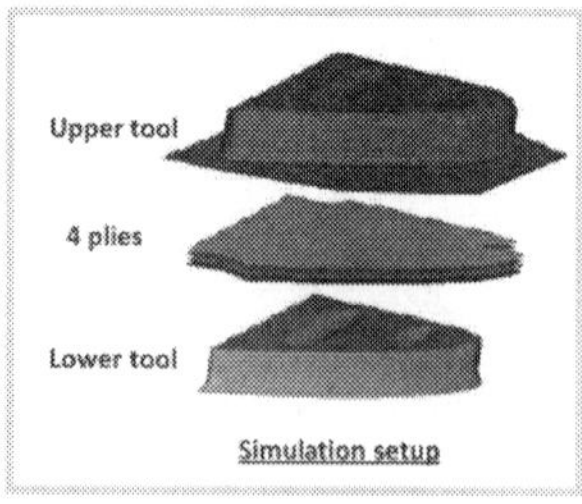

- Manufacturing Simulation can help you:
  - Reduce wrinkling
  - Eliminate bridging
  - Predict thickness of material after forming
  - Deliver optimum flat patterns
  - Determine final fiber orientation

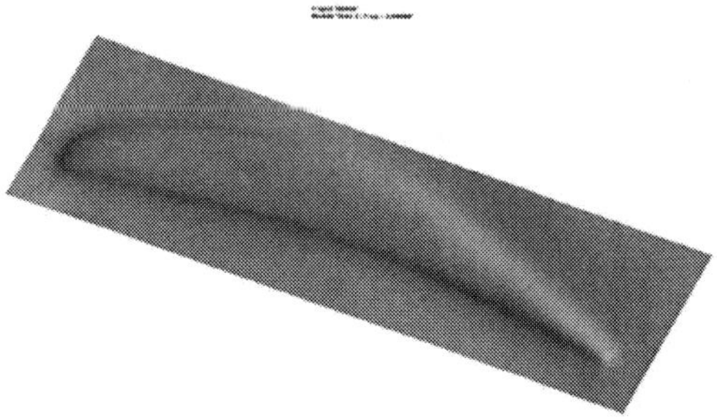

www.esi-group.com

Rubber Pad Forming

Bridging Prediction

get it right®

Fabric / 6 plies / PPS Matrix

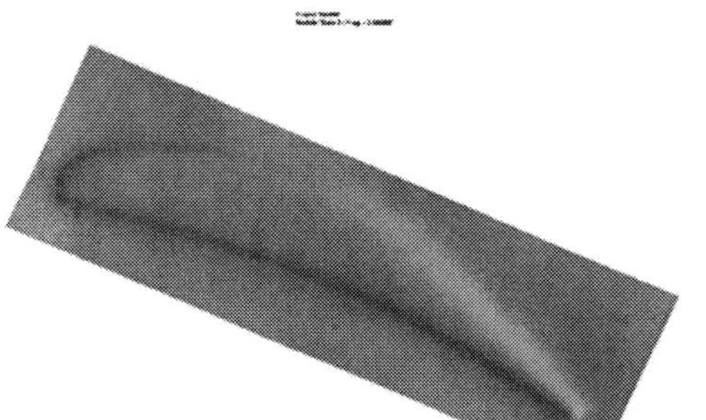

Rubber pad forming simulation

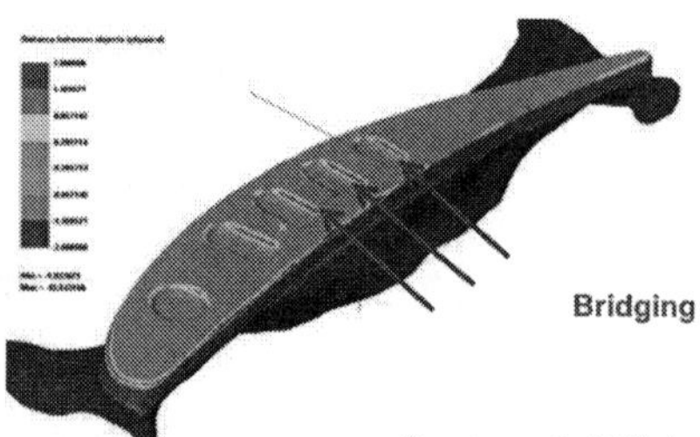

www.esi-group.com

Courtesy: Delft University of Technology

Shearing/Fiber Reorientation

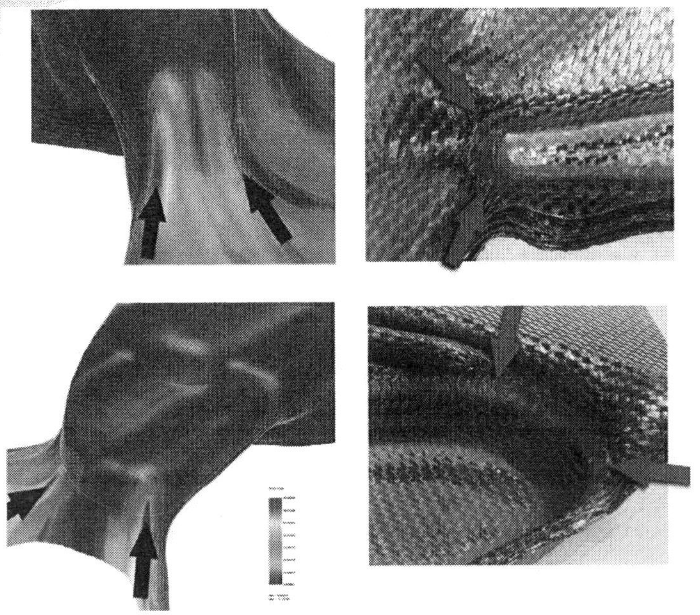

www.esi-group.com

Courtesy: Delft University of Technology

Pre-forming/Draping of Dry-Fabric

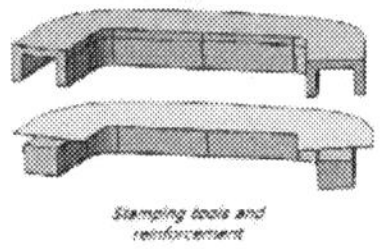

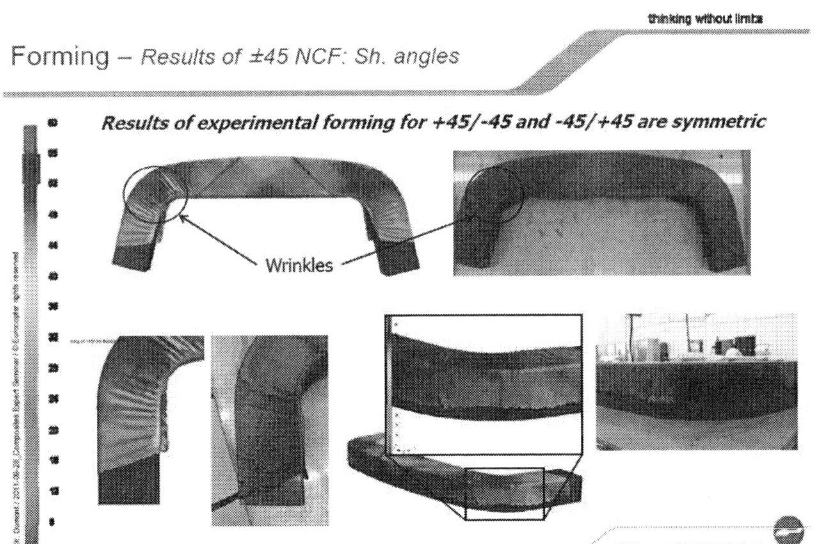

www.esi-group.com

Process Optimization: Forming Speed

Woven fabrics thermoforming: impact of forming speed

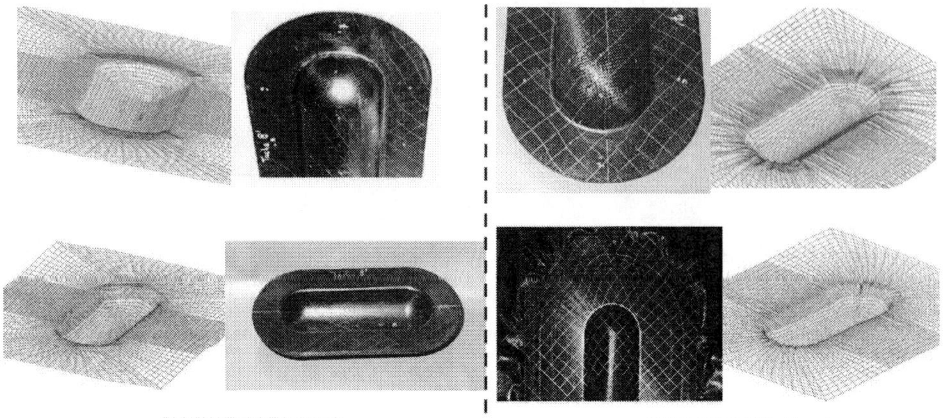

PUNCH VELOCITY= 5.mm/s | PUNCH VELOCITY=40.mm/s

PEI-CETEX; 8 plies

Courtesy: Dassault Aviation

www.esi-group.com

Process Optimization: Tooling Type

Diaphragm forming strategy

Use of 1 diaphragm
=> wrinkling

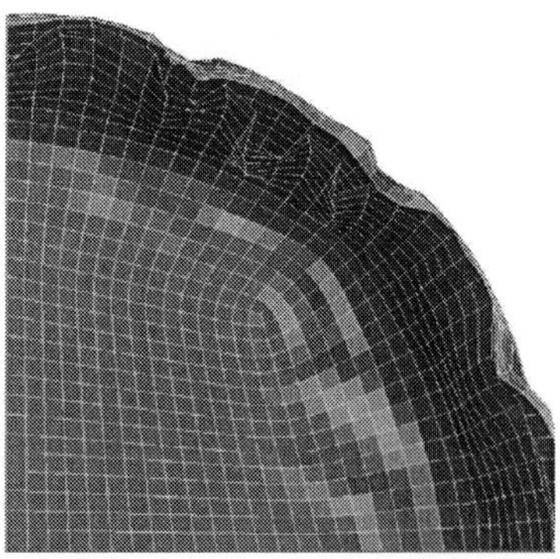

Use of 2 diaphragms
=> No wrinkling

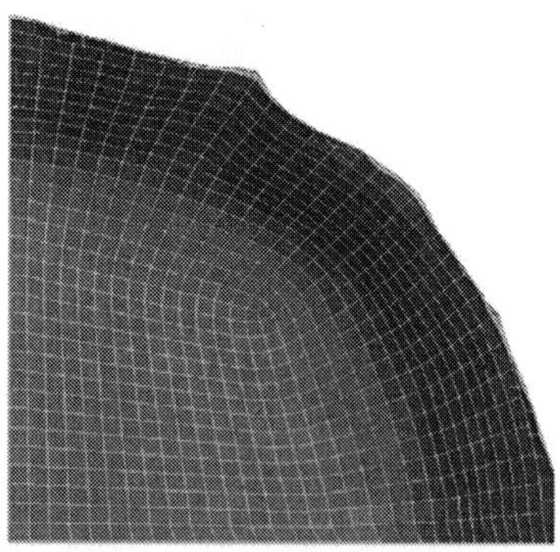

Courtesy: CRC–ACS 2000+

www.esi-group.com

get it right®

Process Optimization: Flat Pattern

J-RIB: 4 plies / thermoforming

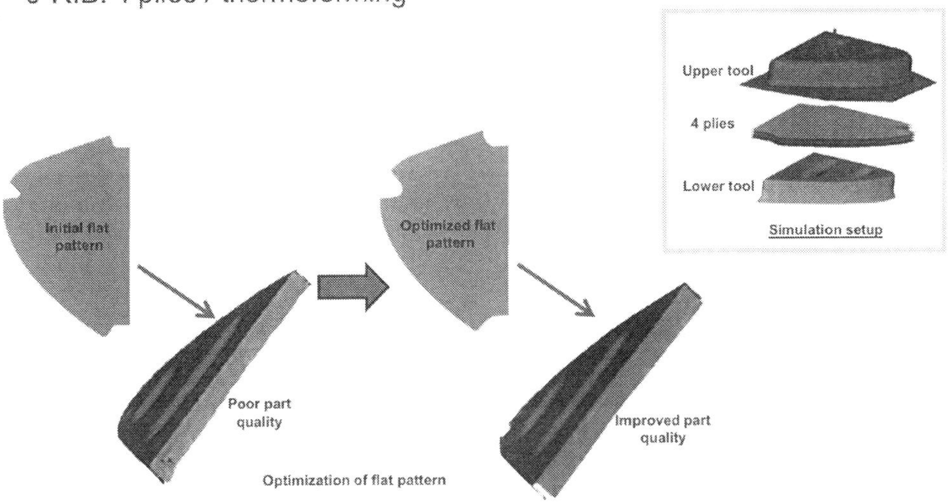

Courtesy: Airbus UK

get it right®

Thermoforming: Link Back to Performance

8 UD Ply CF Wing Rib

Thermoforming Process

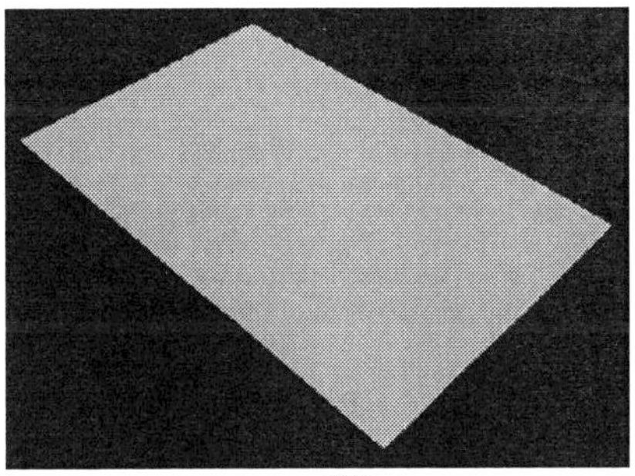

Ply 1

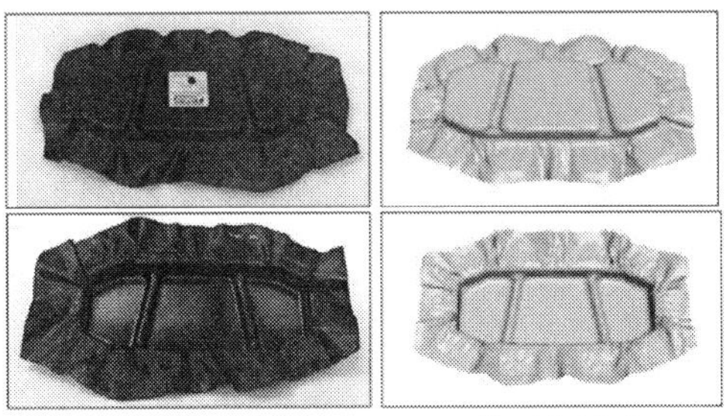

Actual Formed Part ***Simulated Part***

Courtesy: British Aerospace

Thermoforming: Link Back to Performance

8 UD Ply CF Wing Rib

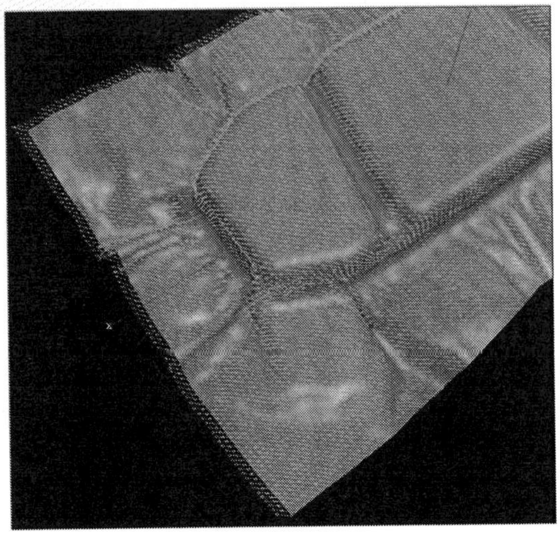

As-Built Fiber Orientation

Computed Fiber Orientation in each ply is critical for actual vs. assumed stiffness

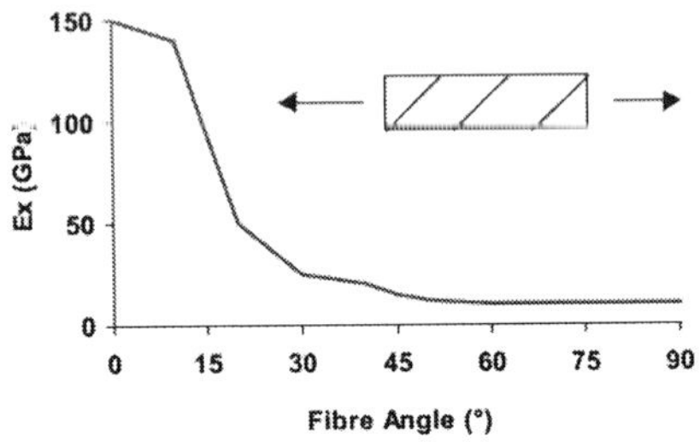

Courtesy: British Aerospace

Liquid Composites Molding Simulation

- Evaluate and optimize
  - Injection strategy (**RTM, VACUUM INFUSION, VARTM**...)
  - Injection pressure and flow rate
  - Injection gates, vents and vacuum ports location
  - Molding temperature
  - Flow media
- Through the prediction of
  - Dry spots
  - Filling and curing times
  - Flow front velocity / Fiber washing
  - Pressure in the mold
- Taking into account
  - Fiber angle variation (permeability variation) of the preform

Rotor Component Infusion Flow Front

Marketing.composites@esi-group.com

Cycle Time Optimization

Injection Strategy

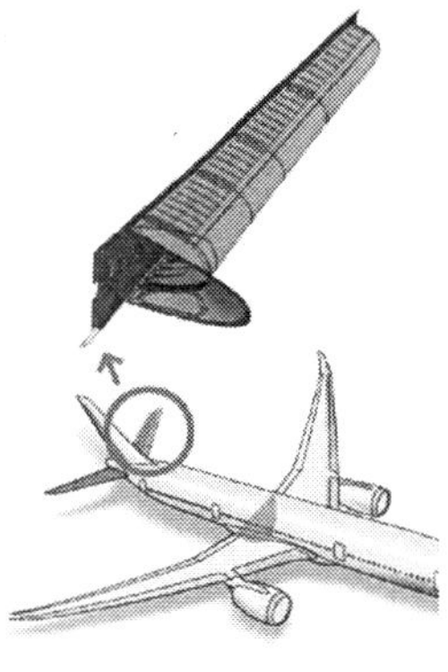

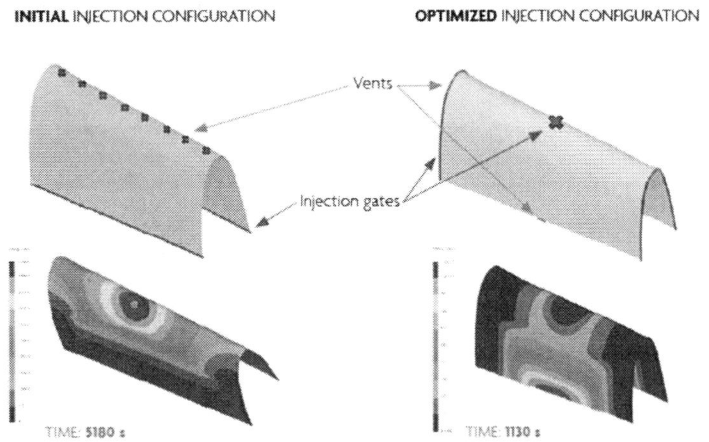

Part Quality Optimization

Reduction/Elimination of Porosities

- Porosity prediction and reduction:
  - Principle: Critical impregnation velocity: $\max\left\|\vec{v}_{front}\right\| \leq \left\|\vec{v}_{crit}\right\|$

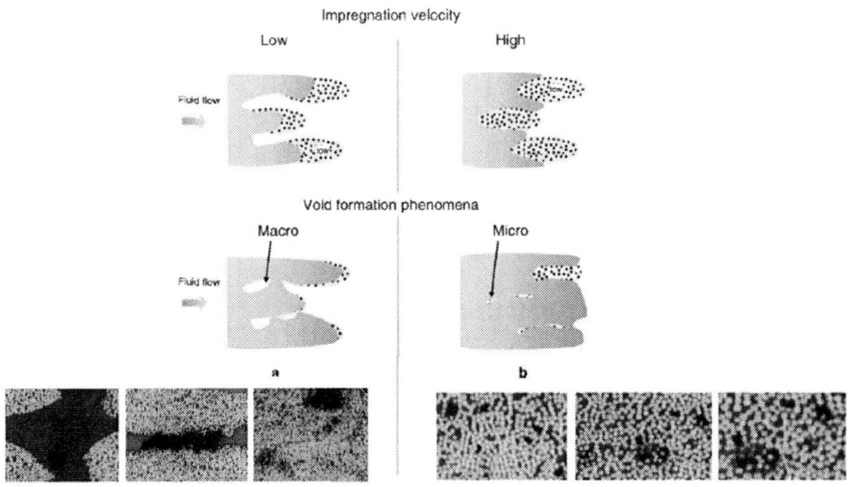

get it right®

Part Quality Optimization
Reduction/Elimination of Porosities

- Porosity prediction and reduction:
  - PAM-RTM input data:

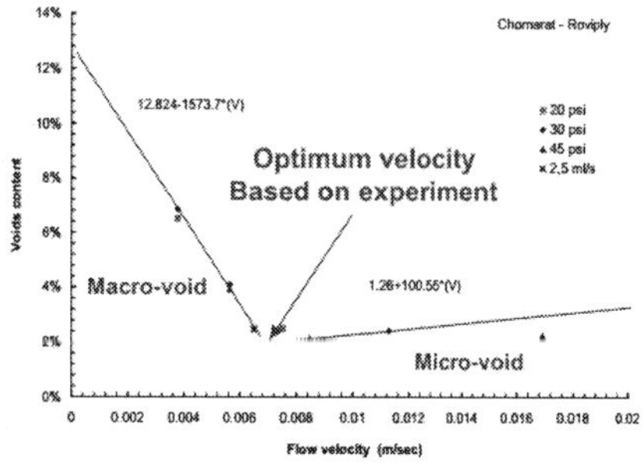

Experimental data

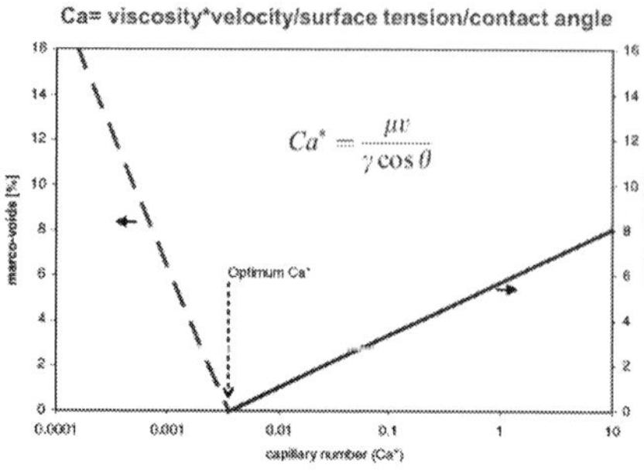

PAM-RTM input curve

get it right®

Part Quality Optimization
Reduction/Elimination of Porosities

- Porosity prediction and reduction:
  - PAM-RTM output: injection flow rate curve

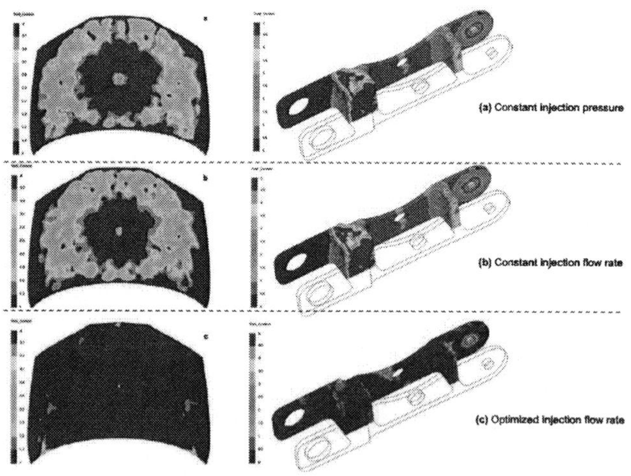

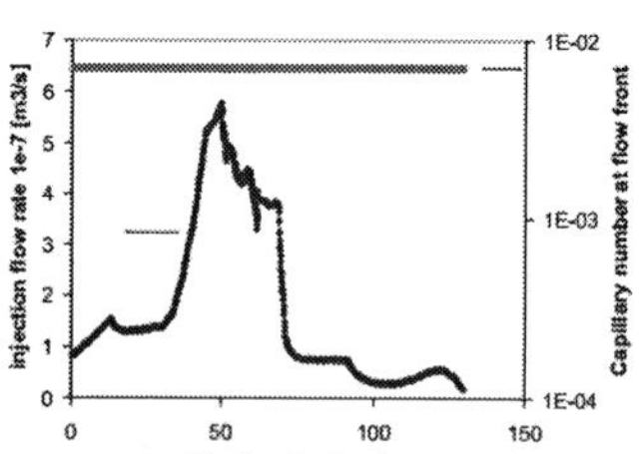

Engineering Research & Flow Technology for Composites

Making tomorrow's engines come true

- CFAN, Texas: International, joint venture Snecma & GE Motors
- **LEAP56**™ *"Leading Edge Aviation Propulsion"*, new engine for a demanding market
- New-generation composite fan blades made by RTM ("Resin Transfer Molding")

JEC Forums (2010)
JEC Composites Show

LEAP56™ engine

Flow simulation

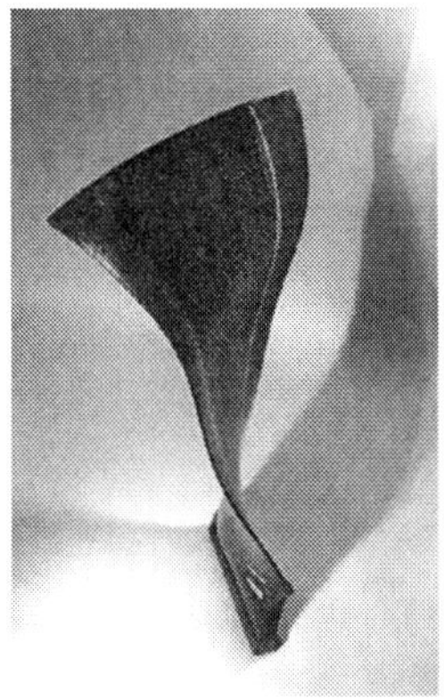

Composite fan blade

ÉCOLE POLYTECHNIQUE MONTRÉAL

bruno.dam

Snecma Groupe SAFRAN

Engineering Research & Flow Technology for Composites

Optimization of the injection process

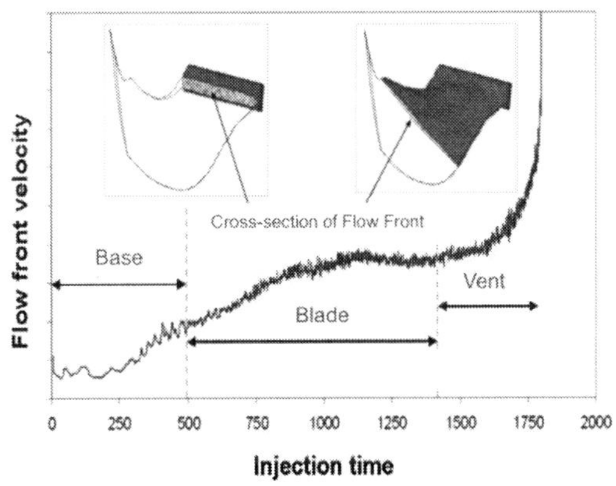

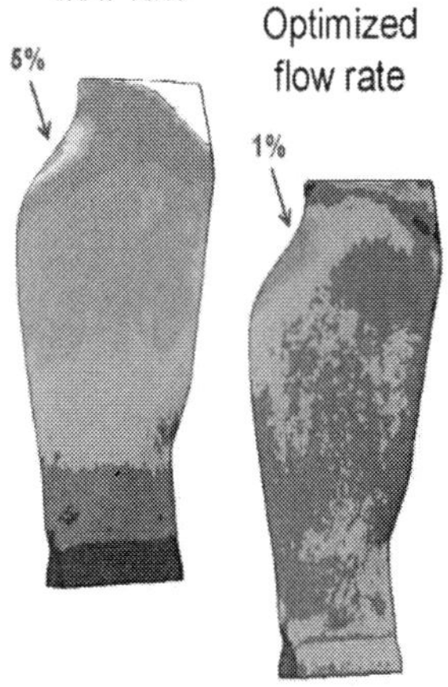

Optimization of injection flow rate to minimize micro/macro-voids formation in resin transfer molded composites

Ruiz, V. Achim, S. Soukane, F. Trochu, J. Bréard

posites Science and Technology 66 (2006) 475-486

Post-Cure Distortion Analysis

- Predict shape distortion of composites parts through a 2 steps computation:
  - Curing step: Thermo-chemical analysis to compute degree of cure and temperature history taking into account the exothermic aspect of the curing process
  - Distortion step: Thermo-chemical stress analysis using temperature and degree of cure history from the curing step
- The stacking definition of the composites part is taken into account
- Thermal and Mechanical interactions with the mold can be considered

www.esi-group.com

 25

Post-Cure Distortion Predictions

Wing Cover

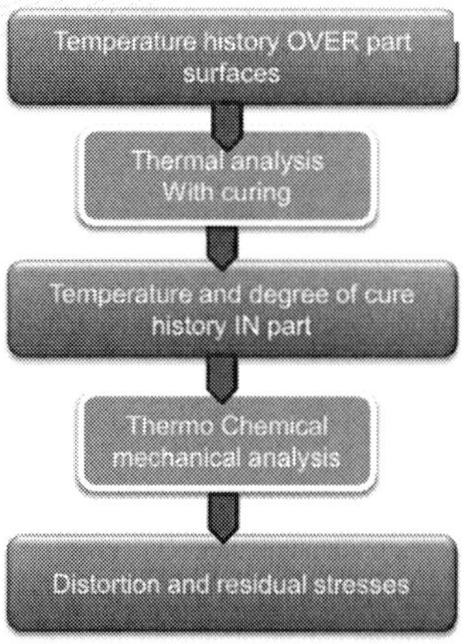

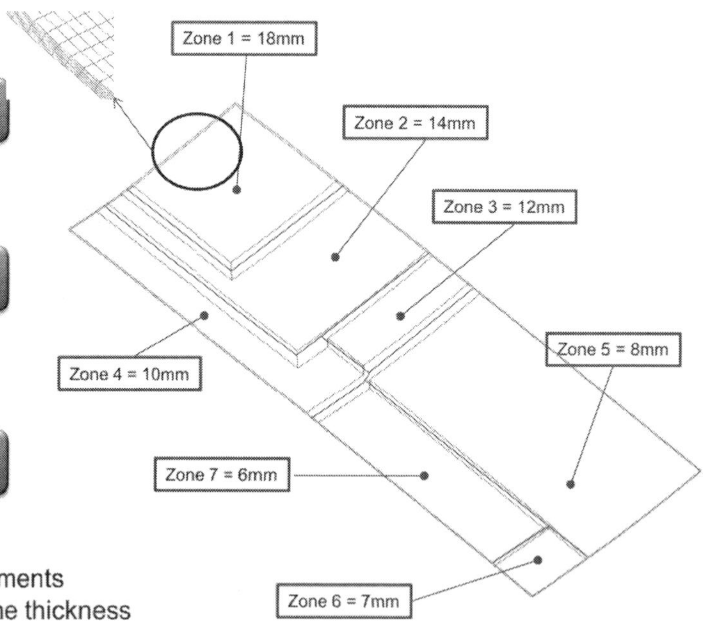

38406 solid elements
6 elements in the thickness

www.esi-group.com

 26

Post-Cure Distortion Predictions
Wing Cover

- Temperature history OVER part surfaces
- Thermal analysis With curing
- Temperature and degree of cure history IN part
- Thermo Chemical mechanical analysis
- Distortion and residual stresses

Thermal cycle applied over part surfaces

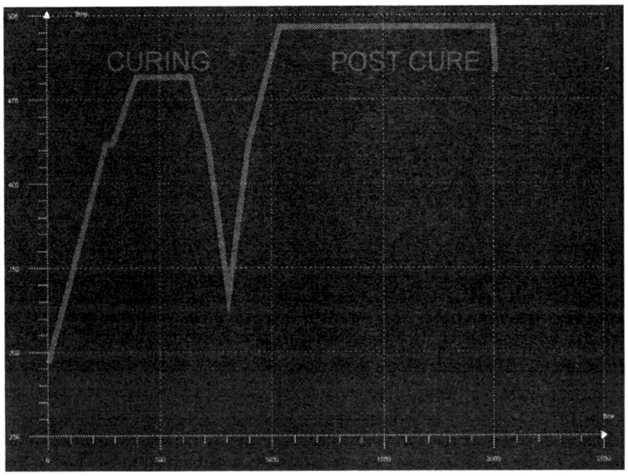

Material, BC and load don't relate to the real part but were defined only for this demo.

www.esi-group.com

Post-Cure Distortion Predictions
Wing Cover

- Temperature history OVER part surfaces
- Thermal analysis With curing
- Temperature and degree of cure history IN part
- Thermo Chemical mechanical analysis
- Distortion and residual stresses

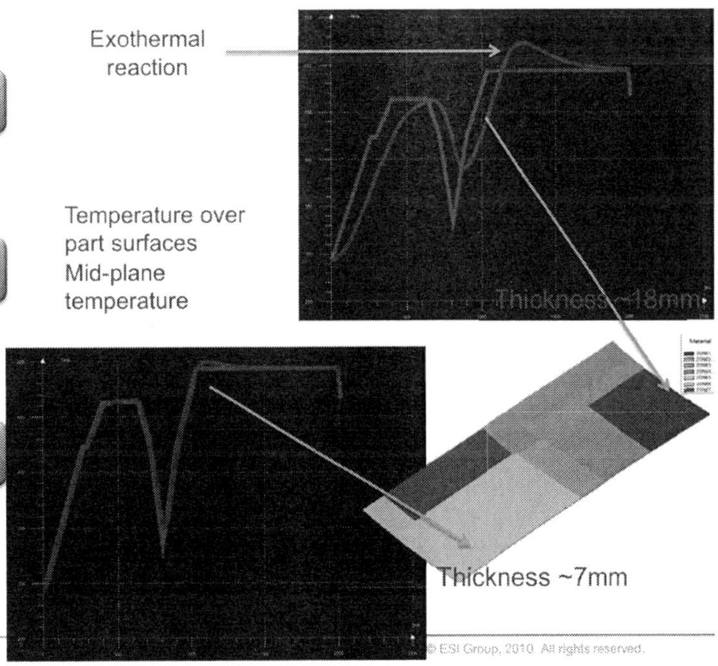

Material, BC and load don't relate to the real part but were defined only for this demo.

www.esi-group.com

get it right®

Post-Cure Distortion Predictions

Wing Cover

- Temperature history OVER part surfaces
- Thermal analysis With curing
- Temperature and degree of cure history IN part
- Thermo Chemical mechanical analysis
- Distortion and residual stresses

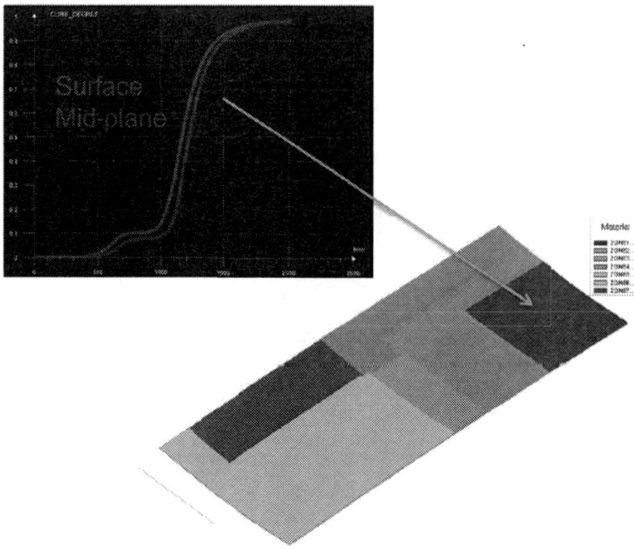

Material, BC and load don't relate to the real part but were defined only for this demo.

get it right®

Post-Cure Distortion Predictions

Wing Cover

- Temperature history OVER part surfaces
- Thermal analysis With curing
- Temperature and degree of cure history IN part
- Thermo Chemical mechanical analysis
- Distortion and residual stresses

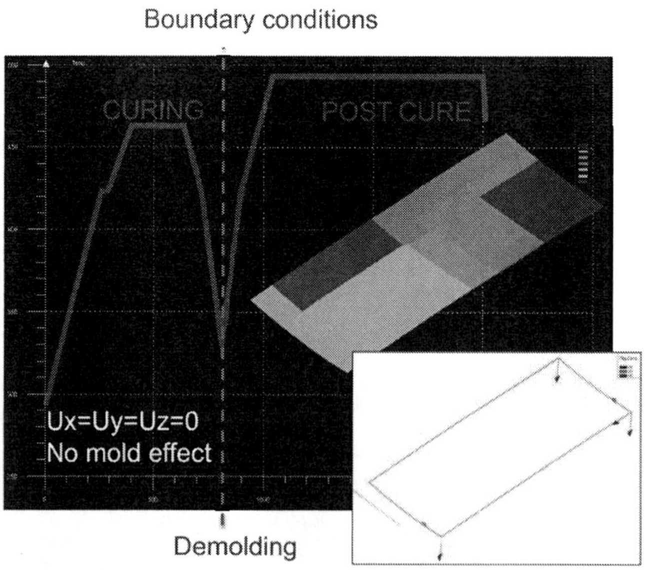

Material, BC and load don't relate to the real part but were defined only for this demo.

www.esi-group.com

Copyright © ESI Group, 2010. All rights reserved.

Post-Cure Distortion Predictions

Wing Cover

get it right®

Geometrical distortion

Temperature history OVER part surfaces

Thermal analysis With curing

Temperature and degree of cure history IN part

Thermo Chemical mechanical analysis

Distortion and residual stresses

Material, BC and load don't relate to the real part but were defined only for this demo.

www.esi-group.com

End-to-End Virtual Prototyping for Composite Materials

get it right®

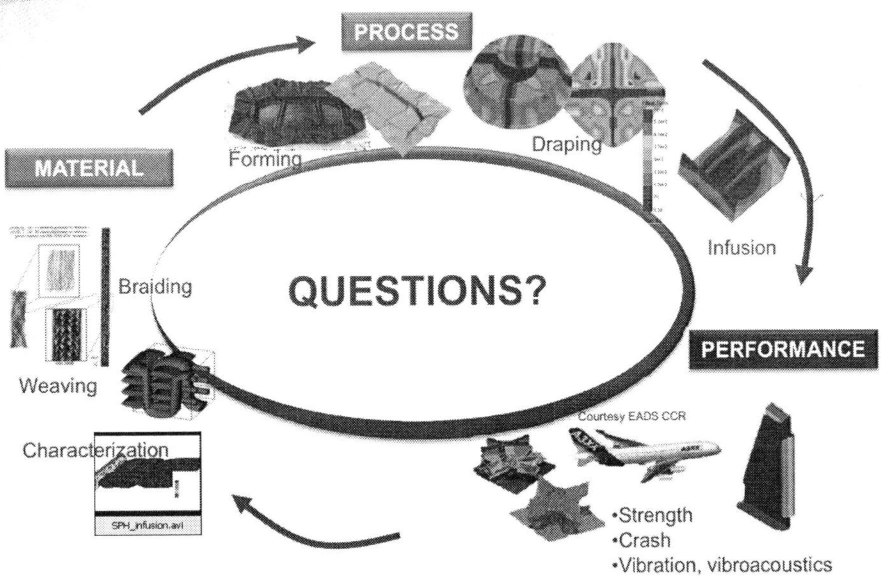

www.esi-group.com

 32

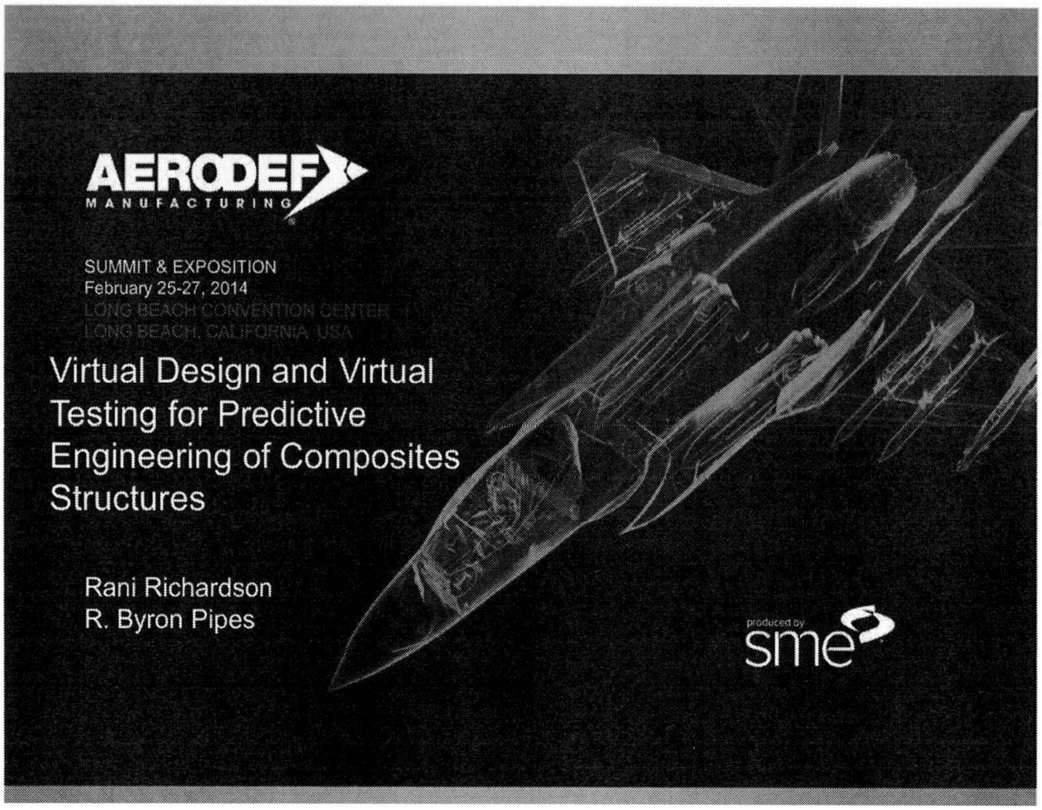
AERODEF
MANUFACTURING
SUMMIT & EXPOSITION
February 25-27, 2014
LONG BEACH CONVENTION CENTER
LONG BEACH, CALIFORNIA USA
Virtual Design and Virtual Testing for Predictive Engineering of Composites Structures
Rani Richardson
R. Byron Pipes
produced by
sme

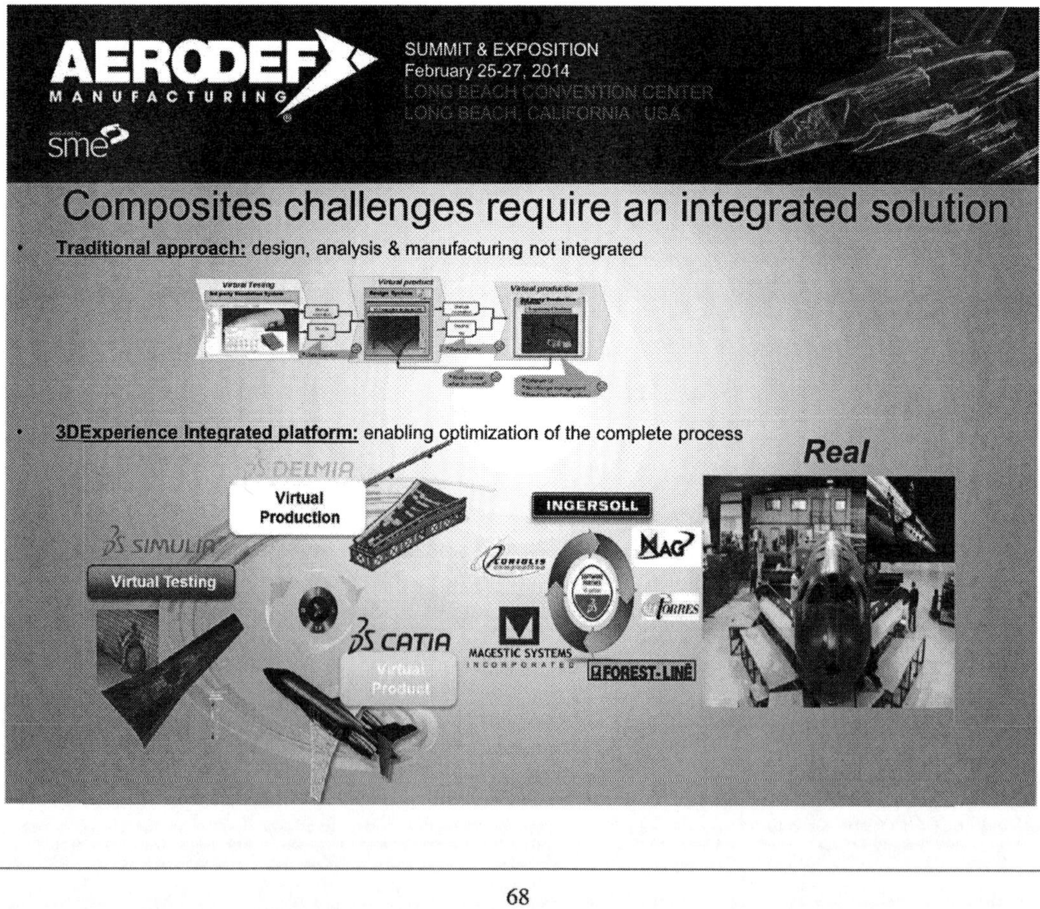
AERODEF
MANUFACTURING
sme
SUMMIT & EXPOSITION
February 25-27, 2014
LONG BEACH CONVENTION CENTER
LONG BEACH, CALIFORNIA USA
Composites challenges require an integrated solution
• Traditional approach: design, analysis & manufacturing not integrated
• 3DExperience Integrated platform: enabling optimization of the complete process
Real
DELMIA
Virtual Production
SIMULIA
Virtual Testing
CATIA
Virtual Product
INGERSOLL
MAG
MAGESTIC SYSTEMS INCORPORATED
FOREST-LINE

AERODEF MANUFACTURING

sme

SUMMIT & EXPOSITION
February 25-27, 2014
LONG BEACH CONVENTION CENTER
LONG BEACH, CALIFORNIA USA

The 3D EXPERIENCE for Composites

- 3D for better communication: See what you mean
- Surfaces are complex, lay-up is complex, 1D or 2D information not enough to represent the reality!
- Need to integrate Design/Simulation/Manufacturing

to 3D plies, virtual manufacturing 3D Finite Element Analysis

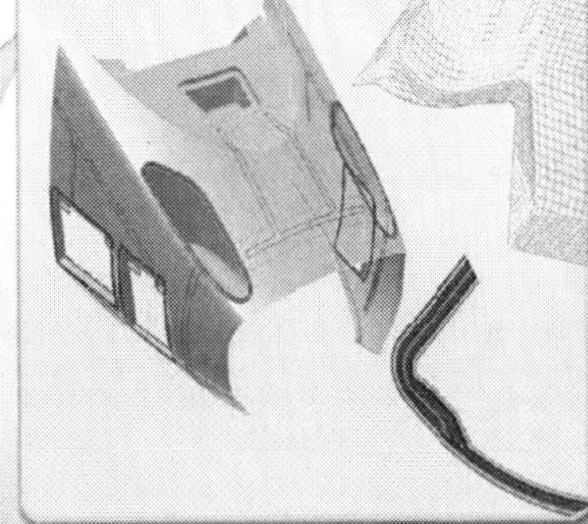

From Excel spreadsheet, drawings Analytical structural analysis…

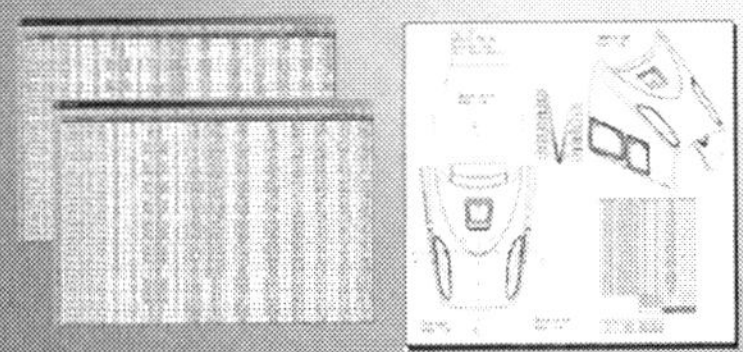

AERODEF MANUFACTURING

sme

SUMMIT & EXPOSITION
February 25-27, 2014
LONG BEACH CONVENTION CENTER
LONG BEACH, CALIFORNIA USA

The 3DEXPERIENCE for Composites Engineering

- Create Detailed 3D Plies from Requirements (Grid, Zones, Solid Slicing, Spreadsheet import)
- Integrating Composites Design rules
- Generates Deliverables : 3D Solid for Digital Mock-up, 2D Drawings, Reports (Ply table Excel)
- Weight & Balance accurate analysis and Cost Estimate

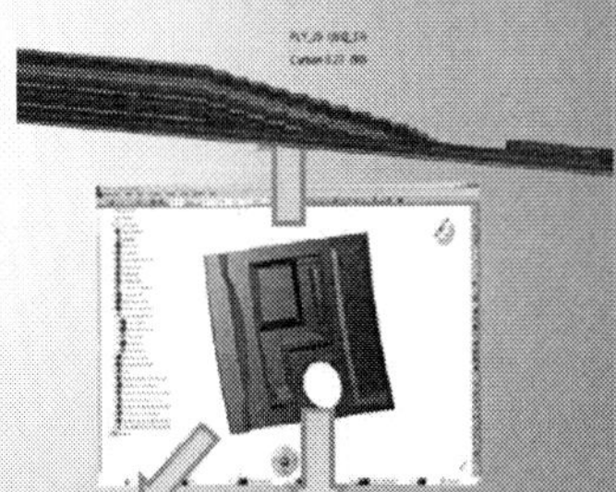

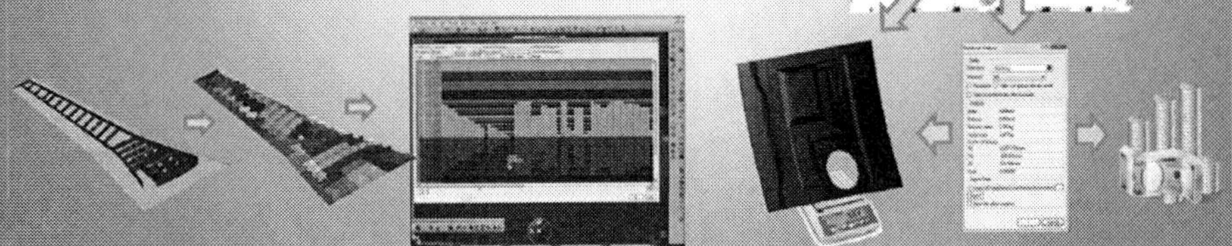

AERODEF MANUFACTURING
SUMMIT & EXPOSITION
February 25-27, 2014
LONG BEACH CONVENTION CENTER
LONG BEACH, CALIFORNIA USA
sme

The 3DEXPERIENCE for Composites Engineering

Prepare and simulate the manufacturing in a 3D virtual environment

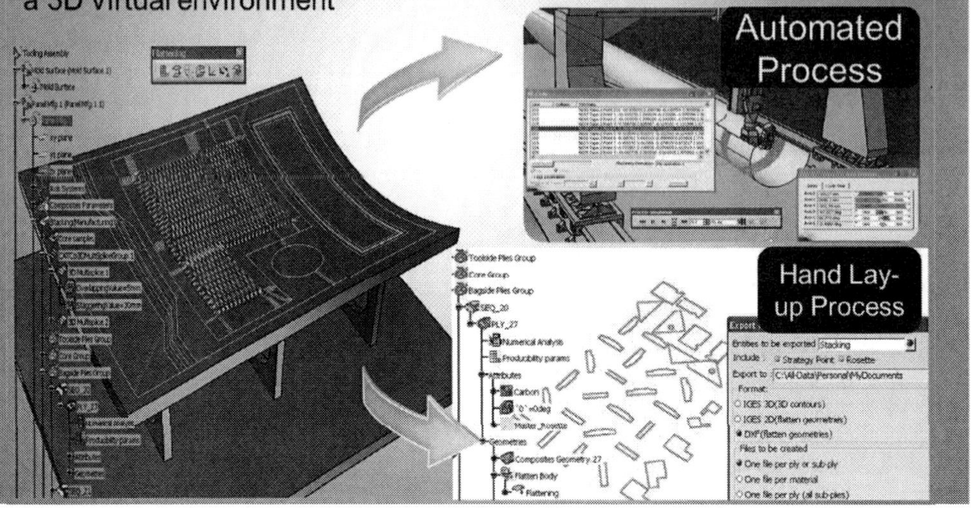

AERODEF MANUFACTURING
SUMMIT & EXPOSITION
February 25-27, 2014
LONG BEACH CONVENTION CENTER
LONG BEACH, CALIFORNIA USA
sme

Polymer Composites

- Polymer composites consist of polymers containing micro and/or nano constituents to produce enhanced multifunctional properties.
- Properties derive from composition and micro/nano structure (constituents, interface and polymer)
- Two primary classes of polymer composites:
  - Light weight structural materials
  - Multifunctional polymer materials with enhanced optical, thermal, electric/dielectric and ionic performance.

AERODEF MANUFACTURING

sme

SUMMIT & EXPOSITION
February 25-27, 2014
LONG BEACH CONVENTION CENTER
LONG BEACH, CALIFORNIA USA

What is the answer?

- Paradigm shift in simulation comprehensiveness: design for manufacturing and performance
  - replace the building block approach with simulation and **test for validation**
  - certify products in a manner that allows for composition and processing to be adjustable **without recertification**
- Combine experiments **curated data, data mining and validated simulation** for *a priori* materials design
- Build the **simulation base in design and manufacturing**
- Understand the **origins of uncertainty** and control them
- Simulation tools can guide understanding of **propagation of uncertainty in design and manufacturing**
- Make interconnected **simulation tools broadly available** for designers, materials suppliers and manufacturers

AERODEF MANUFACTURING

sme

SUMMIT & EXPOSITION
February 25-27, 2014
LONG BEACH CONVENTION CENTER
LONG BEACH, CALIFORNIA USA

Polymer Composites Development

- Reduction in the huge cost of polymer composites **development** is still needed.
- Primary costs are associated with engineering required for product certification (especially for aircraft) via experiment dominated "building-block-approach".
- Material supplier OEM interaction **process too slow and costly**
- Cost reduction via replacement of a portion of experiments by **validated simulations using the MGI approach consisting of physically realistic multi-scale models**, starting at the atomistic level and extending continuously to the macroscopic scale to enable accurate prediction of full scale performance.
- A **simulation suite with individual tools ranked by a technology readiness level system** for measured validation and verification (tool maturity level).

AERODEF MANUFACTURING
sme
SUMMIT & EXPOSITION
February 25-27, 2014
LONG BEACH CONVENTION CENTER
LONG BEACH, CALIFORNIA USA

Polymer Composites Discovery

- Composite performance is limited by current materials, both polymer and reinforcement, therefore material **discovery** is required.
- Taking the MGI approach in the composites will produce the largest impact if efforts are focused on discovery of:
  - new polymer and constituent combinations to produce enhanced performance
  - conventional composites with multifunctional performance ((electrical conductivity)
  - tailorable polymer chemistries to enable enhanced processing
  - materials with improved life-cycle performance
  - recyclable and sustainable systems and methods
  - methods for control of nanoparticle-network meso-structure.

AERODEF MANUFACTURING
sme
SUMMIT & EXPOSITION
February 25-27, 2014
LONG BEACH CONVENTION CENTER
LONG BEACH, CALIFORNIA USA

After 50 years of progress in composites research

- Commercial aircraft are a reality
- Defense aerospace composites are pervasive
- The world-wide failure analysis proved that prediction of strength has been elusive
- Yet we design successfully
- But, we do so with significantly conservative approaches based on experimental tests
- What competency has changed the most In the last 50 years?

AERODEF MANUFACTURING

SUMMIT & EXPOSITION
February 25-27, 2014
LONG BEACH CONVENTION CENTER
LONG BEACH, CALIFORNIA USA

sme

The next decade according to Poursartip

- After more than 40 years of promise, the next decade will see an explosion in the use of composite materials as the major aerospace players, namely Boeing, Airbus and general aviation, have finally fully committed to this technology:
  - **The point of no-return has been crossed for aerospace**
- This commitment means that composites design and manufacturing technology will change dramatically, first in aerospace and then in all sectors as the technology permeates throughout the industry
  - **Automotive and alternative energy markets will follow**
- In ten years, no manufacturing company or material supplier will be untouched:
  - if in composites, they must stay competitive
  - if not in composites, they must ask whether another company can make their product in composites

AERODEF MANUFACTURING

SUMMIT & EXPOSITION
February 25-27, 2014
LONG BEACH CONVENTION CENTER
LONG BEACH, CALIFORNIA USA

sme

Why has it taken so long?

- New materials and processes present significant risks
- Uncertainty is at the heart of the matter
  - The number of possible combinations of polymers, constituents, interface designs and microstructures is huge
  - Edisonian approach to optimization
  - Product and material design is simultaneous
  - Ultra-conservative designs have been pervasive
  - Building block certification approach is very expensive
  - Certification is testing based ($millions/material)
  - Manufacturing is empirical, not science-based
  - Large scale integration of subassemblies to monocoque structures (too large to fail!)
  - Repair and joining technologies not robust

AERODEF MANUFACTURING

SUMMIT & EXPOSITION
February 25-27, 2014
LONG BEACH CONVENTION CENTER
LONG BEACH, CALIFORNIA USA

Simulation-based Engineering

- The foundation tools for composites manufacturing simulation are available, but do not meet the broad range of needs
- The process for developing new manufacturing simulation tools are challenging.
- Accelerated development of design and simulation tools will require new approaches
- The economic incentive to accelerate composites manufacturing technology to a level consistent with mature industries such as automotive will continue unabated.

AERODEF MANUFACTURING

SUMMIT & EXPOSITION
February 25-27, 2014
LONG BEACH CONVENTION CENTER
LONG BEACH, CALIFORNIA USA

What is needed?

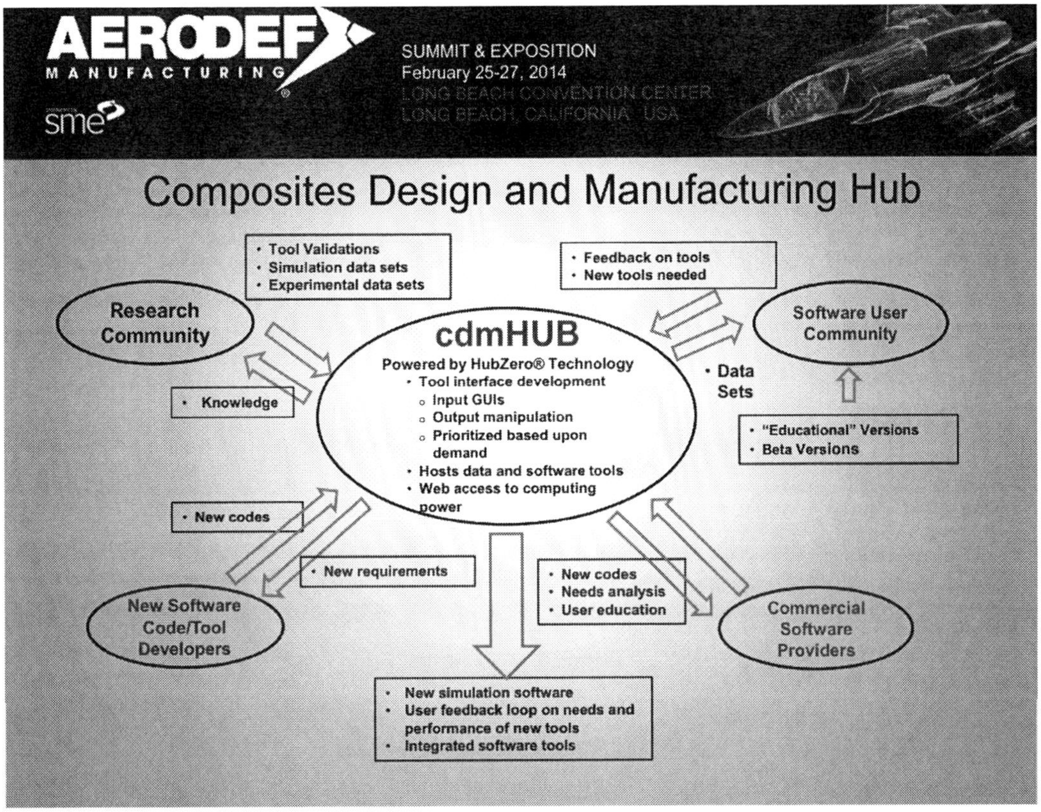

AERODEF MANUFACTURING

sme

SUMMIT & EXPOSITION
February 25-27, 2014
LONG BEACH CONVENTION CENTER
LONG BEACH, CALIFORNIA USA

Contact information:

Rani Richardson
Dassault Systèmes
CATIA Composites Consultant
(201) 675-8361 (Cell)
rani.richardson@3ds.com
www.3ds.com/composites

R. Byron Pipes, NAE, IVA
John Leighton Bray Distinguished Professor of Engineering
Purdue University Neil Armstrong Hall of Engineering
(765) 494-5767
(765) 418-5447 (Cell)
bpipes@purdue.edu
www.cdmHUB.org

LOCKHEED MARTIN

1

Presentation Contents

- C-130 Legacy and Background
- Current and Future Automated drilling and Drill Fastening System
- Automation aiding technological advancements
  - Noteworthy Technologies
  - Active Panel Protection
  - Vision System Part Orientation
  - Vision System Scanning for Hole Locations

2

C-130 Program Legacy

- Longest continuous production run of any military aircraft in history

- In many cases today's C-130's are built with same tooling and processes as the first production models in the 1950's

3

C-130J Automated Drilling

The Evolution of Hand Drilling

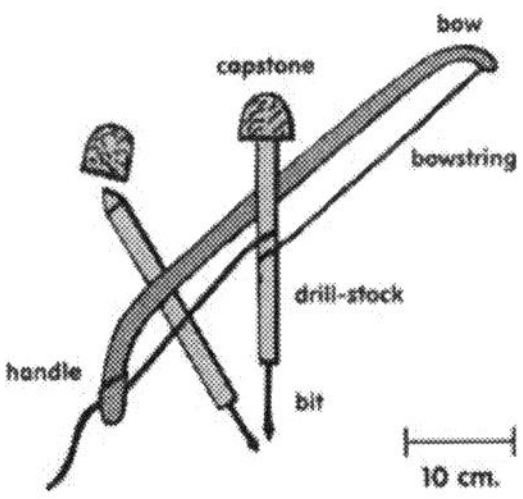

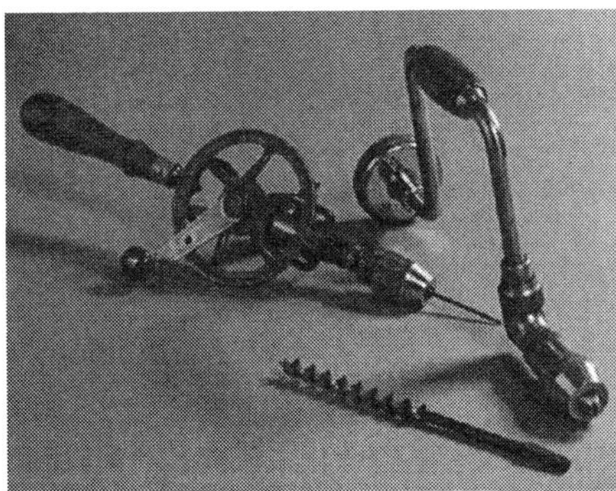

4

C-130J AutoDrill Examples

Wing Beam & Panel Drill Fasten Machine

- 6 Beams & 2 Wing Panels
- ~35,000 Fasteners
- Installs Rivets and Threaded Fasteners Alike

5

C-130J AutoDrill Examples

Cab Top Drilling System

- Drill and Countersink 3,000 Fastener holes
- Rout Interchangeable and Replaceable Window Openings
- Holes normally back-drilled from highly inaccessible frames

6

C-130J AutoDrill Examples

Cargo Ramp Floor

- 1,400 Fasteners
- 1,400 Nutplates
- Assembled like an Onion

7

C-130J AutoDrill Examples

Wing Panel Flexible Track Drilling

- Light-weight modular drilling system
  - Each of 4 components weighs less than 26lb
- Flexible x-axis track vacuum suctioned to wing skins
- Applicable across multiple assemblies

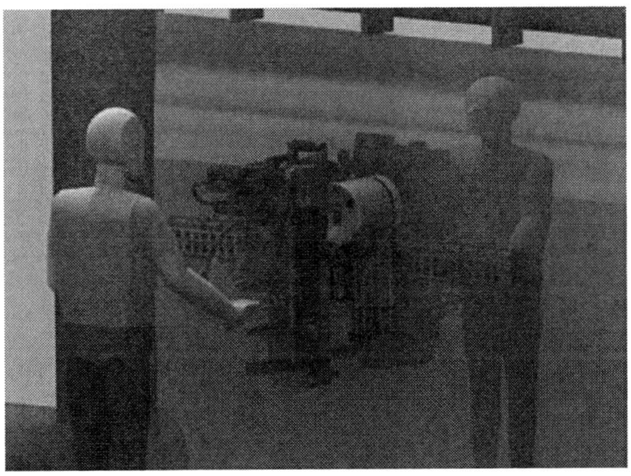

8

C-130J AutoDrill Examples

Cargo Floor Belly Skin Drill

- ~14,500 fastener locations
- Back-drilled from frames
- Stepped up through skins
- Each hole touched 2-3 times

9

Notable Technologies

Nexen Rack and Pinion Drive

- Zero-Backlash roller system
- Long life ~ 30X10^6 cycles

Non-Contact Metrology

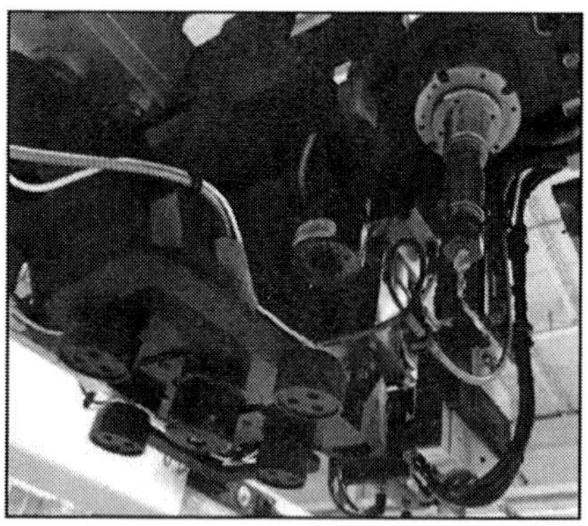

Uses Eddy Currents to determine:

- offset to part
- normality to surface

10

Active Panel Protection

Drill-Riveting System

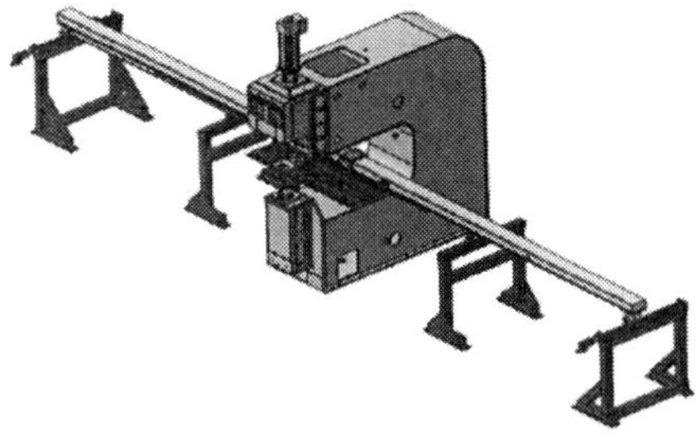

- Drilling & Riveting System for Cargo Floor Panels
- Drills hole, applies sealant, inserts & upsets rivet
- All Electric vs. pneumatic/hydraulic previous generation
- Provides in Process grip length measurement
  - Guarantees correct rivet grip selected
- Active Panel Protection
  - Prevents rivets from being inserted crooked and damaging panel

11

Active Panel Protection

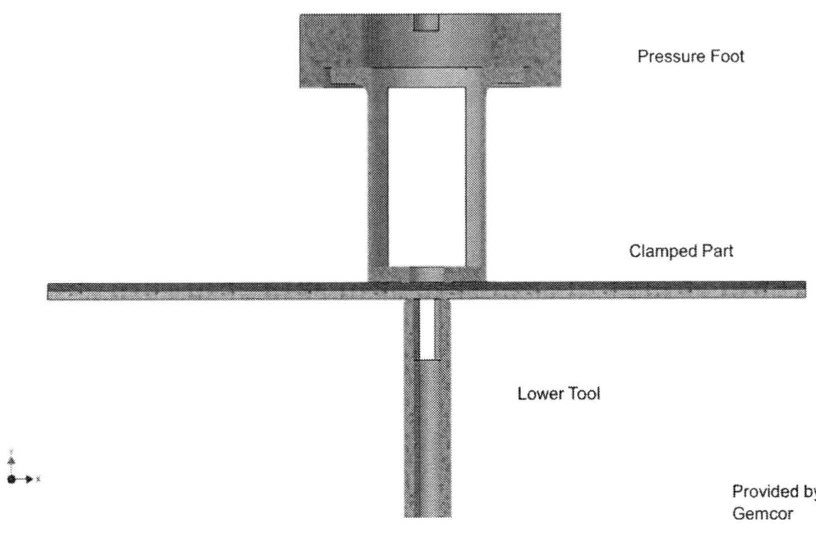

12

Active Panel Protection

Provided by:
Gemcor

13

Active Panel Protection

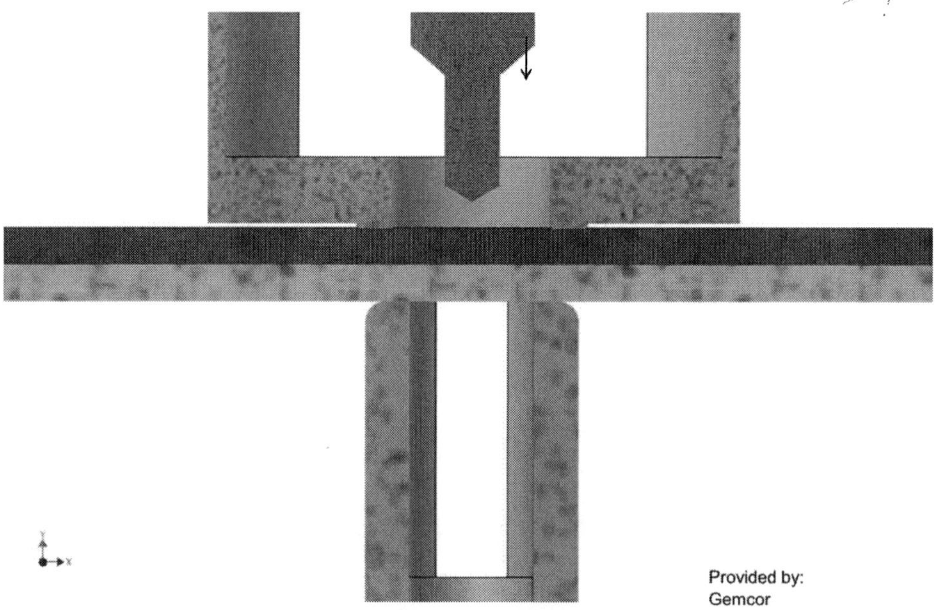

Provided by:
Gemcor

14

Active Panel Protection

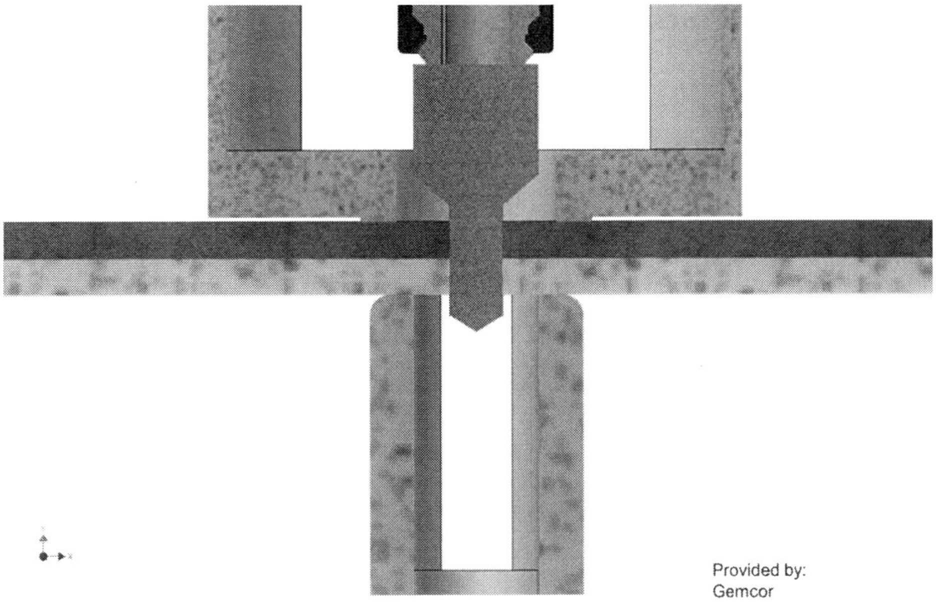

Provided by:
Gemcor

15

Active Panel Protection

Provided by:
Gemcor

16

Active Panel Protection

Provided by:
Gemcor

17

Active Panel Protection

Provided by:
Gemcor

18

Active Panel Protection

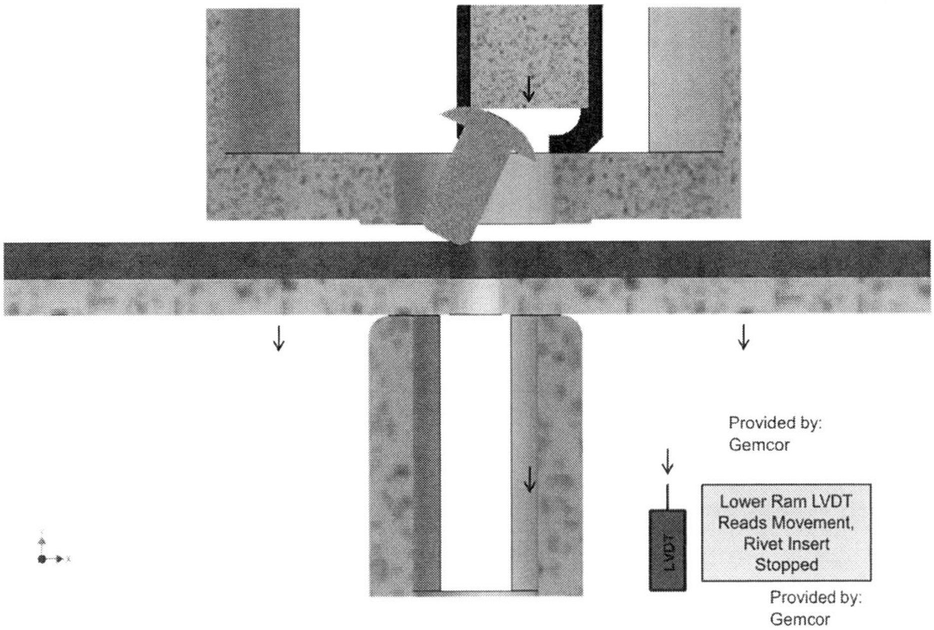

19

Active Panel Protection

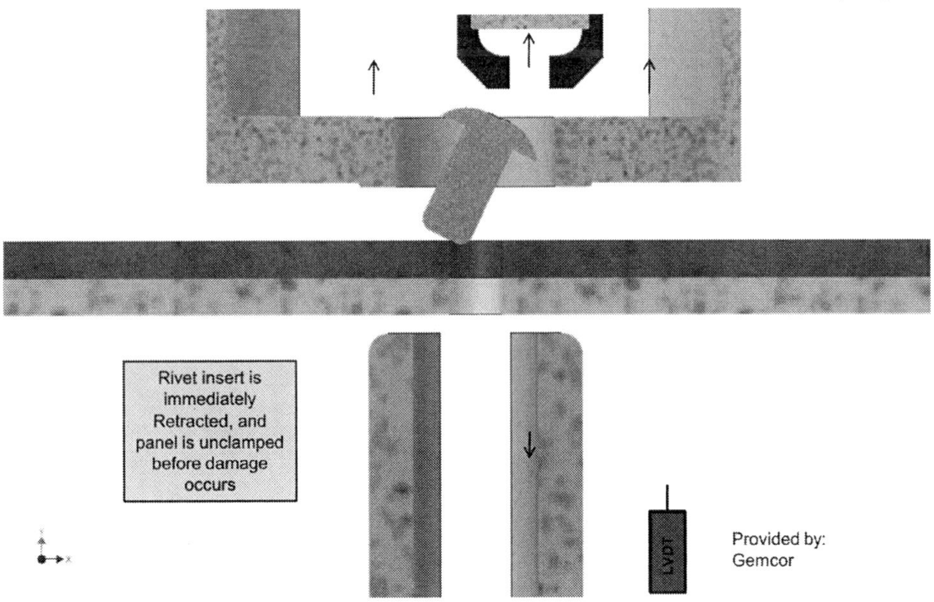

20

Active Panel Protection

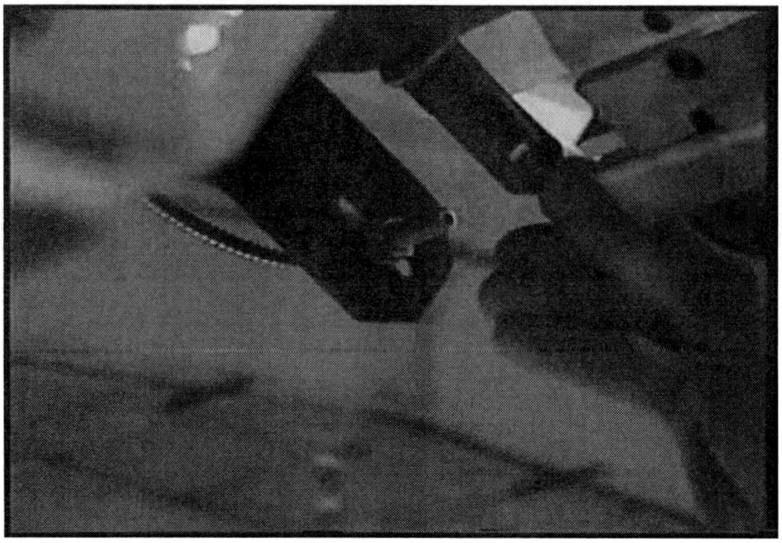

Active Panel Protection (Active Panel Protection.avi)

21

Vision System Part Orientation

Flexible track drilling system

- Drills & countersinks fastener holes in wing panels
- Part Orientation from vision system using index holes (K-holes)
- Can insert temporary fasteners to maintain part clamp-up

22

Vision System Part Orientation

Machine and Rail Installation

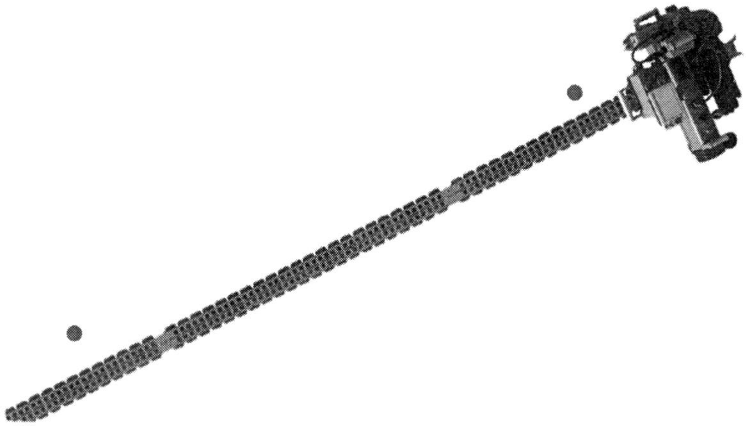

23

Vision System Part Orientation

Locating Index Hole 1

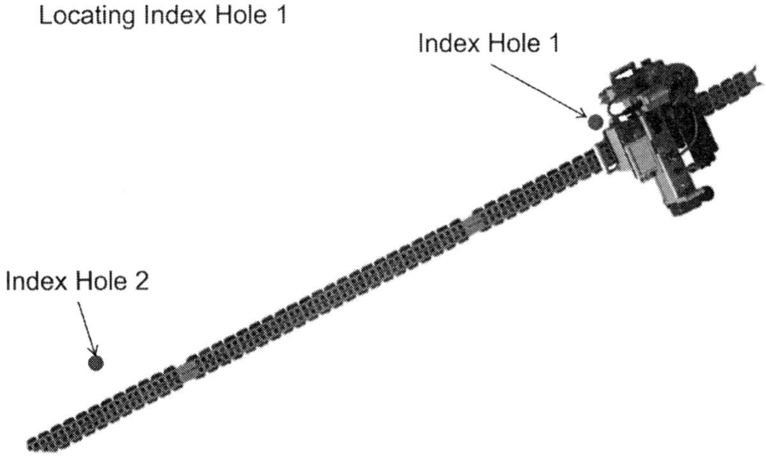

24

Vision System Part Orientation

Locating Index Hole 2

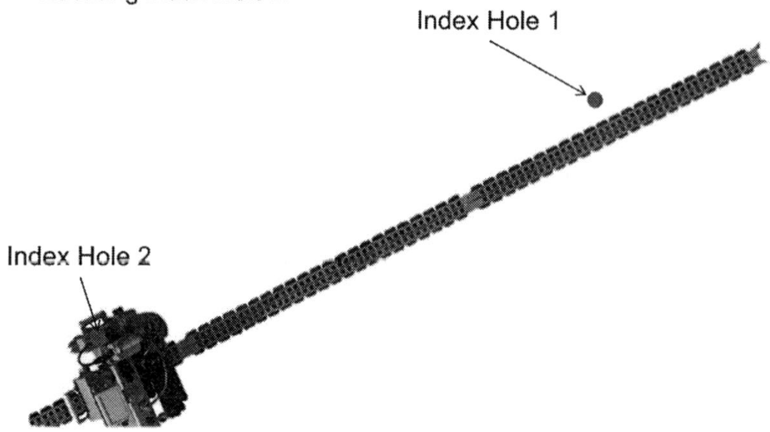

25

Vision System Part Orientation

Program Orientation to Part

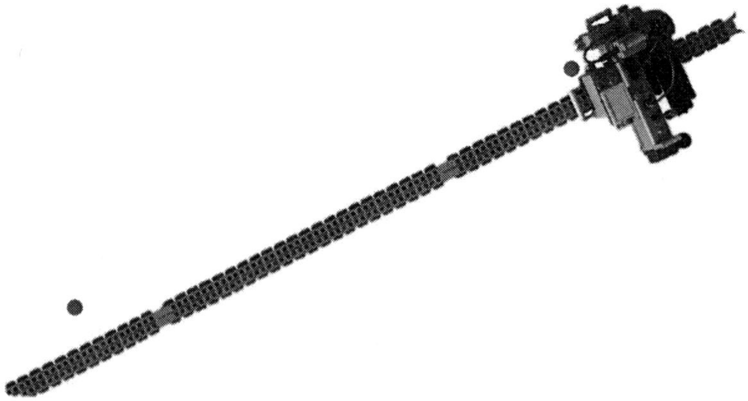

26

Vision System Part Orientation

Drilling Commences

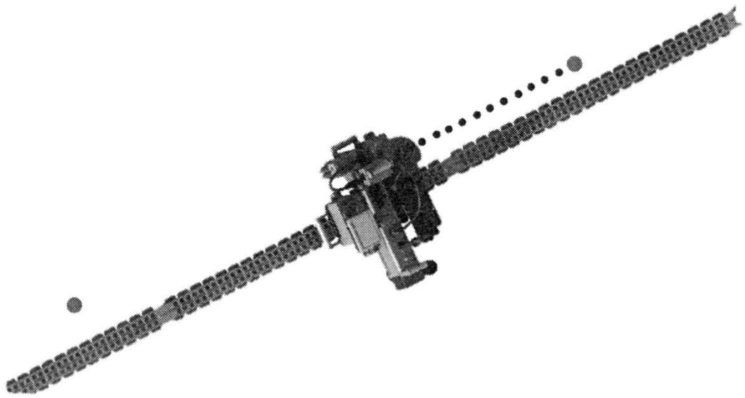

27

Vision System Scanning for Hole Locations

- Frame Sections pre-piloted
- Fastener holes back-drilled from frame sections into overlaying skin

28

Vision System Scanning for Hole Locations

- Vision System Scans pilot holes and records position in machine coordinates
- Skins installed and temporary fastened
- Vision system locates temporary fasteners
- Controller orients part program to match hole locations from scanning and index hole detection

Vision System Camera

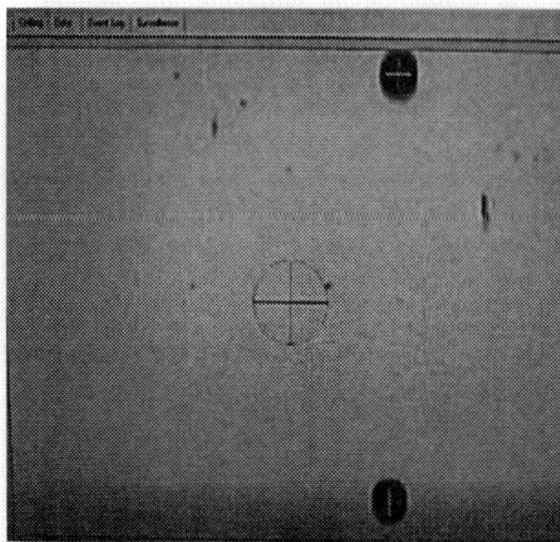

Vision system location capture

29

Vision System Scanning for Hole Locations Video

30

Thank You

Special Thank you to…

- AeroDef Conference coordinators
- Gemcor
- PaR Systems
- West Coast Industries / Mobile Tool Management

31

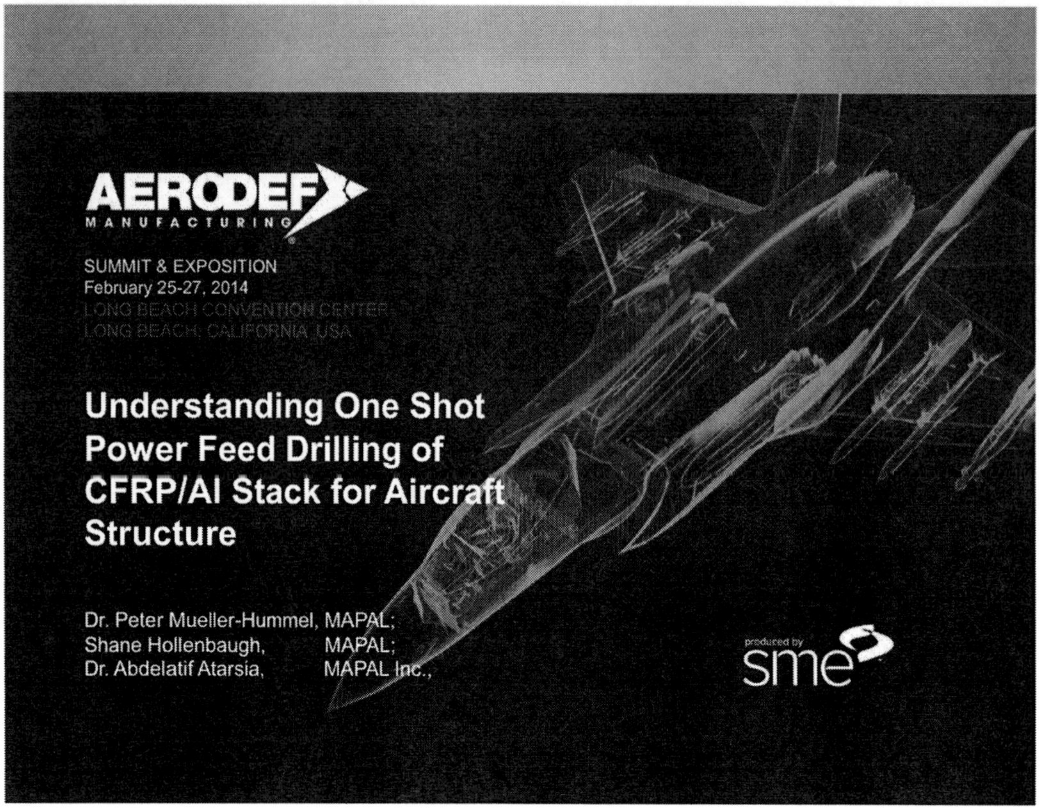
AERODEF MANUFACTURING
SUMMIT & EXPOSITION
February 25-27, 2014
LONG BEACH CONVENTION CENTER
LONG BEACH, CALIFORNIA, USA
Understanding One Shot Power Feed Drilling of CFRP/Al Stack for Aircraft Structure
Dr. Peter Mueller-Hummel, MAPAL;
Shane Hollenbaugh, MAPAL;
Dr. Abdelatif Atarsia, MAPAL Inc.,
produced by sme

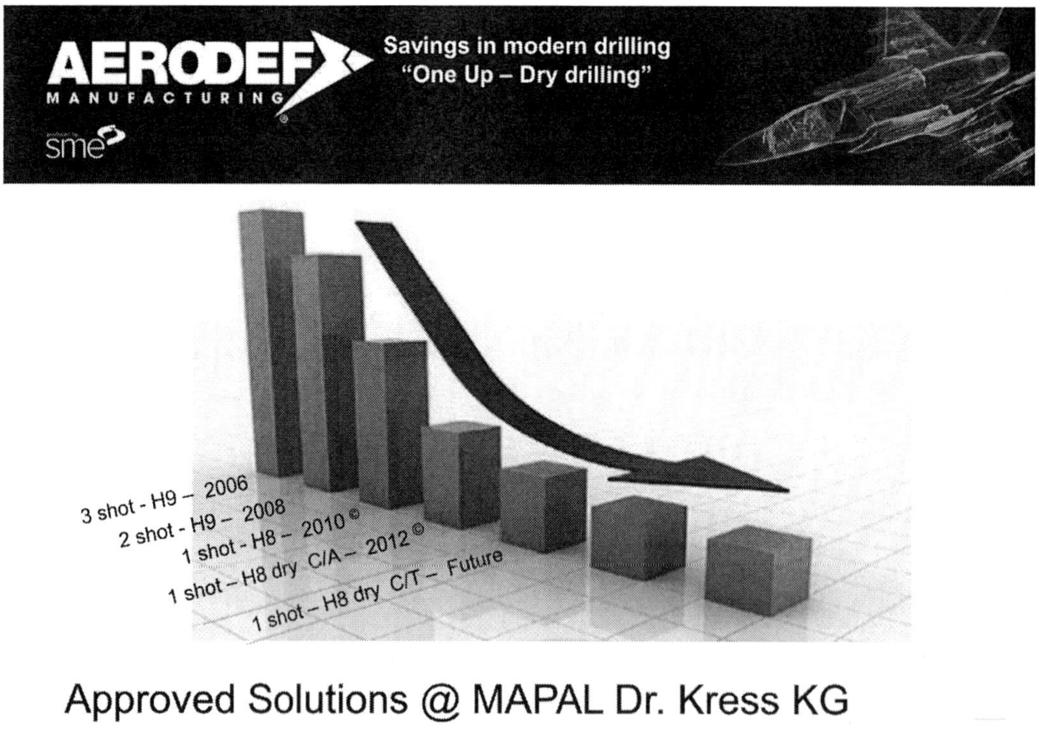
AERODEF MANUFACTURING
sme
Savings in modern drilling
"One Up – Dry drilling"
3 shot - H9 – 2006
2 shot - H9 – 2008
1 shot - H8 – 2010 ©
1 shot – H8 dry C/A – 2012 ©
1 shot – H8 dry C/T – Future
Approved Solutions @ MAPAL Dr. Kress KG

Savings in modern drilling
"One Up – Dry drilling"

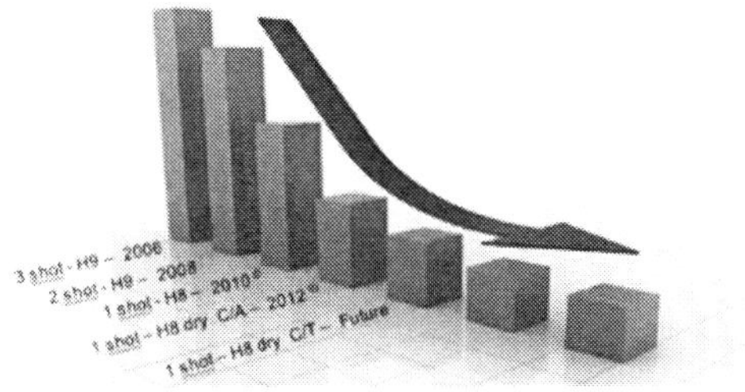

Content

MAPAL Company

Dry Drilling CFRP/Alu

Conclusions

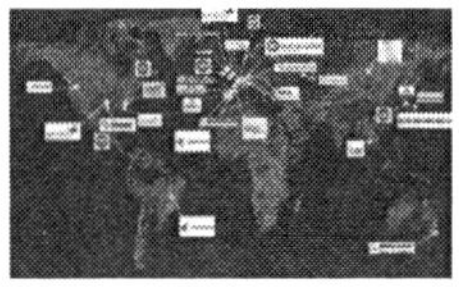

Savings in modern drilling
"One Up – Dry drilling"

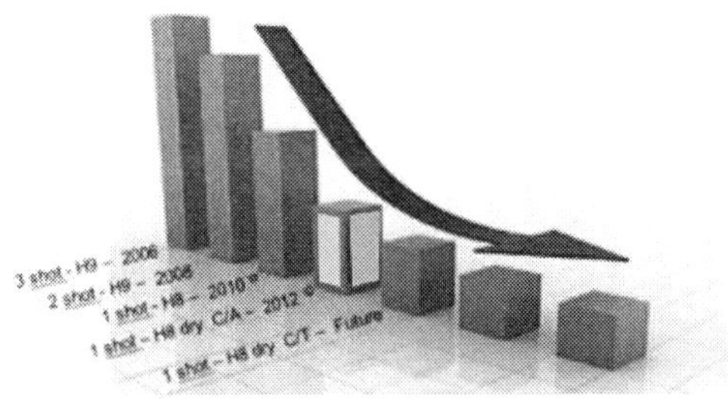

Content

MAPAL Company

Dry Drilling CFRP/Alu

Conclusions

Savings by:
- Avoid MQL
- 50% higher Tool Life
- 20% lower Tool Cost
- Better cpk

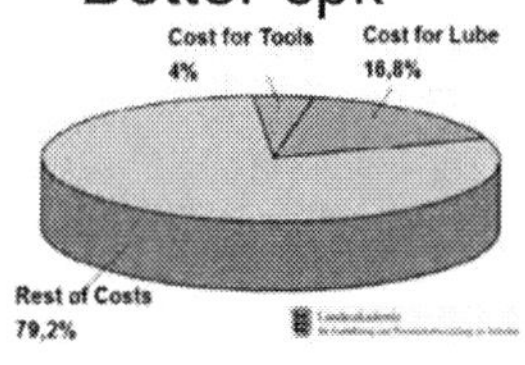

Savings in modern drilling "One Up – Dry drilling"

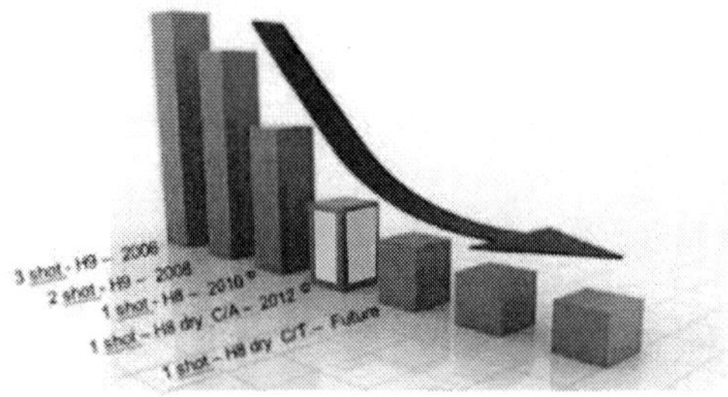

- Content
- MAPAL Company
- Dry Drilling CFRP/Alu
- Conclusions

Mapal is a global company

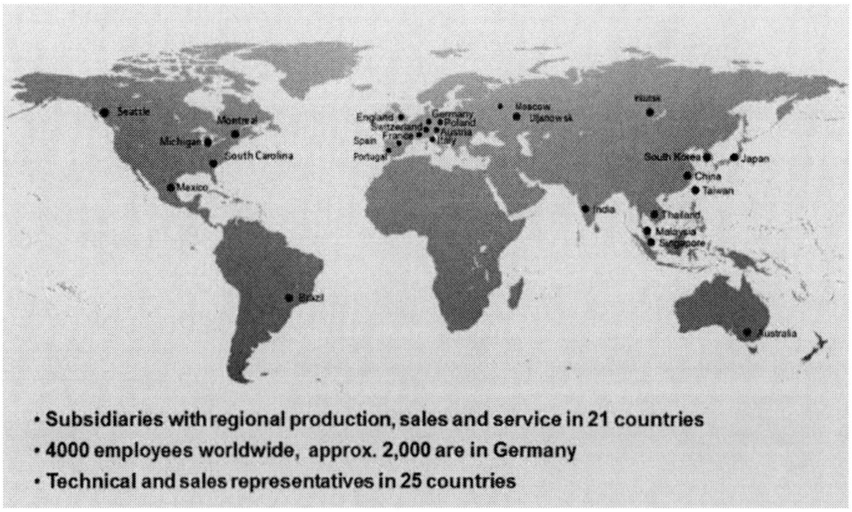

- Subsidiaries with regional production, sales and service in 21 countries
- 4000 employees worldwide, approx. 2,000 are in Germany
- Technical and sales representatives in 25 countries

Research & Development in the Headquarter

Complete measurement and test equipment

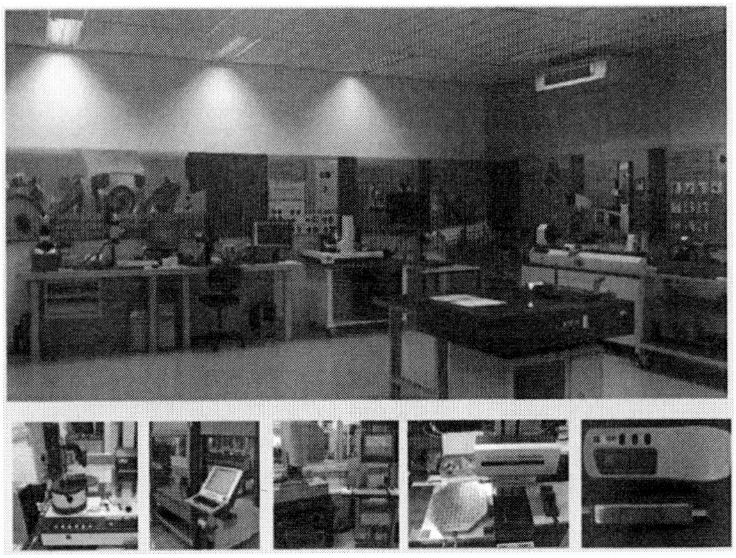

2 shot H9 drilling on Gantry

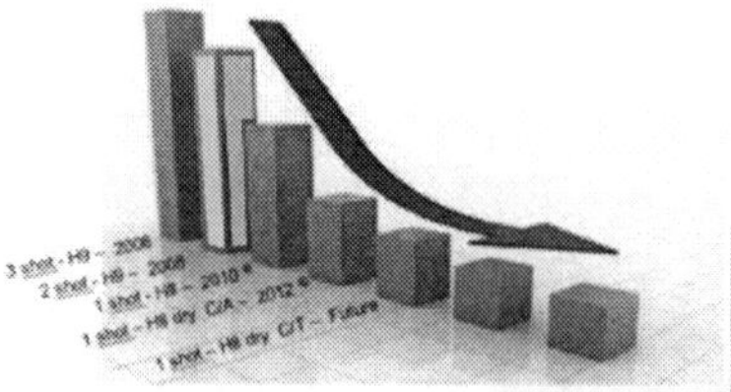

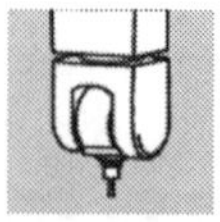

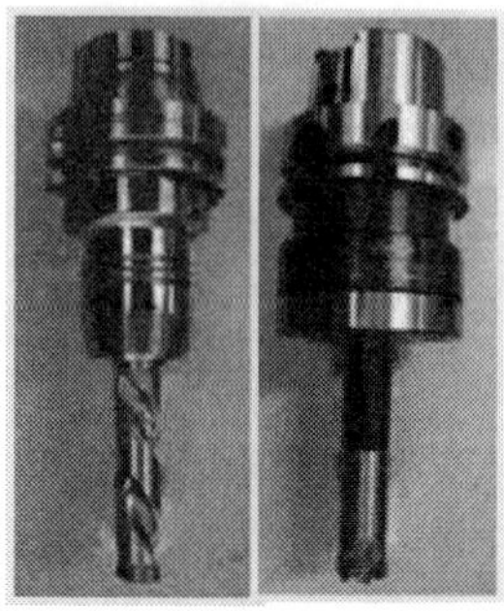

	Speed vc (m/min)	MEGA-Stack-Drill-CT		
		feed f (mm/U)		
		3mm bis 5mm	5mm bis 8mm	8mm bis 12mm
Alu	60-400	0,05-0,2	0,1-0,3	0,15-0,35
CFK	75-150	0,05-0,07	0,06-0,08	0,07-0,1
Titan	10-60	0,02-0,07	0,04-0,15	0,06-0,2

40 holes 200 ho
Tool life

1 shot H8 drilling on Robot

	Speed vc (m/min)	MEGA-Stack-Drill-CT		
		feed f (mm/U)		
		3mm bis 5mm	5mm bis 8mm	8mm bis 12mm
Alu	60-400	0,05-0,2	0,1-0,3	0,15-0,35
CFK	75-150	0,05-0,07	0,06-0,08	0,07-0,1

800holes
Tool life

Alu / Composite / Alu Stack

Dry Drilling H8 one shot on UPA

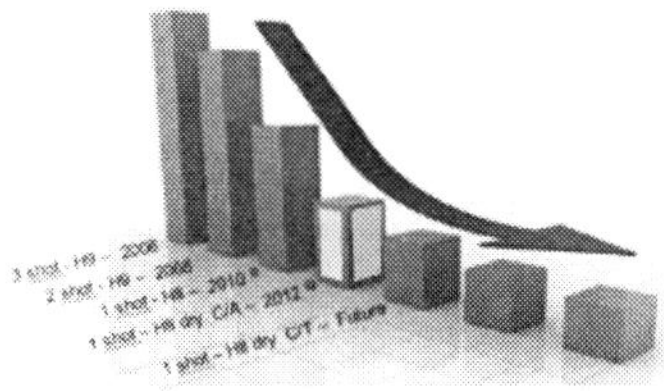

	Speed vc (m/min)	MEGA-Stack-Drill-CT		
		feed f (mm/U)		
		3mm bis 5mm	5mm bis 8mm	8mm bis 12mm
Alu	60-400	0,05-0,2	0,1-0,3	0,15-0,35
CFK	75-150	0,05-0,07	0,06-0,08	0,07-0,1

CFRP Alu (Lubbering)
1 shot

Dry Reaming of TaperLock

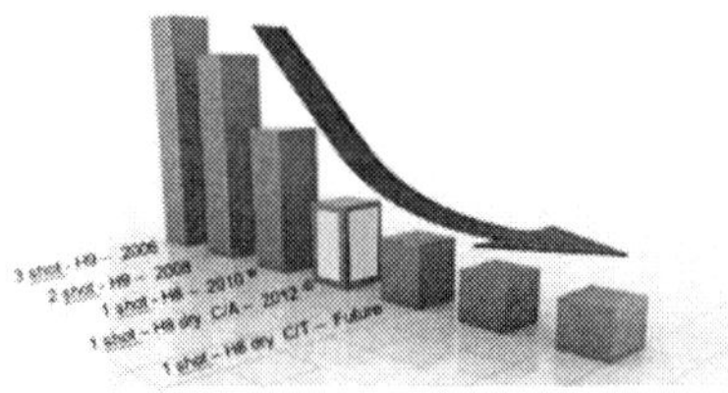

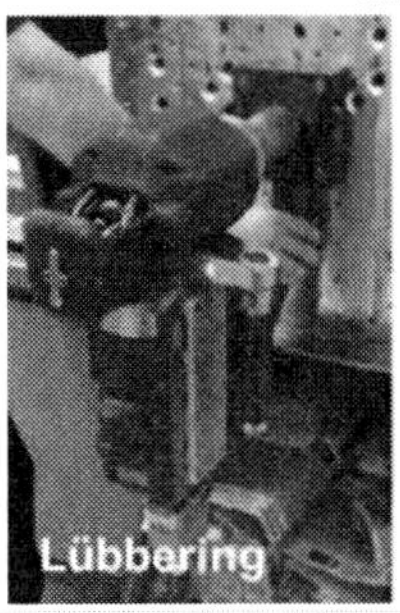

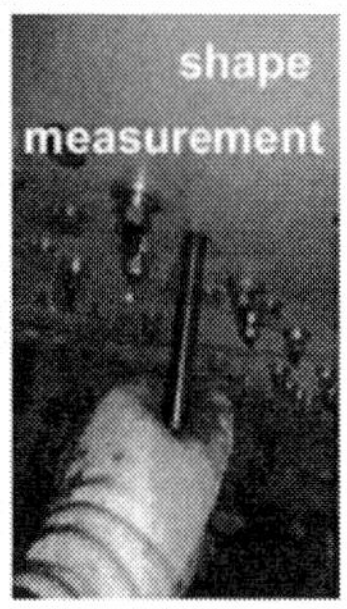

Wing Assy Airbus Broughton ©

Tools for Hand Drilling

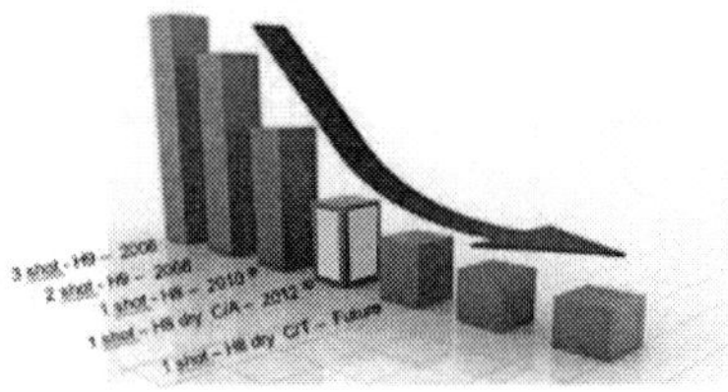

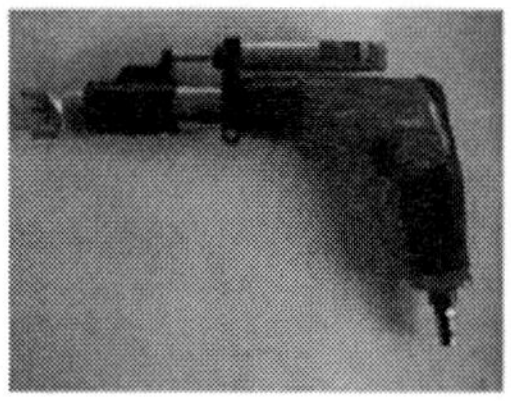

drilling CFRP Titan

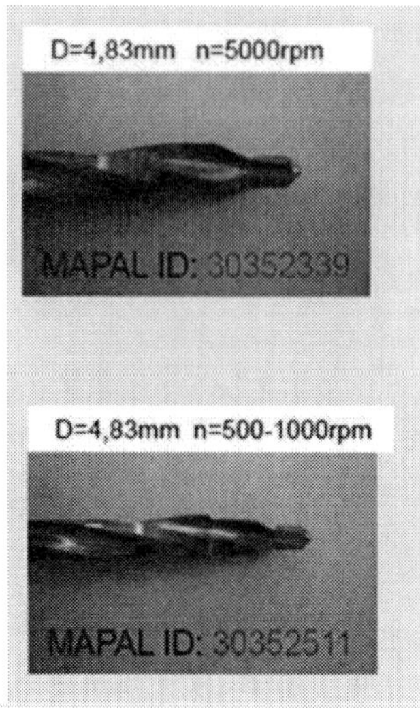

Exit CFK IM-M21E

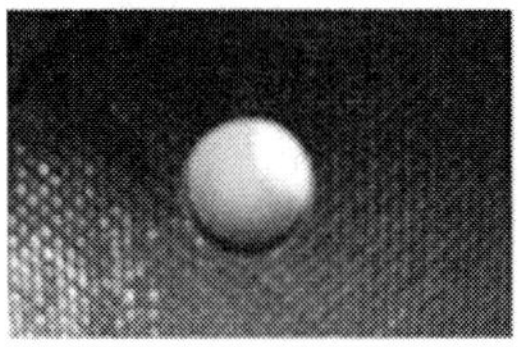

Exit Titan

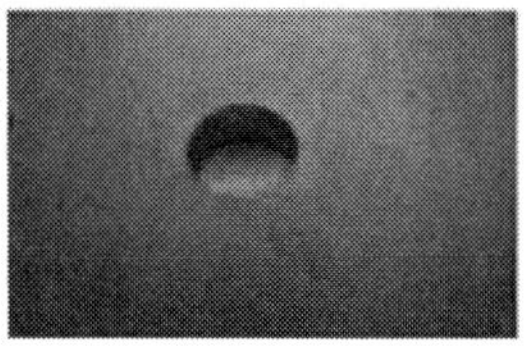

Drilling dry – set up

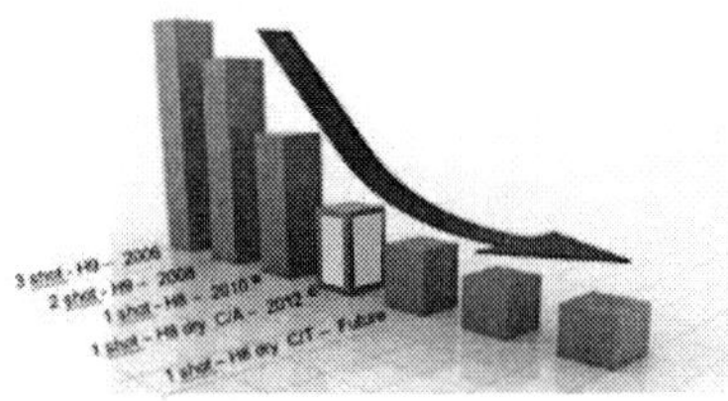

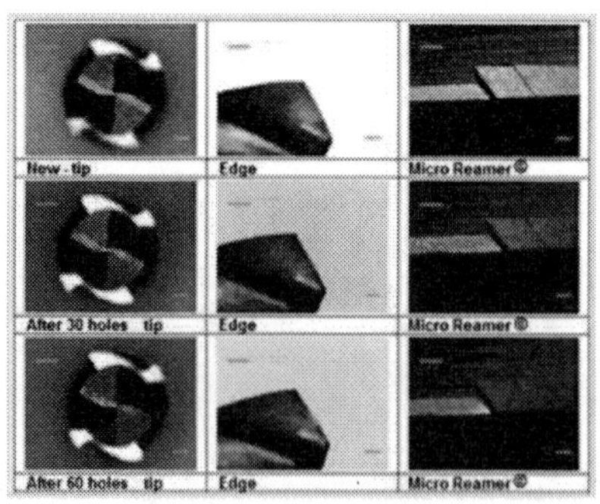

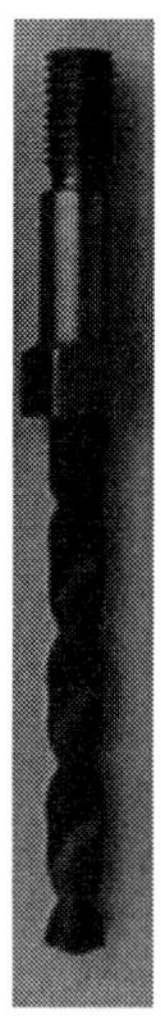

Power feed Luebbering with peck n=2600rpm, feed=0,10mm/rev

Drilling dry - Diameter

Targeting hole size
3/16" + .003" (4,826-4,902)

Average Diameter	4,85948

Tool:	30508265
n =	2600
fu =	0,1
Vc =	40
Ø =	4,86

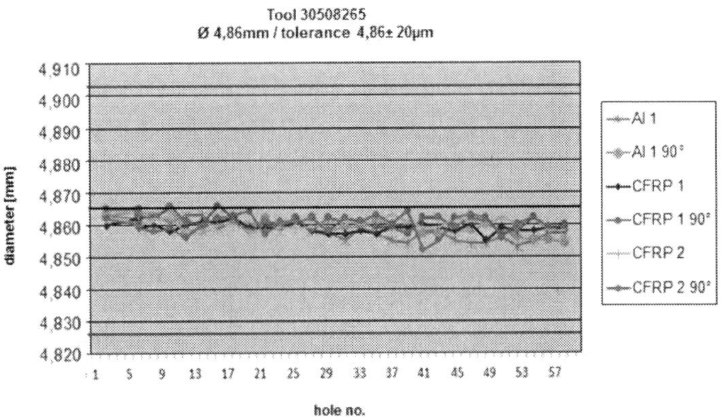

Drilling AL 7075, CFK IM: Stacks:AL-4mm/C-3/4"=19mm and C-3/4" / AL-4mm
One shot, with peck, no MQL / dry

Drilling dry - cpk

Targeting hole size:
3/16" + .003" (4,826-4,902)

Overview of Column Error	Max	4,86300	4,86500	4,86200	4,86600	4,86300	4,86400	
	Min	4,85300	4,85400	4,85500	4,85700	4,85700	4,85200	
	Range	0,01000	0,01100	0,00700	0,00900	0,00600	0,01200	
	Tolerance	0,00900		0,01100		0,00700		
Cpk per material	tolerance	0,04						
	Avg	4,85840	4,85953	4,85900	4,86097	4,86010	4,86007	
	STDEV	0,003158	0,002813	0,001554	0,002470	0,001583	0,002828	
	Cp	2,11	2,37	4,29	2,70	4,21	2,36	**2,11 Min Cp**
	Cpk	1,94	2,31	4,08	2,83	4,23	2,37	**1,94 Min Cpk**
Overall Cpk	STDEV	0,00271						
	AVE	4,85948						
	Cp	**2,46**						
	Cpk	**2,395**						

Drilling AL 7075, CFK IM: Stacks:AL-4mm/C-3/4"=19mm and C-3/4" / AL-4mm
One shot, with peck, no MQL / dry

Drilling dry – Burr height

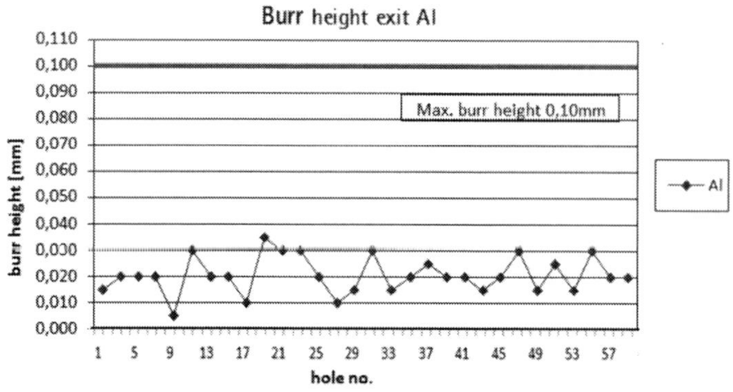

Drilling AL 7075, CFK IM: Stacks:AL-4mm/C-3/4"=19mm and C-3/4" / AL-4mm
One shot, with peck, no MQL / dry

Drilling dry – Burr height

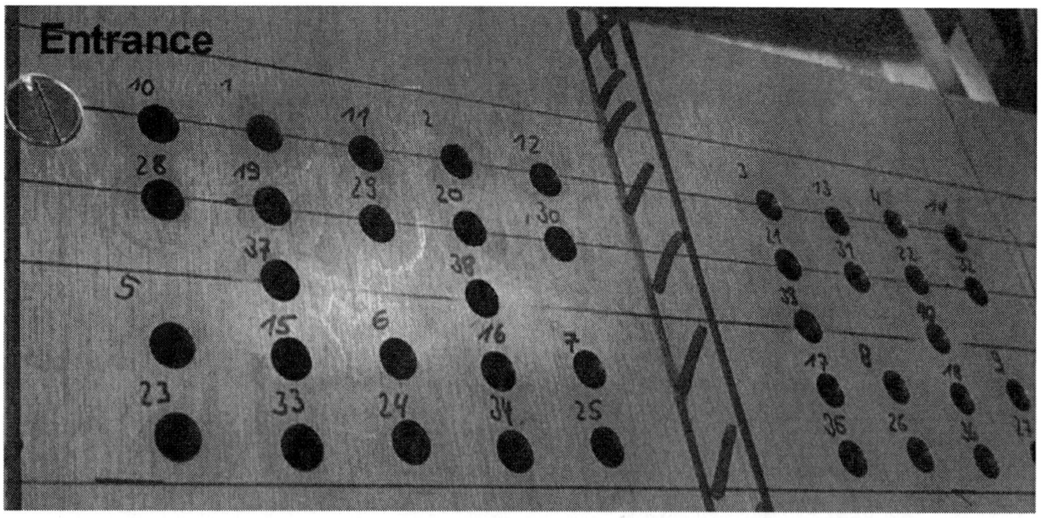

Drilling dry – Burr height

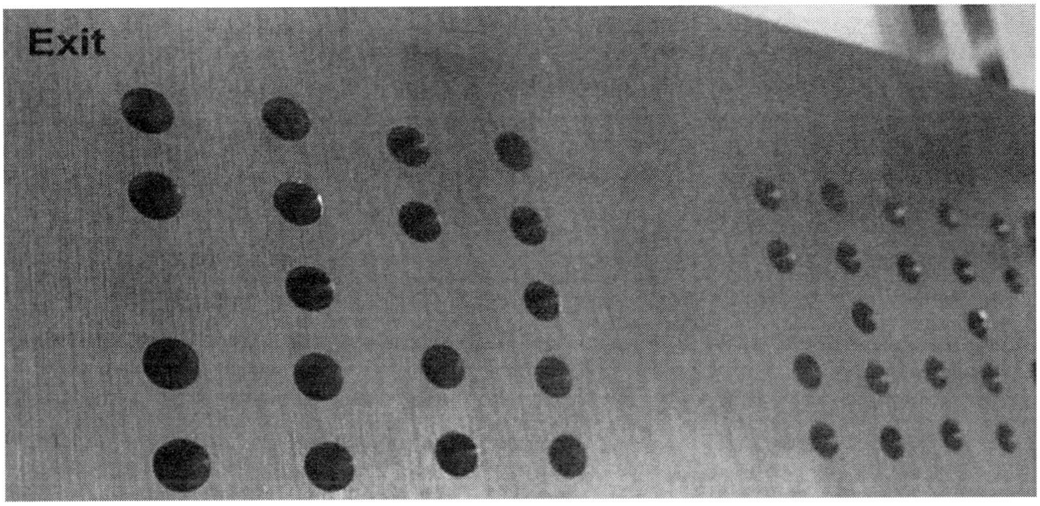

Drilling dry – Hole Quality

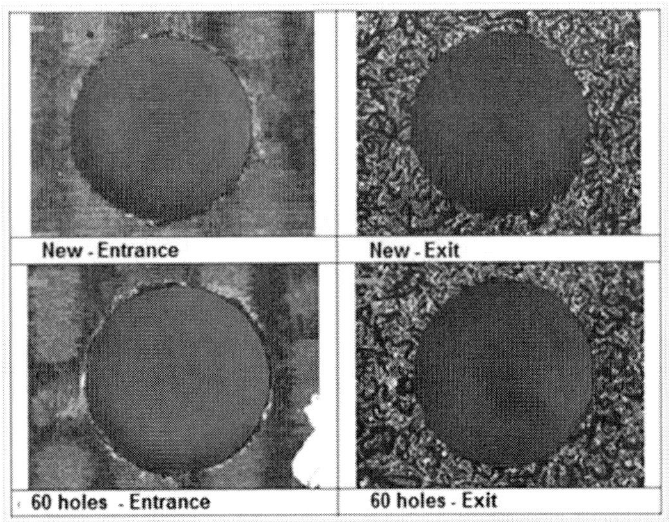

Drilling dry – Temperature Measurment

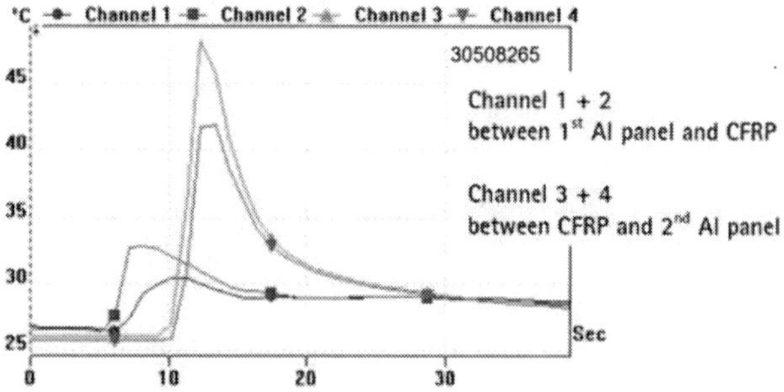

4 Thermo couples
Chanel 1 3mm distance
Chanel 2 closed to the hole

Chanel 3 closed to the hole
Chanel 4 3mm distance

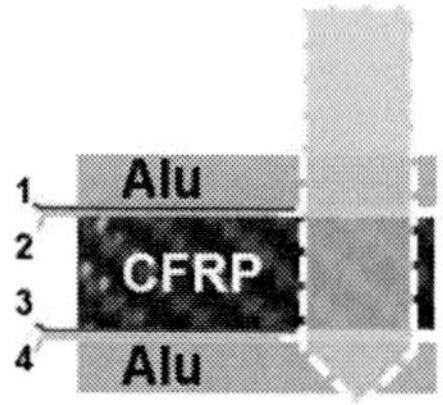

Why we are using the MQL

- MQL is for compensating gaps in cutting tool technology
  - Prevents the build up
  - Lowers the friction
  - Enhance the Chip remov

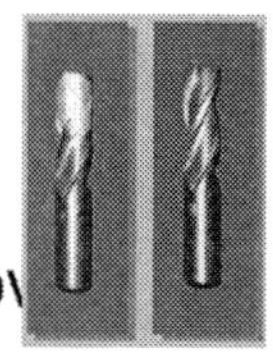

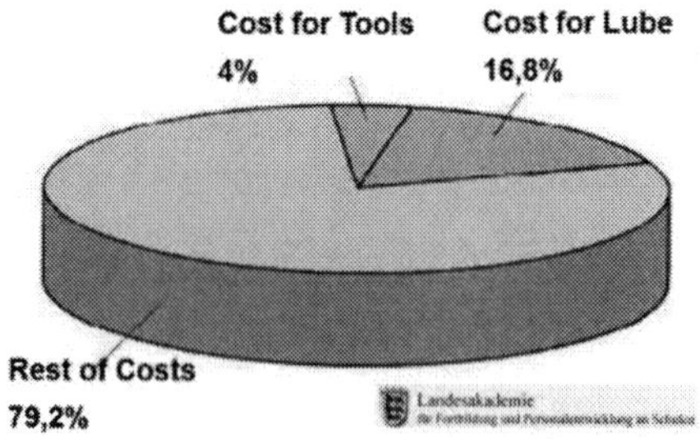

but

- increase in part value? → no
- wellness for the people? → no

Dry Drilling - Benefits and who pays the bill?

- Higher Stability without ic*
- Higher tool life

- Cost for tools with ic*
- No jamming of wet dust

- Improved health and safety

No overspray at the exit
No slip out on jigs

- Improved industrial hygiene
  Lower cost for disposal
  Less cleaning of part, machine, site ...

No shim and coating Adhesion Problems
No strip of coating of jigs and tools

- No limit in tool spiral

Lower burr value

- Lower qualification of the stuff

MQL qualification   No MQL calibration

- Lower cost for equipment
  No Storage, tubes, pumps for the oil

Lower weight of the machine tool

* ic internal coolant holes

Drilling dry – Résumé with UPA

- For Composite / Titanium MQL is still necessary
- For Composite / Aluminum
  - All hole Ø are within the tolerance.
  - All hole burr height are within the tolerance.
  - Ø and burr height is very constant.
  - No critical delaminations are visible.
  - MicoPeck is mandatory

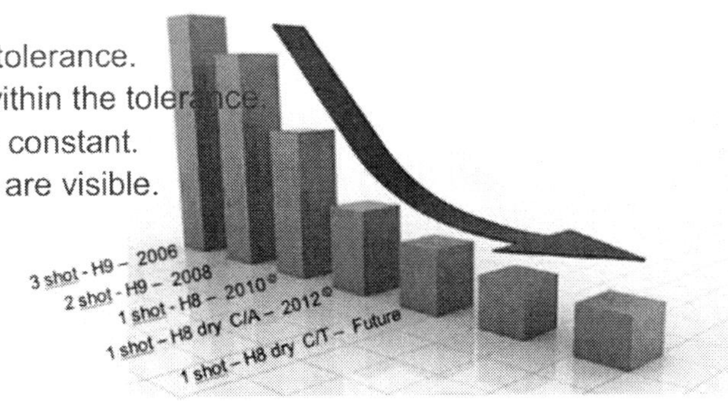

- The savings in dry cutting
  - Better cpk,
  - Less cost of cleaning and heath protection
  - cheaper tools and longer tool life
- makes the capital investment of micro Peck worth

AERODEF MANUFACTURING

SUMMIT & EXPOSITION
February 25-27, 2014
LONG BEACH CONVENTION CENTER
LONG BEACH, CALIFORNIA USA

Integrated CNCs and Robots

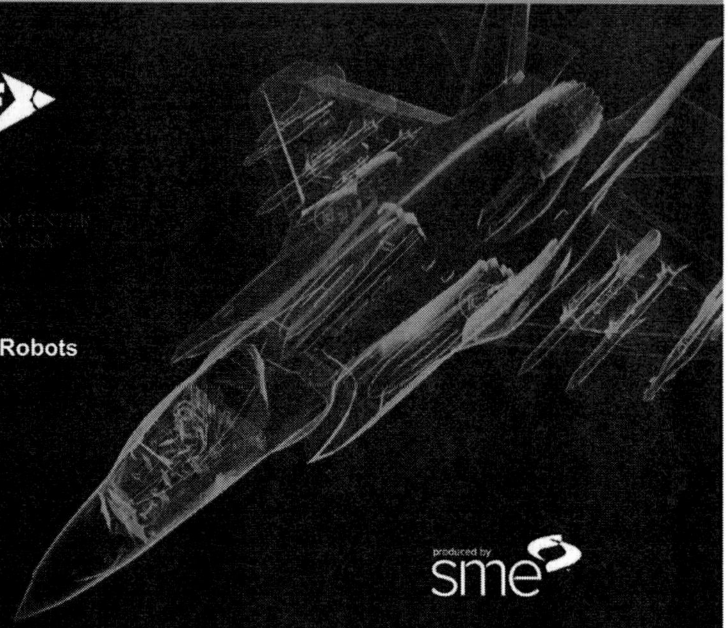

produced by sme

Integrated CNCs and Robots
Objectives

1. What are the drivers?
2. Technology basis
3. CNC control architectures
4. PLM Support
5. Examples

Integrated CNCs and Robots
What are the drivers?

- Lower cost
- Smaller footprint
- Easier installation
- Larger reach
- Access to more sides of work-piece
- Work simultaneously on a single part

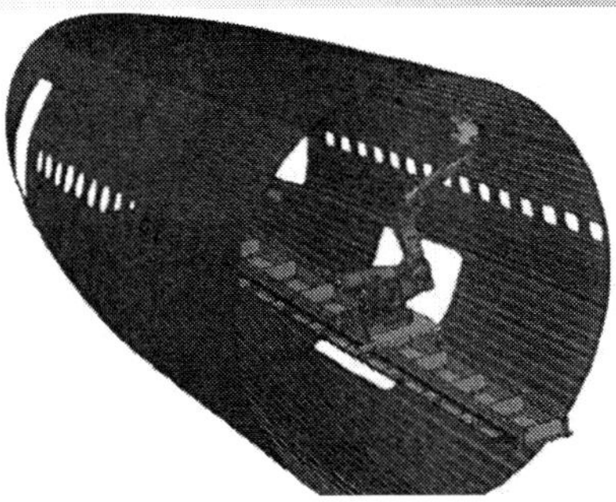

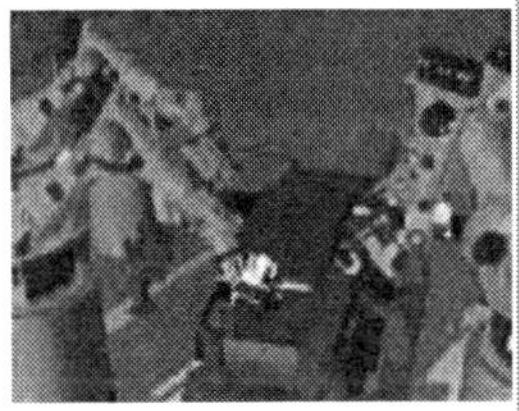

Integrated CNCs and Robots
What are the drivers?

- Tight integration with CNC automation capabilities
- Complex end effectors requiring motion and automation
- Common Programming with CNC machines
- Reduced control complexity

Integrated CNCs and Robots
Technology basis

1. Direct CNC control of serial robot kinematics allows robots to be substituted for some traditional CNC controlled machines
   - Light process forces
     - Drilling and riveting
     - Composites lay-up
     - Waterjet
     - Light milling and trimming
     - Ultrasonic cutting
     - De-burring
     - Inspection
     - Fastening
     - Laser cutting

Integrated CNCs and Robots
Technology basis

2. Robot kinematics can be part of a larger kinematic chain supported by the CNC
   - For example a robot riding on a 3 axis Cartesian system with a rotating workpiece.

- Include end-effector axes
- Entire kinematics chain solved by CNC
- Traditional CNC TCP and orientation programming

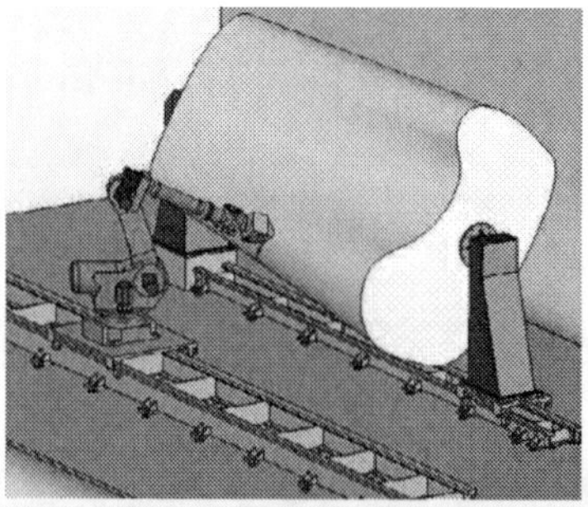

Integrated CNCs and Robots
Technology basis

3. A single CNC can coordinate and synchronize multiple machines
   - For example a Lathe with a loader robot and de-burring robot controlled by a single CNC.

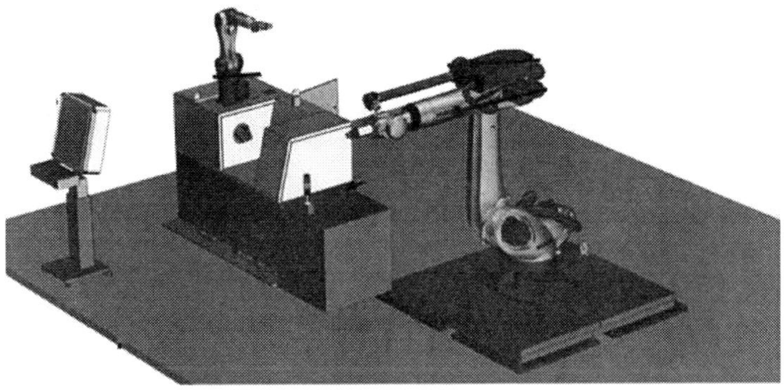

Integrated CNCs and Robots
Multiple Channels / Robots with Single Point of Control

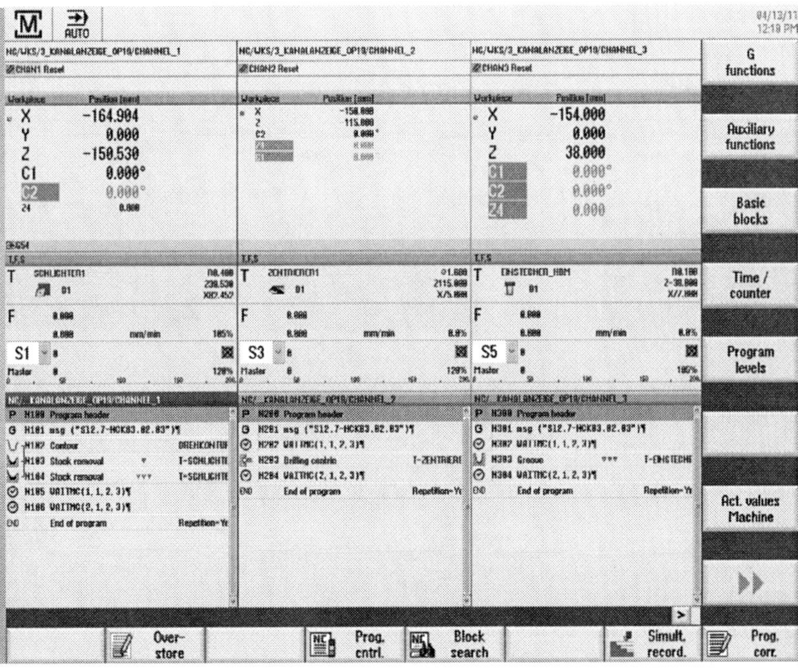

Integrated CNCs and Robots
CNC control architectures

Traditional CNC interface with Robots

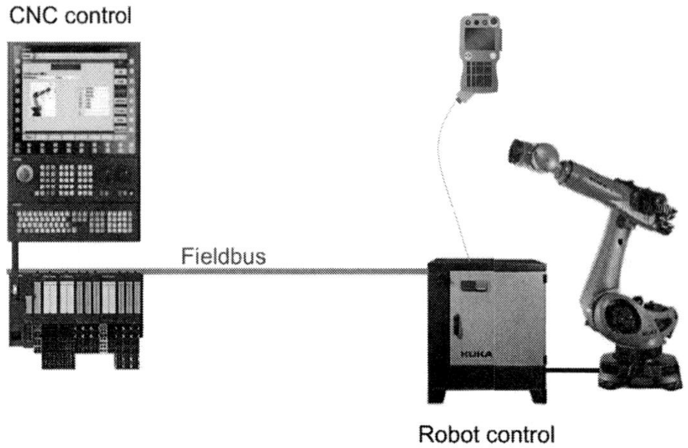

Integrated CNCs and Robots
CNC control architectures

Modern / evolving CNC integration with Robots

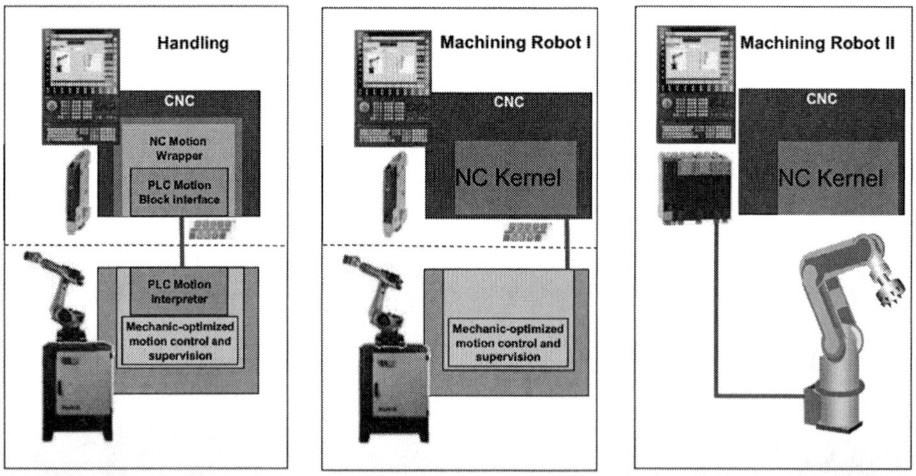

Integrated CNCs and Robots
CNC control architectures

"Handling" architecture

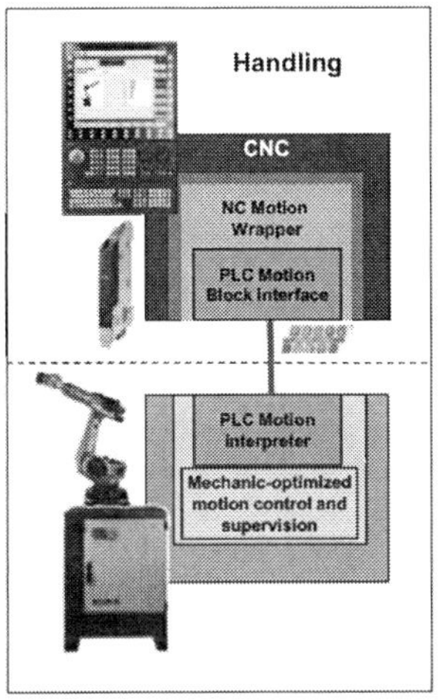

Pros

- Simple Integration into existing CNC architectures
- Programming in CNC language, rather than PLC
- Single point of control and diagnostics information
- Possible automation add-on at end-user site.

Cons

- Programming limitations
- Geometry planning limitations
- Update bandwidth still at PLC/fieldbus rate

Integrated CNCs and Robots
CNC control architectures

"Machining Robot I" architecture

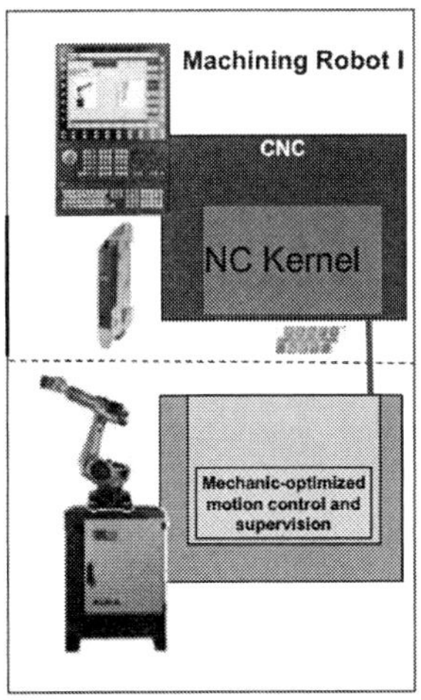

Pros

- Full CNC geometry planning and execution
- Programming in full CNC language
- Single point of control and diagnostics information
- Possible automation add-on at end-user site.
- Robot Safety and compensation provided by robot control

Cons

- Requires CNC control and Robot control
- CNC to Robot interface can be complex

Integrated CNCs and Robots
CNC control architectures

Machining Robot II" architecture

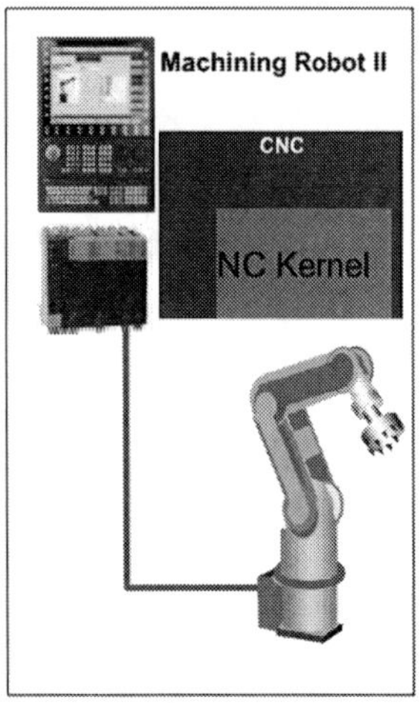

Pros

- Full CNC geometry planning and execution
- Programming in full CNC language
- Single point of control and diagnostics information
- Single control, w/ possibly multiple robots

Cons

- Typically engineered by OEM, not end-user add-on
- Robot Safety and compensation provided by OEM
- Engineering intensive

Integrated CNCs and Robots
PLM Support

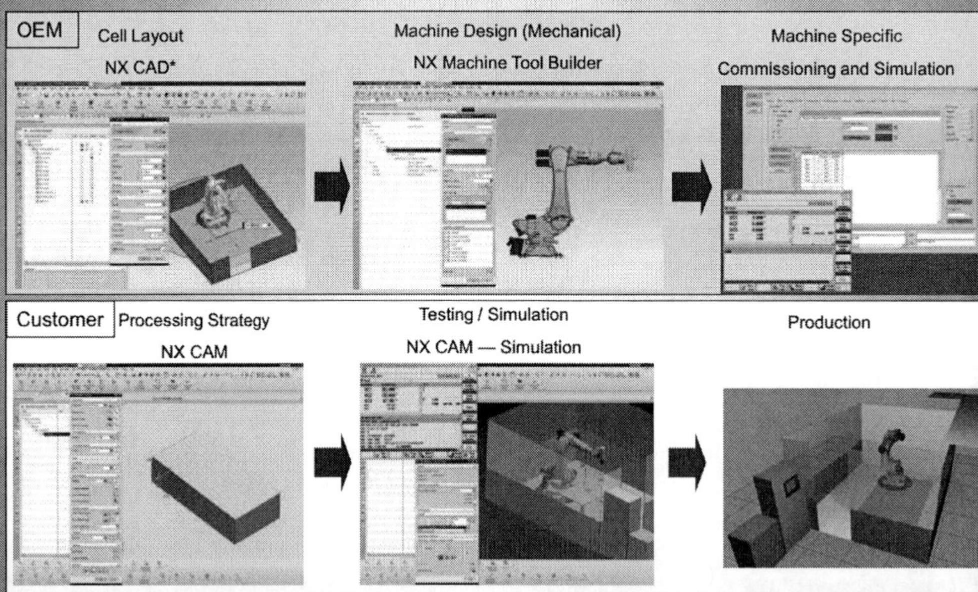

Integrated CNCs and Robots
Example

A lathe with a loader robot and de-burring robot controlled by a single CNC.

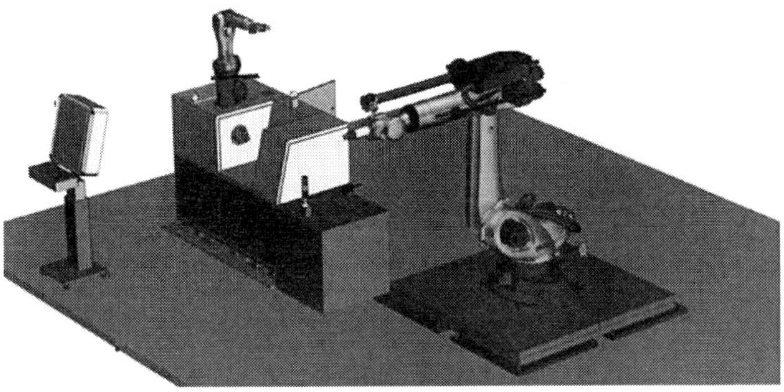

Integrated CNCs and Robots
Advantages

Summarizing Advantages

- Integration of end-effector and other automation
- Extended Kinematics solutions
- Built-in synchronization of multiple robots / machines
- Commonality with existing CNC infrastructure
  - PLM Tools
  - Operator and programmer training / documentation
  - IT and MES systems
  - Backup and Restore
  - Spare parts
- Single Screen and point of control for machine, vision, and automation
- Avoid proprietary robot controllers
- Availability of CNC specialists

AERODEF MANUFACTURING

sme

SUMMIT & EXPOSITION
February 25-27, 2014
LONG BEACH CONVENTION CENTER
LONG BEACH, CALIFORNIA USA

Integrated CNCs and Robots

Roger Hart
R&D Engineering Manager

Siemens Industry, Inc
4710 Columbia Rd.
Lebanon, OH

SIEMENS

tel: +1 (513) 234-9713

Roger.Hart@siemens.com

Advances and New Technologies in Additive Manufacturing of Metals

AeroDef 2014, Long Beach, CA February 25., 2014

Dr. Ian D. Harris
Technology Leader, AW
Founding Director, AMC
614.688.5131
iharris@ewi.org

Additive Manufacturing Consortium
Operated by EWI

EWI®
We Manufacture Innovation

Outline

- **An Overview of Metals Additive Manufacturing Processes**
- **Metals Additive Manufacturing (AM) Developments**
  - Arc Welding,
  - VHP UAM
  - Material development partnership with EOS for PBF-L
- **Additive Manufacturing Consortium (AMC)**
  - Need for collaboration, especially for aerospace design allowables data
  - Response and formation of AMC with 30 partners and members
  - Database for property development for Alloy 625 and DMLS
  - In-process monitoring for NIST/NAMII ($7M)
- **Summary**

EWI
We Manufacture Innovation

Bridging the Gap Between Research and Manufacturing Implementation

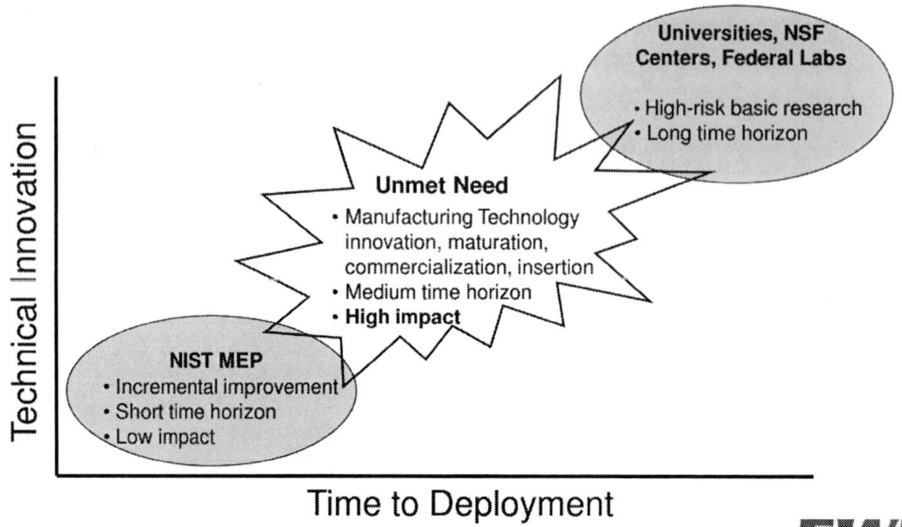

EWI
We Manufacture Innovation

Advanced Manufacturing Technologies at EWI

Innovate, mature, commercialize <u>Materials Joining and Advanced Manufacturing</u> technology for industry

- AcousTech™ Machining
- **Additive Manufacturing**
- Advanced arc welding
- Automation, sensors, controls
- Brazing and soldering
- Dissimilar materials joining
- Friction processing
- Hot forming
- Laser processing
- Nondestructive evaluation
- Numerical modeling and simulation
- Plastic and composite fabrication
- Resistance welding
- Ultrasonic additive manufacturing
- Ultrasonic joining
- Weldability and mechanical testing; metallurgical analysis

EWI
We Manufacture Innovation

AM Process Chain at a 50,000 foot view...

"...allows rapid transformation from "art" (CAD model) to "part" (manufactured product)..."

http://www.commerce.gov/news/fact-sheets/2012/08/15/fact-sheet-additive-manufacturing

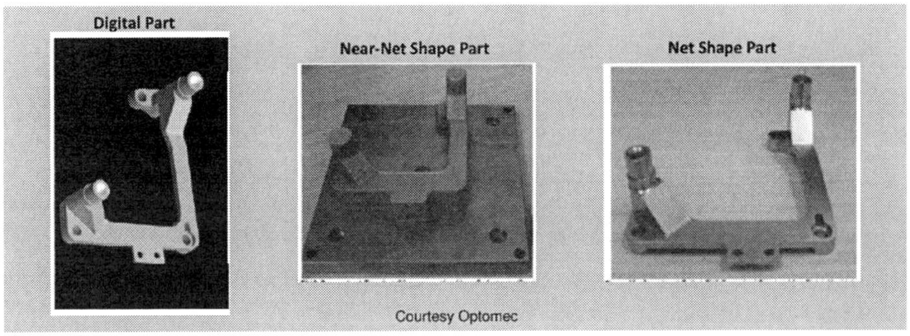

EWI
We Manufacture Innovation

Additive Manufacturing

local melting

CAD model directed CNC toolpaths

Near Net Finish

- **Cost effective for low volume applications (currently)**
- **Enables designs not constrained by conventional manufacturing**
- **CAD to Part in hours to days**
- **Supply chain shifts to on demand**

EWI
We Manufacture Innovation

AM Process Basic Components

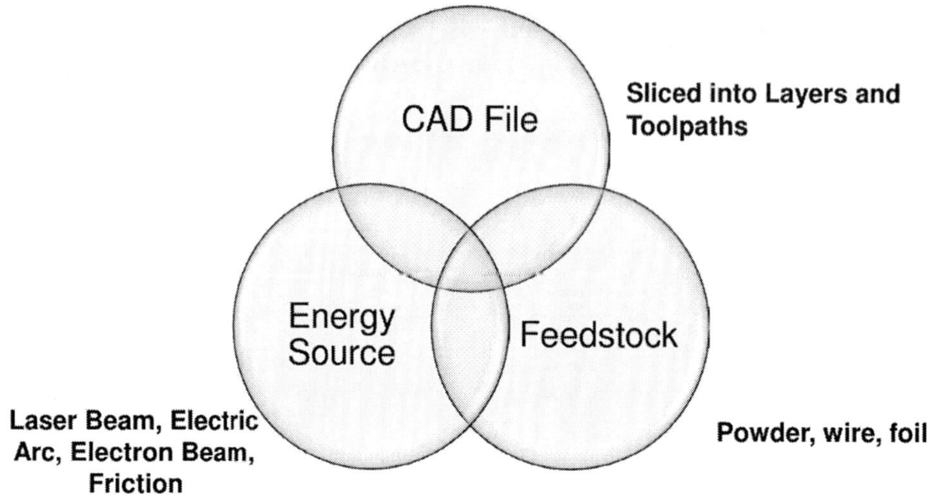

EWI
We Manufacture Innovation

AM...Reality

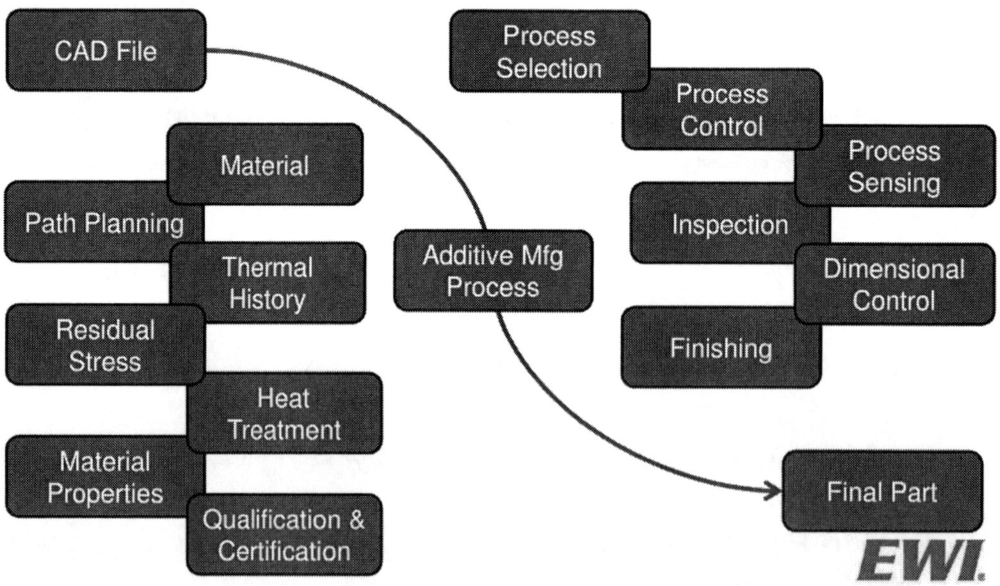

EWI
We Manufacture Innovation

Metals AM – 'Mind to Part'

- AM can eliminate joints between parts by integrating the parts 1-5, 10, 20 or more
- But, the whole thing is a made of 100's or 1000's of beads whether by EB, Laser, or other processes
- Rules of metallurgy apply! – need strong materials input, support
- Develop and optimize through DOE techniques to minimize/eliminate need for HIP ASTM F42 F2924 – highest quality category does not use HIP
- Need to inspect volumetrically
- Room for more productive process 'tools'/toolkit

EWI
We Manufacture Innovation

EOS Direct Metal Laser Sintering (DMLS) M280-400W

- Powder Bed Fusion Additive Manufacturing
  - Uses a fiber laser to fuse powder metal in a reservoir layer by layer to build up final part
- EOS has deployed >400 M270/M280 machines worldwide and is the current market leader
- EWI is a materials development partner with EOS.

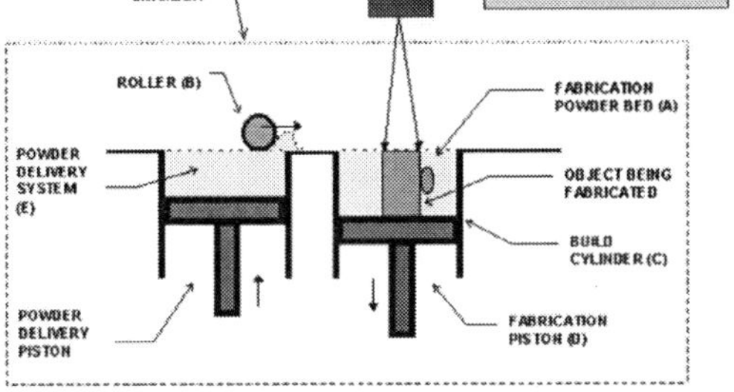

EWI
We Manufacture Innovation

AM is Materials Joining

Manufacturing of complex 3D parts by *joining* successive layers

1-inch PBF-L Cube

675 feet of weld
(Audi R8)

5 miles of weld

3,400 feet of weld

EWI
We Manufacture Innovation

Examples of AM (3D Printing) in the mainstream media

Science Home   Archaeology   Air & Space   Planet Earth   Wild Nature   Natural Science   Dinosaurs

NASA tests 3D-printed rocket engine

Published July 15, 20   FoxNews.com

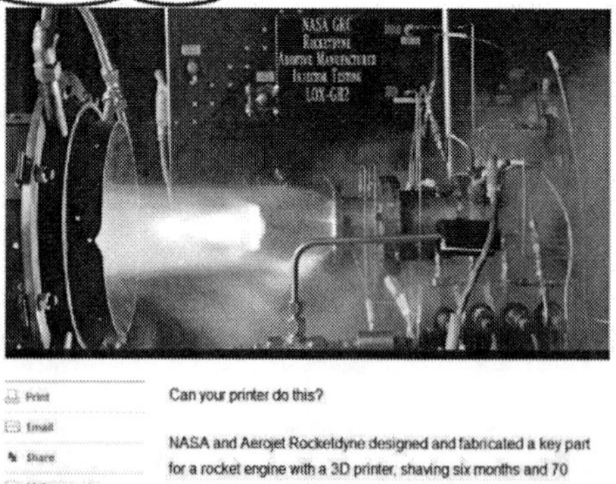

Print
Email
Share
Comments
Like
Tweet

Can your printer do this?

NASA and Aerojet Rocketdyne designed and fabricated a key part for a rocket engine with a 3D printer, shaving six months and 70 percent off the cost of production. A successful test of the component, a rocket injector assembly that delivers liquid oxygen and hydrogen into the engine's combustion chamber, may lead to more efficient manufacturing in the future.

EWI
We Manufacture Innovation

Examples of AM (3D Printing) in the mainstream media

GE's Making 3D Printed Jets

Kyle Maxey posted on May 09, 2013 | 2 Comments

Like 24 Tweet 15 +1 2 Share 9

The prestigious MIT Technology Review has named General Electric's work on 3D printing as one of its '10 Breakthrough Technologies 2013'.

Specifically, GE is developing a metal nozzle for its LEAP jet engine. They will use 3D metal printers to produce the nozzles, which will be lighter in weight due to an advanced design producible only on 3D printers.

Wait, we've heard this story many times before. A company improves their business process by involving 3D printing. What's the big deal?

The big deal here is the volume. GE is MASS PRODUCING these nozzles with 3D printing.

Read More about GE's CEO opinion of 3D printing.

Read More at Fabbaloo

EWI
We Manufacture Innovation

AM Processes for Metals

- Laser and EB powder bed, from companies such as EOS, Concept Laser, SLM, Renishaw, Phenix, and Arcam (EBM) in confined envelope (g/hr)
- EBW freeform fabrication - EB(FFF) (kg/hr)
- Laser powder and wire FFF from companies such as POM, Optomec (LENS) (kg/hr)
- **VHP UAM – high power ultrasonic AM of strip (kg/hr)**
- **Emerging - Arc processes – SMD, MER, GTAW-HW, GMAW-P, PTA (wire and powder) for FFF (kg/hr) Deposition rates from 1 in^3/hr to ~100's in^3/hr**

EWI
We Manufacture Innovation

Metal AM Processes

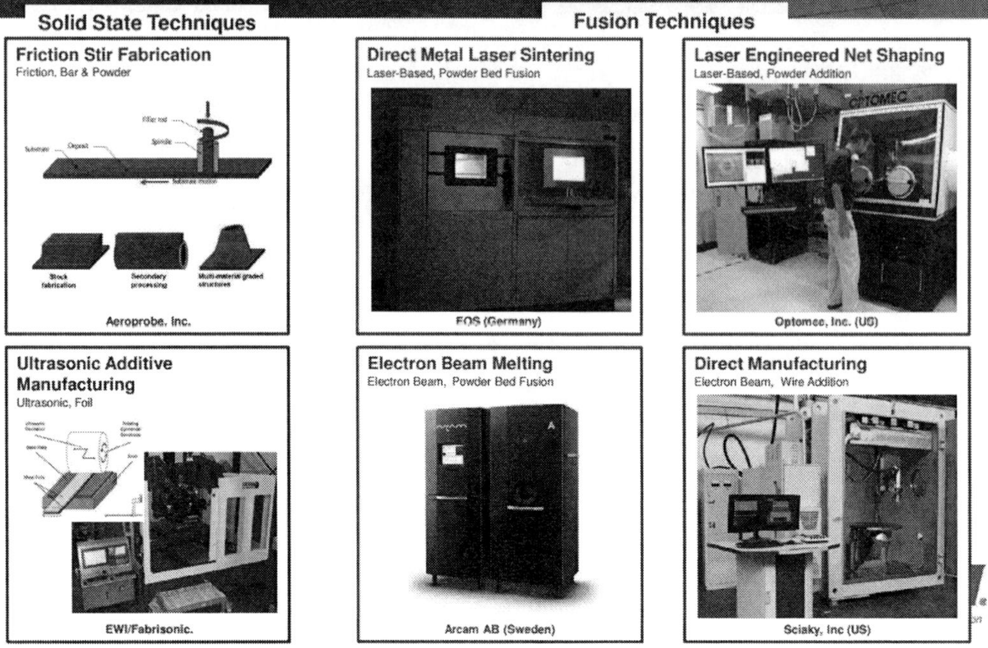

Deposition Rate versus Feature Resolution – More tools

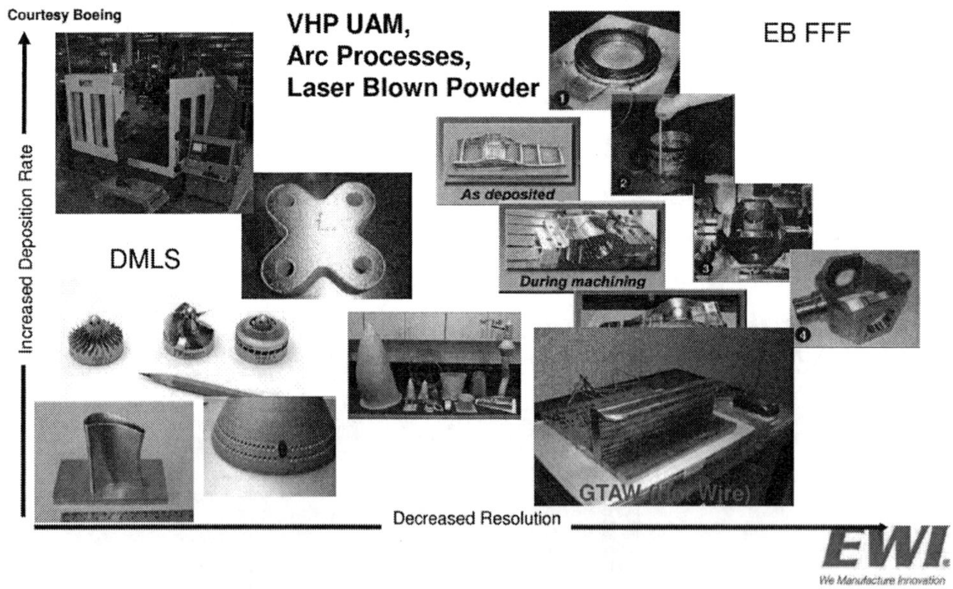

AM is Material Creation

Recognize that AM is the joining of volume elements of material to create a larger functional volume.

LENS deposition of Ti-6Al-4V blocks

- 2943 Individual Tracks
- 490 linear feet of deposit.

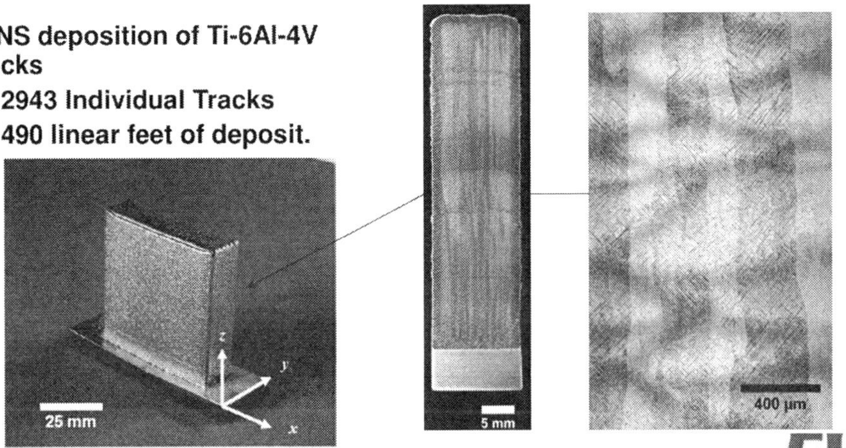

S.M. Kelly, ARL Penn State, 2011

EWI
We Manufacture Innovation

Ti-6-4 Vehicle Control arm with GTAW-HW

- First layer, and completed deposit, 80 lb. (36 kg) part on 36 x 36 in plate (~1m by 1m) – layer-by layer approach – software required for true CAD-part AM
- Aerospace and nuclear qualified GTAW process
- Ubiquitous equipment and potential supply chain

EWI
We Manufacture Innovation

GTAW-HW Distortion Control

- End and side views showing low level of distortion at 4 lbs/hr deposition rate

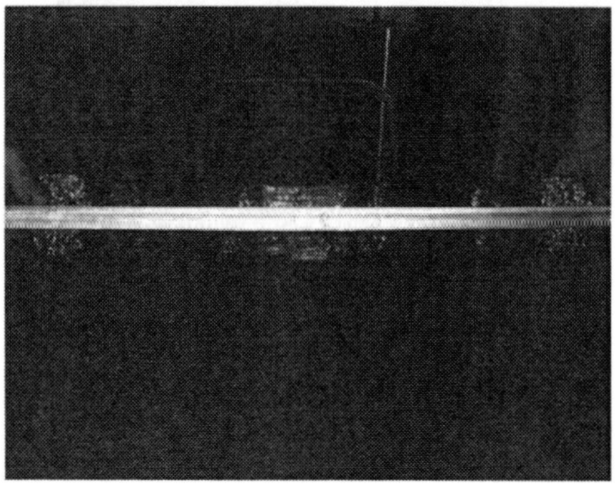

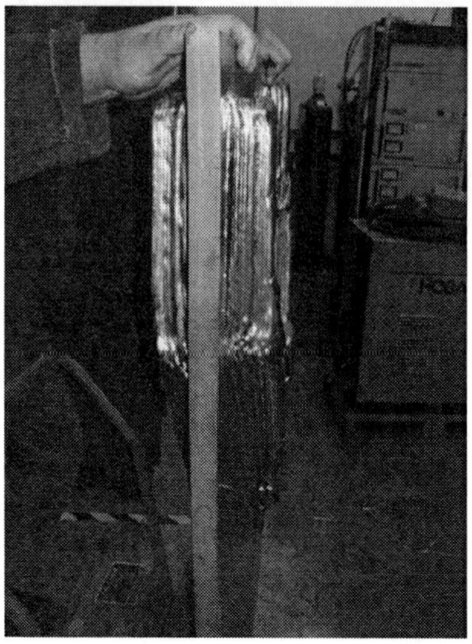

Arc-Based Additive Manufacturing

- Demonstrated arc-based processes for low cost, high deposition rate Titanium additive manufacturing 4-9.5 lbs/hr.
  - GMAW-P
  - RWF-GMAW
  - PAW (Cold Wire)
  - PTA (Powder)
  - GTAW (Hot Wire)

EDX Analysis of Weld Deposit

- 2nd layer of weld metal
- Trail shields to meet low (<0.20%) oxygen requirements
- Deposition rates of 4.5-9.0 lbs/hr demonstrated

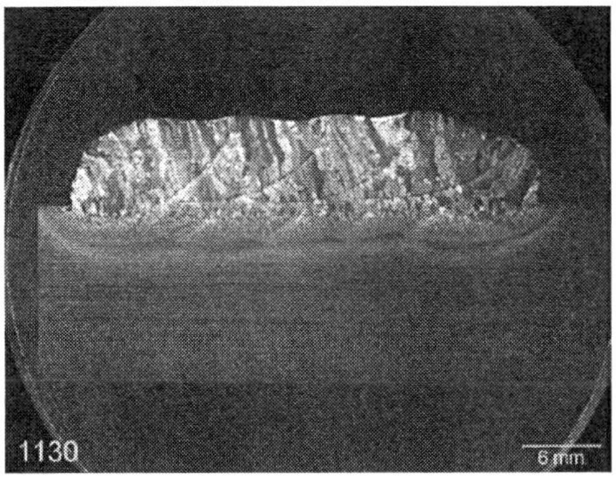

- Element Weight (%)
- Al K 6.09
- Ti K 89.52
- V K 3.91
- Fe K 0.14
- W M 0.34
- Totals 100.00

- Meets Ti-6-4 chemistry requirements with COTS wire

EWI
We Manufacture Innovation

GTAW-HW Work with Ti-6-4 ELI

- Conventional GTAW-HW equipment with trail shield

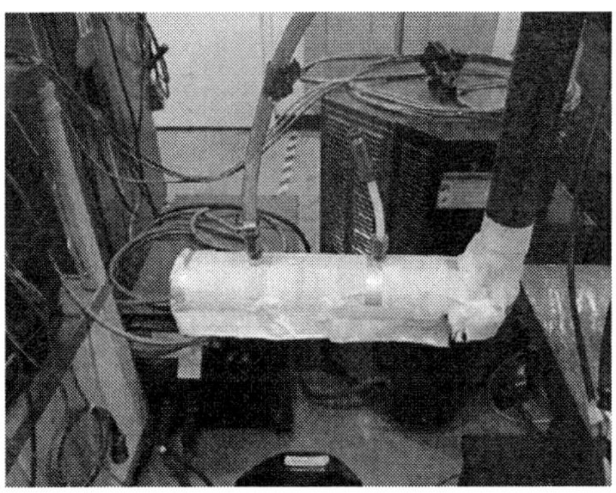

EWI
We Manufacture Innovation

GTAW-HW work with Ti-6-4 ELI

- GTAW-HW at 4kW arc power
- Ti-6-4 ELI (Grade 23) wire 0.045-in diam.

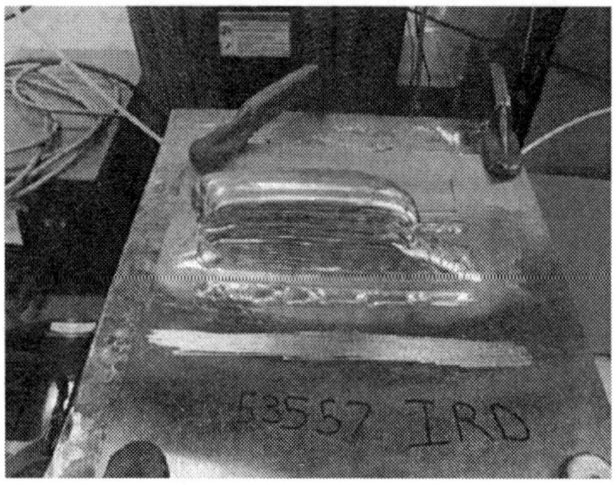

EWI
We Manufacture Innovation

Metallographic Section

- No lack of fusion or porosity discontinuities
- Sound weld metal Equiaxed grain structure – isotropic properties

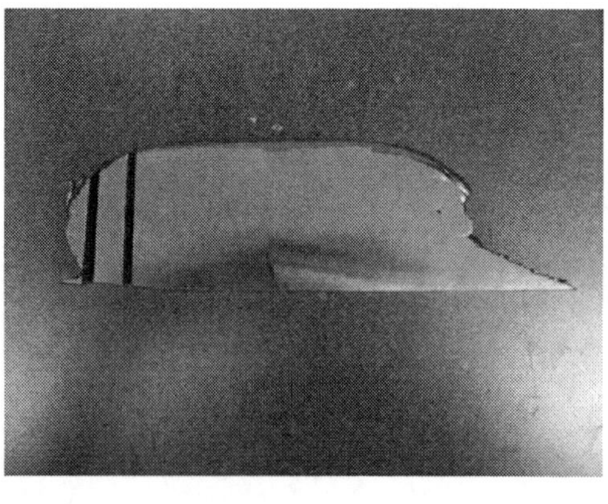

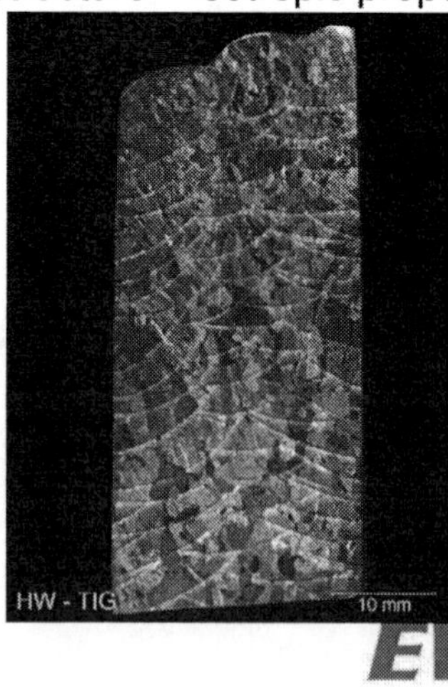

EWI
We Manufacture Innovation

Tensile Test Results

- Tensile and yield strength of the buildup exceeds the requirements of Ti 6-4 ELI listed on MatWeb

Spec ID	Test Results		Matweb	
	UTS (ksi)	Yield Strength (ksi)	UTS (ksi)	Yield Strength (ksi)
H1	137	124.5	125	115
H2	133.4	116.1		
V1	136.3	119.2		
V2	134.6	117.4		

EWI
We Manufacture Innovation

Chemical Analysis

- Leco Furnace Method
- H, N, and O composition of the buildup is much less than the maximum permissible values for Ti 6-4 ELI base metal

Element	Test Result for HW-GTAW Buildup	Matweb Published Requirement for Ti 6-4 ELI
H	0.0013%	≤ 0.0125%
N	0.0078%	≤ 0.030%
O	0.077%	≤ 0.13%

EWI
We Manufacture Innovation

Ultrasonic Additive Mfg (UAM) Commercialization

TECH COLUMBUS InnovationAwards

- New AM process - Increases design flexibility and reduces manufacturing waste enables hybrid material capability
- Enables use of internal channels, sensors, and actuators
- Very high power (9 kW), large envelope (6'x6'x3'), multi-material system for development <u>Wrought strip feedstock</u>
- EWI Formed Fabrisonic LLC, builds/sells systems – 3 sizes

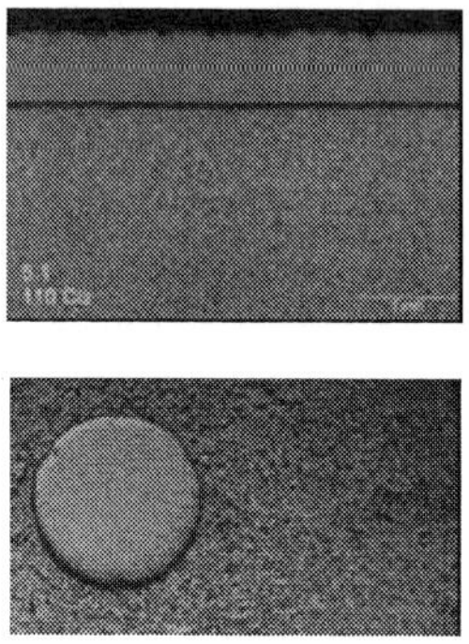

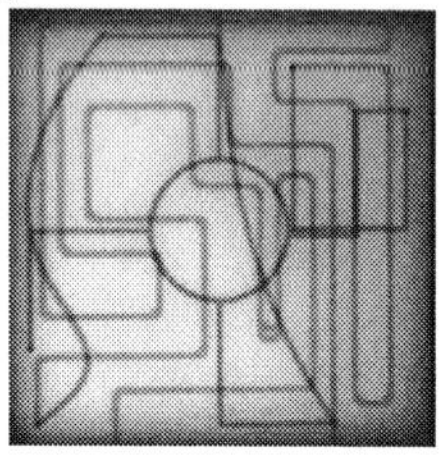

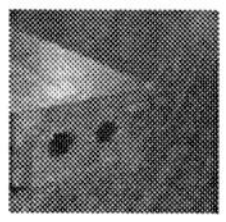

EWI
We Manufacture Innovation

ULTRASONIC ADDITIVE MANUFACTURING

… a new technology – "UAM" creates net-shape solid metal parts using solid state ultrasonic metal welding and CNC contour milling

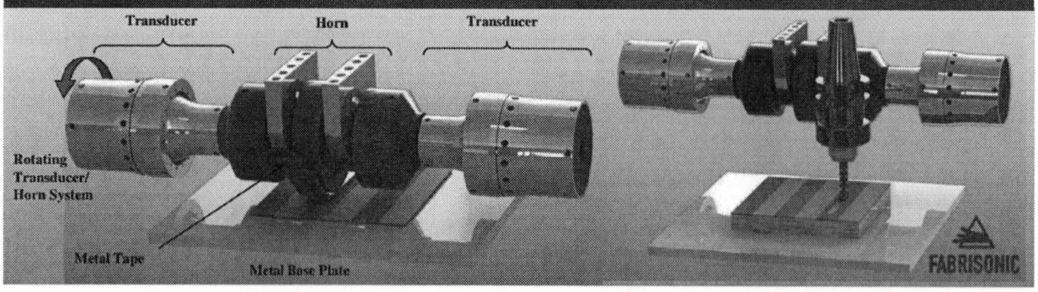

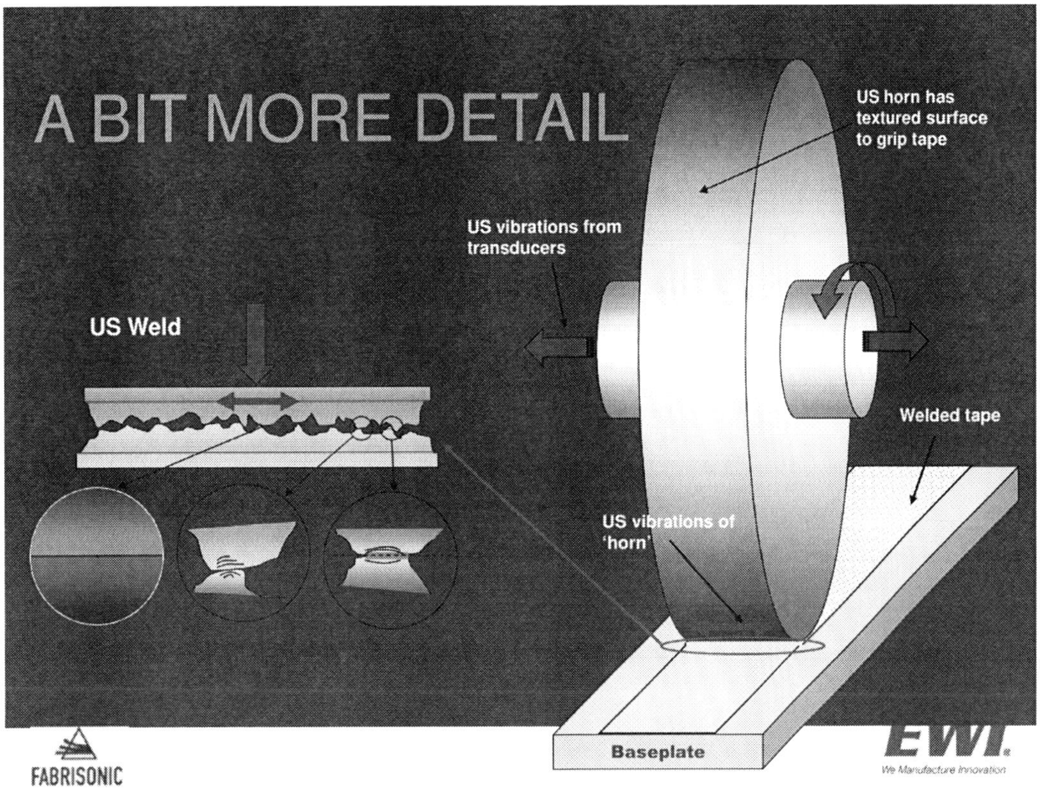
A BIT MORE DETAIL
US horn has textured surface to grip tape
US vibrations from transducers
US Weld
Welded tape
US vibrations of 'horn'
Baseplate
FABRISONIC
EWI
We Manufacture Innovation

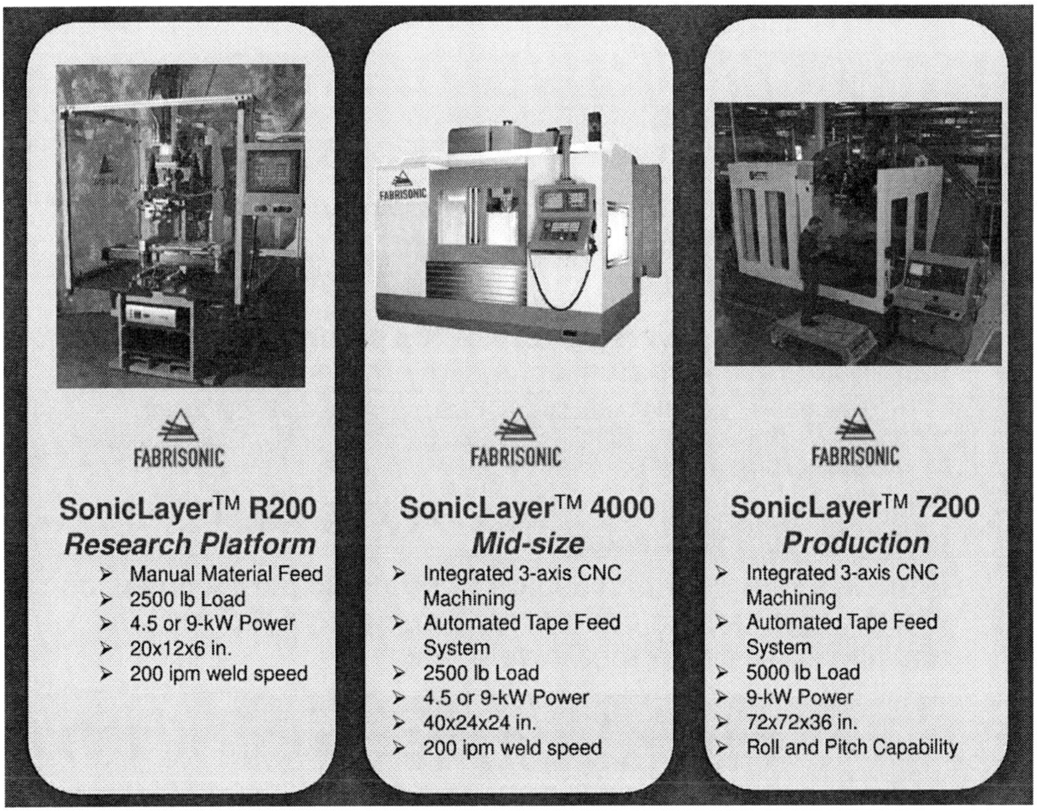
FABRISONIC
SonicLayer™ R200
Research Platform
➢ Manual Material Feed
➢ 2500 lb Load
➢ 4.5 or 9-kW Power
➢ 20x12x6 in.
➢ 200 ipm weld speed
FABRISONIC
SonicLayer™ 4000
Mid-size
➢ Integrated 3-axis CNC Machining
➢ Automated Tape Feed System
➢ 2500 lb Load
➢ 4.5 or 9-kW Power
➢ 40x24x24 in.
➢ 200 ipm weld speed
FABRISONIC
SonicLayer™ 7200
Production
➢ Integrated 3-axis CNC Machining
➢ Automated Tape Feed System
➢ 5000 lb Load
➢ 9-kW Power
➢ 72x72x36 in.
➢ Roll and Pitch Capability

AMC Advancing Manufacturing Readiness – Crossing the 'Valley of Death'

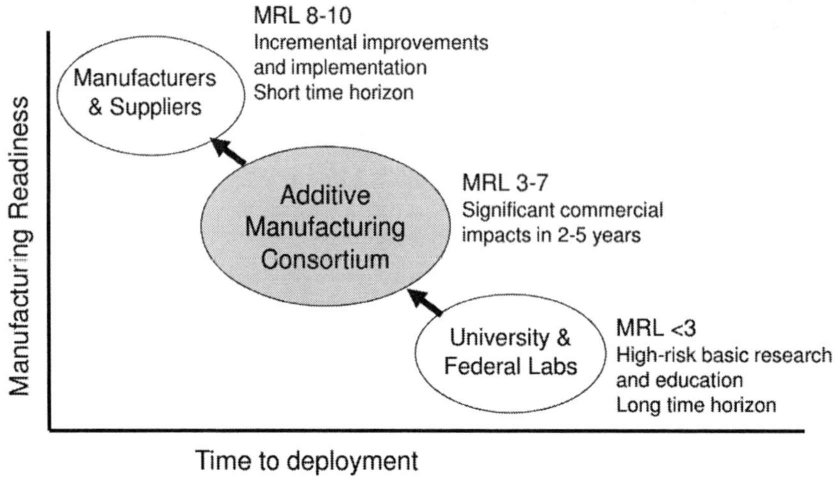

EWI
We Manufacture Innovation

The Additive Manufacturing Consortium

Mission: Accelerate the technology/innovation & advance manufacturing readiness of AM technologies

Goals:

- **Bring together a diverse group of AM practitioners and stakeholders across Academia, Government, and Industry**
  - ~30 members including aerospace OEMs, machine and equipment suppliers, powder metal supplier, service bureau(s), national labs, leading researchers
- **Present example case studies of applying AM approach / principles for manufacturing**
- **Provide a forum for discussion of strategic perspectives on where organizations are currently applying AM principles and how to foster further development**

Database Development is an Urgent Industry Need

EWI
We Manufacture Innovation

AMC – NIST Project and JIP

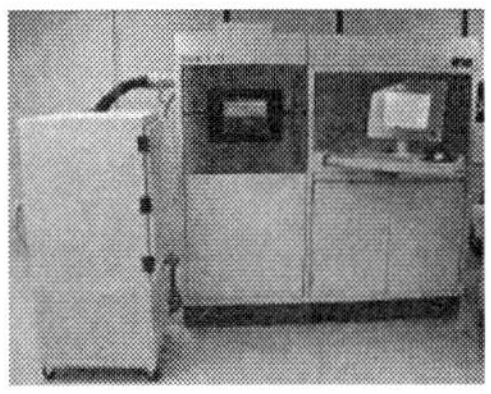

- **NIST Project – DMLS of IN625**
  - IN 625 Laser powder bed with EOS M270
  - $160K, EWI project 53776GTH
  - **Phase 1 Mine available AM property data available from consortium members and other sources globally**
  - **Phase 2 Develop test protocol, identify available testing resources**
  - **Phase 3 Execute test plan, analyze data, create database structure within MMPDS, deploy data, and incorporate feedback.**
- **AMC JIP for DMLS of IN625**
  - IN 625 Laser powder bed with EOS M270
  - $160K, EWI project 53449CPQ (open to more sponsors)
  - Shared Data Set (SDS) with Fixed Process Agreement (FPA) - MMPDS methodology
  - Execute test plan, analyze data, create database structure within MMPDS, deploy data, and incorporate feedback

EWI
We Manufacture Innovation

What we are testing

- **EOS DMLS M270/280 platforms at 200W.**
  - Nickel alloy 625 parameter set.
  - Argon and comparative nitrogen environment
- **Full Heat Treatment: SR, HIP, SA**
- **Coupons are from blocks, not near net shape or individual cylinders.**
- **Machined surface finish.**
- **Properties in two orientations: X-Y and Z.**

Provides a solid foundation for the dataset. Future testing should include effect of thickness, surface finish, recycled powder, other layer thickness, PBF platforms

EWI
We Manufacture Innovation

Outcomes of the NIST JIP Program

- Standard test build to characterize material property data (process agnostic)
- Manufacturing Plan (33 pages)
  - Fixed process agreement: defines requirements
  - Process Control Document: ensures requirements are met and process is documented.
- The goal of the program is to validate the FPA for generating data.

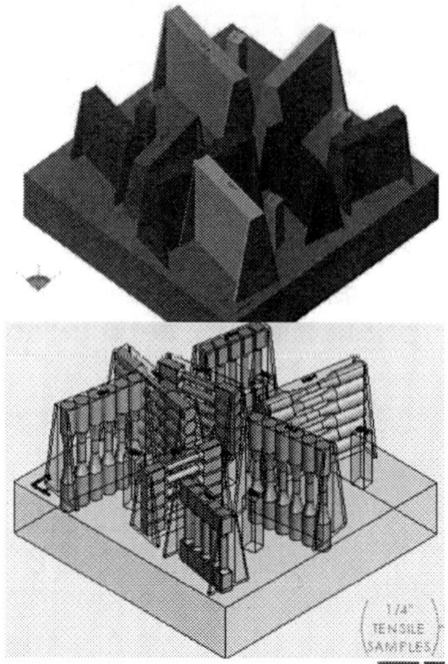

EWI
We Manufacture Innovation

Current Scope

- 1 powder lot, 4 build lots
- 2 Build Locations (NIST and Directed Manufacturing)
- EOS M270 DMLS machines.
- Full HT and NDI
- NADCAP testing
- Metallography
- Goal
  - Supports design allowable generation
  - Codes and Standards development
  - Rapid Qualification strategies.

Material Source		Single Heat Nickel Alloy 625			
Parameter Set		EOS IN625 (M270)			
Environment		Nitrogen		Argon	
Machine		A	B	C	B
Test Temp	Test-Orientation	Build 1 (Lot 1)	Build 2 (Lot 2)	Build 3 (Lot 3)	Build 4 (Lot 4)
RT	Tensile-XY	4	4	4	4
	Tensile-Z	4	4	4	4
	LCF-XY	3	2	3	2
	LCF-Z	2	3	2	3
	HCF-XY	3	2	3	2
	HCF-Z	2	3	2	3
1000°F	Tensile-XY	2	2	2	2
	Tensile-Z	2	2	2	2
	LCF-XY	2	2	2	2
	LCF-Z	2	2	2	2
	HCF-XY	2	2	2	2
	HCF-Z	2	2	2	2

EWI
We Manufacture Innovation

In-Situ Process Monitoring for PBF-L

- EWI/NIST/NAMII project - Development of open architecture PBF-L test bed and application of in-situ monitoring using thermal, acoustic, and optical sensors and HSV, with GE Aviation and EOS $2.25M
- Use of physical sensors and model based in situ process measurement for the detection of material and geometric defects
- Vision - Foundation program for in-situ process monitoring to achieve feed forward process control

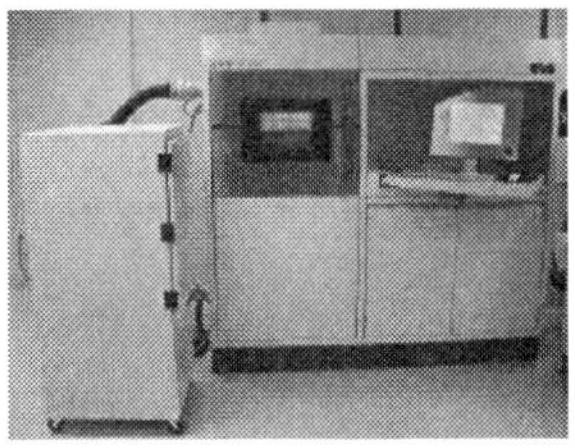

EWI
We Manufacture Innovation

Sensor test bed

- Utilizes ScanLab optics, IPG lasers at EWI

EWI
We Manufacture Innovation

In Process Sensing

- **EWI Leading the 'Informative In-Situ Process Monitoring' Team ($2.2M)**
  - **EWI**: L-PBF/EOS M280, sensor test bed, high speed video, digital imaging, ultrasonic inspection; leverages experience in developing sensing and data fusion strategies for welding.
  - **UNC Charlotte**: Alternative visual techniques leveraged from metrology background
  - **Georgia Tech**: Virtual Defect Simulation
  - **Stratonics**: Thermal sensors
  - **B6Sigma**: Data acquisition
  - **3DS/Paramount**: Off-ramp to Polymer Processing
  - Industrial Partners: **GE Aviation, EOS**

EWI
We Manufacture Innovation

Integrating AM with other processes

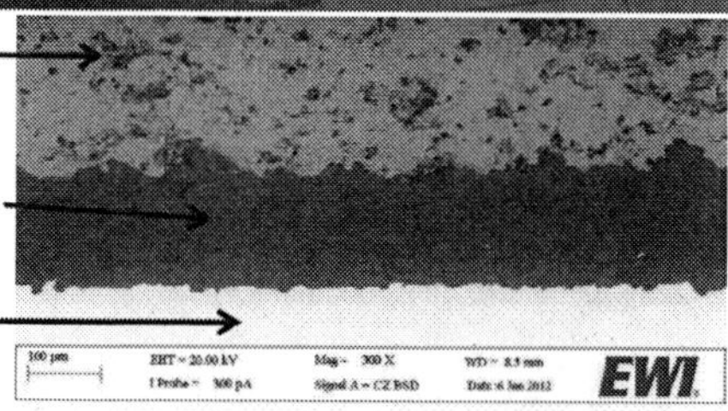

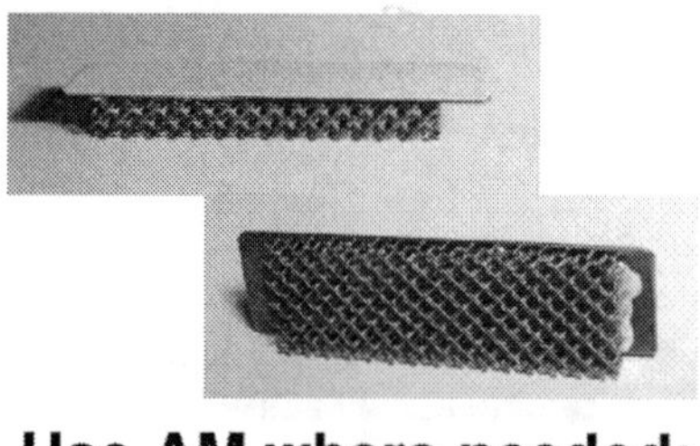

- **Use AM where needed; use conventional processes elsewhere.**

Gratefully Acknowledge Ola Harrysson and Tim Horn at NCSU for lattice structures

EWI
We Manufacture Innovation

Summary

- Metals AM is rapidly developing with a growing suite of technologies providing both complementary and competing attributes
- Develop, optimize and inspect as a weldment!
- A whole new industry and supply chain is in formation for this exciting field
- The Additive Manufacturing Consortium (AMC) offers precompetitive collaboration and works with NAMII/America Makes
- EWI/AMC work together to increase MRL/TRL in the metals AM field
- In-process sensing and monitoring is a key need and being developed

Additive Manufacturing Consortium

EWI

Questions?

Ian D. Harris, Ph. D.
Founding Director, AMC
Technology Leader
614 440 1277
iharris@ewi.org

EWI

THE VALUE OF PERFORMANCE.
NORTHROP GRUMMAN

Surface Finish Enhancement of Ti6Al4V Components Fabricated via Electron Beam Additive Manufacturing

February 25th, 2013

Dr. Eric Fodran
Northrop Grumman Manufacturing Technology Development

NORTHROP GRUMMAN

Ti-6Al-4V Additive Manufacturing (AM): Need

- Manufacturing and design needs of titanium components on developing/ demonstration aircraft, as well as the demands of production platforms, require titanium components with:
  - Improved cost
  - More rapid means of fabrication
  - Increased unitization
  - Reduced buy-to-fly (10-20:1 common)
  - Need for unique manufacturing capability to meet designs

Implementation of AM Components Allows for Rapid, Affordable Manufacturing of Complex Titanium Components

NORTHROP GRUMMAN

Ti-6Al-4V Additive Manufacturing: Challenges

Challenges:

- Current surface finish of E-Beam Additive Manufactured components results in excessively rough surface finish from the standpoint of structural, fatigue rated components
- Current surface finish prohibits baseline surface NDI techniques (penetrant, Eddie current, UT, etc.) for accurate quality assurance.

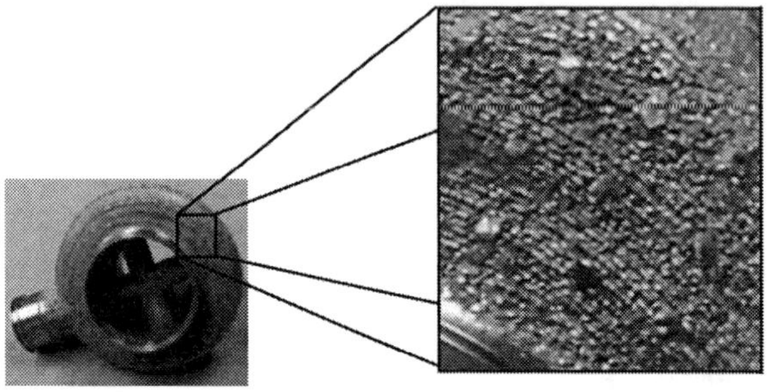

Addressing Key Challenges Will Enable the Beneficial Implementation of Titanium AM Components Providing Rapid, Affordable Manufacturing

DISTRIBUTION STATEMENT A. APPROVED FOR PUBLIC RELEASE

NORTHROP GRUMMAN

Program Objectives

- Optimize surface finish of E-Beam Titanium via optimization AM process and input variables
- Attain surface finish comparable to traditional wrought, subtractive manufacturing methods (124 μin) via AM process parameter modification and subsequent post AM surface finish operations
- Characterize static and dynamic mechanical performance of AM fabricated product, targeting annealed Ti-6Al-4V (AMS 4911) equivalency
- Fabricate full scale demonstration article with targeted 124 μin surface finish

4

DISTRIBUTION STATEMENT A. APPROVED FOR PUBLIC RELEASE

Potential Technology Overall Solution\Approach

- **Fabrication Process Optimization**
  - Detailed evaluation of impact of process parameters on surface finish
    - Powder size\distribution
    - Layer thickness
    - Energy Source
      - Single\multiple beam
      - Power\focus offset
      - Beam speed
- **Post process surface finish application**
  - Abrasive slurry
  - Chemical, chemical assisted polishing
  - Engineered aggregate
- **Properties verification and characterization of surface finish related deleterious effect on performance**

Digital Direct Manufacturing: Program Team

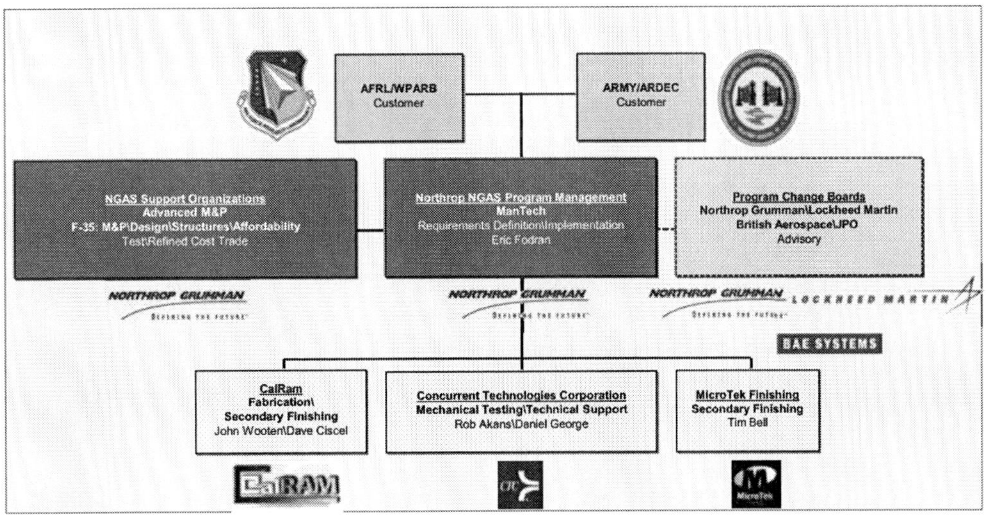

NORTHROP GRUMMAN

Alternative Powder Screening

- **Evaluated Several Alternative Ti-6Al-4V Powders for Potential DOE Inclusion**
  - Targeted attributes of interest
    - Size\distribution
    - Geometry
    - High Apparent Density
    - Low Internal Porosity

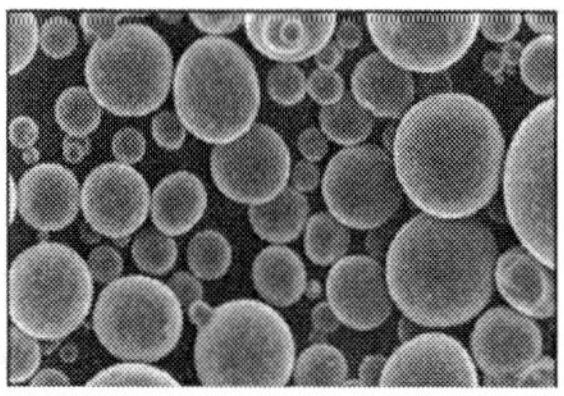

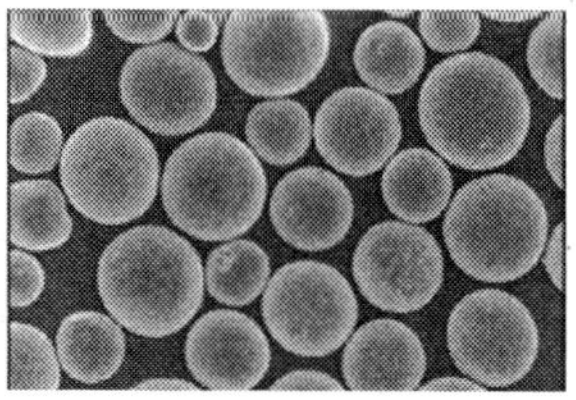

- **Identify Available Candidate for Subsequent Consideration**

7

DISTRIBUTION STATEMENT A. APPROVED FOR PUBLIC RELEASE

NORTHROP GRUMMAN

Candidate Alternative Powder Evaluation :Arcam

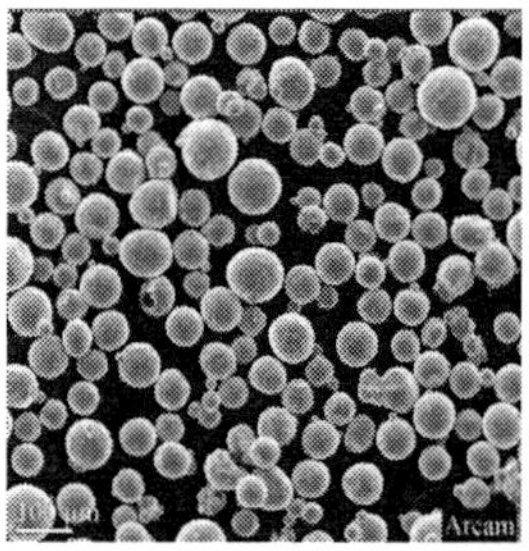

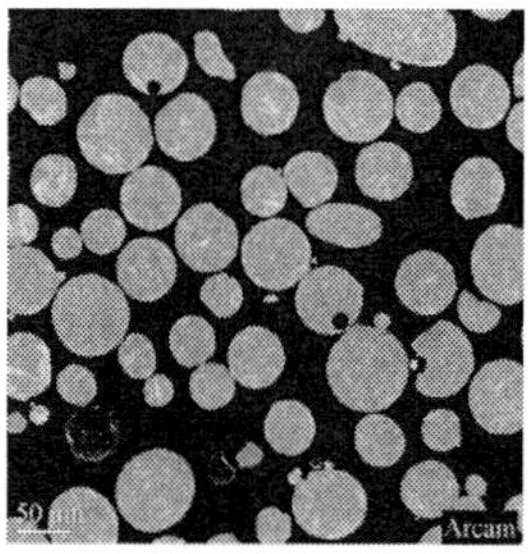

- Broad size distribution
- Good spherodicity
- Significant satellites and internal void density

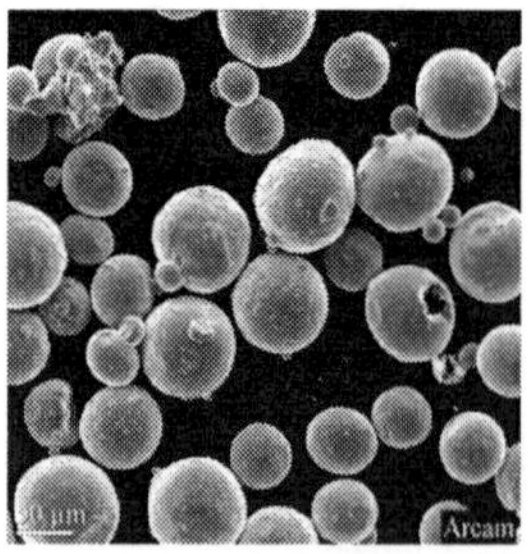

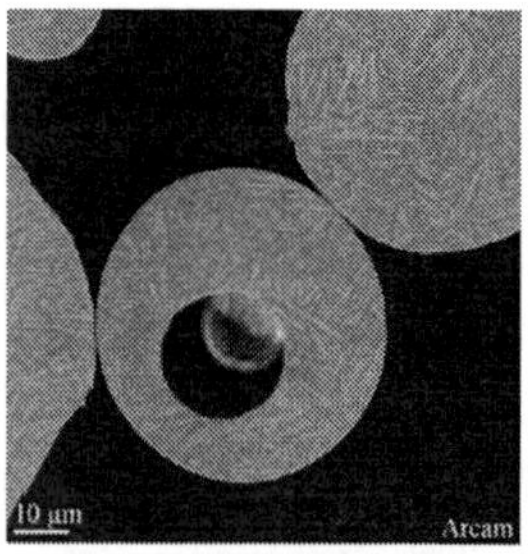

DISTRIBUTION STATEMENT A. APPROVED FOR PUBLIC RELEASE

Candidate Alternative Powder Evaluation :Timet

NORTHROP GRUMMAN

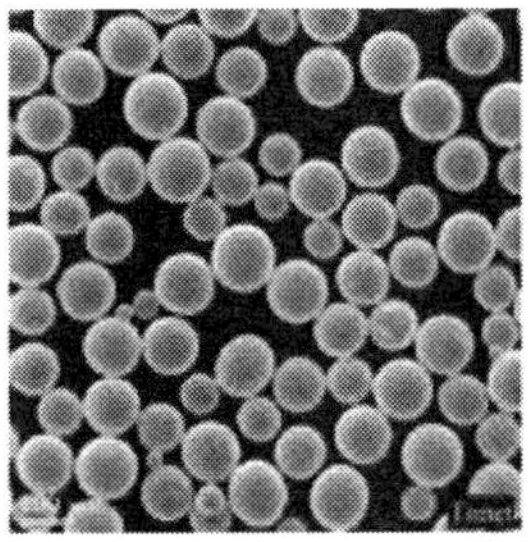

- Moderate size distribution
- Excellent spherodicity
- Minimal satellites and internal void density

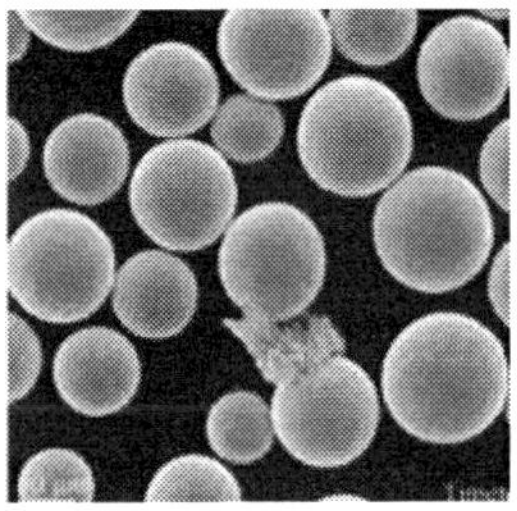

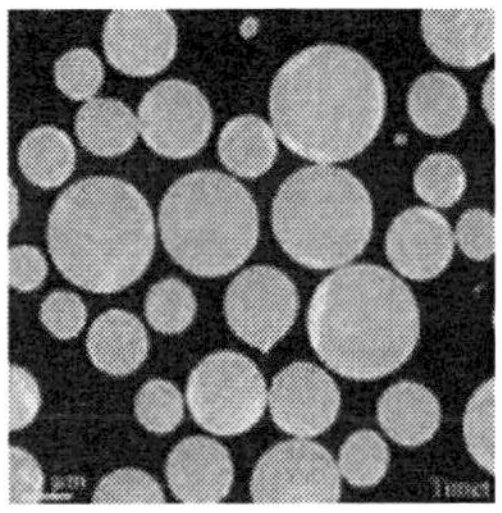

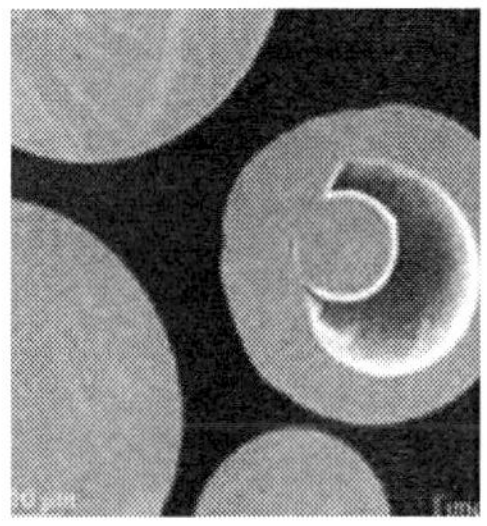

9

DISTRIBUTION STATEMENT A. APPROVED FOR PUBLIC RELEASE

Candidate Alternative Powder Evaluation : Ametek

NORTHROP GRUMMAN

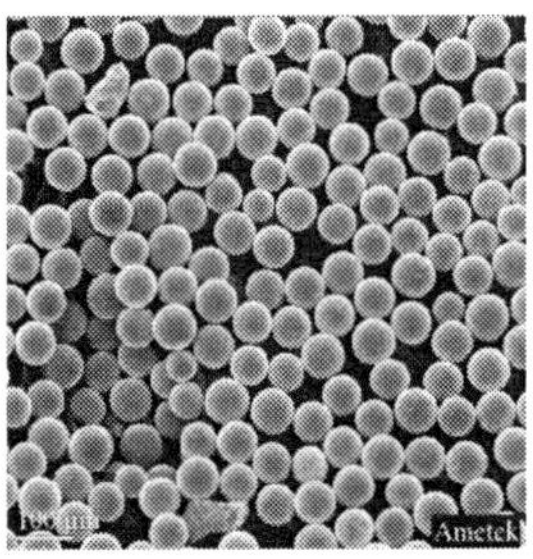

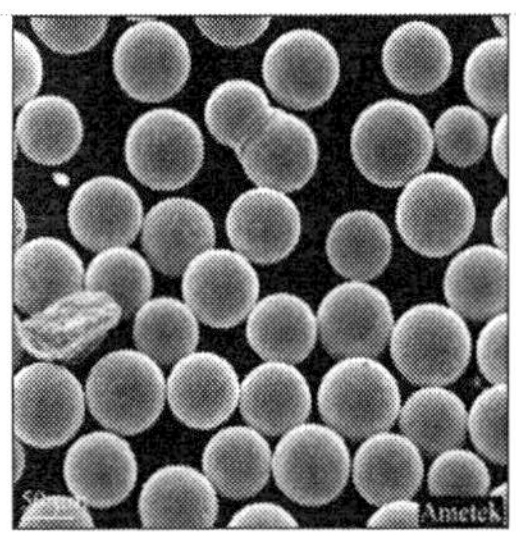

- Narrow size distribution
- Excellent spherodicity
- Minimal satellites and internal void density
- Oxide layer observed on a small population of particles

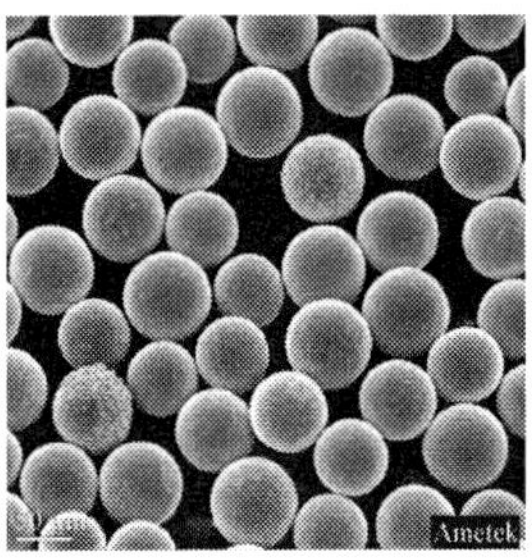

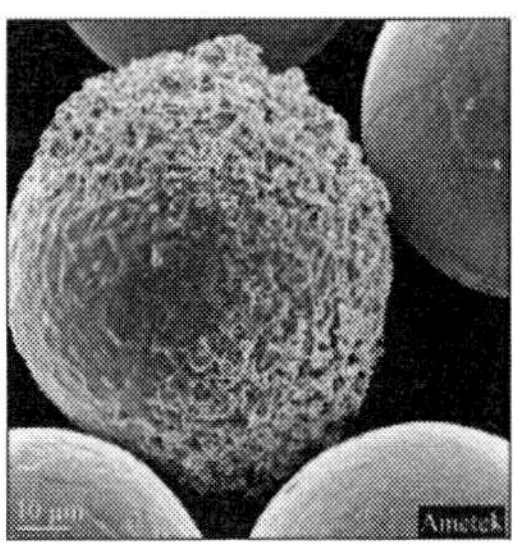

10

DISTRIBUTION STATEMENT A. APPROVED FOR PUBLIC RELEASE

Candidate Alternative Powder Evaluation :FMW

NORTHROP GRUMMAN

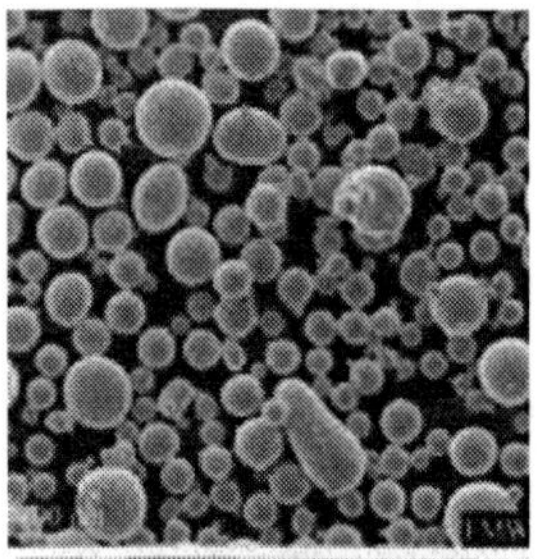

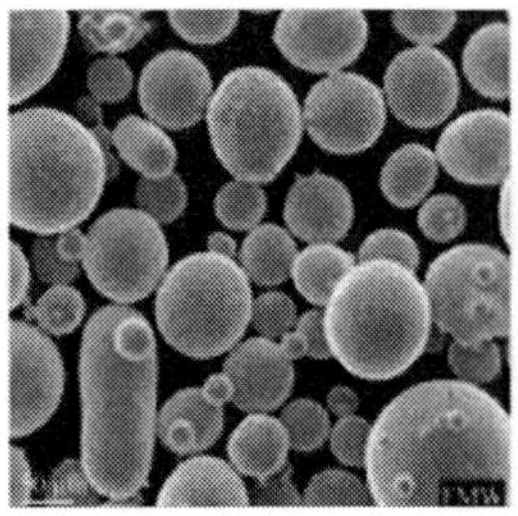

- Moderate size distribution
- Significant density of non-spherical powders
- Moderate satellites and internal void density

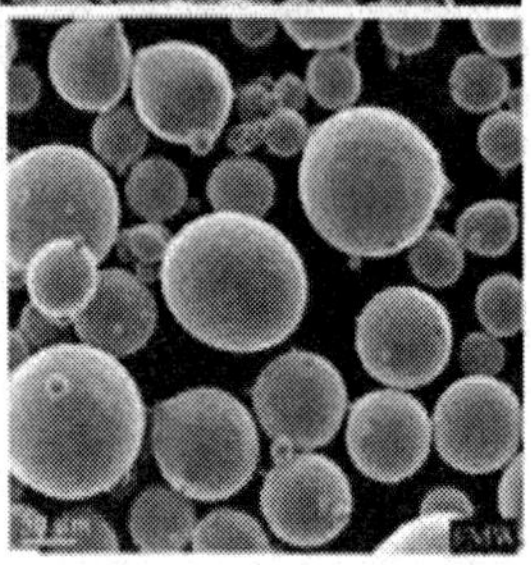

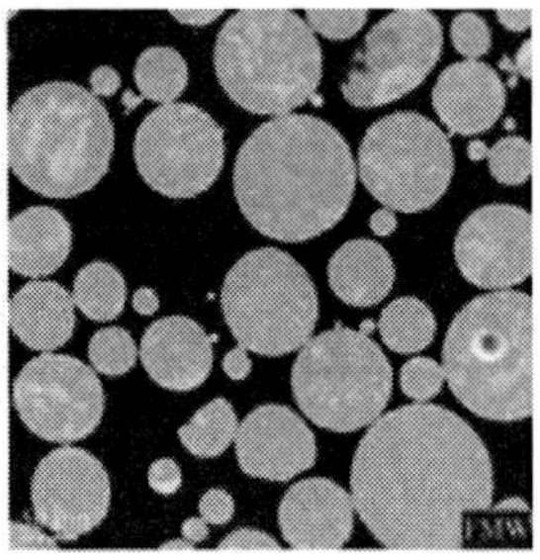

11

DISTRIBUTION STATEMENT A. APPROVED FOR PUBLIC RELEASE

Histogram by Frequency

NORTHROP GRUMMAN

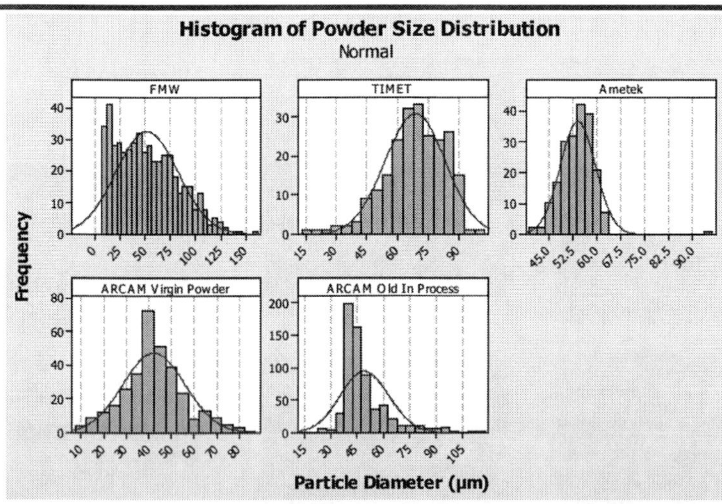

Variable	N	Mean	StDev	Min	Q1	Median	Q3	Max
FMW	503	52.4	30.76	10.61	25.98	48.6	73.48	158.83
TIMET	226	68.555	14.651	16.77	60.402	70.15	79.37	97.78
ARCAM Virgin Powder	329	42.456	13.776	10.61	33.96	41.76	50.03	86
ARCAM Old In Process	654	49.032	13.78	15.91	41.08	44.68	51.48	117.03
Ametek	203	54.012	5.530	39.680	50.590	54.60	57.610	96.190

DISTRIBUTION STATEMENT A. APPROVED FOR PUBLIC RELEASE

Alternative Powder Screening

NORTHROP GRUMMAN

Supplier	Max Particle Size (Micron)	Min Particle Size (Micron)	Internal Void	Satellites	Non-Spherical	Composition Verification
Timet	98	17	very few	none	very few	Yes
Ametek	96	40	very few	none	very few	Yes
FMW	158	11	moderate	moderate	moderate	Yes
ARCAM	117	10	significant	significant	moderate	Yes

- **Arcam - Baseline gas atomized powder**
  - Broad size distribution, high density of satellite and internal voids
- **Timet PM – PREP (Plasma Rotating Electrode Process)**
  - Narrow distribution, fewest degree of non-spherical particluates\defects, "tunable" size distribution by rod rotation speed (~$400/kg)
- **Ametek - Low Pressure Plasma Spray spheriodization of low cost, non-spherical powder from the hydride-dehydride process**
  - Low cost and narrowest size distribution (~$250/kg)
- **Other GA powders considered were, but were not procured:**
  - ATI (previously Crucible Research) – Potential challenges with quality (ceramic contamination from atomization nozzle) and cost – small capacity plant
  - FMW – Unable to meet CalRAM PO requirements – particle size distribution

EBM AM In-Process Surface Finish Optimization and Characterization

NORTHROP GRUMMAN

- **Developed and executed comprehensive DOE to evaluate EBM AM process parameters and identify parameters for most refined surface finish**
  - Targeted Attributes
    - Layer thickness
    - Powder size\distribution
    - Energy source\beam characteristics
      - Qty of beams
      - Beam size\focus beam offset
      - Beam velocity
      - Preheat duration
- **Coupons with representative geometric features (pocket depth, fillet radii, etc.) to be fabricated**

Design of Experiment Development and Execution

NORTHROP GRUMMAN

Quantity of Beams	Beam Speed	Powder	Layer Thickness	Focus Offset
25	6	Arcam	25 Micron	0 mA
25	6	Timet	40 Micron	1.5 mA
25	6	Ametek	50 Micron	3 mA
25	6	Arcam	25 Micron	1.5 mA
25	6	Timet	40 Micron	3 mA
25	6	Ametek	50 Micron	0 mA
25	2	Arcam	25 Micron	3 mA
25	2	Timet	40 Micron	0 mA
25	2	Ametek	50 Micron	1.5 mA
25	2	Arcam	40 Micron	3 mA
25	2	Timet	50 Micron	0 mA
25	2	Ametek	25 Micron	1.5 mA
25	2	Arcam	40 Micron	3 mA
25	2	Timet	50 Micron	0 mA
25	2	Ametek	25 Micron	1.5 mA
100	6	Arcam	40 Micron	0 mA
100	6	Timet	50 Micron	1.5 mA
100	6	Ametek	25 Micron	3 mA
100	6	Arcam	40 Micron	1.5 mA
100	6	Timet	50 Micron	3 mA
100	6	Ametek	25 Micron	0 mA
100	6	Arcam	50 Micron	1.5 mA
100	6	Timet	25 Micron	3 mA
100	6	Ametek	40 Micron	0 mA
100	2	Arcam	50 Micron	1.5 mA
100	2	Timet	25 Micron	3 mA
100	2	Ametek	40 Micron	0 mA
100	2	Arcam	50 Micron	3 mA
100	2	Timet	25 Micron	0 mA
100	2	Ametek	40 Micron	1.5 mA
100	2	Arcam	50 Micron	0 mA
100	2	Timet	25 Micron	1.5 mA
100	2	Ametek	40 Micron	3 mA

- 5 Factor DOE developed with up to 3 levels within each factor
- Results in 34 unique sets of process parameters
- 9 experimental builds with 4 coupons within each build to execute all experimental combinations

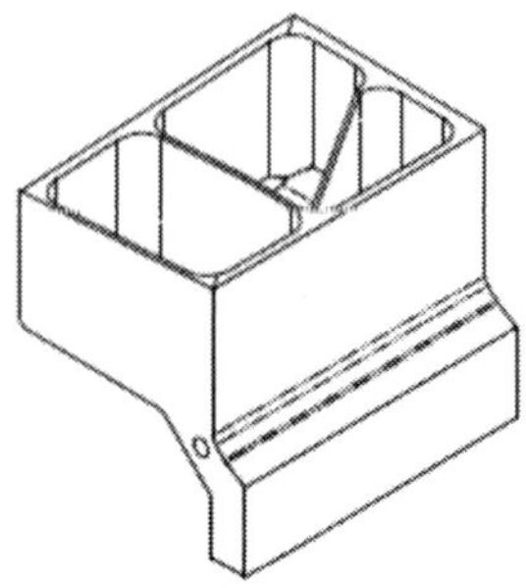

Ti DDM MODEL

Design of Experiment Development and Execution

NORTHROP GRUMMAN

BUILD # 1

Row #	Powder	Layer Thickness	Coupon	Quantity of Beams	Beam Speed	F/O
23	Arcam	50	A	100	6	1.5
26	Arcam	50	B	100	2	1.5
29	Arcam	50	C	100	2	3
32	Arcam	50	D	100	2	0

BUILD # 2

Row #	Powder	Layer Thickness	Coupon	Quantity of Beams	Beam Speed	F/O
11	Arcam	40	A	25	2	3
14	Arcam	40	B	25	2	3
17	Arcam	40	C	100	6	0
20	Arcam	40	D	100	6	1.5

BUILD # 3

Row #	Powder	Layer Thickness	Coupon	Quantity of Beams	Beam Speed	F/O
2	Arcam	25	A	25	6	0
5	Arcam	25	B	25	6	1.5
8	Arcam	25	C	25	2	3
XX	Arcam	25	D	100	6	3

BUILD # 4

Row #	Powder	Layer Thickness	Coupon	Quantity of Beams	Beam Speed	F/O
18	Timet	50	A	100	6	1.5
21	Timet	50	B	100	6	3
15	Timet	50	C	25	2	0
12	Timet	50	D	25	2	0

BUILD # 5

Row #	Powder	Layer Thickness	Coupon	Quantity of Beams	Beam Speed	F/O
3	Timet	40	A	25	6	1.5
6	Timet	40	B	25	6	3
9	Timet	40	C	25	2	0
XX	Timet	40	D	100	6	3

BUILD # 6

Row #	Powder	Layer Thickness	Coupon	Quantity of Beams	Beam Speed	F/O
24	Timet	25	A	100	6	3
27	Timet	25	B	100	2	3
30	Timet	25	C	100	2	0
33	Timet	25	D	100	2	1.5

BUILD # 7

Row #	Powder	Layer Thickness	Coupon	Quantity of Beams	Beam Speed	F/O
4	Ametek	50	A	25	6	3
7	Ametek	50	B	25	6	0
10	Ametek	50	C	25	2	1.5
XX	Ametek	50	D	40	6	3

BUILD # 8

Row #	Powder	Layer Thickness	Coupon	Quantity of Beams	Beam Speed	F/O
25	Ametek	40	A	100	6	0
28	Ametek	40	B	100	2	0
31	Ametek	40	C	100	2	1.5
34	Ametek	40	D	100	2	3

BUILD # 9

Row #	Powder	Layer Thickness	Coupon	Quantity of Beams	Beam Speed	F/O
13	Ametek	25	A	25	2	1.5
16	Ametek	25	B	25	2	1.5
19	Ametek	25	C	100	6	3
22	Ametek	25	D	100	6	0

DISTRIBUTION STATEMENT A. APPROVED FOR PUBLIC RELEASE

DOE Coupon Design

NORTHROP GRUMMAN

- Representative geometric features for structural components within air and space vehicle program integrated into coupon design
  - Wall\stiffener\floor thickness
  - Fillet\corner radius
  - Stiffener height
  - Pocket size
  - Rib\stiffener separation

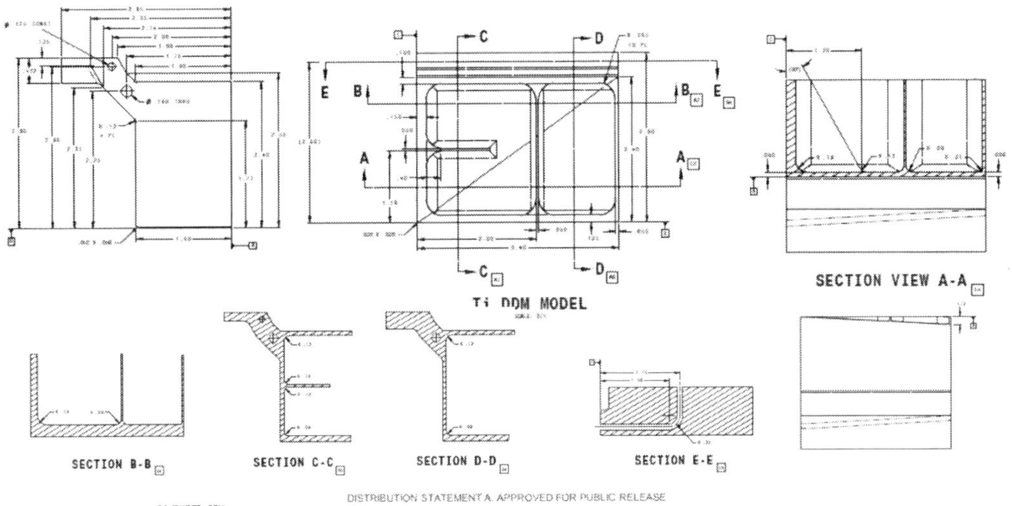

DISTRIBUTION STATEMENT A. APPROVED FOR PUBLIC RELEASE

Surface Finish Evaluation: Unique Locations

NORTHROP GRUMMAN

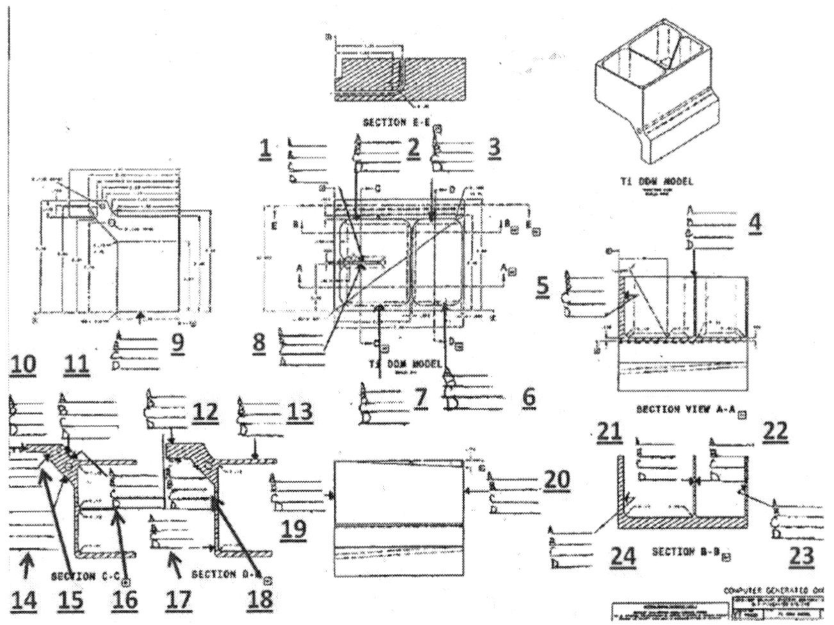

18

DISTRIBUTION STATEMENT A. APPROVED FOR PUBLIC RELEASE

MiniTab Statistical Analysis: Main Effects Plots

NORTHROP GRUMMAN

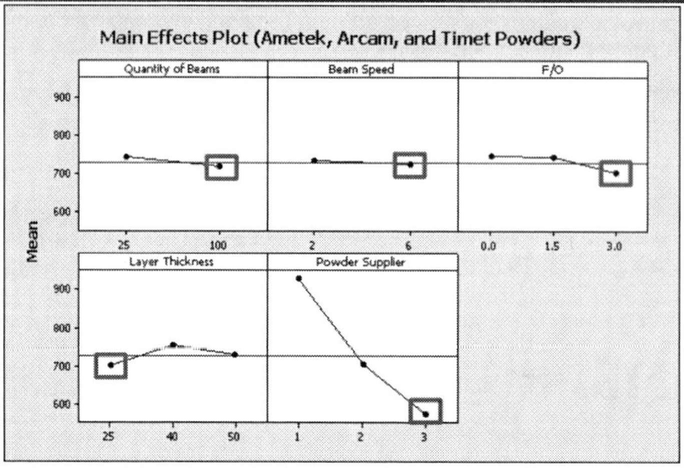

- Optimal Machine Parameters for Lowest Average Surface Roughness
  - Timet
  - Beam Speed: 6
  - F/O: 3
  - Quantity of Beams: 100
  - Layer Thickness: 25

EBM AM Mechanical Properties Assessment

NORTHROP GRUMMAN

- **Characterization of AM fabricated coupons post surface finish optimization DOE**
  - Static and dynamic properties assessment of as AM fabricated surfaces
    - Tensile
    - Fatigue\Stress-Life Curve Development

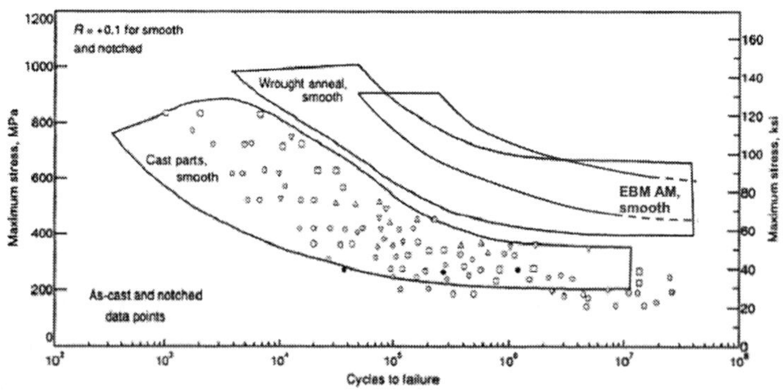

Data points for notched castings are replotted here.
Source: *Metals Handbook*, Vol 3, 9th ed., *Properties and Selection: Stainless Steels, Tool Materials and Special-Purpose Metals*, American Society for Metals, 1980, p411. Most data points are based on $K_t = 3.0$, but some are for smooth specimens ($K_t = 1$). Each symbol represents data from a different source.

Metallography: Transverse

NORTHROP GRUMMAN

- No change in microstructural constituents between legacy Arcam processes and optimized
- Only refinement in overall microstructure observed

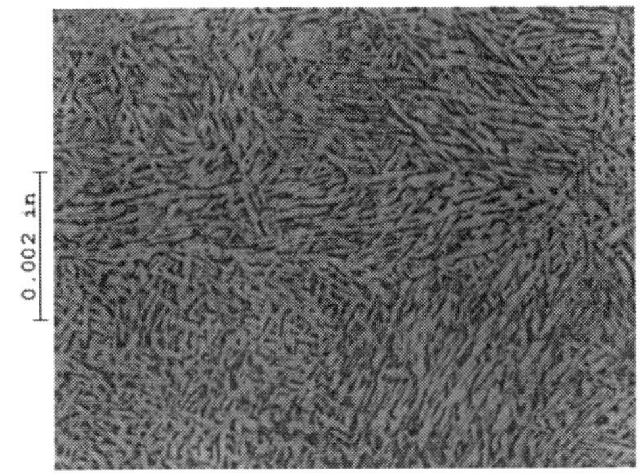

Legacy Arcam

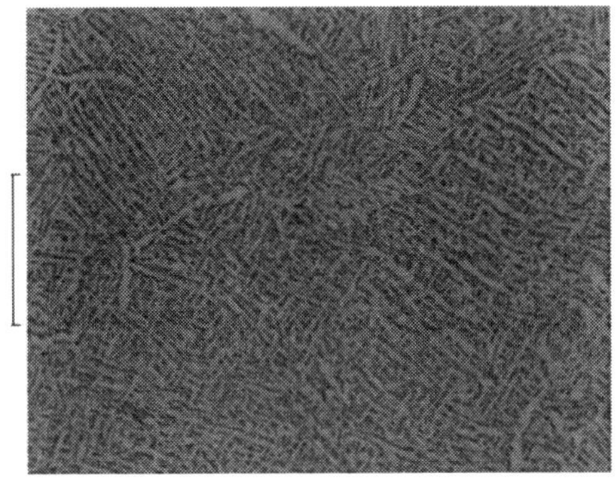

Optimized AM Processes

21

Metallography: Longitudinal

NORTHROP GRUMMAN

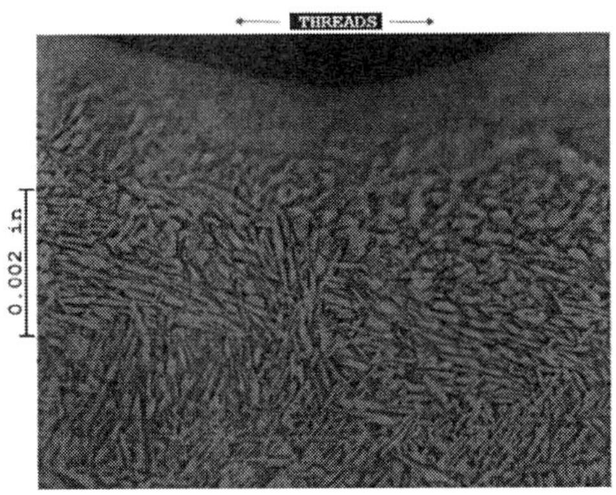

Legacy Arcam

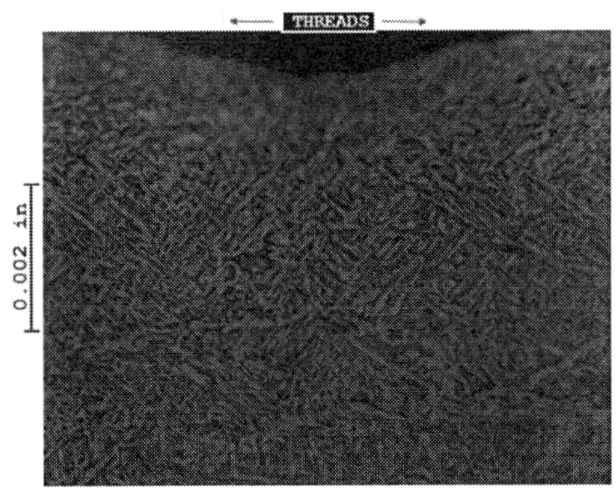

Optimized AM Processes

22

EBM AM Mechanical Properties Assessment: Fatigue Properties

NORTHROP GRUMMAN

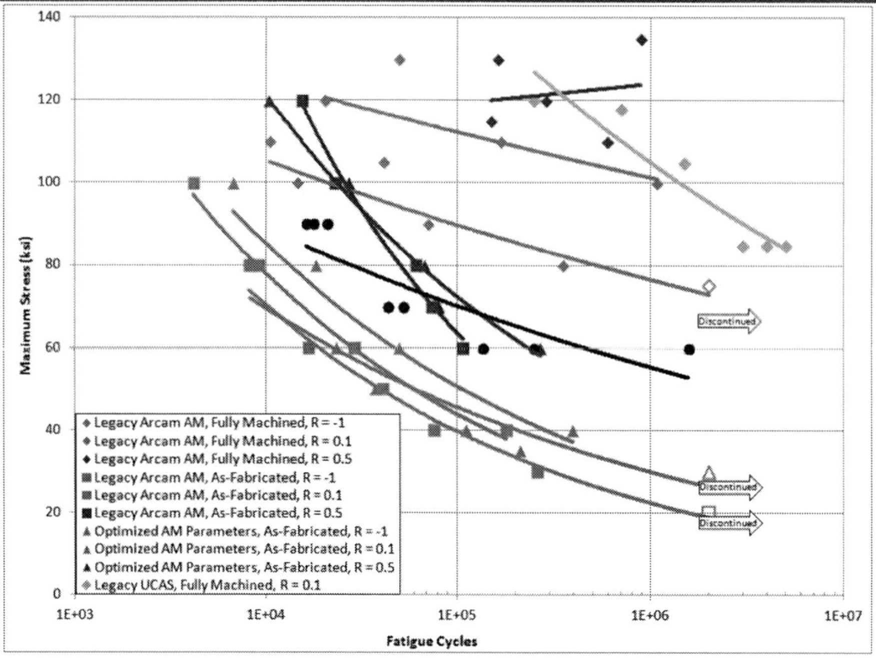

23

DISTRIBUTION STATEMENT A. APPROVED FOR PUBLIC RELEASE

EBM AM Mechanical Properties Assessment: Fatigue Properties

NORTHROP GRUMMAN

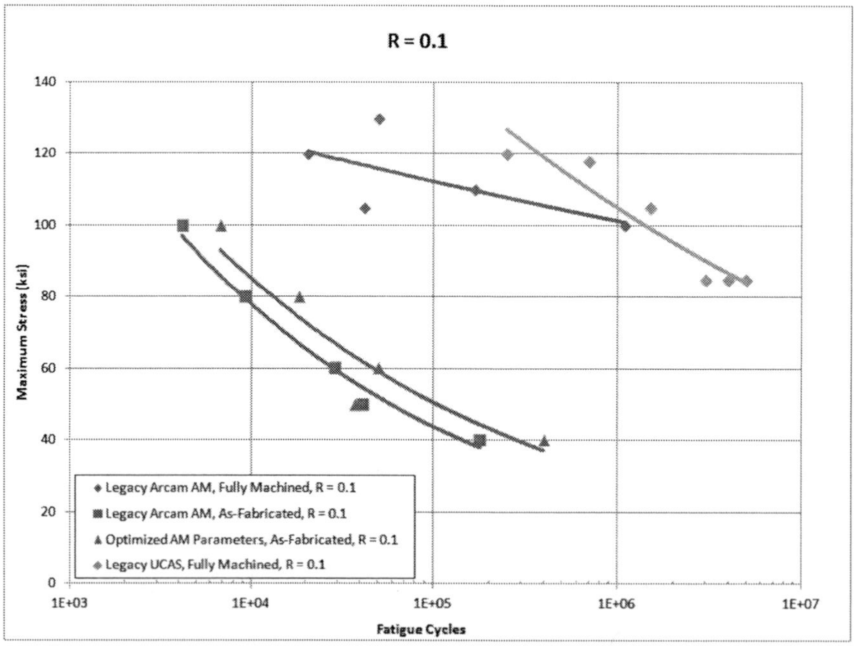

24

DISTRIBUTION STATEMENT A. APPROVED FOR PUBLIC RELEASE

Surface Finish Evaluation, Post-AM Processing

NORTHROP GRUMMAN

- Optimized process parameter static and dynamic performance assessment in work
- Post- AM surface finish process evaluation in work
  - Improvement in overall surface finish:
    - Legacy baseline average of **850-950 µin**
    - Achieving **500-600 µin** with DOE optimized process parameters
    - Achieving **150-250 µin** with post-AM processing

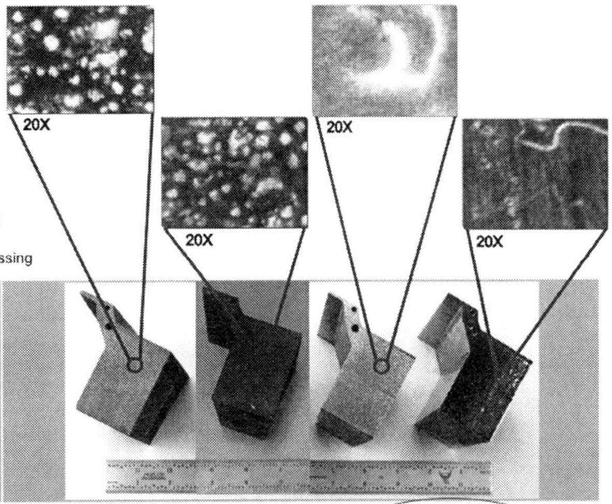

Surface Finish Evaluation Coupons

MicroTek Finishing Selected as the Most Promissing Post-AM Finishing Process for the Mechanical Properties Validation

Surface Finish Process		As-EBM Built, Timet 50 micron layers	Chemically-Assisted Aggregate	Chem Mill + Chemically-Assisted Aggregate	Engineered Aggregate
Supplier		None	REM	Ducommun* + REM	MicroTek
Example Result (R_a)	Min.	173 µin	191 µin	139 µin	27 µin
	Mean	599 µin	515 µin	271 µin	227 µin

DISTRIBUTION STATEMENT A. APPROVED FOR PUBLIC RELEASE

* Chem mill did not trade favorably as a stand-alone post-process

Remaining Program Tasks

NORTHROP GRUMMAN

- Completion of Mechanical Properties Evaluation of AM Fabricated and Secondary Finished Coupons
  - Static and dynamic properties assessment
    - Tensile
    - Fatigue\Stress-Life Curve Development
- AM Fabrication of Full-Scale Component

26

DISTRIBUTION STATEMENT A. APPROVED FOR PUBLIC RELEASE

THE VALUE OF PERFORMANCE.

NORTHROP GRUMMAN

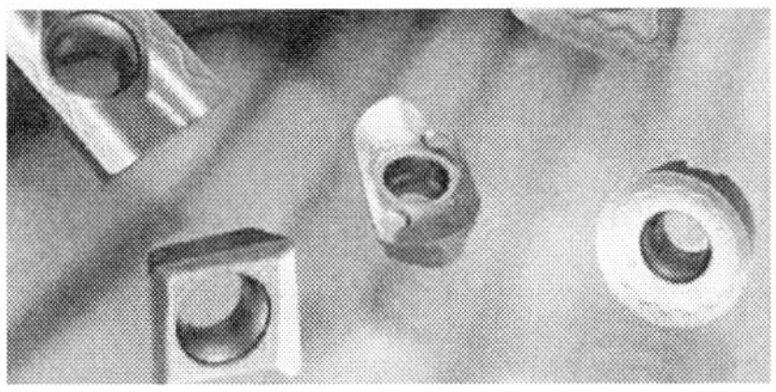

CERATIZIT

Synthesis, structure, and behaviour of a new CVD- TiB_2 coating with extraordinary properties for high performance applications

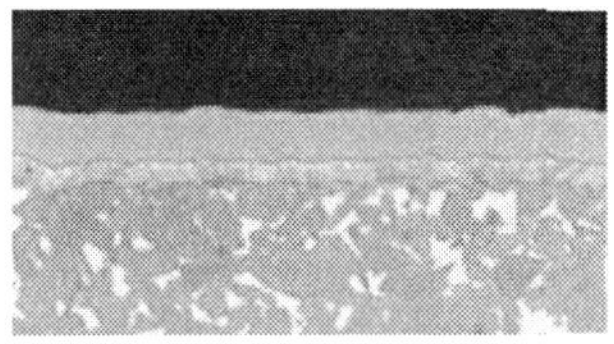

hard material matters

OUTLINE

CERATIZIT

- Introduction
- Demands
- Process
- Properties
- Performance
- Conclusion

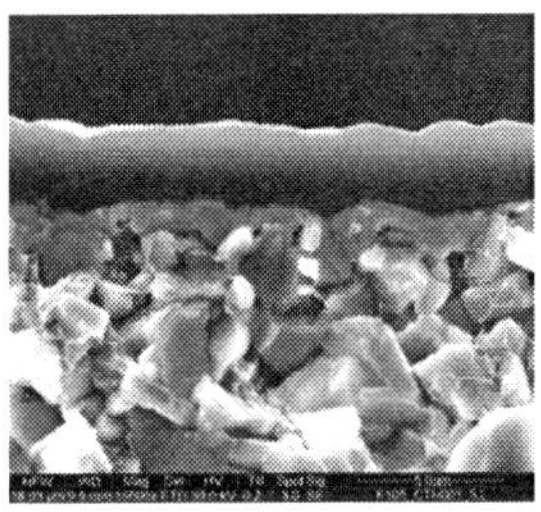

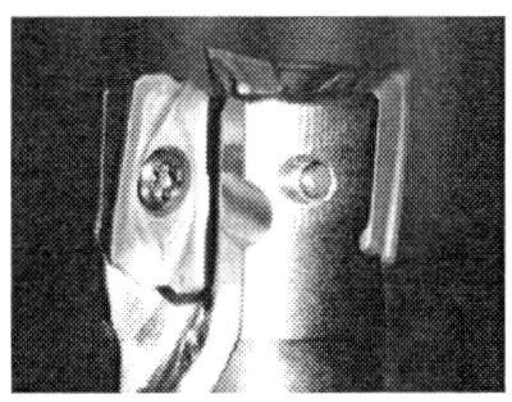

hard material matters

INTRODUCTION

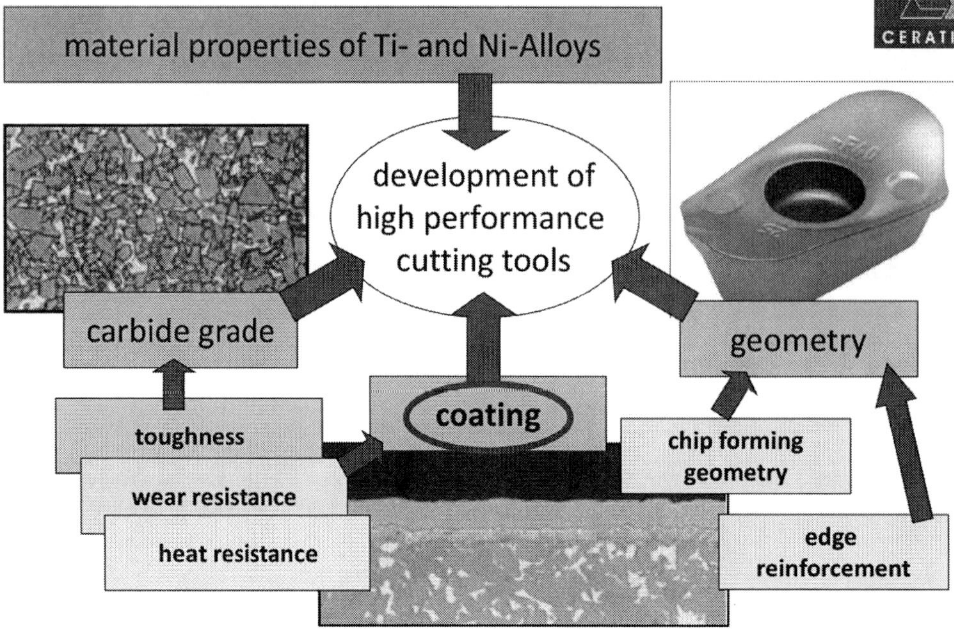

hard material matters

DEMANDS

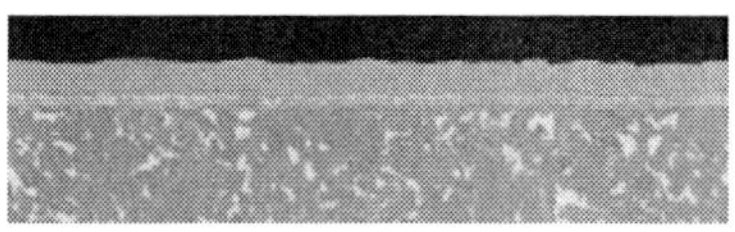

hard to machine materials require specific behaviour of coating

- **highest hardness**
  - best wear resistance
  - sufficient toughness
- **temperature barrier**
  - low heat conductivity into substrate
  - low friction value
- **diffusion resistance**
  - significantly reduced crater wear
  - high chemical stability

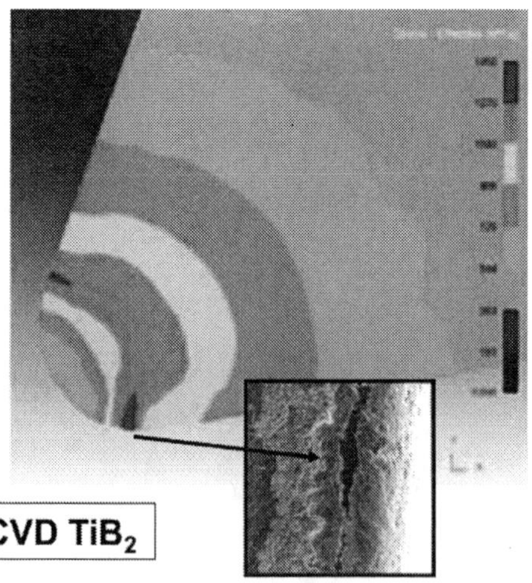

targeted coating system → CVD TiB_2

hard material matters

PROCESS

„state of the art PVD coating process“

coating structure

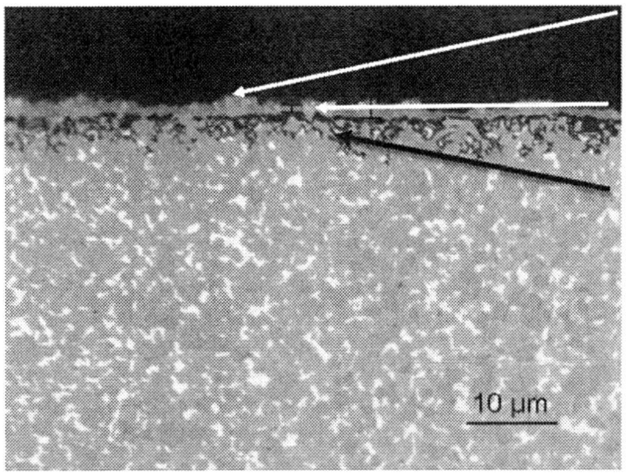

uneven layer

binder-diffusion

Boron/W/Co diffusion + h-phase

unconrolled process

PROCESS

„new developed coating process“

coating structure

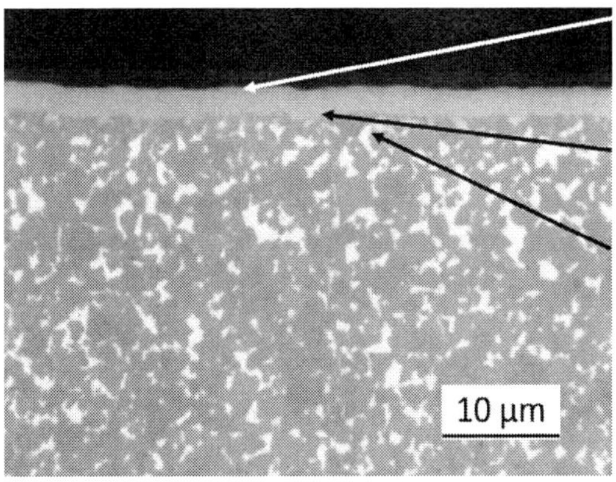

even, smooth
and uniform layer

perfect interface

no negative influence on
substrate

controlled process

PROPERTIES
hardness [nano indentation method]

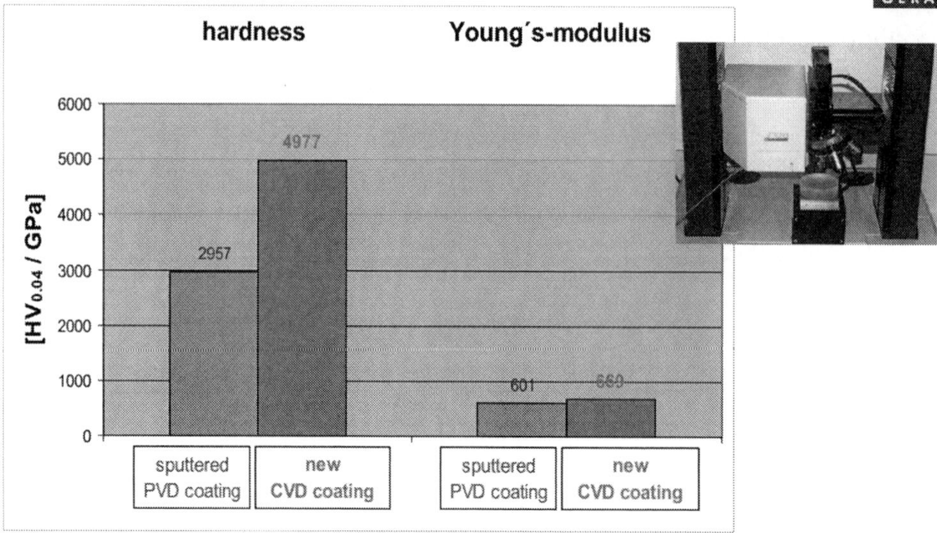

>60% higher hardness of the new CVD TiB_2 coating compared to a sputtered PVD TiB_2 coating

PROPERTIES
morphology and structure [SEM-investigation]

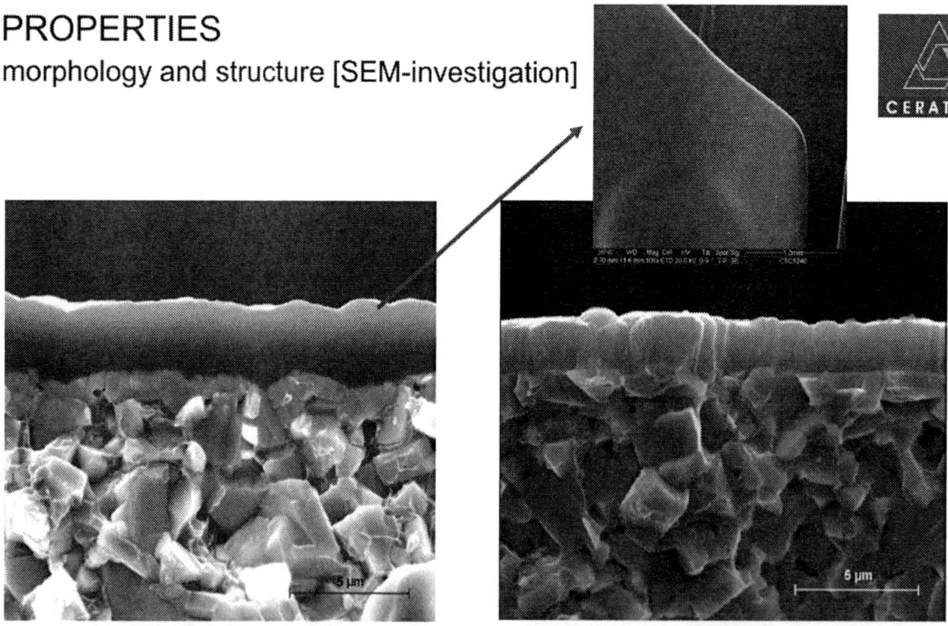

CVD TiB_2 coating **PVD TiB_2 coating**

PERFORMANCE

machining trial at the CERATIZIT lab

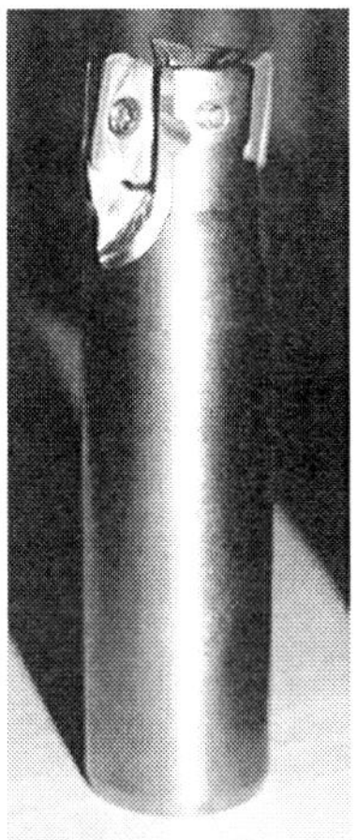

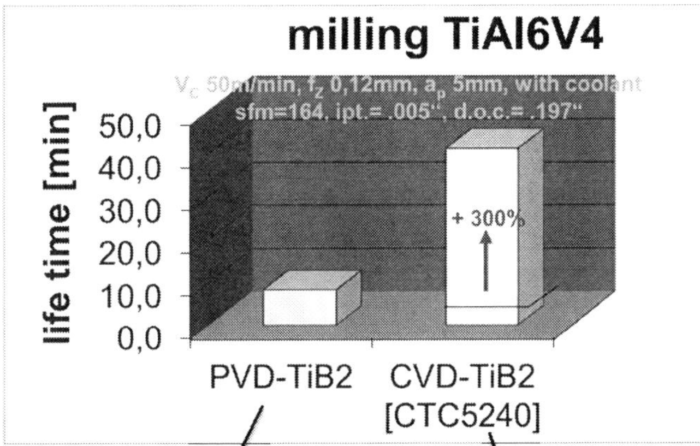

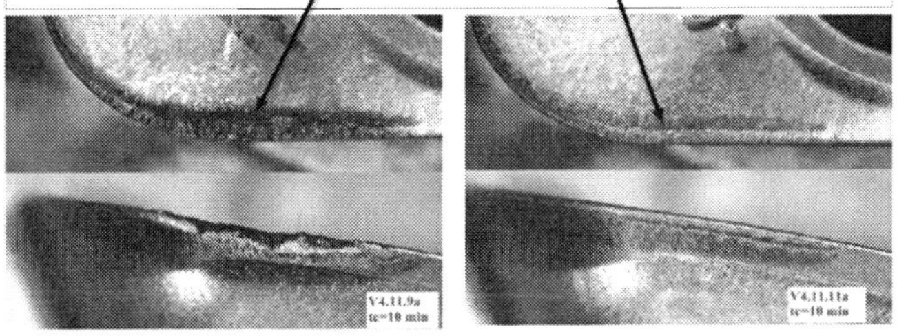

New CVD – TiB_2

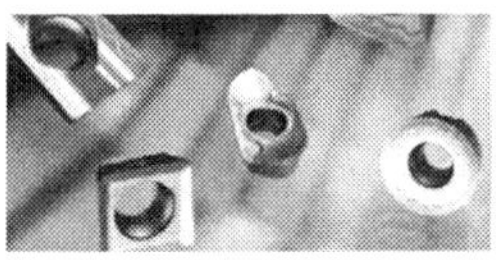

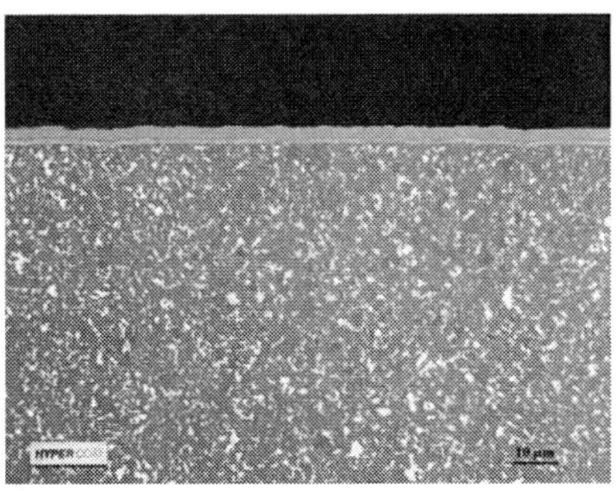

CONCLUSION

- new approach for a high wear resistant coating
- efficient CVD hot wall process usable
- very smooth surface compared to conventional PVD and CVD coatings
- ultra fine grained structure (< 50 nm)
- uniform, regular morphology without textures
- good adhesion on substrate material (prooved by scratch test)

New solution for high performance applications on difficult to machine materials – CTC5240

CERATIZIT **TiB_2** HYPERCOAT coating and high performance machining of Titanium alloys

CERATIZIT

High performance machining of Ti- alloys

hard material matters

XDKT 11T3… long edge profile cutter
MaxiMill C245.43.R.03, dia.=1.7" , 3 flutes

CERATIZIT

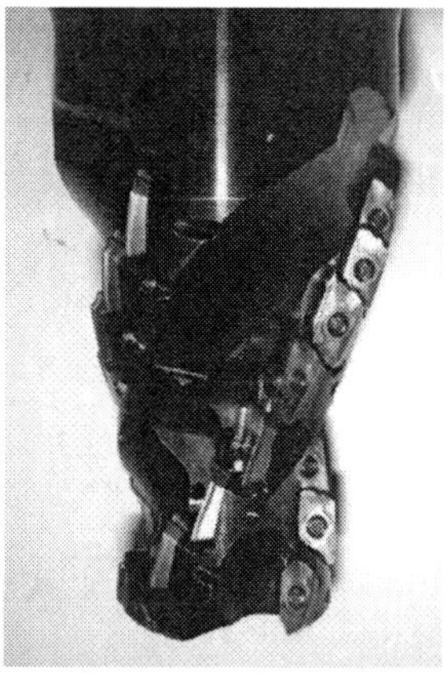

Tool life:
354.3" way= 112min.

Result: a drop of cycle time of 30min./ pce.

hard material matters

CERATIZIT TiB_2 HYPERCOAT coating and high performance machining of Titanium alloys

CERATIZIT

Insert: XDKT 11T308ER-F40
XDKT 11T332ER-F40

Tool: C245.40.R.03-11-32-47

Material: TiAl6V4

Cutting data:

- 120 sfm.
- .005 ipt
- d.o.c. = 1.4"
- w.o.c.= $^2/_3$" – 1 ½"
- Tool life: 630" = 140min. & 2 components

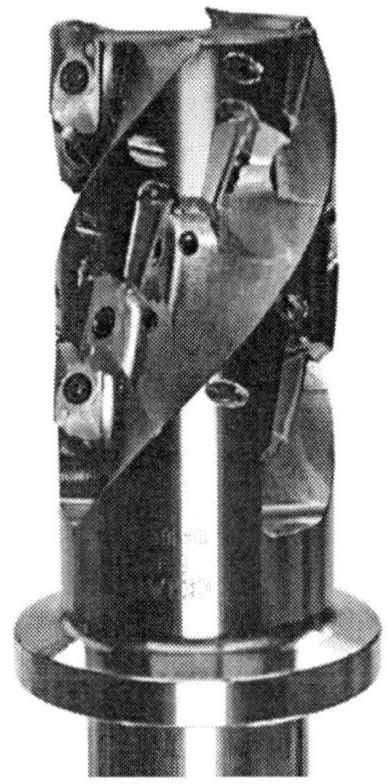

hard material matters

CERATIZIT TiB_2 HYPERCOAT coating and high performance machining of Titanium alloys

CERATIZIT

XDKT11T308ER- F40 CTC5240 & wear

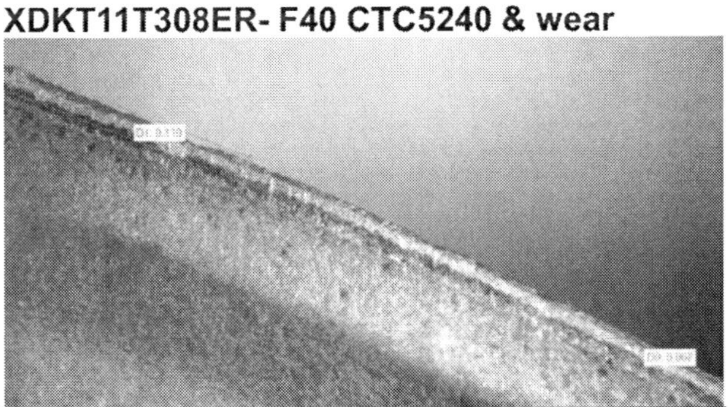

wear patterns after

70min. = 1 component

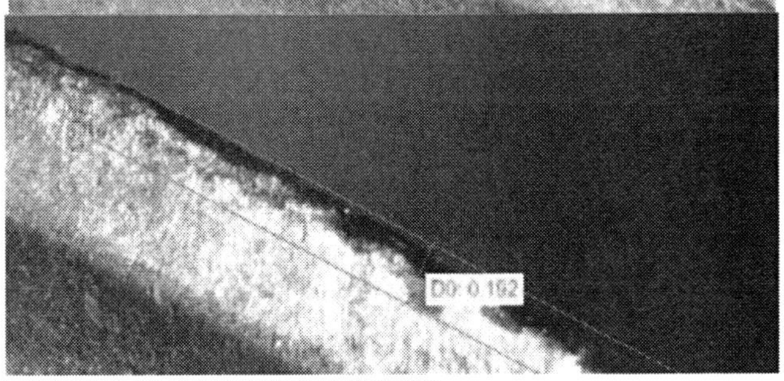

wear patterns after

140min. = 2 components

hard material matters

CERATIZIT TiB_2 HYPERCOAT coating and high performance machining of Titanium alloys

CERATIZIT

Machine: Starrag Heckert STC 1000
Taper: SK50

Cutting data: English & metric

247.7 sfm
.006" ipt.
9.4" ipm.
4.134" d.o.c.
.866" w.o.c.
49.8 ci/min. MRR!!!
30 min. Tool life!!!

Vc= 75.5m/min.
fz= .16mm
Vf= 240mm/min.
ae= 22mm
ap= 105mm

Bench mark
@ Starrag Heckert!!!

hard material matters

CERATIZIT TiB_2 HYPERCOAT coating and high performance machining of Titanium alloys

CERATIZIT

recommended cutting data (for TiAl6V4):

		30	65	100	130	165	200	230	265	300	330	365	400
v_C	sfm												
		.002	.003	.004	.005	.006	.007	.008	.009				
f_Z	ipt												
		.04	.08	.120	.160	.200	.240	.280	.320	.360	.400	.440	
d.o.c	inch												

hard material matters

CERATIZIT TiB_2 HYPERCOAT coating and high performance machining of Titanium alloys

Insert: XOLX 120410ER-F40 CTC5240

Tool: AHFC.52.R.05-12/ High Feed Tool

Material: TiAl6V4

Cutting data:

150 sfm.

.35 ipt.

d.o.c. = .03"

w.o.c.= 100%

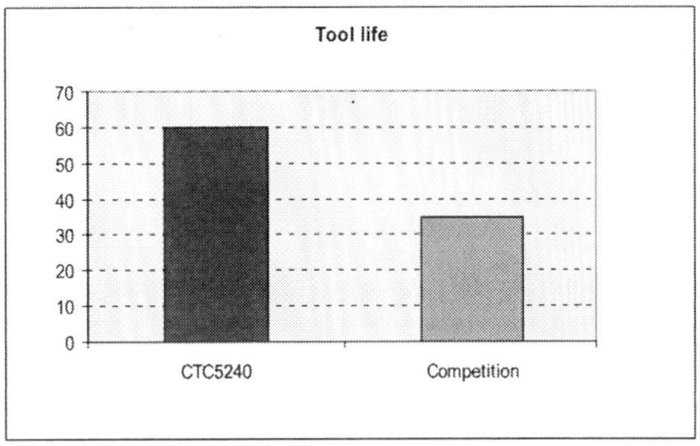

Perfect tool to plunge in Titanium!

hard material matters

CERATIZIT TiB_2 HYPERCOAT coating and high performance machining of Titanium alloys

recommended cutting data (TiAl6V4):

sfm.	30	65	100	130	165	200	230	265	300	330	365	400				
ipt.	.0012	.0016	.020	.0240	.0280	.0320	.0360	.0400	.0440	.0480	.0520	.0560	.0600	.0640	.0680	.0720
d.o.c.	.020	.0240	.0280	.0320	.0360	.0400	.0440	.0480	.0520	.0560	.0600	.0640	.0680	.0720	.0760	.0800

hard material matters

CERATIZIT **TiB_2** HYPERCOAT coating and high performance machining of Titanium alloys

CERATIZIT

Thanks'!

hard material matters

FANUC
FA AMERICA

FANUC Aerospace

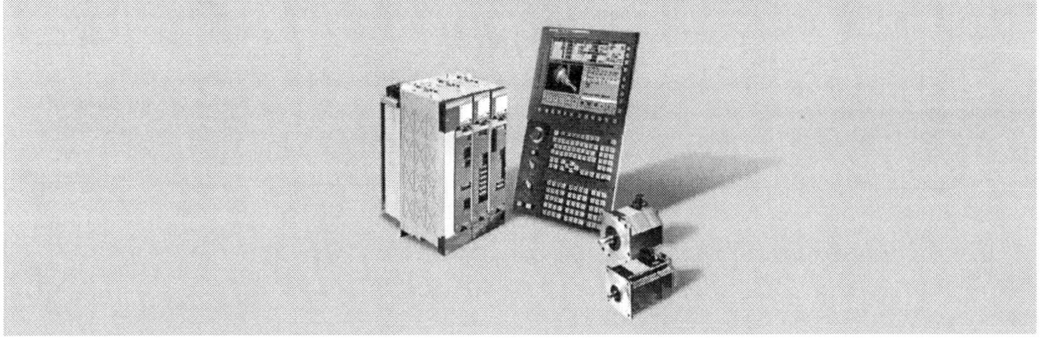

FANUC

FANUC
FA AMERICA

- 350 Employees in North America
  - 120 Service Technicians
  - 30 Engineers
- 65% Market share (installed NA)
  - Closest competitor – 14%
- Monthly Production
  - 30,000 CNC's
  - 200,000 Servo Motors
- Fully automated production
  - 1,750 FANUC Robots used

It's pronounced "FAN-uck"

888-FANUC-US

FANUC FA America

FANUC
FA AMERICA

Service & Support where you need it

34 Service Center & Satellite Locations

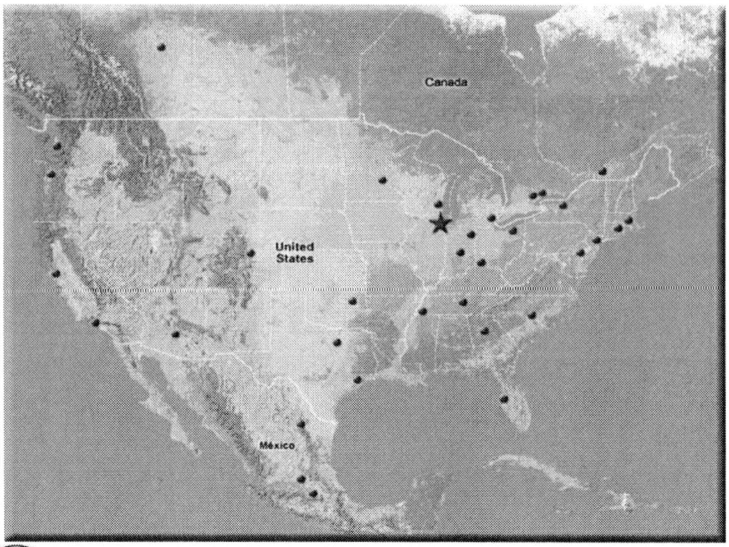

UNITED STATES

★ Chicago Headquarters

Atlanta	Memphis
Boston	Milwaukee
Charlotte	Minneapolis
Cincinnati	Nashville
Cleveland	New Jersey
Dallas	Philadelphia
Denver	Phoenix
Detroit	Portland
Fort Wayne	Rochester
Hartford	San Francisco
Houston	Seattle
Indianapolis	Tampa
Los Angeles	Tulsa

CANADA

Edmonton	Montréal
Guelph	Toronto

MEXICO

Mexico City	Monterrey
Querétaro	

888-FANUC-US

FANUC Support

FANUC
FA AMERICA

- 1-888-FANUC US
  - Any support issue, any time, anywhere
  - Add-on Options
  - Break / Fix Phone Support or On-site Filed Service
  - Replacement Parts
- Mark Lohmeier
  - Aerospace Program Manager
  - New Technologies, Manufacturing Efficiency Improvement, New Machine Capital Investment or Machine Retrofits

888-FANUC-US

Contact Information

FANUC FA AMERICA

Call Toll-free

1-888 FANUC US

(1-888-326-8287)

and Select from the Following Departments

- 1 – Part Sales
- 2 – Service / Technical Support & CS24 After Hours Service
- 3 – PCB & Motor Repair
- 4 – Technical Training
- 5 – Technical Sales & Options
- 6 – Accounts Receivables & Billing
- 7 – Service Contract Sales
- 8 – CNC Express – CNC Sales
- 9 – Repeat the Menu

888-FANUC-US

In Production 24/7

FANUC FA AMERICA

Unequaled reliability

- Measured MTBF as Complete CNC Systems
- What Makes-Up a CNC System :
  - CNC Control
  - LCD Display and Operator Interface Panel
  - I/O Modules and I/O Rack
  - Servo and Spindle Amplifiers
  - Spindle and Servo Motors
- Failures Caused by Machine Tool Builders & End-Users -- Excluded from the calculation
- Failures With Unclear Origin -- Included in Calculation, Even If Caused by Machine Tool Builder or End-User
  - *Ask competition for their measured failure rates*
- Average Failure Rate -- Less Than 0.01 Failure Per System per Month
  - Less Than 1 Failure every 10 Years
- ***MTBF of up to 27.8 years on the new CNC systems***

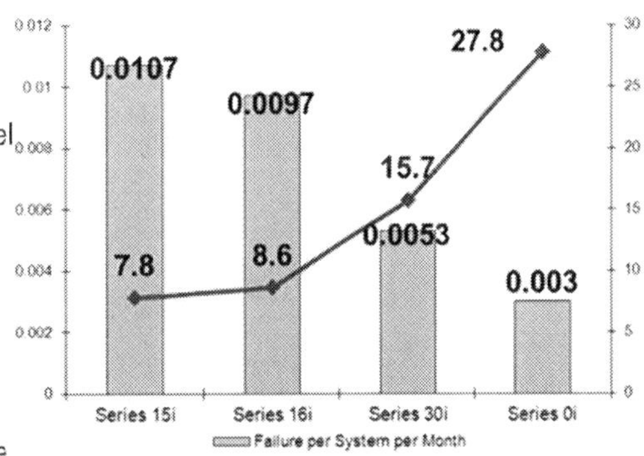

888-FANUC-US

FANUC Long Term Commitment

FANUC
FA AMERICA

TOTAL SUPPORT

CNC Support for *25 years* After Production Stops

- Average machine life…
  - 20 years
- Support & Repair Parts 25 Years

FANUC CNC's Never Become "Obsolete"

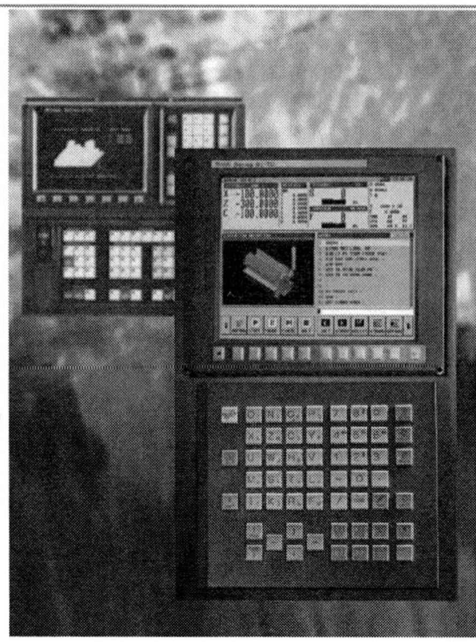

888-FANUC-US

Long Term Commitment

FANUC
FA AMERICA

- After mass production has stopped, FANUC puts enough components in stock to cover at least 25 years of availability
  - If the component could not be made available any more, a sub-circuit would be developed to provide an identical function

Memory component

Equivalent sub-circuit

888-FANUC-US

Fanuc Product Line

FANUC
FA AMERICA

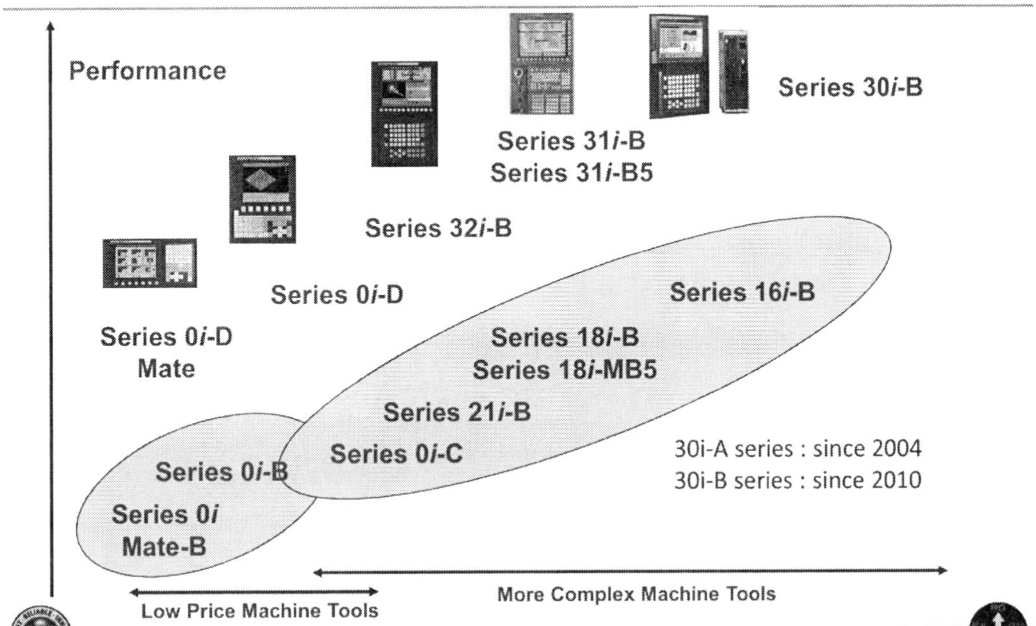

888-FANUC-US

FREE Services

FANUC
FA AMERICA

- Maintenance Workshop (4 hours)
  - Recommended preventive maintenance
  - Recommended spare parts
  - Diagnostic overview
- Programming Workshop (2-4 hours)
  - 5 axis programming features
  - Optimal programming techniques
- Operator Workshop (2 hours)
  - Basic operations (program execution)
  - Basic operational diagnostics

888-FANUC-US

≫Free Services

FANUC
FA AMERICA

- **Parameter Comparison**
  - ≫ Evaluate existing machines
  - ≫ Recommendations for consistent, optimal programming
- **Machine Inventory Analysis**
  - ≫ Evaluate existing machines
  - ≫ Recommendations for spare parts and preventive maintenance
  - ≫ Provides catalog of entire facility inventory
- **Machine specification and runoff assistance**
  - ≫ Machine specification assistance
  - ≫ Recommendations for optimal machine verification

888-FANUC-US

FANUC
FA AMERICA

Aerospace Technology

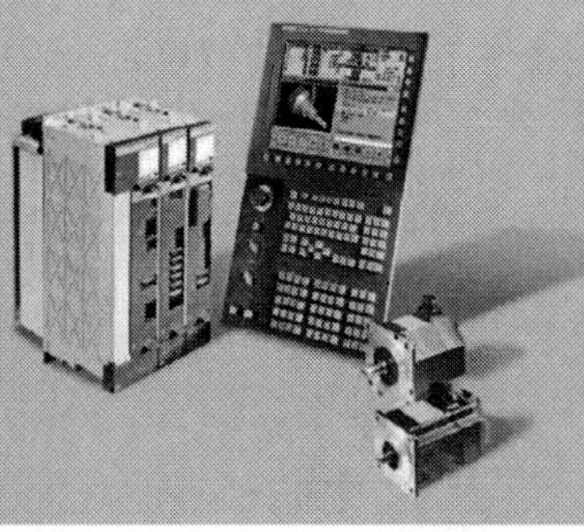

“5 axis”

FANUC
FA AMERICA

- “5 axis” motion is any contouring motion that includes a contouring capable rotary axis.
  - 4, 5, 3+2, 3+1 are all “5 axis”
  - Basic concepts apply to “3 axis”

Typical Process

FANUC
FA AMERICA

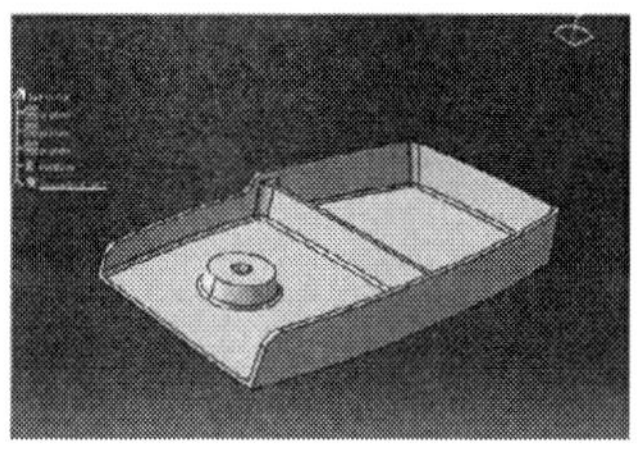

- CAD
  - Part model
- CAM
  - Dot to dot approximation map
- Post-Processor
  - Formats the dots into specific control/Machine language
- CNC
  - Connects the dots

Property of FANUC FA America © All Rights Reserved
888-FANUC-US

Connecting the Dots

FANUC
FA AMERICA

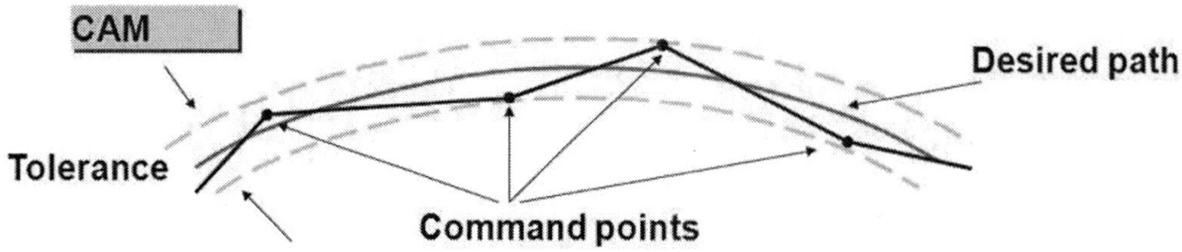

- Connecting the dots means rough, inaccurate motion
- Common solutions:
  - "Detuning"
  - Flooding with points

4, 5+ Axes

FANUC
FA AMERICA

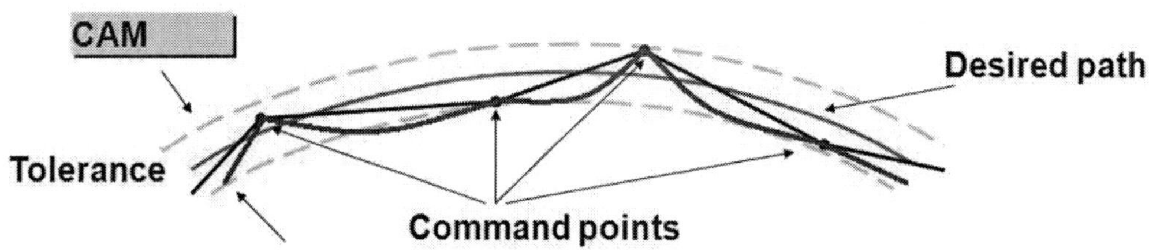

- "5 axis" problem significantly worse
- Tool tip "scallops"
  - Often mistaken for faceting

Part Coordinate Programming

FANUC FA AMERICA

888-FANUC-US

Workpiece Setting Error Correction

FANUC FA AMERICA

Example:G54.4 P1 (WSEC)

- (Use Tilted Working Plane G68.x for local operations)

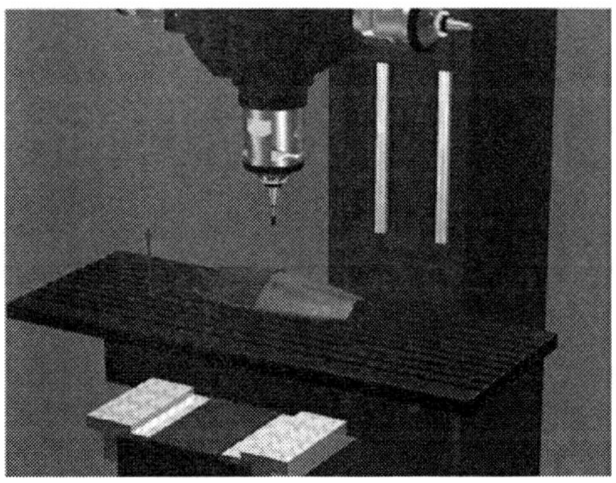

Example: Benefits

- Lower Labor Costs – Higher Productivity
- Improved Flexibility
- Simplified Programming

888-FANUC-US

Tilted Working Plane (G68.x)

FANUC
FA AMERICA

- Create Local Coordinate System for Drilling, etc...

Property of FANUC FA America © All Rights Reserved

888-FANUC-US

Tool Center Point

FANUC
FA AMERICA

- **Pivot Point Programming**
  - G01 X_ Y_ Z_ B_ C_ F_
  - Program individual axes to achieve tool position, orientation, and speed
  - Post handles kinematics & tool

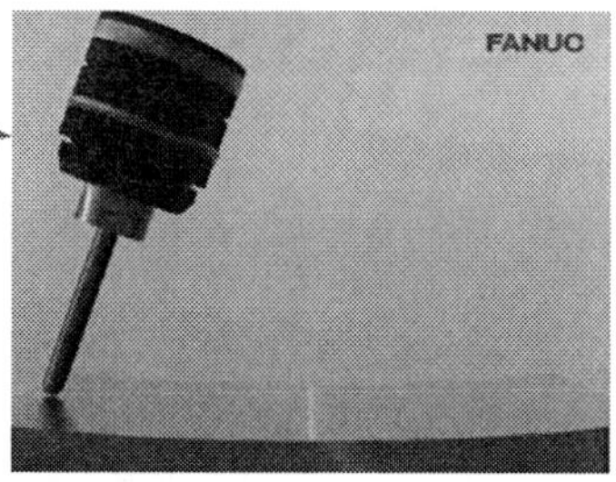

- **Tool Center Point Programming**
  - G43.4 H1 G01 X_ Y_ Z_ B_ C_ F_
  - Program cutting point position, tool orientation, and speed
  - CNC handles kinematics & tool

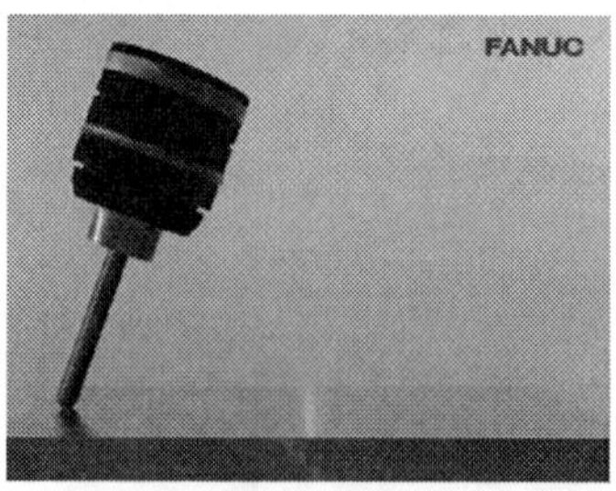

Property of FANUC FA America © All Rights Reserved

888-FANUC-US

Side Mill Cutting

FANUC
FA AMERICA

- **Without Tool Posture Control**
  - Tool scalloping

- **With Tool Posture Control**
  - Tool follows exact path

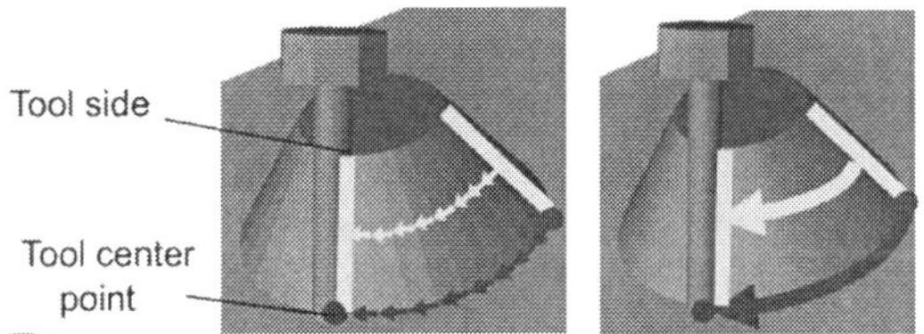

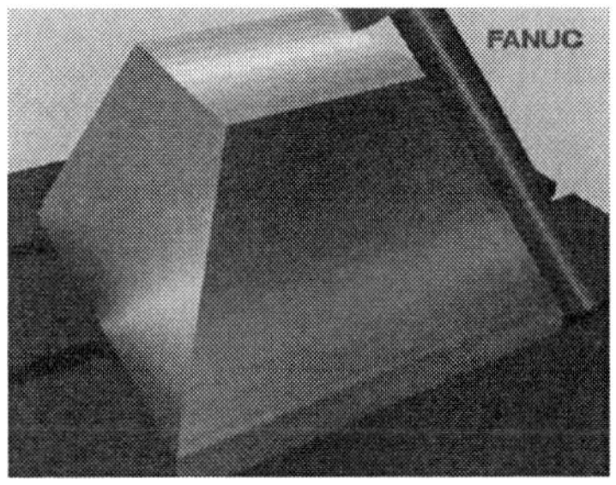

888-FANUC-US

Lookahead - AiCC

FANUC
FA AMERICA

- **Lookahead is equivalent to car headlights at night**
  - **Lets the driver know what's coming so they can adjust speed accordingly.**
- **FANUC AiCC provides 600 or 1,000 block lookahead**

888-FANUC-US

Spline Interpolation

FANUC
FA AMERICA

- Real-time lookahead generates spline based on standard G-code program

- Benefits
  - Better accuracy
  - Highly tuned servos
  - Shorter cycle times
  - Improved machine and tool life
  - Achieve higher feedrates

888-FANUC-US

Process Specific Settings

FANUC
FA AMERICA

- Configure different profiles for different processes
  - Example: FANUC Machining Condition Selection Function
  - R1 – R10

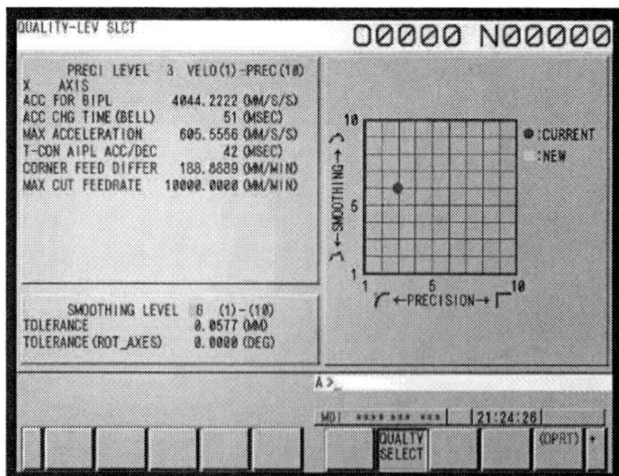

- Process specific
  - Roughing
  - Finishing
- Material specific
  - Hard metal
  - Soft metal

888-FANUC-US

Tool Orientation

FANUC
FA AMERICA

- **Type I: Rotation Axes Angles**
  - Command rotary axes angles directly
  - G43.4
  - E.g. A-30.0 C-110.0

- **Type 2: Vectors**
  - Command tool orientation using CAM vectors
  - G43.5
  - E.g. I0.470 j-0.171 k0.866

This technique requires minimal post processor logic.

888-FANUC-US

Simplified Post-Processor

FANUC
FA AMERICA

- With the proper features and CNC configuration, the post-processor simply reformats CAM output.
  - CAM: GOTO/5.20, 3.14, 9.20, 0.42, 0.81, 0.13
  - G CODE: G1 X5.20 Y3.14 Z9.20 I0.42 J0.81 K0.13
- Benefits
  - Simplification of post-processor
  - Geometry machine independent
  - Better accuracy
  - Easier long-term program support
  - Greater flexibility / faster model to manufacture

888-FANUC-US

MMS – May 2012

FANUC
FA AMERICA

888-FANUC-US

"High Speed Smooth TCP"

FANUC
FA AMERICA

- Includes several options
  - Tool Center Point (TCP)
  - Tool Posture Control (TPC)
  - Tool tip and tool vector smoothing**
  - Tool orientation smoothing**
  - Fairing**

888-FANUC-US

"Fairing"

FANUC
FA AMERICA

- Creates path through the approximated points – accurately creating the original, pre-CAM curve.

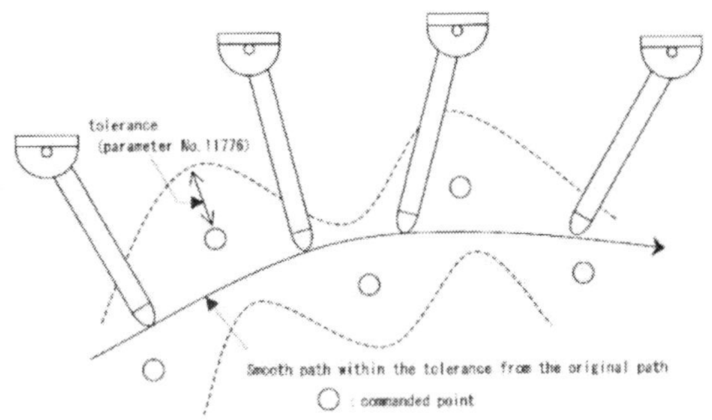

Spline Interpolation

FANUC
FA AMERICA

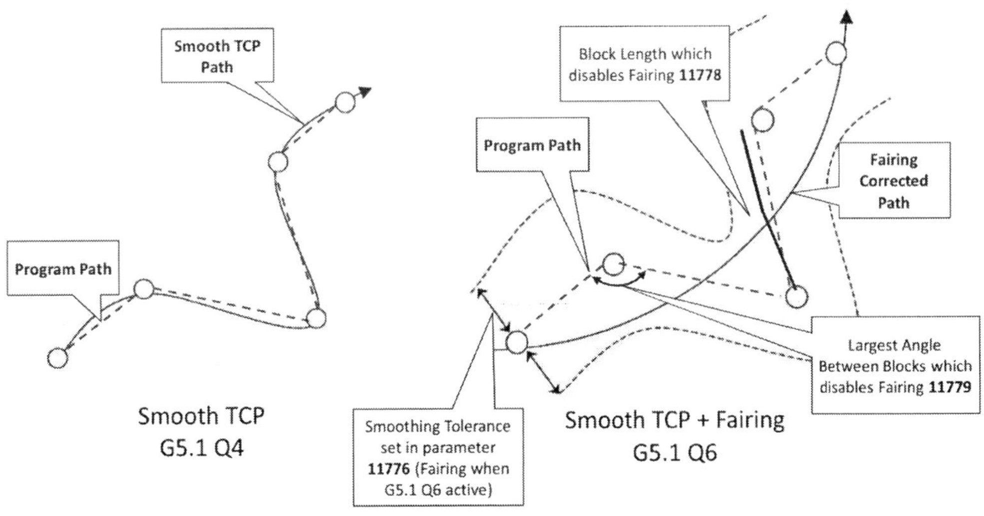

High Speed Smooth TCP

FANUC
FA AMERICA

Without Smooth TCP

- Stripes on machined surface

With High Speed Smooth TCP

- Very smooth machined surface

888-FANUC-US

Modular Programming

FANUC
FA AMERICA

- Basic program architecture:
  - Locate part
  - Feature activation macro
  - Machine independent geometry
  - Feature de-activation macro
- Benefits
  - Simple future modification
  - Long term program support
  - Easy to add new features
  - Easy to move programs to other machines

888-FANUC-US

Modular Programming (example)

FANUC
FA AMERICA

```
O0001
;
G65 P9004 (Probing Macro to setup G54.4
P1 values);
;
G65 P9006 (Activation Macro);
;
G00 X0.0 Y0.0 (Part geometry);
Z1.0;
G01 Z-1.0 F100.;
X10;
Y10.;
X0;
Y0;
G00 Z1.0;
;
G65 P9008 (Deactivation Macro);
;
M30;
```

```
O9006 (Activation Macro);
G54.4 P0;
G80 G40 G69;
G90 G49 G70 G94 G17 G64;
G54;
;
G54.4 P1; (WSEC)
G5.1 Q1 R1.; (AiCC + MCSF)
G43.4 H1 P3.; (HSSTCP)
G5.1 Q6.; (FAIRING)
;
M99;
```

```
O9008 (Deactivation Macro)
G40;
G5.1 Q4;
G49;
G5.1 Q0.;
;
G91 G28 Z0;
G90;
M99;
```


888-FANUC-US

Benefits

FANUC
FA AMERICA

- Immediate Benefits
  - Part Surface Quality Improvement
  - Part Accuracy Improvement
  - Cycle Time Reduction
- Long Term Benefits
  - Increased Tool Life
  - Increased Machine Life
  - “Art to Part” Process Simplification
    - » Cost Reduction – Increased Flexibility
  - Program is Machine Independent

888-FANUC-US

Interference Zones

FANUC
FA AMERICA

888-FANUC-US

Presenter

FANUC
FA AMERICA

Mark Lohmeier: Aerospace Program Manager for Aero – Engine and Energy Vertical Markets at FANUC FA America.

Mark's career includes over 34 years' in the Machine Tool Industry as Field Service Engineer, Installer of CNC Retrofits, a CNC Application Engineer, and Sales of New Machines, and Machine Tool Retrofitting and Rebuilding.

Mark is a Graduate of the University of Cincinnati in Business Development and Marketing, coupled with a Degree in Medical Electronics from United Electronics Institute, Louisville, KY.

Thank you for your time today!!

FANUC
FA AMERICA
Mark Lohmeier
Program Manager – Aero Engine & Energy
FANUC FA America
O:513-336-0808
Mark.lohmeier@fanucfa.com
www.fanucfa.com 888-FANUC-US

888-FANUC-US

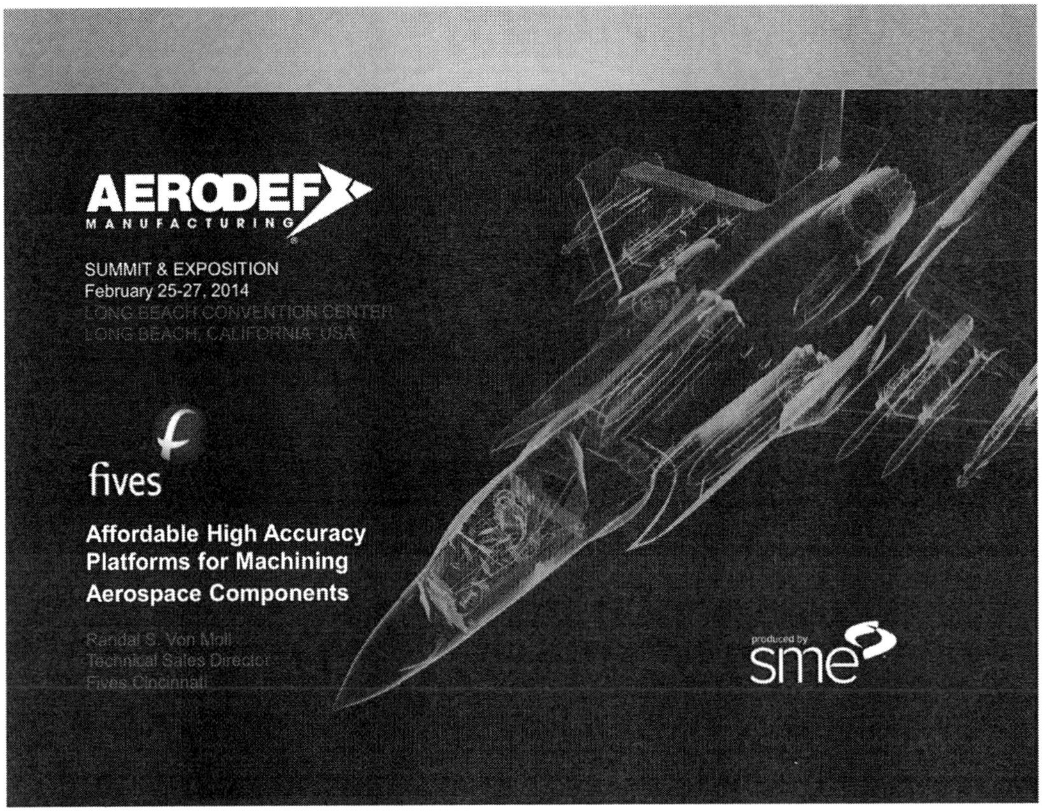
AERODEF
MANUFACTURING
SUMMIT & EXPOSITION
February 25-27, 2014
LONG BEACH CONVENTION CENTER
LONG BEACH, CALIFORNIA USA
fives
Affordable High Accuracy
Platforms for Machining
Aerospace Components
Randal S. Von Moll
Technical Sales Director
Fives Cincinnati
produced by
sme

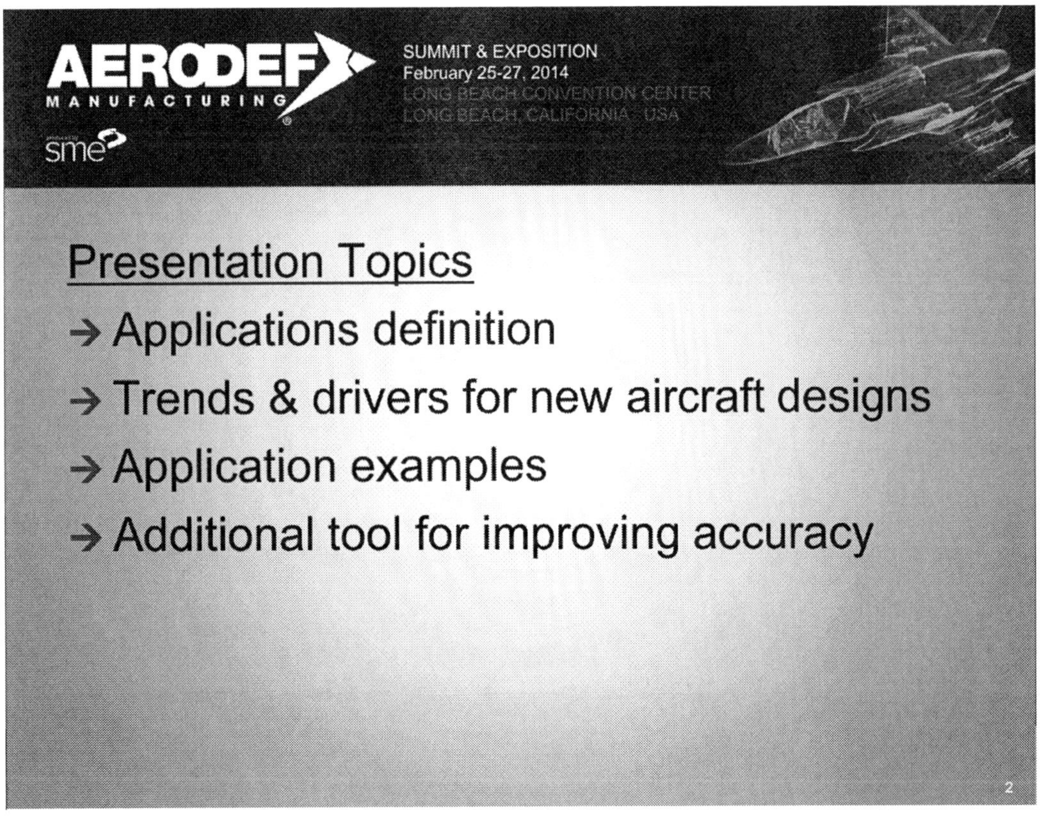
AERODEF
MANUFACTURING
sme
SUMMIT & EXPOSITION
February 25-27, 2014
LONG BEACH CONVENTION CENTER
LONG BEACH, CALIFORNIA USA
Presentation Topics
→ Applications definition
→ Trends & drivers for new aircraft designs
→ Application examples
→ Additional tool for improving accuracy
2

Affordable high accuracy platforms for machining of Aerospace components:

Applications include:

- → Carbon and glass fiber composite parts
- → Composite sandwich structures
- → Drilling
- → Precision countersinking
- → Milling & trimming
- → On-machine inspection
- → Aluminum profiling

3

AERODEF MANUFACTURING
sme
SUMMIT & EXPOSITION
February 25-27, 2014
LONG BEACH CONVENTION CENTER
LONG BEACH, CALIFORNIA USA

Cured Composites Aircraft Parts

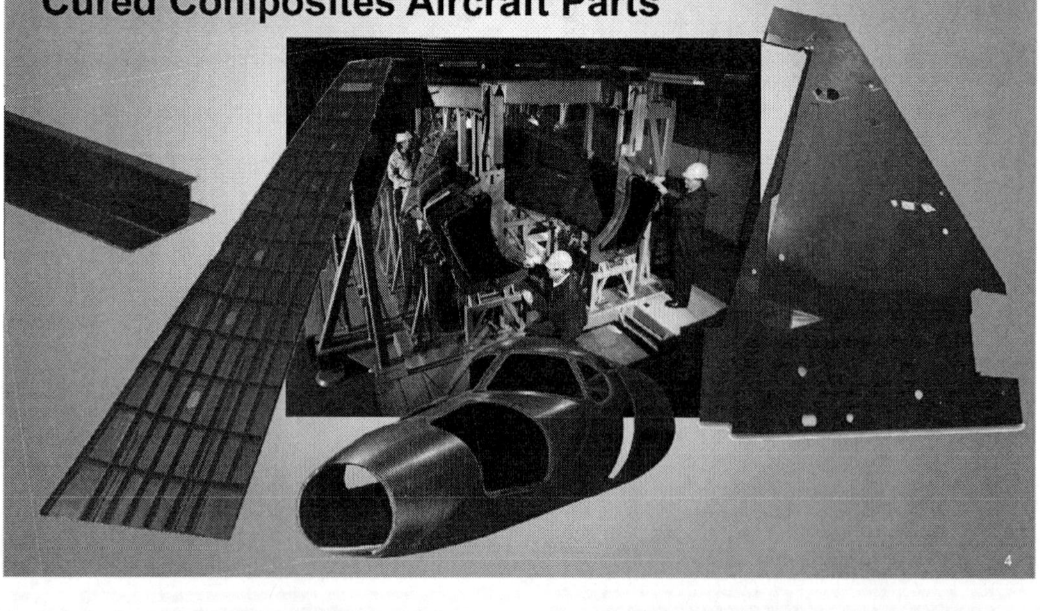

4

AERODEF MANUFACTURING

sme

SUMMIT & EXPOSITION
February 25-27, 2014
LONG BEACH CONVENTION CENTER
LONG BEACH, CALIFORNIA USA

Industry Trends & Drivers for Composite Parts

- New airframe designs have increased volume of composite detail parts and assemblies
  - Current/Future Defense
    - V22 Osprey
    - F-35 A–B–C variants
    - Global Hawk
    - UCAS (X-47B)
    - LRS-Bomber
  - Commercial (B787; A350; CSERIES)
- **Composites parts require machining**

5

AERODEF MANUFACTURING

sme

SUMMIT & EXPOSITION
February 25-27, 2014
LONG BEACH CONVENTION CENTER
LONG BEACH, CALIFORNIA USA

Determinate Assembly

- Digital Thread
- Sophisticated 3D modeling software
- Digital detail parts designs
  - Digital data drives machine tool cutter paths
- The finished machined parts, by design and digital manufacturing process:
  - Fit together without further fitting, shimming, or secondary machining processes
  - Simpler/less assembly tooling required
  - Interchangeable/replaceable parts

6

Affordable high accuracy platforms for machining of Aerospace components:

Application Examples

AERODEF MANUFACTURING
sme
SUMMIT & EXPOSITION
February 25-27, 2014
LONG BEACH CONVENTION CENTER
LONG BEACH, CALIFORNIA USA

Boeing – Frederickson WA
→ (3) E mill - Extrusion Mill
→ Program: Commercial jets

<u>E mill</u>
→ Long extrusion-like composite stringers
→ Profile mill; drill; gauge reduction; cut-off
→ 3-axis horizontal machining center
→ 4th and 5th axes provided by rotary head and tailstock

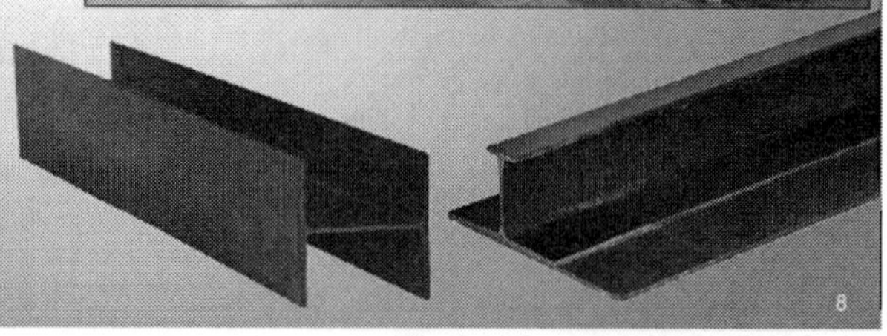

AERODEF MANUFACTURING

SME

SUMMIT & EXPOSITION
February 25-27, 2014
LONG BEACH CONVENTION CENTER
LONG BEACH, CALIFORNIA · USA

Lockheed Martin – Ft. Worth TX

- → (7) Cincinnati Auto Drill Gantries
- → Wing Drill & Forward Fuselage Systems
- → Program: **F-35**

Single and Dual Gantry – Autodrill

- → Precision drilling, countersinking, & milling
- → 5-Axis B/C head with drilling quill
- → Pressure foot position feedback for precision counter sink depth

9

AERODEF MANUFACTURING

SME

SUMMIT & EXPOSITION
February 25-27, 2014
LONG BEACH CONVENTION CENTER
LONG BEACH, CALIFORNIA · USA

Lockheed Martin – Marietta, GA

- → (**2**) Cincinnati Auto Drill Gantries
- → Wing Drilling Systems
- → Program: **F-35**

Single Gantry – Wing Autodrill

- → Precision drill, countersink
- → 5-Axis B/C head with drilling quill
- → Pressure foot position feedback for precision counter sink depth
- → Dynamic Ram Compensation
- → Hole measurement system for hole diameter and countersink depth

*Some covers and other guarding removed

10

AERODEF MANUFACTURING

sme

SUMMIT & EXPOSITION
February 25-27, 2014
LONG BEACH CONVENTION CENTER
LONG BEACH, CALIFORNIA, USA

Boeing – Charleston SC

→ (**1**) FTR Autodrill
→ Program: **787** Fuselage Barrels

Traveling Column Autodrill System

→ Profile mill, cutout mill, & drill
→ Floor-type boring mill with ram
→ 5-Axis B/C head with drilling quill axis
→ Pressure foot position feedback for precision counter sink depth
→ **Dynamic Ram Compensation**
→ Arranged for MQC (Minimal Cryogenic Cooling) cutter cooling technology

*Some covers and other guarding removed

11

AERODEF MANUFACTURING

sme

SUMMIT & EXPOSITION
February 25-27, 2014
LONG BEACH CONVENTION CENTER
LONG BEACH, CALIFORNIA, USA

Boeing Tukwila, WA
Advanced Development Center

→(**2**) Cincinnati 5-Axis Routers
→B2; F22; other classified programs

→MasterCarver 5-Axis Routers
→Trimming and drilling of composite panels
→Wet & Dry Machining capabilities
→High efficiency vacuum dust collection system

12

AERODEF MANUFACTURING

sme

SUMMIT & EXPOSITION
February 25-27, 2014
LONG BEACH CONVENTION CENTER
LONG BEACH, CALIFORNIA, USA

Major aircraft engine builder - USA

- → PMT - Precision Mill & Trim Gantry
- → Carbon composite & composite sandwich structures
- → Program: Large commercial aircraft **nacelle components**

PMT – Precision Mill & Trim

- → High Rail gantry configuration
- → 4m x 7m pallet shuttle system
  - › Offline load of parts and tooling
- → Precision drill, countersink & mill/trim
- → Patent Pending dust/chip collection system for dry machining
  - › Enclosure includes flexible roof cover

*Some covers and other guarding removed

13

AERODEF MANUFACTURING

sme

SUMMIT & EXPOSITION
February 25-27, 2014
LONG BEACH CONVENTION CENTER
LONG BEACH, CALIFORNIA, USA

Triumph – Hawthorne CA

- →(**2**) PMT - Precision Mill & Trim Gantry
- →Aluminum panels
- →Program: **747**; others

PMT – Precision Mill & Trim

- →Low Rail gantry configuration
- →Precision drill, trim, cutout
- →Flexible tooling – Pogo's
- →Patent pending chip collection system for dry machining

14

Affordable high accuracy platforms for machining of Aerospace components:

Additional tool for improving accuracy:
VEC - Volumetric Error Compensation

AERODEF MANUFACTURING
sme
SUMMIT & EXPOSITION
February 25-27, 2014
LONG BEACH CONVENTION CENTER
LONG BEACH, CALIFORNIA USA

VEC - Volumetric Error Compensation

- → Service event & Software
- → Determines existing geometric errors
- → Derives needed geometric comps
- → Loads VEC comp table into the CNC
- → VEC comp file works with the Siemens 840D real-time compile cycle
- → Rapid Calibration Time
  - › Single setup for 100's points
  - › Few hours vs. weeks
- → ~300% improvement - average & maximum volumetric errors

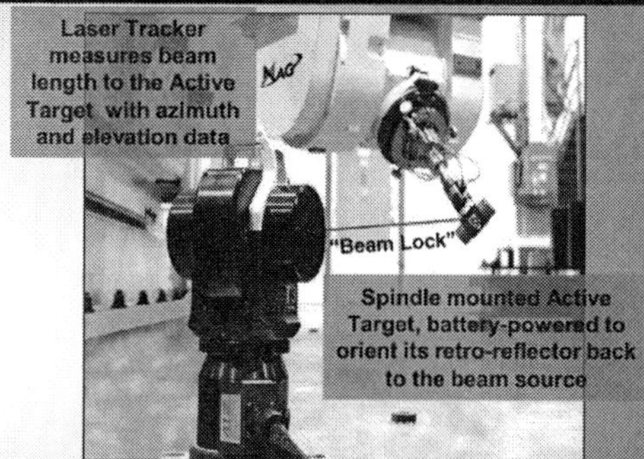

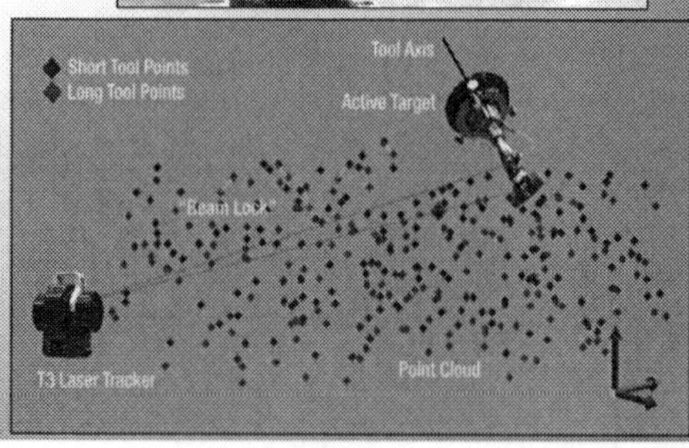

AERODEF MANUFACTURING

sme

SUMMIT & EXPOSITION
February 25-27, 2014
LONG BEACH CONVENTION CENTER
LONG BEACH, CALIFORNIA USA

Conclusions

→ Applications: composites & Sandwich structures

→ Trends & drivers for new aircraft designs

→ Application examples – modular affordable solutions for tight tolerance work

→ Additional tool for improving accuracy - VEC

17

---

fives

Metal Cutting | Composites

Cincinnati | Giddings & Lewis | Forest - Liné | Liné Machines

Thank you

E-Drill Fastener Removal System

David Siedal

Technical Fellow

Lockheed Martin Aeronautics

E-Drill Fastener Removal System

Presentation Overview

- What is the E-Drill System?
- How Does It Work?
- Why Is Lockheed Martin Interested?
- What Type Of Testing Was Done To Evaluate The System?
- Where Is Lockheed Martin Using the E-Drill?
- Future Plans

AERODEF MANUFACTURING
February 25-27, 2014 - SME

E-Drill Fastener Removal System

- Developed By Perfect Point, Huntington Beach California
- A New And Portable Way To Quickly Remove Fasteners Without Drilling

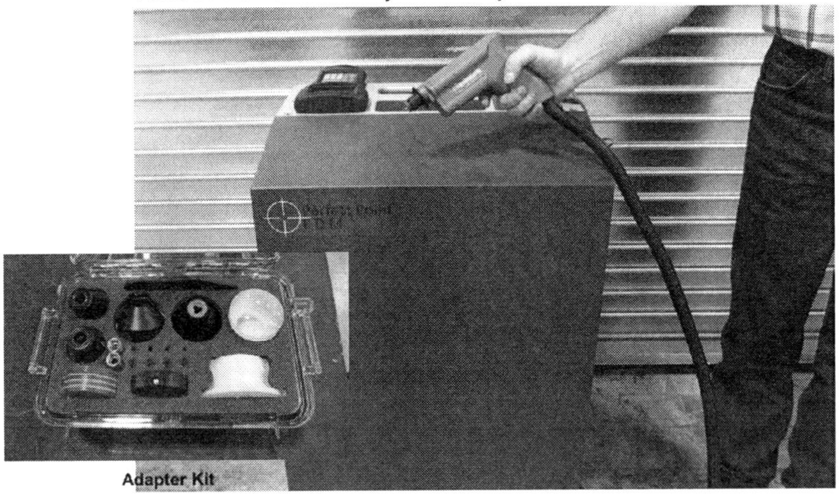

AERODEF MANUFACTURING
February 25-27, 2014 - SME

E-Drill - What Is It And How Does It Work?

1) Electrode Burns A Hollow Cylindrical Groove Into The Fastener
2) Fastener Shank Is [redacted]king With A Center Punch

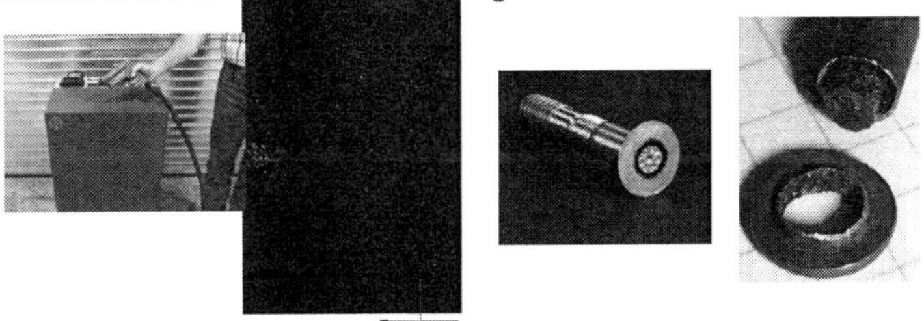

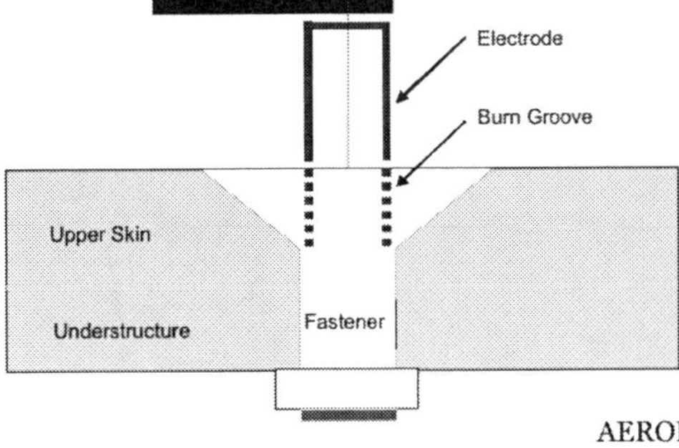

AERODEF MANUFACTURING
February 25-27, 2014 - SME

E-Drill Testing At Lockheed Martin-Aeronautics

- System Durability And Stability
- Process Control And Repeatability
- Operator Interface
- Electrode Life
- Routine Service Requirements
- Extended Service Maintenance Tasks

5

Lockheed Martin Laboratory Testing

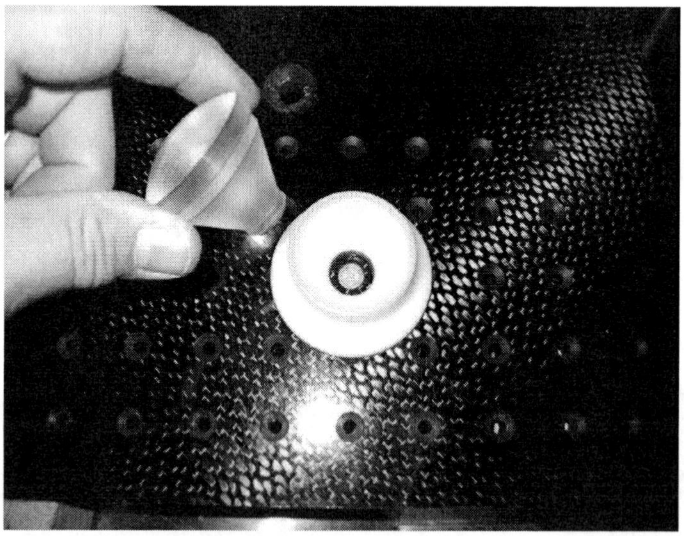

Fastener Locating Reticle

6

AERODEF MANUFACTURING
February 25-27, 2014 - SME

Flush Fastener Locator With Reticle Removed

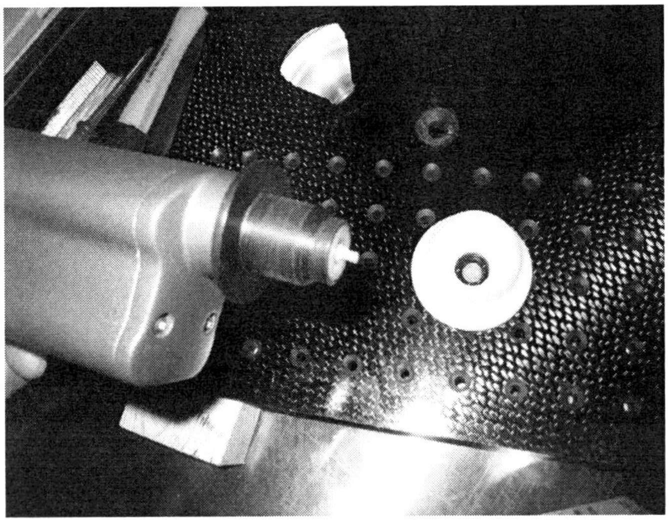

AERODEF MANUFACTURING
February 25-27, 2014 - SME

7

Handheld Unit Cutting Groove Into Fastener

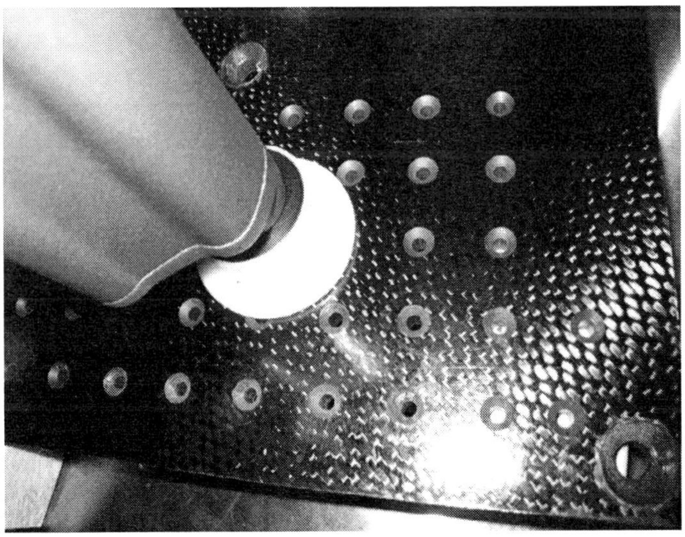

AERODEF MANUFACTURING
February 25-27, 2014 - SME

8

Properly Burned Fastener

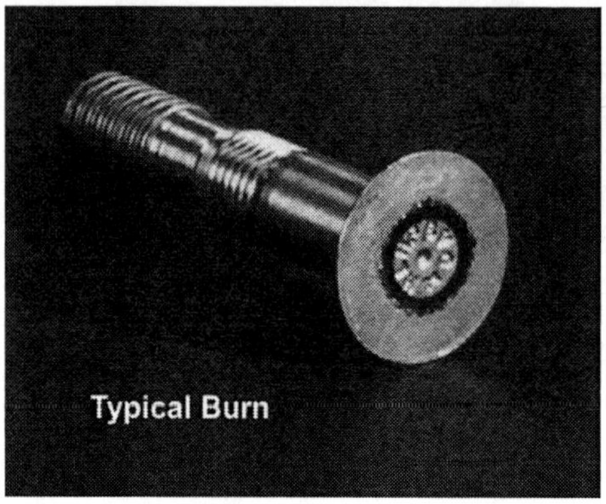

Punched Out Fastener

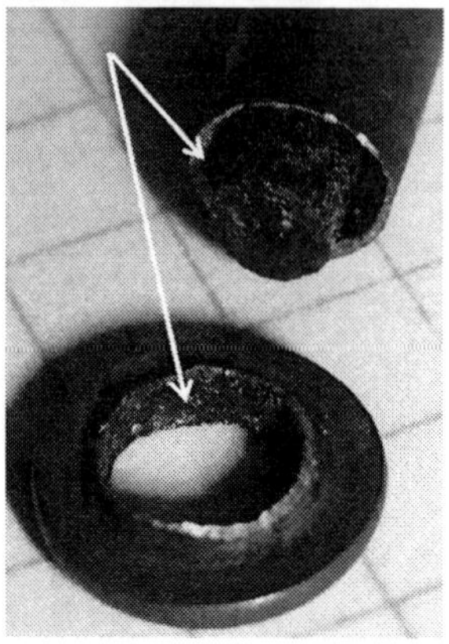

AERODEF MANUFACTURING
February 25-27, 2014 - SME

9

Early Laboratory Test Results Of E-Drill

- Over All The Equipment Performed As Advertised And Was Reliable
- The E-Drill Process Demonstrated The Potential To Save Time And Money
- Many Variables Affect The Process
- Locating The Hand Tool Precisely Over The Center Of The Fastener Is Critical
- Consistent Centering Of The Fastener With The Prototype Bombsight Locator Is Difficult
- Fastener Burn-Through Occurred Intermittently And Needed Further Investigation
- Fastener Burn-Through In Aluminum Is Not A Major Issue
- Fastener Burn-Through In Fiberglass Panels Much Less Damaging
- Fastener Burn-Through In Carbon Fiber Composites Is A Major Issue
- Machining Enhancements To The Electrodes Were Needed
- Electrode Life Was Not As Long As Expected
- First Prototype Unit Not UL or Canadian Standards Association Certified
- Most Lockheed Martin Applications Require Very Precise Fastener Alignment
- A Clean Conductive Fastener Surface Is Needed For The E-Drill System

AERODEF MANUFACTURING
February 25-27, 2014 - SME

©2014 Lockheed Martin Corporation

10

Fastener Burn-Through

AERODEF MANUFACTURING
February 25-27, 2014 - SME

Fastener Burn-Through Sources Of Error

- Electrode Not Centered On Fastener
- Wrong Electrode Used
- Electrode Improperly Machined Or Bent
- Depth Of Cut Too Deep
- Hand Held Unit Positioned Off Angle To Fastener
- Improper Machine Settings
- Operator Error

AERODEF MANUFACTURING
February 25-27, 2014 - SME

Electrode Burn Groove Analysis

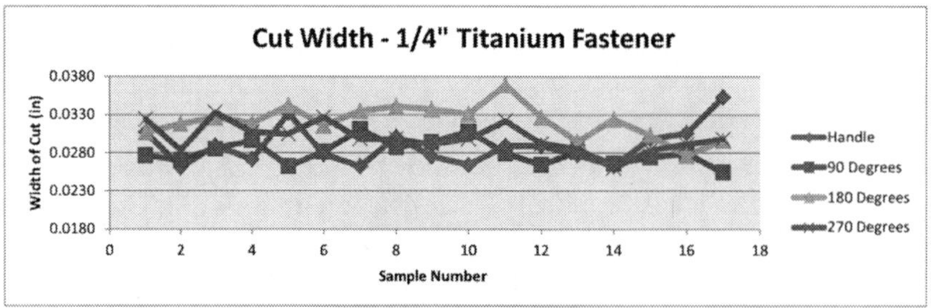

Burn Groove Cross-Section Analysis

Vacuum Fastener Locator Fixture Development

Two Approaches Investigated During Centering Fixture Development

1) Commercial Off The Shelf Fixtures With CCD Camera
2) Custom Designed And Built Fixtures With CCD Camera

Modified Newport Microscope Stage Mechanical Locator

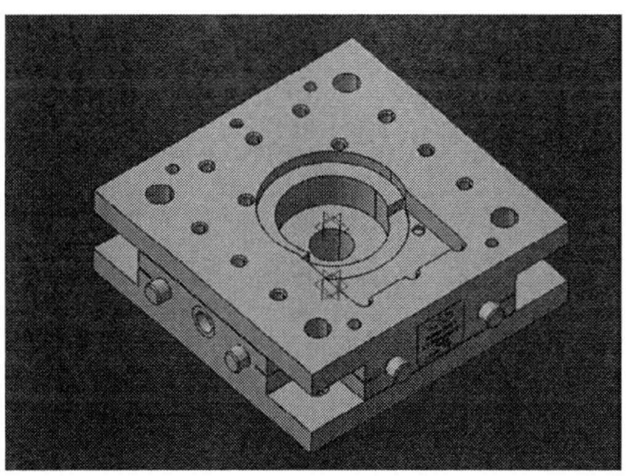

Lockheed Martin Prototype Fastener Locating Fixture

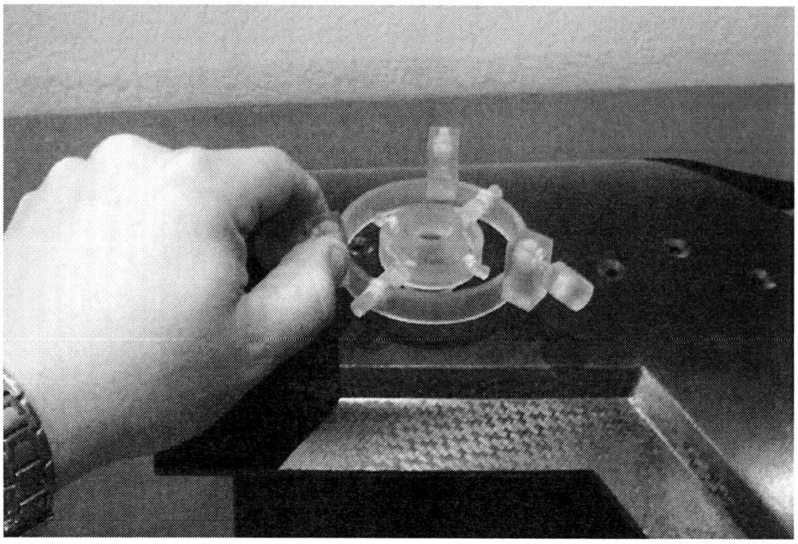

AERODEF MANUFACTURING
February 25-27, 2014 - SME

17

Lockheed Martin Vacuum Locating Fixture Components

AERODEF MANUFACTURING
February 25-27, 2014 - SME

18

Camera Image

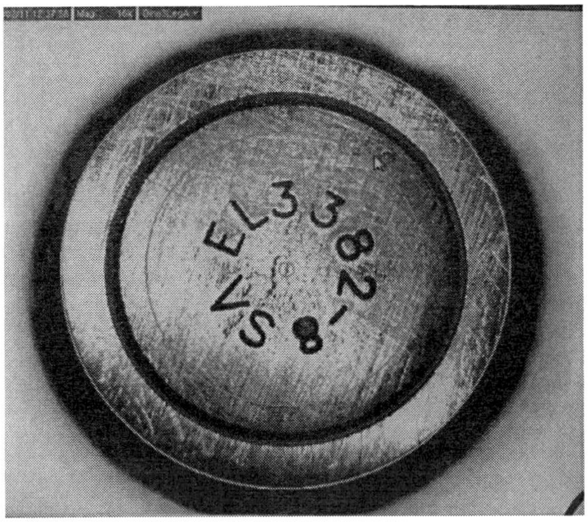

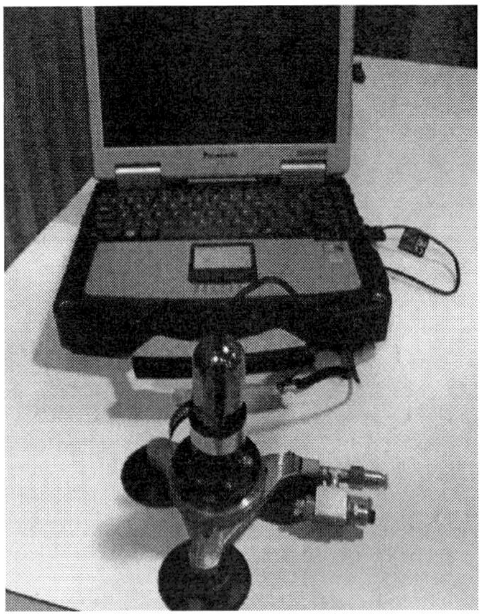

AERODEF MANUFACTURING
February 25-27, 2014 - SME

Latest Perfect Point Offset Vacuum Locating Fixture

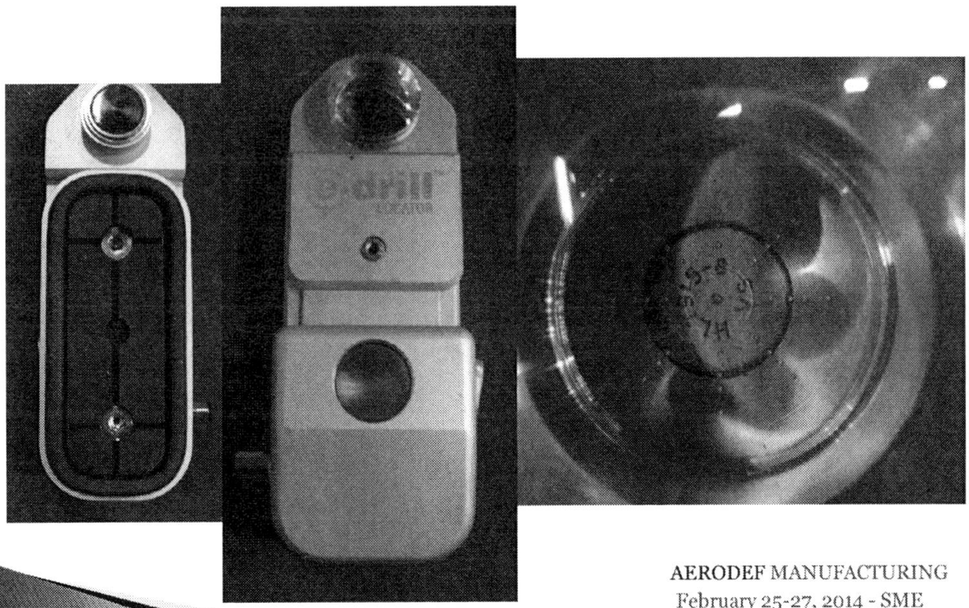

AERODEF MANUFACTURING
February 25-27, 2014 - SME

Fastener Burn-Through Test Panels

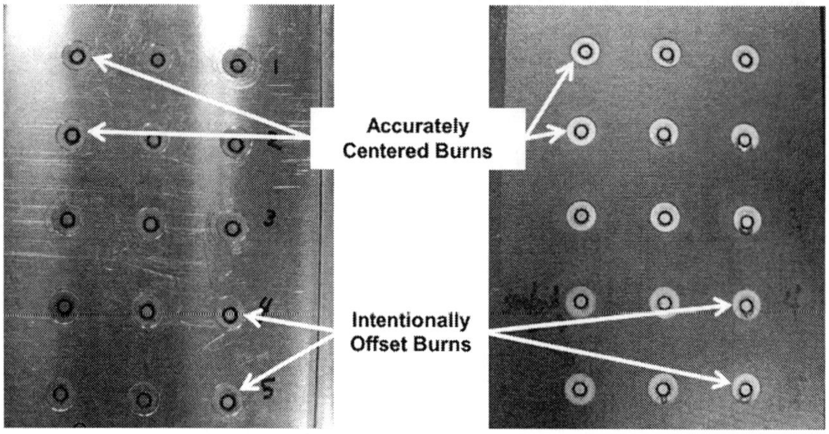

Aluminum Skin **Composite Skin**

Aluminum Skin Fastener Burn-Through Testing

Minimal Damage To The Part Occurs
When The Burn Groove Penetrates The Fastener In Aluminum

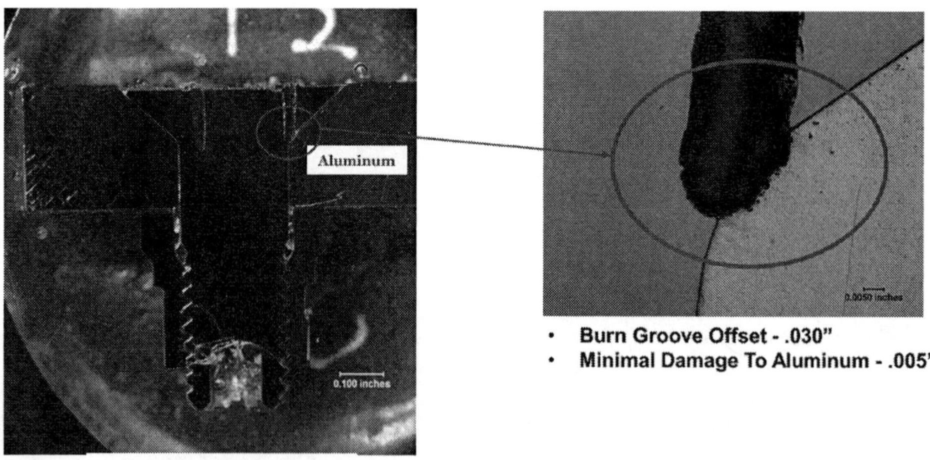

- **Burn Groove Offset - .030"**
- **Minimal Damage To Aluminum - .005"**

Fastener (1/4" Diameter)

Thermal Testing With Temperature Sensitive Paint Applied To Fastener Shank

No Thermal Damage To The Part Is Apparent
When The Burn Groove Is Contained Within The Fastener Shank
(Less Than 275 Degrees Fahrenheit)

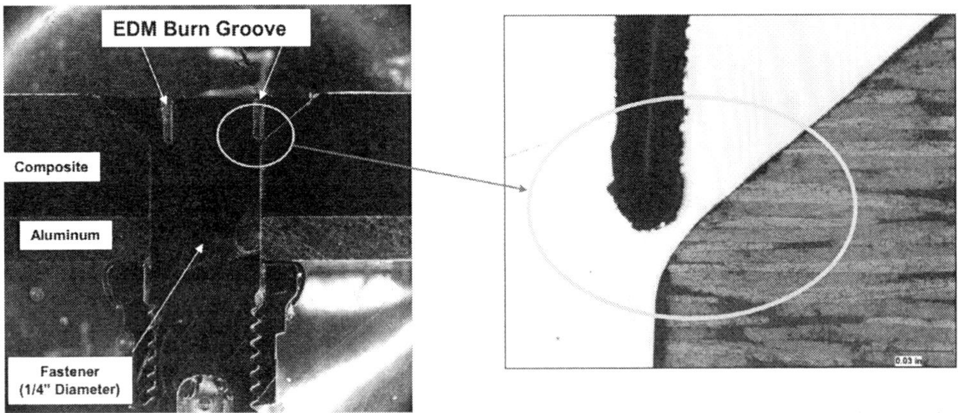

AERODEF MANUFACTURING
February 25-27, 2014 - SME

Fastener Burn-Through Composite Test Panel

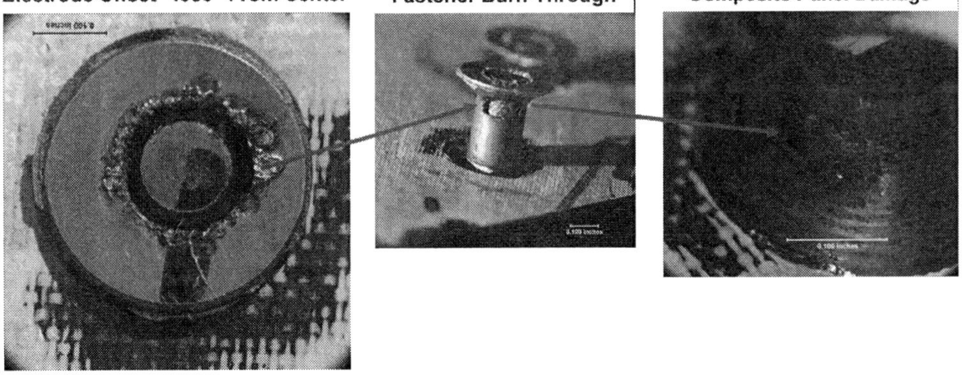

AERODEF MANUFACTURING
February 25-27, 2014 - SME

Composite Panel Fastener Burn-Through Testing

Significant Damage To The Part Occurs
When The Burn Groove Penetrates The Side Of The Fastener In Composites

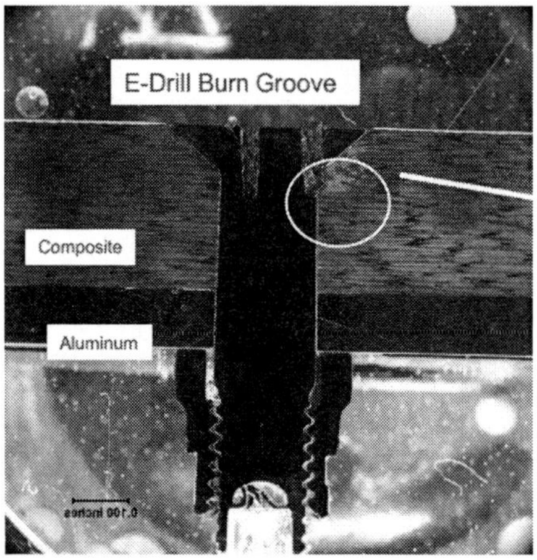

Fastener (3/16" Diameter)

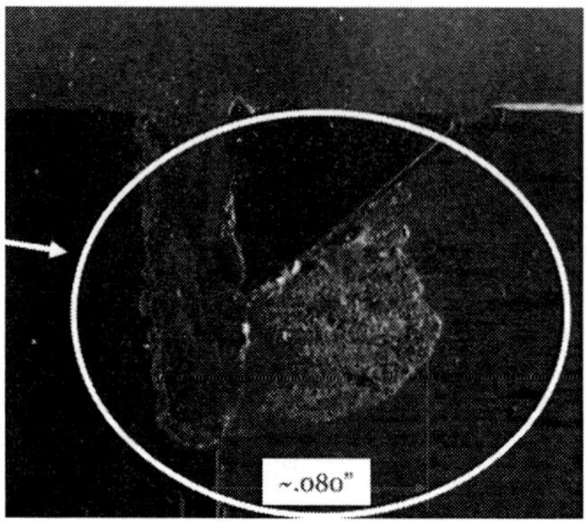

Electrode Intentionally Offset From Center By .030"

AERODEF MANUFACTURING
February 25-27, 2014 - SME

25

Properly Burned Fasteners

AERODEF MANUFACTURING
February 25-27, 2014 - SME

26

Separated Fasteners After E-Drill Processing

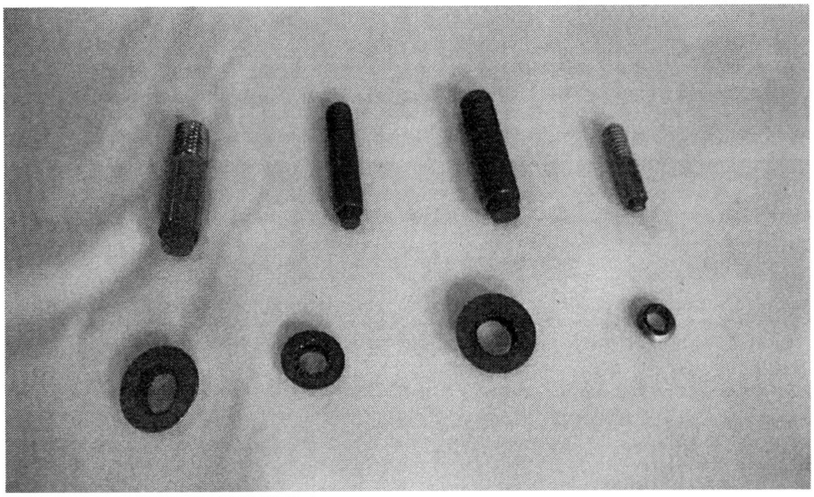

AERODEF MANUFACTURING
February 25-27, 2014 - SME

E-Drill System Status

- Perfect Point Continues To Improve System Performance With Numerous Refinements
- Flush Fastener Burn-Through Dramatically Reduced When Used With Improved Mechanical Locating Fixture And Camera
- Electrode Machining Accuracy And Repeatability Significantly Improved
- Hand Held Gun Improved In Several Areas
- Electrode Life Significantly Improved
- Canadian Standards Association (CSA) Certification Obtained
- Ongoing Collaboration Between Lockheed Martin And Perfect Point On New Applications And Equipment

AERODEF MANUFACTURING
February 25-27, 2014 - SME

Thank You

AERODEF MANUFACTURING
February 25-27, 2014 - SME

29

Composite Drilling Seminar

Feb 25th 2014

Presented by

Scott Daggett – Nat'l Aerospace Specialist

OSG TAP & DIE, INC.

Agenda

- Overview of Today's Composites
- Overview of Drill Geometries
- Common Quality Issues
- Experiment Overview & Results
- Summary of Experiment

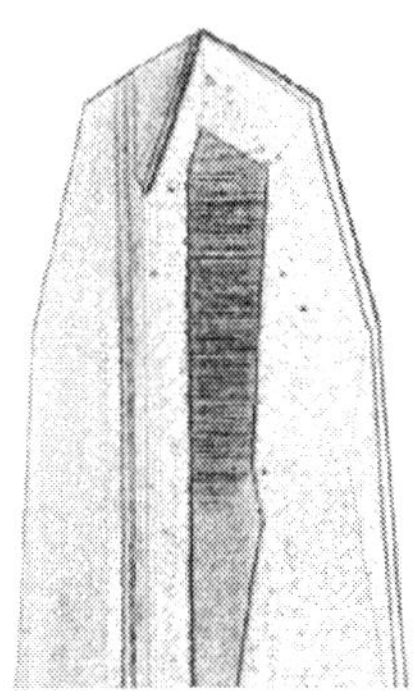

OSG TAP & DIE, INC.

Today's Composites

OSG TAP & DIE, INC.

Good v. Bad Holes

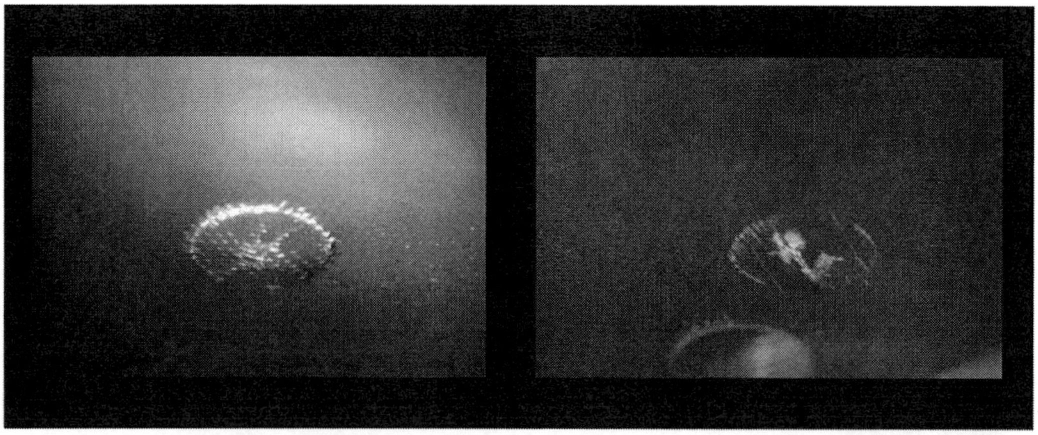

OSG TAP & DIE, INC.

Quality Issues

- Surface Delamination
- Fiber Rolling
- Fiber Breakout
- Burr
- Splintering/Fraying
- Exposed / Un-Cut Fibers
- Hole Expansion
- Heat Degradation
- Surface Roughness

OSG TAP & DIE INC.

Common Quality Issues

Damage of Surface Layers

Delamination

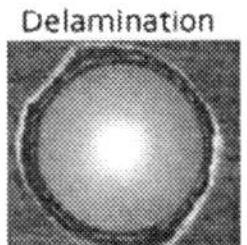

Uncut Fibers

Fiber Pullout

Edge Chipping

Surface Roughness

R_a, R_z, RMS, R_{MAX}

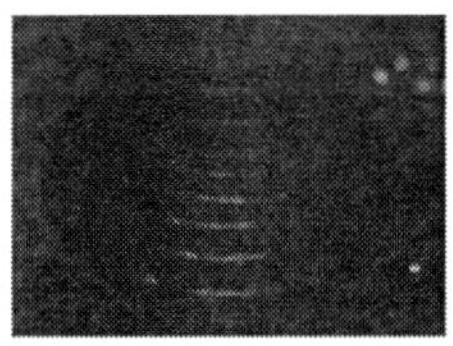

Damage of Peripheral Zone

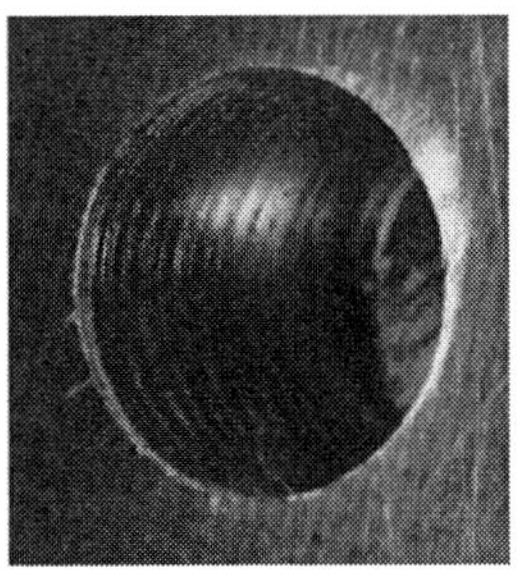

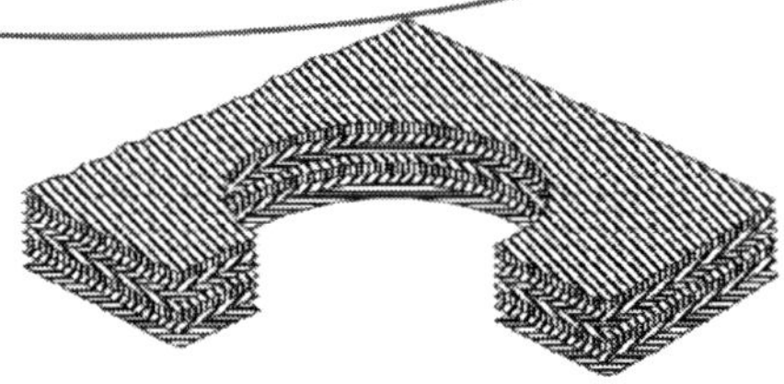

Roundness Error

$f_k = (d_{max} - d_{min})/2$

Dimensional Error

$\Delta D = d\text{-}D$

Crack Formation

OSG TAP & DIE INC.

Uncut Fibers

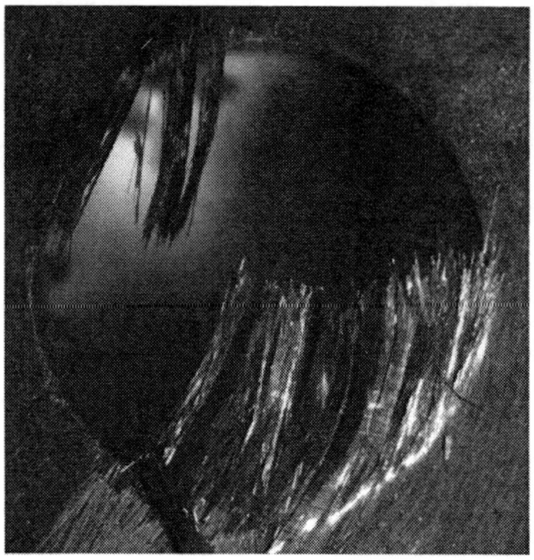

OSG TAP & DIE, INC.

Fiber Orientation in Drilling

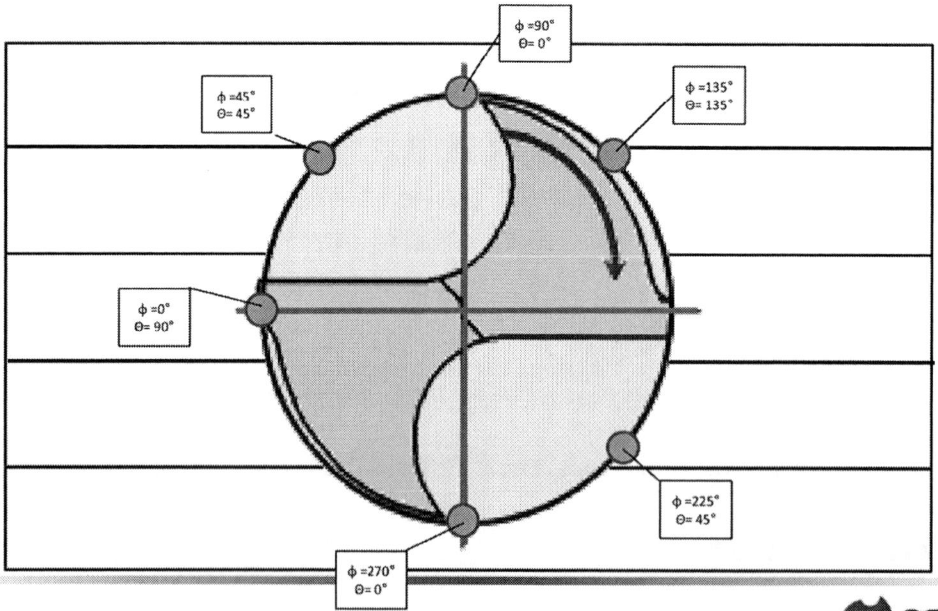

Source: Machining of Polymer Composites – Sheikh-Ahmad

OSG TAP & DIE, INC.

Thrust Force & Torque Curve

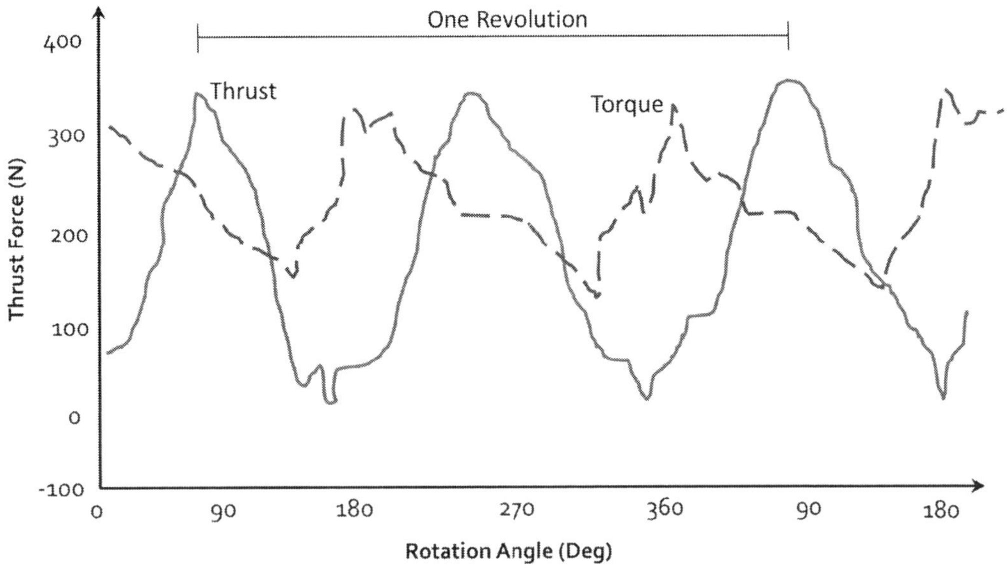

Source: Machining of Polymer Composites – Sheikh-Ahmad

OSG TAP & DIE, INC.

Characteristics of Uncut Fibers

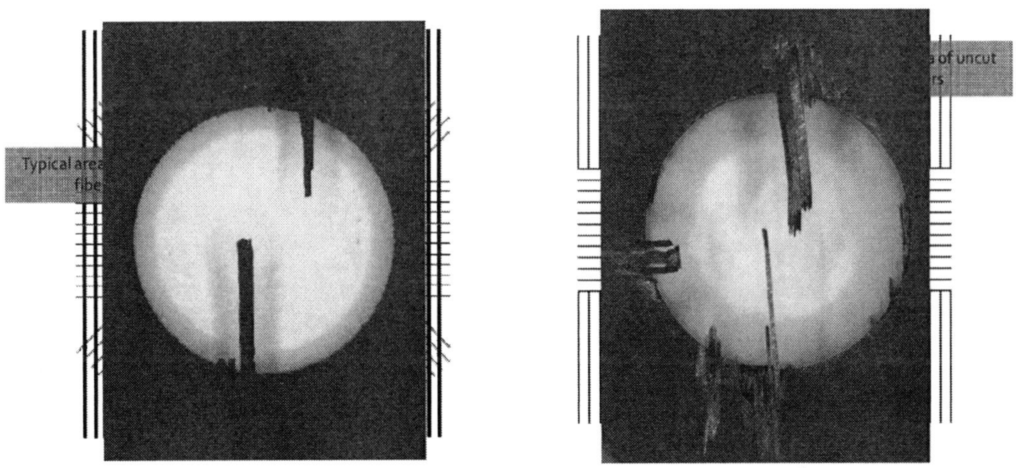

Unidirectional Fibers

Woven Fibers

OSG TAP & DIE, INC.

Fiber Pullout

OSG TAP & DIE, INC.

Hole Shrinkage

Internal Damage

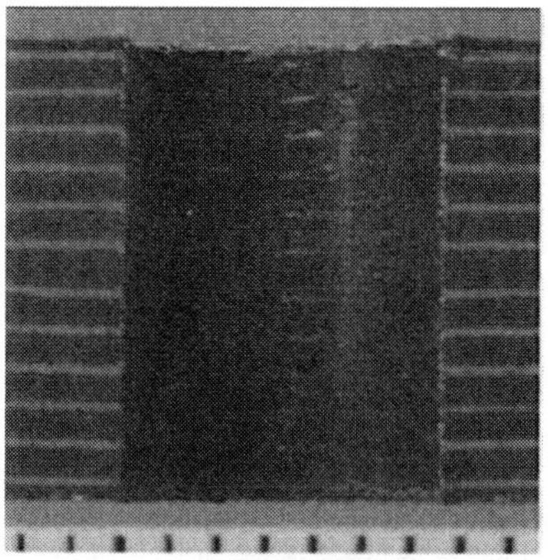

Acceptable

OSG TAP & DIE, INC.

Hole Shrinkage

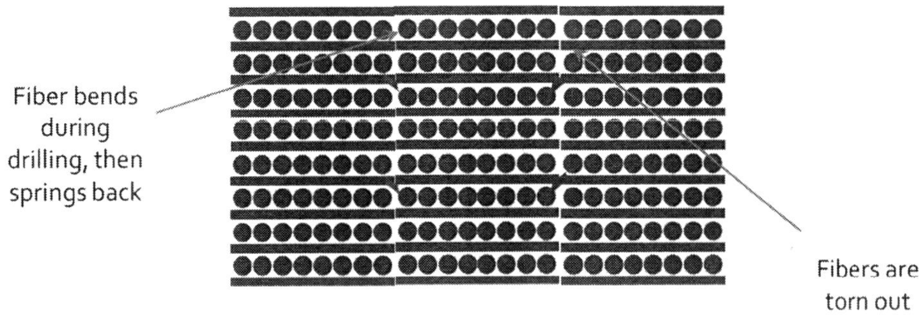

Chipping

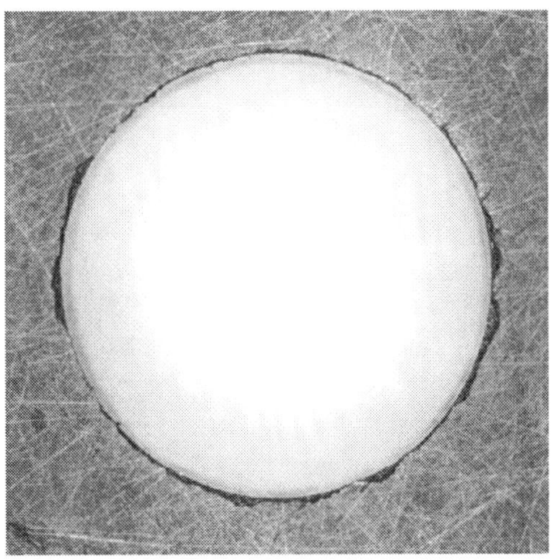

Delamination

Types of Delamination

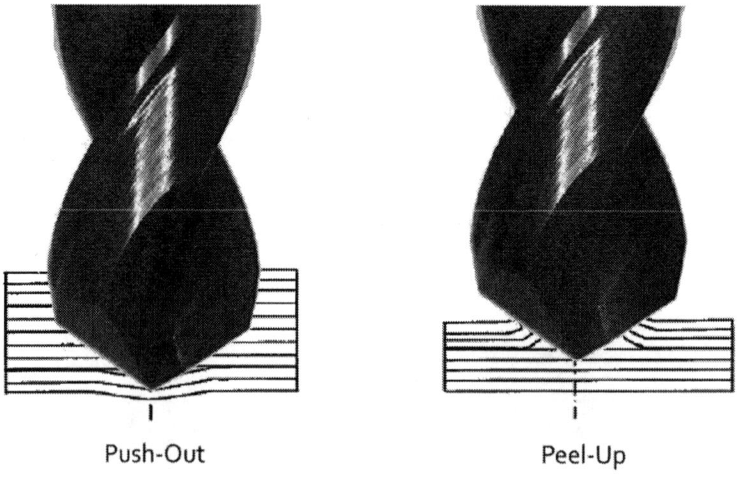

Delamination

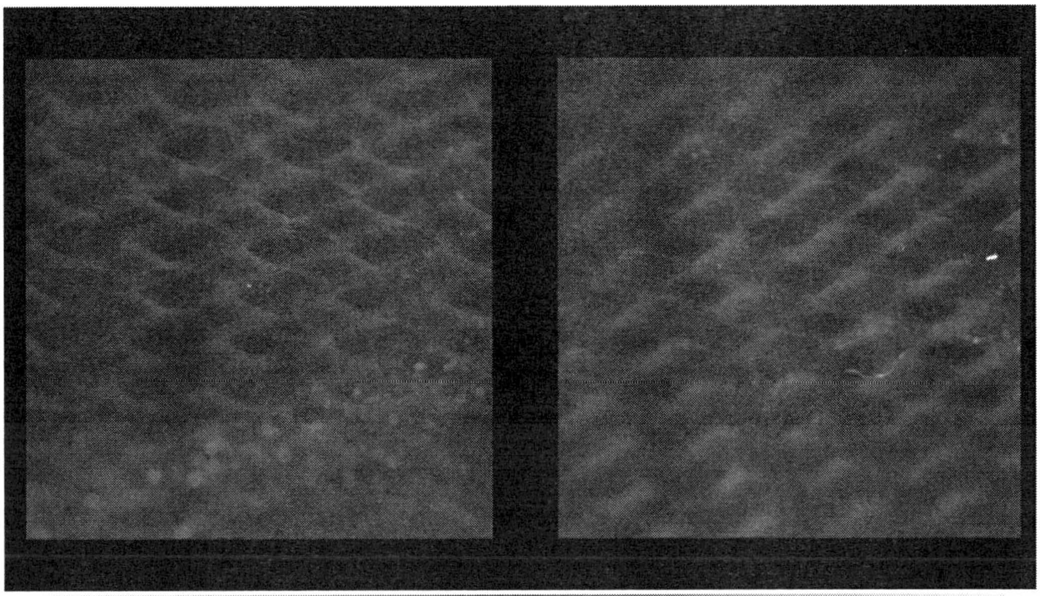

OSG
TAP & DIE, INC.

6 Major Features

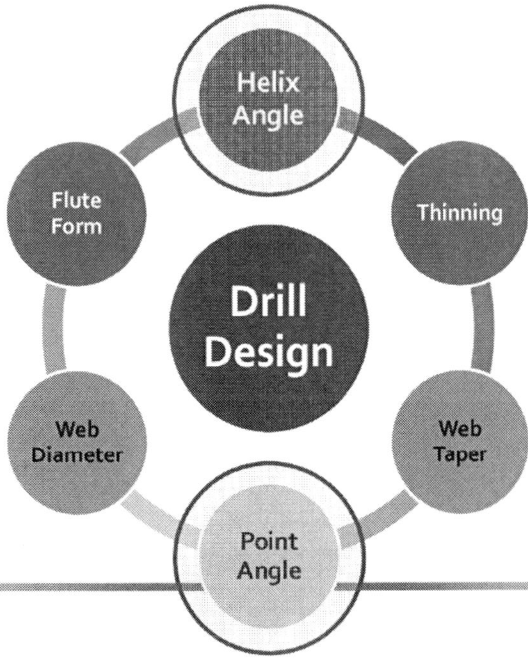

OSG
TAP & DIE, INC.

Helix Angles

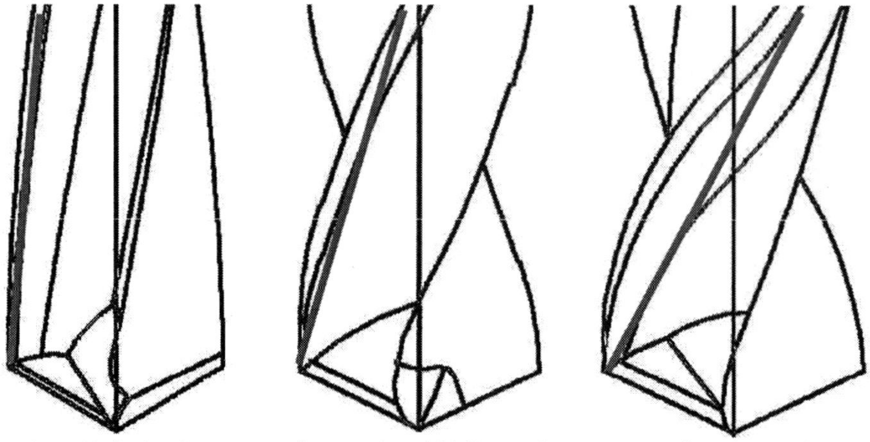

Low Helix Angle
Example: 12°

Conventional Helix Angle
Example: 30°

High Helix Angle
Example: 40°

Geometry Related Quality Issues

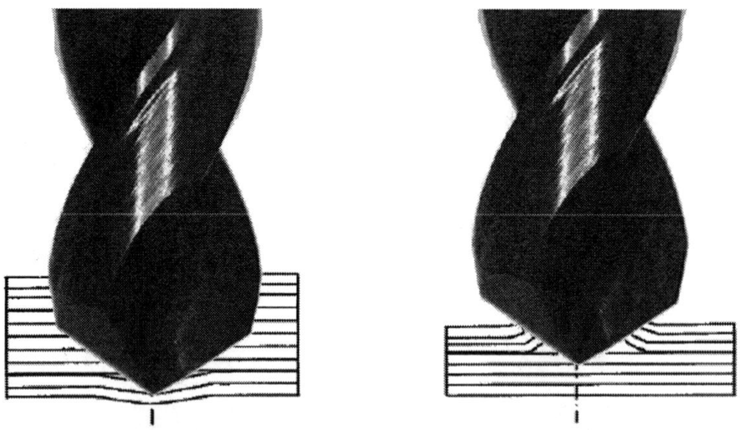

Push-Out

Peel-Up

Types of Delamination

Point Angles

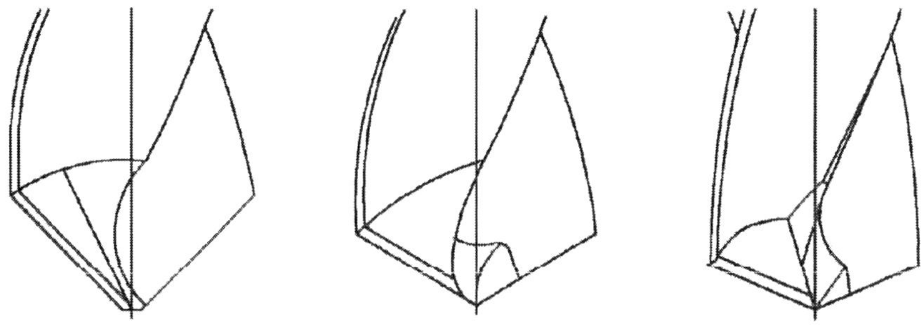

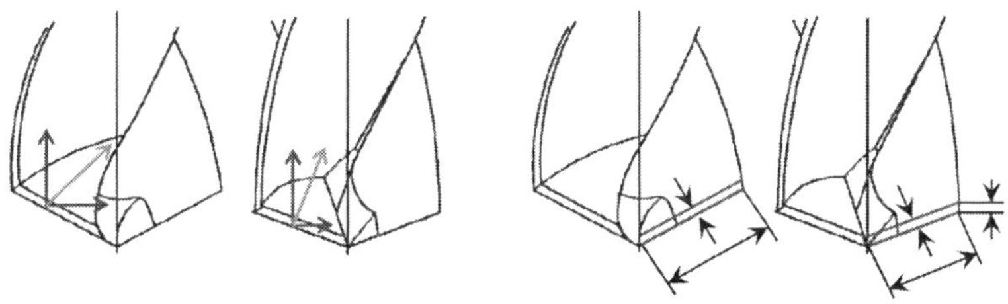

Effect of Triple & Double Angles

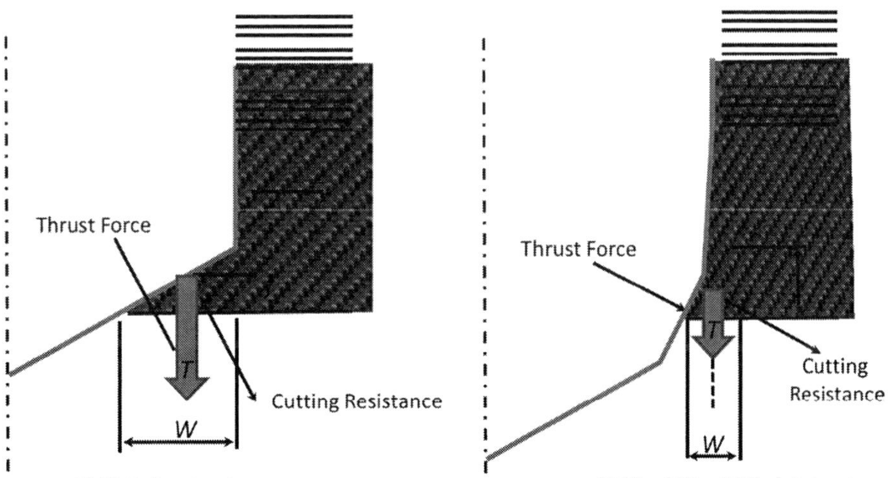

120° Point Angle
Conventional Drill

120° x 30° x 6° Point Angle
Triple Angle

Fiber Orientation in Drilling

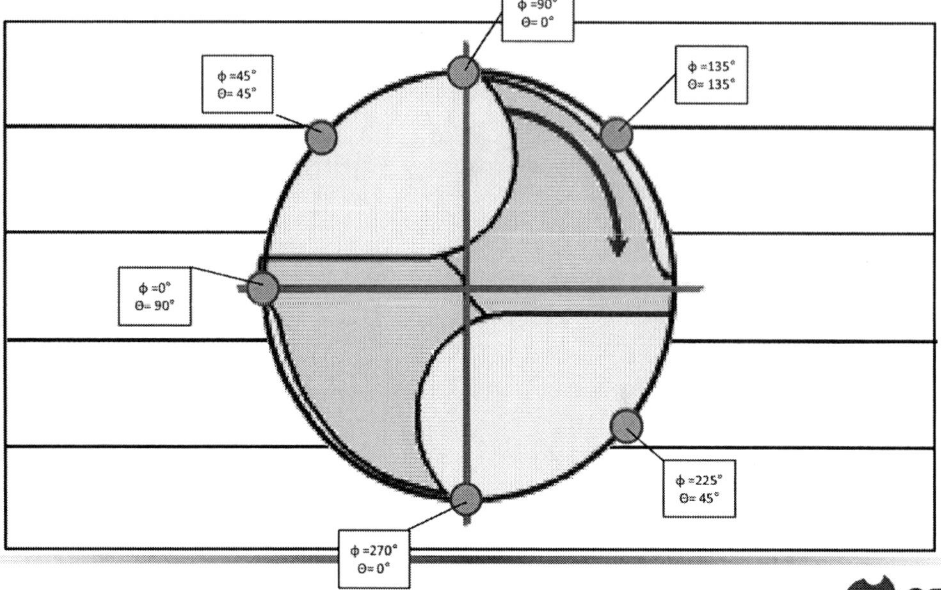

Characteristics of Uncut Fibers

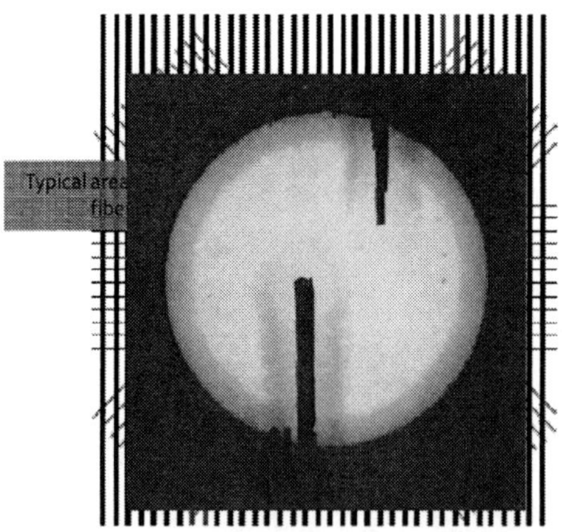

Unidirectional Fibers

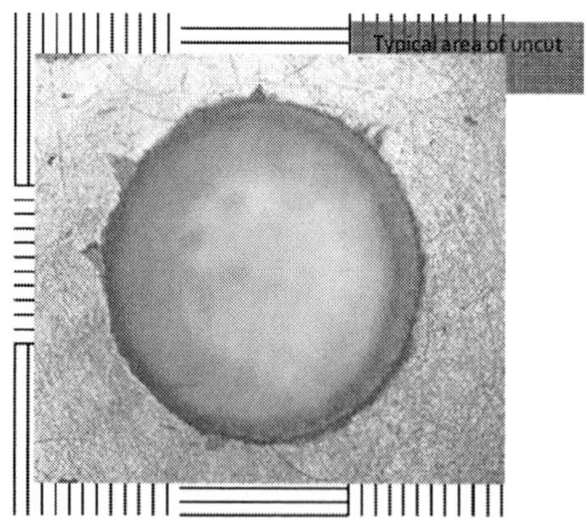

Woven Fibers

Experiment Overview

To test six drill designs in four different composite laminates to analyze effect of each drill geometry

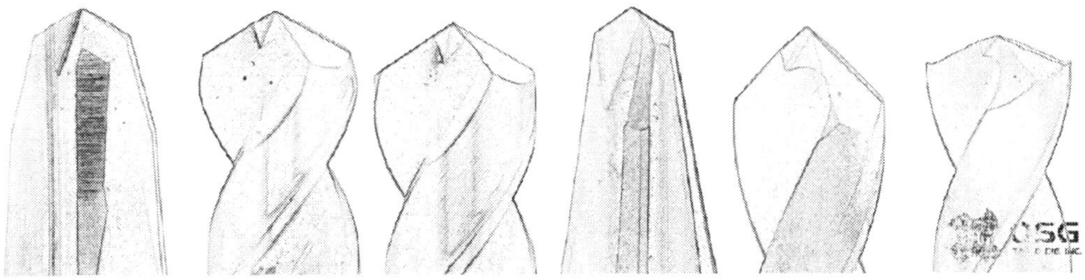

Experiment Overview-Drills

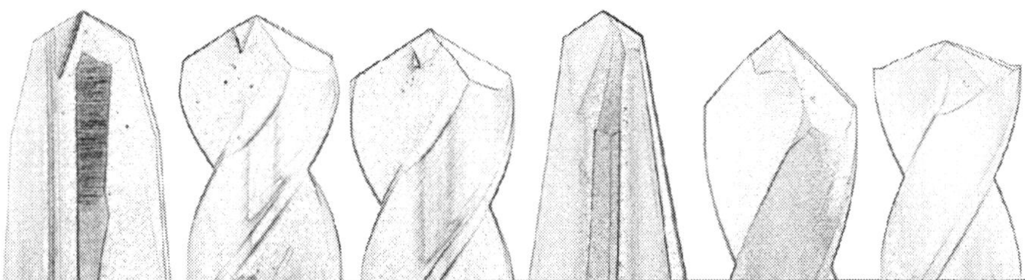

Drill Name	Multi Taper Drill 0 Helix	Short Taper High Helix	120° Point High Helix	Long Taper Low Helix	90° Point Standard Helix	Flat Brad Standard Helix
Diameter	0.251″	0.251″	0.251″	0.251″	0.251″	0.251″
Point Angle	3 Angles	2 Angles	120°	2 Angles	90°	140°
Helix Angle	0°	40°	40°	8°	30°	30°
Coating	Diamond	Diamond	Diamond	Diamond	Diamond	Diamond

Test Parameters

200 SFM – 3,043 RPM

Low – 0.001 / Med – 0.002 / High – 0.003 IPR

OSG TAP & DIE, INC.

Composite #1

Entrance Surface

Exit Surface

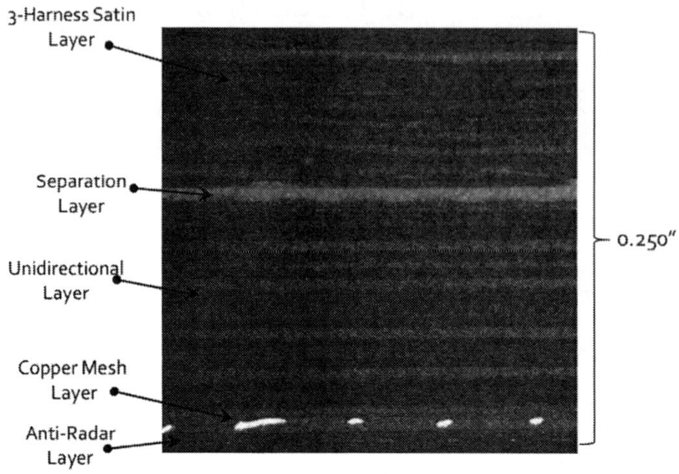

3-Harness Satin Weave with Cu Mesh & Anti-Radar Layer

OSG TAP & DIE INC.

Summary of Experiment

OSG TAP & DIE INC.

Composite #1	Low Feed - Hole #3		Medium Feed - Hole #6		High Feed - Hole #9	
	Entrance	Exit	Entrance	Exit	Entrance	Exit
Multi-Taper Drill	Good	High Delamination	Good	Med. Delamination	Good	Low Delamination
Short Taper	Fiber Pullout	High Delamination	Fiber Pull Out Uncut Fibers	Low Delamination	Fiber Pullout Uncut Fibers	Good
120° Point	Good	Good	Fiber Pullout Uncut Fibers	Good	Fiber Pullout	Good
Long Taper	Good	High Delamination	Uncut Fibers	Med. Delamination	Good	Low Delamination
90° Point	Good	Good	Fiber Pullout	Good	Fiber Pullout	Good
Flat Brad	Good	Med. Delamination	Good	Low Delamination	Fiber Pullout	Chipping Low Delamination

✓ Best O OK ⊘ No Good	Composite #1		Comments
	Entrance	Exit	
Multi-Taper Drill	✓	O	Entrance: Excellent Exit: Delamination / Worst at Low Feed
Short Taper	⊘	O	Entrance: Fiber Pull Up & Pull Out Exit: Delamination at Low & Med. Only
120° Point	✓	✓	Entrance: Excellent / High Feed shows Pull Up Exit: Excellent
Long Taper	✓	O	Entrance: Excellent, Small Uncut Fibers Exit: Delamination / Worst at Low Feed
90° Point	✓	✓	Entrance: Excellent / High Feed shows Pull Up Exit: Excellent
Flat Brad	O	⊘	Entrance: Good / Fiber Pull Out at High Feed Exit: Delamination at Low Feed / Chipping at High

This composite was found to be best drilled by 120 & 90 drills
Long taper drills generate too much chatter when drilling at low feeds

OSG TAP & DIE, INC.

Composite #2

Entrance Surface

Smooth Top Surface

Exit Surface

Fiberglass Surface

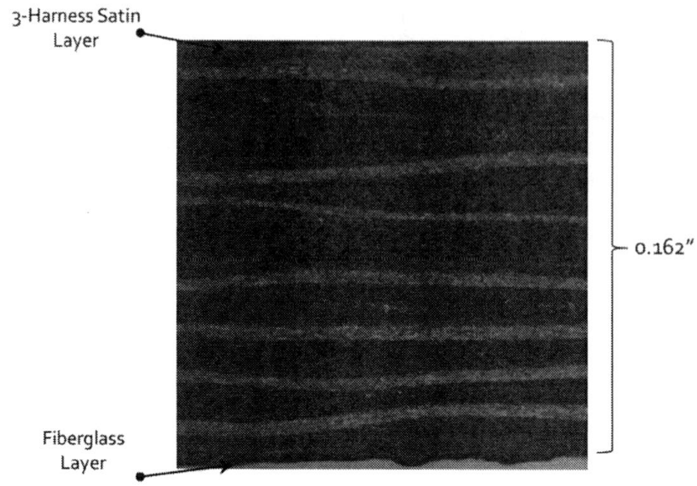

3-Harness Satin Weave with Fiberglass Exit

OSG TAP & DIE INC.

Composite #2	Low Feed - Hole #3		Medium Feed - Hole #6		High Feed - Hole #9	
	Entrance	Exit	Entrance	Exit	Entrance	Exit
Multi-Taper Drill	Low Delamination	Good	Low Delamination	Good	Small Uncut Fibers Low Delamination	Good
Short Taper	Fiber Pull Out Uncut Fibers	Uncut Fibers Low Delamination	Fiber Pull Out Uncut Fibers	Uncut Fibers Low Delamination	Fiber Pull Out Uncut Fibers	Uncut Fibers Low Delamination
120° Point	Fiber Pull Out Uncut Fibers Delamination	Good	Fiber Pull Out Uncut Fibers Delamination	Good	Fiber Pull Out Uncut Fibers Delamination	Good Low Delamination
Long Taper	Low Delamination	Good	Low Delamination	Good	Low Delamination	Small Uncut Fibers
90° Point	Fiber Pull Out Uncut Fibers	Uncut Fibers	Fiber Pull Out Uncut Fibers Low Delamination	Good	Fiber Pull Out Uncut Fibers Delamination	Good
Flat Brad	Delamination	Severe Delamination	Fiber Pull Out Uncut Fibers	Severe Delamination	Fiber Pull Out Uncut Fibers	Severe Delamination

✓ Best / O OK / ⊘ No Good	Composite #2		Comments
	Entrance	Exit	
Multi-Taper Drill	✓	✓	Entrance: Good, Fiber Peel Up Present Exit: Excellent
Short Taper	O	O	Entrance: At Low Feed Quality is OK Exit: At Low Feed Quality is OK
120° Point	⊘	✓	Entrance: Delamination at All Feeds Exit: Excellent
Long Taper	✓	✓	Entrance: Good, Fiber Peel Up Present Exit: Excellent, small uncut fibers at High Feed
90° Point	O	✓	Entrance: Good quality at low to med feed rates Exit: Excellent
Flat Brad	⊘	⊘	Entrance: Exit: Severe Delamination at All Feed Rates

Long taper drills were found to be the best, but surface delamination is concerning.
Due to weak laminate bond

OSG TAP & DIE INC.

Composite #3

Entrance Surface

Copper Mesh Surface

Exit Surface

Fiberglass Surface

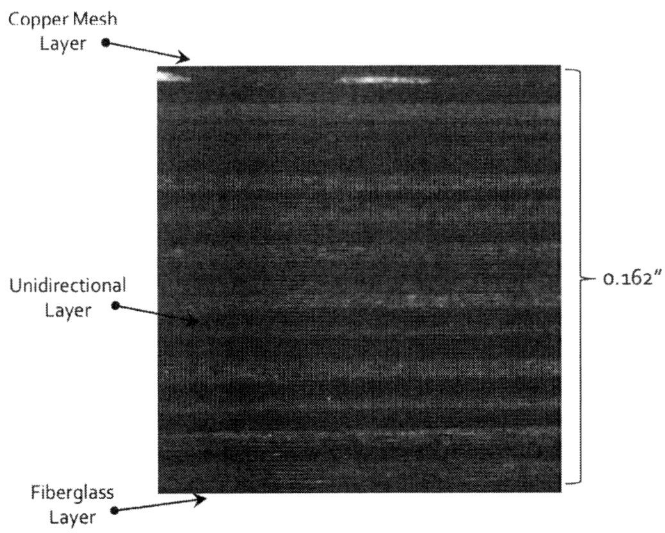

Unidirectional w/ Cu Mesh

OSG TAP & DIE INC.

Composite #3	Low Feed - Hole #3		Medium Feed - Hole #6		High Feed - Hole #9	
	Entrance	Exit	Entrance	Exit	Entrance	Exit
Multi-Taper Drill	Mesh Pull Up Uncut Mesh	Small Uncut Fiber Low Delamination	Mesh Pull Up Uncut Mesh	Small Uncut Fiber Low Delamination	Mesh Pull Up Uncut Mesh	Small Uncut Fiber Low Delamination
Short Taper	Mesh Pull Up Uncut Mesh	Uncut Fiber Delamination	Mesh Pull Up Uncut Mesh	Delamination	Mesh Pull Up Uncut Mesh	Delamination
120° Point	Good	Good	Uncut Mesh	Good	Mesh Pull Up Uncut Mesh	Good
Long Taper	Good	Low Delamination	Good	Low Delamination	Good	Uncut Fibers Low Delamination
90° Point	Uncut Mesh	Good	Uncut Mesh	Good	Uncut Mesh	Good
Flat Brad	Mesh Pull Up Uncut Mesh	Severe Delamination	Mesh Pull Up Uncut Mesh	Severe Delamination	Mesh Pull Up Uncut Mesh	Severe Delamination

✓ Best ○ OK ⃠ No Good	Composite #3		Comments
	Entrance	Exit	
Multi-Taper Drill	⃠	○	Entrance: Light Peel Up w/ Uncut Cu Mesh Exit: Small Uncut fibers with Low Delamination
Short Taper	⃠	⃠	Entrance: Severe Uncut Cu Mesh Exit: Delamination at All Feeds
120° Point	○	✓	Entrance: Low Feed is Excellent / High Feed - X Exit: Excellent
Long Taper	✓	○	Entrance: Excellent Exit: Low Delamination at All Feeds
90° Point	⃠	✓	Entrance: Uncut Cu Mesh All Feeds Exit: Excellent
Flat Brad	⃠	⃠	Entrance: Delamination, Uncut Cu Mesh Exit: Severe Delamination

Overall best drill is 120 point, but ONLY at low feeds
Long Taper drill is good but low amount of delamination is present on exit of all holes

Composite #4

Entrance Surface

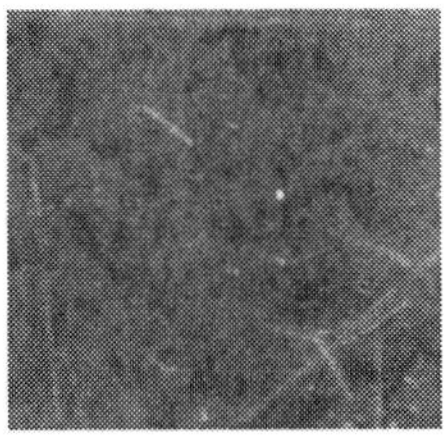

Exit Surface

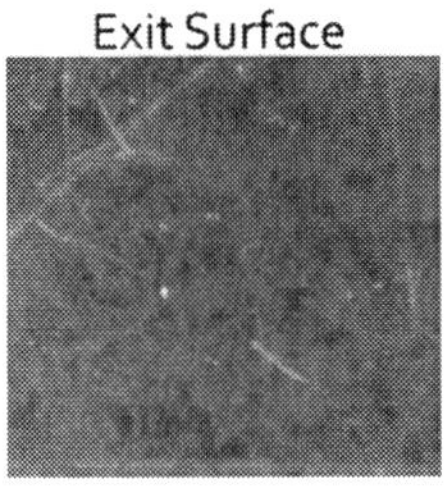

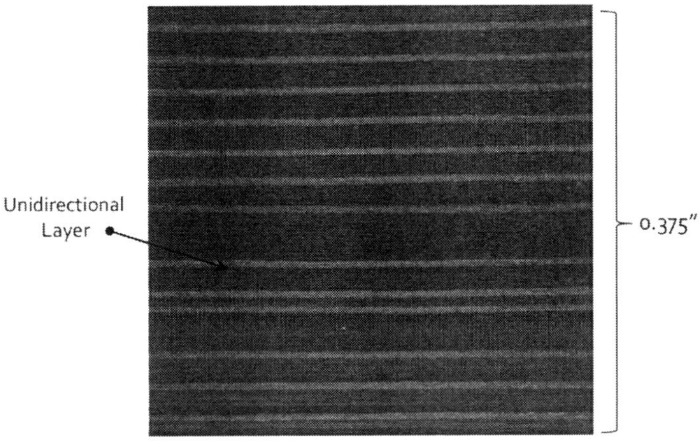

Unidirectional

OSG TAP & DIE, INC.

Composite #4	Low Feed - Hole #3		Medium Feed - Hole #6		High Feed - Hole #9	
	Entrance	Exit	Entrance	Exit	Entrance	Exit
Multi-Taper Drill	Uncut Fibers	Delamination Uncut Fibers	Uncut Fibers	Delamination Uncut Fibers	Uncut Fibers	Delamination Uncut Fibers
Short Taper	Peel Up	Delamination Uncut Fibers	Peel Up	Delamination Uncut Fibers	Peel Up	Delamination Uncut Fibers
120° Point	Good	Delamination Uncut Fibers	Peel Up Uncut Fibers	Delamination Uncut Fibers	Peel Up Uncut Fibers	Delamination Uncut Fibers
Long Taper	Good	Delamination Uncut Fibers	Uncut Fibers	Delamination Uncut Fibers	Uncut Fibers	Delamination Uncut Fibers
90° Point	Good	Delamination Uncut Fibers	Peel Up	Delamination Uncut Fibers	Peel Up	Delamination Uncut Fibers
Flat Brad	Peel Up	Delamination Uncut Fibers	Peel Up Uncut Fibers	Delamination Uncut Fibers	Peel Up Uncut Fibers	Delamination Uncut Fibers

✓ Best O OK ⊘ No Good	Composite #4		Comments
	Entrance	Exit	
Multi-Taper Drill 	O	⊘	Entrance: Low Feed / Low Uncut Fibers Exit: Delamination at All Feed Rates
Short Taper 	O	⊘	Entrance: Low Feed / Low Uncut Fibers Exit: Delamination at All Feed Rates
120° Point 	O	⊘	Entrance: Low Feed / Low Uncut Fibers Exit: Delamination at All Feed Rates
Long Taper 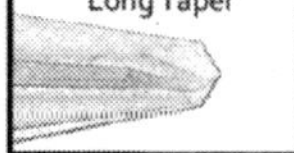	O	⊘	Entrance: Low Feed / Low Uncut Fibers Exit: Delamination at All Feed Rates
90° Point 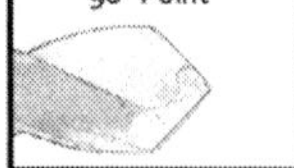	✓	⊘	Entrance: Good Hole at Low Feed Exit: Delamination at All Feed Rates
Flat Brad 	O	⊘	Entrance: Low Feed / Low Uncut Fibers Exit: Delamination at All Feed Rates

Extremely tough laminate to drill
Delamination present at all holes but low feed with 90 Point drill was best of worst

OSG TAP & DIE, INC.

	Composite #1		Composite #2		Composite #3		Composite #4	
	Entry	Exit	Entry	Exit	Entry	Exit	Entry	Exit
	✓	O	✓	✓	⊘	O	O	⊘
	⊘	O	O	O	⊘	⊘	O	⊘
	✓	✓	⊘	✓	O	✓	O	⊘
	✓	O	✓	✓	✓	O	O	⊘
	✓	✓	O	✓	⊘	✓	✓	⊘
	O	⊘	⊘	⊘	⊘	⊘	O	⊘

Summary of Experiment

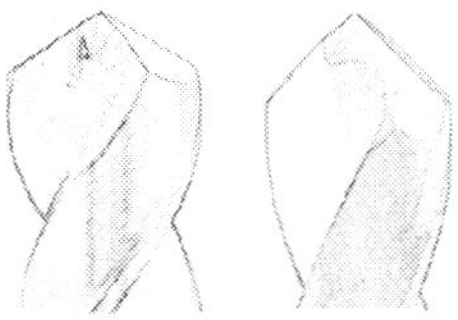

90° & 120° Point Angle drills are very effective in a wide range of composites

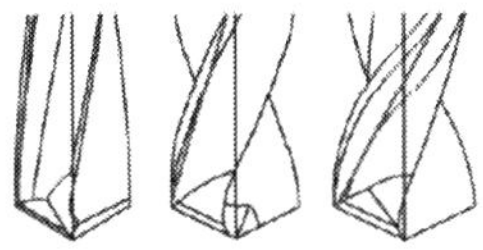

Degree of helix angle is very sensitive at entry of composite

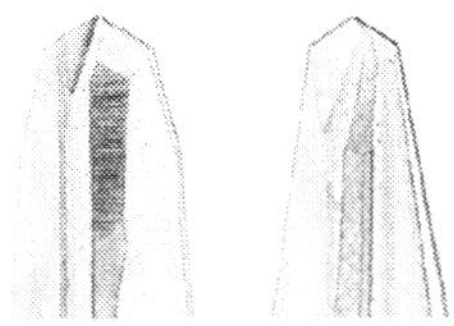

Long taper drills very sensitive to feed rates, performing better at higher feeds

OSG TAP & DIE INC.

Summary of Experiment

Know your composite laminate

All laminates & applications are unique

Supply laminates to your cutting tool maker

Partner with a manufacturer and distributor who can help you

OSG TAP & DIE INC.

Thank you

OSG
TAP & DIE, INC.

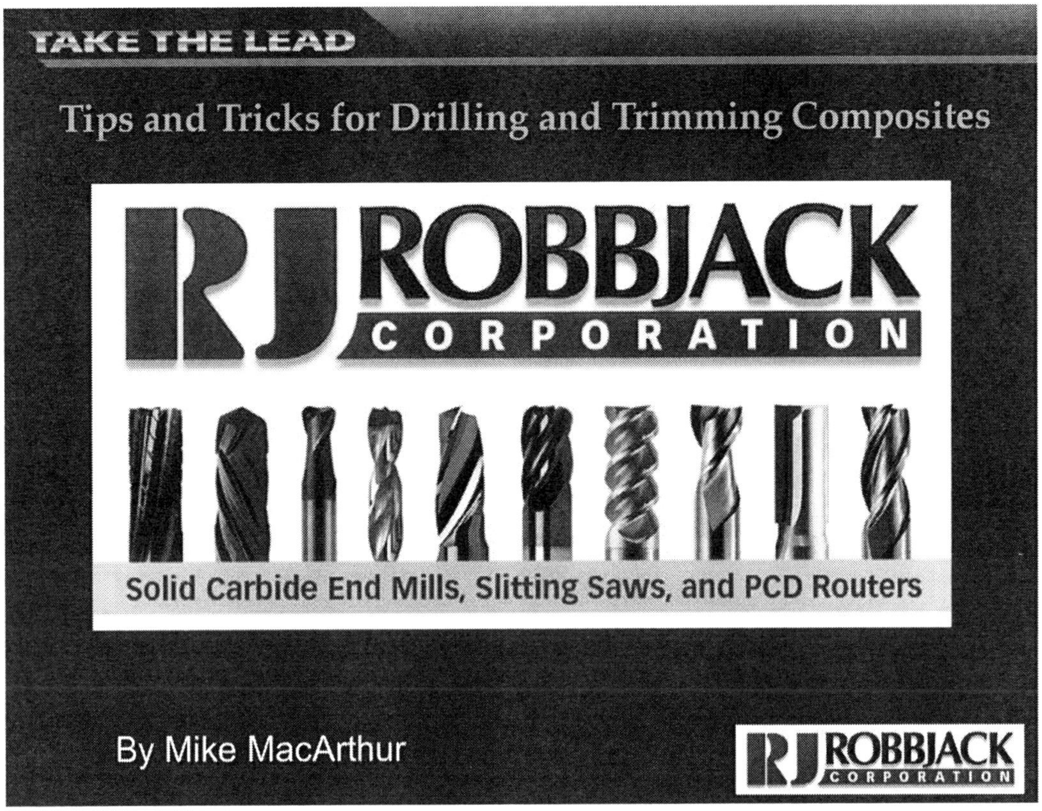
TAKE THE LEAD
Tips and Tricks for Drilling and Trimming Composites
RJ ROBBJACK
CORPORATION
Solid Carbide End Mills, Slitting Saws, and PCD Routers
By Mike MacArthur
RJ ROBBJACK
CORPORATION

TAKE THE LEAD
Cutting Carbon Fiber Right?
Random Fiber
Prepreg
Intermediate Layers
Vacuum Bag

TAKE THE LEAD

Tips for Fixturing

- Rigidity is important
  - Tool Life
  - Consistency
- Vacuum
  - Closely Matches Part
- Tips and Tricks
  - Tape unsupported areas
  - Sand Bags

RJ ROBBJACK CORPORATION

TAKE THE LEAD

Choosing the Best Drill

Carbide Drills

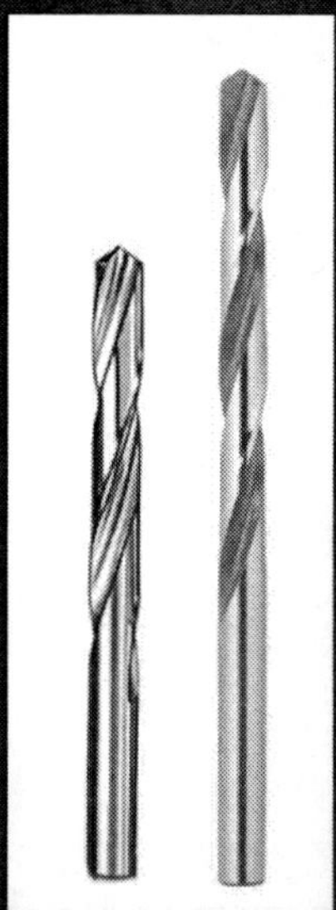

Diamond Coated Drills

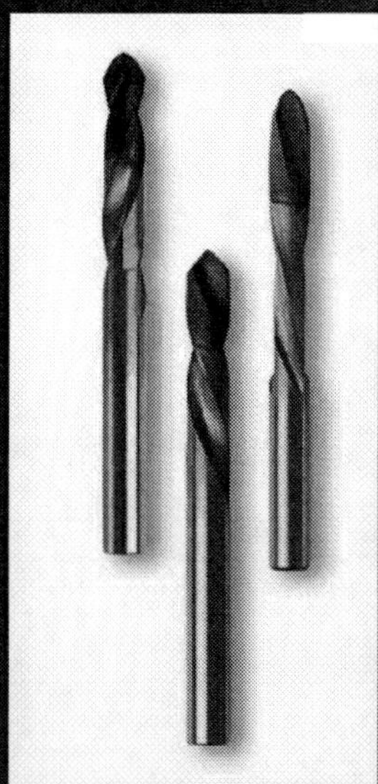

PCD Drills

TAKE THE LEAD
Choosing the Best Drill
Diamond Coated Drills
Used for Low Plastic Content CFRP
New Tip Geometries to Eliminate Delimitation
Great Tool life
Disadvantages
-Limited Carbide Grades
- 6% Co
-No Re-sharpening
RJ ROBBJACK CORPORATION

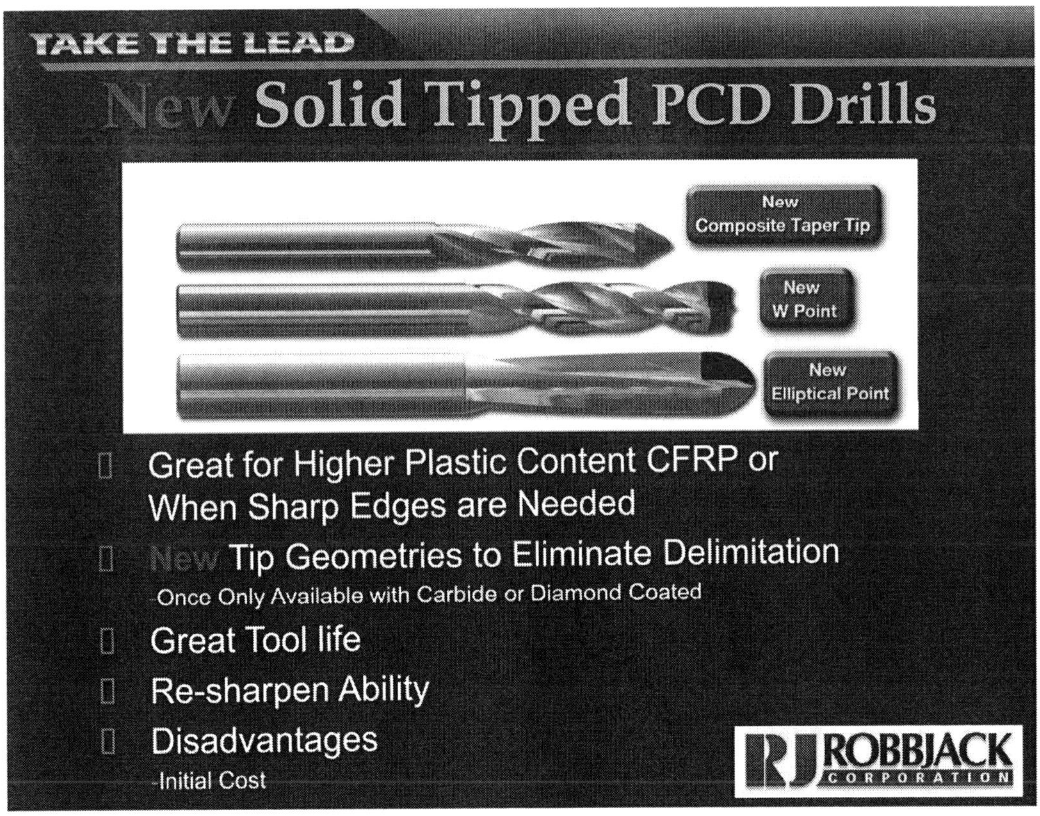
TAKE THE LEAD
New Solid Tipped PCD Drills
New Composite Taper Tip
New W Point
New Elliptical Point
Great for Higher Plastic Content CFRP or When Sharp Edges are Needed
New Tip Geometries to Eliminate Delimitation
-Once Only Available with Carbide or Diamond Coated
Great Tool life
Re-sharpen Ability
Disadvantages
-Initial Cost
RJ ROBBJACK CORPORATION

TAKE THE LEAD

PCD Drill Video

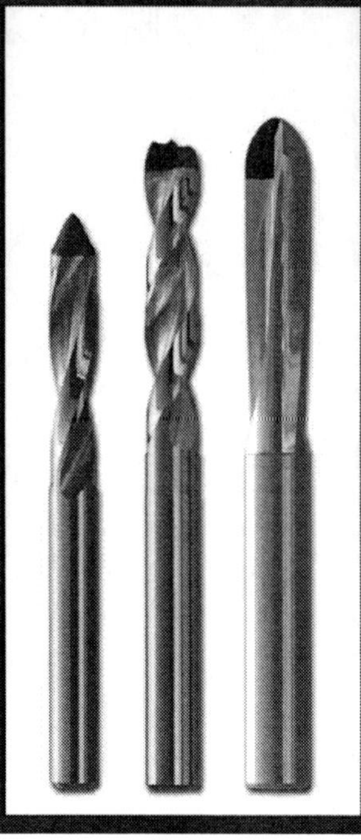

1/4" Dia. Double Angle PCD Drill

233 SFM
3,560 RPM
21.4 inch per min. (543 mm/min)
.006" per rev

ROBBJACK CORPORATION

TAKE THE LEAD

PCD Drill Video

ROBBJACK CORPORATION

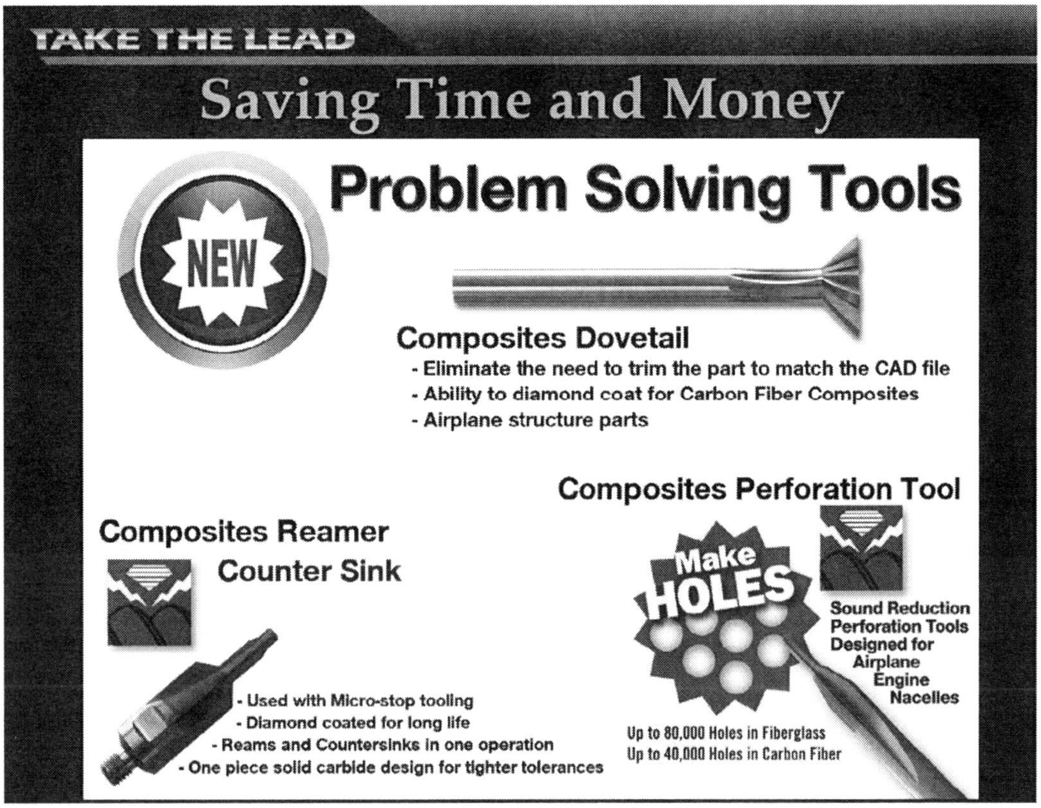
TAKE THE LEAD
Saving Time and Money
Problem Solving Tools
NEW
Composites Dovetail
- Eliminate the need to trim the part to match the CAD file
- Ability to diamond coat for Carbon Fiber Composites
- Airplane structure parts
Composites Perforation Tool
Composites Reamer
Counter Sink
Make HOLES
Sound Reduction Perforation Tools Designed for Airplane Engine Nacelles
- Used with Micro-stop tooling
- Diamond coated for long life
- Reams and Countersinks in one operation
- One piece solid carbide design for tighter tolerances
Up to 80,000 Holes in Fiberglass
Up to 40,000 Holes in Carbon Fiber

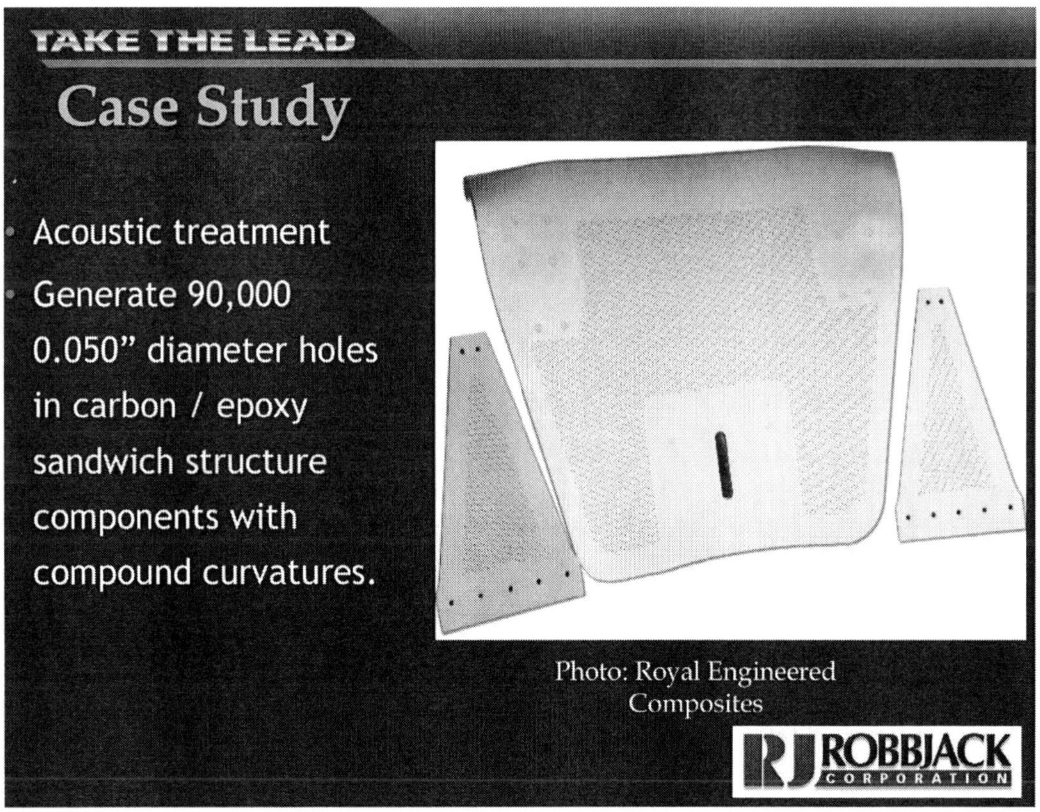
TAKE THE LEAD
Case Study
• Acoustic treatment
• Generate 90,000 0.050" diameter holes in carbon / epoxy sandwich structure components with compound curvatures.
Photo: Royal Engineered Composites
RJ ROBBJACK CORPORATION

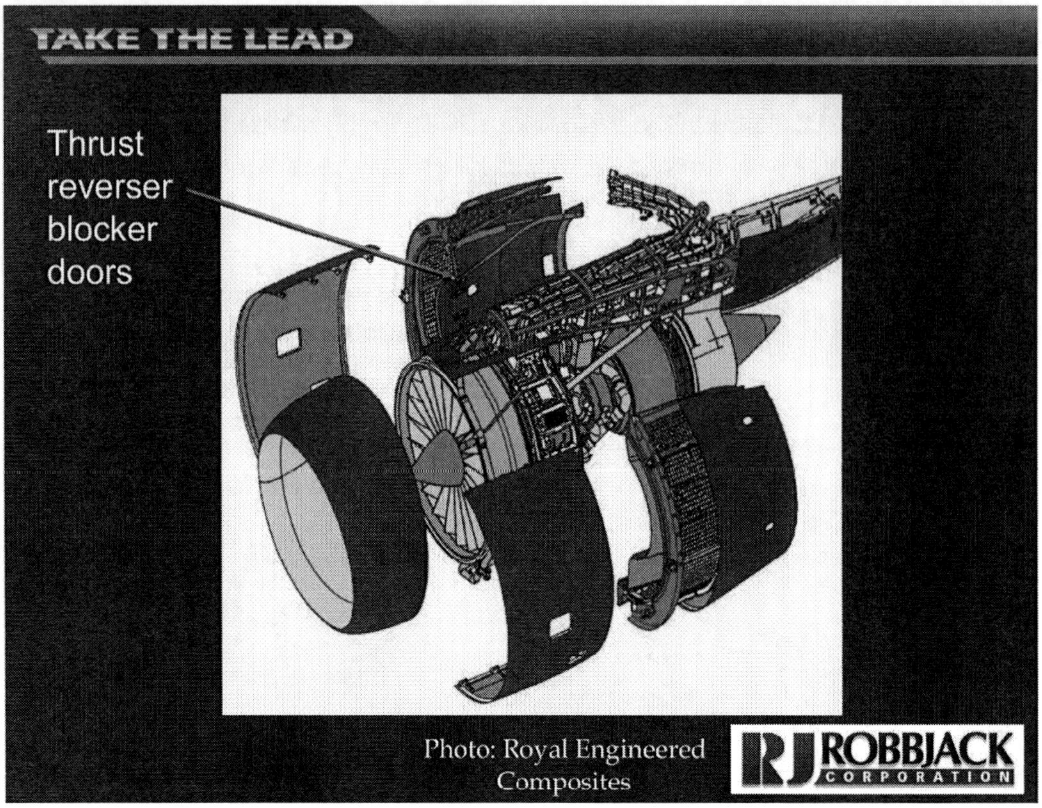
TAKE THE LEAD
Thrust reverser blocker doors
Photo: Royal Engineered Composites
RJ ROBBJACK CORPORATION

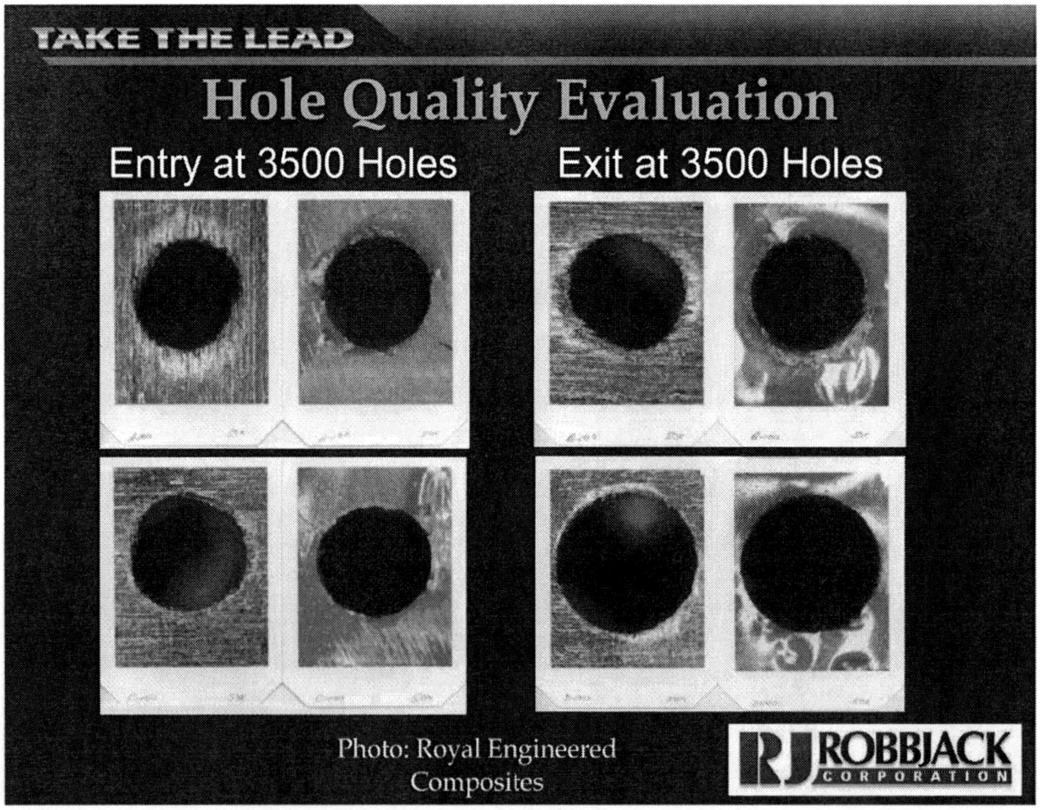
TAKE THE LEAD
Hole Quality Evaluation
Entry at 3500 Holes
Exit at 3500 Holes
Photo: Royal Engineered Composites
RJ ROBBJACK CORPORATION

TAKE THE LEAD

Hole Quality Evaluation

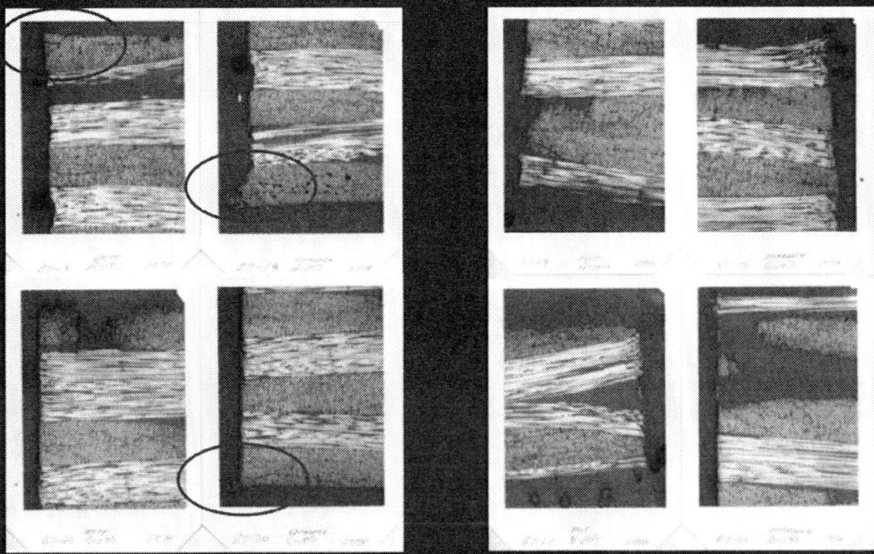

Photo: Royal Engineered Composites

RJ ROBBJACK CORPORATION

TAKE THE LEAD

Conclusions for Acoustical Panels

- Critical
  - CVD coating
  - Adhesion
  - Spindle power
  - Tool handling
    - Non contact measurement
    - Installation in the collet
- Important
  - Tool geometry
  - Chip load
  - Programming
  - (RJ Tests 40,000 RPM and 32.5 ipm)

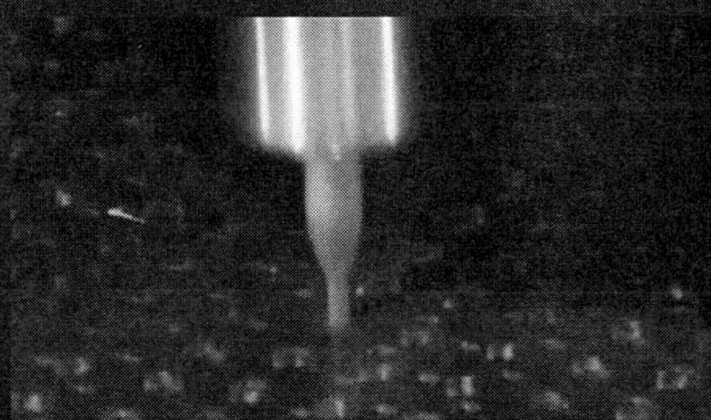

Photo: Royal Engineered Composites

RJ ROBBJACK CORPORATION

TAKE THE LEAD

Choosing Cutting Tools Trimming

Solid Carbide Routers

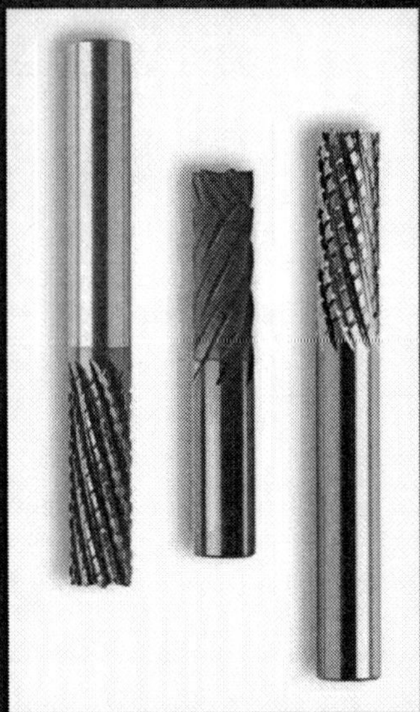

- Applications
  - Un-rigid Fixturing or part design
  - Used as a Rougher for PCD
    - Trimming
    - Pocketing
- Advantages
  - Geometry
  - Forms
  - Availability
  - Cost?
- Disadvantages
  - Poor Tool Life

ROBBJACK CORPORATION

TAKE THE LEAD

Choosing Cutting Tools Trimming

Diamond Coated Routers

- Applications
  - Low Plastic CFRP Materials
  - Trimming
  - Interpolation of holes
  - pocketing
- Advantages
  - Geometry
  - Forms
  - Tool Life
- Disadvantages
  - Limits on Carbide Grade
  - Shock Resistance at cutting edge

ROBBJACK CORPORATION

TAKE THE LEAD

Cutting Tools Trimming

PCD Router

- Applications
  - Rigid Fixturing
    - Trimming
    - Interpolation of holes
    - Pocketing
  - Un-rigid Fixturing
    - Finishing
- Advantages
  - Cut to net shape in one pass
  - Shock Resistance at Cutting Edge
  - Tool Life
- Disadvantages
  - Initial cost

ROBB JACK CORPORATION

TAKE THE LEAD

PCD Router Video

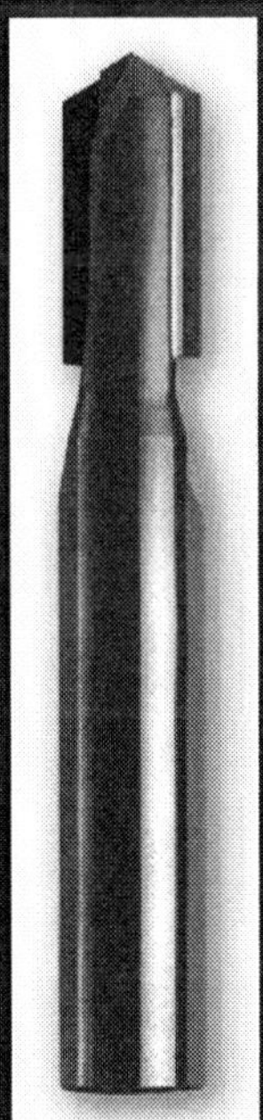

3/8" Dia. PCD Router (CPCD-203-12)

982 SFM
10,000 RPM
80 inches per minute (2032 mm/min)
.008" per rev

ROBB JACK CORPORATION

TAKE THE LEAD

PCD Router Video

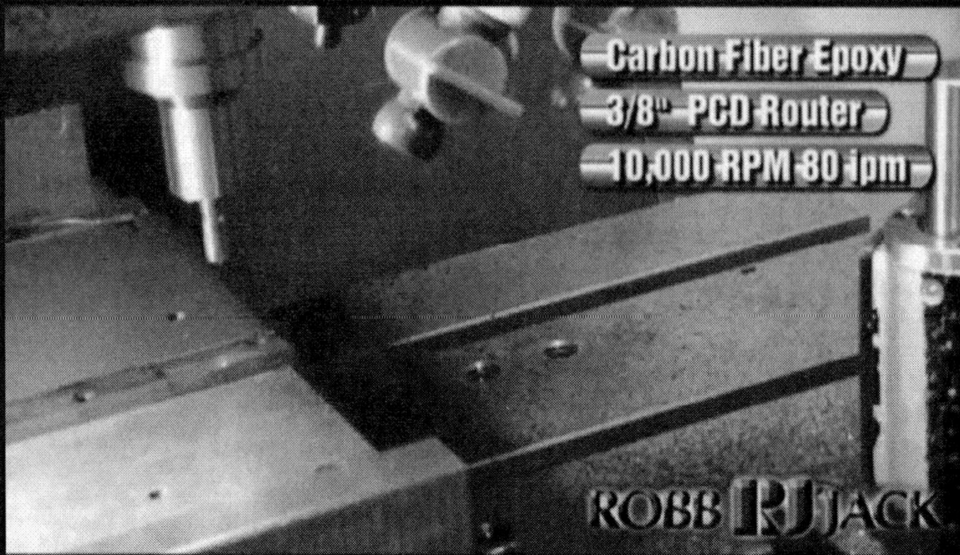

ROBB RJ JACK CORPORATION

TAKE THE LEAD

Tool Life

- Proper RPM (800-1000 SFM)
  - Reduces heat
    - Reduces thermal failure
  - Increases abrasion resistance.
  - Reduces chipping
- Proper Chip Load per Tooth
  - Gets the heat out in the chip
  - Reduces the amount of time in part
  - Do not rub the tool
- Cut full depth up too 2x tool diameter.
- Cut to net shape minimize finish passes

ROBB RJ JACK CORPORATION

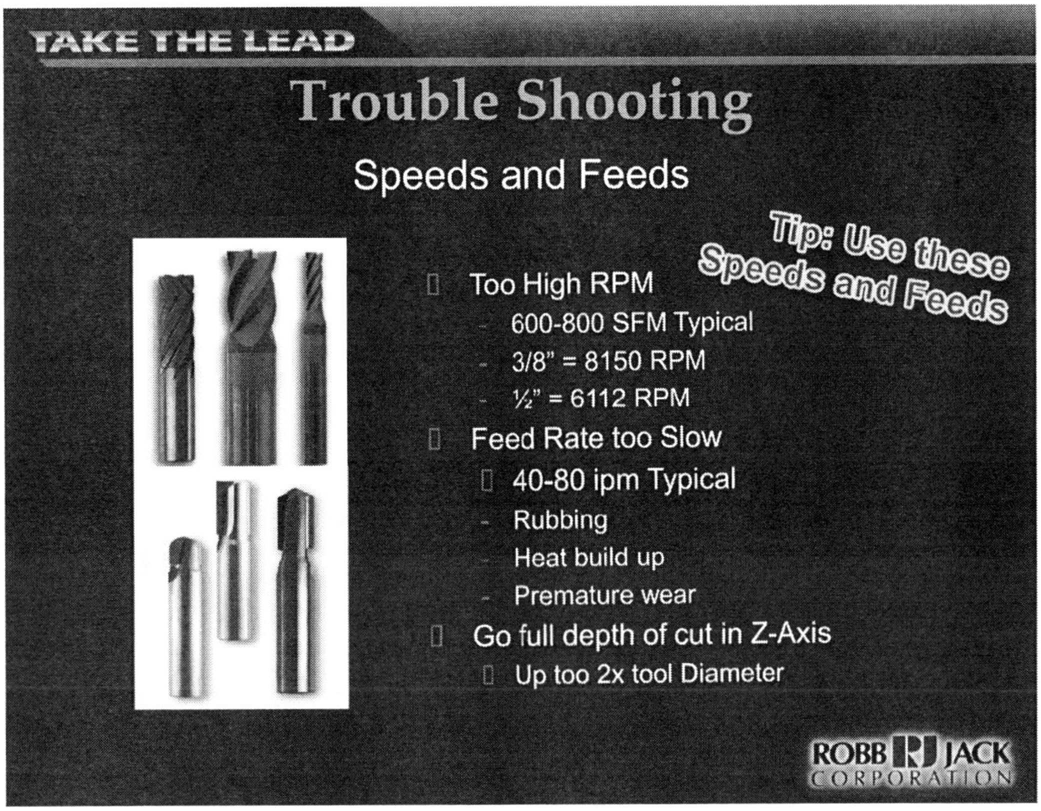
TAKE THE LEAD
Trouble Shooting
Speeds and Feeds
Tip: Use these Speeds and Feeds
Too High RPM
600-800 SFM Typical
3/8" = 8150 RPM
½" = 6112 RPM
Feed Rate too Slow
40-80 ipm Typical
Rubbing
Heat build up
Premature wear
Go full depth of cut in Z-Axis
Up too 2x tool Diameter
ROBB JACK CORPORATION

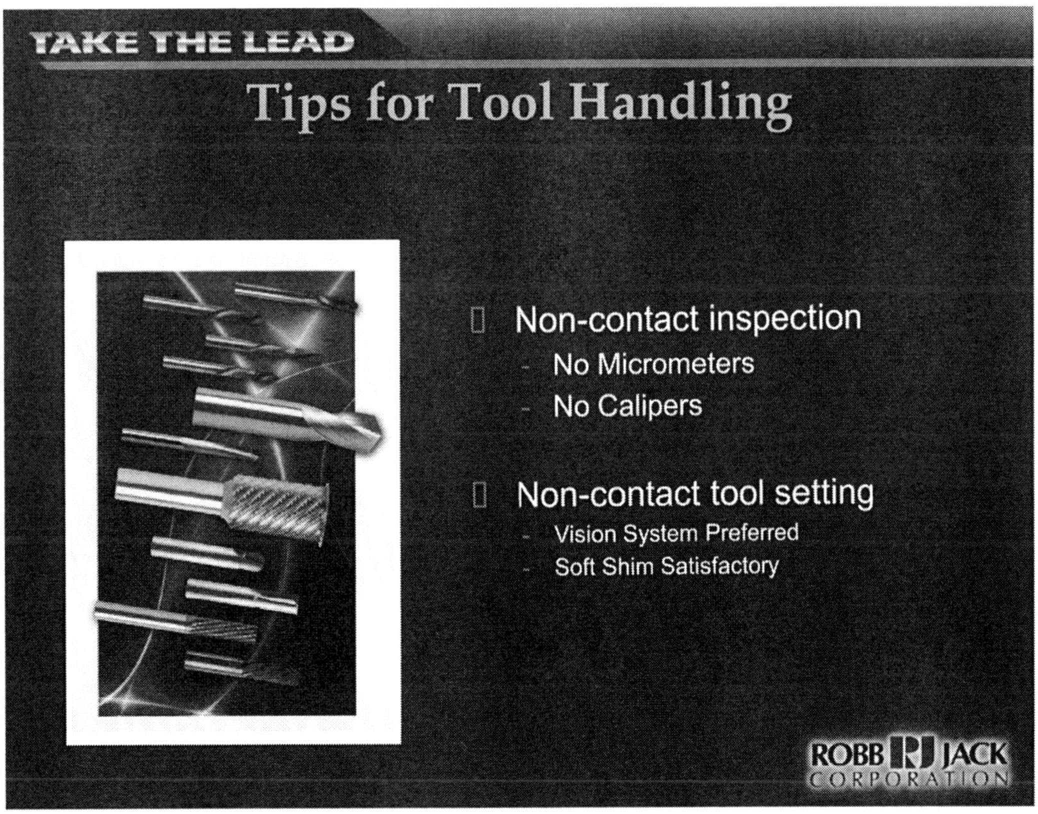
TAKE THE LEAD
Tips for Tool Handling
Non-contact inspection
No Micrometers
No Calipers
Non-contact tool setting
Vision System Preferred
Soft Shim Satisfactory
ROBB JACK CORPORATION

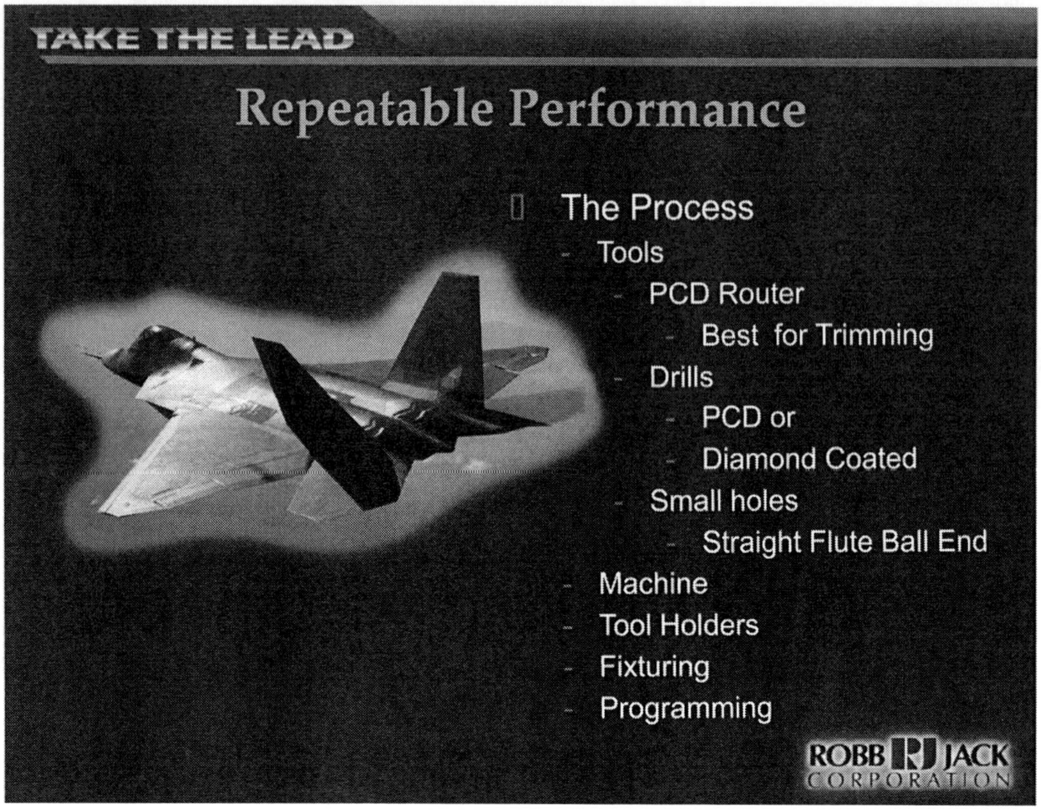
TAKE THE LEAD
Repeatable Performance
The Process
- Tools
- PCD Router
- Best for Trimming
- Drills
- PCD or
- Diamond Coated
- Small holes
- Straight Flute Ball End
- Machine
- Tool Holders
- Fixturing
- Programming
ROBB JACK CORPORATION

Cryogenic Machining And Supply Chain Insertion

AeroDef Manufacturing Summit And Exhibition

Feb 25, 2014

LOCKHEED MARTIN

Jeff Langevin
Manufacturing Technology
Lockheed Martin Aeronautics

Jason Huff
Supply Chain Innovation
Lockheed Martin Aeronautics

1

Agenda

- Titanium And The JSF
- Cryogenic Machining Development
- Advantages
- Supply Chain Opportunities
- Supplier Technology Insertion Capability
- Questions & Answers

2

Titanium And The JSF

	CTOL	STOVL	CV
Titanium Make-Up Of Aircraft (By Weight)	6 %	6%	11%
Amount of Titanium Impacted By Technology (By Weight)	85%	85%	91%
Potential "High Performance" Parts	61	52	85
Weight of Aircraft Impacted By Technology (By Weight)	5 %	5 %	10 %

3

Titanium And The JSF

Titanium Advantages

- High strength-to-weight ratio
- Corrosion resistance
- Fatigue / fracture resistance
- High temperature resistance

Machining Efficiency Obstacles

- Poor thermal conductor
- Elasticity
- Heat doesn't dissipate through part
- Rubbing instead of cutting

Result

- Heat build up at cutting edge
- Premature tool wear
- Slower cutting speeds
- Short tool life
- Poor quality

4

Putting Out The Fire

Standard machining coolants are:
- Costly to maintain and discard
- Unsafe
- Thermally inefficient
- Environmentally unfriendly
- Potentially corrosive
- Of controversial lubrication value

The ideal cutting tool cooling system:
- Extends tool life
- Eliminates worker health issues
- Reduces disposal cost
- Reduces maintenance costs
- Is non-corrosive and "green"
- Requires no operator intervention
- Decreases overall manufacturing costs

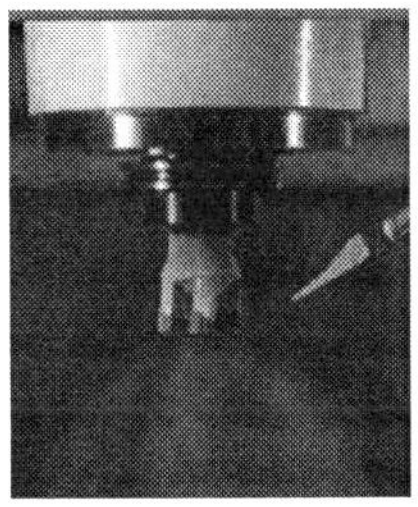

5

Cryogenic Machining Development

Small Business Innovative Research (SBIR)

- Creare, Inc. – Technology Developers
- Manufacturing Technology – Technical Support
- 5ME– Commercialization Partner
- F-35 Joint Program Office (JPO)
- U.S. Navy

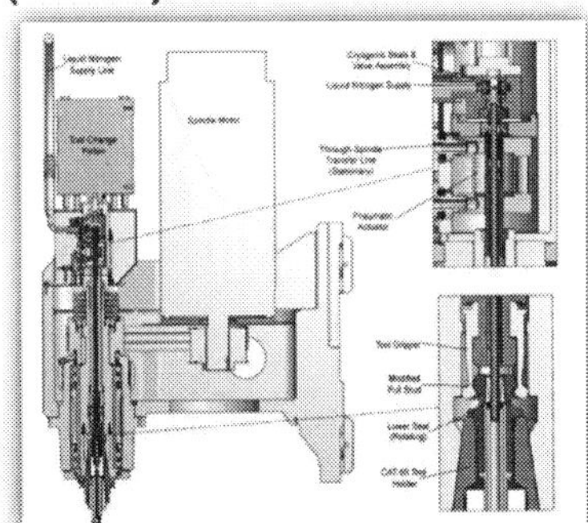

Solution

- Concentrated cooling at the cutting edge
- Through the spindle (LN_2) cooling
  - Direct delivery to cutting edge
  - Indirect delivery to cutting inserts

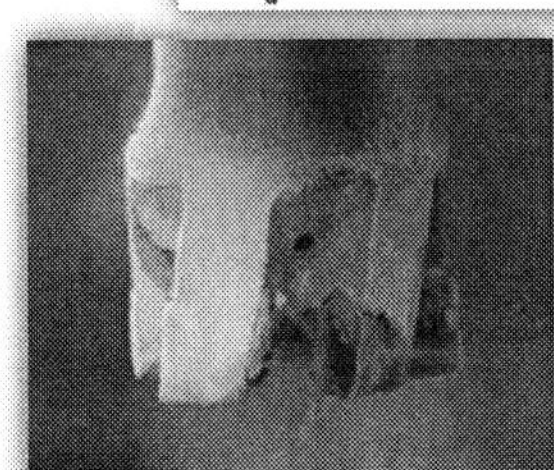

6

Cryogenic Milling
For Titanium Performance

Machining Performance

- Up to 10x cutting tool life
- 2x material removal rate (MRR)
  - Plunge
  - Pocket Edge Milling
- Quality
  - Reduced material contamination
  - Bur reduction

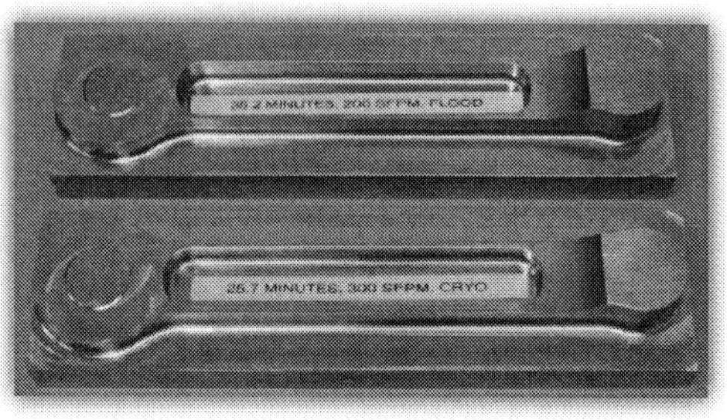

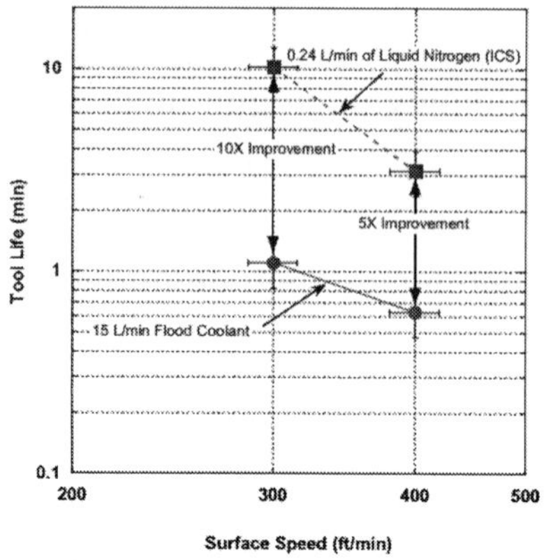

7

Demonstration

Integrated Process

Machining Trials

8

Going Green Is ~~Cool~~

Green and Economical

- Nitrogen is not a greenhouse gas
- Eliminates flood coolant management costs
- Eliminates secondary cleaning processes
- Liquid nitrogen is inexpensive ~ $0.25/L

Health and Safety

- Safe non-slip coolant approach for large walk-on machine tables
- Eliminates hazards related to aerosolized flood coolants
- Easily safeguarded with an oxygen sensor

9

Development To Insertion

10

Supply Chain Insertion

Supply Chain Affordability Challenge

- **Business as usual cannot achieve return for USG**
- **Develop capability to transition new disruptive technologies into supply chain**

Manufacturing Technology Focus

- **Lower cost, faster return on investment**
- **Changing aircraft costly, risky and time consuming**

Huge Opportunity

- **Aeronautics cost approximately 30%**
- **Supply Chain cost approximately 70%**

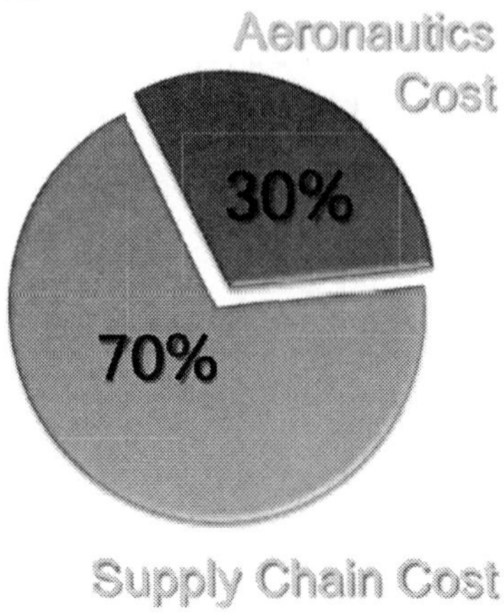

11

New Tool for Supply Chain Management

1. **Wrestle Supply Chain**
   - Fight for Value
   - Procurement *(Contracting and Negotiation)*
   - Mature Capability
2. **Optimize Supply Chain**
   - Eliminate Wasted Value
   - Lean Six-Sigma
   - Mature Capability
3. **Change Supply Chain (NEW)**
   - Create and Leverage New Value
   - Supplier Technology Insertion Capability *(STIC)*
   - Cryogenic Machining is Pilot

Enabler

12

Supplier Technology Insertion Capability (STIC)

- ManTech
- SBIR
- Tech Co.
- R&D Orgs

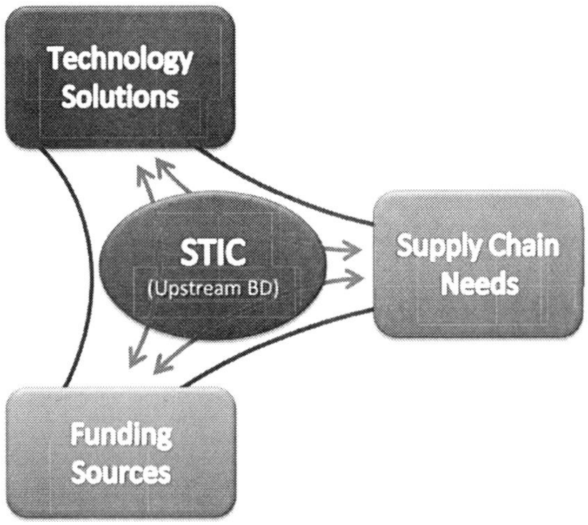

- Affordability

- Government
- Private
- Corporate

13

STIC Pilot Cryogenic Machining

Value Proposition

- **Faster machining speeds** (up to 2x)
- **Less tool changes** (up to 10x tool life)
- **No coolant cleanup**

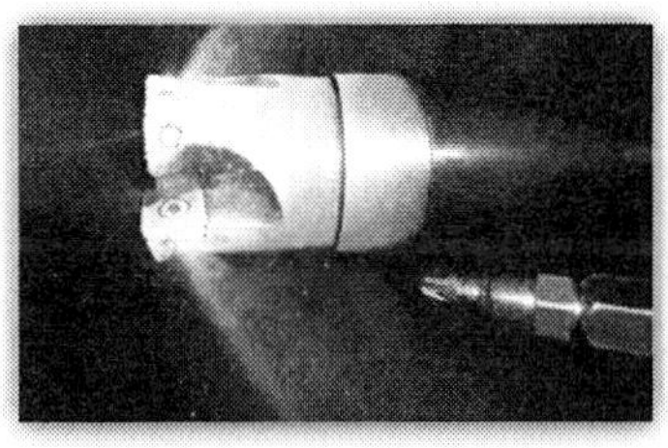

Cryogenic Cooling

Insertion Strategy

- **Establish low cost supply chain partner**
- **Leverage low cost point to expand benefit in supply base**

Traditional Cooling

14

Insertion Plan

Broker Insertion - *Complete*

- **Supply Chain Partner**
  - ***Over 400 suppliers solicited***
  - ***Small privately owned machine shop***
- **Funding**
  - ***Air Force Commercialization Readiness Program (CRP) – 6 months***
  - ***Mentor-protégé – 3 years***

Supply Base Development Plan

- **Phase 1 – Establish Low Cost Production** – *In Progress*
  - ***Define technology value and risk***
- **Phase 2 – Expand Benefit in Supply Base** – *In Planning*
  - ***Expand parts produced***
  - ***Identify additional suppliers for technology***
- **Phase 3 – Fully Seeded – Technology used by two or more suppliers**

15

STIC Opportunity Portfolio

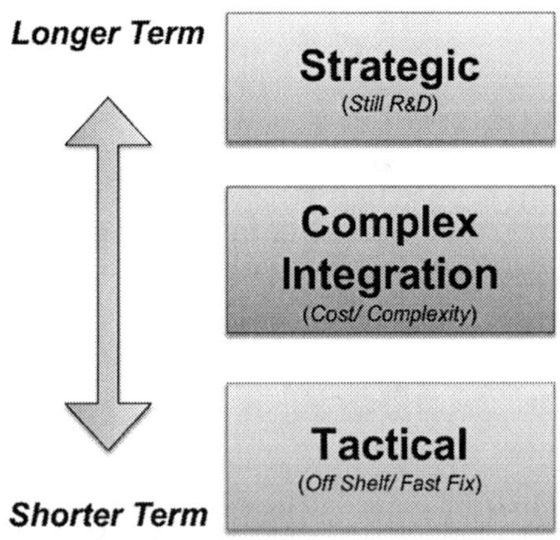

- **Cryogenic Machining**
- **Electron Beam Direct Mfg. (EBDM)**
- **Laser Assisted Consolidation (LAC)**

- **Closed Volume Molding (CVM)**
- **Non-Contact Metrology (NCM)**
- **Optical Projection (OP)**

- **ManTech Developed Tooling**

16

Insertion Challenges

Supplier Partner Identification

Lockheed is a Prime

Slow System and Dynamic Markets

Culture, Culture, Culture

17

Discussion

LOCKHEED MARTIN

We never forget who we're working for™

MANUFACTURING TECHNOLOGY

18

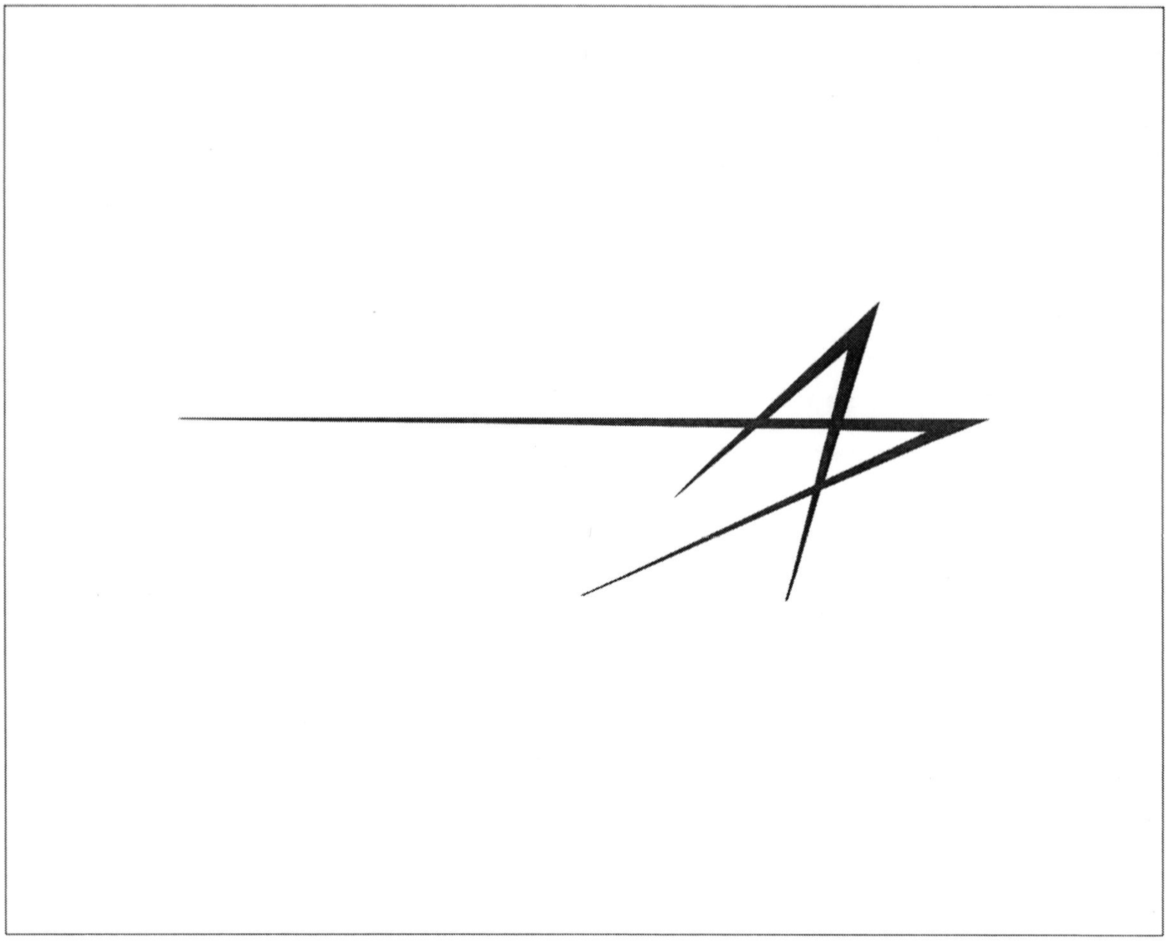

AERODEF MANUFACTURING

SUMMIT & EXPOSITION
February 25-27, 2014
LONG BEACH CONVENTION CENTER
LONG BEACH, CALIFORNIA USA

Planning and Designing a Lean Manufacturing Facility

"You Only Get One Chance to Do It Right"

PRESENTED BY: Chris Williams

AERODEF MANUFACTURING

SUMMIT & EXPOSITION
February 25-27, 2014
LONG BEACH CONVENTION CENTER
LONG BEACH, CALIFORNIA USA

Introduction

- Lean Six Sigma (LSS) has historically been seen as quality and efficiency tool used on the manufacturing floor
- Manufacturing companies, including aerospace, have begun to incorporate LSS methodologies into the design/ renovation of their facilities
  - Reduce Square Footage
  - Reduce Capital Construction Cost
  - Increase Efficiency
  - Product Flow
  - Leverage Integration
  - Leverage New Technologies

SUMMIT & EXPOSITION
February 25-27, 2014
LONG BEACH CONVENTION CENTER
LONG BEACH, CALIFORNIA, USA

Start with the End in Mind

The Value Stream

- People Flow
- Material Flow
- Material Management
- Production of Goods
- Management of Finished Goods
- Information Management
- Waste/Dunnage Management

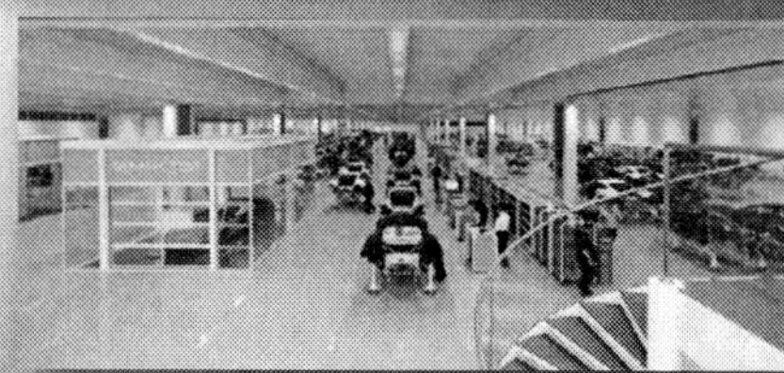

SUMMIT & EXPOSITION
February 25-27, 2014
LONG BEACH CONVENTION CENTER
LONG BEACH, CALIFORNIA, USA

The Process

- Data Collection
  - Cycle Times
  - Work Schedules
  - Rate Demand
  - Takt Time
- Create Process Maps
  - VSM
  - Swim Lane
  - Spaghetti
- Collect Customer Requirements
  - Financial Requirements
  - Facility Goals
  - Quality and Efficiency Targets
  - Compliance & Permitting

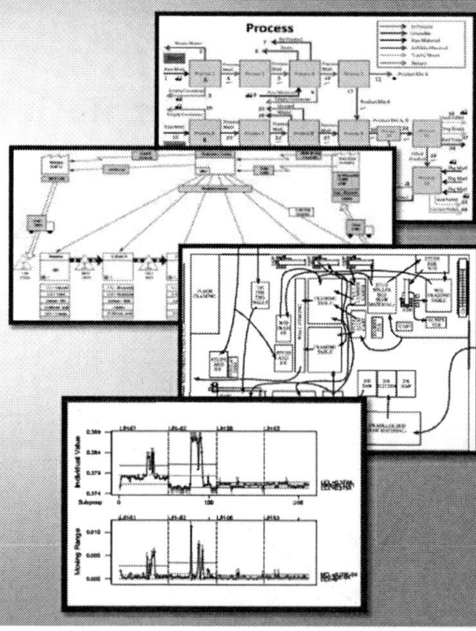

AERODEF MANUFACTURING
presented by sme

SUMMIT & EXPOSITION
February 25-27, 2014
LONG BEACH CONVENTION CENTER
LONG BEACH, CALIFORNIA USA

The Process

- Collect Facility & Process Requirements
- Select a Site
  - Cause & Effect Matrix
  - Pugh Matrix
- Design the Building or Space Around the Process
- Generate conceptual Layouts Based on Data
- Select Layout and Simulate

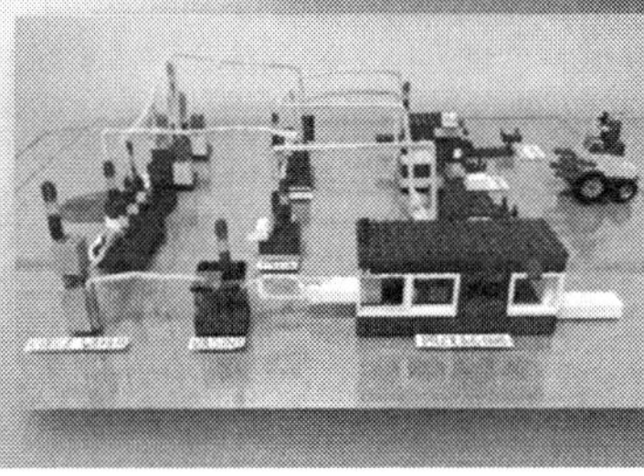

AERODEF MANUFACTURING
presented by sme

SUMMIT & EXPOSITION
February 25-27, 2014
LONG BEACH CONVENTION CENTER
LONG BEACH, CALIFORNIA USA

Iterative Design Process

Technology

Process Mapping

Data Analysis & Concepts

Capacity Studies & Simulation

Manufacturing Layouts

Building Layouts, & Simulation

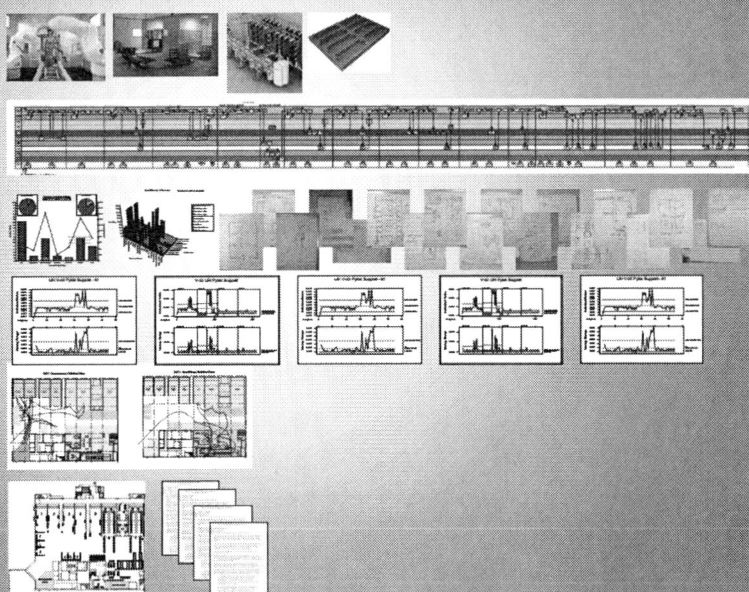

AERODEF MANUFACTURING

sme

SUMMIT & EXPOSITION
February 25-27, 2014
LONG BEACH CONVENTION CENTER
LONG BEACH, CALIFORNIA USA

Aerospace Coating Facility

Process

- Teamed with Industrial & Manufacturing Engineers
- Mapped the Process
- Improved the Process Flow
- Generated the Process Layout
- Integrated technologies in New Ways
- Developed the Building Layout & Design

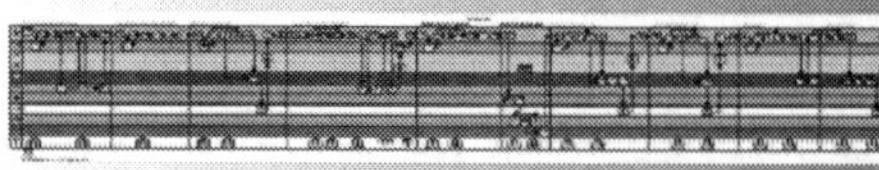

Results

- Reduced Facility Footprint by 20% from Original Concept
- Determined a Phased Build Plan to for Production
- Reduced anticipated Capital Spend by 25%

AERODEF MANUFACTURING

sme

SUMMIT & EXPOSITION
February 25-27, 2014
LONG BEACH CONVENTION CENTER
LONG BEACH, CALIFORNIA USA

High Speed Manufacturer

Process

- Teamed with Equipment Engineers
- Mapped the Process
- Created Manufacturing Cells
- Developed Single Piece Flow
- Generated the Process Layout
- Integrated Automated Material Handling
- Developed the Building Layout & Design
- Simulated the Process

Results

- Reduced Facility Footprint by 10% from Original Concept
- Determined a Phased Build Plan to Support Increased Production
- Reduced Cycle Time from 9 Days to 1 Day

AERODEF MANUFACTURING
sme
SUMMIT & EXPOSITION
February 25-27, 2014
LONG BEACH CONVENTION CENTER
LONG BEACH, CALIFORNIA USA

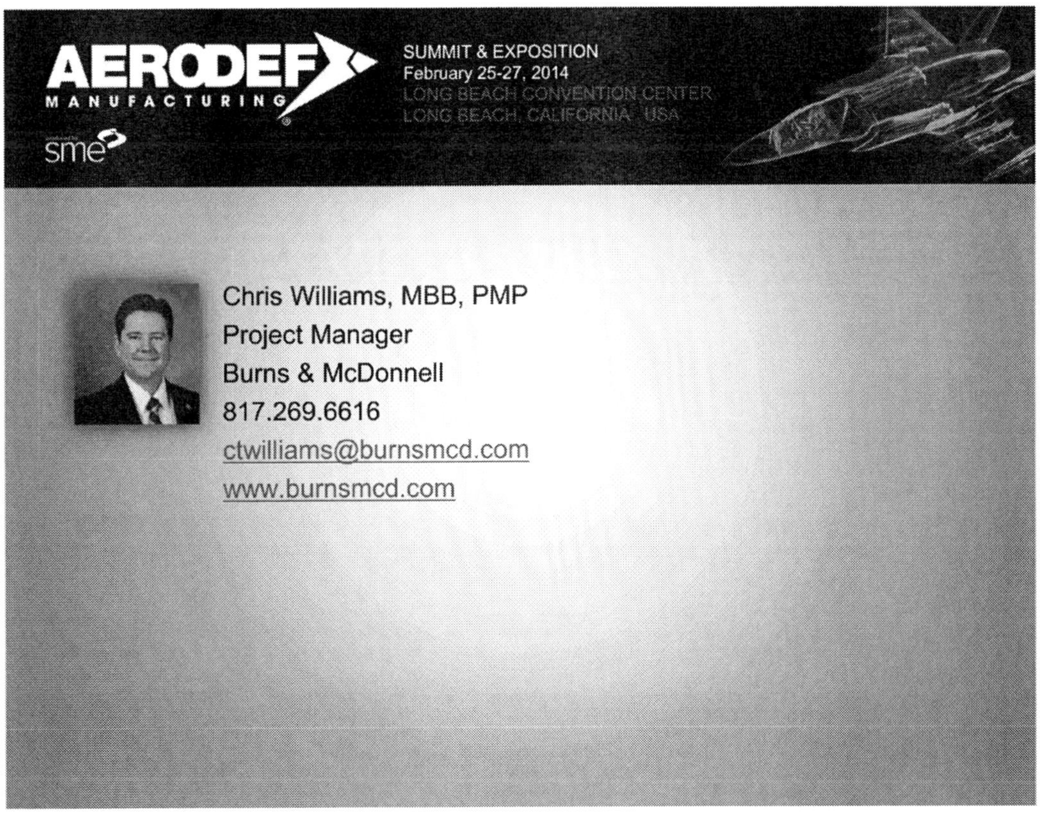
AERODEF MANUFACTURING
sme
SUMMIT & EXPOSITION
February 25-27, 2014
LONG BEACH CONVENTION CENTER
LONG BEACH, CALIFORNIA USA
Chris Williams, MBB, PMP
Project Manager
Burns & McDonnell
817.269.6616
ctwilliams@burnsmcd.com
www.burnsmcd.com

Minimizing Energy Risk
in Manufacturing
BY: DR. DARREN MCKNIGHT
ZF ENERGY DEVELOPMENT
AERODEF
MANUFACTURING
FEBRUARY 2014
ZF ENERGY
DEVELOPMENT LLC

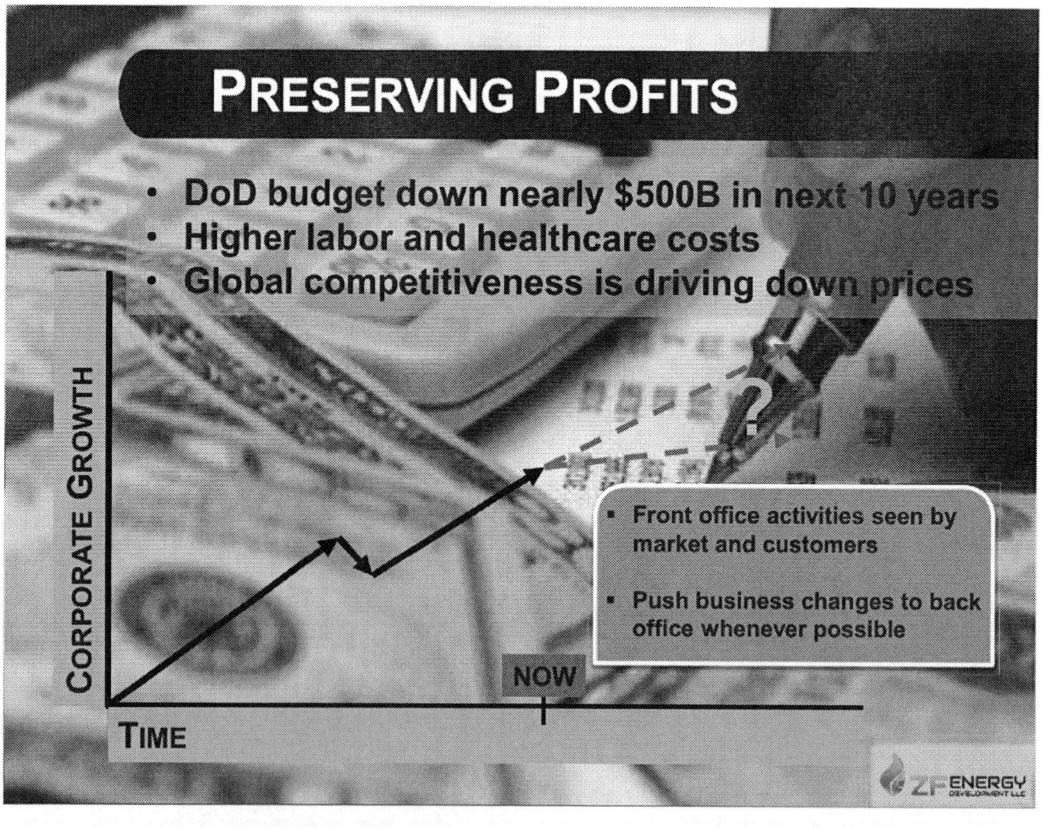

PRESERVING PROFITS
• DoD budget down nearly $500B in next 10 years
• Higher labor and healthcare costs
• Global competitiveness is driving down prices
CORPORATE GROWTH
?
▪ Front office activities seen by market and customers
▪ Push business changes to back office whenever possible
NOW
TIME
ZF ENERGY

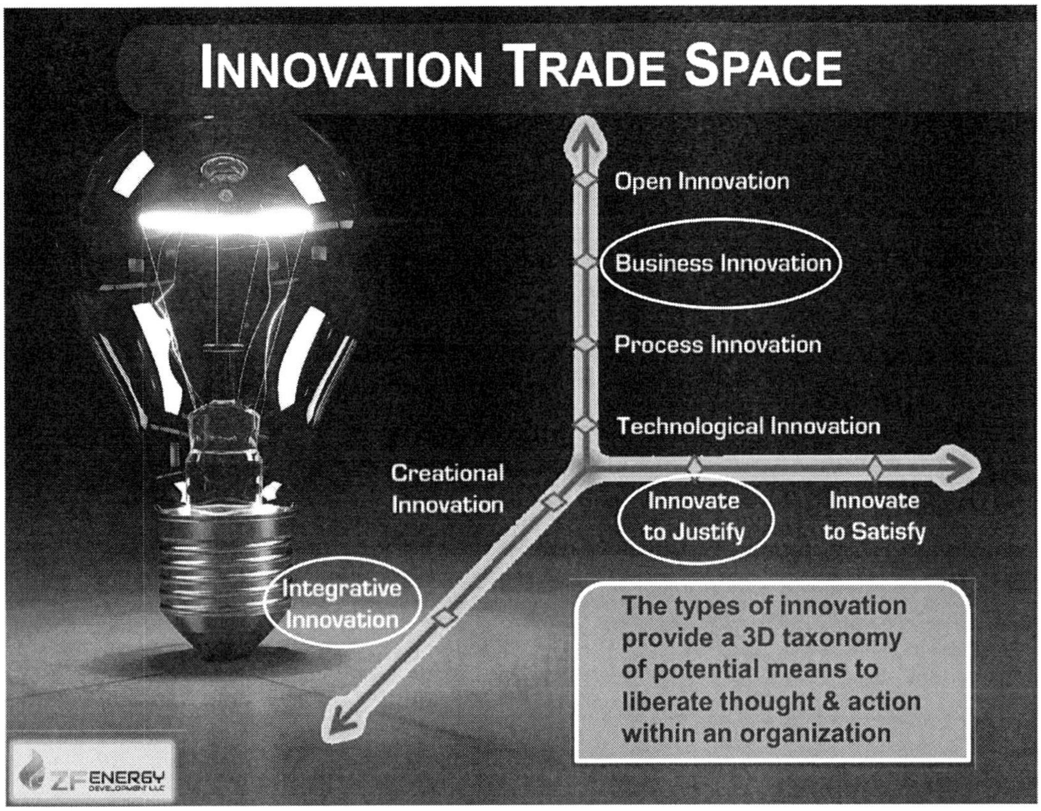

INNOVATION TRADE SPACE
Open Innovation
Business Innovation
Process Innovation
Technological Innovation
Creational Innovation
Innovate to Justify
Innovate to Satisfy
Integrative Innovation
The types of innovation provide a 3D taxonomy of potential means to liberate thought & action within an organization
ZF ENERGY DEVELOPMENT LLC

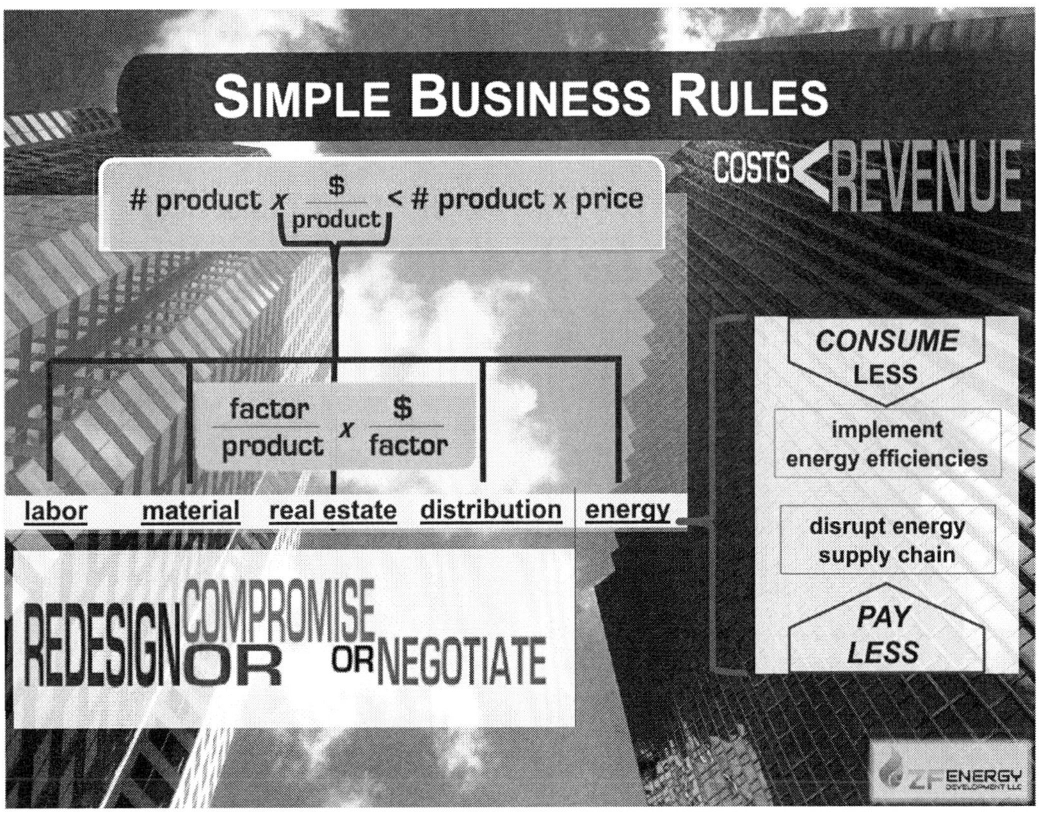

SIMPLE BUSINESS RULES
COSTS<REVENUE
product x $/product < # product x price
factor/product x $/factor
labor
material
real estate
distribution
energy
REDESIGN OR COMPROMISE OR NEGOTIATE
CONSUME LESS
implement energy efficiencies
disrupt energy supply chain
PAY LESS
ZF ENERGY DEVELOPMENT LLC

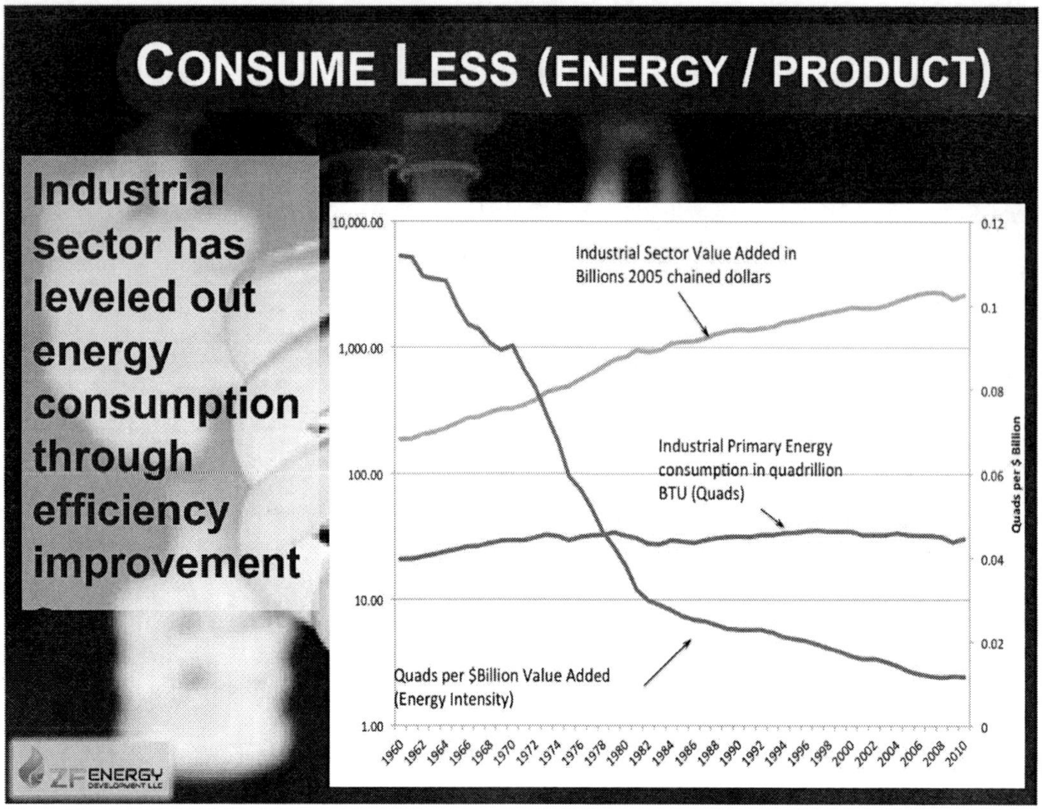

CONSUME LESS (ENERGY / PRODUCT)
Industrial sector has leveled out energy consumption through efficiency improvement
Industrial Sector Value Added in Billions 2005 chained dollars
Industrial Primary Energy consumption in quadrillion BTU (Quads)
Quads per $Billion Value Added (Energy Intensity)
Quads per $ Billion
10,000.00
1,000.00
100.00
10.00
1.00
0.12
0.1
0.08
0.06
0.04
0.02
0
1960 1962 1964 1966 1968 1970 1972 1974 1976 1978 1980 1982 1984 1986 1988 1990 1992 1994 1996 1998 2000 2002 2004 2006 2008 2010
ZF ENERGY DEVELOPMENT LLC

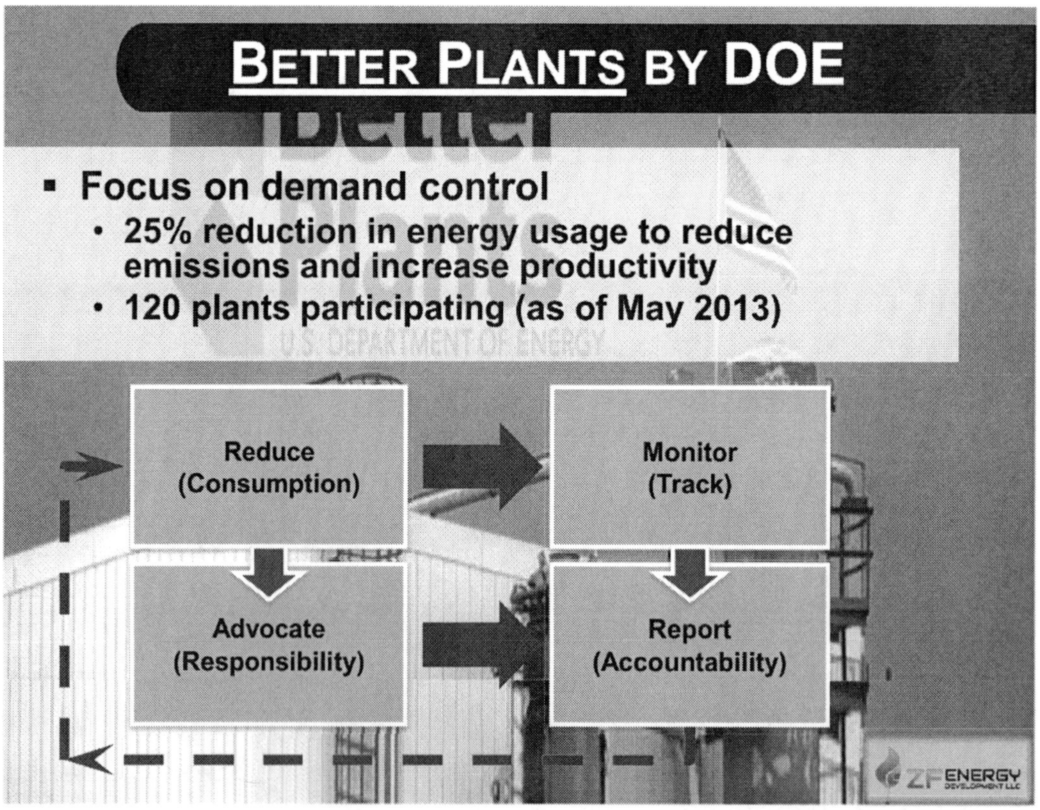

BETTER PLANTS BY DOE
Focus on demand control
25% reduction in energy usage to reduce emissions and increase productivity
120 plants participating (as of May 2013)
U.S. DEPARTMENT OF ENERGY
Reduce (Consumption)
Monitor (Track)
Advocate (Responsibility)
Report (Accountability)
ZF ENERGY DEVELOPMENT LLC

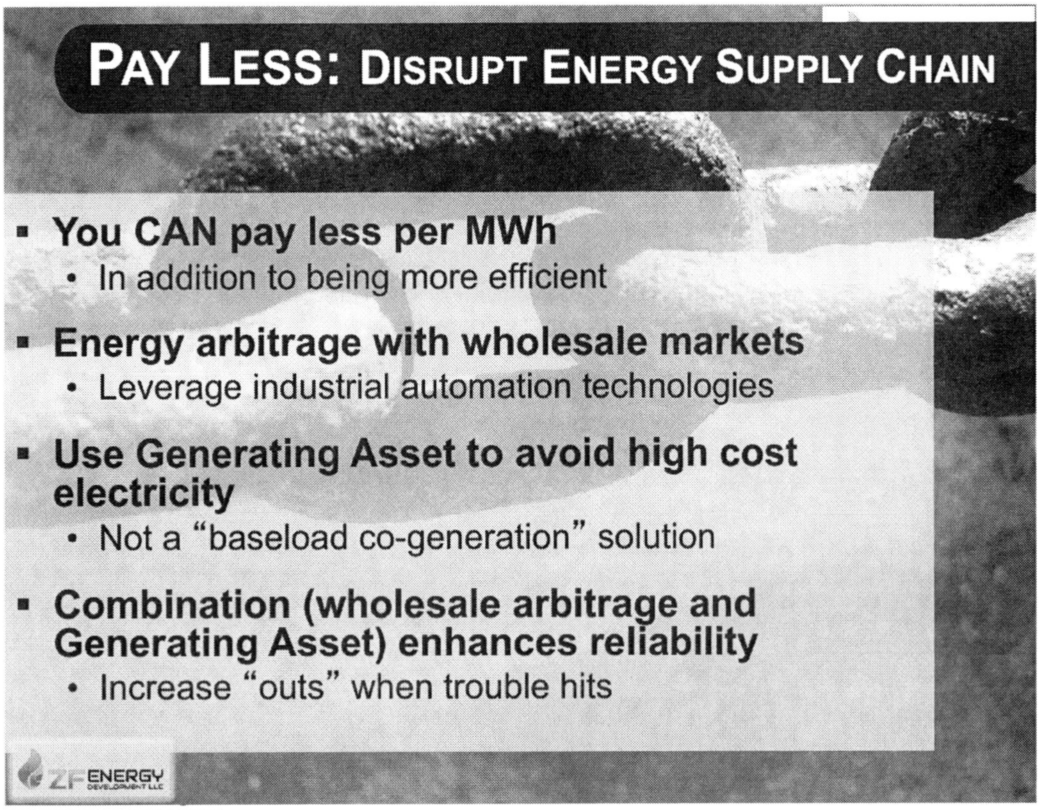
PAY LESS: DISRUPT ENERGY SUPPLY CHAIN
You CAN pay less per MWh
In addition to being more efficient
Energy arbitrage with wholesale markets
Leverage industrial automation technologies
Use Generating Asset to avoid high cost electricity
Not a "baseload co-generation" solution
Combination (wholesale arbitrage and Generating Asset) enhances reliability
Increase "outs" when trouble hits
ZF ENERGY DEVELOPMENT LLC

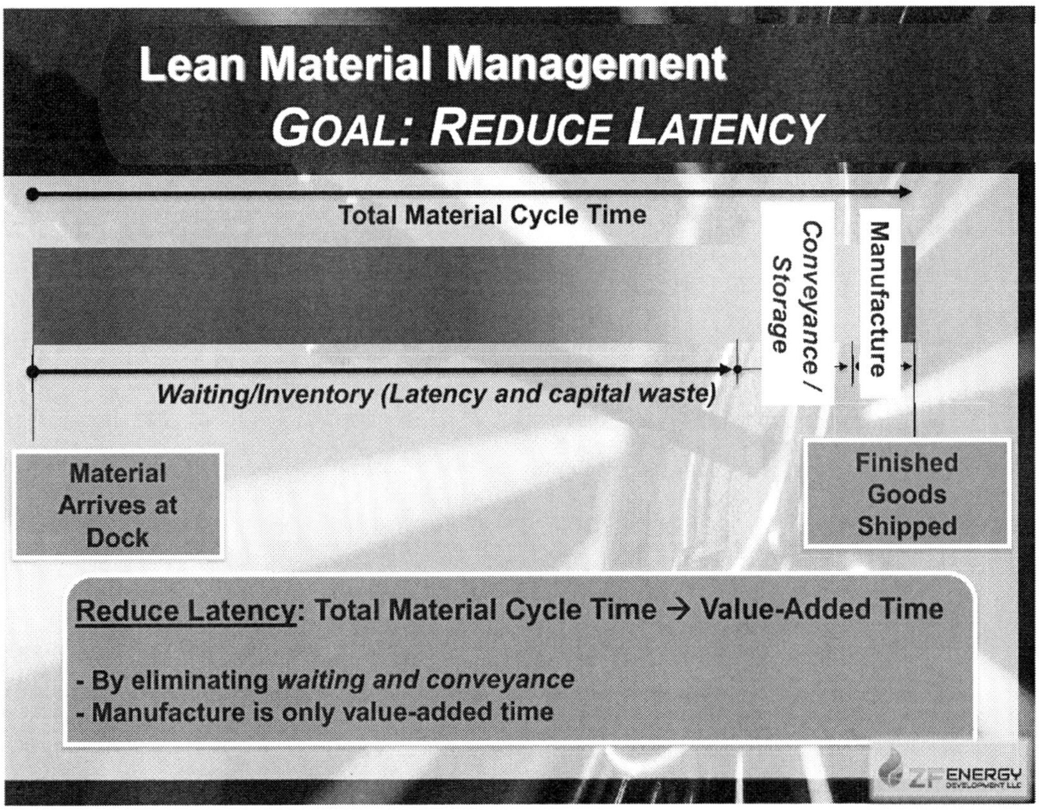
Lean Material Management
GOAL: REDUCE LATENCY
Total Material Cycle Time
Conveyance / Storage
Manufacture
Waiting/Inventory (Latency and capital waste)
Material Arrives at Dock
Finished Goods Shipped
Reduce Latency: Total Material Cycle Time → Value-Added Time
- By eliminating waiting and conveyance
- Manufacture is only value-added time
ZF ENERGY DEVELOPMENT LLC

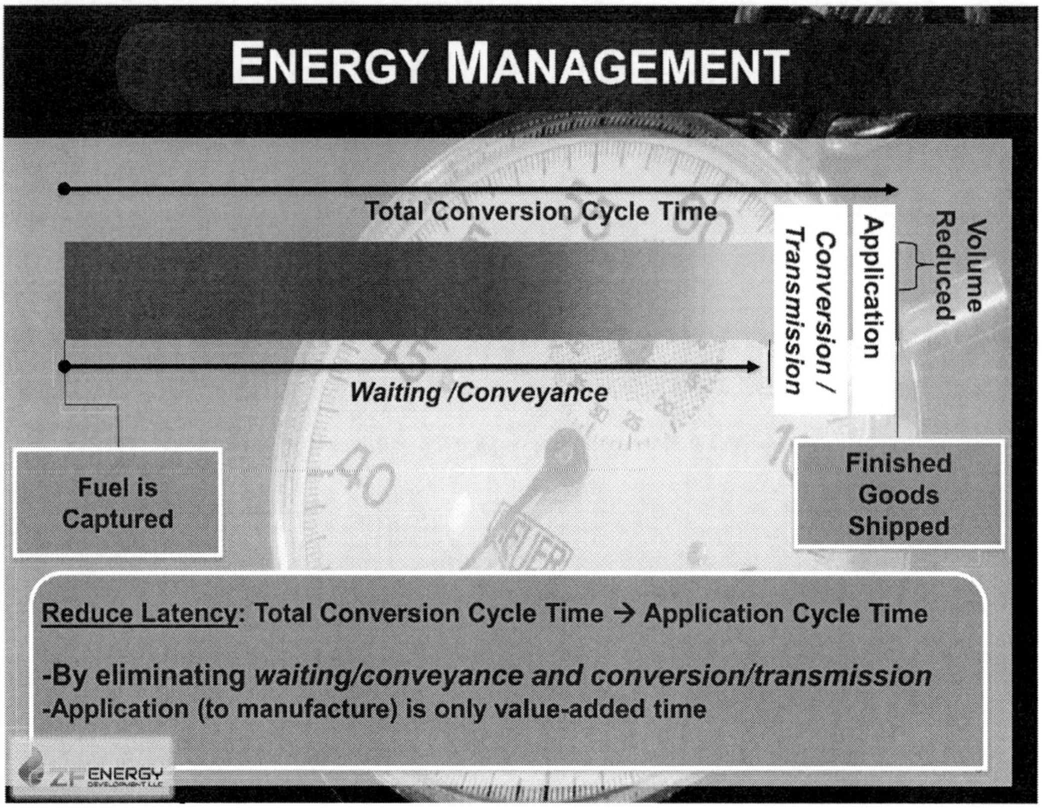
ENERGY MANAGEMENT
Total Conversion Cycle Time
Volume Reduced
Application
Conversion / Transmission
Waiting /Conveyance
Fuel is Captured
Finished Goods Shipped
Reduce Latency: Total Conversion Cycle Time → Application Cycle Time
-By eliminating waiting/conveyance and conversion/transmission
-Application (to manufacture) is only value-added time
ZP ENERGY DEVELOPMENT LLC

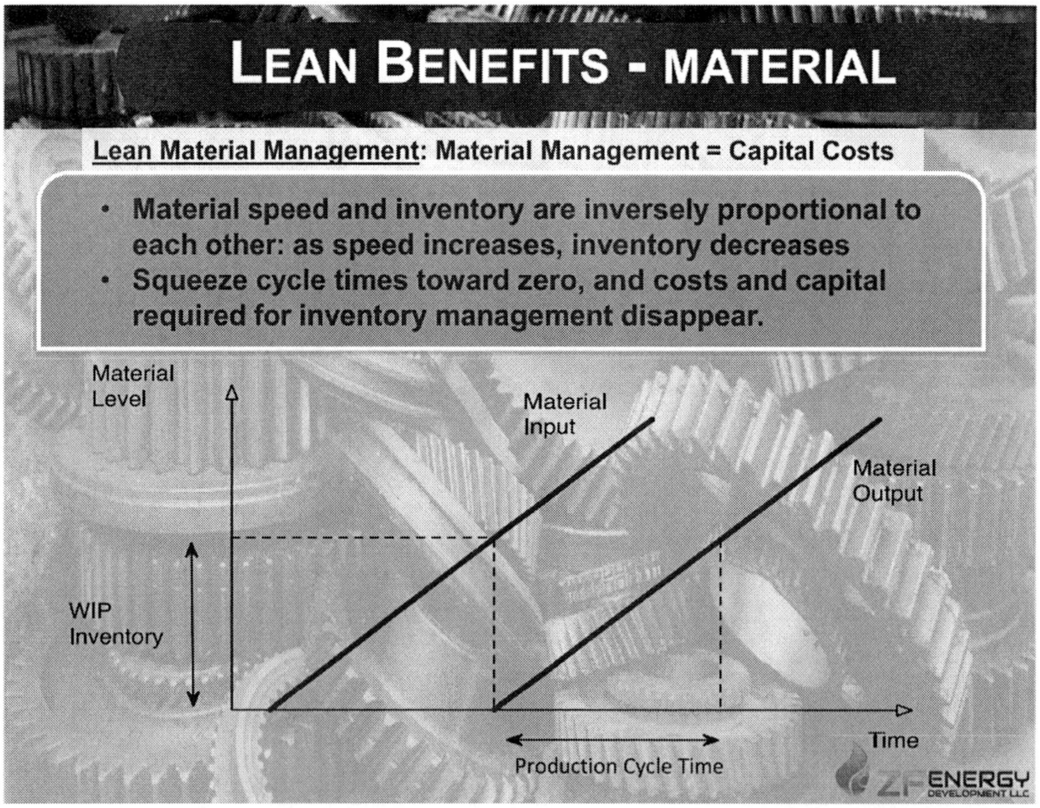
LEAN BENEFITS - MATERIAL
Lean Material Management: Material Management = Capital Costs
• Material speed and inventory are inversely proportional to each other: as speed increases, inventory decreases
• Squeeze cycle times toward zero, and costs and capital required for inventory management disappear.
Material Level
Material Input
Material Output
WIP Inventory
Time
Production Cycle Time
ZP ENERGY DEVELOPMENT LLC

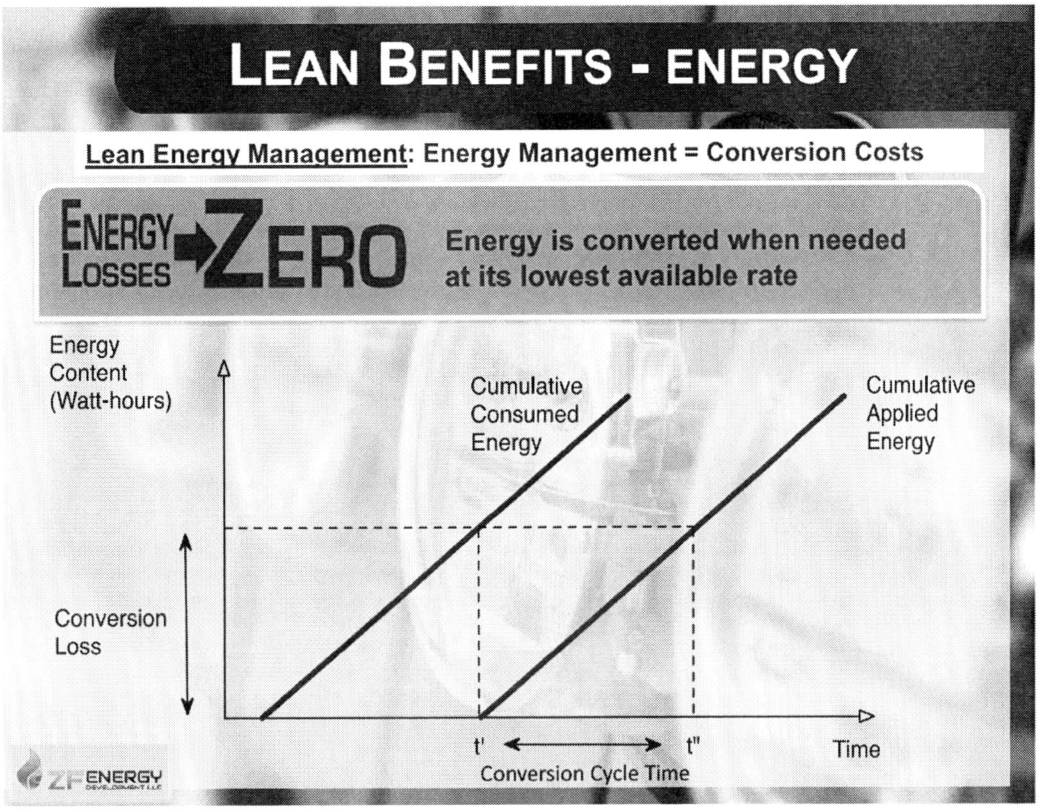
LEAN BENEFITS - ENERGY
Lean Energy Management: Energy Management = Conversion Costs
ENERGY LOSSES ZERO
Energy is converted when needed at its lowest available rate
Energy Content (Watt-hours)
Cumulative Consumed Energy
Cumulative Applied Energy
Conversion Loss
t'
t"
Time
Conversion Cycle Time
ZF ENERGY

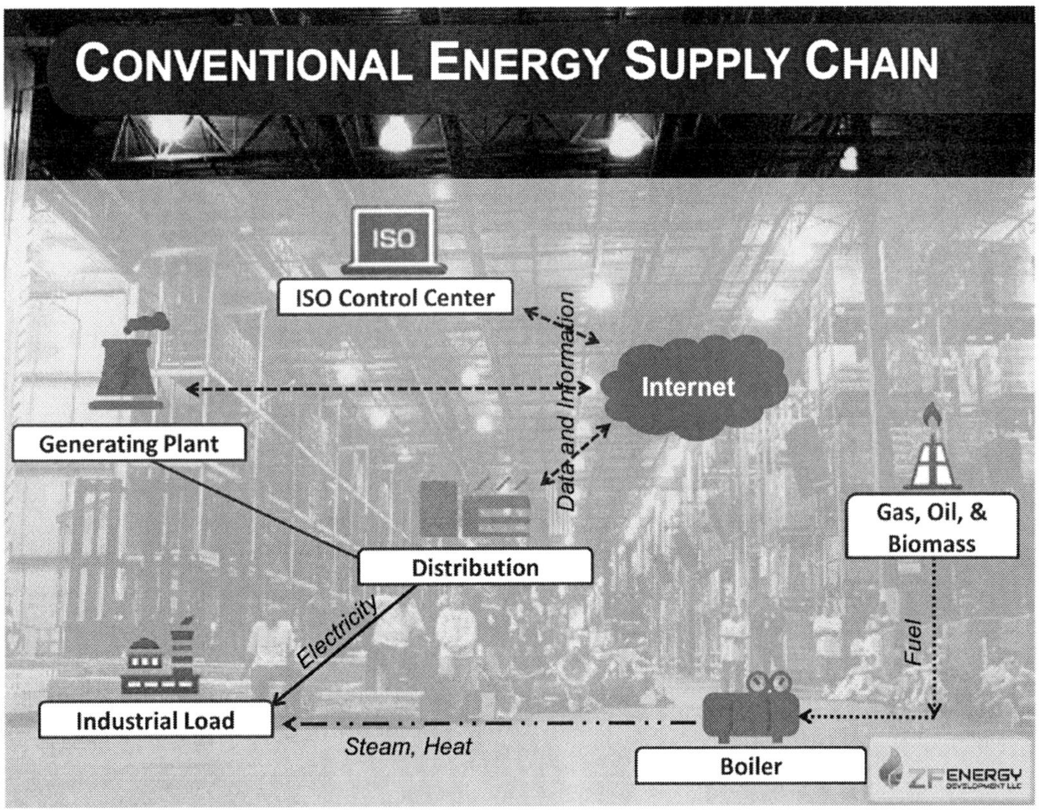
CONVENTIONAL ENERGY SUPPLY CHAIN
ISO
ISO Control Center
Data and Information
Internet
Generating Plant
Gas, Oil, & Biomass
Distribution
Electricity
Fuel
Industrial Load
Steam, Heat
Boiler
ZF ENERGY

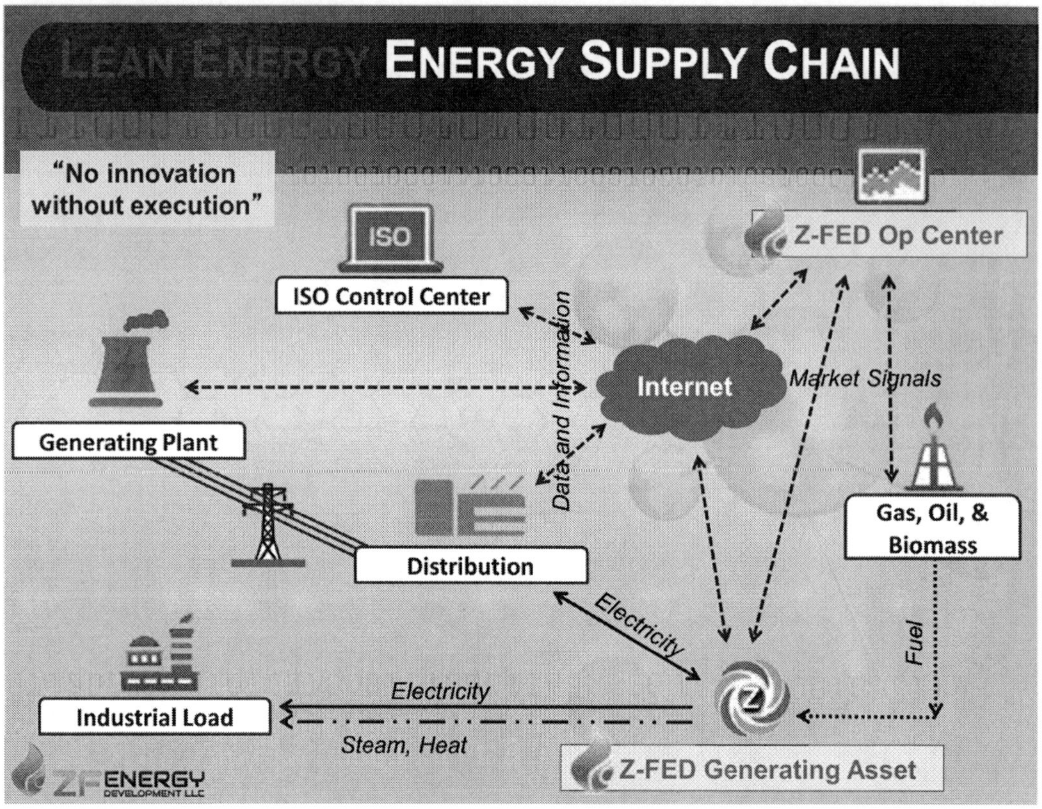

Energy Productivity Metrics

- **Lower Cost per MWh**
  - ISO NE: average \$190/MWh retail can be replaced with natural gas co-gen at \$70-90/MWh and wholesale rates are \$60/MWh

- **Energy Density Proportional to Cost Reduction Potential**
  - Higher consumption per square meter means lower unit cost potential
  - Tailor to industrial energy segment
  - Hospital energy efficiency thrust calling for \$2/ft^2 while most are at \$5-6/ft^2

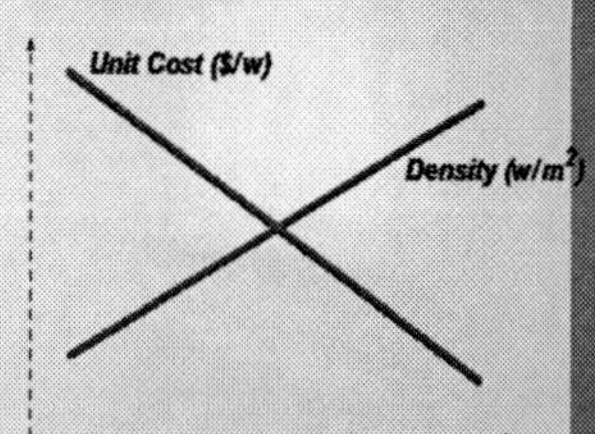

- **Energy productivity (energy consumed per value added component) reflects enhancements to the bottomline**
  - Emphasize actions that go straight to profit

14

ZF ENERGY DEVELOPMENT LLC

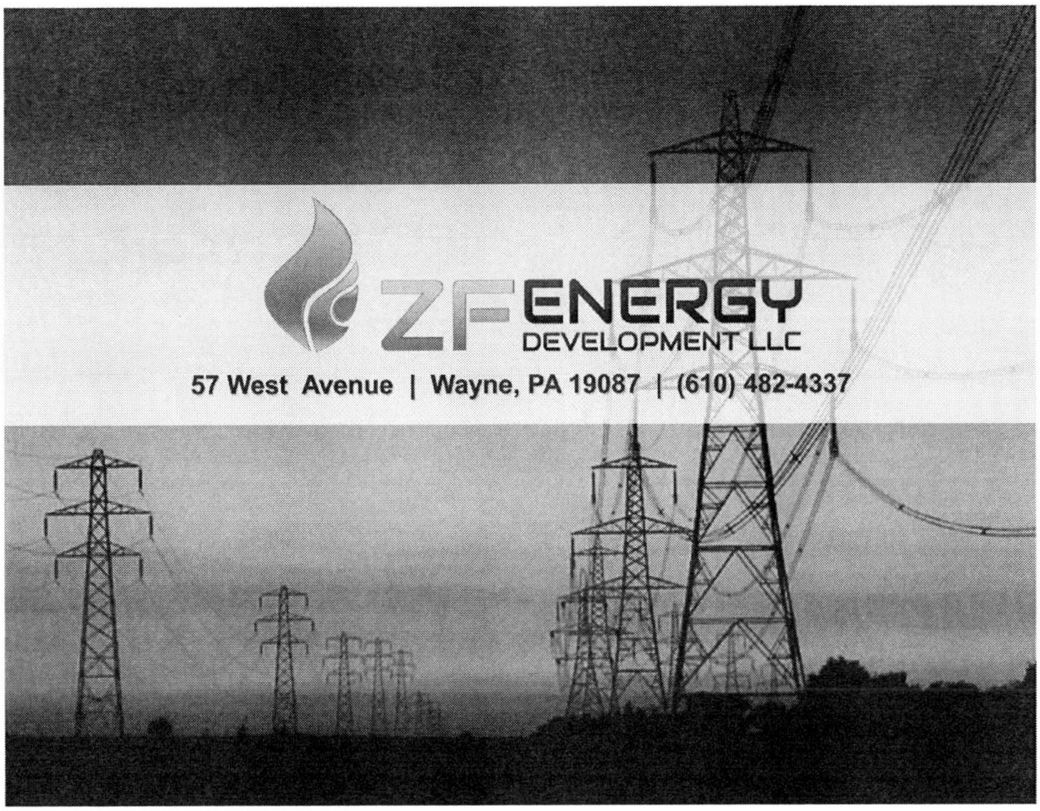
ZF ENERGY
DEVELOPMENT LLC
57 West Avenue | Wayne, PA 19087 | (610) 482-4337

Automated CCFX Inspection
2014 Aerodef Conference

AERODEF MANUFACTURING produced by sme

THE VALUE OF PERFORMANCE.
NORTHROP GRUMMAN

Northrop Grumman Aerospace Systems

February 25, 2014

Presenter: Chad Victorino
ManTech Development Engineer

Contributors: Dale Perkins (Manufacturing Engineering), Bill Louder (JWST Test Engineering), Jason Bither (Mgr El Segundo Site ManTech Engineering), Todd Szallay (Mgr NGAS ManTech Engineering)

Outline

NORTHROP GRUMMAN

- Why Automated CCFX Inspection?
- Hardware & Software Requirements
- Background on Vector Measurement and Kellering Approach
- Overview of Vector-Based Models
- Engineering Rework Activities
- Driving First Time Quality
- Digital Thread
- Applications of Automated CCFX Inspection
- Questions

1

Why Automated CCFX Inspection?

NORTHROP GRUMMAN

- **Issue:**
  - Fabricating complex composite parts with stringent profile requirements are frequently subjected to lengthy measurement times and routinely subjected to lengthy repairs
- **Automated CCFX Inspection:**
  - Advancements in inspection methods and technologies have enabled unique opportunities to decrease inspection and repair processing time by utilizing digital thread while decreasing overall program cost. First time quality inspection measurements can be leveraged to continuously optimize processes and yields.

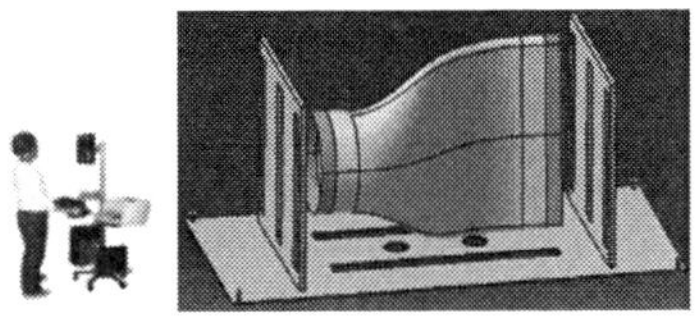

Methodology applied to any composite part with profile requirements driving first time quality and reducing engineering related rework costs

2

Hardware & Software Requirements

NORTHROP GRUMMAN

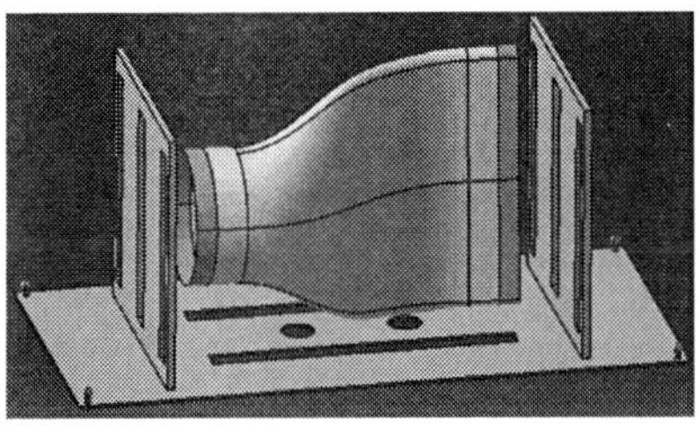

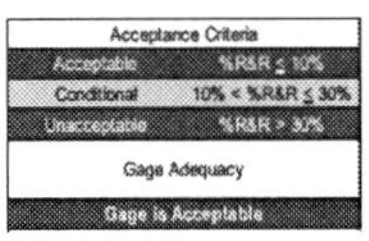

Acceptance Criteria	
Acceptable	%R&R ≤ 10%
Conditional	10% < %R&R ≤ 30%
Unacceptable	%R&R > 30%
Gage Adequacy	
Gage is Acceptable	

- **Composite Check Fixture (CCFX)**
  - Tooling coordinate system can be referenced into with metrology device using tooling spheres
  - Tool Usage Instructions (TUI) describes in detail, process for loading composite part onto fixture to ensure repeatability of measurements across shipsets
- **Metrology Device**
  - References into a tooling coordinate system using tooling spheres
  - Meets Gage R&R Requirements
- **Metrology Software**
  - Generates seek and destroy targets using vector based measurements
  - Customized automated analysis reports through development with software developer

3

Vector Based Measurement

NORTHROP GRUMMAN

- A vector based model is created when a complex, multiple intersecting curved surface is measured after being machined using the milling technique of Kellering
- The vector based measurement model duplicates the cutting tool's normal vector as opposed to the cutting tool's centerline axis vector
- This vector based modeling technique eliminates an induced cosine measurement error and produces a more accurate dimensional feature measurement value result

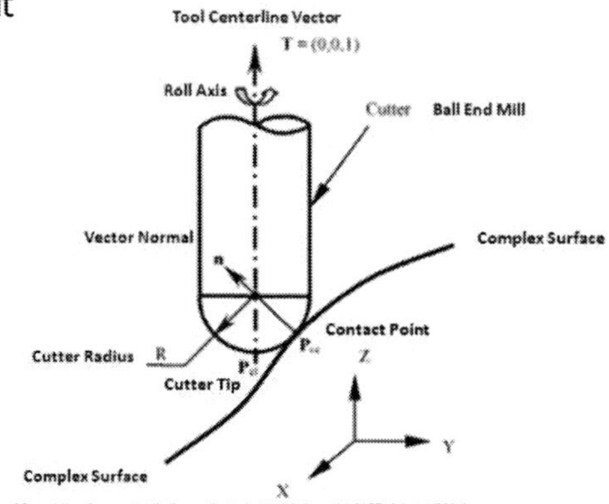

4

Vector Based Models

NORTHROP GRUMMAN

2 Types of Vector-Based Models:

CATIA Cutter Contact Point Model (.CATPart File Output)

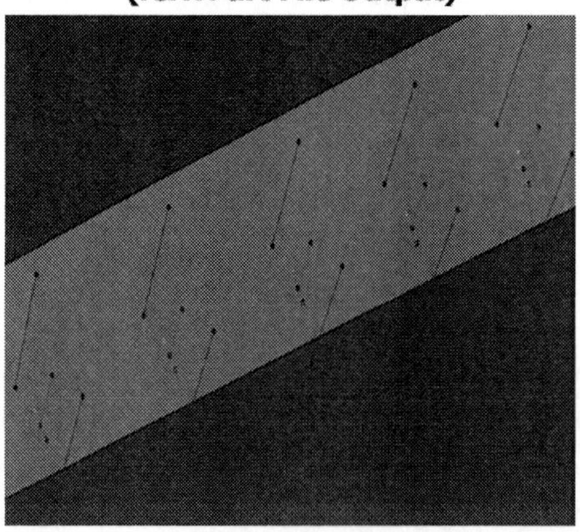

"Seek & Destroy" **Targets Model (.mcx File Output)**

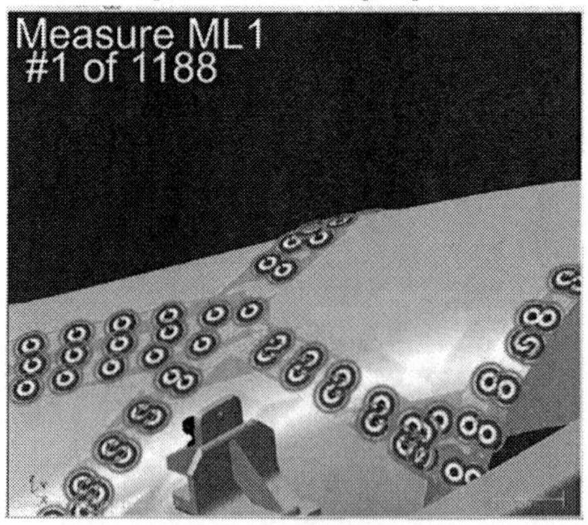

5

Engineering Rework Activities

NORTHROP GRUMMAN

- **Manufacturing Engineering - Sanding**
  - After surface profile data is analyzed, determines the over tolerance conditions and communicates how much and where to sand through PowerPoint work instructions.

- **Design Engineering – Sacrificial Patch Repair**
  - Takes raw under tolerance surface profile data analyzed to the nominal surface of the model and communicates the ply stack sequences, orientations, shapes and locations through PowerPoint work instructions.

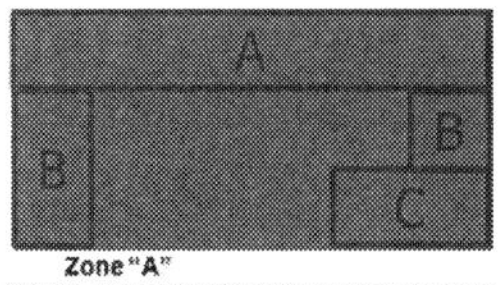

Lay	Angle	Thick	Material Name
1	0	0.007	Fabric
2	45	0.007	Fabric
3	0	0.004	Barrier

- **Stress Engineering – Liquid Shim Allowables**
  - Takes raw surface profile data at specific locations analyzed to the nominal surface of the mating model to develop a gap map for determining liquid shim allowables.

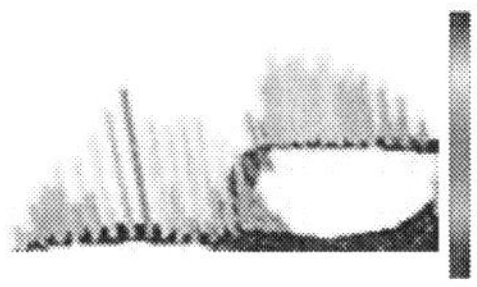

Customized automated analysis reports available to reduce processing time for various Engineering rework activities.

6

Approved for public release; distribution unlimited. NGAS Case 14-0187 dated 1/30/14.

Driving First Time Quality

NORTHROP GRUMMAN

- **Iterative Design Changes**
  - Design Engineering takes all surface profile data from past shipsets and based on that information issues an engineering change to optimize the sacrificial ply with the goal of reducing rework activities
    - Collecting data and analyzing trends across multiple shipsets can be cumbersome and time intensive
- **3D Virtual SPC Heat Map**
  - Provides a vehicle for mapping SPC data visually across multiple shipsets allowing Design Engineers to quickly pin-point areas frequently out of tolerance
  - Greatly reduced turn around time on an iterative design changes result in an overall reduction in rework activities across multiple shipsets

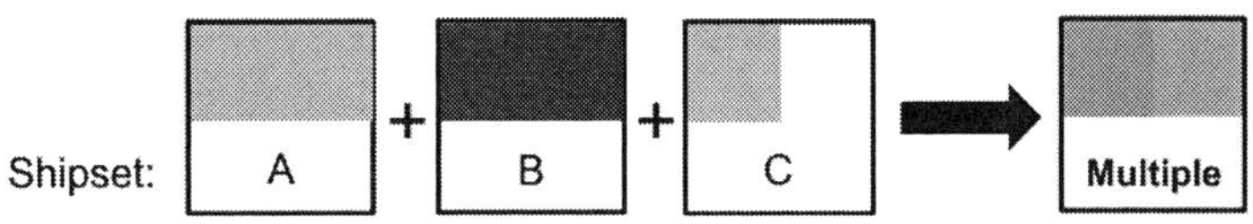

3D Visual SPC Heat Map allows for data driven design changes to be efficiently generated

7

Approved for public release; distribution unlimited. NGAS Case 14-0187 dated 1/30/14.

Digital Thread & the Primary Goal of Automated CCFX Inspection

NORTHROP GRUMMAN

- **Government Definition of Digital Thread:**

  The Digital Thread is the creation and use of a digital surrogate of a material system to allow dynamic, real-time assessment of the system's current and future capabilities to inform decisions across all acquisition phases

- **How does Automated CCFX Inspection Help Achieve Digital Thread Goals?:**
  - The same inspection points are used repeatedly from shipset to shipset satisfying requirements for collection of Statistical Process Control (SPC) data as required by most government contracts.
    - Surface profile metrology data can be captured in most industry standard SPC software through .csv files
  - Apply consistent rework decisions from shipset to shipset based on engineering logic captured in automated analysis reports.
    - Reduces turn around time on engineering rework activities
  - 3D Virtual SPC Heat Map
    - Facilitates quick iterative design change to stop rework from happening in the first place

8

Additional Applications of Automated CCFX Inspection

NORTHROP GRUMMAN

- **Automated Repair Kits**
  - Use CAD and/or metrology software to take the automated patch repair and translate it into a file understood by the ply cutting machine for automated cutting of patch repair ply necessary to build back into within tolerance.
  - Collect surface profile data on mating surface prior to installation to ensure proper gap requirements are met and shipping with the part pre-built shim kits for ease of integration into subassembly
- **Thickness Measurements**
  - Utilize vector based measurements to determine part thickness by using the CMM to give an as-built cure tool condition to more accurately determine part thickness.
- **Data for validation of Springback Prediction Software**
  - Utilize vector based measurements to determine part thickness by using the CMM to give an as-built cure tool condition to more accurately determine part thickness

9

Automated CCFX Inspection has many applications to be leveraged.

NORTHROP GRUMMAN

Questions?

THE VALUE OF PERFORMANCE.

NORTHROP GRUMMAN

AERODEF MANUFACTURING

SUMMIT & EXPOSITION
February 25-27, 2014
LONG BEACH CONVENTION CENTER
LONG BEACH, CALIFORNIA, USA
Finishing & Coatings

Handheld Measurement System for Multi-Layer Coatings Inspection

Presented by
Gary Ostrowski
Engineering Fellow
Northrop Grumman
Aerospace Systems

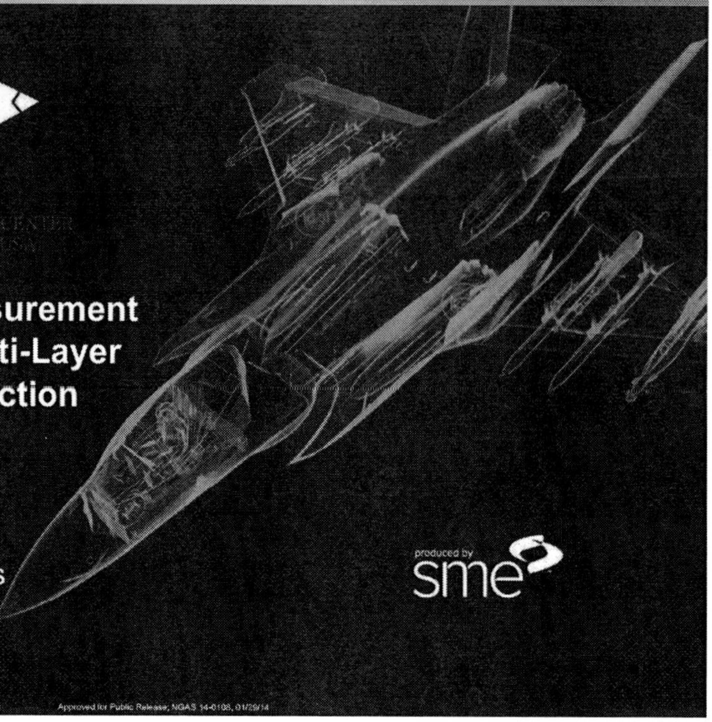

produced by sme

AERODEF MANUFACTURING
sme

SUMMIT & EXPOSITION
February 25-27, 2014
LONG BEACH CONVENTION CENTER
LONG BEACH, CALIFORNIA, USA

Objectives of Presentation

- The Need for Multi-layer Coatings Measurement Tools in Manufacturing
- Unique Nature of the Problem
- Solutions Investigated
- Technical Overview of Our Solution
- Tech Development Path
- Tech Maturation Level Evaluation
- Path Forward

The Core Team

- Air Force Research Laboratory AFRL/RXCA Materials and Mfg. Directorate
  - –Juan Calzada - NDE for LO System Program Mgr.
  - –Bryan C Foos (ctr.), Ph.D. - Materials State Awareness & Supportability
- JENTEK Sensors, Inc
  - –Yanko Sheiretov, Ph.D. - Sr. Vice President, Product Development
  - –Zachary M. Thomas, Ph.D., Esq., Vice President, Operations
- Northrop Grumman Aerospace Systems
  - –Andy Jakubek, Engineer / Project Mgr., Manufacturing Technology
  - –Gary Ostrowski, Engineering Fellow, Manufacturing Technology

The Need for Multi-Layer Coatings Measurement Tools in Manufacturing

- Coatings: sprayed elastomers
  - Highly filled, precise thickness
- Layers: 3 or more disparate coatings
- Production environment needs:
  - Rapid NDT thickness verification
  - Measurement unaffected by cure
  - Single operator, ergonomic
  - Repetitive measurement
  - Close quarters
  - Auto data recordation

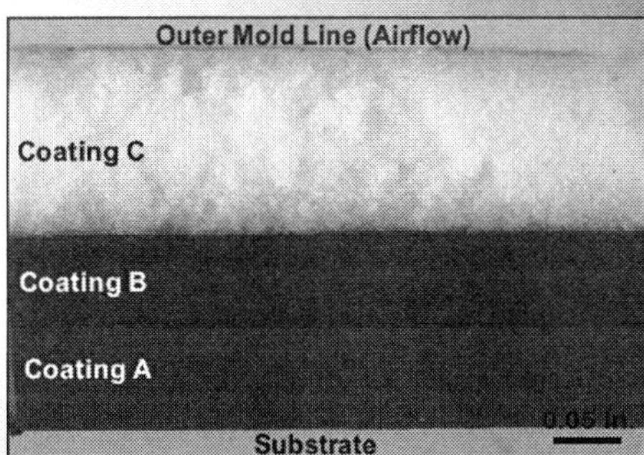

Typical Coating Stack

The Need for Multi-Layer Coatings Measurement Tools in Manufacturing

- Baseline pin displacement
- Labor intensive
- Inadequate precision, resolution
- Inadequate repeatability, reproducibility
- Manual data recording
- Return to exact location
- Requires operator subtraction

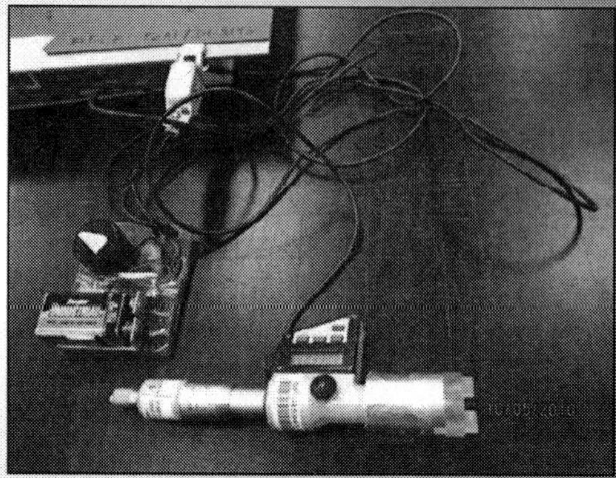

Baseline Mechanical Pin Measurement

Unique Nature of Problem

- 3+ highly disparate materials
- Multiple fillers and binders
- Varied substrates
  - Metallic and non-metallic
- Robotic and manual spray
- Extended coating cure times
- Critical path schedules
- Depot maintenance
- Field repairs
- Manual use
- Robotic adaptation

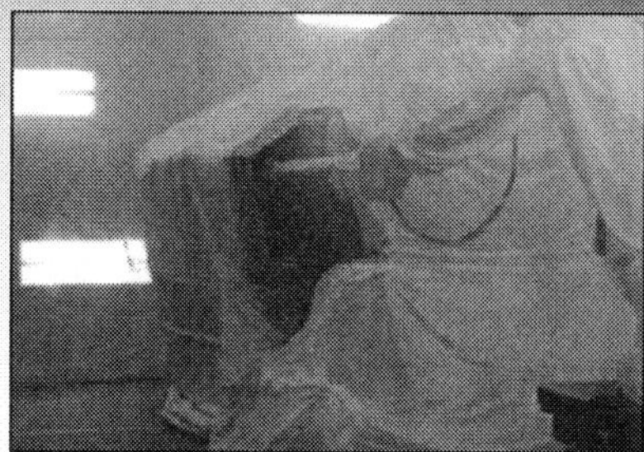

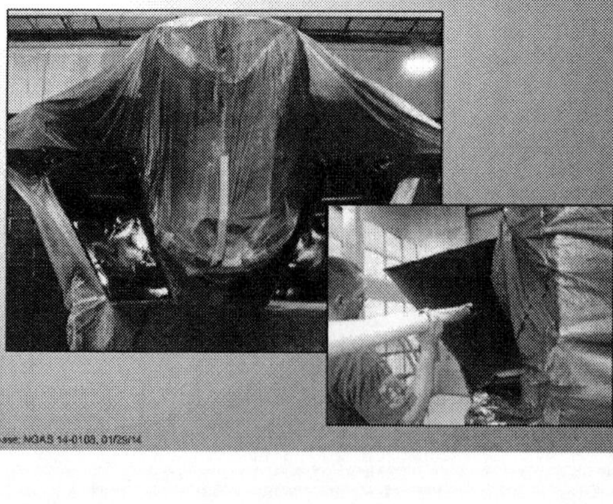

Solutions Investigated

- Measurement solutions exist for thick coatings of *two or fewer* disparate layers
  - Terahertz
  - Eddy current
  - Ultrasonic
  - Standard impedance
  - Automated / robotic / manual
  - In-process and post-process
  - Direct mechanical thickness
  - Electrical properties
- None currently available for *three or greater* disparate layers
  - Various substrates
  - Various fillers and binders

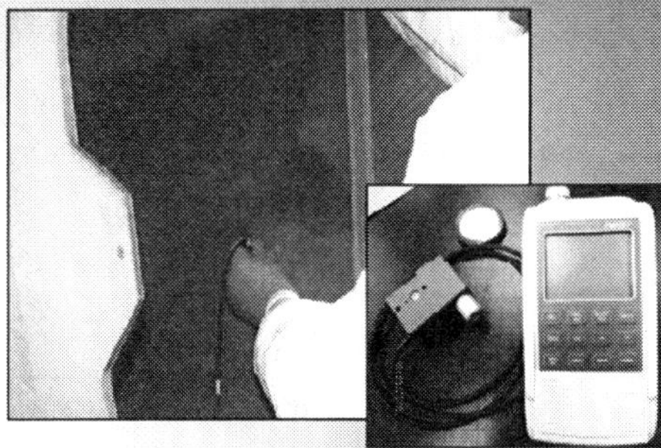

Approved for Public Release; NGAS 14-0108, 01/29/14

Solutions Investigated

- Various promising NDT technologies investigated
  - Ultrasonic sensors
  - Laser sensors
  - Laser and ultrasonic combinations
  - Eddy current sensor arrangements
  - Impedance devices
- 3+ disparate layers presented significant challenges
- Sensor arrangements could not consistently differentiate boundaries

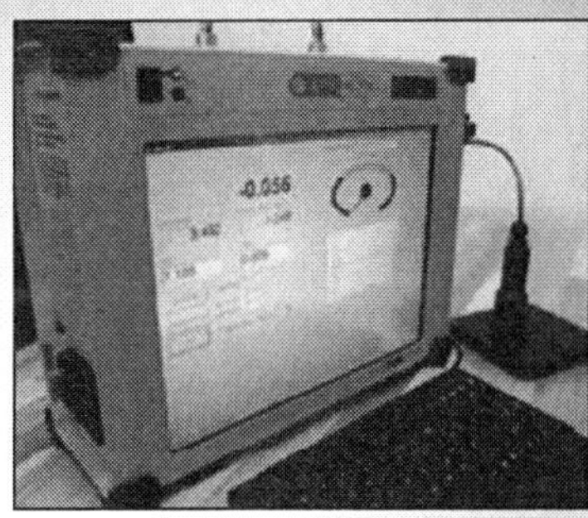

Approved for Public Release; NGAS 14-0108, 01/29/14

JENTEK Solution

Sensors: MWM®-Arrays & MWM®-Rosettes

- Paradigm shift in sensor design (first priority is predictable response based on physics-based modeling)

Next Generation 8200 GridStation® System

- 10x signal-to-noise improvement
- Very low frequencies (deep penetration)
- 19-119 fully parallel impedance channels
- Substantially increased scan speeds

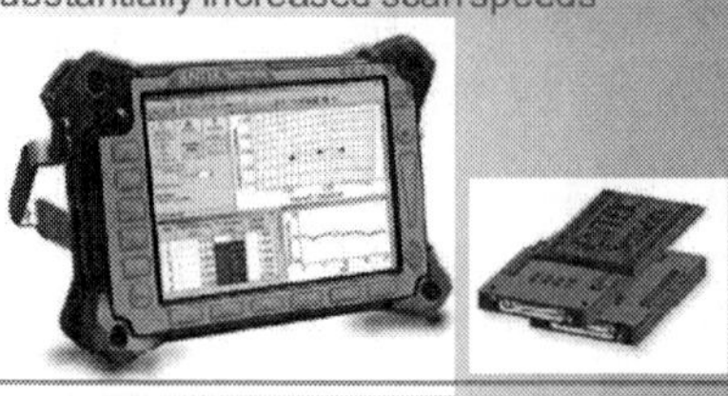

GridStation® Software using Hyperlattices™

- Rapid, autonomous data analysis for decision support in NDT and SHM – Performs multivariate inverse method using precomputed databases

Fuzzy-Hyperlattices

- Rapid uncertainty and risk assessment using precomputed databases

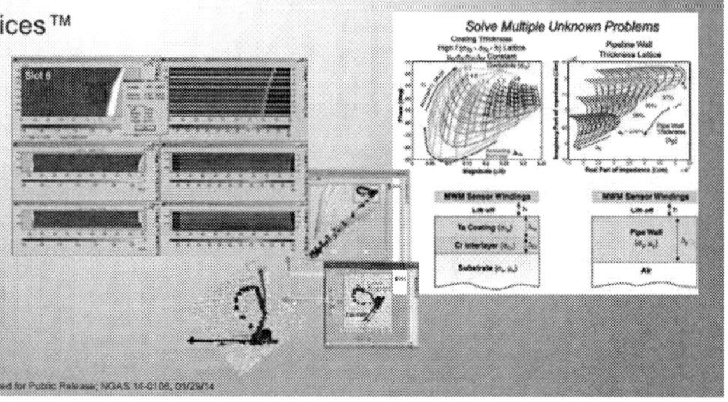

Approved for Public Release; NGAS 14-0106, 01/29/14

JENTEK Solution

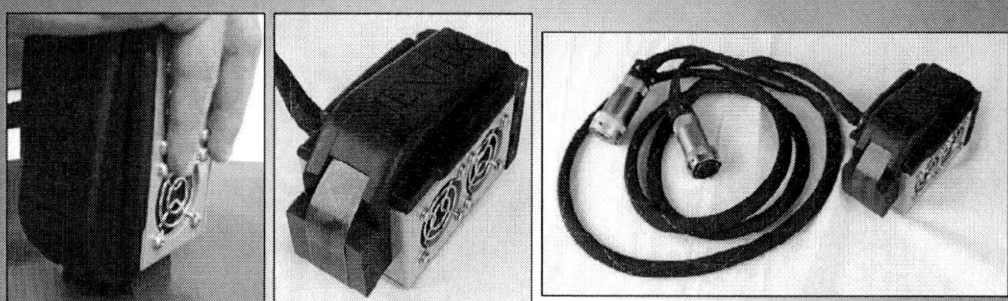

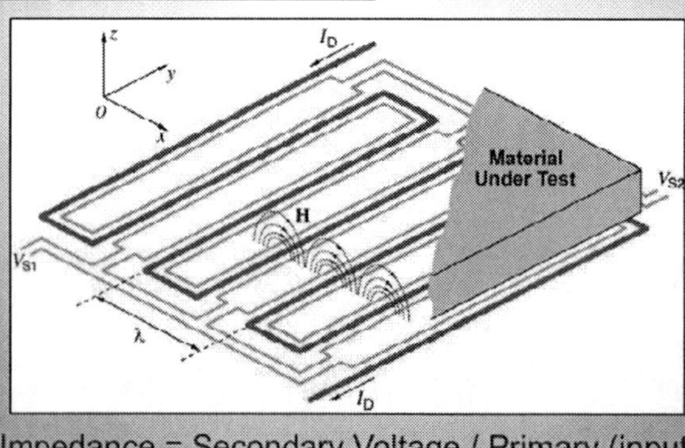

Transfer Impedance = Secondary Voltage / Primary (input) Current

Approved for Public Release; NGAS 14-0106, 01/29/14

JENTEK Product Evolution

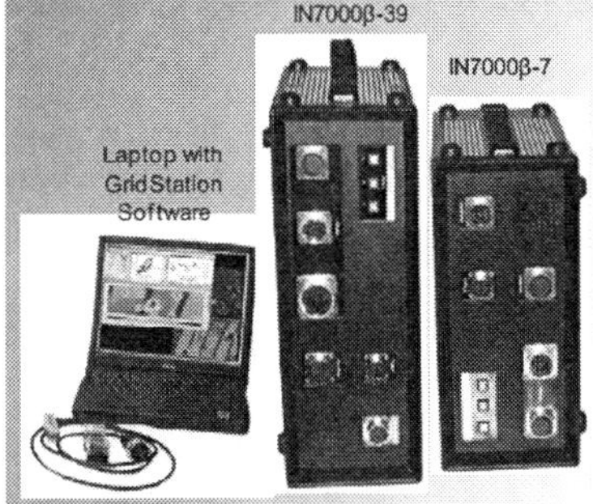

IN7000β Series Systems
- Size 12"*w* x 15"*l* x 5.5"*h* (-39)
- Weight: 30 lbs (-39)
- Supports 7 or 39 channels
- Requires external display & computer (laptop or desktop)

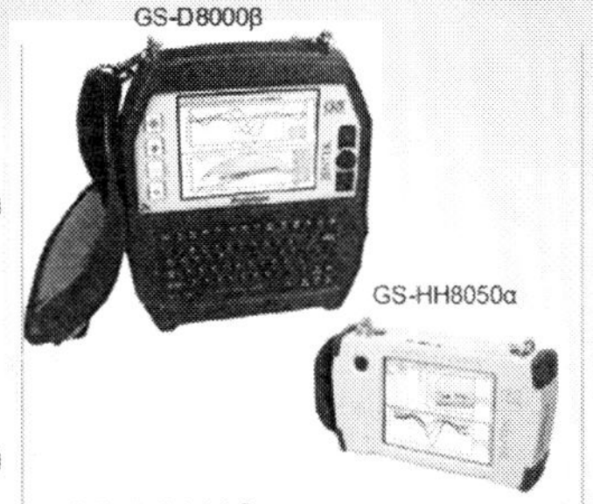

GS-D8000β
- Size: 10"*w* x 11"*l* x 5"*h* (-39)
- Weight: 11 lbs (-39)
- Supports 7, 15, or 39 channels
- Integrated computer
- External power required

GS-HH 8050α (HandHeld)
- Size: 3.5"*w* x 10"*l* x 5"*h*
- Weight: <5 lbs, inc. battery
- Supports 7 channels
- Touch screen interface
- Integrated computer

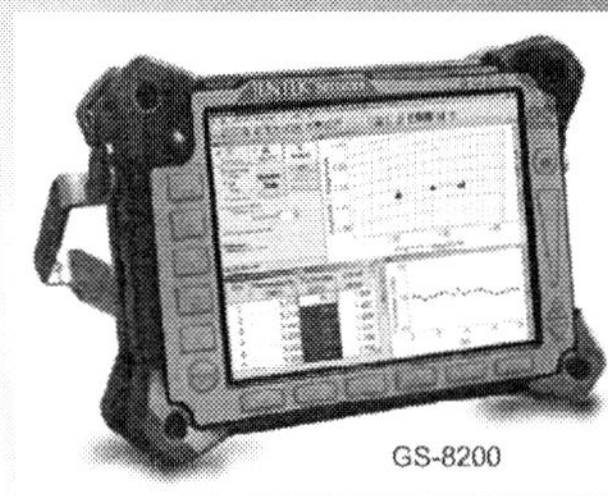

GS-8200α
- Size: 4.3"*w* x 14.7"*l* x 10.9"*h* (-19)
- Weight: 8 lbs
- Supports 19 – 119 channels
- 100x faster than GS-D8000β
- 10x improved signal-to-noise (at same scan rate)
- DC–20MHz operating frequency
- Multi-touch screen
- Integrated computer

Approved for Public Release; NGAS 14-0108, 01/29/14

Initial Technical Development Path

- Initially determined 3+ layer measurement feasibility
  - First system to succeed
  - Advanced from MRL 4 to MRL 5
  - Lab environment
  - Prototype sensors
  - Free film materials
  - Multiple substrates
  - Gage R&R work
  - Algorithm Development
  - Ergonomic challenges
  - Sensor durability

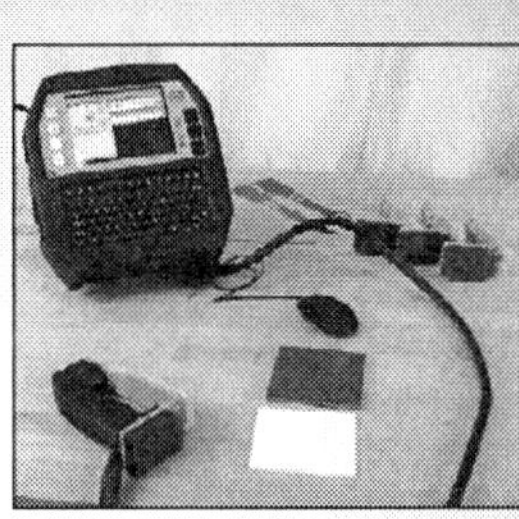

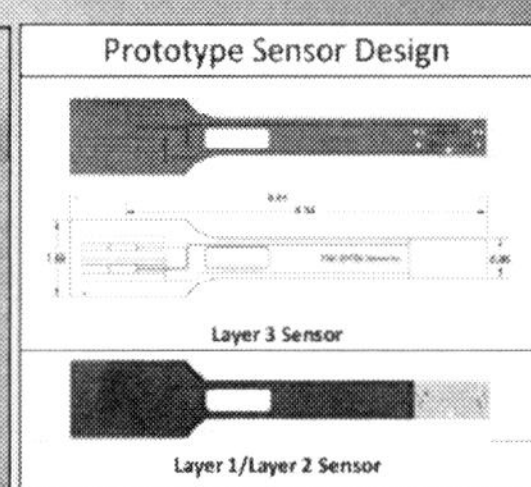

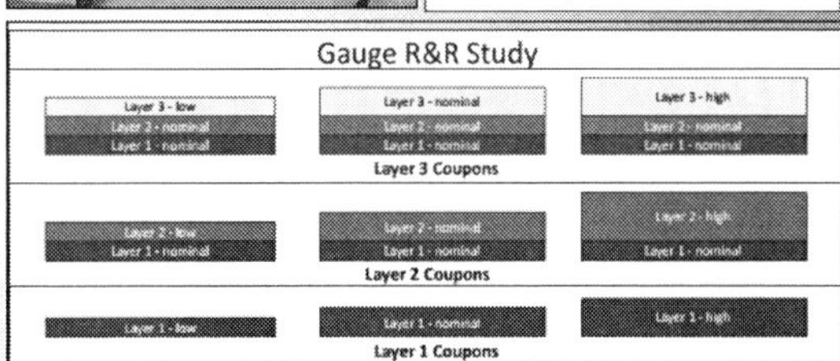

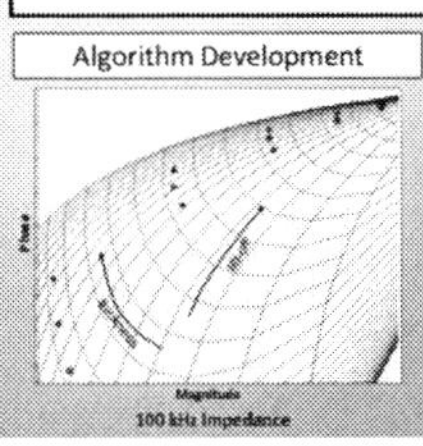

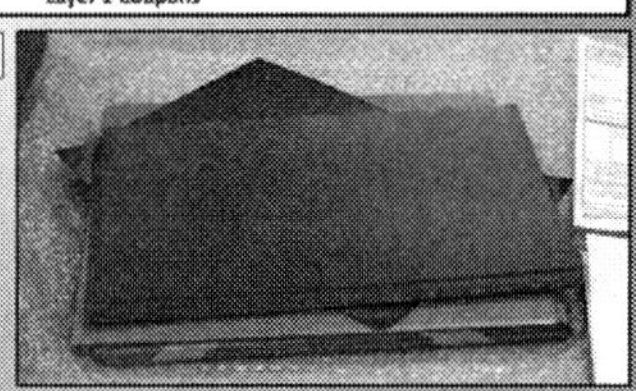

Approved for Public Release; NGAS 14-0108, 01/29/14

Evolution of Tech Development Path

- Expanded use, multiple applications
- Advanced materials evolution
- Free and adhered films
- Various substrates
- Advanced substrate films
- Multiple stack arrangements
- Multiple resin systems
- Multiple fillers and densities
- Complex curvatures
- Confined enclosures
- Vertical, horizontal, inverted
- Substructure variations
- OML features (gaps, fasteners, etc.)
- Production environments

Current Tech Development Path

- Production trials
  - Palmdale, CA 11/13 – 2/14
- Production personnel
- Production personnel training
- Technician evaluation
- Paint booth environment
- Ergonomic factors
- All attitude use
- Safety reviews
- Rugged sensor evaluation
- Snap-in sensor cartridge
- Cable / plug optimization
- Gage R&R – new stacks

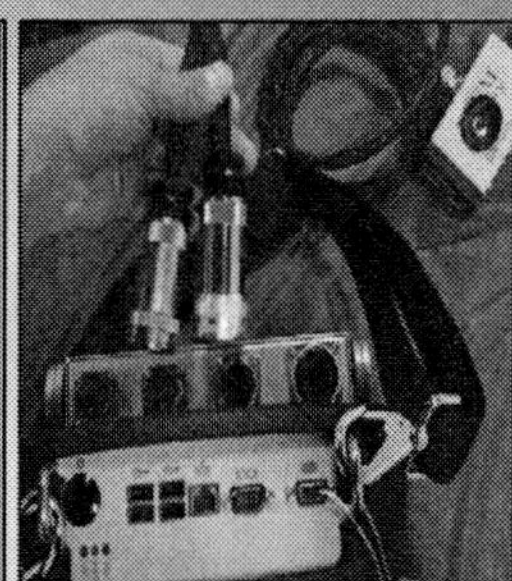

Tech Maturation Level Evaluation

- Initial production trial feedback mostly good
- Ease of use generally good
- Consistent measurement results
- Calibrations performed
- Multiple operators trained by engineer
- Multiple operators used equipment
- Confined space limitations
- Smaller, lighter probe design desired
- Concave radius limitations
- Prefer static spot measurement
  - Probe must be moved to scan
  - High contour & overhead challenge
- Precision, resolution, repeatability, reproducibility look acceptable
  - Controlled, repeated trials required
- Data analysis and comparison to witness panels are pending

Path Forward

- Current MRL Level is 6: ***Prototype System in a Production Relevant Environment***
- Continue to evaluate within the production environment
- Provide feedback for design optimization
  - Ease of use
  - Ergonomics and safety
  - Optimize sensor size and mass for greater applicability
- Repeat production environment evaluations after design optimizations
- Evaluate production environment precision, resolution, repeatability, reproducibility through data analysis and Gage R&R
- Advance to MRL 7 and 8
  - MRL 7: System in a production representative environment
  - MRL 8: Ready for LRIP
- Sensor design evolution with advancing resin and filler packages
- Gage R&R on new coating system
  - "Stack B"

Summary

- Thickness measurement of 3+ layer coatings is difficult
- Typical sensor probe designs can't differentiate layers
- JENTEK's Meandering Winding Magnetometer (MWM®) can
- Production trials underway to refine MWM® and advance MRL
- MRL 6 demonstrated
- Working toward MRL 7 and 8

The AFRL, JENTEK, and Northrop Grumman Aerospace Systems team is effectively working toward the development and production implementation of MWM®

AERODEF MANUFACTURING

SUMMIT & EXPOSITION
February 25-27, 2014
LONG BEACH CONVENTION CENTER
LONG BEACH, CALIFORNIA USA

Intelligent Sealant Application System ISAS™

produced by sme

Presented by
C. Y. Ingham

SMRC SYSTEMS & MATERIALS RESEARCH CORPORATION

AERODEF MANUFACTURING

sme

SUMMIT & EXPOSITION
February 25-27, 2014
LONG BEACH CONVENTION CENTER
LONG BEACH, CALIFORNIA USA

- Objectives
  - Describe the difficulty of fuel area sealing
  - Estimate weight savings vs. current methods
  - Describe proper use of ISAS Kaps
  - Describe rapid proficiency curve
  - Describe benefits of implementing ISAS Kaps

AERODEF MANUFACTURING

SUMMIT & EXPOSITION
February 25-27, 2014
LONG BEACH CONVENTION CENTER
LONG BEACH, CALIFORNIA USA

- Applying sealant is an art
  - No standard set of tools
    - Brushes
    - Nozzles
    - Cups
    - Spatulas
    - Caps
  - Most are customized by individual sealer
  - Takes a long time to become proficient
  - Can be slow, tedious process

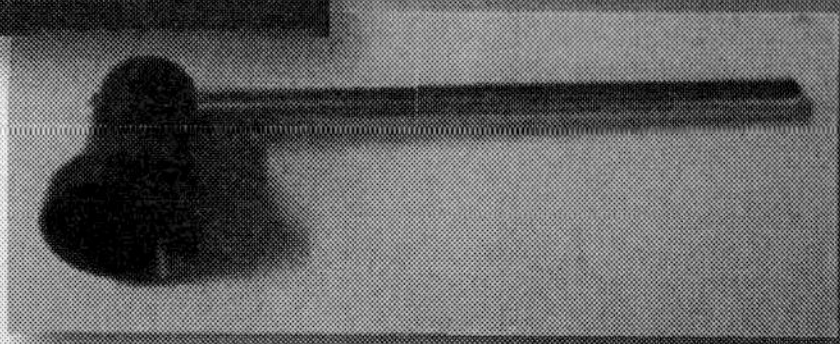

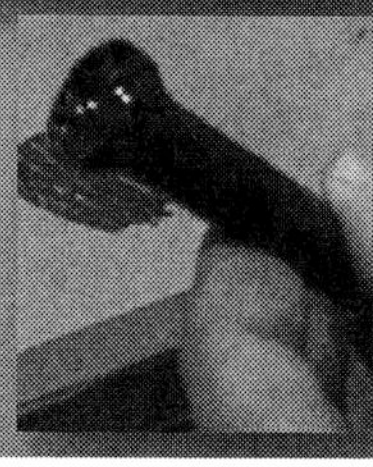

AERODEF MANUFACTURING

SUMMIT & EXPOSITION
February 25-27, 2014
LONG BEACH CONVENTION CENTER
LONG BEACH, CALIFORNIA USA

- Seal areas are difficult to reach
  - In tight corners
  - Behind other structures
  - Crowded nutplate spacing

AERODEF MANUFACTURING
sme
SUMMIT & EXPOSITION
February 25-27, 2014
LONG BEACH CONVENTION CENTER
LONG BEACH, CALIFORNIA USA

- Blind areas
  - Use of mirrors required to see
- Tight tolerances
- Exclusion areas
  - Upper portion of enclosed dome nutplate

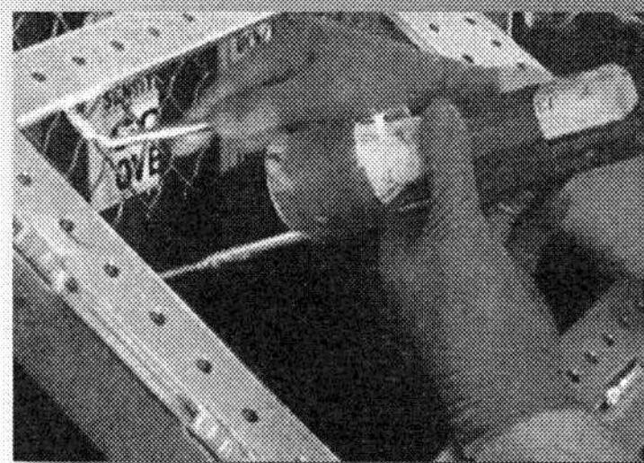

AERODEF MANUFACTURING
sme
SUMMIT & EXPOSITION
February 25-27, 2014
LONG BEACH CONVENTION CENTER
LONG BEACH, CALIFORNIA USA

- ISAS is born
  - Air Force SBIR program launched in 2011
  - Currently in Phase II
- Primary focus is backside fastener seals
  - Enclosed dome nutplates
  - Hi-Lock, Hi-Tigue, Eddie Bolts

AERODEF MANUFACTURING
sme

SUMMIT & EXPOSITION
February 25-27, 2014
LONG BEACH CONVENTION CENTER
LONG BEACH, CALIFORNIA, USA

- SMRC designed nozzles
  - Unable to achieve desired results
- SMRC designed molds (Kaps)
  - Easy, high quality seals

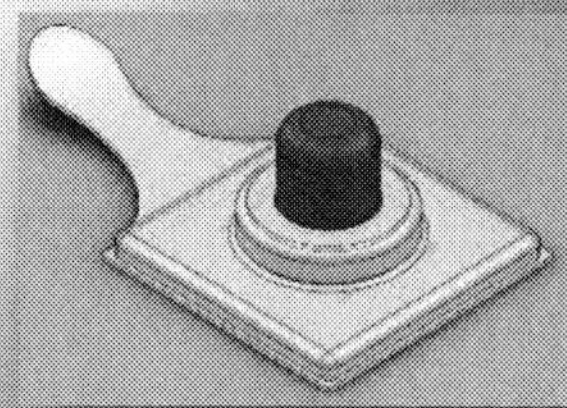

SMRC

AERODEF MANUFACTURING
sme

SUMMIT & EXPOSITION
February 25-27, 2014
LONG BEACH CONVENTION CENTER
LONG BEACH, CALIFORNIA, USA

- Kap features (Enc. Dome Nutplate)
  - Consistent sealant volume & profile
  - 360° squeeze-out verification
  - Exclusion of sealant from upper dome
  - Colored transparent plastic
    - Quick, easy selection per fastener diameter
    - Stands out against structure (FOD prevention)

AERODEF MANUFACTURING
sme

SUMMIT & EXPOSITION
February 25-27, 2014
LONG BEACH CONVENTION CENTER
LONG BEACH, CALIFORNIA USA

- Conventional Kap features (Bolt & Nut)
  - Consistent sealant volume & profile
  - 360° squeeze-out verification
  - Pin for alignment
  - Colored transparent plastic
    - Quick, easy selection per fastener diameter
    - Stands out against structure (FOD prevention)

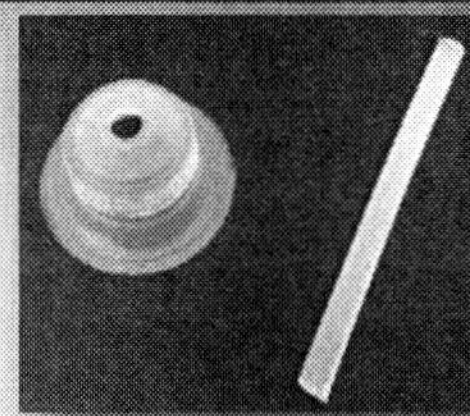

SMRC

AERODEF MANUFACTURING
sme

SUMMIT & EXPOSITION
February 25-27, 2014
LONG BEACH CONVENTION CENTER
LONG BEACH, CALIFORNIA USA

- Current generation enclosed dome nutplate ISAS Kap
  - Sliding cylinder for alignment, and ease of filling
  - Improved Kap profile
    - minimization of sealant
    - smooth transitions
    - feathered edges
    - adherence to specified dimensions

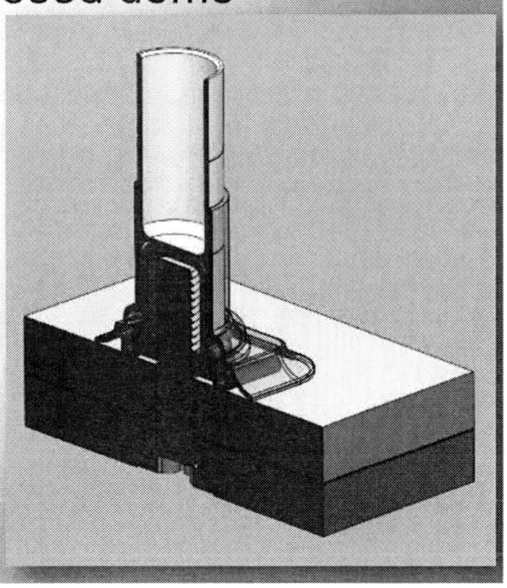

AERODEF MANUFACTURING
sme
SUMMIT & EXPOSITION
February 25-27, 2014
LONG BEACH CONVENTION CENTER
LONG BEACH, CALIFORNIA USA

ISAS Kap Seal Procedure

- Assemble Kap
  - Slide plunger into Kap until plunger is flush with Kap top as shown (Kap shown inverted)
  - With Kap facing cavity side up, add sealant to "Fill Line"

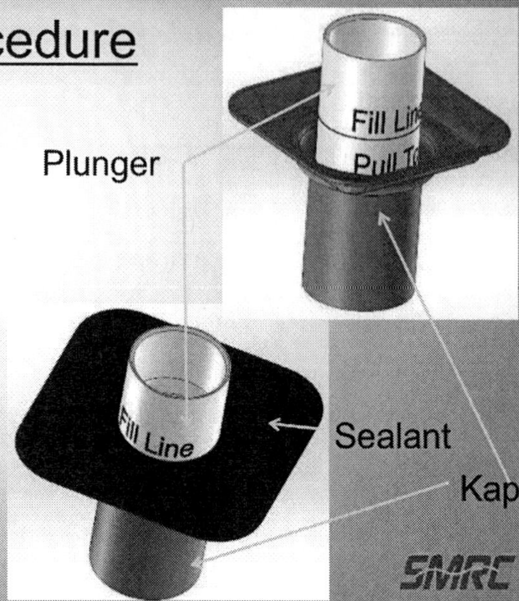

SMRC

AERODEF MANUFACTURING
sme
SUMMIT & EXPOSITION
February 25-27, 2014
LONG BEACH CONVENTION CENTER
LONG BEACH, CALIFORNIA USA

ISAS Kap Seal Procedure

- Align
  - Invert and position plunger on nutplate dome to align
  - Push plunger down until it stops
- Position
  - Slide Kap down plunger until bottom is flush with substrate surface
  - Maintain orientation with nutplate base
  - If "nonstandard" nutplate, pull plunger back to "Pull Line"
  - 360 degree sealant squeeze-out should be visible

SMRC

AERODEF MANUFACTURING sme

SUMMIT & EXPOSITION
February 25-27, 2014
LONG BEACH CONVENTION CENTER
LONG BEACH, CALIFORNIA USA

- Current generation conventional ISAS Kap
  - Pursuing sliding alignment pin and integral pin designs
  - Achieving 30% weight reduction versus current process
  - Procedure similar to above

AERODEF MANUFACTURING sme

SUMMIT & EXPOSITION
February 25-27, 2014
LONG BEACH CONVENTION CENTER
LONG BEACH, CALIFORNIA USA

- Analysis of Variance (ANOVA)
  - ¼" Eddie Bolt seals meet all dimensional restrictions in the specification with a 95% confidence level (meeting nominal thicknesses ranges)
    - Average 'A' dimension in spec.
    - Average 'B' dimension in spec.
    - Average 'D' dimension in spec.
  - Weight savings of >30%
  - Average weight of 0.63 grams versus ~1.01 gram in current production settings
  - Newest design for Enclosed Dome Nutplates have yet to be ANOVA evaluated

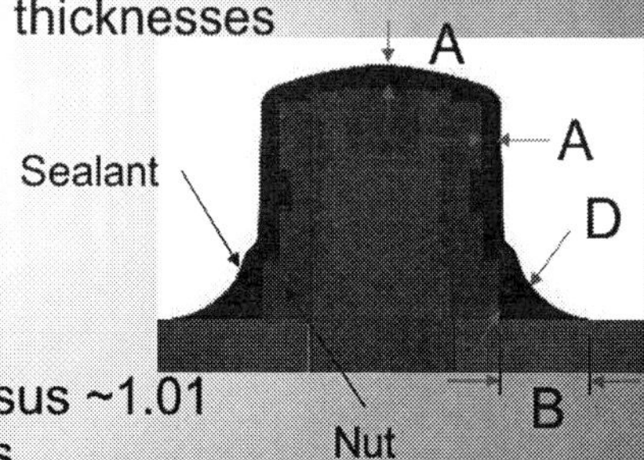

SMRC

AERODEF MANUFACTURING
sme
SUMMIT & EXPOSITION
February 25-27, 2014
LONG BEACH CONVENTION CENTER
LONG BEACH, CALIFORNIA USA

- Analysis of Variance (ANOVA)

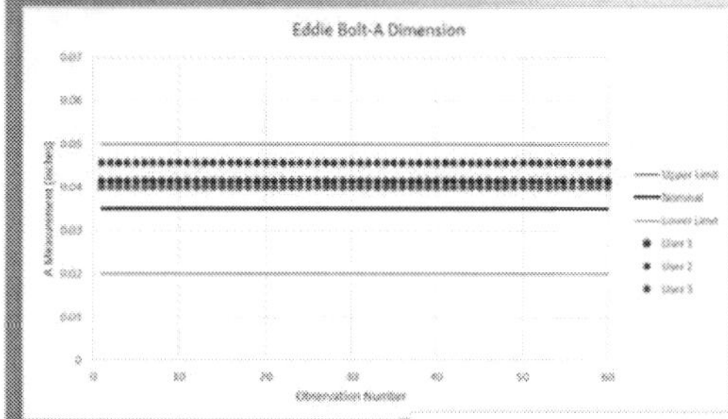

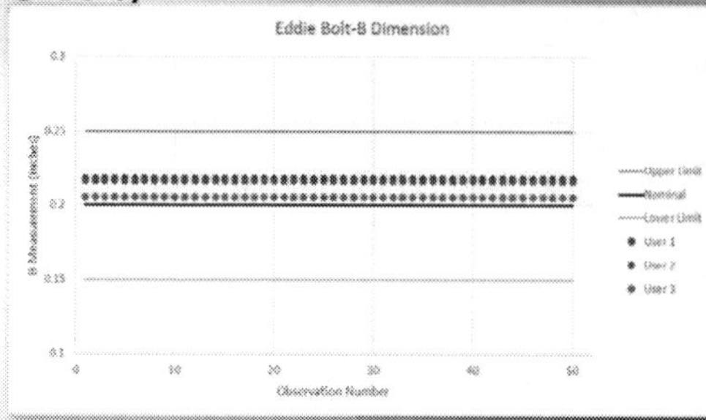

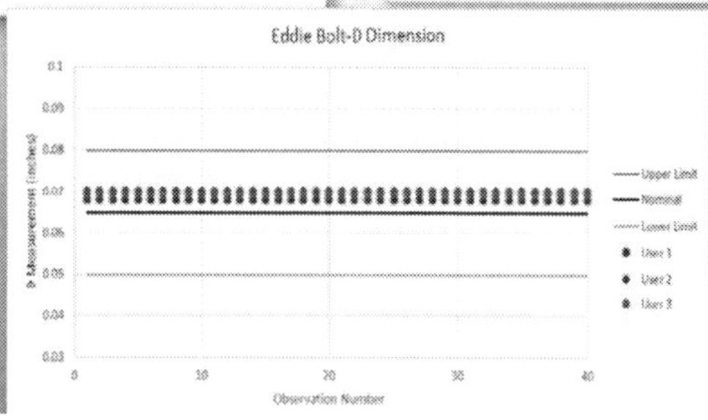

SMRC

AERODEF MANUFACTURING
sme
SUMMIT & EXPOSITION
February 25-27, 2014
LONG BEACH CONVENTION CENTER
LONG BEACH, CALIFORNIA USA

- Ease of inspection
  - Consistency of the ISAS seals makes inspection easier and quicker
  - Seals look the same every time
  - Voids easily detected

ISAS Dome Nutplates

AERODEF MANUFACTURING

sme

SUMMIT & EXPOSITION
February 25-27, 2014
LONG BEACH CONVENTION CENTER
LONG BEACH, CALIFORNIA USA

- Summary
  - Fuel sealing is very difficult
  - Weight savings of >30%
  - ISAS Kaps easy to use
  - Proficiency gained quickly
  - Benefits include the above and…
    - Consistency
    - Ease of inspection
    - Flexibility

AERODEF MANUFACTURING

sme

SUMMIT & EXPOSITION
February 25-27, 2014
LONG BEACH CONVENTION CENTER
LONG BEACH, CALIFORNIA USA

Thank you for your support!

FA8650-13-C-5187

AIR FORCE RESEARCH LABORATORY

Contact Information:

Systems and Materials Research Corporation
11525 Stonehollow Drive, A120, Austin, TX 78758
Office: (512) 535-7791 • Fax: (512) 206-4966
www.systemsandmaterials.com

Malcolm D. Prouty, Ph.D.
President
Cell: (512) 968-4750
malcolmprouty@systemsandmaterials.com

C.Y. Ingham
Senior Automation Specialist
Cell: (512) 788-2005
cyingham@systemsandmaterials.com

Chelsea Boundy
Senior Chemist
Cell: (512) 680-7386
chelseaboundy@systemsandmaterials.com

SMRC

PICOMETRIX
an API company

In-Process Monitoring of Specialty Coatings

David Zimdars, Ph.D.* — Manager of Terahertz Research and Development, Picometrix LLC, an Advanced Photonix Company

Owen Cupp, MSEE — Program Manager, Northrop Grumman Aerospace Systems

Miles Honkawa — Material & Process Engineer, Northrop Grumman Aerospace Systems

Juan Calzada — NDE for Low Observable System Program Manager, Air Force Research Laboratory, AFRL/RXCA

*Presenter
Contact: Picometrix, LLC., 2925 Boardwalk Dr., Ann Arbor, MI 48104, USA
Tel: 734-864-5639 Fax: 734-998-3474, dzimdars@picometrix.com

NORTHROP GRUMMAN

AEROBOTIX

Funded under Phase II SBIR Topic AF071-117, Contract FA8650-08-C-5006

AERODEF
Long Beach, CA
February 25, 2014 2:10-2:40 pm

Approved for public release: 2/4/14.
Subject: In-Process Measurement of Specialty Coatings (Briefing Charts)
Case Number 88ABW-2014-0359

PICOMETRIX
an API company

Overview

2

PICOMETRIX
an API company

Challenge

- Aircraft may be robotically spray painted with thick multi-layer specialty coatings.
- Proper performance of the aircraft requires that the thickness of the coatings be controlled to a tight tolerance.
- Coatings take several days to apply and cure.
- Traditional non-destructive thickness gauges (e.g. UT and Eddy current) require contact to the surface and are inappropriate for wet coatings.
- Measurement of witness coupons (by pin gauge) may not represent the high and low coating buildup over complex curved surfaces.

3

PICOMETRIX
an API company

Opportunity

- Currently, the verification of coating thickness must wait until the coating has fully cured. The thickness testing is performed manually.
- A sensor technology that *can measure the wet coating concurrent with the spray process* may be used to predict the final cure state thickness.
- Predicting the cured final thickness may help minimize costly corrective actions (e.g. sanding over-built areas or recoating under-built areas).

4

PICOMETRIX
an API company

Solution

- Compact Robot Mounted Pulsed THz Sensor
  - Non-Contact
  - Stand-Off
  - In-Process
  - Sees Through Plume
  - Measures Through Thick Opaque Coatings
  - Wet or Cured

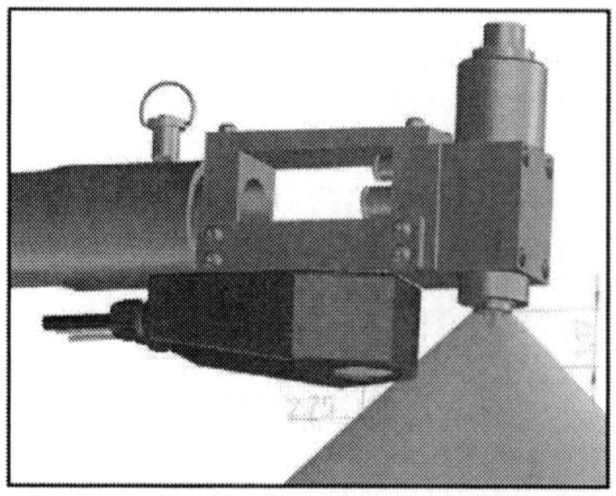

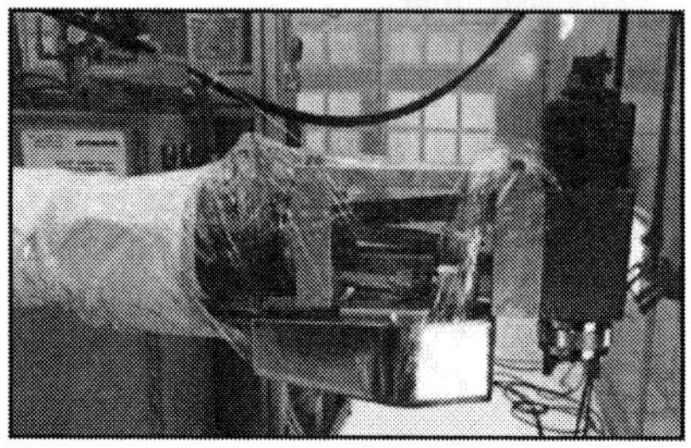

5

PICOMETRIX
an API company

Background of Picometrix Time Domain Pulsed Terahertz Instrumentation

6

PICOMETRIX
an API company

Picometrix, LLC

- Formed in 1992 in Ann Arbor, Michigan
- **Subsidiary of Advanced Photonix, Inc. since 2005**
  - **NYSE: AMEX listed (API)**
  - Leading supplier of TD-THz instrumentation and optical receivers utilizing Si, GaAs and InGaAs
  - Industrial, military, homeland security, medical and telecom markets
- Facility 50,000 sq. ft. with four TD-THz application labs
- **THz product brand transitioning to Advanced Photonix (API)**

7

T-Ray 5000® Modular Product Line

PICOMETRIX
an API company

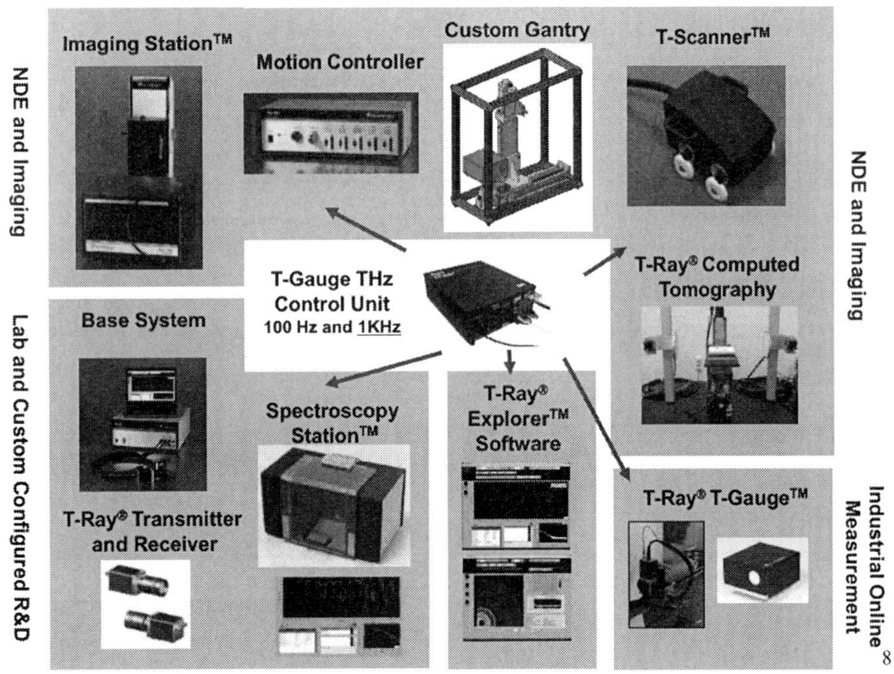

8

T-Gauge ® (Model 521X series) Terahertz Control Unit (TCU)

PICOMETRIX
an API company

Industrially Hardened

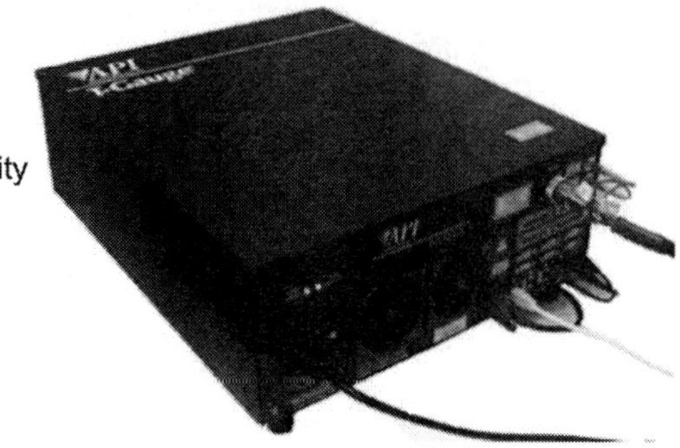

- Operating Environmental Conditions
  - 0 to +50 C, 20 - 90 % relative humidity (non-condensing)
- Mechanical Operating
  - Vibration 5 -500 Hz, 0.006 in.p-p displacement, 1 g max
  - Shock +/- 15 g pk, ½ sine, 11ms
- AC Power requirements 90 - 264 VRMS 47 - 63 Hz, 550W max.
- Overall weight 40 lbs
- Emissions EN 55011 Class A
- FCC Part 15 Class A
- FDA CDRH Class 3R Laser Product
- UL Certified UL 61010A-1

9

Pulsed THz Aerospace Applications

PICOMETRIX
an API company

Terahertz Control Unit

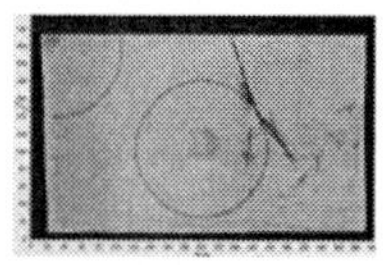

Radome delamination and water intrusion; Hill AFB

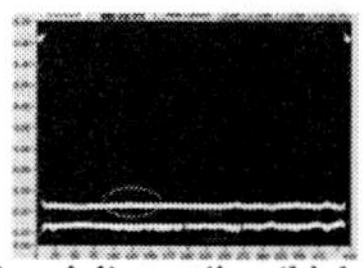

Specialty coating thickness measurement; with WPAFB, Northrop Grumman

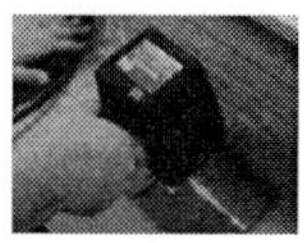

Sub-surface step and gap; with WPAFB, Lockheed Martin

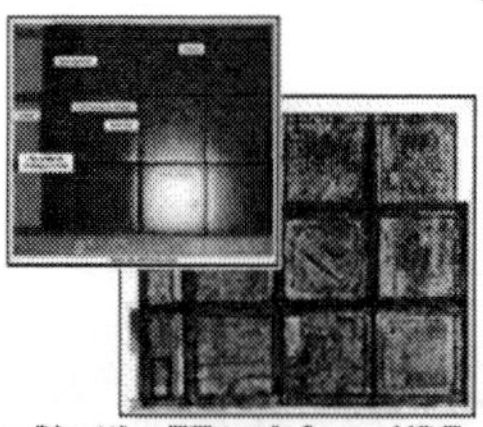

Space Shuttle: ET tank foam NDE; Orbiter TPS Tiles-hidden corrosion detection; NASA

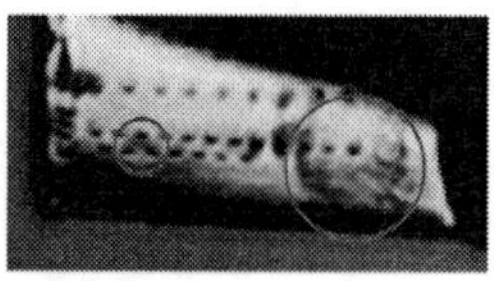

Aeroturbine thermal barrier coating measurement; with Navy STTR, GE

10

PICOMETRIX
an API company

Pulsed THz Measurements

11

Features and Benefits of Pulsed THz Measurements

PICOMETRIX
an API company

- Technology for non-destructive testing
- Electromagnetic Impulses → Non Contact
- Penetrates Thick Dielectrics
- High Dynamic Range → Coatings
- Waveform record like ultrasound "A-Scan"
- Sub-picosecond → High Resolution,
- Near single-cycle → <0.001 in. (1 mil)
- Non ionizing → Safe
- Low power
- High speed → Suitable for Spray Robot
- Remote compact sensors → In Process Monitoring

12

Sensors Emit and Detect THz Pulses

PICOMETRIX® an API company

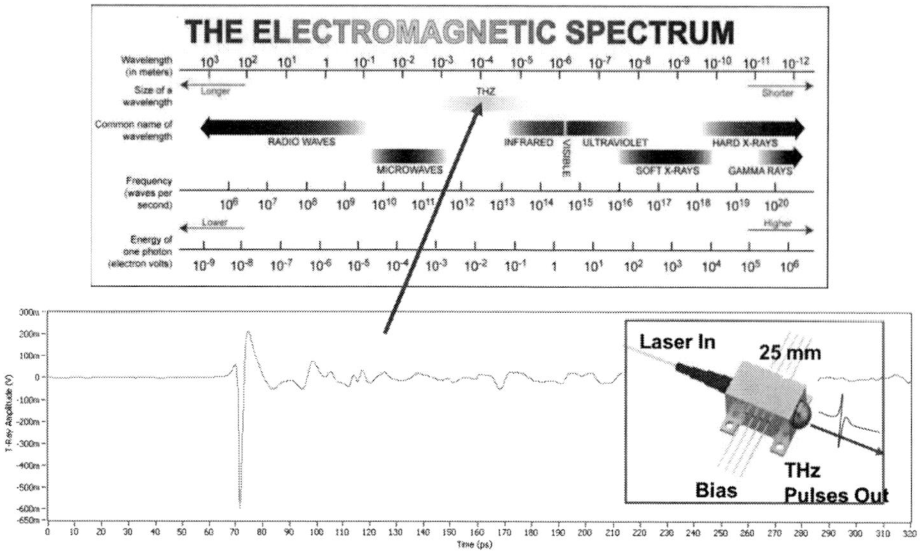

Raw Pulsed THz Waveform

THz Transmitter

13

Coating Measurement

PICOMETRIX® an API company

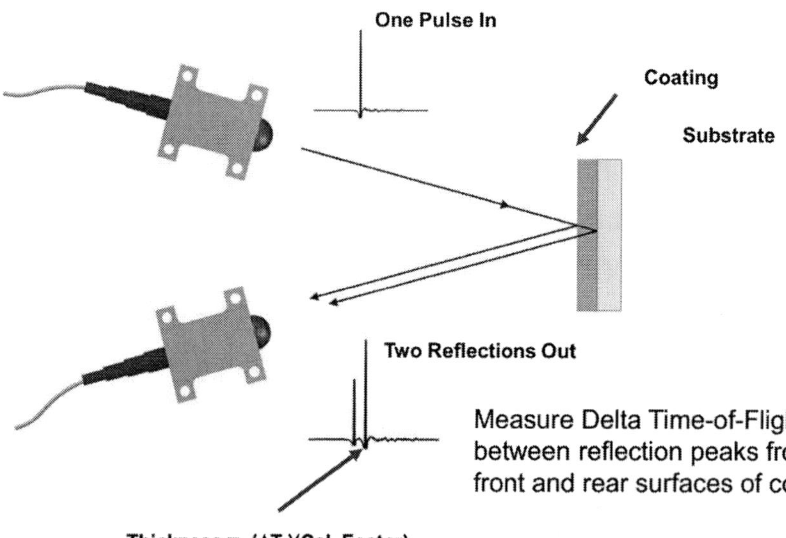

(Concept shown pitch-catch for clarity – implemented monostatic)

Measure Delta Time-of-Flight (ToF) between reflection peaks from front and rear surfaces of coating.

This Delta time value (ps) is directly correlated to coating thickness.

14

THz B-Scan
Single Form Coating Thickness

PICOMETRIX
an API company

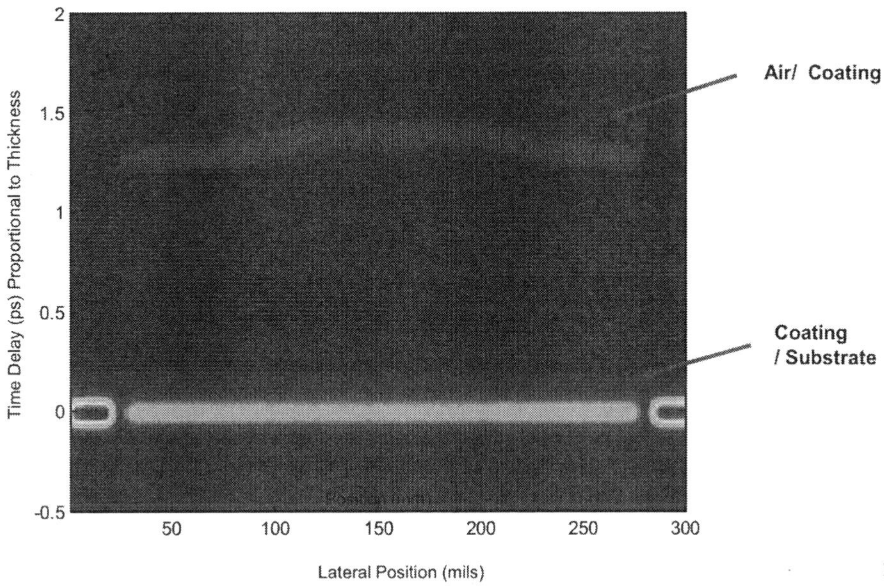

15

PICOMETRIX
an API company

Spray-Robot Mounted THz Sensor

16

Sensor Features

PICOMETRIX® an API company

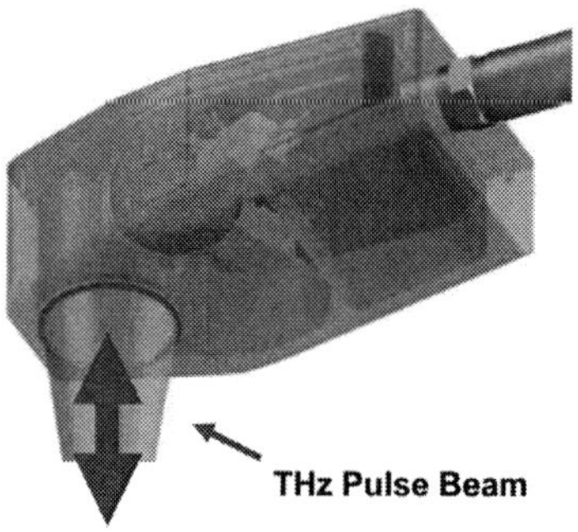

Features

- Compact for mounting on existing robots
  - Size: 1.9 in. high, 2.9 in. wide, 5.1 in. long
  - Weight ~3.5 lbs
  - Mount using three 1/4-20 tapped holes
- **Class I Div. 1 Design**
  - **Stainless steel non-sparking body**
  - **Head contains <3% aluminum**
  - **Fully sealed to IP6 (withstand water jet)**
  - **Electronics voltage and current limited**
- Solvent Resistant
  - Teflon coated HDPE lens
  - Buna-N O-rings and cable gland seals
  - Nylon 6/6 cable glands
- Sensor Output beam
  - Centered in body
  - 0.89 in. from end of body
  - Accepts Picometrix standard 1.5 in diameter lenses.

17

Sensor Mounted on Boom

PICOMETRIX® an API company

Parameter	Value
Lens f/#	3.3
Focal distance	4.5 inches
Spot size 0.25 THz	5.0 mm
Rayleigh range @ 0.25 THz	33 mm
Bandwidth	• • • •
Resolution	• • • •
Polarization	Axis rot 23.5°
Transmitter	Std bowtie

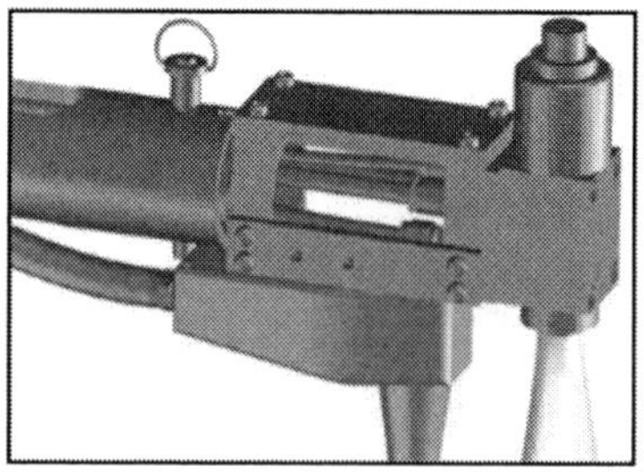

• Legacy paint booths, robot, and spray paths not are specifically designed to accommodate in process sensors
• Sensor housing must be small and light enough to not impact aircraft at any point or alter spray path
•THz sensor "sees through" spray plume, pajamas, and light collected overspray
•Sensor must be mounted close to spray location yet avoid collection sufficient to drip

18

Northrop Grumman Paint-Booth Installation

PICOMETRIX
an API company

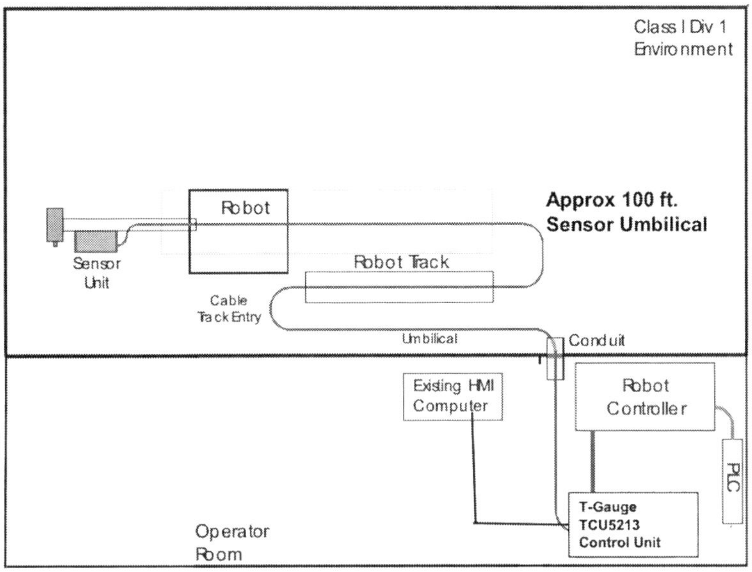

19

Intrinsically Safe Class I Div 1 Interconnection

PICOMETRIX
an API company

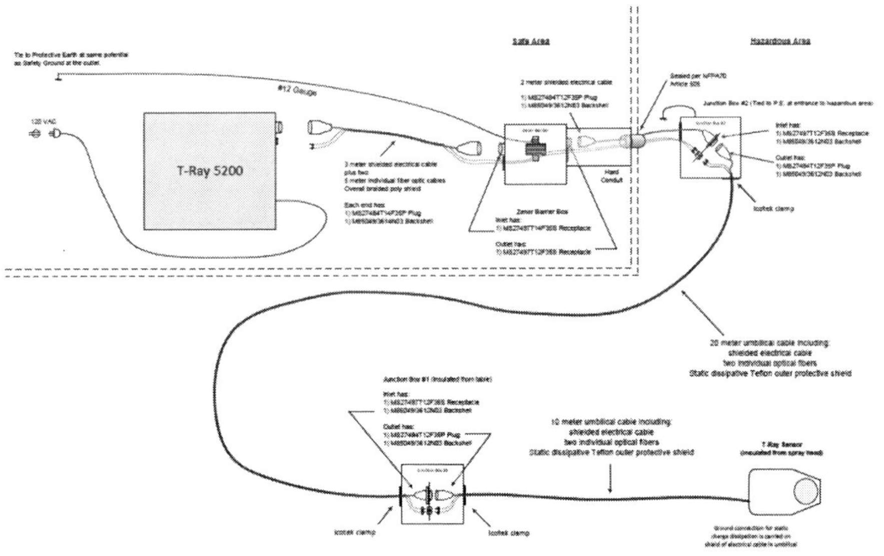

20

PICOMETRIX
an API company

Demonstration Measurements

21

Coupon Spray Pattern

PICOMETRIX
an API company

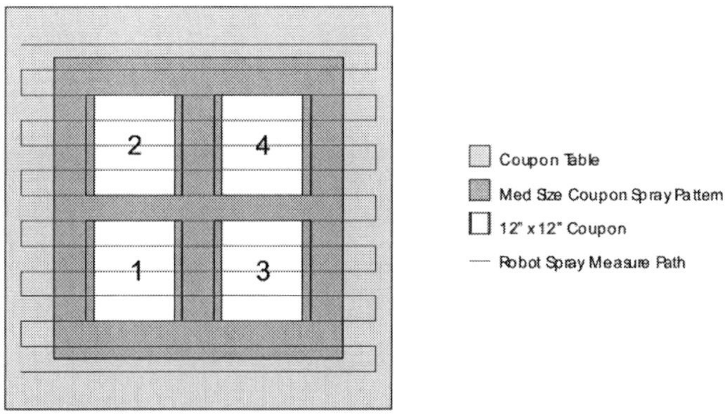

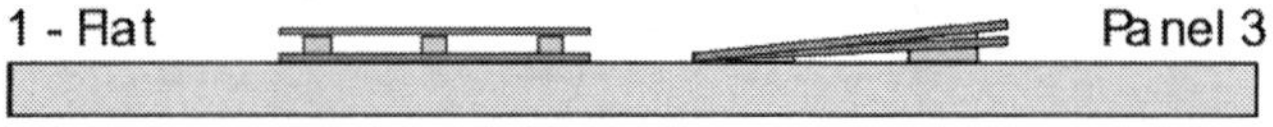

22

THz B-Scans
15 Applications of Form

PICOMETRIX
an API company

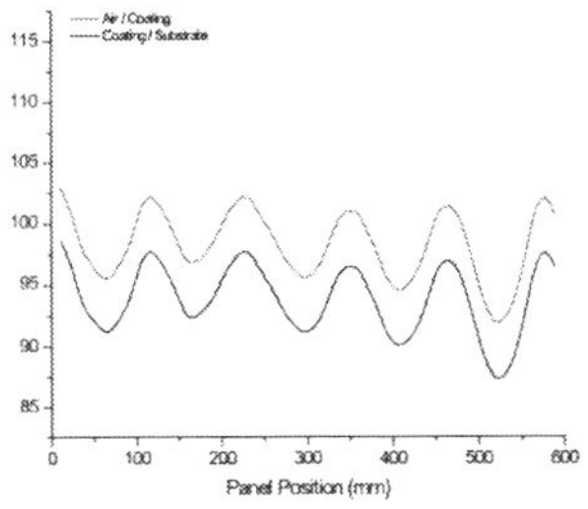

3 Passes

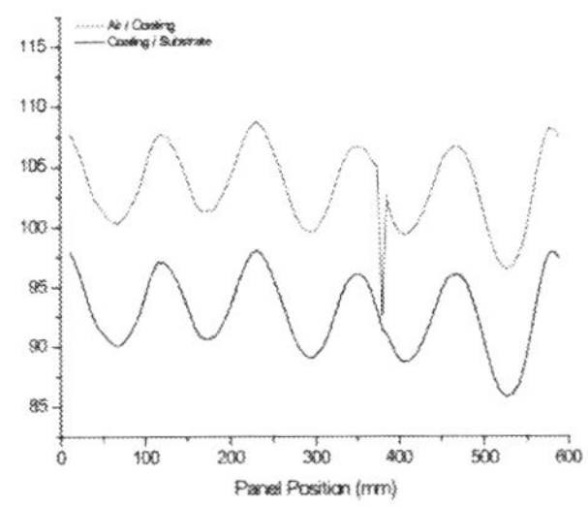

9 Passes

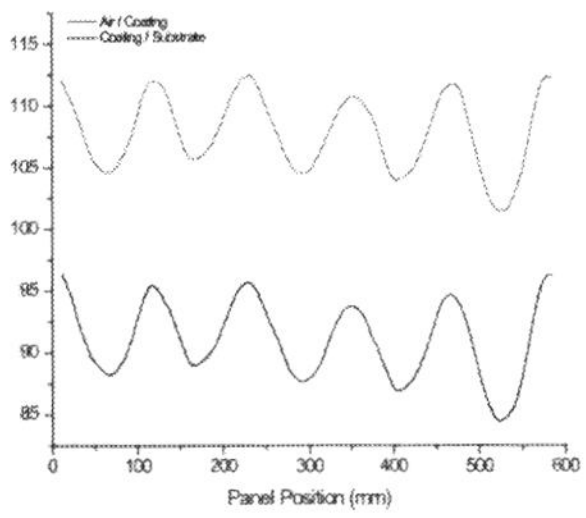

15 Passes

- Data illustrates sinusoidal vertical motion of robot boom
- Display exaggerates motion due to high value for speed of light = > actual displacement is **only a few mils**
- Coating thickness measurement is the difference => not affected by motion.

23

Panel Tilt Does Not Affect Difference in Time of Flight

PICOMETRIX
an API company

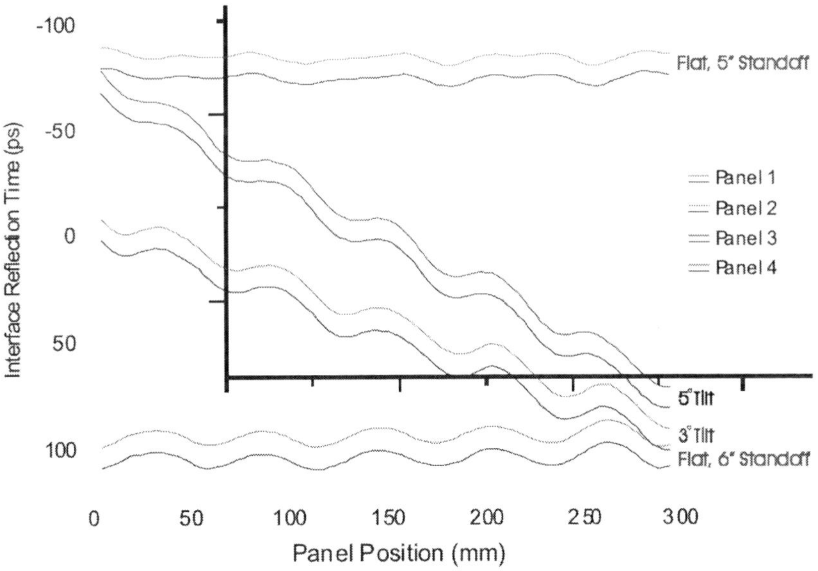

24

Lateral Scan
15 Applications of Form

PICOMETRIX® an API company

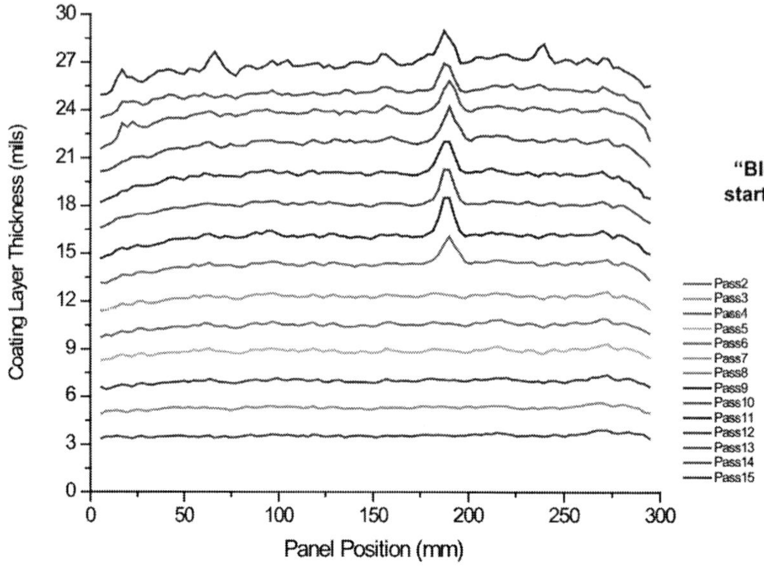

"Blob" appears starting at Pass 8

25

Example Form Application Profile

PICOMETRIX® an API company

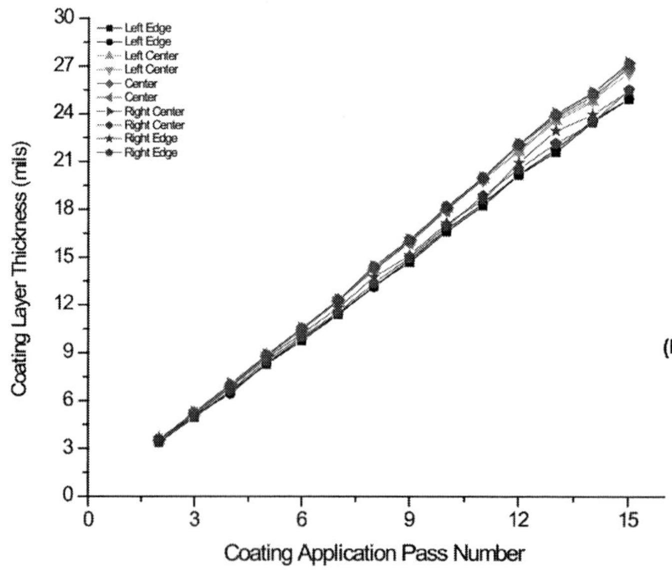

Build slightly Thinner at edges (Blue, Black traces)

26

Very Thick Coating Profile

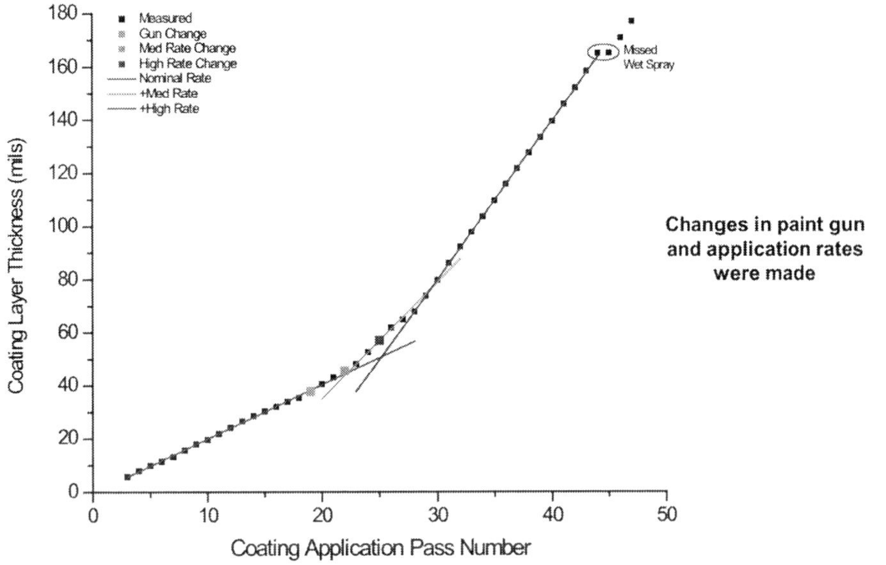

27

Detection of Under-Thickness

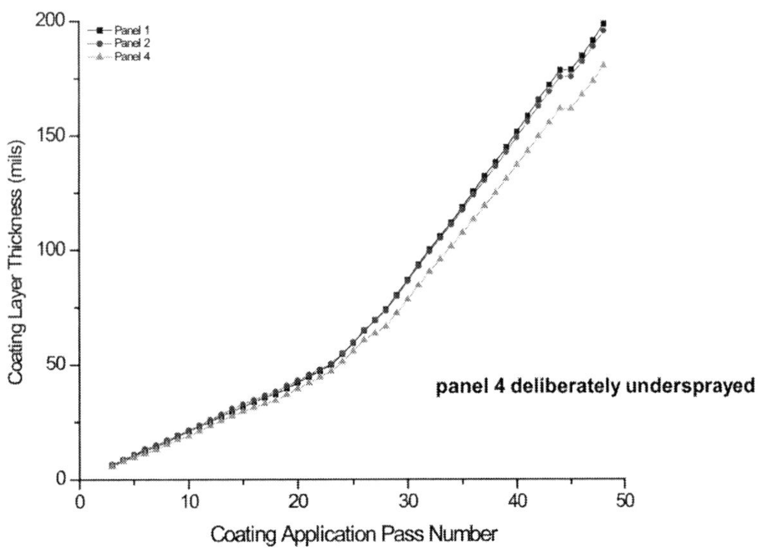

28

Monitor Cure Kinetics: Prediction Factors

PICOMETRIX
an API company

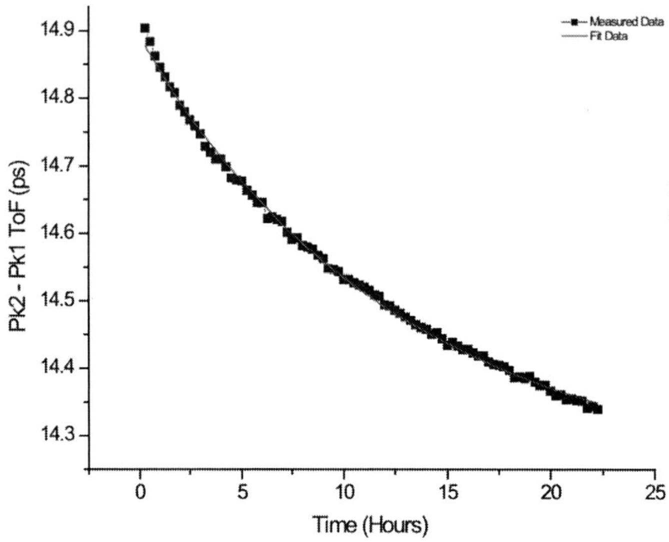

3.5% change over 22 hours

29

Insignificant Dependence on Spray Environment (Cured Prediction Factors)

PICOMETRIX
an API company

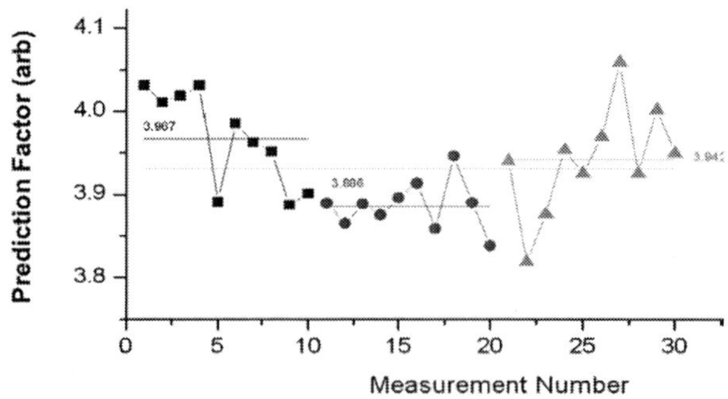

Nominal Environment
Temperature: 75 ± 5°F
Humidity: 60 ± 5% RH
Low Humidity, Low Temperature
Temperature: 65 ± 5°F
Humidity: 45 ± 5% RH
High Humidity, Low Temperature
Temperature: 65 ± 5°F
Humidity: 70 ± 5% RH

Calibration tightly grouped regardless of environment
< 1% total error for single form if environment ignored

(Also, no significant difference between fresh and >8 month old material)

30

PICOMETRIX
an API company

Concluding Remarks

31

PICOMETRIX
an API company

Transition of Technology

- Measurements have been demonstrated on airframes of interest.
- Technology is transitioning for production and process improvement validation testing.
  - Engineering data collection
  - Evaluation of ROI
  - Retained data for process improvement

32

Measurement Modes

PICOMETRIX an API company

- Mode 2 – Spray Off
  - Robot drives sensor to series of airframe OML Q/A measurement points using custom path.
  - Displays and records predicted thickness of coating.
  - Shows areas low, in tolerance and high.
- Mode 1 – Spray On
  - Sensor watches while production spray path runs.
  - Sensor reports thickness at the locations it is pointing along the spray path.
  - Displays and records predicted thickness of coating (if you were to stop now).
  - Predicts number of passes to completion.

33

PICOMETRIX an API company

Thank You

Acknowledgements:

The authors would like to thank Keith Pfeiffer of Aerobotix; Madeleine Low and Roy Thompson of Northrop Grumman Aerospace Systems; and Jeff White of Picometrix for current and previous efforts on this project.

34

Automated Composite Structure Inspection System (ACSIS) for Fiber Placement

AeroDef Manufacturing Summit and Exposition

Air Force Research Lab – Amber Davis
NCDMM – Jim Fisher
Ingersoll Machine Tools – Chris Esser

Defense-Wide Manufacturing Science & Technology (DMS&T) Program

February 25, 2014

INGERSOLL Machine Tools

Distribution Statement C: Distribution authorized to US Government agencies and their contractors dated December 18, 2013. Other requests for this document shall be referred to AFRL/RXMP, WPAFB OH 45433-7739.

Project Team Chart

Funded through Defense Wide Manufacturing Science and Technology (DMS&T) Program – 2012 IBIF

Managed by Air Force Research Laboratory (AFRL)
Amber Davis – Materials Engineer, PM for Effort

NCDMM NATIONAL CENTER FOR DEFENSE MANUFACTURING AND MACHINING

Prime Contractor: National Center for Defense Manufacturing and Machining (NCDMM)
Jim Fisher – Director of Operations, PM
Randy Gilmore – Chief Technology Officer
Brian Schmidt – Project Engineer

INGERSOLL Machine Tools

Subcontractor: Ingersoll Machine Tools
Dr. Tino Oldani – President & CEO
Chris Esser – Project Technical Lead
Clarissa Hennings – Business Development Manager

Distribution Statement C: Distribution authorized to US Government agencies and their contractors dated December 18, 2013. Other requests for this document should be referred to AFRL/RXMP, WPAFB OH 45433-7739.

2

Industry Advisory Board

ATK	Bell Helicopter	General Dynamics
Barrett Milenski	Paul Oldroyd	Paul Hauwiller
HITCO Carbon Composites: Les Cohen	Jeff Hendrix	Lockheed Martin: Bob Koon, Keely Povlitz
Northrop Grumman: Brent Meredith, Dan Worsham, Doug Decker, Jason Leon	Renegade Materials Corporation: Laura Gray	SCRA: Danna Kelley
Sikorsky: Tom Carstensen	U.S. Army: Jon Schuck	Chris Coughlin

Distribution Statement C: Distribution authorized to US Government agencies and their contractors dated December 18, 2013. Other requests for this document should be referred to AFRL/RXMP, WPAFB OH 45433-7739.

3

Ingersoll AFPM

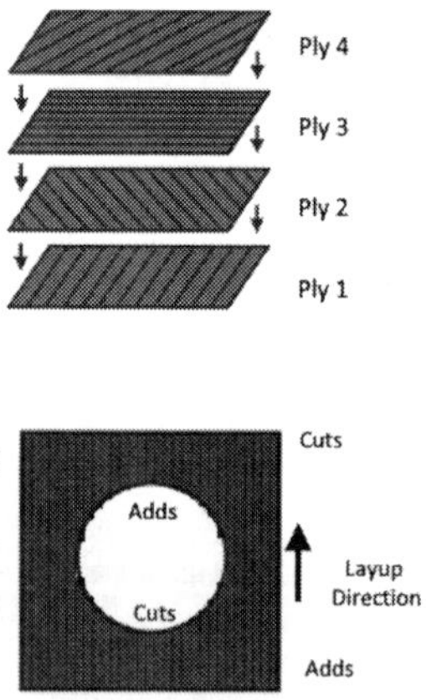

Distribution Statement C: Distribution authorized to US Government agencies and their contractors dated December 18, 2013. Other requests for this document should be referred to AFRL/RXMP, WPAFB OH 45433-7739.

4

Composite Structures

- AFP is typically used for aerospace structures, where tight tolerances and high quality are essential
- As such, defects in the composite material during layup must be identified and repaired

5

Problem

- Even though machines are reliable, potential problems such as gaps, twisted tows, wrinkles, etc. can occur following layup which requires stringent inspection.
- Current inspection methods require the machine to be interrupted to allow 100% manual, visual inspection and measurements to be performed.

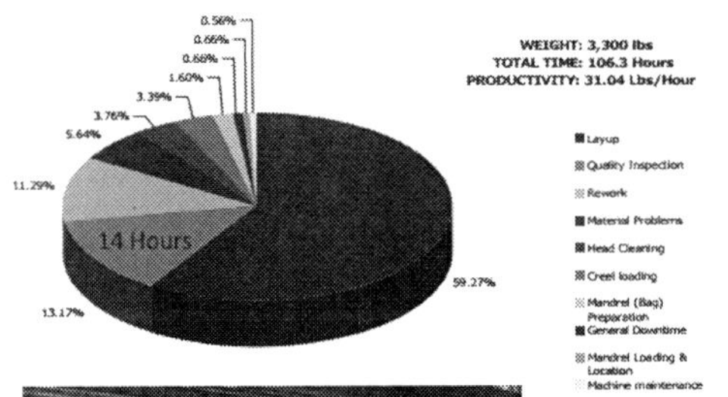

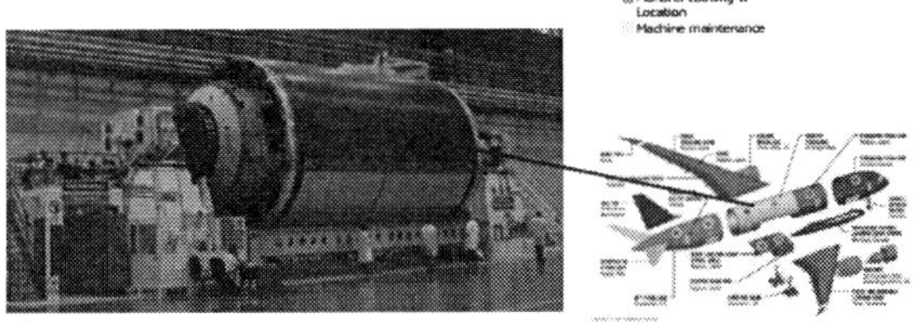

Manual Inspection Limitations

- Inspection must occur while AFPM is stopped
  - Reduces production rate
- Defect detection depends on inspector alertness
- Defect logs (if any) are manually generated and maintained
  - Limited detail and information
  - Process can continue without logging defect
- High overhead cost of keeping inspectors onsite

6

Objective

- Develop a system to automate the process of inspecting fiber placement for fabricating low cost composite structures.
  - Reduce inspection time
  - Increase reliability of defect detection
  - Maintain detailed electronic manufacturing records for aerospace certifications
  - Reduce paper for quality part tracking through digital collection of data

Automate Inspection using Imaging System Technology

Range of Defects →

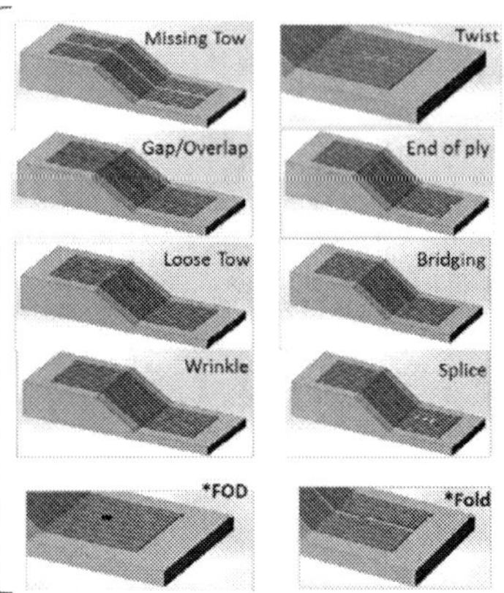

Distribution Statement C: Distribution authorized to US Government agencies and their contractors dated December 18, 2013. Other requests for this document should be referred to AFRL/RXMP, WPAFB OH 45433-7739.

7

Simulated Aerospace Part

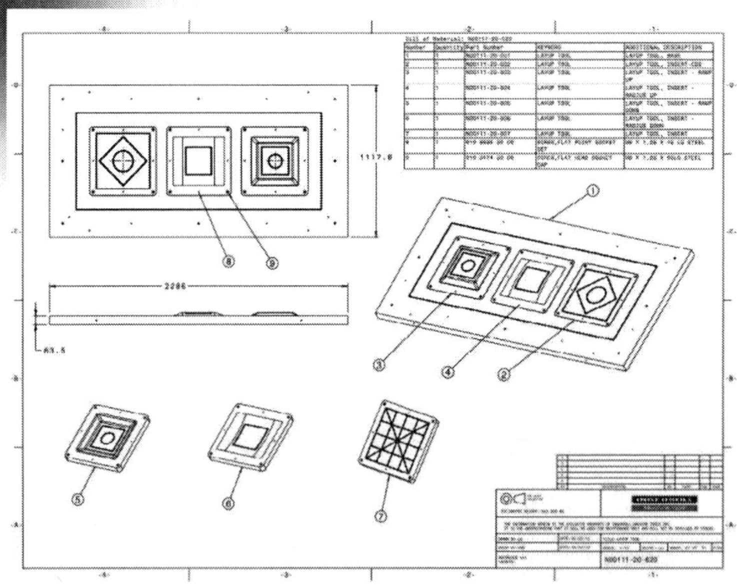

Part designed to have geometry that would be found on a typical aerospace structure

Distribution Statement C: Distribution authorized to US Government agencies and their contractors dated December 18, 2013. Other requests for this document should be referred to AFRL/RXMP, WPAFB OH 45433-7739.

8

Inspection Hardware Mounting

Imaging System Selections

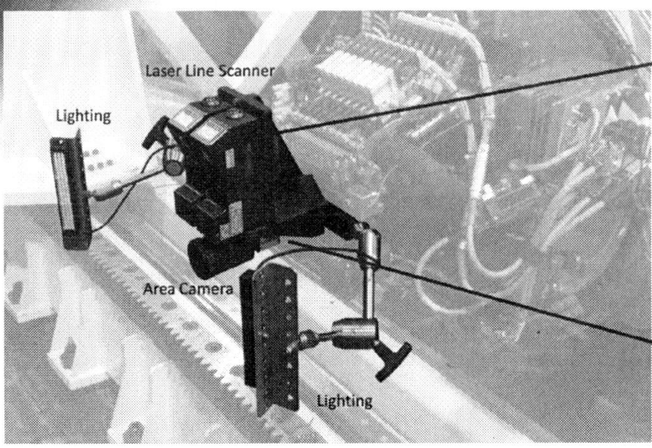

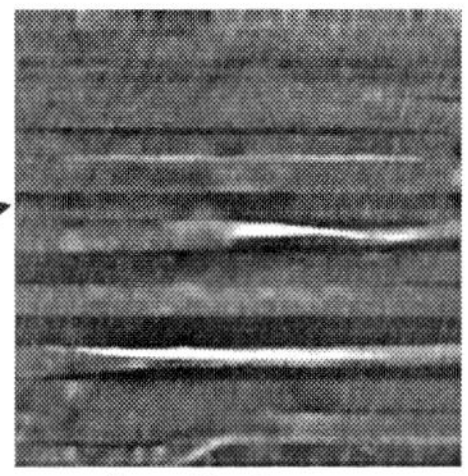

Laser Line Scanner

Area Camera

Defect Detection - Camera

Laser Scanner Data Gathering

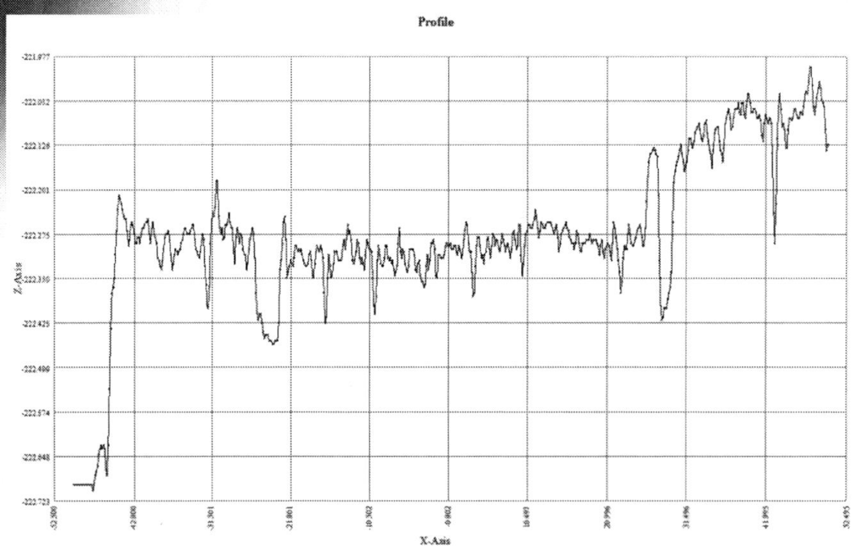

Laser Scanner Data Gathering

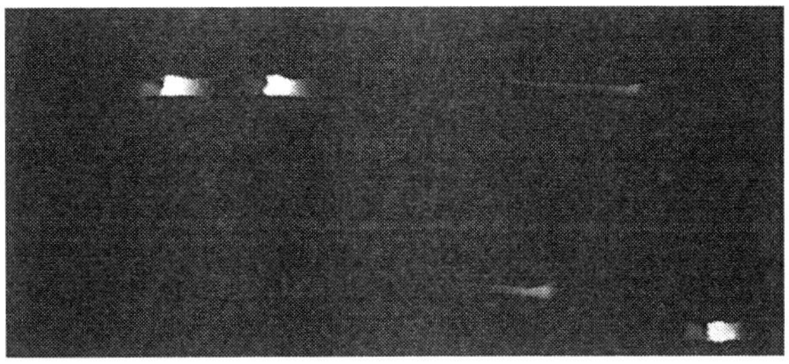

Individual Profiles

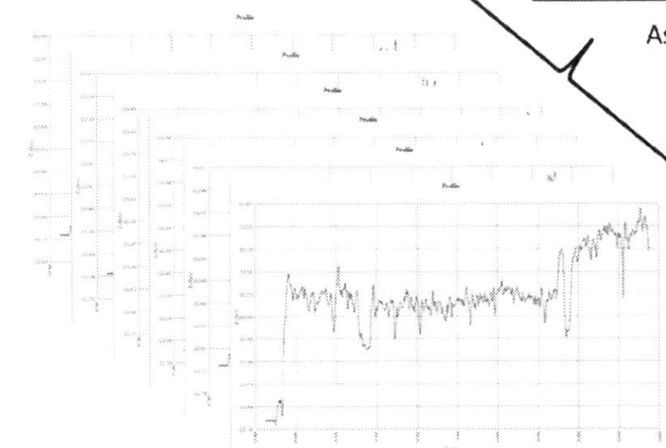

Assembled Profiles form height 'image'

Stack profiles

Distribution Statement C: Distribution authorized to US Government agencies and their contractors dated December 18, 2013. Other requests for this document should be referred to AFRL/RXMP, WPAFB OH 45433-7739.

13

Defect Detection – Raw Image

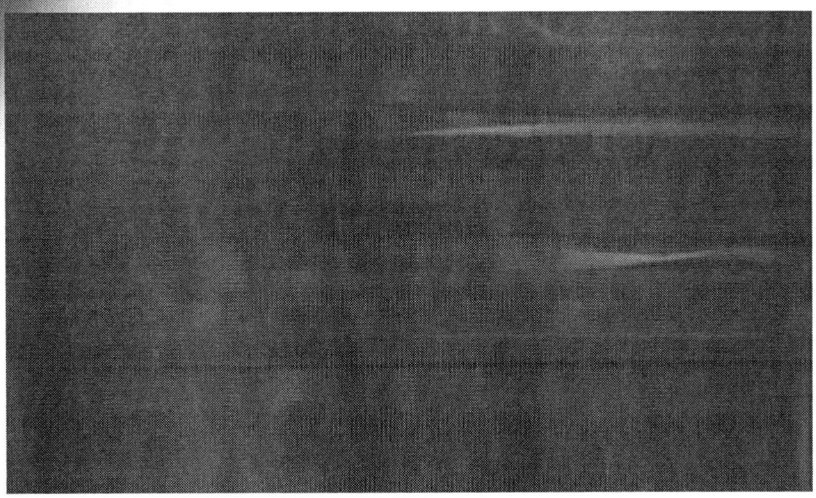

Raw scan of a flat plate. 16 x ¼" tows. 13 plies thick

Distribution Statement C: Distribution authorized to US Government agencies and their contractors dated December 18, 2013. Other requests for this document should be referred to AFRL/RXMP, WPAFB OH 45433-7739.

14

Defect Detection – Ply Noise

Perform filtering to eliminate the noise introduced by layers below the top ply

Defect Detection – Filtered

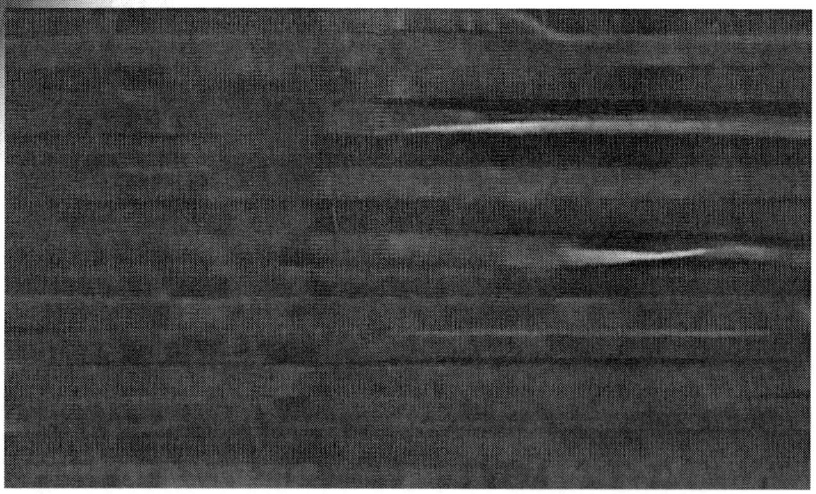

Perform further filtering to bring out features in the top ply

Defect Detection – Contrast

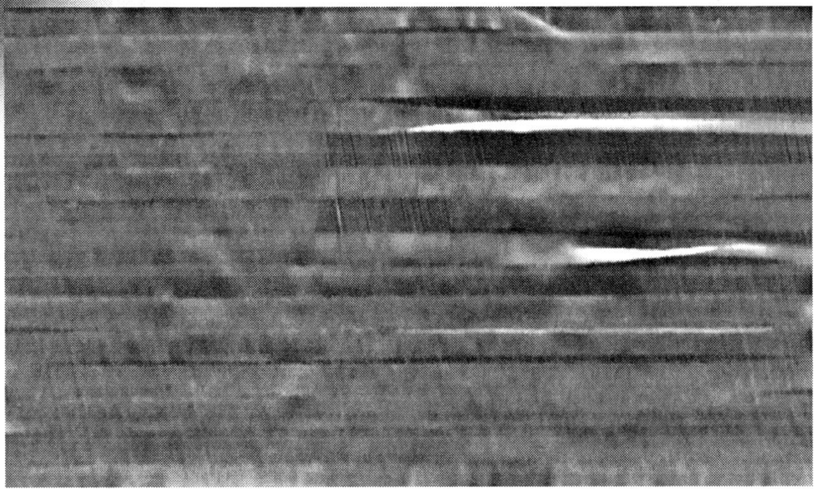

Increase the contrast of the image to bring out the details of the ply

Defect Detection – Areas of Interest

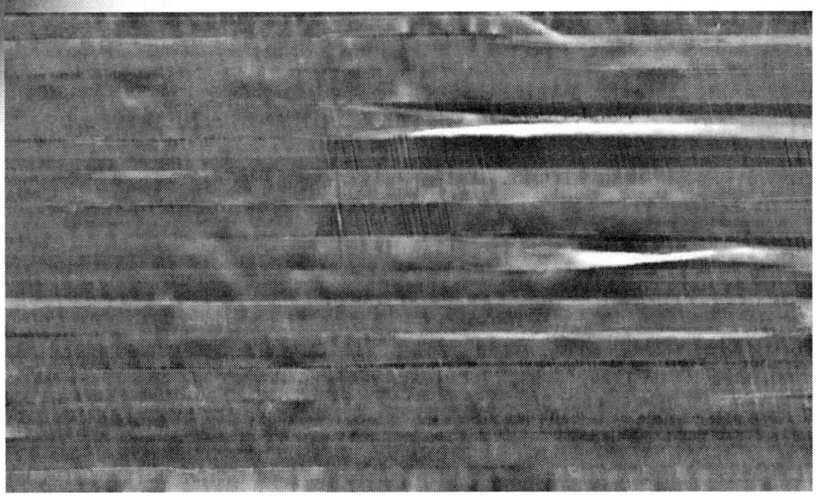

Find areas that could be defects, areas of interest

Defect Detection – Identification

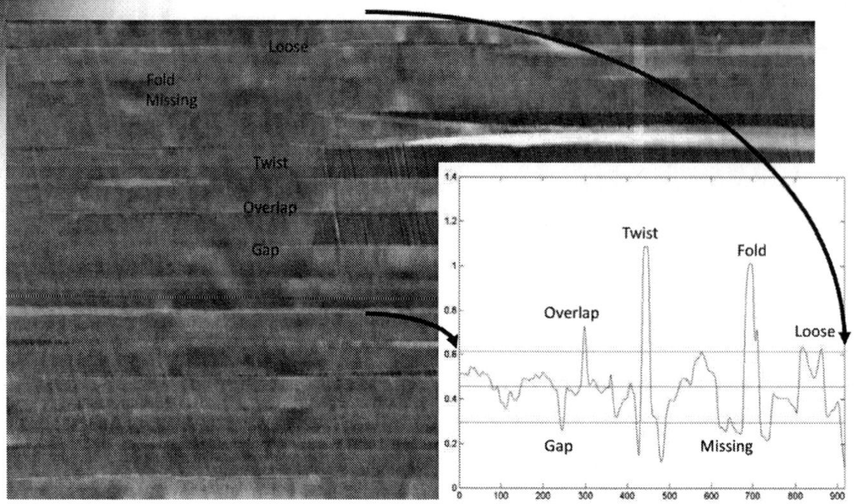

With areas of interest, perform vertical scan for defect identification

Defect Log

A log will be kept of each defect found

- Part ID
- Course
- Time Stamp
- Defect Type
- XYZ Position
- Defect Measurements
- Ply
- Repair Action

Ply List | Detected Defects

ID	Project ID	Sync Time Stamp	Defect Type	Actual Type	Course Number	Defect Num	X Position	Y Position	Z Position	Repair Action	User ID	Inspector ID
[illegible]	1	8:33:45	Twisted Tow	Twisted Tow	3	1	1	1	.8	Repaired	1	1
3	1	8:35:01	Add/Cut Positio	Add/Cut Positi	11	1	-16.5	14.5	0	Disregarded	1	1
4	1	8:35:12	Splice	Splice	11	2	-34	12	0	Disregarded	1	1
5	2	9:20:13	FOD	FOD	2	1	-18	6	1	Repaired	1	1
6	2	9:22:27	Missing Tow	Add/Cut Positi	13	1	-30.5	-6.5	1	Repaired	1	1
7	3	10:10:10	Bridging	Not A Defect	1	1	20	7	0	Not A Defect	2	3
8	3	10:11:12	Bridging	Bridging	6	1	20	0	0	Repaired	2	3
9	3	10:12:55	Loose Tow	Loose Tow	17	1	28.5	-4	1	Repaired	2	3
10	3	10:13:03	Wrinkle	Wrinkle	17	2	-17	-4.5	1	Repaired	2	3
11	3	10:14:48	Splice	Twisted Tow	21	1	-10	12	0	Rapaired	2	3
12	4	16:22:22	Gap	Gap	2	1	0	12	0	Disregarded	2	3
13	4	16:22:45	Splice	Splice	2	2	-10	-12	0	Repaired	2	3
14	4	16:24:13	Wrinkle	Wrinkle	14	1	23	12	0	Repaired	2	3
15	4	16:24:16	FOD	FOD	14	2	0	0	0	Repaired	2	3
16	4	16:24:21	FOD	FOD	14	3	-14	-12	0	Repaired	2	3
17	4	16:24:29	Twisted Tow	Twisted Tow	14	4	-1	-1	.8	Repaired	2	3
18	4	16:27:12	Add/Cut Positio	Add/Cust Posit	17	1	-1.5	-14	0	Repaired	2	3
19	4	16:27:32	Wrinkle	Twisted Tow	17	2	-2	2	.8	Repaired	2	3
20	4	16:29:15	Splice	Not A Defect	21	1	34	12	0	Not A Defect	2	3
21	4	16:33:59	Fold	Fold	24	1	35.5	-5	0	Repaired	2	3
(New)												

Operator Interface

The Operator GUI is the central hub where the machine operator interacts with the Inspection System. Operator will be able to:

- Log repairs conducted
- Control laser projection
- View past defects
- Control the inspection process

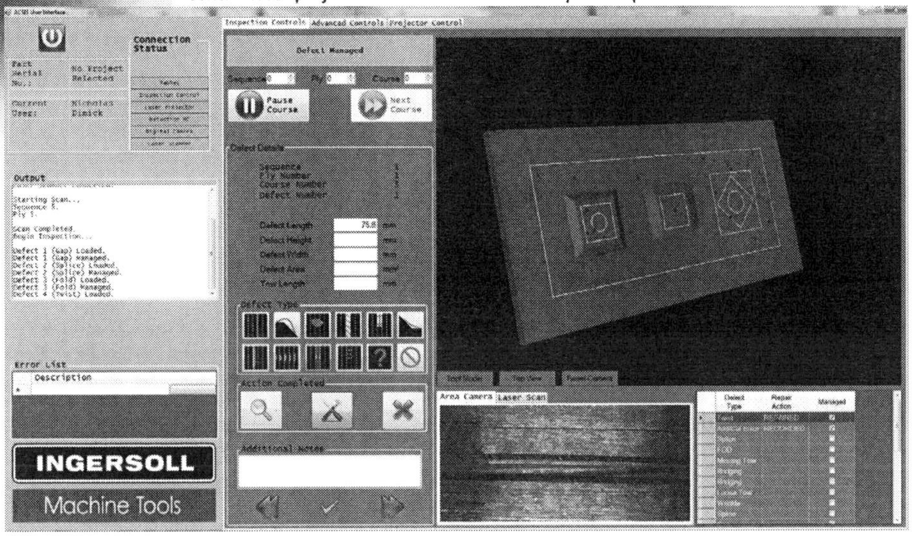

Distribution Statement C: Distribution authorized to US Government agencies and their contractors dated December 18, 2013. Other requests for this document should be referred to AFRL/RXMP, WPAFB OH 45433-7739.

21

Defect Location Projection

Defect location will be displayed to operator by a laser projector

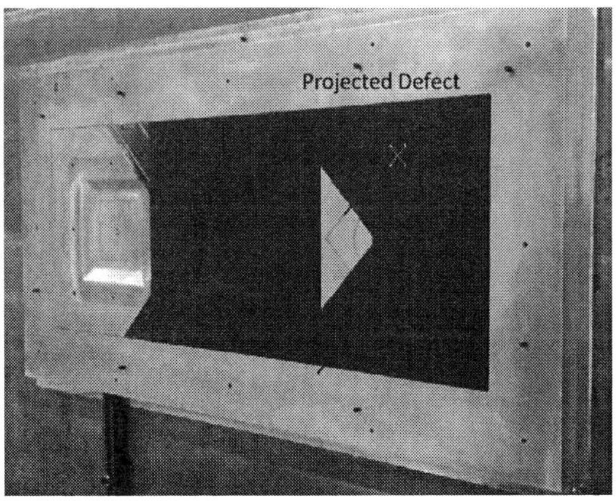

Distribution Statement C: Distribution authorized to US Government agencies and their contractors dated December 18, 2013. Other requests for this document should be referred to AFRL/RXMP, WPAFB OH 45433-7739.

22

Wireless Pendant

A wireless pendant is also being developed to allow the operator to view defect information about the ply, control the laser projector, and take notes about defects while moving freely around the part

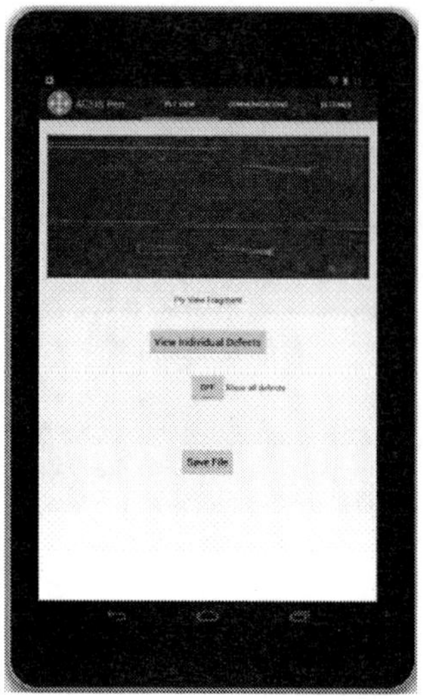

Distribution Statement C: Distribution authorized to US Government agencies and their contractors dated December 18, 2013. Other requests for this document should be referred to AFRL/RXMP, WPAFB OH 45433-7739.

Repair Workflow

AFP will layup a ply

With the ply completed, the inspection system will scan the entire ply

A list of all defects is created. The operator will move to the part to begin repair.

The operator will repair a defect

The operator will log the defect as managed

All defects have been managed

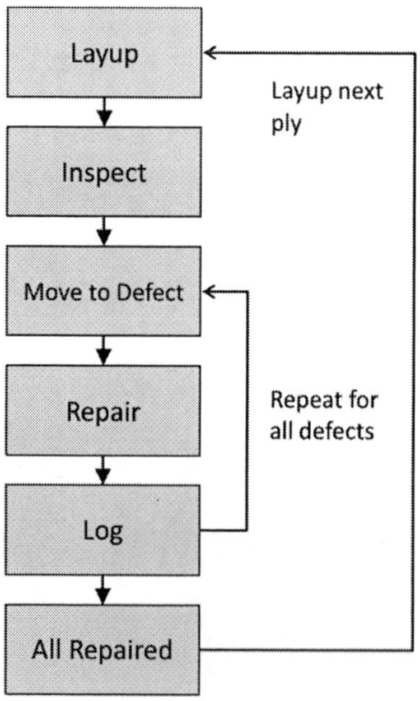

Distribution Statement C: Distribution authorized to US Government agencies and their contractors dated December 18, 2013. Other requests for this document should be referred to AFRL/RXMP, WPAFB OH 45433-7739.

Defect Cloud

Defect information will be used to overlay defects onto a 3D representation of the part. This information can be used to identify patterns in defect locations

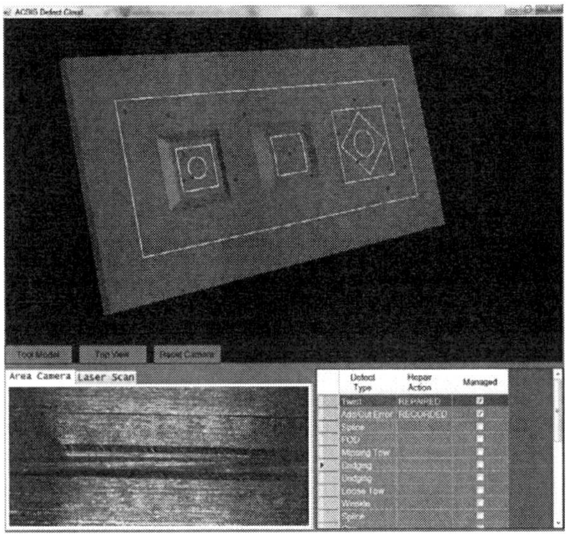

Current State of ACSIS

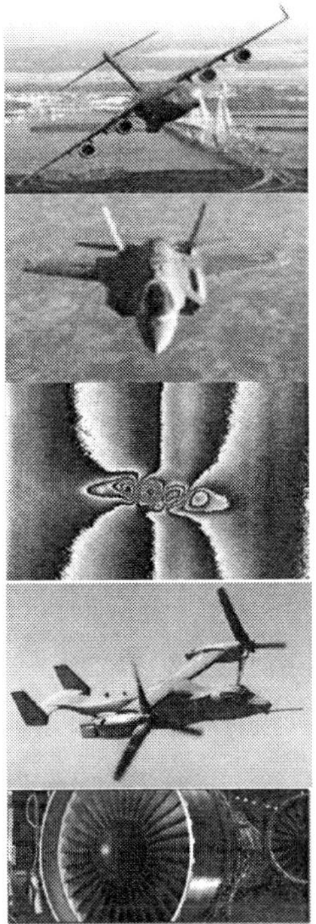

Shearography Nondestructive Testing of Aerospace Structures

Presented By
Flynn Spears
Strategic Sales
Laser Technology, Inc

February 26th 2014

Laser Technology Inc.
Aerospace NDE Systems www.LaserNDT.com

Shearography Nondestructive Testing

Technical Input for this presentation provided by

Laser Technology Inc.
Mr. John Newman
President
Laser Technology, Inc

Laser Technology Inc.
Aerospace NDE Systems www.LaserNDT.com

What is Shearography NDT?

- Vibration resistant digital imaging laser interferometer
- Measures changes of test part surface to < 5nm.
- Images material defects when combined with an engineered stress.
- Can be done with portable and large production systems.

Features:

- Non-Intrusive, Real Time NDT
- Electronic Images in JPEG, TIF or PMF format
- Detects and measures:
  - Disbonds
  - Delaminations
  - Core damage
  - Impact Damage
  - Porosity
  - Wrinkled Fibers

Why is Shearography NDT Important?

- Shearography NDT delivers large increase in aerospace composite manufacturing productivity at a reduced cost.
- Tests parts in excess of 3X to 4X times faster than UT C-Scan and X-Ray.
- Non-Contact, Non-Contaminating part inspection for porosity, disbonds, delamination, core and impact damage
- Test parts during build up- allowing repair or scrapping part at lowest possible cost.
- Offers unique engineering solutions for advanced materials and structures.

Laser Technology Inc.
Aerospace NDE Systems Copyright © Laser Technology Inc. 2010 www.LaserNDT.com

Shearography NDT Theory

How it works...

Laser Technology Inc.
Aerospace NDE Systems *Copyright © Laser Technology Inc. 2010* *www.LaserNDT.com*

Laser Methods Application

- **Industrial Applications**
  - Measuring
  - Straight Line Marking
  - Laser Peening of Metal, Hardening, Cutting
  - Alignment
  - Non-Destructive Testing
- **Medical Applications**
  - Laser Scalpels
  - Ablation
  - Photo Chemical Reactions
  - Eye Surgery
- **Military Applications**
  - Range Finding
  - Target Acquisition
  - Directed Energy Weapons
  - Non-Lethal Weapons
  - Isotope Separation
- **Consumer Products**
  - CD and DVD Players and Recorders
  - Laser Printers
  - Bar Code Scanner
  - Laser Pointers

Laser Technology Inc.
Aerospace NDE Systems *Copyright © Laser Technology Inc. 2010* *www.LaserNDT.com*

Laser Wavelengths

- Wavelength Classification
  - Ultraviolet 180 Nm - 400 Nm
  - Visible 400 Nm - 700 Nm
  - Infrared 700 Nm - 1 mm

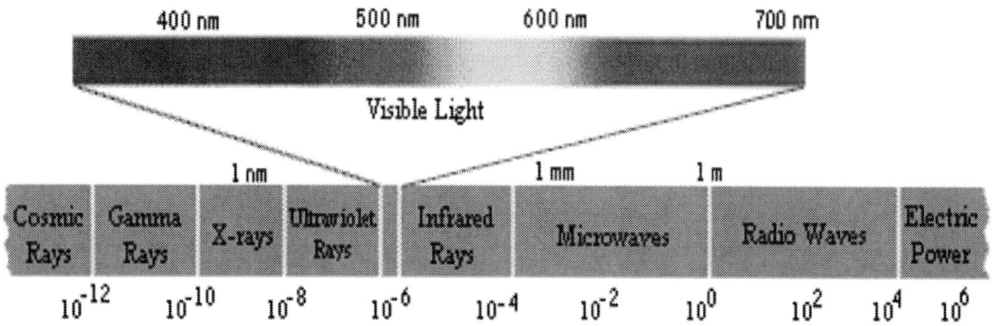

Portable Shearography Systems are Class 3a (3r)

Laser Technology Inc.

Aerospace NDE Systems

7 *www.LaserNDT.com*

Shearography NDT System Schematic Diagram

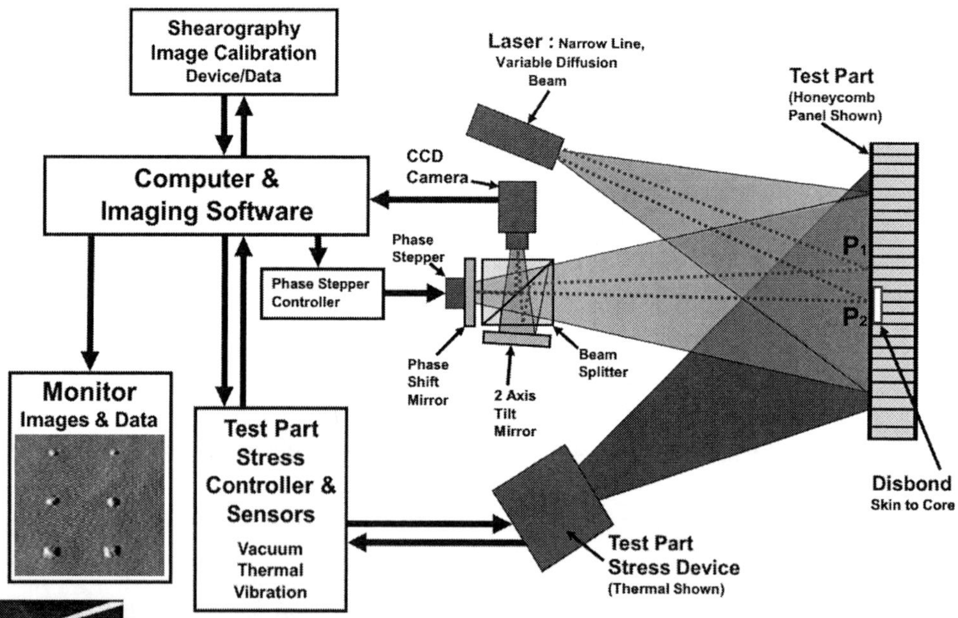

Laser Technology Inc.

Aerospace NDE Systems

www.LaserNDT.com

CCD Pixel voltage output changes between two strain states caused by a phase shift Δ due to the presence of an disbond or other anomaly.

Phase Difference Strain State 1 = Φ
Phase Difference Strain State 2 = Φ+Δ

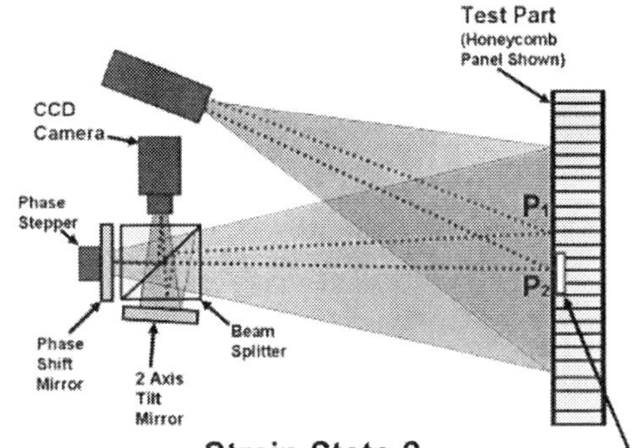

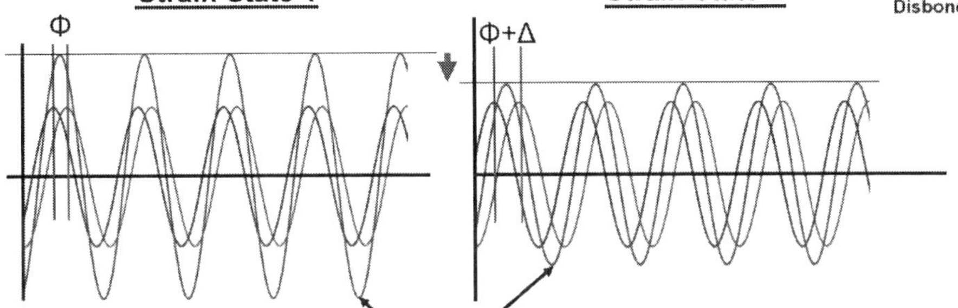

Laser Technology Inc.
Aerospace NDE Systems

www.LaserNDT.com

Images of the Part with Changing Strain States

Images at Strain States 1 & 2 are Combined for a Phase Map which is then "unwrapped"

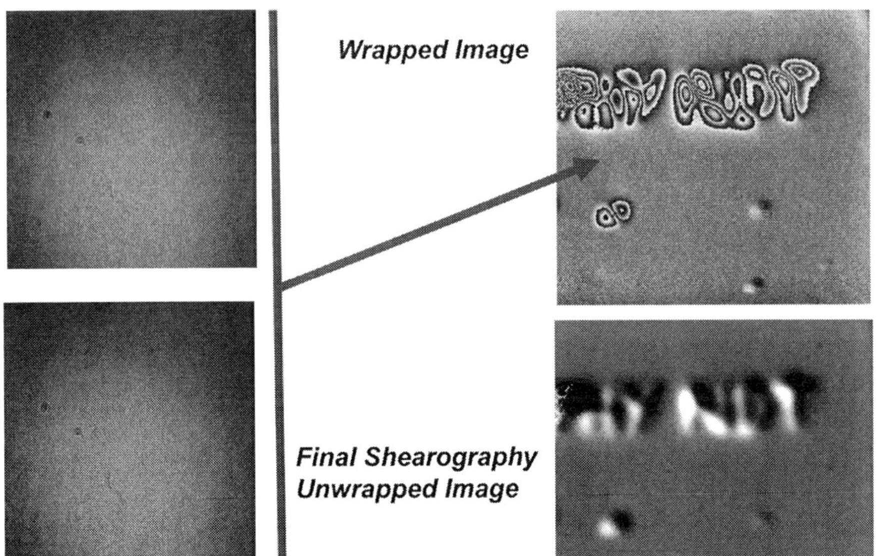

Laser Technology Inc
Aerospace NDE Systems

www.LaserNDT.com

Fringe Presentation

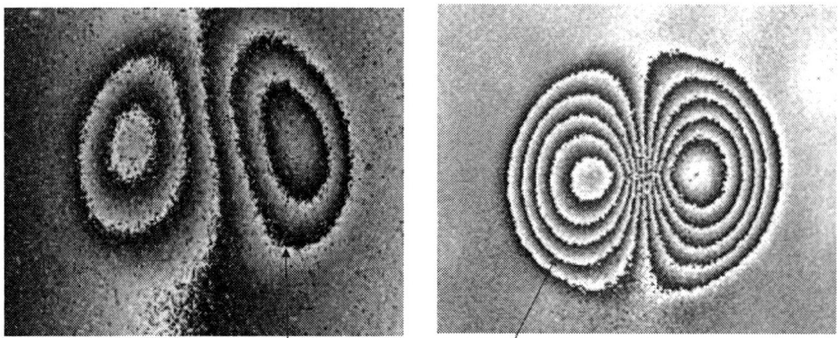

Each fringe is a line of Iso-Strain

1 fringe = 0.26 um (Microns) = ½ Wave Length of Light

Signal Interpretation

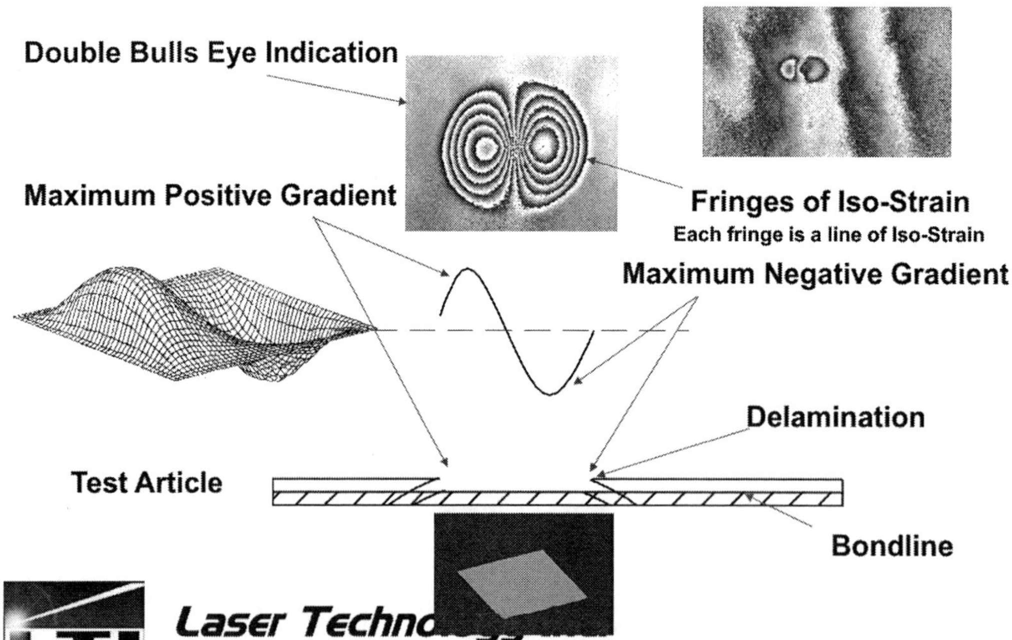

Practical Effect of Shear Vector

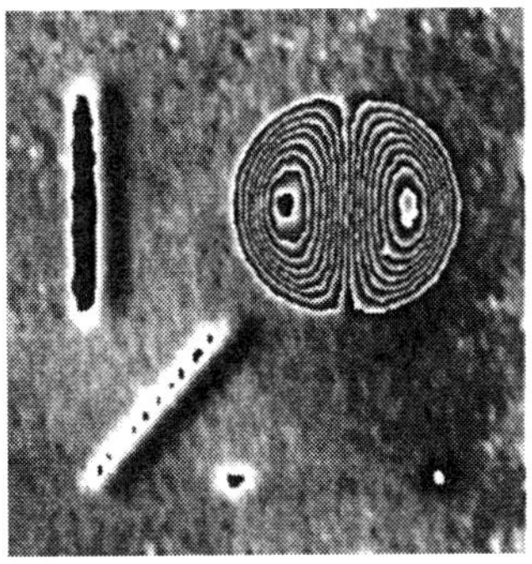

Horizontal Shearing

Vertical defect visible but
Only end points of horizontal
Horizontal indication visible.

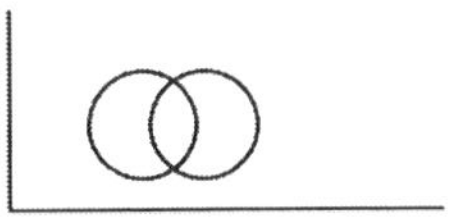

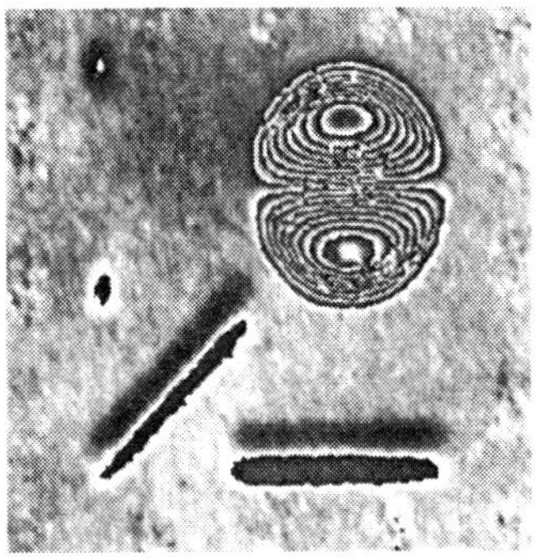

Vertical Shearing

Horizontal defect visible
but only end points of
vertical indication visible.

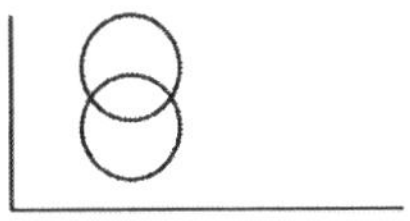

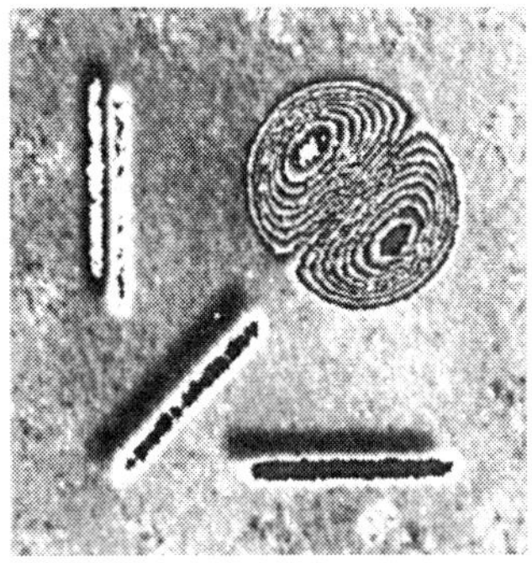

Shearing at 45 degrees

Both horizontal and vertical
Indications visible.

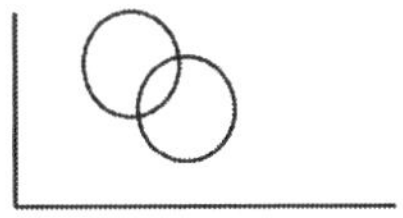

Laser Technology Inc.
Aerospace NDE Systems www.LaserNDT.com

Stressing Methods…

Laser Technology Inc.
Aerospace NDE Systems Copyright © Laser Technology Inc. 2010 www.LaserNDT.com

Methods of Stressing

- Thermal Excitation
- Pressure Excitation
- Vacuum Excitation
- Acoustic Excitation
- Vibration Excitation
- Mechanical loading
- Microwave

Laser Technology Inc.
Aerospace NDE Systems *www.LaserNDT.com*

Shearography Test Results By Stress Method

Thermal Shearography
Aluminum Honeycomb Panel

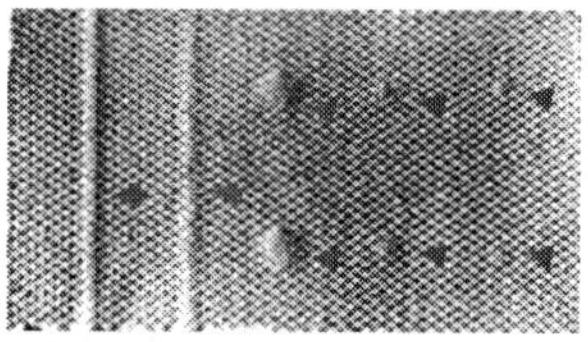

Pressure Shearography
COPV

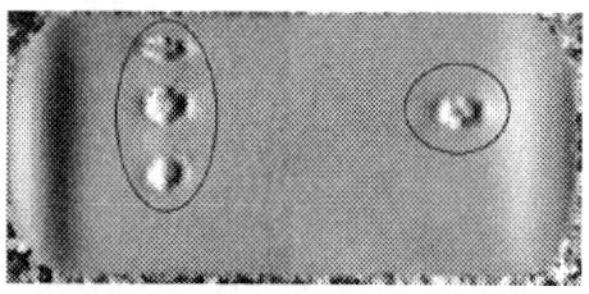

Vacuum Shearography
Composite/Nomex Honeycomb

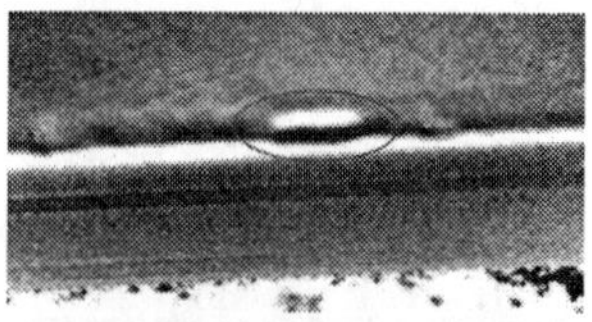

Acoustic Shearography
Foam Cryogenic Fuel Tank TPS

Laser Technology Inc.
Aerospace NDE Systems *www.LaserNDT.com*

Thermal Stressing

- **Stressing Mechanism:**
  - Hot Air Gun
  - Quartz Heat Lamps
- **Theory:**
  - Materials without anomalies display a uniform thermal expansion when heated. Anomalies at the surface or subsurface cause changes to the thermal expansion.
- **Detects:**
  - Damage, Disbonds, Delaminations

Laser Technology Inc.
Aerospace NDE Systems

www.LaserNDT.com

Thermal Stressing

THERMAL	*Degrees Temperature Change*	*1 to 30 ° F*	*<u>Laminate Panels</u>* *Impact Damage, Delamination, Wrinkled Fibers, Porosity, Embedded Foreign Material, Repairs* *<u>Sandwich Panel</u>* *Impact Damage, Skin-to-Core Disbond, Damaged Core, Foam-Foam Disbonds, Metal Core-Skin Disbonds, Repairs Kissing Disbonds, Skin Delaminations, Water* *<u>Resin Transfer Molded Composites</u>* *Resin Lean Areas, Porosity, Engine Stators, Vanes, Composite Fan Blades, Erosion Strip Bond* *<u>Cracks</u>* *Steel, Aluminum, Composites*

Laser Technology Inc.
Aerospace NDE Systems
Copyright © Laser Technology Inc. 2010
www.LaserNDT.com

Shearography Test Results By Stress Method

Thermal Shearography
Aluminum Honeycomb Panel

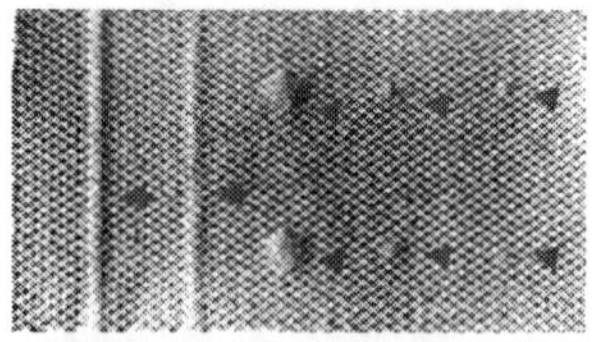

Pressure Shearography
COPV

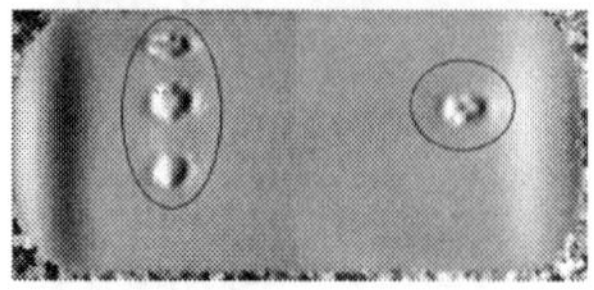

Vacuum Shearography
Composite/Nomex Honeycomb

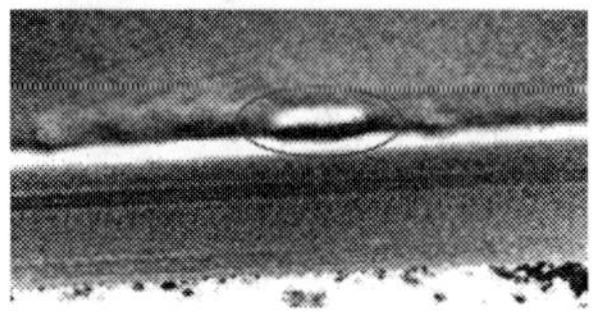

Acoustic Shearography
Foam Cryogenic Fuel Tank TPS

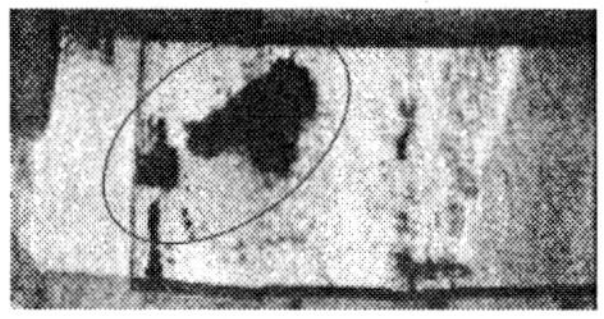

Pressure Stressing

- **Stressing Mechanism:**
  - Internal air or water pressure

- **Theory:**
  - Deforms surface due to structural anomaly.

- **Detects:**
  - Damage, Disbonds, Delaminations, Fiber bridging

Laser Technology Inc.
Aerospace NDE Systems *Copyright © Laser Technology Inc. 2010* *www.LaserNDT.com*

Pressure Stressing

PRESSURE	*PSI, KG/CM² Pressure Change*	*0.01 to 5,000+ PSID*	***COPV & Composite Rocket Motors*** *Impact Damage, Liner-Composite Disbonds, Porosity, Fiber Bridging* ***Pressure Vessels & Heat Transfer Structures*** *Metal Pressure Tanks, Liquid Propellant Rocket Engine Exit Cones, Thrust Ramps, Piping, Space Vented Core Honeycomb* ***Pressure Containing Structures*** *Pumps, Manifolds, Hollow Turbine Blades*

Laser Technology Inc.

Aerospace NDE Systems *www.LaserNDT.com*

Technical Approach: Shearography Detection of Strain Concentrations Due to Damage

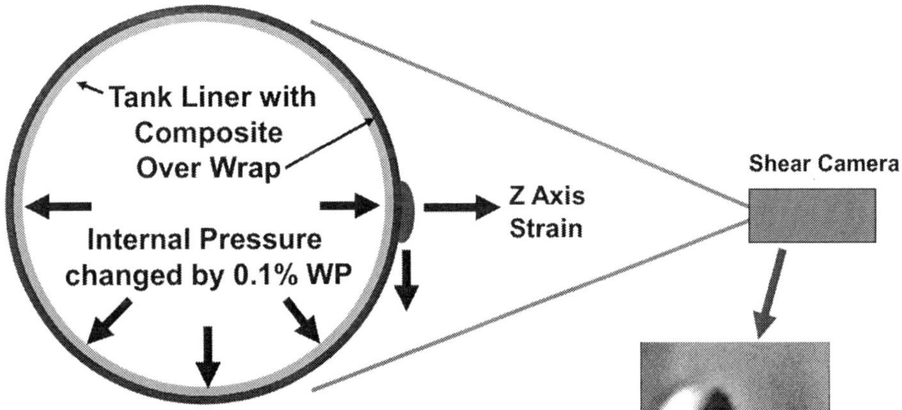

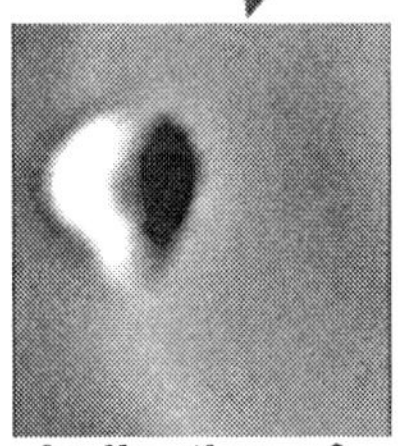

Indication of Impact Damage

- Increased pressure increases hoop and longitudinal strain.
- Z axis component of strain concentrations are detected by the shearography camera.
- As pressure increases, strain rates at damage sites are greater than good areas.

Laser Technology Inc.

Aerospace NDE Systems *www.LaserNDT.com*

18 Inch COPV Shearography

Impact Damage Test

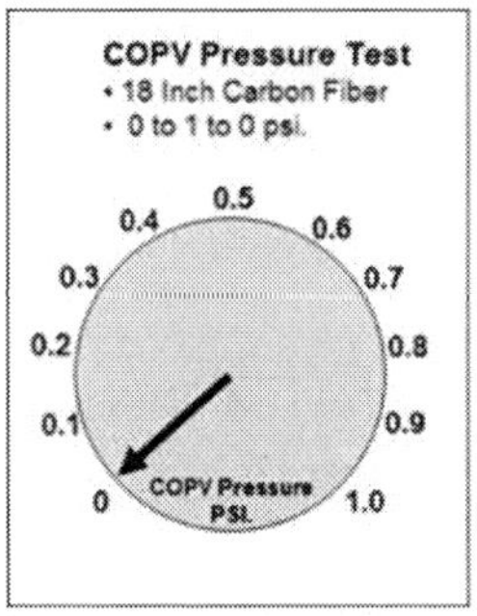

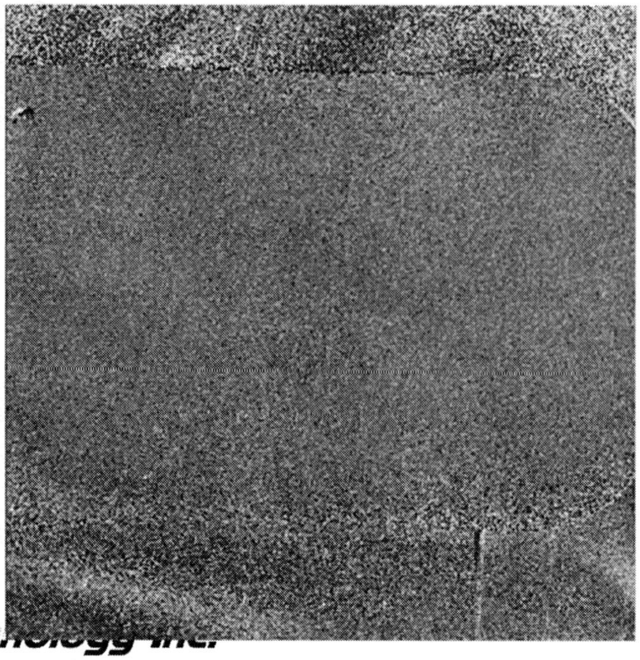

Aerospace NDE Systems *Copyright © Laser Technology Inc. 2010* *www.LaserNDT.com*

Shearography Test Results By Stress Method

Thermal Shearography
Aluminum Honeycomb Panel

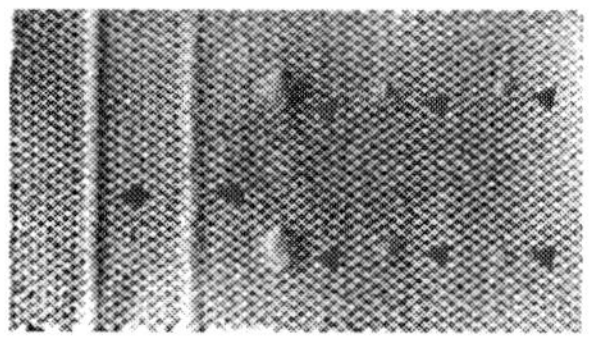

Pressure Shearography
COPV

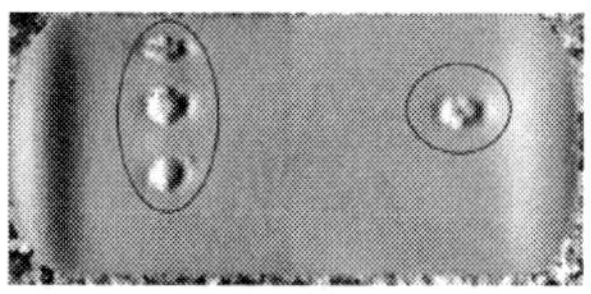

Vacuum Shearography
Composite/Nomex Honeycomb

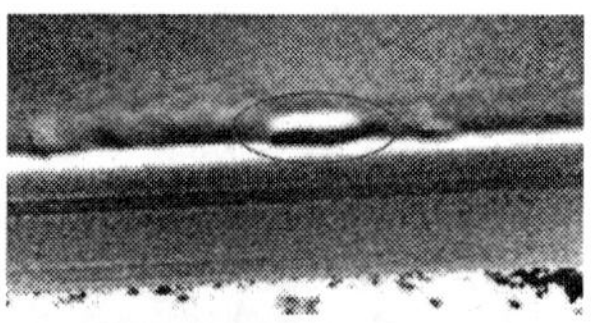

Acoustic Shearography
Foam Cryogenic Fuel Tank TPS

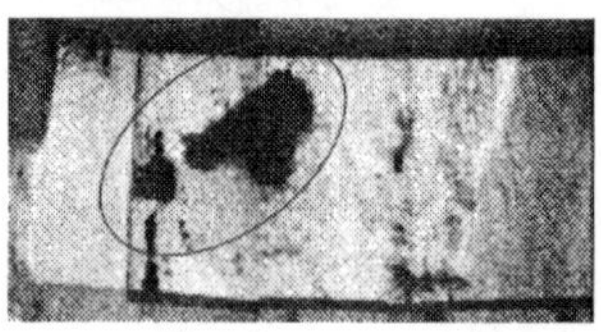

Laser Technology Inc.

Aerospace NDE Systems *Copyright © Laser Technology Inc. 2010* *www.LaserNDT.com*

Vacuum Stressing

- Stressing Mechanism:
  - Test Chamber
  - Vacuum Window
  - Portable Vacuum Shearography System
- Theory:
  - Reduced external air pressure on structure allows subsurface anomalies to expand and deform the surface of the part.
- Detects:
  - Damage, Disbonds, Delaminations, Crushed Core

Vacuum Stressing

PARTIAL VACUUM	*PSI, KG/CM² Pressure Reduction*	*-0.02 to -7.0 PSID.*	*Elastomers* *Coatings, Low Observables, Rubber and Plastic Voids, Disbonds, Tires, Solid Rocket Motor Liners, Rubber to Substrate Bond, Cork to Substrate Bond* *Sandwich Panels* *Impact Damage, Voids, Disbonds, Radomes, Aircraft Control Surfaces, Flaps, Air Brakes, Helicopter Blades, Turbine Engine Ducts, Laminated Wood Structures* *COPV* *COPV Fiber Bridging, Liner Disbonds*

Laser Technology Inc.
Aerospace NDE Systems Copyright © Laser Technology Inc. 2010 www.LaserNDT.com

Portable Vacuum Shearography System

Near Side and Far Side Disbonds Detected

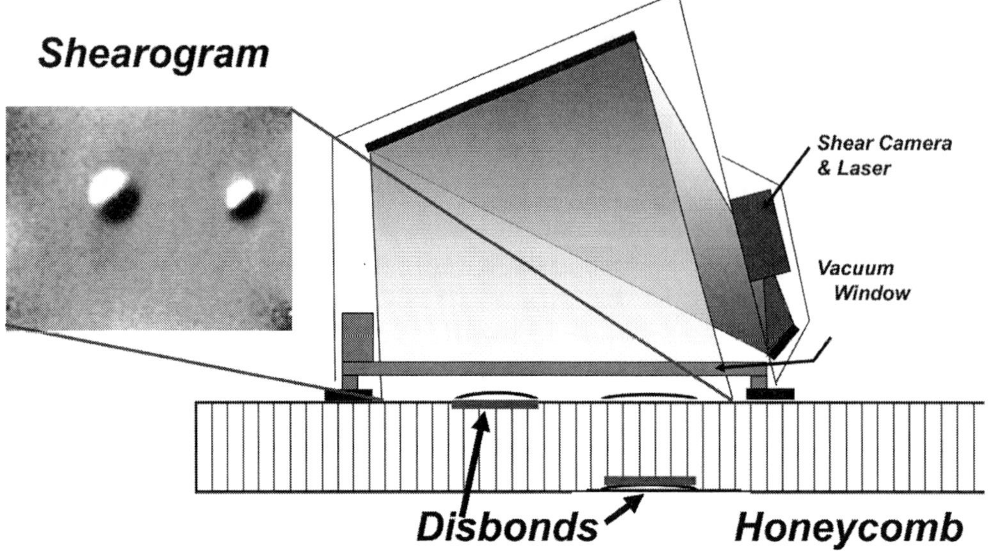

Laser Technology Inc.

Aerospace NDE Systems *www.LaserNDT.com*

Honeycomb Ultrasonic NDT STD.

Tested with Real-Time Shearography Vacuum Stress
Pressure Drop during test = Ambient to -0.5 psid

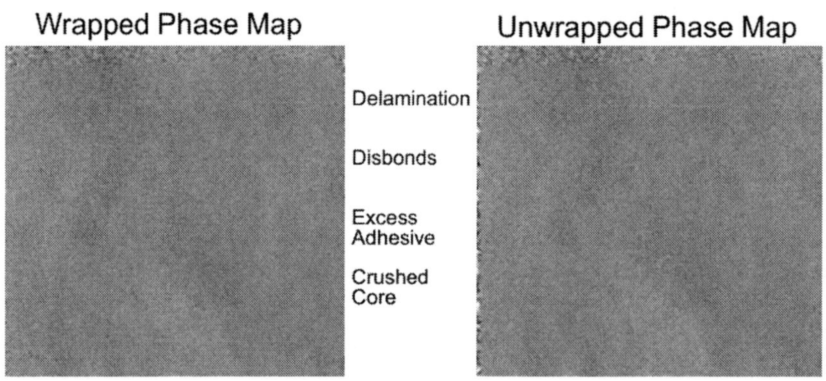

Laser Technology Inc.

Aerospace NDE Systems *Copyright © Laser Technology Inc. 2010* *www.LaserNDT.com*

Shearography Test Results By Stress Method

Thermal Shearography
Aluminum Honeycomb Panel

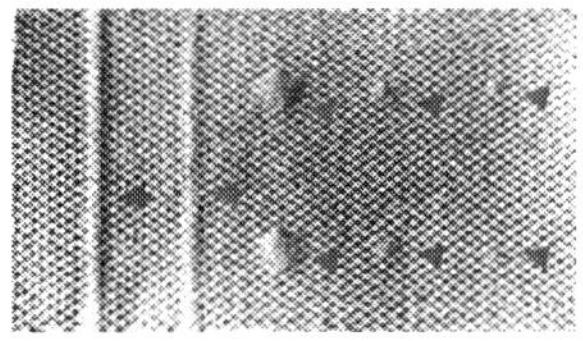

Pressure Shearography
COPV

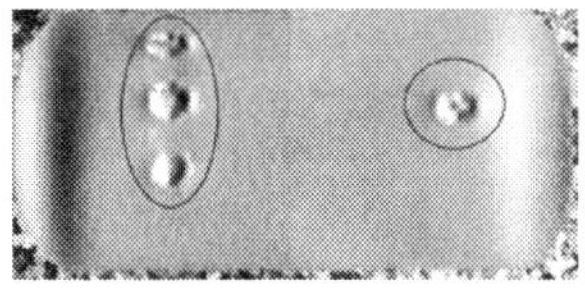

Vacuum Shearography
Composite/Nomex Honeycomb

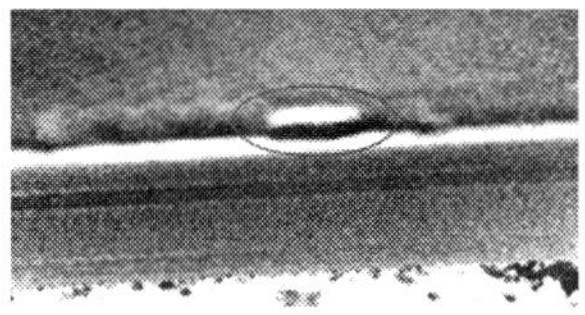

Acoustic Shearography
Foam Cryogenic Fuel Tank TPS

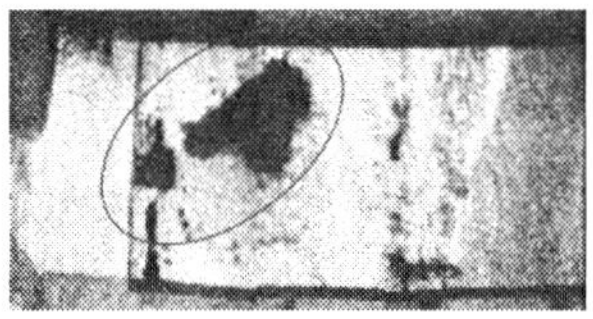

Laser Technology Inc.
Aerospace NDE Systems www.LaserNDT.com

Vibration Stressing

- Air Coupled Acoustic Stress
  - Non Contact Stressing Method
  - Low Density Materials and Structures
- Mechanically Coupled Vibration Stress
  - Vacuum or mechanical coupled exciter
  - High Density Materials and Structures
- Theory:
  - Broad frequency bandwidth excitation excites more compliant disbonded material in complex resonant modes.
- Detects:
  - Disbonds, Delaminations, Damage

Laser Technology Inc.
Aerospace NDE Systems Copyright © Laser Technology Inc. 2010 www.LaserNDT.com

Vibration Stressing

VIBRATION	*AIR COUPLED* *0.5 to 20kHz.*	*90 to 125dB*	*Foam Rocket Thermal Protection Systems* *Damage, Disbonds, Delamination, Cracks* *Light Weight Honeycomb* *Spacecraft Solar Panels, Solar Cell Bond*
	MECHANICALY COUPLED *4.0 to 250kHz.*	*80 to 100db*	*Metal Honeycomb* *Metal Honeycomb Turbine Engine Fan Cases, Fan Blade Erosion Strip bond, Metal to Metal Bonded Panels and Honeycomb*

Acoustic Shearography vs. AirScan UT

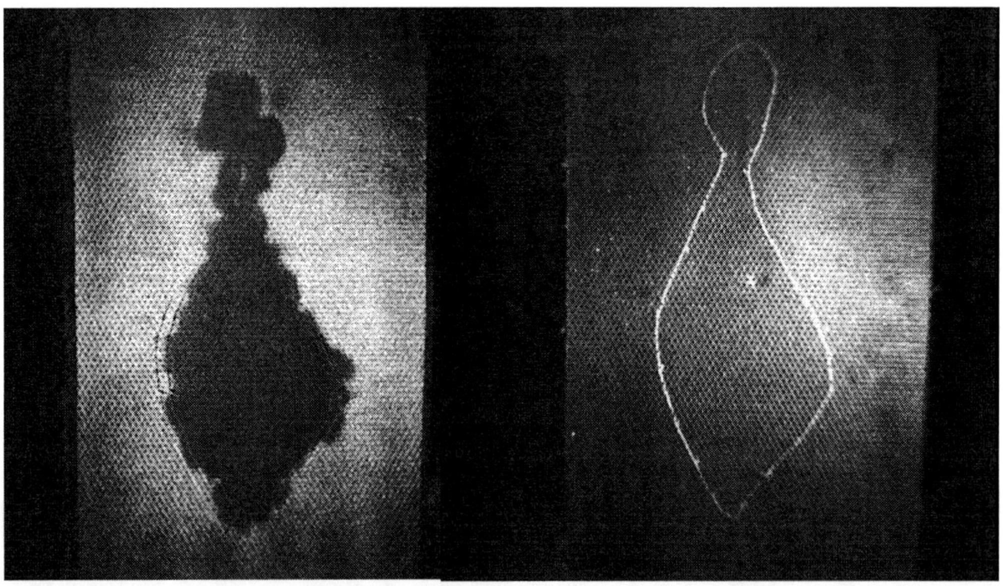

Acoustic Sheargram (26 x 18 in.) *Airscan UT*

Vibration Stressing

Nacelle - Perforated Honeycomb

Live Image | *MECAD Shearogram*

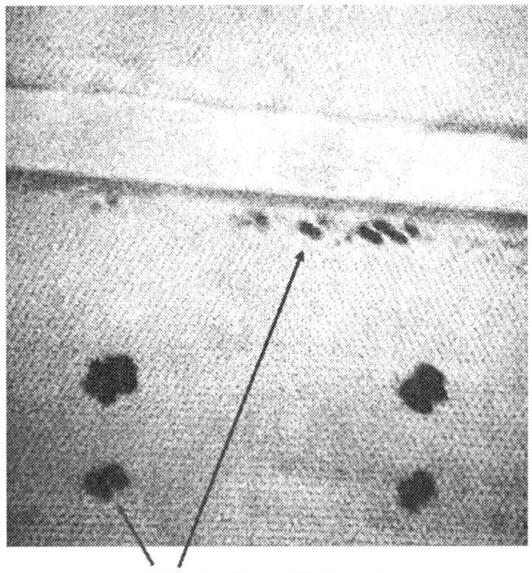

16 x 16 inches

Laser Technology Inc.

Aerospace NDE Systems *www.LaserNDT.com*

Shearography NDT Results on Aerospace Materials Standards and Structures

Laser Technology Inc.

Aerospace NDE Systems *www.LaserNDT.com*

Shearography NDT Applications

- Sandwich Panels:
  - Metal and Organic Honeycomb
  - Foam Core
- Composite Laminate Panels
- Complex Molded Composite Parts
- Resin Transfer Molded Parts
- Composite Over Wrapped Pressure Vessels (COPV)
- On-Vehicle Inspection
- Large Test Object Inspection

LTI **Laser Technology Inc.**
Aerospace NDE Systems www.LaserNDT.com

Shearography NDT of Composite Laminate Panels

- *Shearography is effective for the inspection of carbon fiber laminate panels with single side access however, imaging and heating the panels from both sides simultaneously doubles throughput but also allows imaging through the full volume thickness of the panel.*

- ***Thermal Shearography is the optimal technique for solid laminates to image delaminations, disbonds, voids and porosity.***

- ***Defect detection limit is nonlinear, but for depths of 0.25 inches and less, equal to 1.0 times defect diameter. Access to both sides of a panel gives a detection limit of 0.125 inches in 0.25 inch thick carbon fiber panels.***

LTI **Laser Technology Inc.**
Aerospace NDE Systems Copyright © Laser Technology Inc. 2010 www.LaserNDT.com

Honeycomb Panel
Carbon Fiber Face Sheet With Aluminum Core

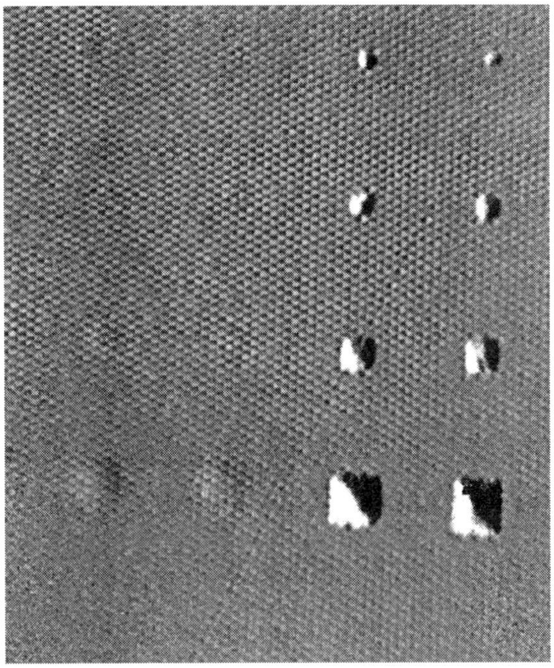

- All near side face sheet to core disbonds detected with thermal shearography.
- Two side inspection recommended.

Inspection Time/Side = _5 seconds_

Panel Dim. Shown = 16 x 16 inches

Laser Technology Inc.

Aerospace NDE Systems www.LaserNDT.com

25 Ply Carbon Fiber Laminate Panel With Teflon Inserts

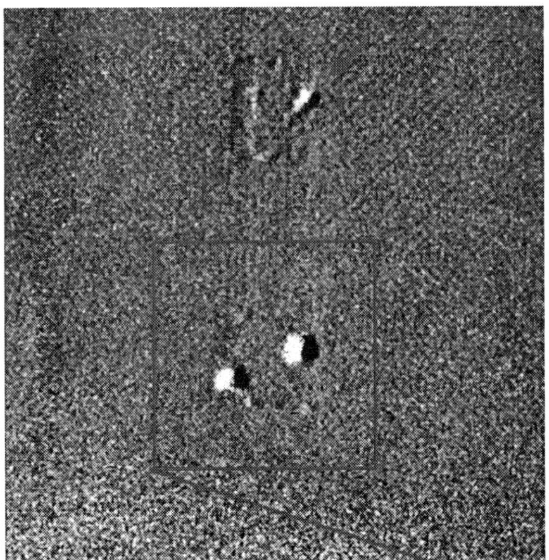

- The single sheet Teflon inserts intended to simulate disbonds instead bonded in (left)
- Thermal shearography shows presence of the Teflon strips and several actual disbonds (below).

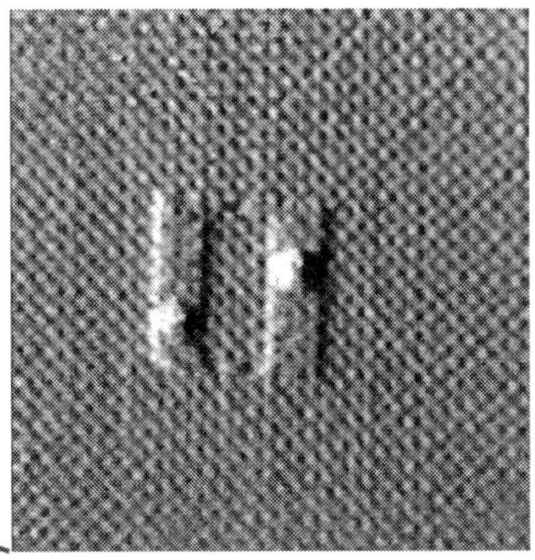

Laser Technology Inc.

Aerospace NDE Systems Copyright © Laser Technology Inc. 2010 www.LaserNDT.com

Panel 6.72 mm. Thick

Resin Starved Areas (irregular pattern left and bottom)

Laser Technology Inc.
Aerospace NDE Systems *www.LaserNDT.com*

Panel variable ply depth

7 defects detected

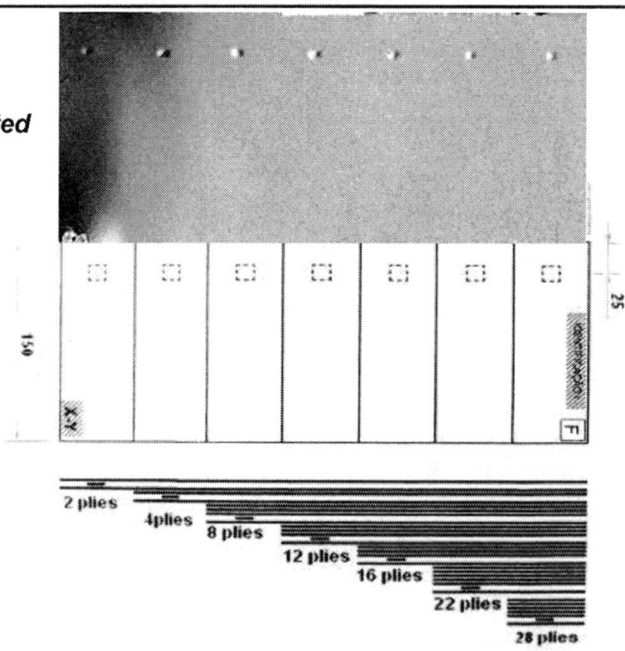

Aerospace NDE Systems *www.LaserNDT.com*

Thermal Shearography Inspection – Honeycomb Panel

Panel Size: 12 x 12 inches ***Construction:*** Carbon Fiber 0.040 in., Nomex core
Inserts Sizes: 0.375, 0.500, 0.250 x 1.000 inches

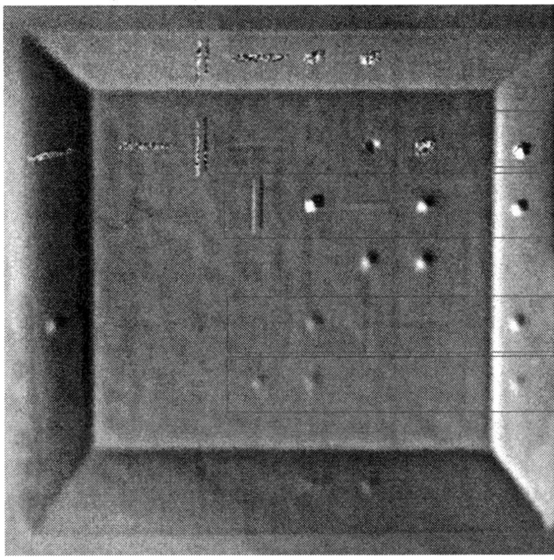

- **Teflon Inserts at face sheet bond, above and below film adhesive**

0.500 above adhesive

0.500 below adhesive

0.375 above adhesive

0.375 below adhesive

Shearography Imaging of Wrinkled Fibers

- **Fiber wrinkles substantially reduce panel stiffness and in-plane tensile strength.**
- **Shearography can detect fiber wrinkles from a single side allowing inspection of foam cored wind turbine blade skins.**

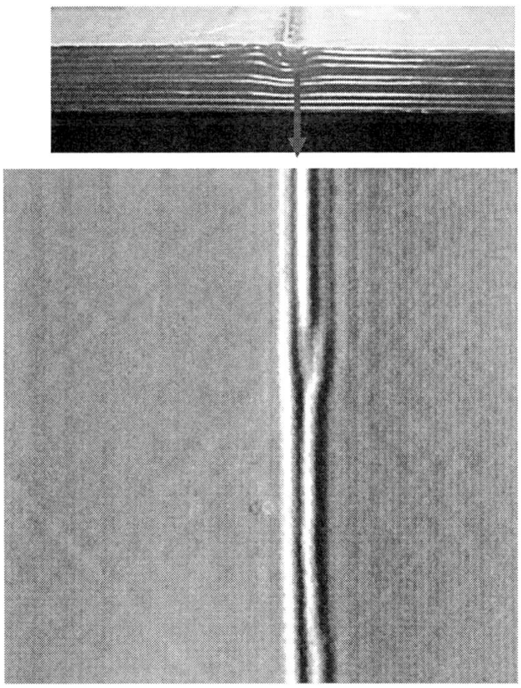

Above: Cross section of carbon fiber laminate panel with fiber wrinkles

Below: Shearography image of the panel face.

Typical Composite Inspection

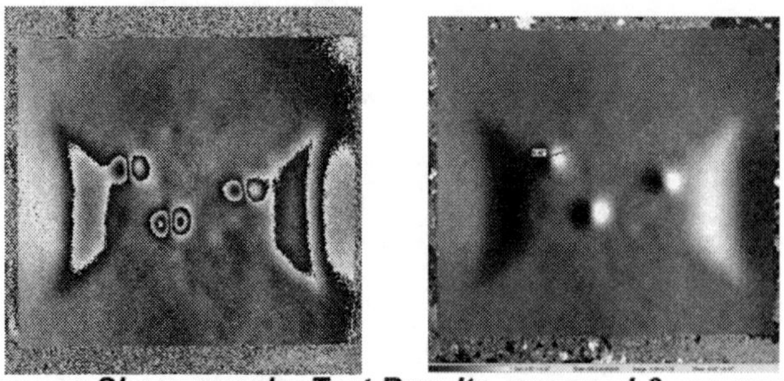

Vacuum Shearography Test Results, wrapped & unwrapped

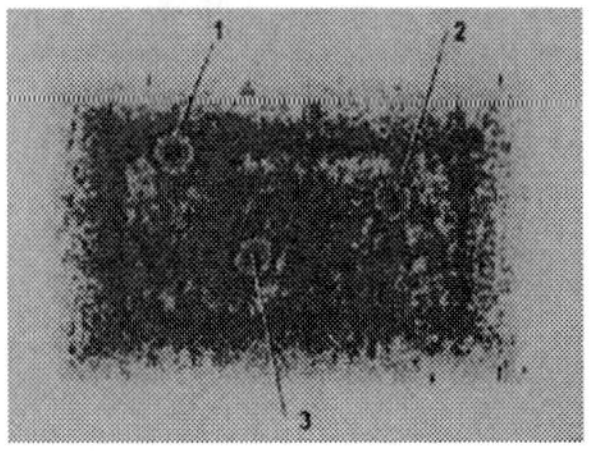

UT C-Scan Test of Panel

- *Shearography inspection of a foam core composite panel using vacuum stressing.*
- *Laser Shearography inspection time 10 seconds per side.*
- *"C-Scan of panel took about 1 hour 10 minutes to produce using through transmission at 10 MHz.*
- *Indications show weakness in panel caused by programmed skin to core separation.*

Laser Technology Inc.
Aerospace NDE Systems

www.LaserNDT.com

Roha Cell Foam Core Flap

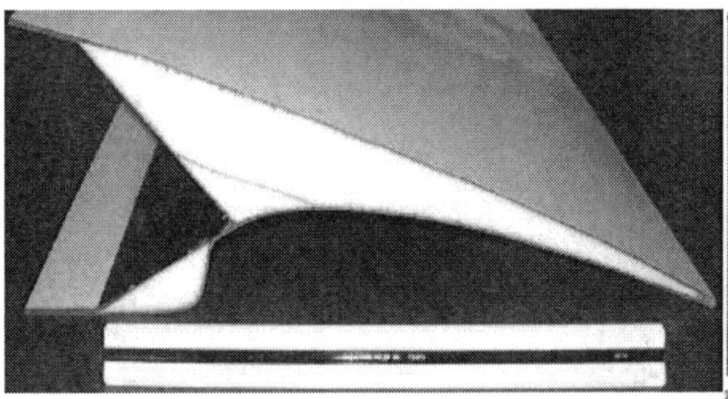

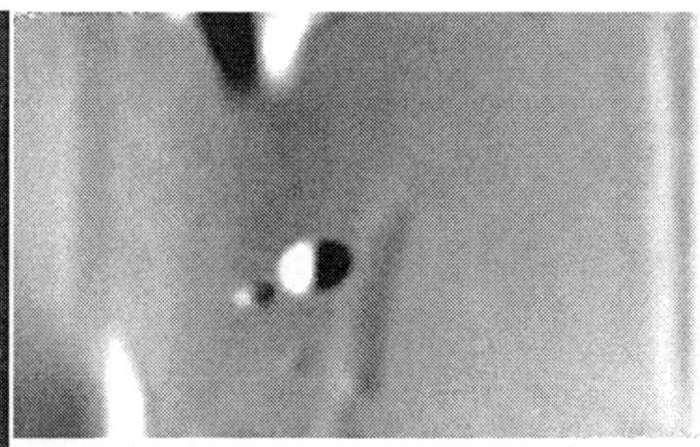

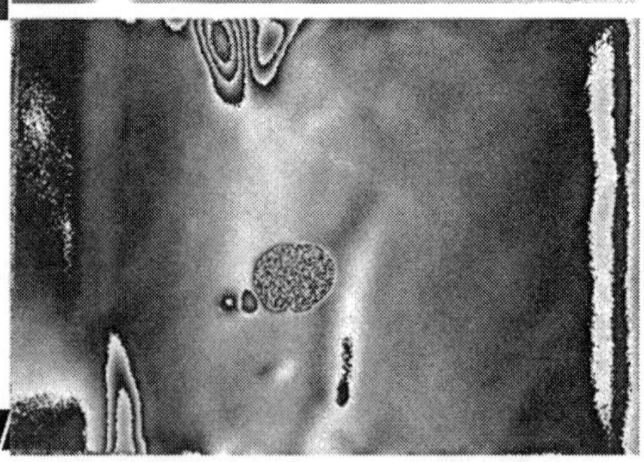

Un-wrapped Phase Map

Wrapped Phase Map

Disbonds and impact damage

Laser Technology Inc.
Aerospace NDE Systems

www.LaserNDT.com

A-6 Radome

30% inspected in 1 minute w/Thermal Stress

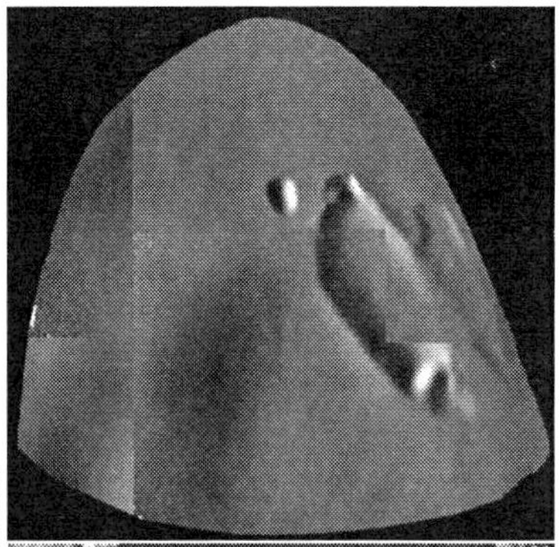

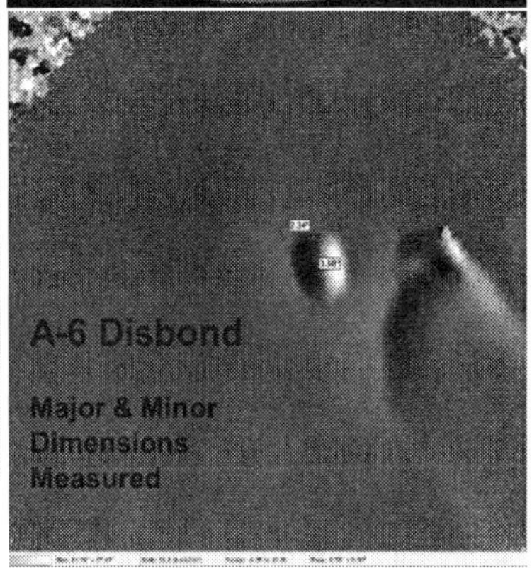

Laser Technology Inc.

Aerospace NDE Systems Copyright © Laser Technology Inc. 2010 www.LaserNDT.com

Shearography NDT of Military Aircraft F-22 Production Radomes

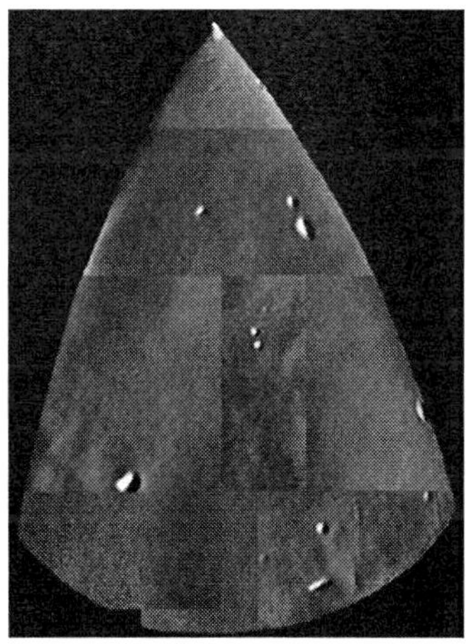

Laser Technology Inc.

Aerospace NDE Systems Copyright © Laser Technology Inc. 2010 www.LaserNDT.com

LTI On-Structure or Aircraft Shearography Testing

Airbus A319
Rudder Inspection

- Vacuum Attach to surface
- Use anywhere
- Up to 14 sq. meters/hour

Laser Technology Inc.
Aerospace NDE Systems

www.LaserNDT.com

Near / Far Side Disbonds Around Potted Fasteners Near Rudder Tip

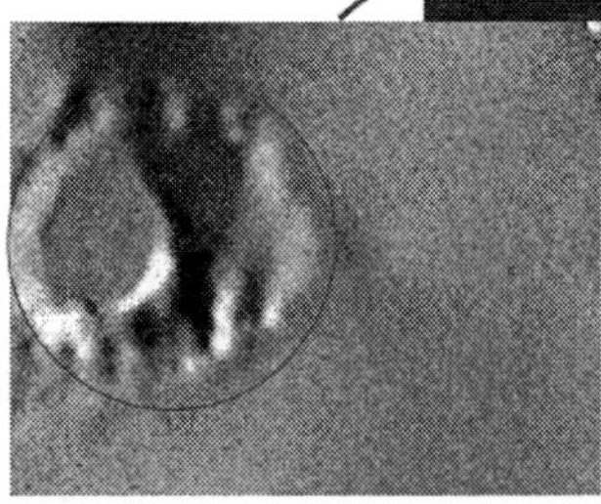

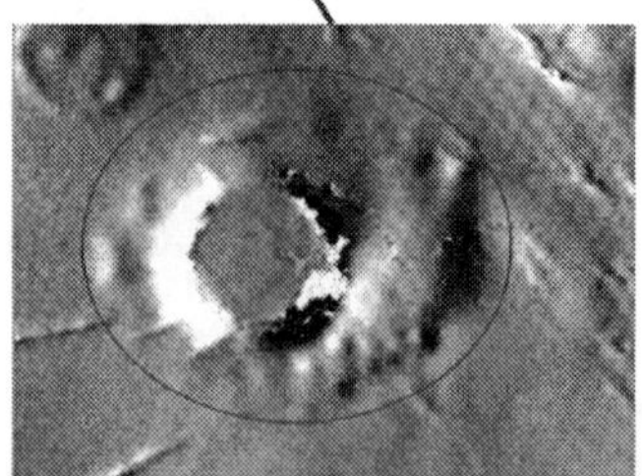

Laser Technology Inc.
Aerospace NDE Systems

www.LaserNDT.com

Impact Tool Drop Damage to Solid Laminate Panel

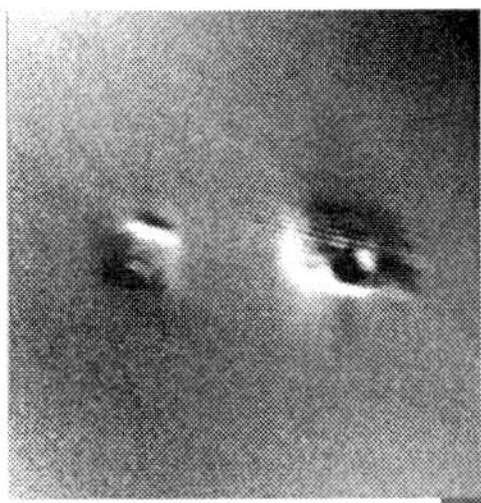

8x8 inches

- Non-Visible Damage easily detected.
- Broken fibers
- Delamination
- Cracking matrix
- Results Correlate with UT C-Scan

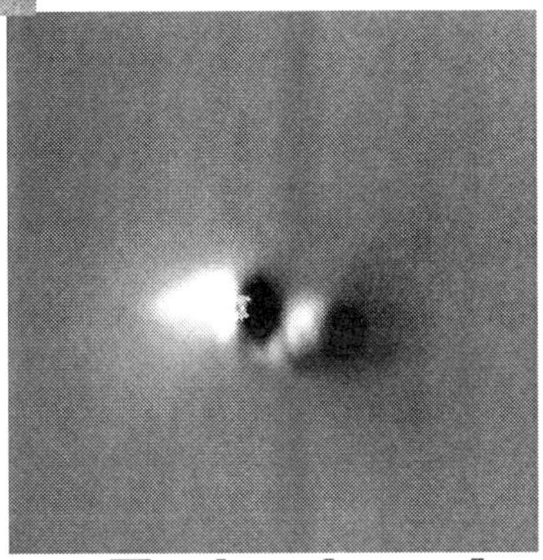

Wrapped Phase Map

Laser Technology Inc.
Aerospace NDE Systems

www.LaserNDT.com

Repaired Aluminum Honeycomb Aircraft Control Surfaces

- Extensive repairs make conventional UT sometimes very difficult or impossible.
- Shearography tests show disbonds and core damage without the need for couplant and reduced inspection time.

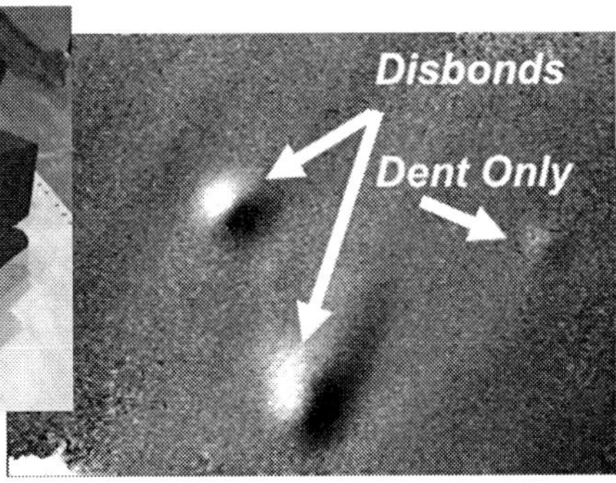

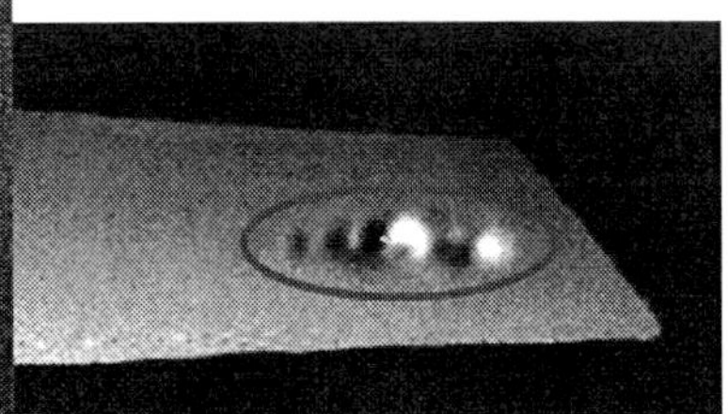

F-16 Leading Edge Flap Disbond inboard end

Laser Technology Inc.
Aerospace NDE Systems
Copyright © Laser Technology Inc. 2010
www.LaserNDT.com

Rolls Royce Tay - Fan Case Inspection

- ***Fan Seals***
- ***NDT of Anti-Ice Panels***
- ***Honeycomb Fan Case Structure***

Shearogram of TAY
Fan seal and ice panel

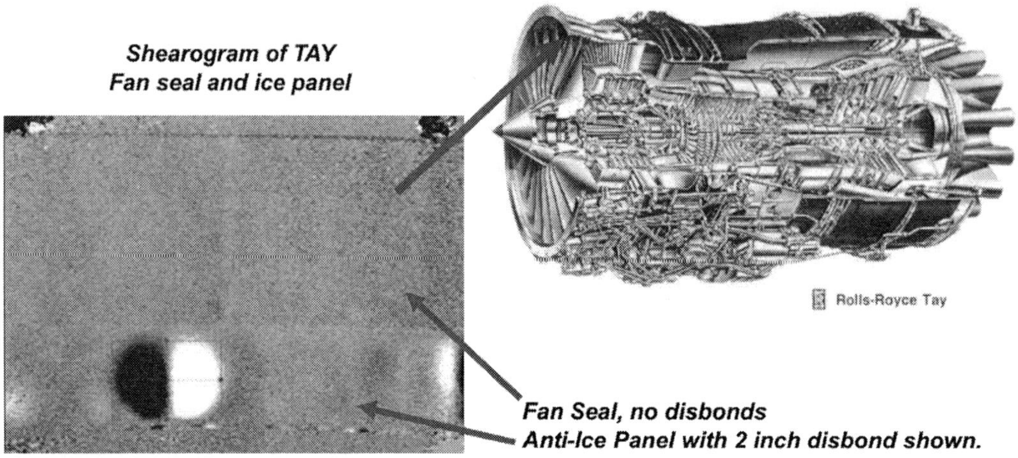

Laser Technology Inc.
Aerospace NDE Systems *www.LaserNDT.com*

Trent 900 Aft Fan Case Septumated Ti Honeycomb

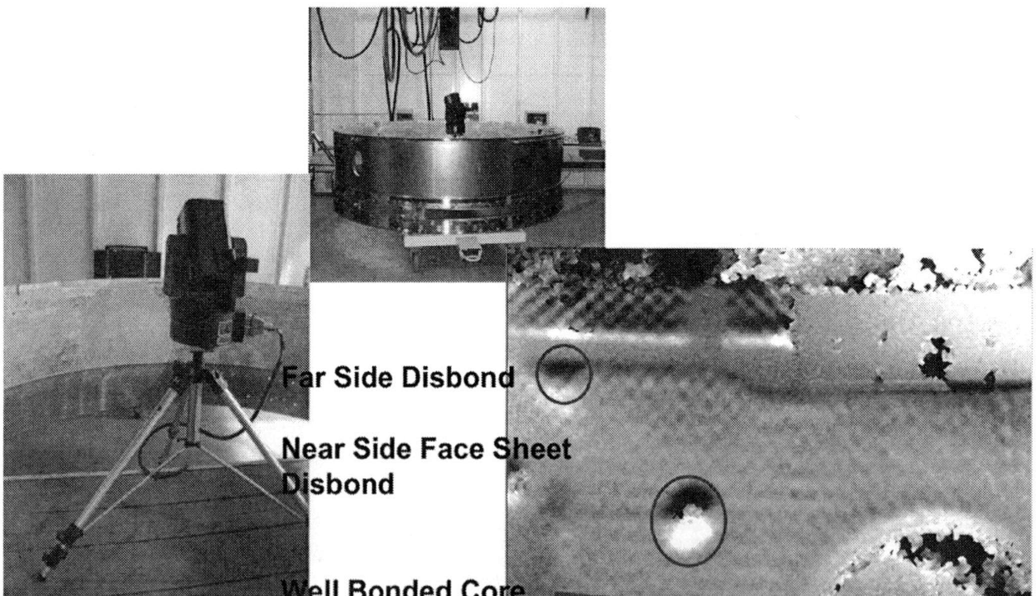

Laser Technology Inc.
Aerospace NDE Systems *www.LaserNDT.com*

Trent 900 Aft Fan Case Septumated Ti Honeycomb

Comparison Between

Thermal Shearography | *Ultrasonic C-Scan*

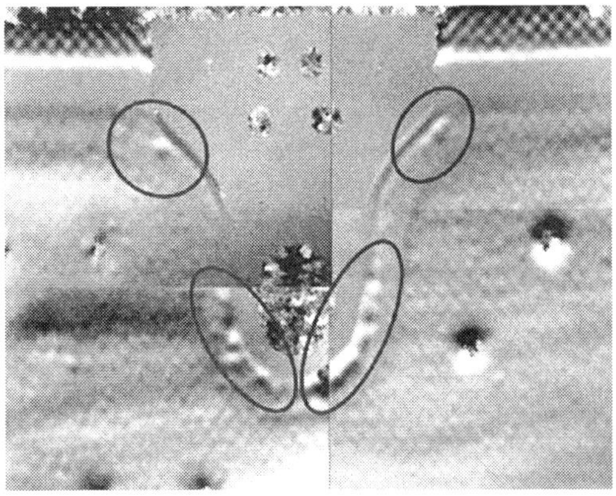

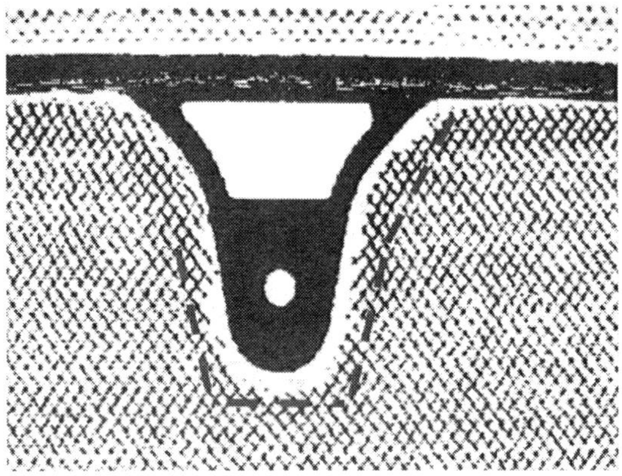

Laser Technology Inc.

Aerospace NDE Systems

www.LaserNDT.com

Engine Duct Shearography Thermal Inspection

ID Indications

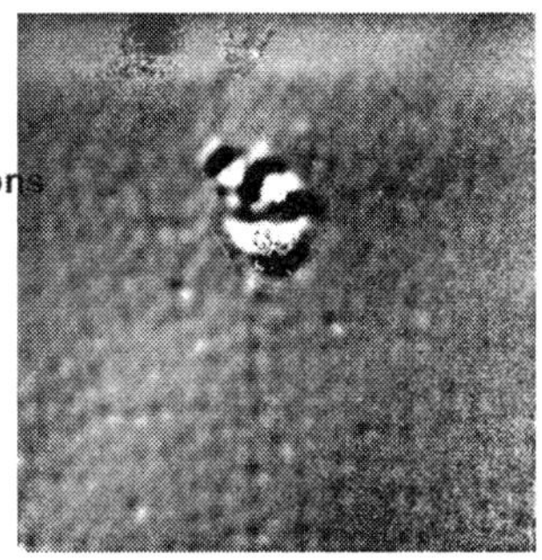

ID Indications

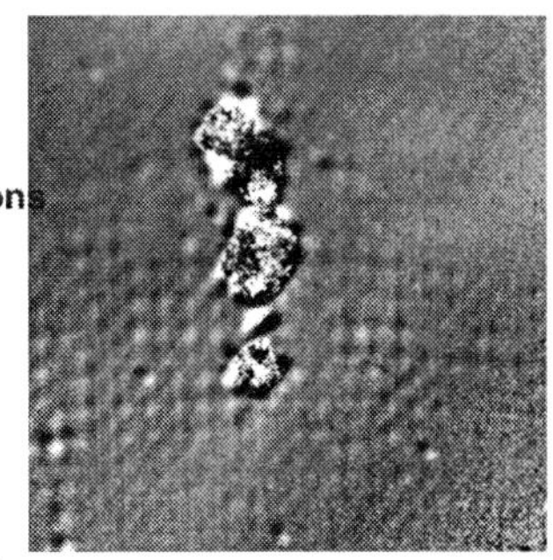

Laser Technology Inc.

Aerospace NDE Systems

www.LaserNDT.com

LTI-2100M testing a Pratt & Whitney Canada

Duct section with thermal shearography

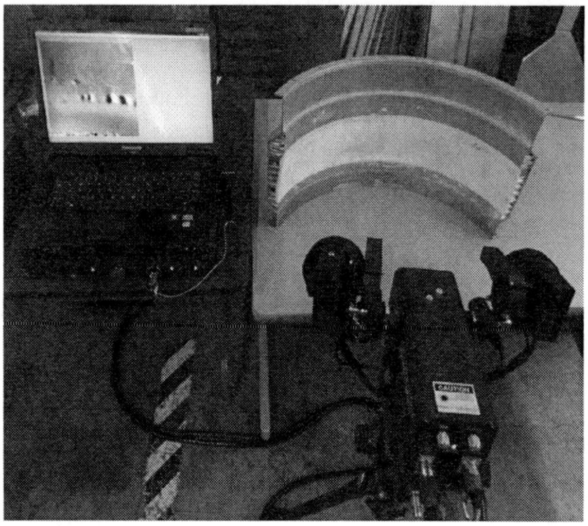

Laser Technology Inc.

Aerospace NDE Systems

www.LaserNDT.com

Disbonds Aluminum Face Sheet/Honeycomb core detected with thermal shearography

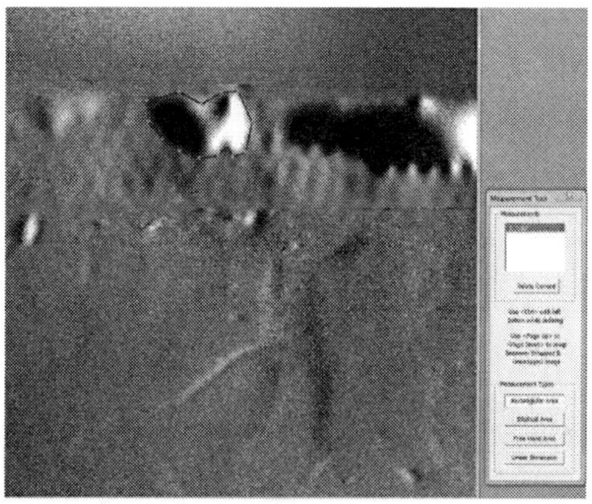

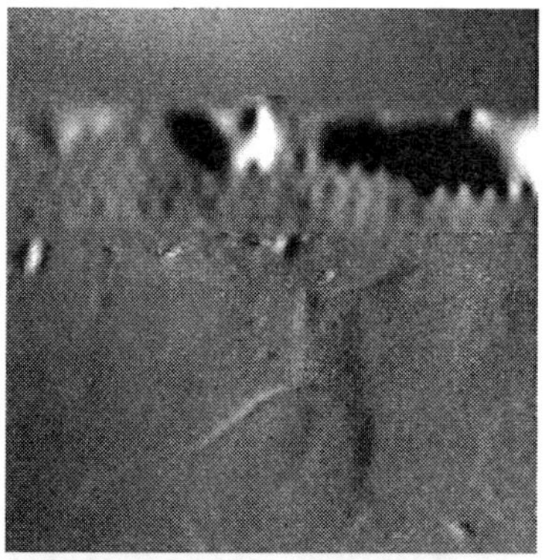

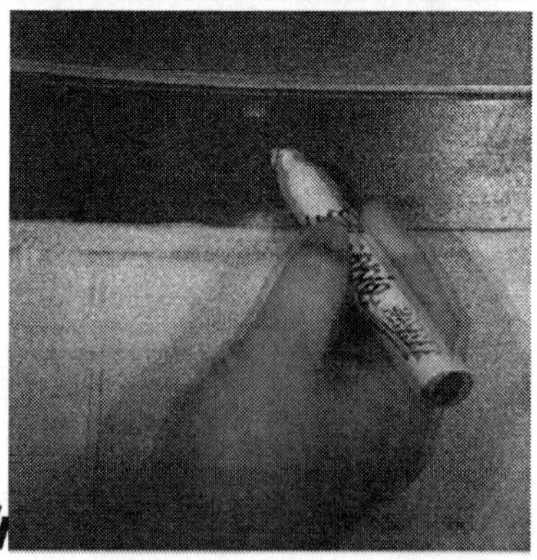

Laser Technology Inc.

Aerospace NDE Systems

www.LaserNDT.com

C-17 Duct Component

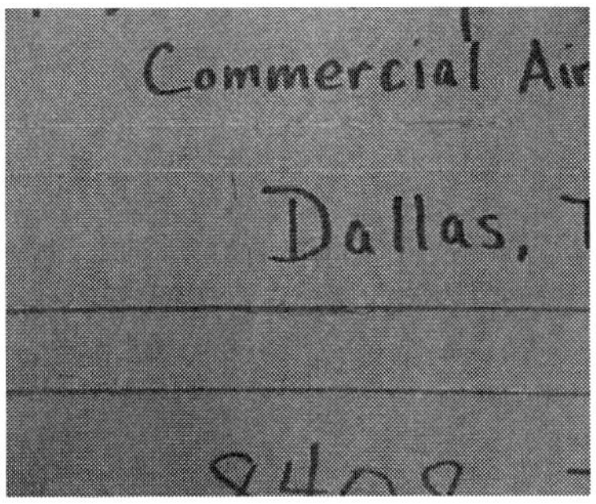

Disbond under 0.050 in. thick doubler

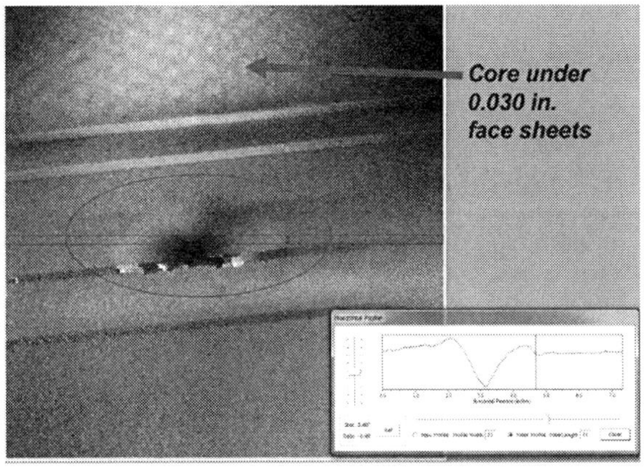

Line scan showing disbond signal

Laser Technology Inc.

Aerospace NDE Systems

www.LaserNDT.com

Aircraft Repairs

- Metal Honeycomb
- Composite Laminates
- Composite Honeycomb

Laser Technology Inc.

Aerospace NDE Systems

Copyright © Laser Technology Inc. 2010

www.LaserNDT.com

Engineered Composite Laminate Repair NDI With Thermal Shearography

Step 1:
Capture Initial Shearography Image Of Repair At Ambient Temperature, Then Warm With Infrared Radiation

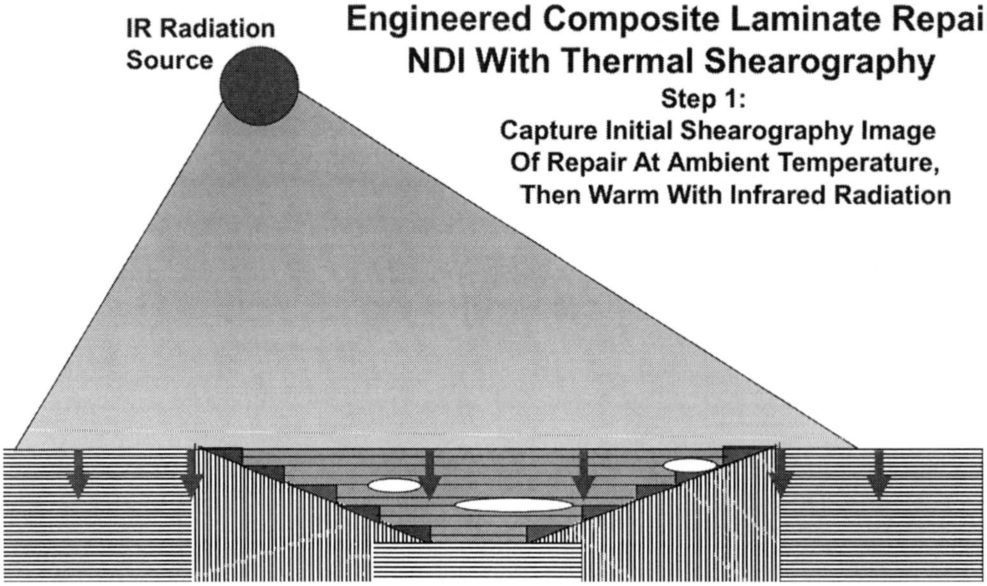

Laser Technology Inc.

Aerospace NDE Systems *Copyright © Laser Technology Inc. 2010* *www.LaserNDT.com*

Engineered Composite Laminate Repair

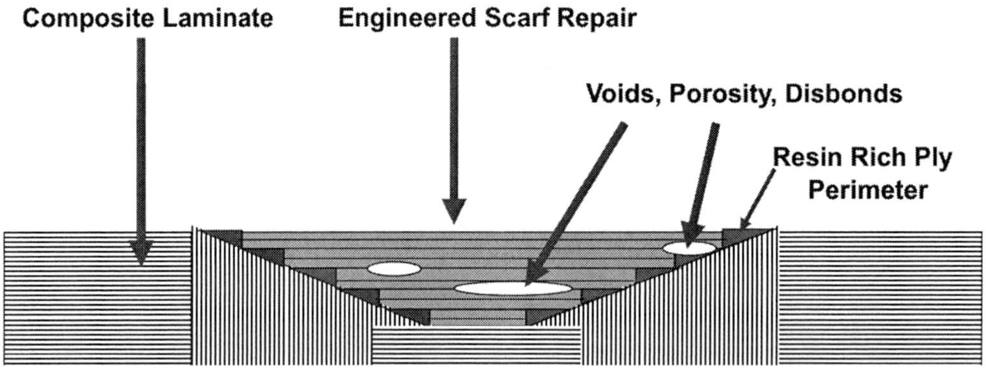

Laser Technology Inc.

Aerospace NDE Systems *Copyright © Laser Technology Inc. 2010* *www.LaserNDT.com*

IR Radiation Source

Engineered Composite Laminate Repair NDI With Thermal Shearography

Step 2:

- Heat diffuses through composite material.
- Uniform material thermally expands (U).
- Areas with discontinuities, voids, porosity and disbonds have greater thermal expansion (D).

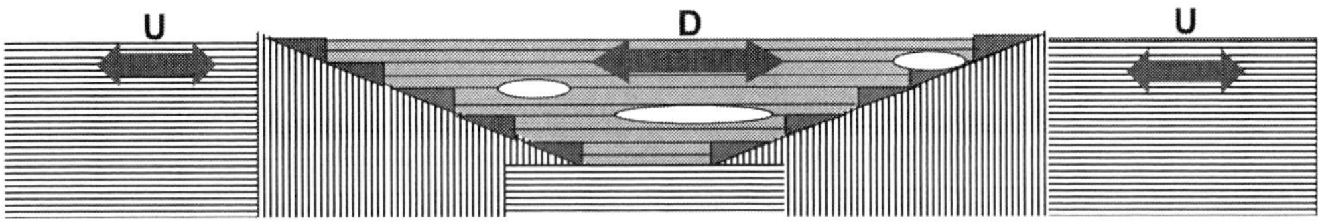

Laser Technology Inc.

Aerospace NDE Systems *www.LaserNDT.com*

IR Radiation Source

Engineered Composite Laminate Repair NDI With Thermal Shearography

Step 3:

- Areas with discontinuities, voids, porosity and disbonds have greater thermal expansion causing material to deform locally out-of-plane.
- Second shearography image data capture with surface deformed, revealing defects.

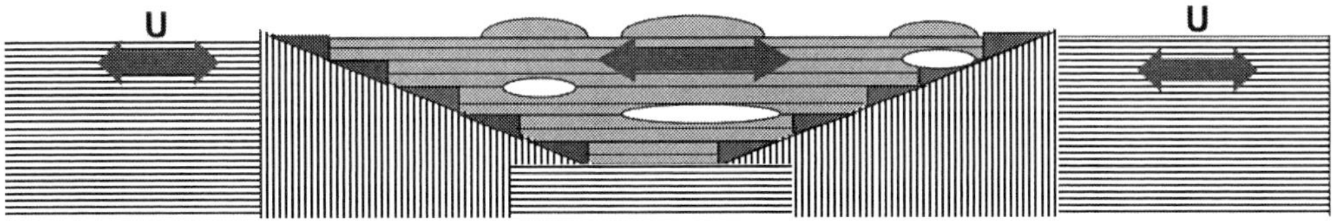

Laser Technology Inc.

Aerospace NDE Systems *www.LaserNDT.com*

External Fuel Tank: Composite Repair NDT

- *Thermal Stress*
- *25 second test time*

Laser Technology Inc.

Aerospace NDE Systems *www.LaserNDT.com*

AWACS Rotodome Shearography NDE

- Detects and measures Impact Damage & Disbonds
- Differentiates between damage and repairs

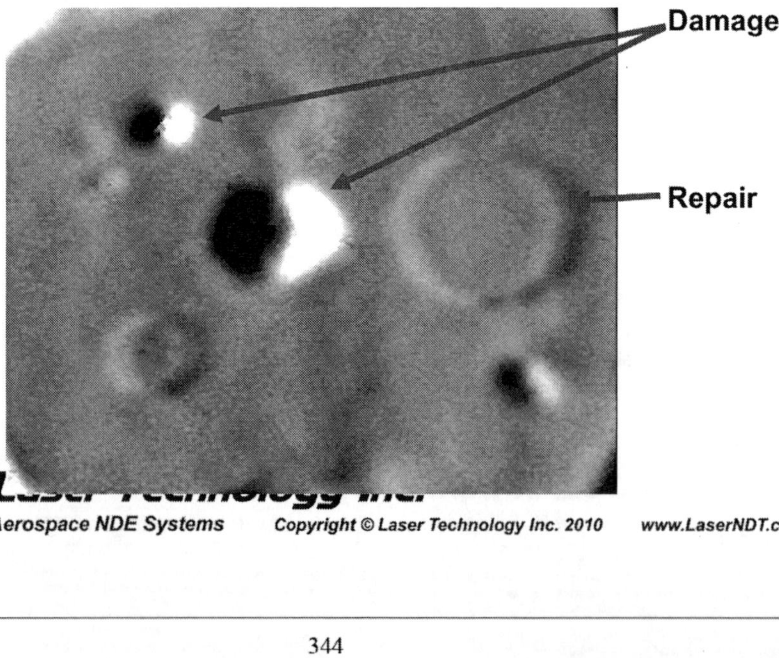

Aerospace NDE Systems *www.LaserNDT.com*

Helicopter Blades

- Metal Honeycomb
- Composite Honeycomb/Foam Core
- Erosion Strip Bond
- Heater Wires: Bond & Lay-up

Bell 212 Blade with Milled Flat Bottom Holes

Thermal Shearography Results showing easy detection of 0.375 inch Disbonds.
Field of View is 18 x 17 inches covering the full cord of honeycomb cored area.

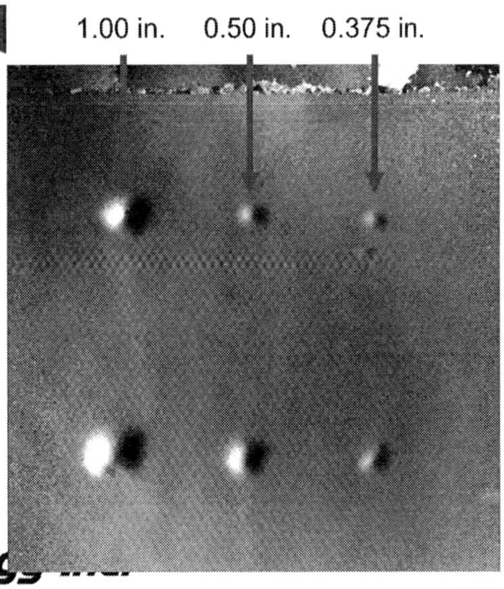

MD-500 Main Rotor Blade Lead Edge Disbonds

String of voids along MD-500 Leading edge.
Image Inspection Time: 5 seconds

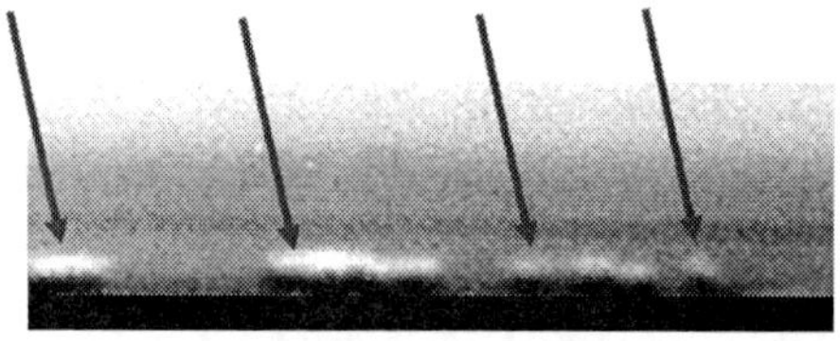

End of long void along MD-500 Leading edge
makes a slight jog.

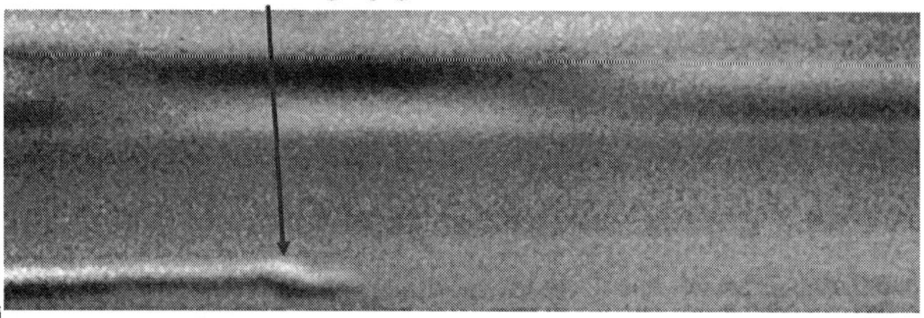

Laser Technology Inc.
Aerospace NDE Systems

www.LaserNDT.com

Inspection of Black Hawk Helicopter Rotor Blade

Trailing Edge

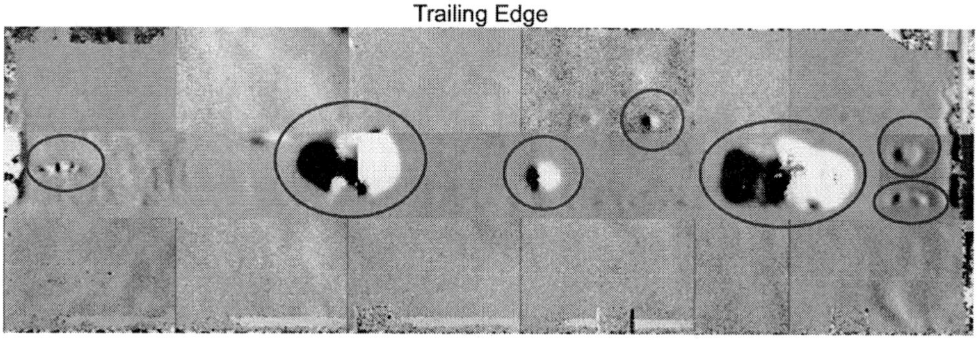

Leading Edge

Blade Test Time per blade for 2 Channel system

- Blade Dimensions (typical): 24 inch span x 280
- Typical Field of View: 12 x 18 inches
- Test Image Array: 2 x 32 images per blade = 64 images/blade
- 15 seconds per sector, 16 minutes/blade + 5 minutes load/unload

Laser Technology Inc.
Aerospace NDE Systems

www.LaserNDT.com

6 Meter Blade with Nomex core tested both sides in 7 minutes

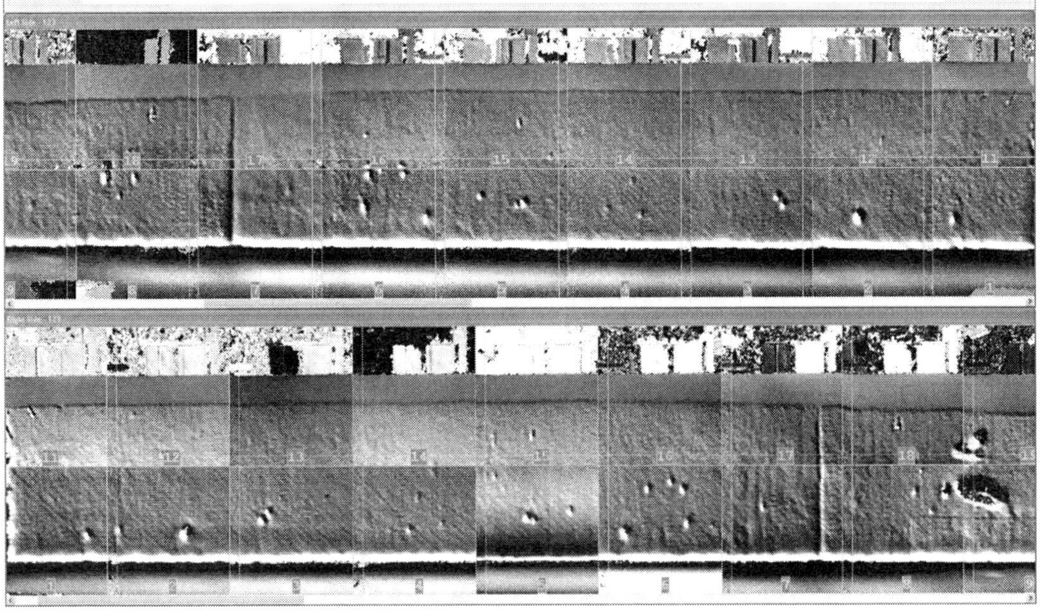

Laser Technology Inc.
Aerospace NDE Systems *Copyright © Laser Technology Inc. 2010* *www.LaserNDT.com*

Production Shearography Systems

Thermal/Vacuum/Acoustic Shearography

Laser Technology Inc.
Aerospace NDE Systems *Copyright © Laser Technology Inc. 2010* *www.LaserNDT.com*

The LTI-5100HD HD All Mode Shearography System

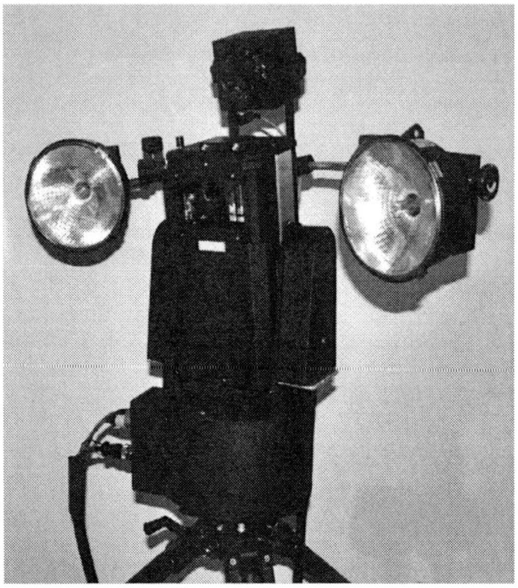

- Highest Resolution and Fastest Frame Rate
- High Power Single Frequency Laser
- Real-Time Imaging
- Powerful Image Analysis Tools
- Integrated, Programmable Stress Controls

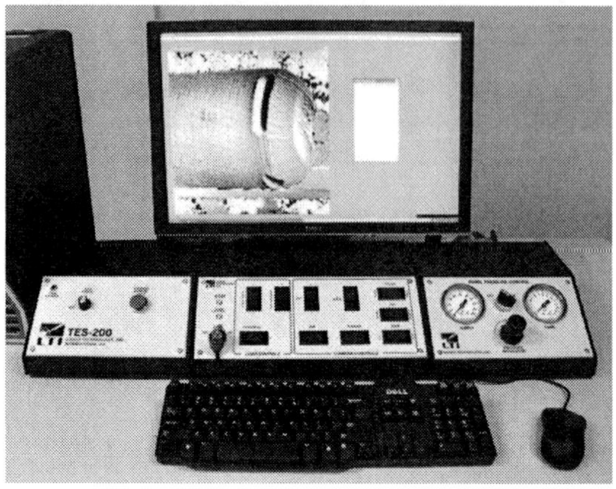

Laser Technology Inc.
Aerospace NDE Systems *www.LaserNDT.com*

LTI Portable Thermal/Vacuum Shearography Systems

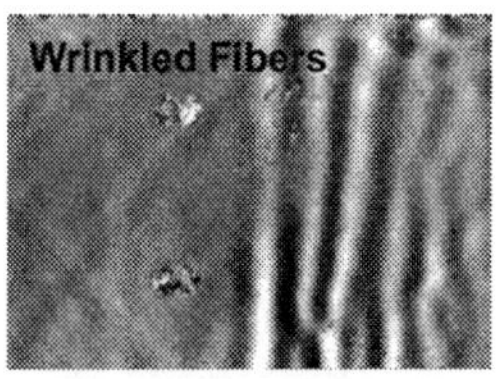

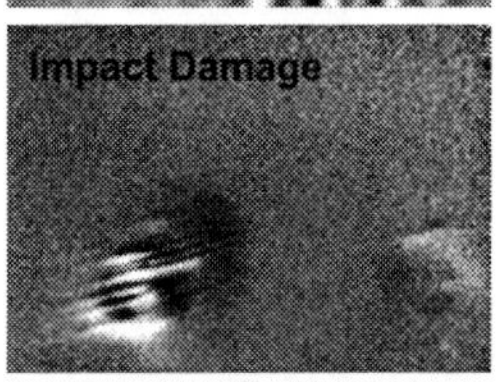

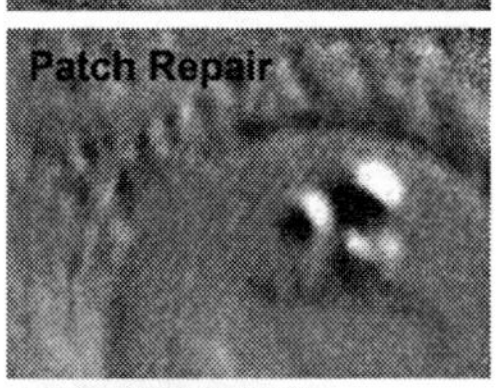

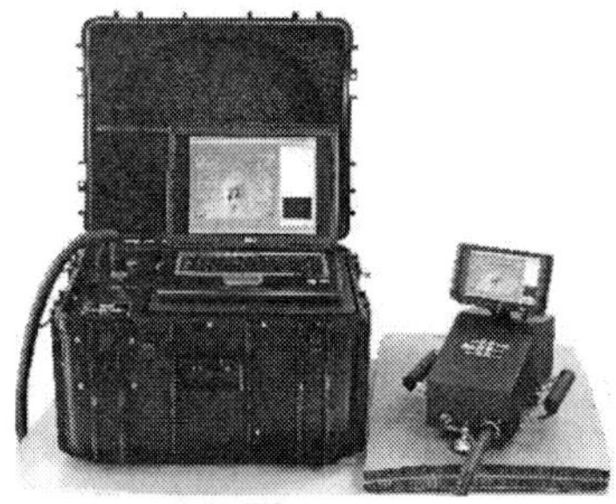

Laser Technology Inc.
Aerospace NDE Systems *Copyright © Laser Technology Inc. 2010* *www.LaserNDT.com*

LTI-9000 Production Shearography Systems

Shearography Scan Gantry

- Multi Axis Motion
- Teach Learn
- Image Stitch
- Auto Scan

Shearography NDT….

Training….

Certification….

Specifications….

Laser Technology Inc.
Aerospace NDE Systems Copyright © Laser Technology Inc. 2010 www.LaserNDT.com

Shearography NDT Certification

Shearography Certification is included in:

AIA NAS 410 rev3
ASNT SNT-TC1A Level I, II, III
EN 4179
ASTM E2581-07

General Publications/Specs

AITM-Airbus Test Method For Inspection Processes
BSS-7057 Boeing Specification Support Standard

Laser Technology Inc.
Aerospace NDE Systems *www.LaserNDT.com*

Summary

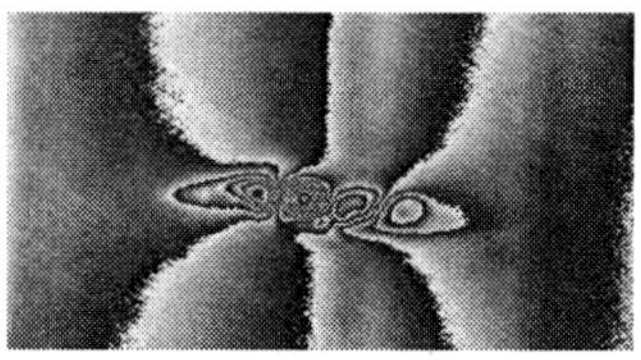

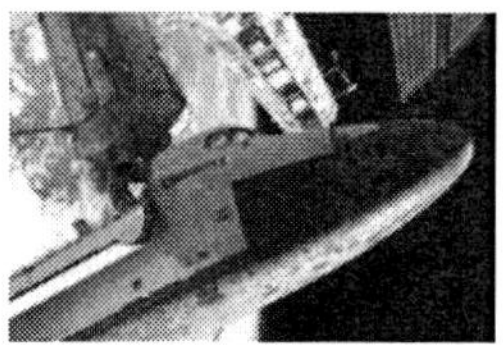

1. Shearography NDT is cost effective, mature and widely accepted NDT technology.
2. Shearography NDT offers much faster inspection over traditional NDT methods for composites, honeycomb and foam core panels and structures.
3. Development of new industry NDT std's is critical for acceptance of new technology and for enhanced reliability and safety in new industries.
4. The cost of systems over the years based on improved electronics and software have come in line with newer flaw detection instruments like Phased Array and Module based systems.

Laser Technology Inc.
Aerospace NDE Systems *Copyright © Laser Technology Inc. 2010* *www.LaserNDT.com*

Shearography Nondestructive Testing

Thank you!

Laser Technology Inc.
1055 W. Germantown Pike
Norristown, PA 19403 USA

Tel: **610-631-5043**

4046 S. 158th LN
Seattle, WA 98188 USA

Tel: **206 246-4244**

Laser Technology Inc.
Aerospace NDE Systems *www.LaserNDT.com*

Surface preparation, contact angle measurements, and adhesive bond performance

Giles Dillingham
Brighton Technologies Group, Inc.

Objectives

- understand and apply principles of wetting measurements to ensure quality and consistency of surface preparation processes for bonding and coating operations
- understand the fundamental relationships between surface preparation and adhesion of adhesives and coatings
- understand the sensitivity of prepared surfaces and the performance of adhesive bonds or coatings on these surfaces to environmental factors such as storage and handling conditions

Topics

- Surface chemistry, surface morphology, and adhesion
- Wetting measurements as quantitative predictors of failure mode and strength
- Examples

Creation and destruction of a bonded structure

- Liquid adhesive or coating establishes contact with substrate
  - Secondary interactions predominate: covalent bond formation is usually absent
- Phase change creates an interface between two rigid materials of similar modulus in intimate contact
- During failure, large scale viscoelastic deformation is absent
  - Local viscoelasticity in crack tip process zone

How do we *predict* adhesion?

- Strength of a bonded structure is determined by:
  - G_{subst}: cohesive toughness of substrate
  - G_{adh}: cohesive toughness of adhesive, ink or coating
  - G_{int}: interfacial fracture toughness
- G_{int} is usually the limiting factor
- Predicting adhesion usually reduces to predicting G_{int}

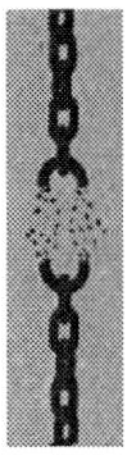

Contact angles and energy of adhesion

- Work of adhesion: energy consumed in creating an interface

$$W_A = \gamma_l + \gamma_s - \gamma_{sl}$$

- Directly measurable by adhesive surface tension and contact angle

$$W_A = \gamma_l(1+\cos\theta)$$

Energy of adhesion as part of interfacial fracture toughness

- Two processes are involved in breaking an established interface:
  - *dewetting* from surface (energy = W_A)
  - *viscoelastic deformation* (energy = f(T,v); T, v = temperature, velocity)
- Total energy required *(toughness):*

$$G_{int} = W_A\,(1+f(T,v))$$

Toughness and W_A

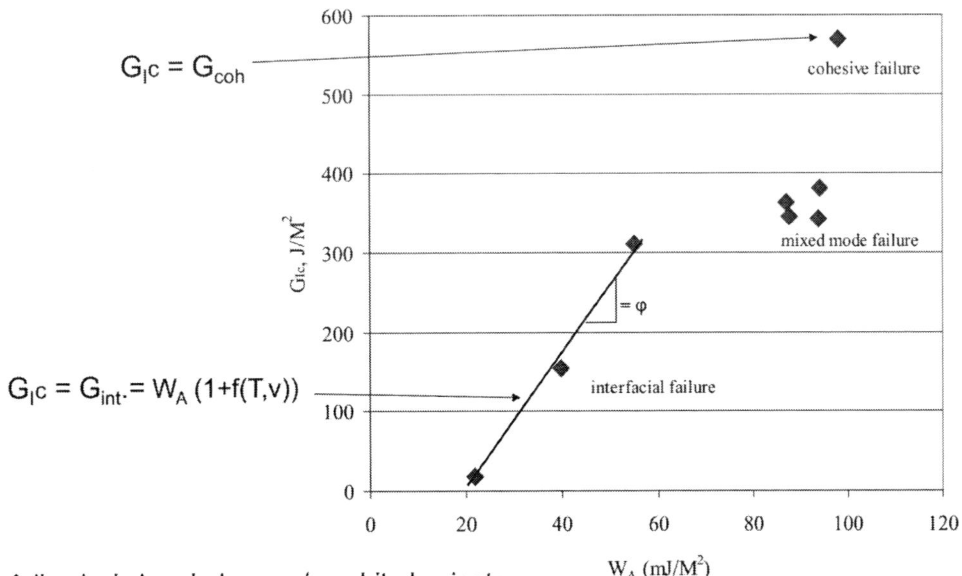

Adhesively bonded epoxy/graphite laminate

Prediction of toughness in manufacturing and in the field

- W_A is a critical parameter that determines practical adhesion
- How does it relate to θ_{adh}?
- How can we evaluate W_A in manufacturing or repair situations?

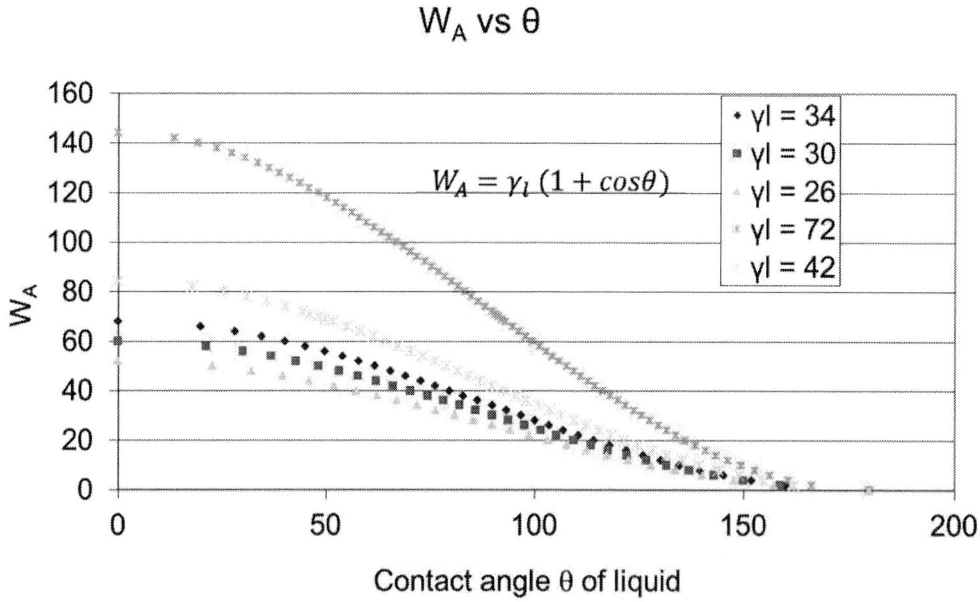

- W_A of high γ_l liquids are most sensitive to substrate surface energy
- θ <~30° maximize W_A for all adhesives

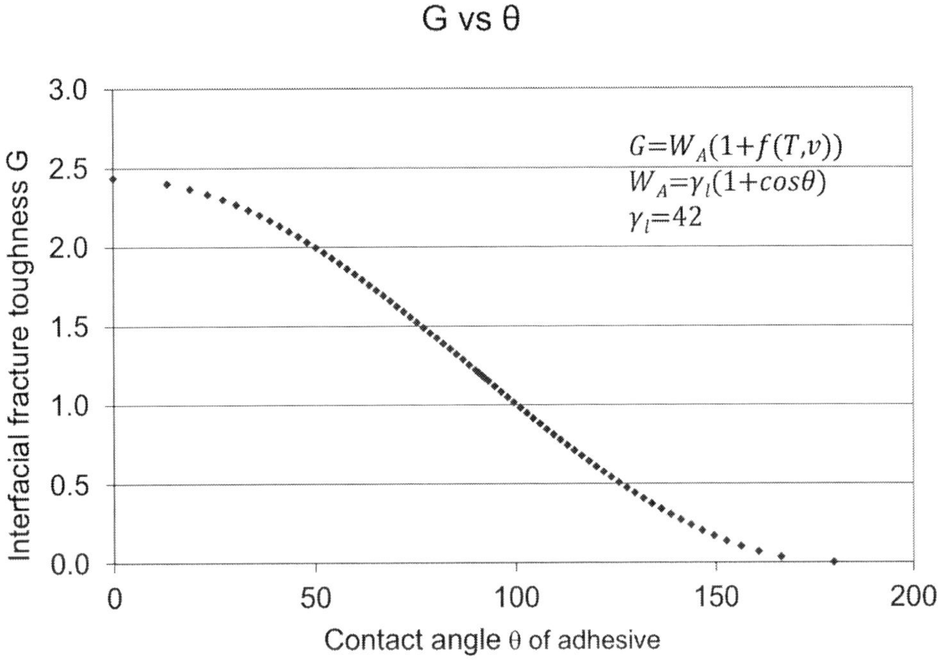

θ <~30° should correspond to maximum G for all adhesives

Prediction of toughness in manufacturing and in the field

- Calculation of W_A requires knowledge of θ_{adh} and γ_{adh}
- θ_{H2O} is a very convenient surrogate liquid
  - Very sensitive to subtle changes in substrate surface energy
  - Particularly sensitive to polar functional groups on surface
    - These groups are very important for adhesion of most adhesives

$G_I c$ vs θ_{H2O}

composite /composite adhesive joints

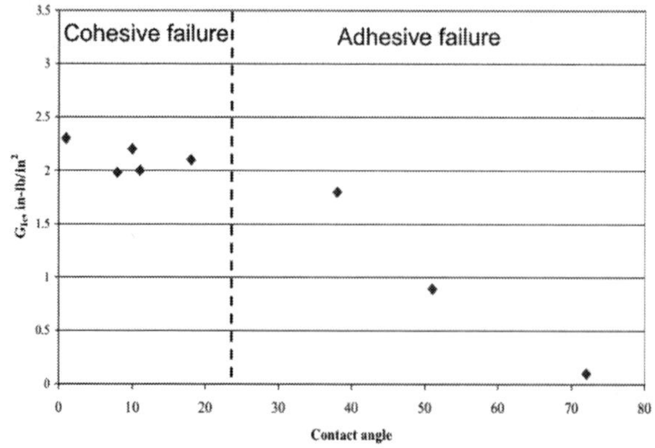

$G_I c$ vs θ_{epoxy}

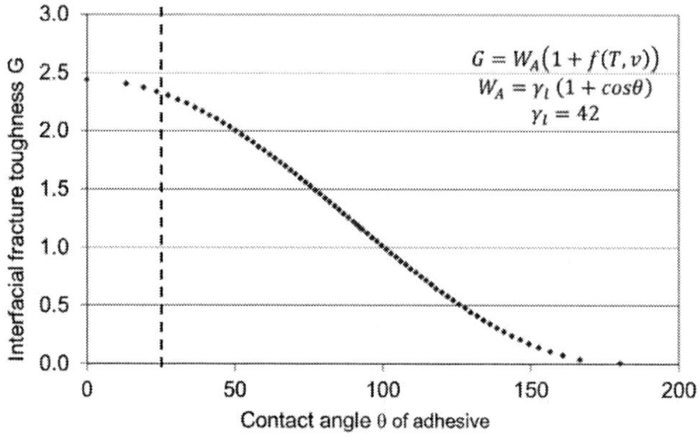

Contact angle methods for QA: ballistic deposition

- Small droplet (<2 µl) of probe fluid is created on surface from a pulsed stream of nanoliter-size drops
- Contact angle is calculated from drop diameter & volume

$$\frac{d^3}{V} = \frac{24\sin^3\theta}{\pi(2-3\cos\theta+\cos^3\theta)}$$

- Correlates to θ_R; γ_s^p

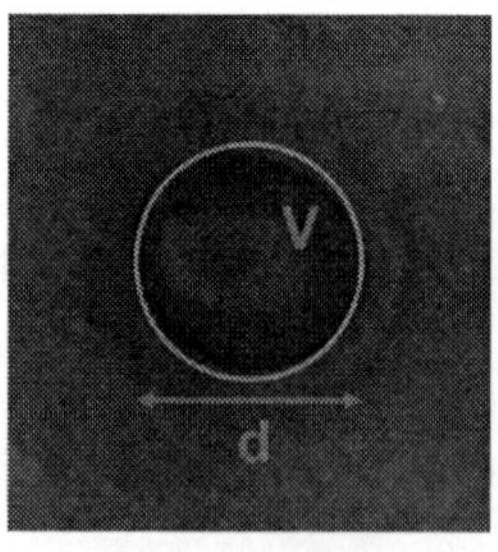

Classes of surface treatments for adhesive bonding of aerospace materials

- Mechanical
  - Abrasive paper
  - Grit blast
  - Peel ply
- Chemical
  - Plasma
  - Corona
  - UV

Mechanical treatments

- Most common approach
- Can function extremely well:
  - Removes contaminants
  - Increases wetting by adhesive
  - Permits mechanical interlocking
- Can create chemically reactive sites for primary bond formation

Grit blasting

cracking

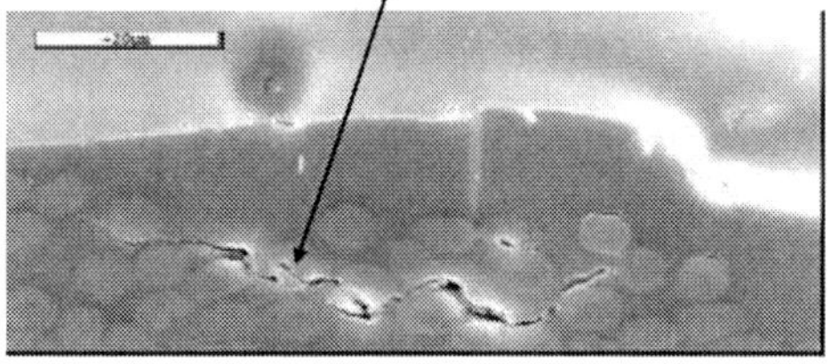

undisturbed resin

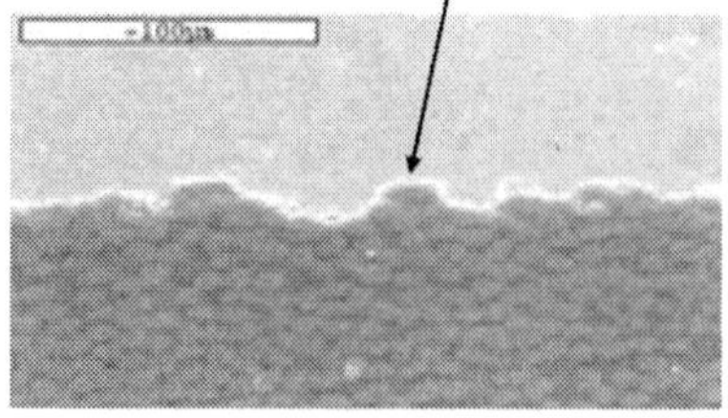

Blasting parameters: 80-grit garnet, 5"/s gun speed, 4" nozzle to surface, 0.375" path spacing.

Blasting parameters: 80-grit garnet, 6"/s gun speed, 6" nozzle to surface, 0.5" path spacing.

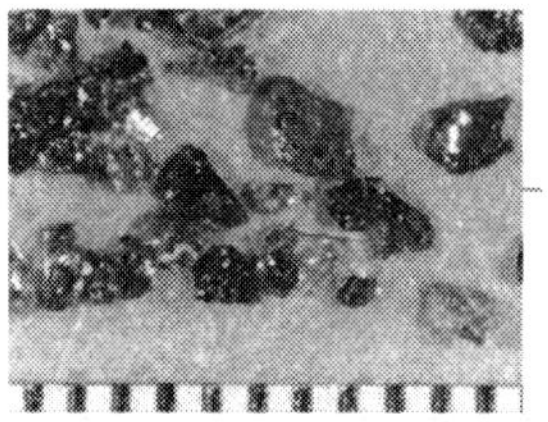

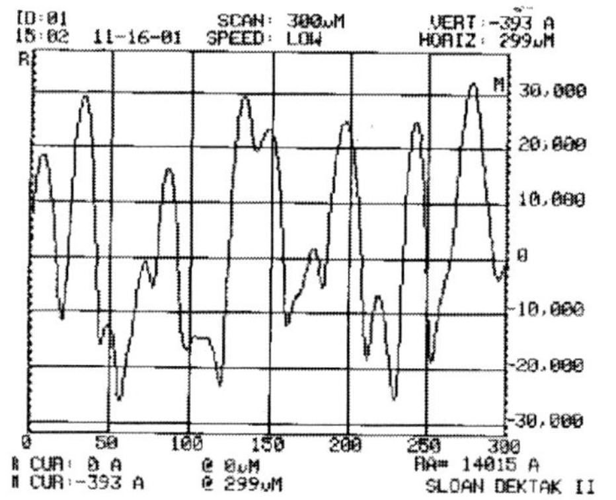

Sample 45-1

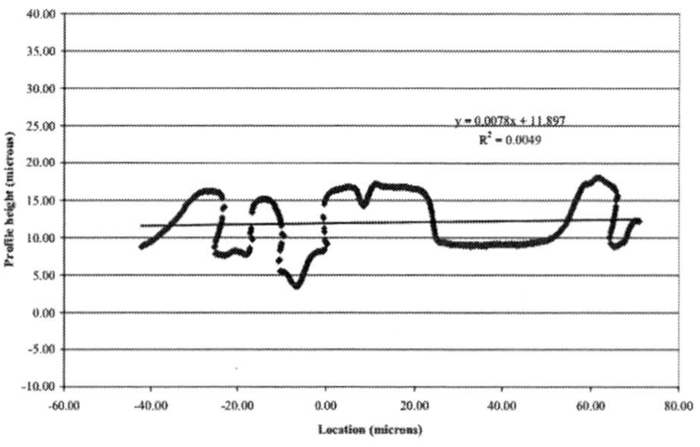

Ballistic H_2O contact angles obtained using Surface Analyst™

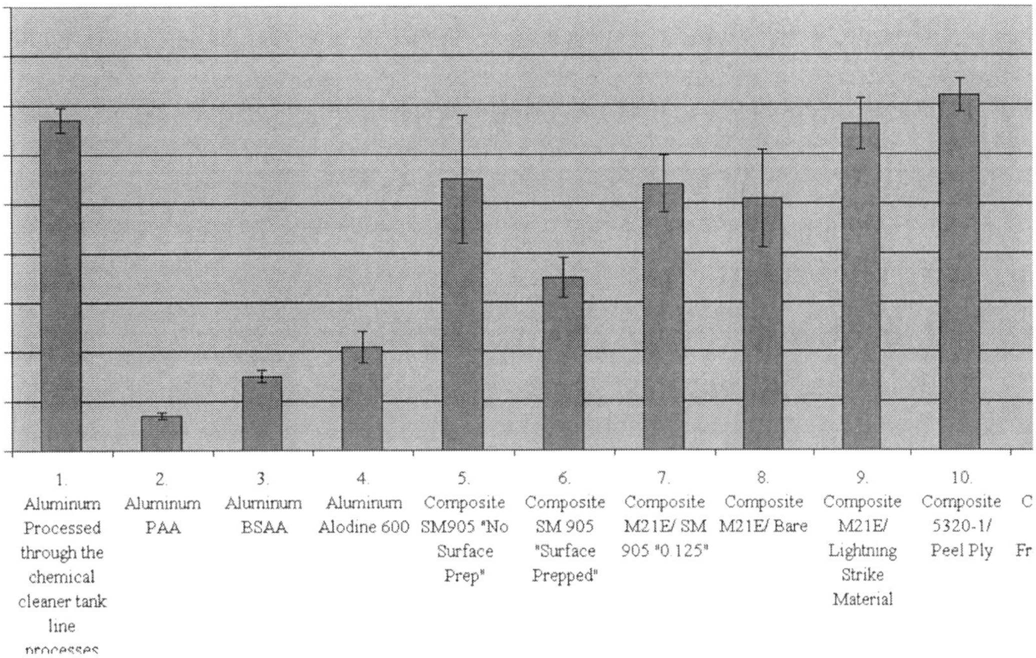

Peel ply

- Protective payer to be removed immediately prior to application of adhesive
- Fractures resin surface during removal to creating a rough and clean substrate
- Can be a source of contamination

Example: Effects of sanding and solvent wiping on peel ply surfaces

- Various panels received from customer
- Peel ply removed; contact angle measurements obtained as a function of
  - Aging time
    - 5 min, 30 min, 8 hr, 12 hr, 24 hr
  - Solvent wiping
    - IPA, acetone
  - Sanding
    - 36 grit, 80 grit

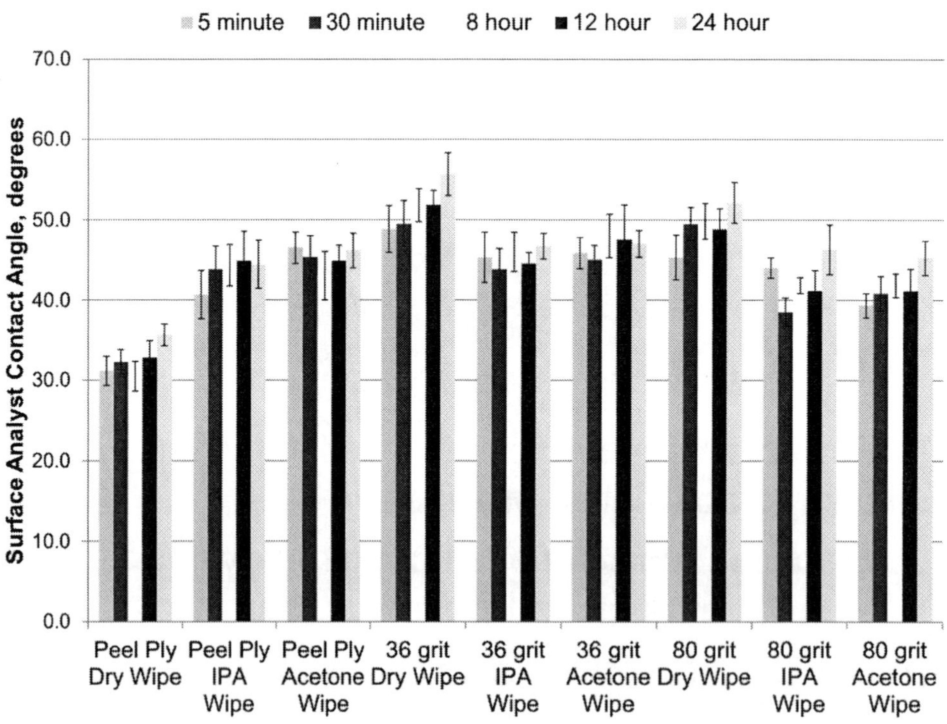

Conclusions

- Cleanest surface is created by peel ply removal
  - Solvent wiping reduces surface energy
- Sanding with Al_2O_3 abrasive paper from McMaster Carr reduces surface energy
  - Solvent wiping of abraded surfaces increases surface energy
  - Results consistent with contamination from abrasive paper
- Aging effects are minimal

Example: effect of contaminated peel ply

- Polyester was scoured, then contaminated via various dilutions of a siloxane solution, dried and heat set (65 s @ 188 °C)
- 10 ply carbon fiber/epoxy prepreg laminated w/ scoured or contaminated peel ply prior to curing @ 179° C
- Adhesives:
  - 3M AF555
  - Cytec MB1515-3
  - Henkel EA9657
- X-ray photoelectron spectra obtained after peel ply removal
- Contact angles obtained from 2μl ballistically deposited water droplets using BTG Surface Analyst™ probe tool.
- DCB Mode I fracture tests (ASTM D5528) for Mode I fracture toughness and failure mode

Silicone contamination of laminate surface by peel ply

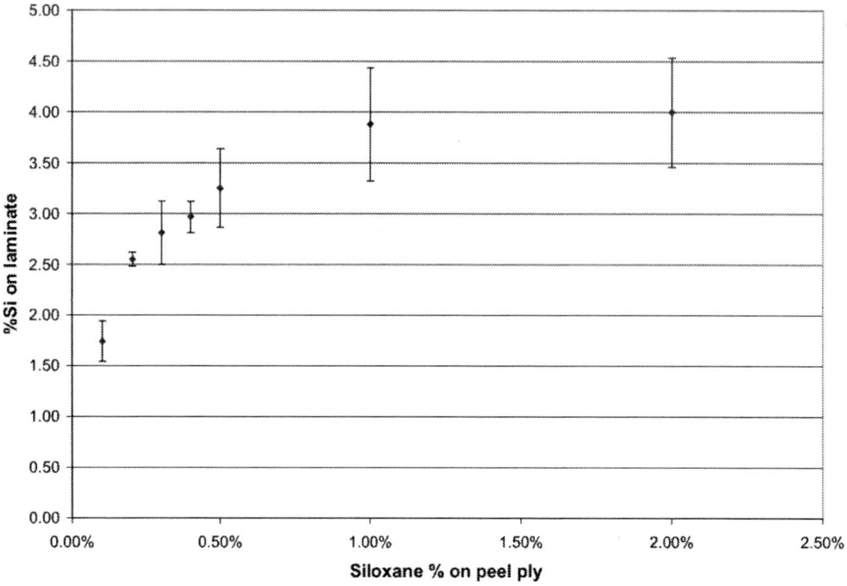

G_{Ic} vs At % Si

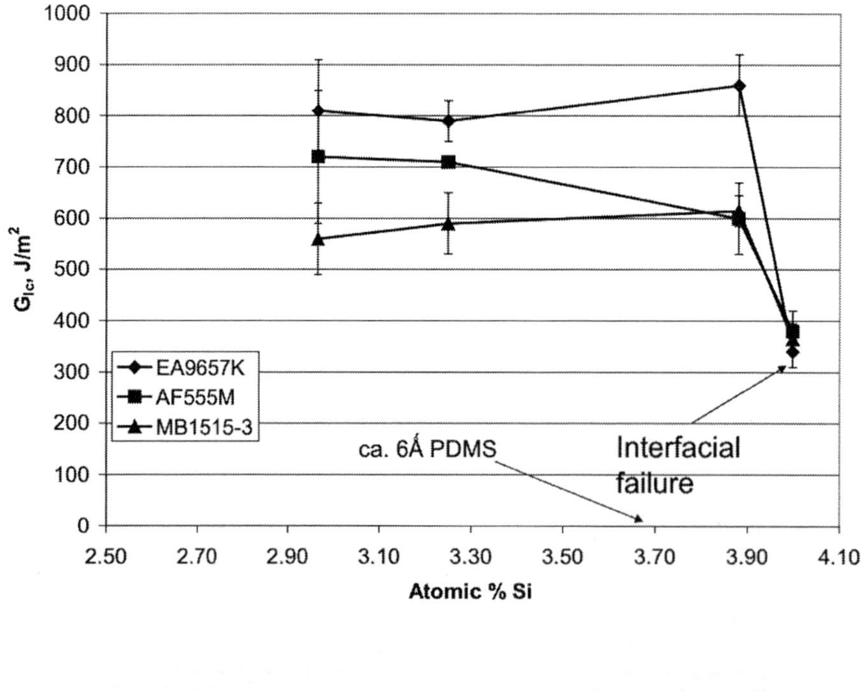

Ballistic water contact angle vs At.% silcon

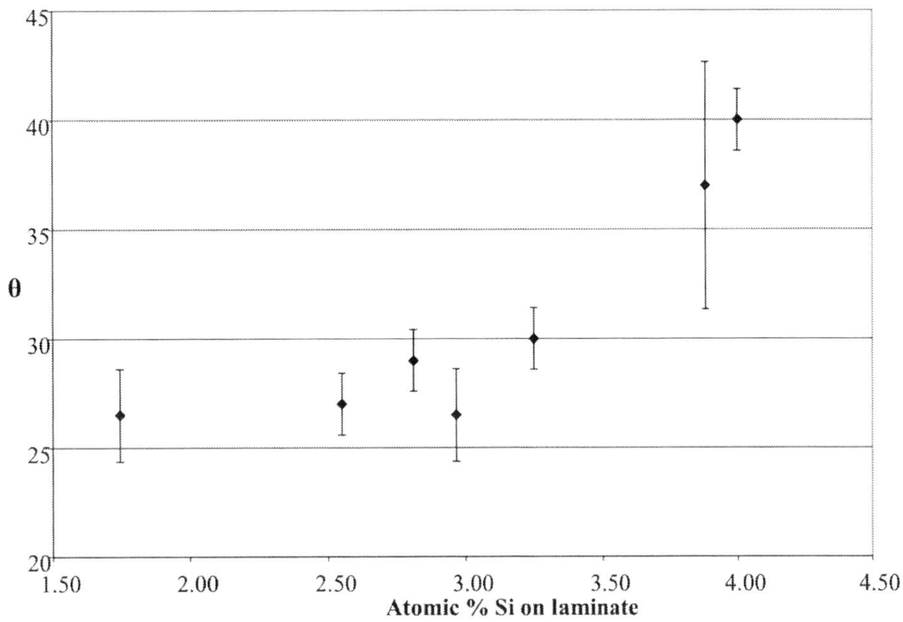

G_{Ic} vs ballistic water contact angle

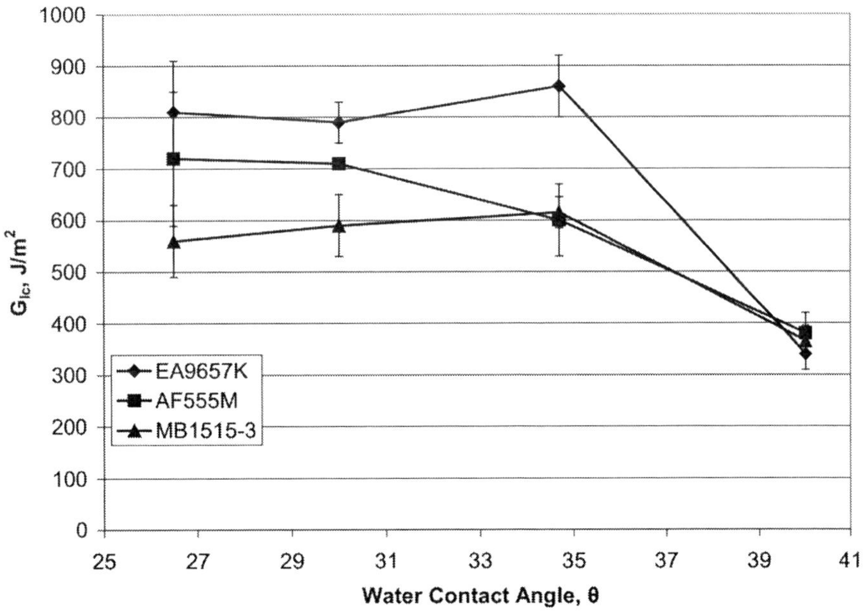

Conclusions

- Transfer of siloxane to laminates from contaminated peel ply can degrade adhesive bond performance.
- These adhesives tolerated the equivalent of a 6 Å siloxane layer before fracture toughness and failure mode were affected.
- Contact angles of ballistically deposited water drops readily discriminated between benign and detrimental levels of siloxane on macroscopically rough laminate surfaces.

Example: quality assurance of surface preparation for bonded nutplates

- Bonded nutplates allow use of threaded fasteners for panel attachment
- Can be 1000's of nutplates per aircraft
- Consistency of surface preparation is critical
- Variables:
  - Optimum surface treatment method
  - Aging after treatment
  - Contamination

Bonded nutplate failures

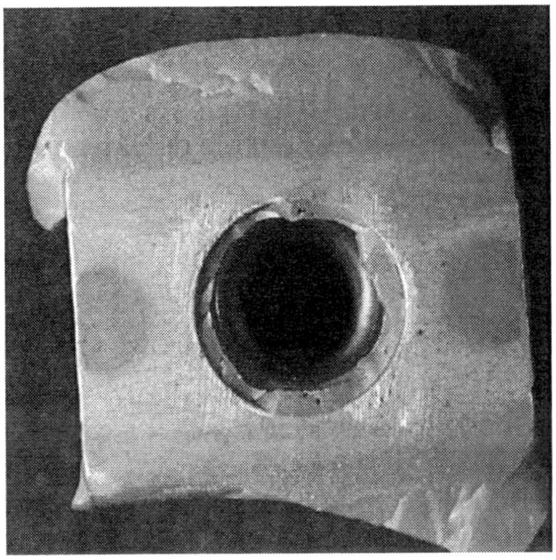

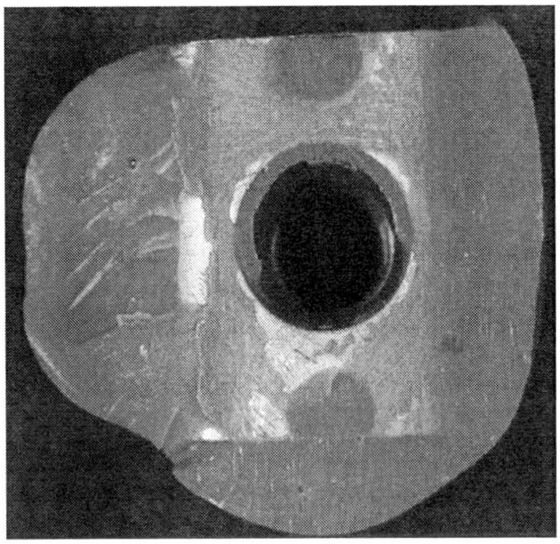

Interfacial failures seen in production

Bonded nutplates

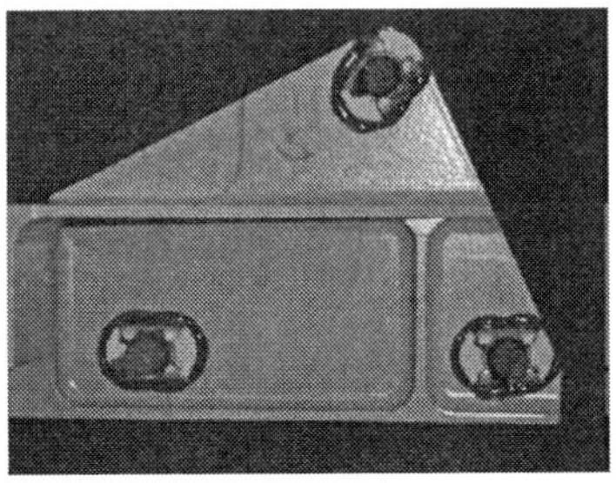

Titanium: effect of aging after surface treatment

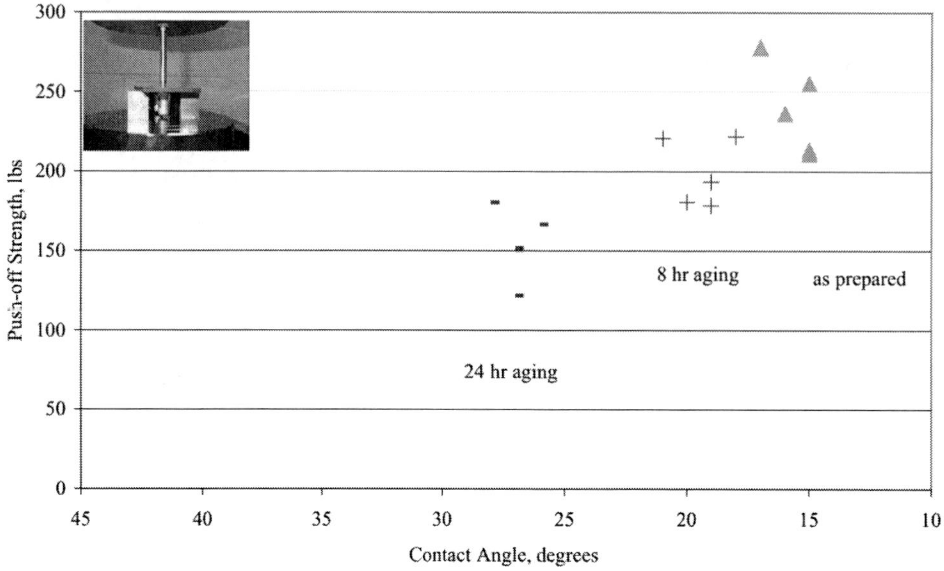

Composite: effect of aging after surface treatment

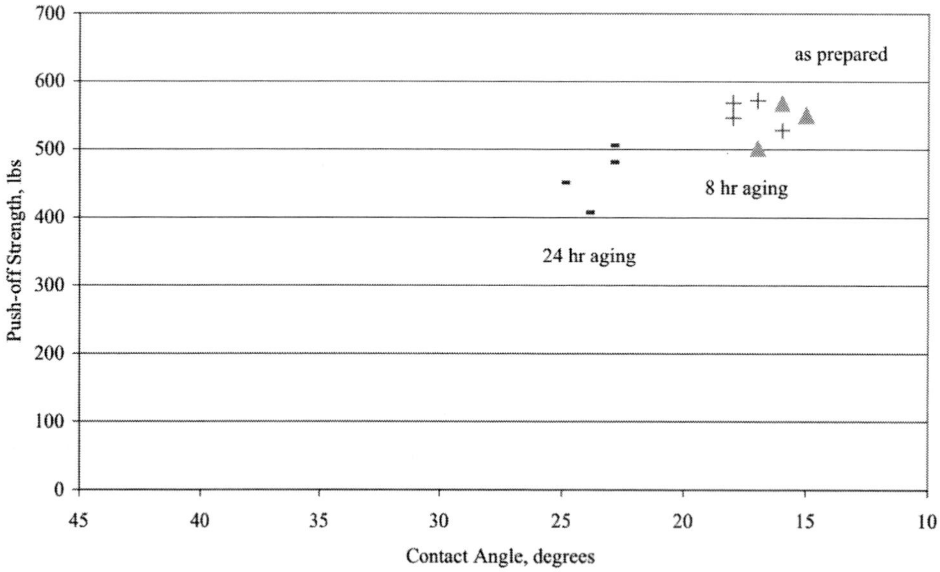

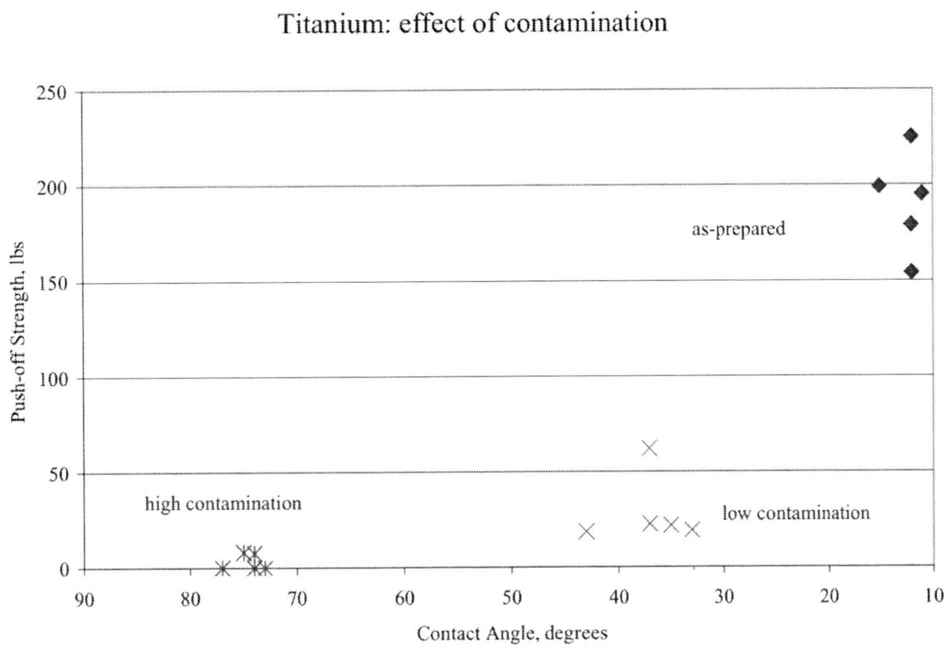
Titanium: effect of contamination
Push-off Strength, lbs
Contact Angle, degrees
as-prepared
high contamination
low contamination
250
200
150
100
50
0
90
80
70
60
50
40
30
20
10

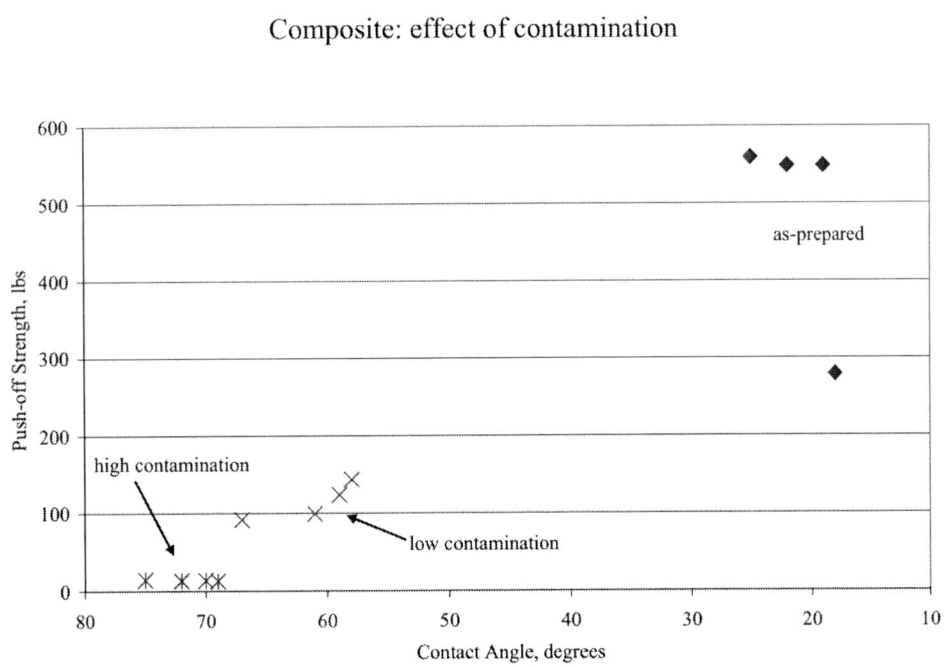
Composite: effect of contamination
Push-off Strength, lbs
Contact Angle, degrees
as-prepared
high contamination
low contamination
600
500
400
300
200
100
0
80
70
60
50
40
30
20
10

Adhesive bond performance depends on:

- Surface morphology
  - Millimeter to submicron scale roughness:
    - Permits mechanical interlocking between adhesive and substrate
    - Increases interfacial area
    - Changes stress state at interface
    - Improves tolerance to contaminants
- Surface chemistry
  - Fundamental source of adhesion between two phases
- Wetting measurements are sensitive to both

What can contact angles tell us about surface preparation?

- Contact angles provide a sensitive index of surface preparation quality and consistency
  - Quantitative relationships between wetting and adhesion developed for PSA's is applicable to rigid structural adhesives
  - With appropriate instrumentation, contact angle data can form the core of a quality assurance program for surface preparation
- θ_{H20} <~30° corresponds to maximum W_A and high adhesion

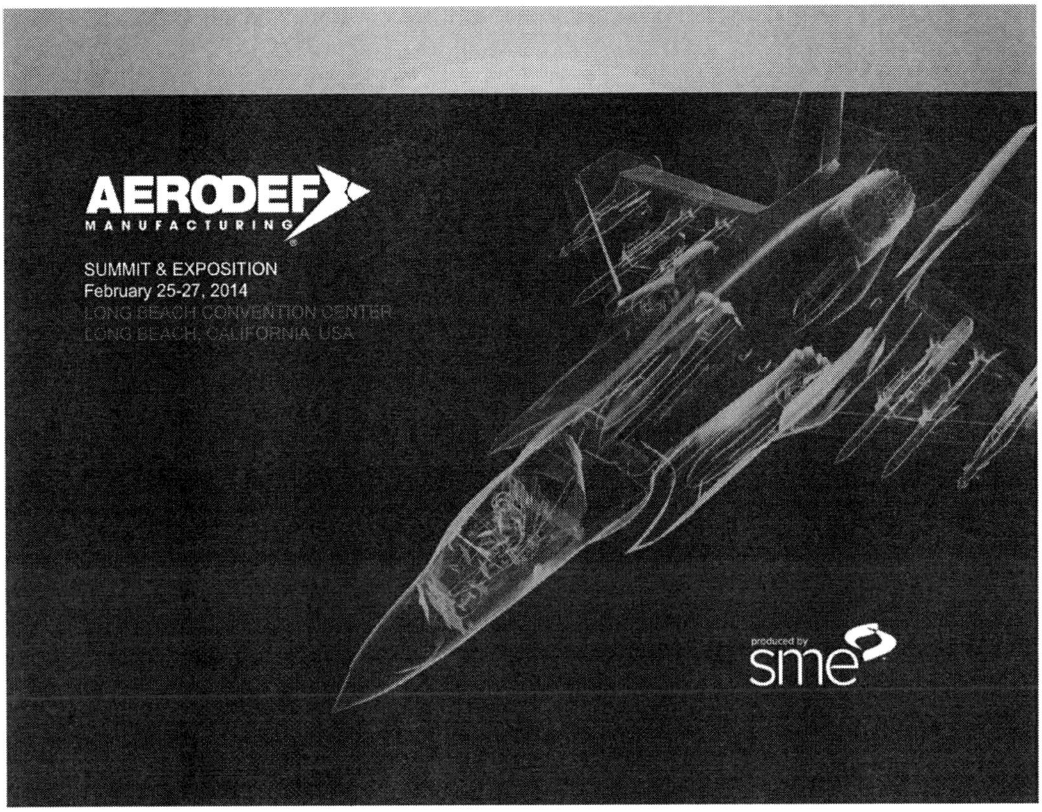
AERODEF
MANUFACTURING
SUMMIT & EXPOSITION
February 25-27, 2014
LONG BEACH CONVENTION CENTER
LONG BEACH, CALIFORNIA, USA
produced by
sme

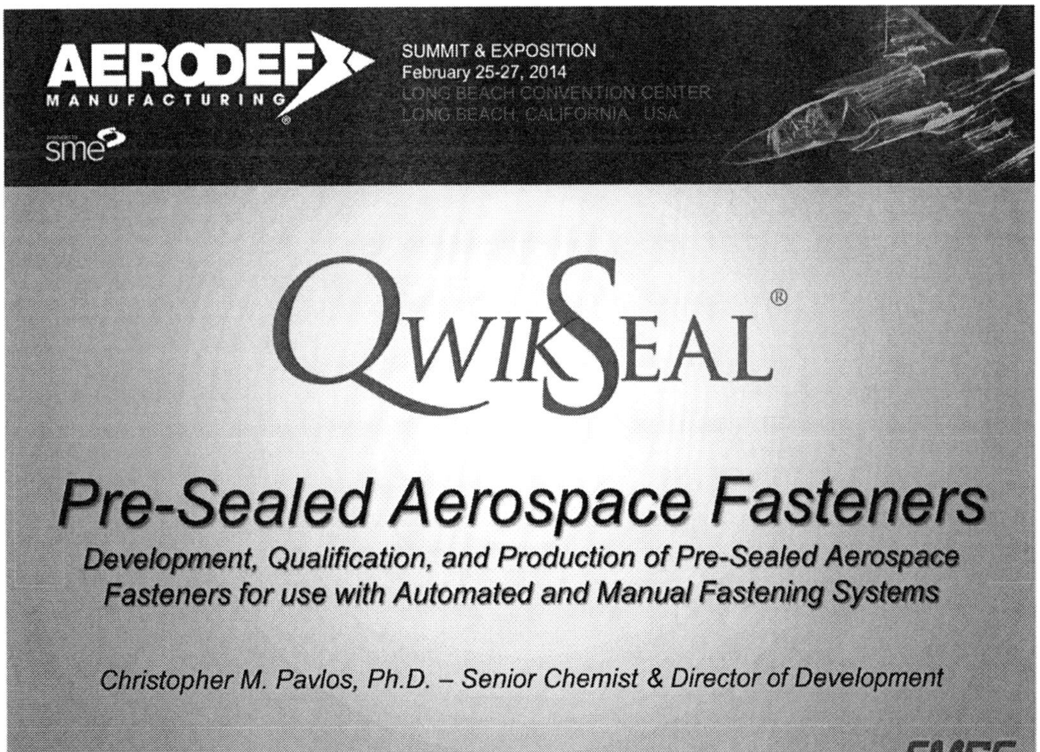
AERODEF
MANUFACTURING
sme
SUMMIT & EXPOSITION
February 25-27, 2014
LONG BEACH CONVENTION CENTER
LONG BEACH, CALIFORNIA, USA
QWIKSEAL®
Pre-Sealed Aerospace Fasteners
Development, Qualification, and Production of Pre-Sealed Aerospace Fasteners for use with Automated and Manual Fastening Systems
Christopher M. Pavlos, Ph.D. – Senior Chemist & Director of Development
SMRC

AERODEF MANUFACTURING

sme

SUMMIT & EXPOSITION
February 25-27, 2014
LONG BEACH CONVENTION CENTER
LONG BEACH, CALIFORNIA USA

Overview

- *Qwik*Seal Background
  - Problem
  - *Qwik*Seal Solution
  - Sealant
  - Barrier Coat
- Current Capabilities
  - Technology Snapshot
  - Fastener Production Capability
- SMRC Internal Testing
  - Installation Methods
  - Barrier Coat Extrusion/Sealant Coverage
  - Preload Retention
- Qualification Program (ESTCP)
  - Project Team
  - Qualification Process
  - Planned Testing
- Future Work

AERODEF MANUFACTURING

sme

SUMMIT & EXPOSITION
February 25-27, 2014
LONG BEACH CONVENTION CENTER
LONG BEACH, CALIFORNIA USA

Problem: "Wet-Installation"

- Mix or thaw sealant
  - *Limited shelf and work life*
- Apply to each fastener
  - *Operator dependent variation*
- Install fastener
- Clean up residual sealant with solvent wipes
  - *Excessive hazmat waste disposal*

This process is messy, time-consuming, and inconsistent

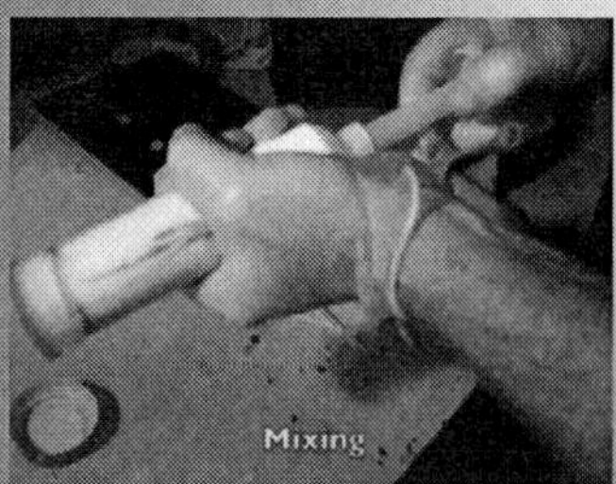

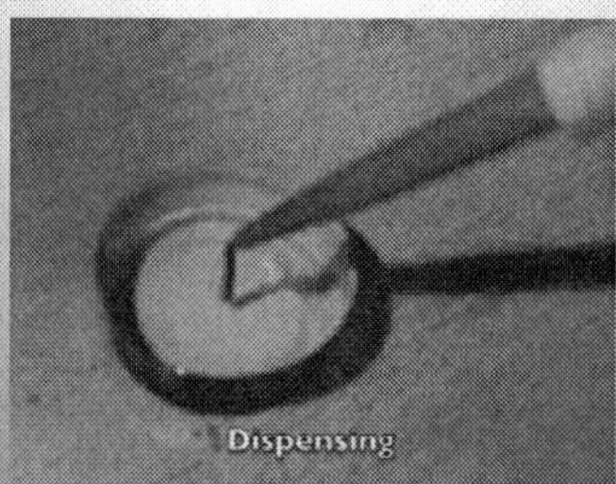

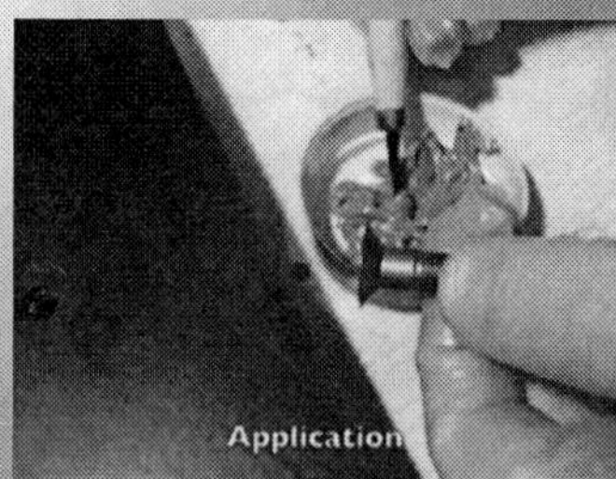

The fastener sealing process must be taken out of the critical path!

AERODEF MANUFACTURING
SUMMIT & EXPOSITION
February 25-27, 2014
LONG BEACH CONVENTION CENTER
LONG BEACH, CALIFORNIA USA
sme

Problem: Excessive Cleanup

- Conventional application of sealant to the panel can result in contamination of fastener threads, necessitating time consuming cleanup
- *Qwik*Seal fasteners can eliminate this problem resulting in significant labor savings

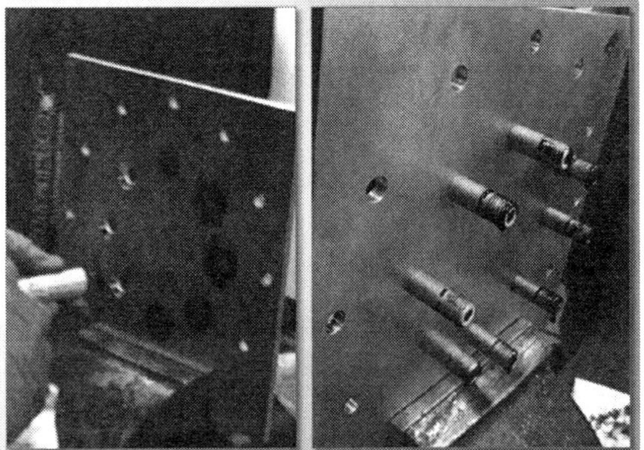

AERODEF MANUFACTURING
SUMMIT & EXPOSITION
February 25-27, 2014
LONG BEACH CONVENTION CENTER
LONG BEACH, CALIFORNIA USA
sme

The *Qwik*Seal Solution

- *Qwik*Seal sealant is a moisture-curable variant of polythioether-based aerospace sealant developed by SMRC and PPG Aerospace – PRC-DeSoto
- A pre-determined amount of *Qwik*Seal sealant is applied to the fastener and quickly encapsulated with a fast-curing barrier coat
- The barrier coat excludes moisture from the system, preventing sealant cure and thus extending the shelf-life of the fastener
- Upon installation, the barrier coat ruptures exposing the sealant to ambient moisture and initiating cure

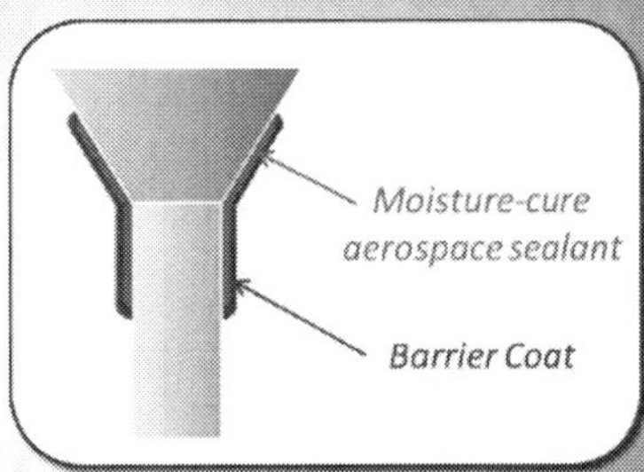

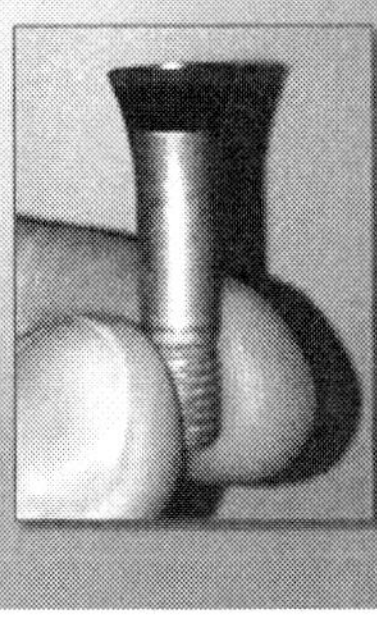

AERODEF MANUFACTURING
sme
SUMMIT & EXPOSITION
February 25-27, 2014
LONG BEACH CONVENTION CENTER
LONG BEACH, CALIFORNIA USA

*Qwik*Seal Sealant

- AMS 3277 compliant polythioether resin modified with moisture-sensitive cure-blocking agent.
- Cure-blocked sealant is stable in dry conditions, but hydrolyzes in ambient air to regain reactivity.

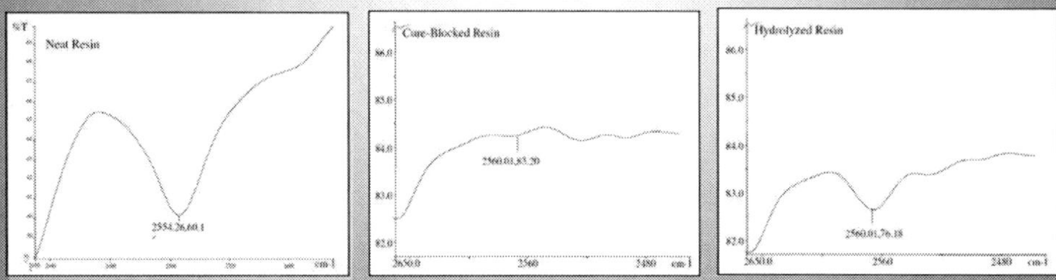

AERODEF MANUFACTURING
sme
SUMMIT & EXPOSITION
February 25-27, 2014
LONG BEACH CONVENTION CENTER
LONG BEACH, CALIFORNIA USA

*Qwik*Seal Barrier Coat

- Barrier coat encapsulates sealant slowing moisture permeation and adding protection against handling
- Extruded from countersink in one piece upon fastener installation.
  - Very durable – provides excellent handling
  - Anticipated long shelf-life (15-30 days, currently under development)

AERODEF MANUFACTURING

sme

SUMMIT & EXPOSITION
February 25-27, 2014
LONG BEACH CONVENTION CENTER
LONG BEACH, CALIFORNIA USA

Technology Snapshot

- *Qwik*Seal cure-blocking chemistry developed by SMRC
  - Sealant formulation and scale-up transitioned to PPG Aerospace – PRC-DeSoto
- Barrier coat characteristics developed by SMRC
  - Material composition and application hardware finalized
  - Profile and thickness – fastener dependent – continually optimizing new types
- Prototype fastener production machine acquired
  - Dispense hardware finalized by SMRC – continually optimizing parameters for new fastener types
  - Transferred to PPG's Valencia, CA Application Support Center (ASC) in March 2013
- Currently undergoing qualification under ESTCP Dem/Val program

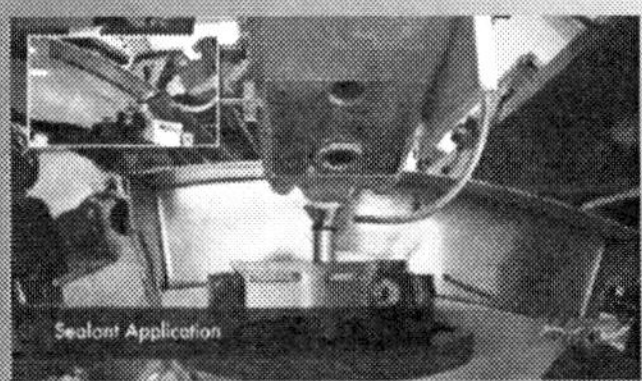

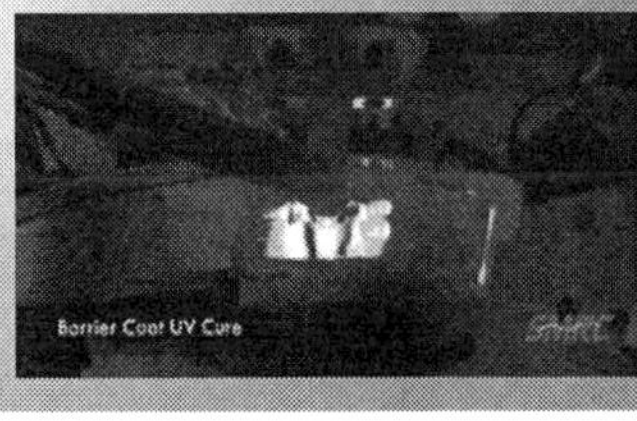

AERODEF MANUFACTURING

sme

SUMMIT & EXPOSITION
February 25-27, 2014
LONG BEACH CONVENTION CENTER
LONG BEACH, CALIFORNIA USA

Sealant Application

- *Qwik*Seal production machine has capability to apply sealant in a variety of positions
- Sealant profile will be dictated by individual applications
- New valve recently acquired to decrease cycle time and increase profile consistency

AERODEF MANUFACTURING
sme

SUMMIT & EXPOSITION
February 25-27, 2014
LONG BEACH CONVENTION CENTER
LONG BEACH, CALIFORNIA USA

Barrier Coat Application

- SMRC Proprietary Material
- UV-Curable
- Spray valve recently integrated onto *Qwik*Seal machine.
- Notching introduces a pre-determined failure point and ensures complete and uniform extrusion
  - Currently being done by hand
  - Capability will be integrated onto *Qwik*Seal production machine

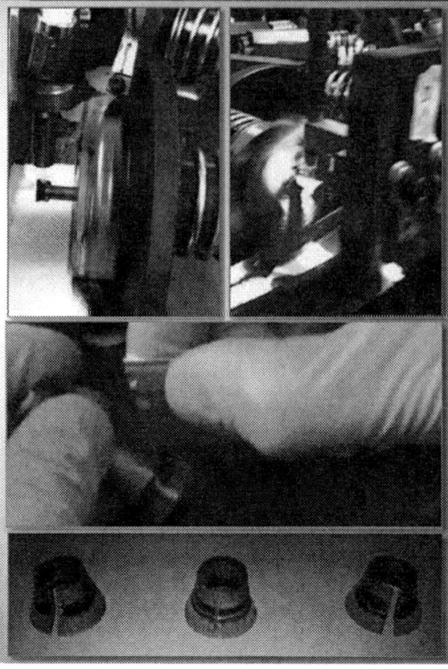

AERODEF MANUFACTURING
sme

SUMMIT & EXPOSITION
February 25-27, 2014
LONG BEACH CONVENTION CENTER
LONG BEACH, CALIFORNIA USA

Fastener Production (Prototype Machine)

- Current prototype machine cycle time ~11 s/fastener
- Next generation expected to significantly speed production rate

AERODEF MANUFACTURING

produced by sme

SUMMIT & EXPOSITION
February 25-27, 2014
LONG BEACH CONVENTION CENTER
LONG BEACH, CALIFORNIA USA

Pneumatic Installation

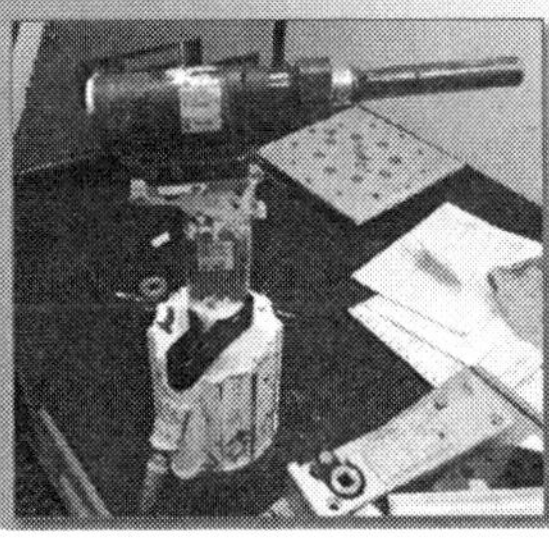

Extrusion/Sealant Coverage

Preload Retention

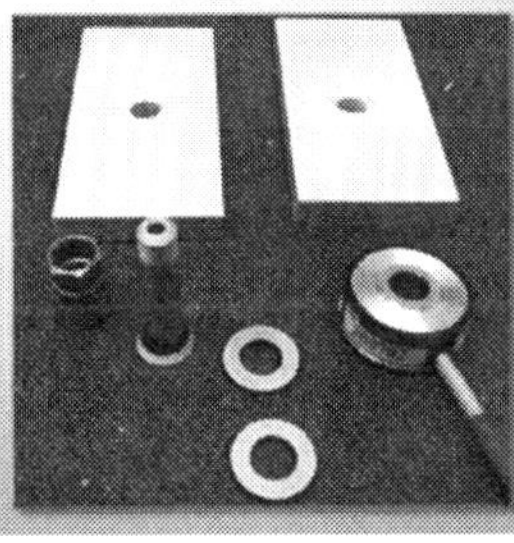

AERODEF MANUFACTURING

produced by sme

SUMMIT & EXPOSITION
February 25-27, 2014
LONG BEACH CONVENTION CENTER
LONG BEACH, CALIFORNIA USA

Push-through Installation

- Driven into panel with pneumatic hammer
- Optimized for ¼" - $^{7}/_{16}$" shear head fasteners (other sizes currently under development)

AERODEF MANUFACTURING
sme

SUMMIT & EXPOSITION
February 25-27, 2014
LONG BEACH CONVENTION CENTER
LONG BEACH, CALIFORNIA USA

Pull-through Installation

- Lockbolt fasteners installed by pulling from the back side of the panel with a pneumatic tool
- Fasteners must be fully seated in one smooth motion to avoid barrier coat becoming trapped
- When the proper stroke length is used, complete extrusion of barrier coat is consistently observed
- ¼" - $^{7}/_{16}$" shear head fasteners optimized to date – other sizes currently under development

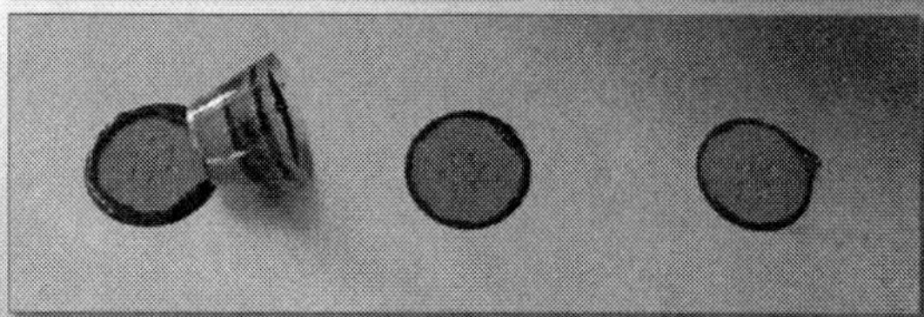

AERODEF MANUFACTURING
sme

SUMMIT & EXPOSITION
February 25-27, 2014
LONG BEACH CONVENTION CENTER
LONG BEACH, CALIFORNIA USA

Preload Retention Test

In aircraft and other manufacturing processes, fasteners are tightened to achieve a certain amount of preload stress, and preload retention is critical to ensuring long term joint integrity.

- When sealing a fastener with a viscous sealant, a certain amount of the preload stress can be lost due to sealant creep out of the countersink hole during cure.

In the case of *Qwik*Seal, small pieces of barrier coat trapped inside the countersink could migrate out of the hole before sealant cure and serve as an additional source of preload loss.

- This is not expected to happen if extrusion of barrier coat is complete.

AERODEF MANUFACTURING

sme

SUMMIT & EXPOSITION
February 25-27, 2014
LONG BEACH CONVENTION CENTER
LONG BEACH, CALIFORNIA, USA

Preload Retention Test

Equipment

- Hi-Lok fasteners and collars
- Al Panels (2)
- Compression Load Cell
- Hardened washers
- DAQ/10 V Power supply
- J-Type Thermocouple
- *Qwik*Seal and control fasteners

Procedure

- Assemble components as shown
- Collar shears off when requisite torque is achieved
- Collect data until sealant has fully cured

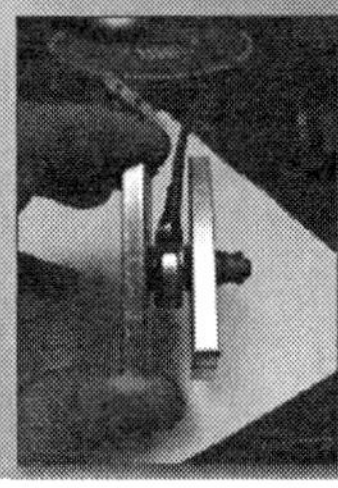

AERODEF MANUFACTURING

sme

SUMMIT & EXPOSITION
February 25-27, 2014
LONG BEACH CONVENTION CENTER
LONG BEACH, CALIFORNIA, USA

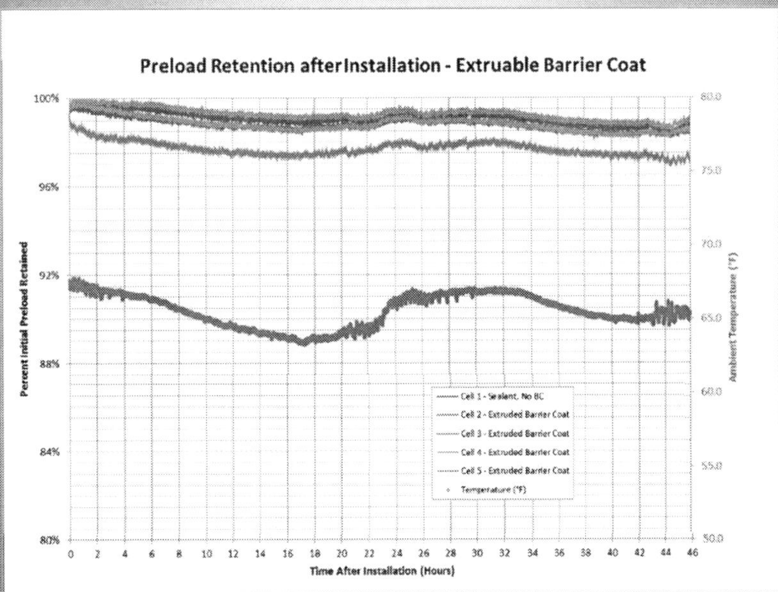

AERODEF MANUFACTURING
sme
SUMMIT & EXPOSITION
February 25-27, 2014
LONG BEACH CONVENTION CENTER
LONG BEACH, CALIFORNIA USA

Aerospace Materials Qualification Program (SAE G9)

- Qualification has been broken into 3 subtasks:
  - Develop specification, test/qualify sealant, qualify fasteners
- Draft specification is being developed between AFRL, NAVAIR, PPG Aerospace – PRC-DeSoto, and SMRC
  - Joint test plan agreed on and submitted (March 2013)
  - Specification to be discussed at G9 Spring 2014 meeting
- Sealant testing anticipated to begin Q1 2014
  - Sealant testing will be performed by UDRI
- On-fastener testing and qualification will begin Q2 2014
  - Testing will be conducted at AFRL, UDRI, and SMRC

AERODEF MANUFACTURING
sme
SUMMIT & EXPOSITION
February 25-27, 2014
LONG BEACH CONVENTION CENTER
LONG BEACH, CALIFORNIA USA

Qualification Program Team

SMRC
- Christopher Pavlos (PI)
- Malcolm Prouty
- Chelsea Boundy
- Keris Ward

PPG
- Thom Dolezal (PM)
- Bruce Virnelson
- Bill Keller
- John Sands

AFRL
- Al Fletcher (PI)
- Michelle Barga

NAVAIR
- Diane Kleinschmidt

UDRI
- Kate Holley

AERODEF MANUFACTURING
sme

SUMMIT & EXPOSITION
February 25-27, 2014
LONG BEACH CONVENTION CENTER
LONG BEACH, CALIFORNIA USA

Sealant Testing

- Cure characteristics
- Hydrolytic Stability
- Resistance to Hydrocarbons
- Tensile Strength
- Peel Strength
- Hardness
- Corrosion Resistance

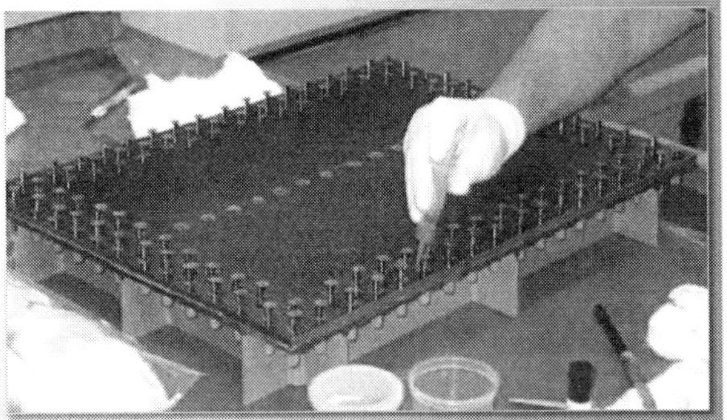

Fastener Testing

- Preload Retention
- Sealability (puffer box)
- Corrosion
- Shelf-life
- Cure-time

AERODEF MANUFACTURING
sme

SUMMIT & EXPOSITION
February 25-27, 2014
LONG BEACH CONVENTION CENTER
LONG BEACH, CALIFORNIA USA

Future Work

- Determine shelf-life (anticipate 2-4 week min)
- Design and build 2nd generation production machine
  - High throughput rate (~ 20 fasteners/min)
  - More sophisticated QA
- Scale materials production processes
- Production trials with potential customers

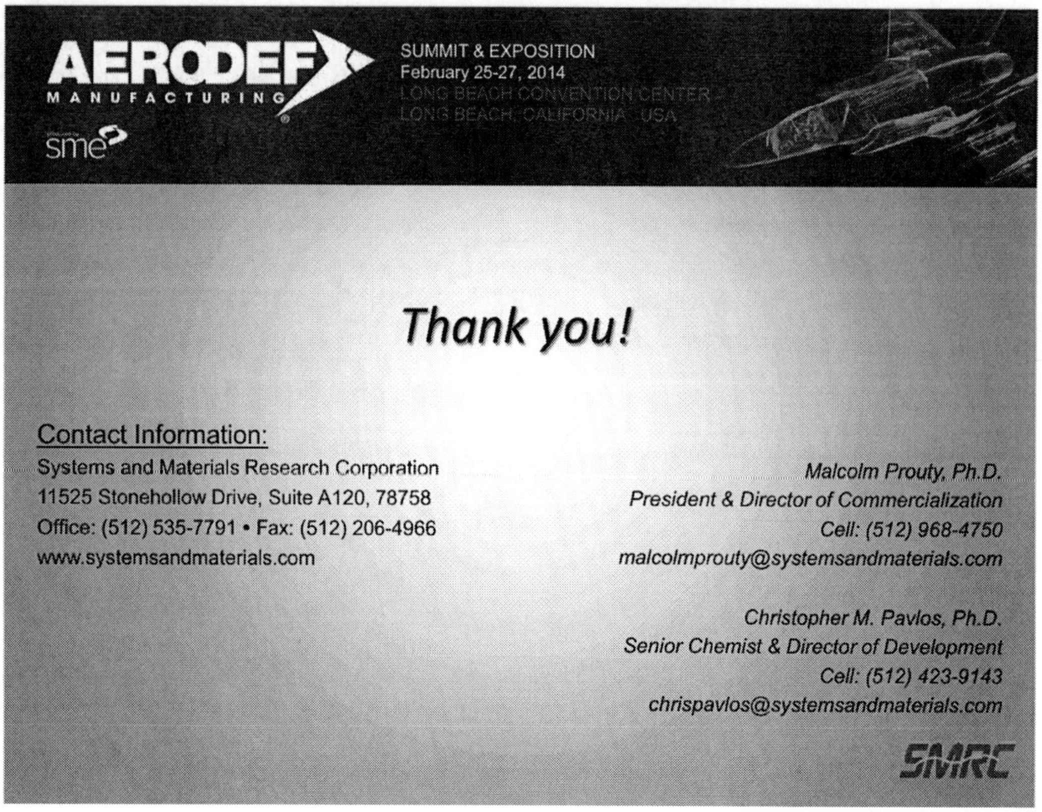
AERODEF MANUFACTURING
SUMMIT & EXPOSITION
February 25-27, 2014
LONG BEACH CONVENTION CENTER
LONG BEACH, CALIFORNIA USA
sme
Thank you!
Contact Information:
Systems and Materials Research Corporation
11525 Stonehollow Drive, Suite A120, 78758
Office: (512) 535-7791 • Fax: (512) 206-4966
www.systemsandmaterials.com
Malcolm Prouty, Ph.D.
President & Director of Commercialization
Cell: (512) 968-4750
malcolmprouty@systemsandmaterials.com
Christopher M. Pavlos, Ph.D.
Senior Chemist & Director of Development
Cell: (512) 423-9143
chrispavlos@systemsandmaterials.com
SMRC

Non-Contact Metrology Applications Development

Fastener Flushness Inspection

Chris Barrow

February 26, 2014

LOCKHEED MARTIN

Non-Contact Metrology Applications Development

OBJECTIVE: Investigate Advancements In Non-Contact Metrology Technology Which Could Address F-35 Application Issues

- Decrease Span
- Increase Quality

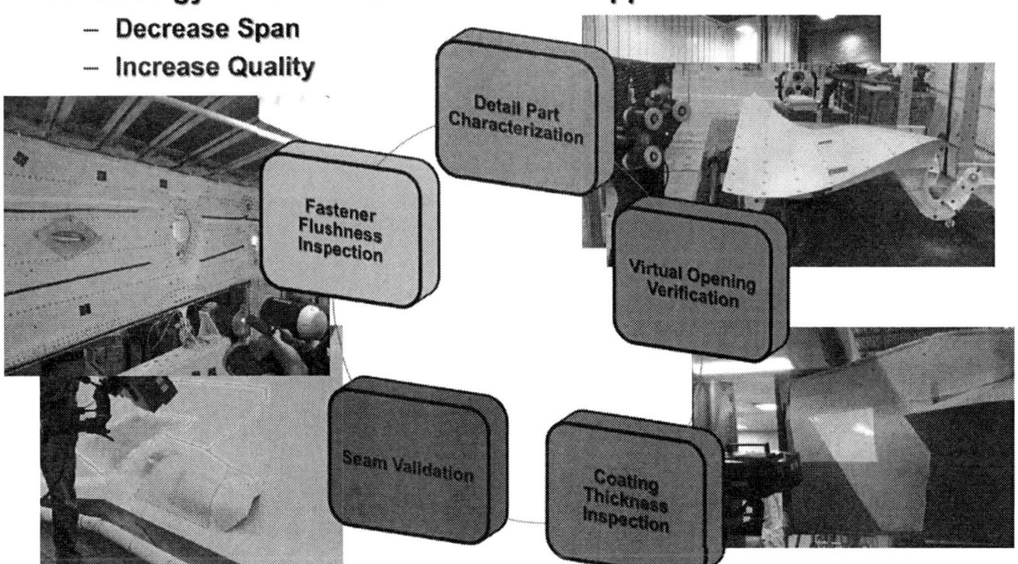

2

Fastener Flushness Issue

- **Requirement:** Inspect Sub-Flush Condition Of All Fasteners Installed On F-35 Exterior Surfaces

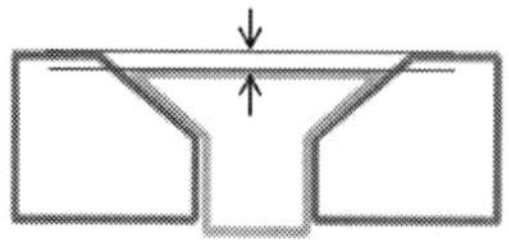

Countersink With Sub-Flush Fastener

- **Baseline Process:** Handheld Fowler Flushness Gauge

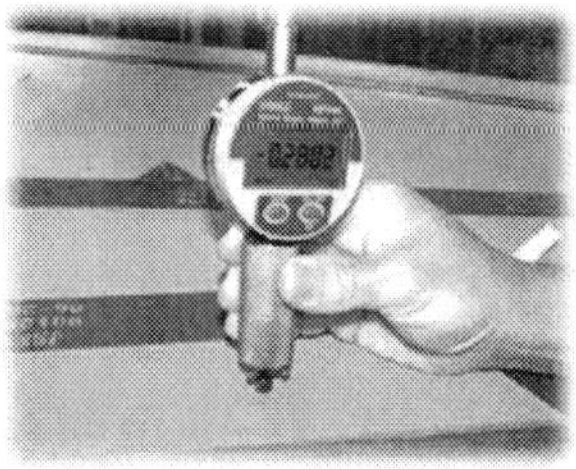

- **Issues**
  - Measurement Quantity
  - Non-Repeatable
  - Limited Data Collection

3

Development Plan

4

Industry Survey

- **RFI Conference Conducted To Solicit Technology Solutions For Application Issues**
  - Multiple Representatives From 10+ Metrology Suppliers In Attendance
  - Reviewed Application Needs Including:
    - Feature Types
    - Feature Condition
    - Environmental Concerns
- **Team Reviewed Supplier Proposals And Downselected Tech For Further Evaluation**

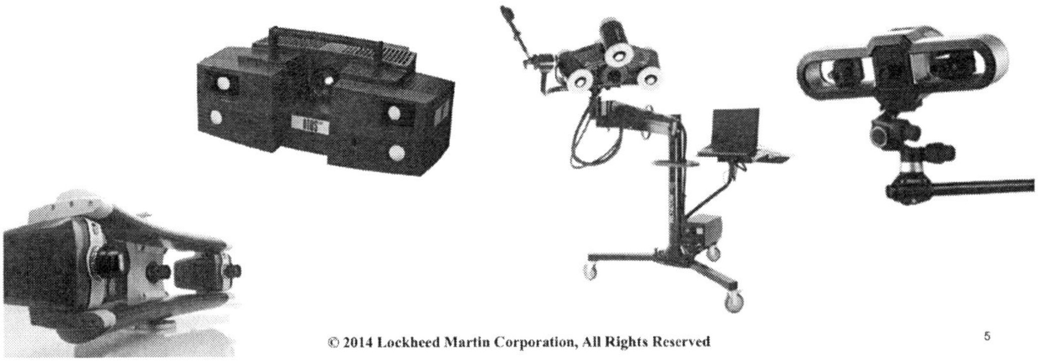

Sub-Scale Evaluation

- **Goal:** Evaluate the Accuracy/Repeatability Of Multiple Systems & Downselect Optimal For Further Development
- **Article:** Calibrated Spotface Panel - 21 “Fasteners”

Selected Technology

- **Cognitens WLS400M**
  - **Among Best Tested In Accuracy/Repeatability**
  - **Least Affected By Vibration**
    - *Can Be Handheld*
    - *Unaffected By Mechanics On Aircraft*
  - **Development Needs Identified → Further Automation Of Data Interrogation/Reporting**

HEXAGON
METROLOGY

Process Overview

Prepare → Measure → Analyze → Report

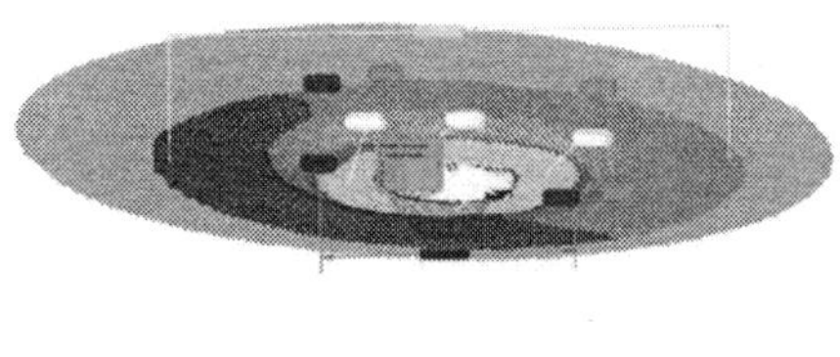

Algorithm Logic

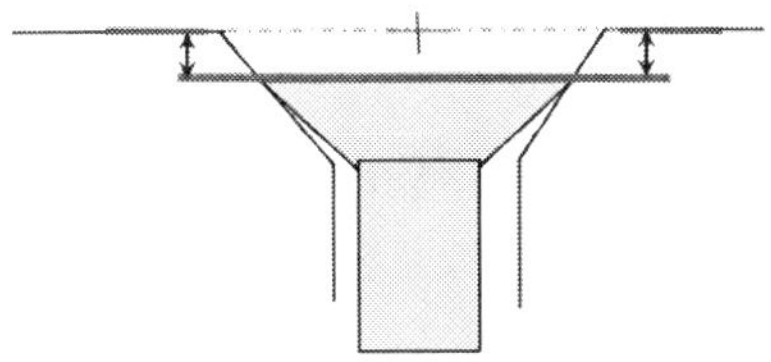

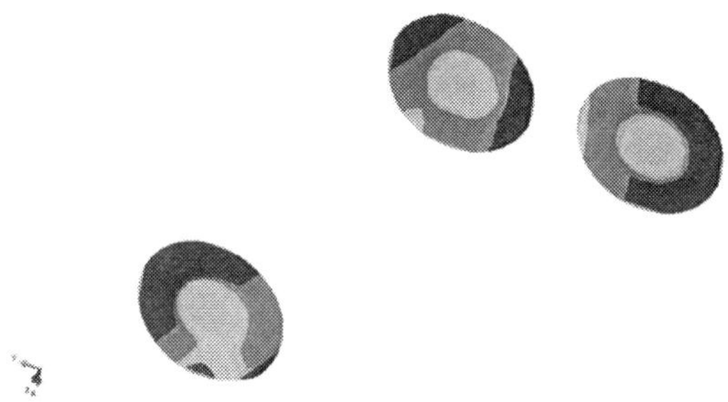

9

Full Scale Evaluation

- **Goal:** Validate Technology & Process On Aircraft

- **Article:** F-35 Lower Wing Skin

- **Result**
  - Passing Gage R&R On Aircraft!
  - Significant Number Fasteners Shown As "Proud"
  - Visual Inspection Of Skin Showed No Proud Fasteners → Bad Correlation

10

Root Cause Analysis

- **Investigation Showed "Proud" Fasteners Primarily Match Two Conditions**
  1. Fasteners Located In Ramp Areas
  2. Tilted Fasteners

- **Explainable Conditions With Current Algorithm**

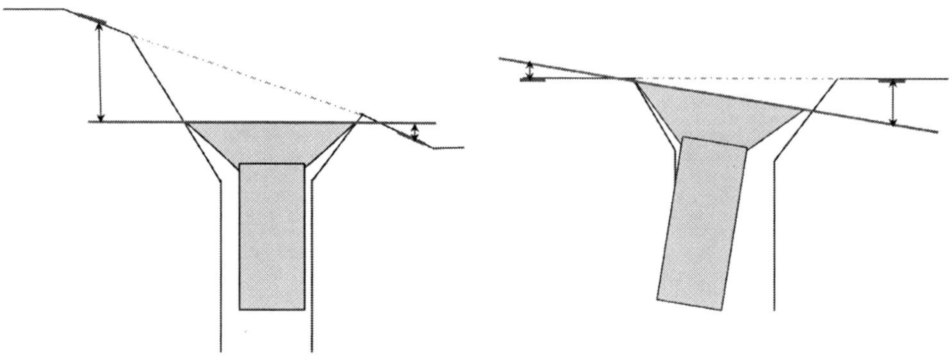

Potential Solutions

- **Fasteners On Ramps**
  - Create Surface Ring Closer To Fastener Head → Minimize Effect Of Ramps

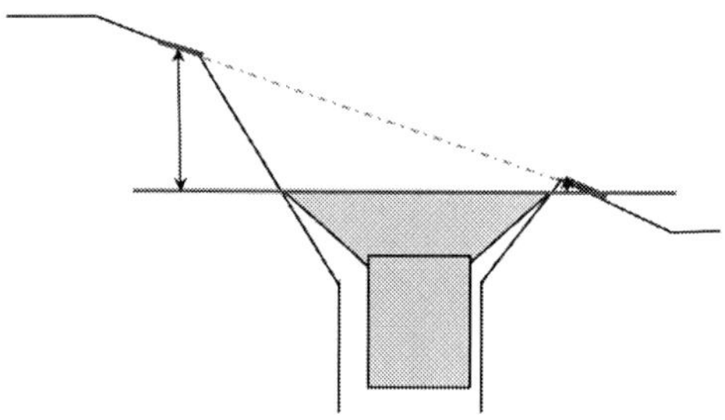

- **Tilted Fasteners**
  - Flag Out Of Angularity Fasteners

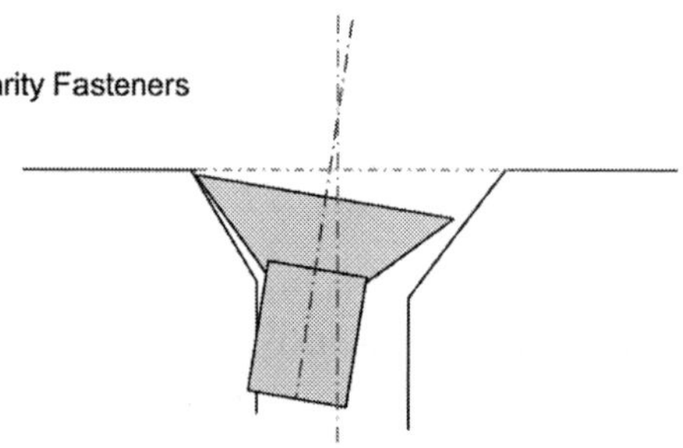

Next Steps

- **Fastener Flushness Inspection**
  - Refine Algorithms
  - Conduct Production Trial To Deploy Process On Floor And Prove Decreased Span
  - Collect SPC Data To Improve Quality Of Manufacturing Processes
  - Expand Deployment To Other Build Stations

- **Develop/Integrate Solutions For Other Application Issues**

13

THE VALUE OF PERFORMANCE.
NORTHROP GRUMMAN

Maturation of High Temperature Laser Sintering (LS) Technologies and Infrastructure for Air and Space Vehicles

AeroDef 2014

Long Beach, CA

Pedro A. Gonzalez

America Makes

Team Roles and Responsibilities

NORTHROP GRUMMAN

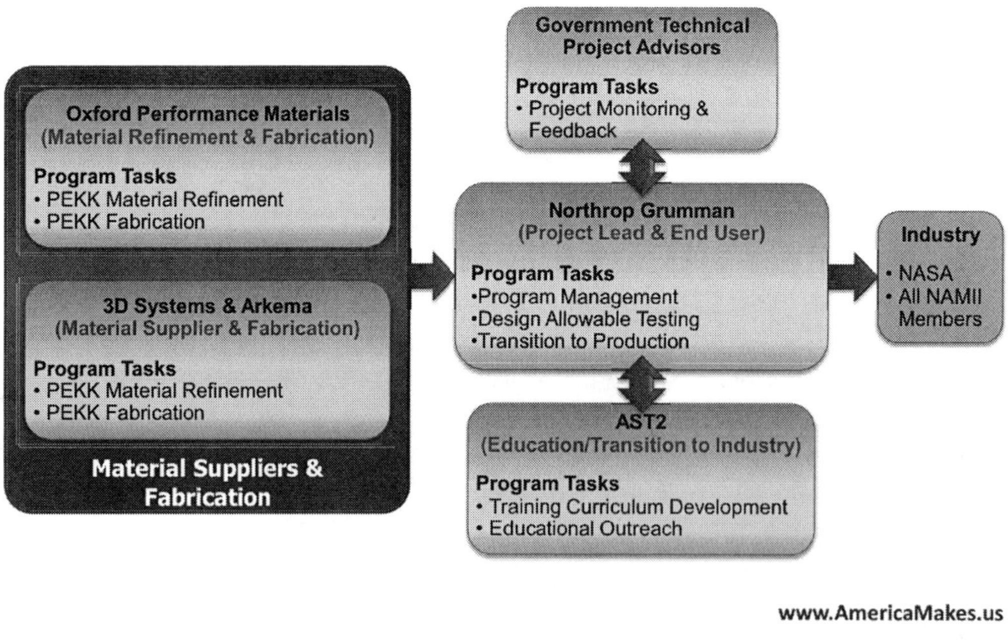

www.AmericaMakes.us

Schedule

NORTHROP GRUMMAN

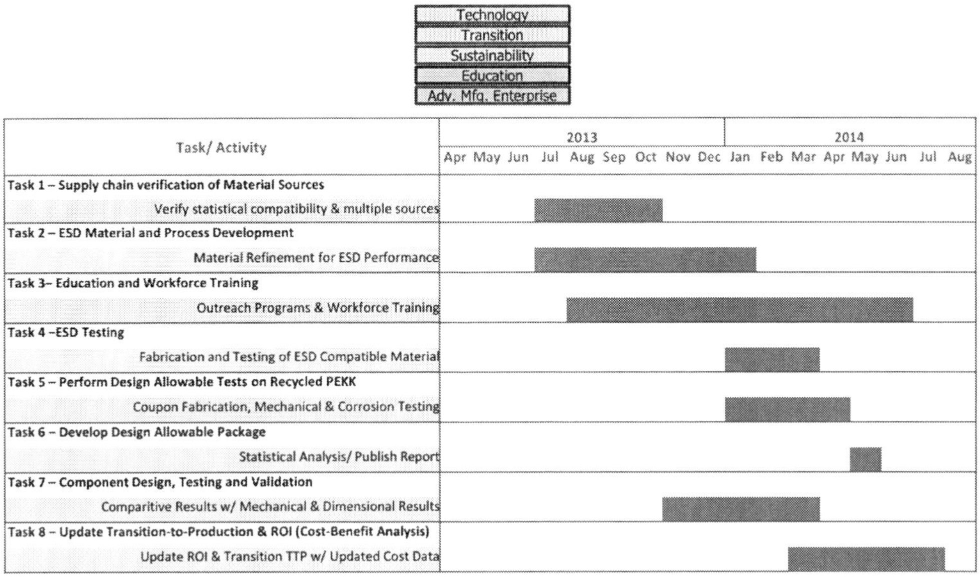

www.AmericaMakes.us

Program Technical Progress

NORTHROP GRUMMAN

- PEKK Test Matrix Developed
  - Create ESD PEKK B-Basis & Neat PEKK S-Basis Data
  - 2,467 Coupons to be fabricated & tested
  - 6 different temperatures
  - Soaked samples
    - (Hydraulic Fluids, Fuel)
  - Corrosion testing
  - Space environment specific tests
    - Extreme Cold, Outgassing, Thermal Conductivity, etc

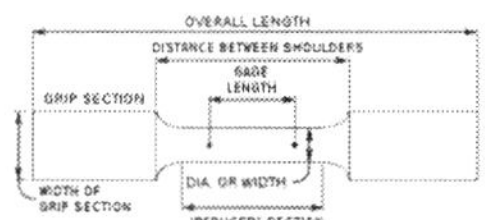

Typical Coupon

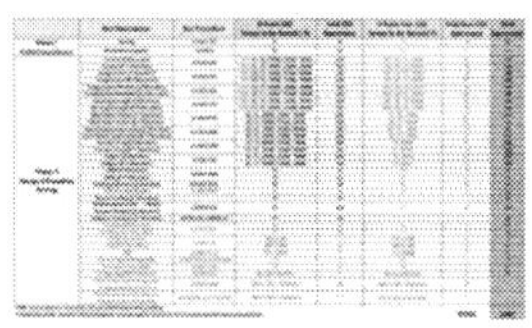

PEKK Test Matrix

www.AmericaMakes.us

Program Technical Progress

NORTHROP GRUMMAN

- Initial Material Screening
  - Baseline material characteristics

- OPM Coupons were placed in build chamber and distributed in four quadrants

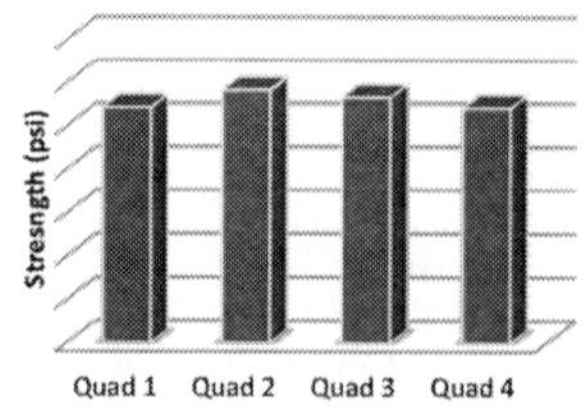

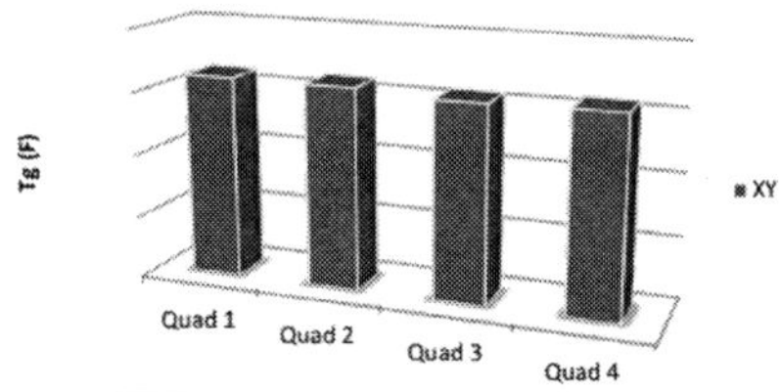

www.AmericaMakes.us

Program Technical Progress

NORTHROP GRUMMAN

- Density Cubes
  - Cubes were examined for porosity & density

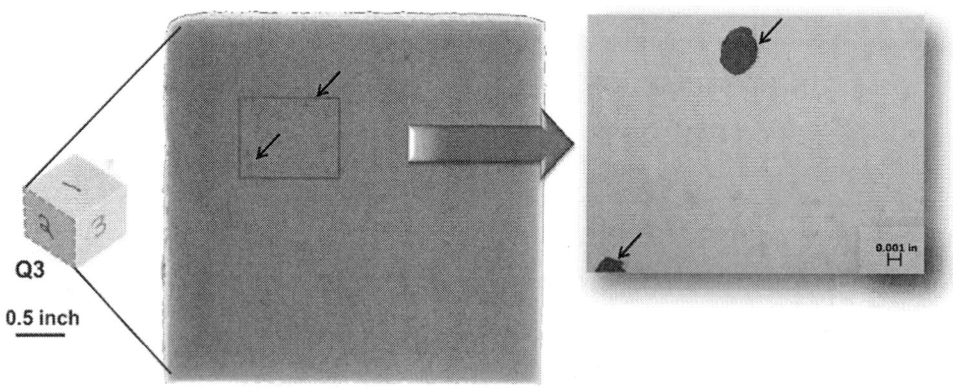

www.AmericaMakes.us

Transition & Education Plans

NORTHROP GRUMMAN

- Transition to Production
  - Various Northrop Grumman Aerospace platforms are being considered for implementation
  - Data available to America Makes members for implementation
- Educational Outreach & Workforce Training
  - Discovery & Inventorcloud Plans Being Developed
    - Educate K-12 students on Additive Manufacturing
  - Assemble AM Body of Knowledge for Workforce Training
  - Create Mobile AM Exhibit to spread the AM word

www.AmericaMakes.us

THE VALUE OF *PERFORMANCE.*

NORTHROP GRUMMAN

America Makes

8

Recent Trends in Metal Additive Manufacturing

Presented At AeroDef Manufacturing Conference
Long Beach, California USA

February 26, 2014

Doug Czaplicki
Honeywell Aerospace

Honeywell

Agenda

Honeywell

- Machine Developments
- Materials
- In-Process Monitoring
- Supply Chain

Machine Developments

Honeywell

- Metal additive manufacturing machines have been rapidly evolving

- Similar traits for many current machines
  - Laser-based systems with single 200 or 400 W laser
  - ~9.5" cube build volume

- Focus for recently announced machines:
  - Higher power lasers, multiple lasers
  - Larger volume
  - Production
  - Modularity

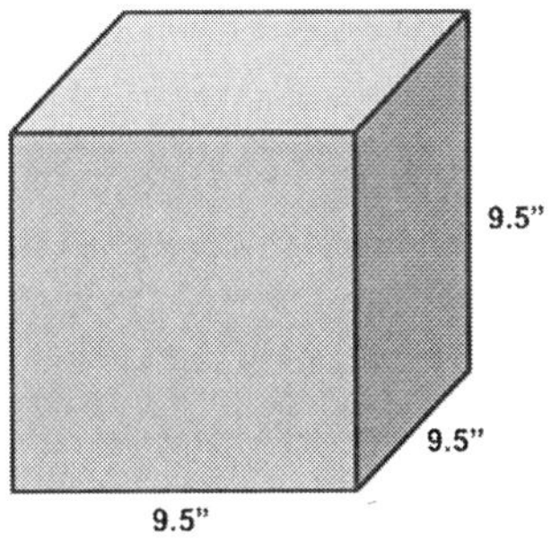

Machine Developments

Honeywell

- Build volumes are increasing
  - 3 of the major vendors now offer machines with build volumes larger than 9.5" cube

  - EOS M400: 15.75" x 15.75" x 15.75" (x,y,z)
  - SLM 500 HL: 19.5" x 11" x 12.75"
  - Concept X 1000R: 24.75" x 15.75" x 19.5"

Reference: Google image search

Machine Developments

Honeywell

- Laser power is trending towards 1 kW
  - Higher laser power = larger layer thickness & faster builds
    - 50 µm vs. 100+ µm layers
    - Trade-off between speed and resolution
- Multi-laser platforms
  - Promise of increased production rates
  - Increased complexity of build parameters that can be utilized
  - Potential maintenance issues? Higher lifetime costs?
  - Calibration issues between separate lasers?
- Production
  - Machine builders are adding in-process monitoring equipment
    - Aerospace has a need to verify part quality
  - Dual recoater blades
    - Small decrease in time between layers
    - Assume 2 sec. savings, 50 µm layer size, ~4" build height (~2000 layers) = **66 minutes saved**

5 HONEYWELL - CONFIDENTIAL

Machine Developments

Honeywell

- Modularity
  - Similar to pallet system for CNC machine centers
  - Separate chambers – one for building, one for post-processing
    - Turntable
    - Swappable frames
  - Decreased machine idle time

6 HONEYWELL - CONFIDENTIAL

Materials

Honeywell

- Many common alloys used in aerospace are available
  - Cobalt Chromium (CoCr)
  - Nickel: Alloys 625 & 718
  - Iron: 15-5, 17-4, maraging steel, 316
  - Titanium: Ti 6-4, CP Ti
  - Aluminum: AlSi10Mg, AlSi12, 6061
- New alloys being qualified by machine manufacturers
  - Hastelloy X (weldable nickel alloy with good oxidation resistance)

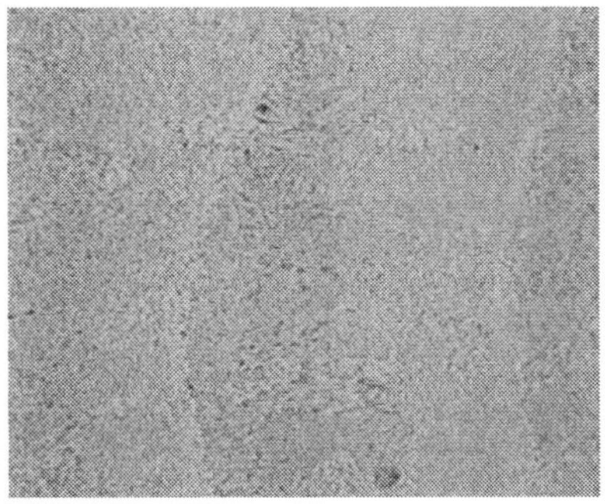

Micrograph of DMLS AlSi10Mg. Magnification: 400X

- Other alloys are being qualified while some companies & suppliers are experimenting on their own

Materials

Honeywell

- Lack of available mechanical property data is a large issue with regards to aerospace
  - Materials engineers need to understand behavior
  - Designers need to data to create a component
  - Engine applications have stricter requirements than most other applications
  - Cost to qualify a material is extremely high

Tensile

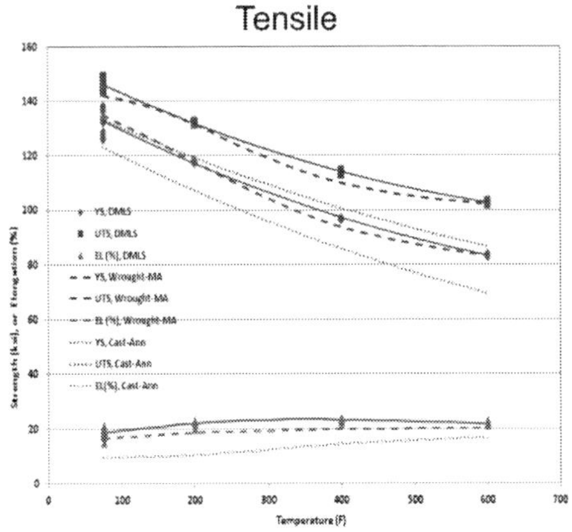

Materials

Honeywell

- Industry needs a methodology to combine and distribute property data for common alloys
  - Similar to MIL-HDBK-5 / MMPDS
  - Standard alloys built using standard parameters

- Increased data availability = faster innovation and adoption

In-Process Monitoring

Honeywell

- Ensuring part quality is critical for aerospace applications
  - NDT is useful but occurs after a part is built

- Additive offers a chance to verify quality during build process

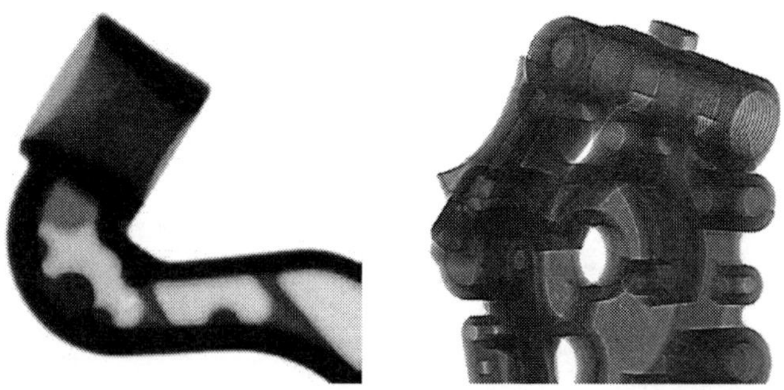

X-ray film of separate castings

Reference: (L) Google image search, (R) http://www.eidosolutions.it/eido_edf_en.htm

In-Process Monitoring

Honeywell

- Weld pool monitoring
  - Capture temperature and shape of weld pool continuously
  - IR camera / pyrometer to view the platform
  - Photo diode / CMOS for looking through the laser
- Monitor oxygen content
  - Detect chamber leaks (increased oxides in material)
  - Critical for reactive metals
- Camera to review platform for defects
  - Poor recoating, abnormal powder particles

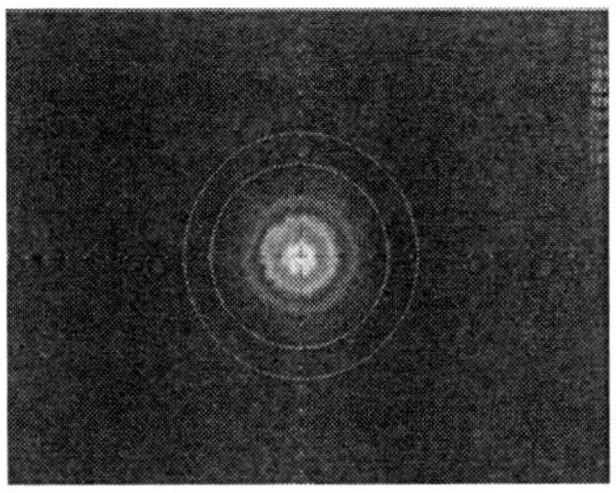

Weld pool monitoring software

- In-process monitoring could minimize need for NDE afterwards
  - Correlate data to any defects found and feedback into build process
  - Concern: data overload

11 HONEYWELL - CONFIDENTIAL

Supply Chain

Honeywell

Pros

- Suppliers continue to gain experience & confidence
  - Many have procured additional machines
  - Exciting designs have been produced
- Acquisitions have occurred
  - AM supplier gains support of a larger company
    - MTI in 2012, Directed Manufacturing in 2014
  - Lends credence to AM as a viable technology

Reference: http://www.mmsonline.com/blog/post/obstacles-to-the-adoption-of-additive-manufacturing

12 HONEYWELL - CONFIDENTIAL

Supply Chain

Honeywell

Cons

- Acquisitions have caused some uncertainty
- Additive suppliers are still geared towards prototyping
- Few suppliers have ISO 9001 & AS9100 certification
  - Necessary to make production components
- Fewer suppliers have full-time materials engineers on staff
  - Fundamental understanding of AM processes are impossible without this expertise
  - Critical to further unlocking of AM's potential

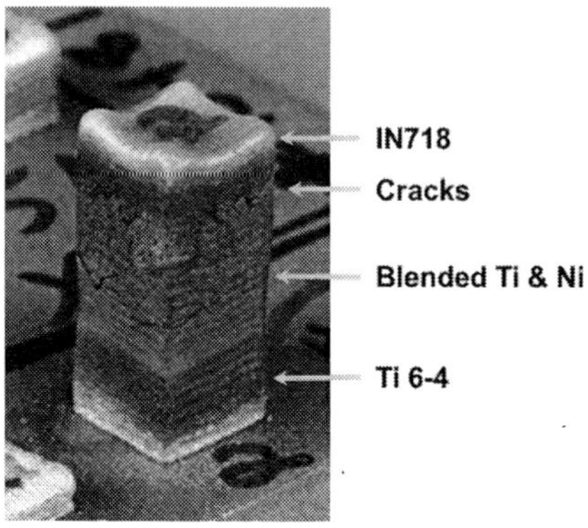

Reference: Development of Nickel-Titanium Graded Composition Components, M. Domack, Rapid Prototyping Journal, Vol. 11 Iss: 1, p. 41 - 51

13 HONEYWELL - CONFIDENTIAL

Supply Chain

Honeywell

Likely Trends Going Forward

- Additional acquisitions
- Evolving business models
  - Suppliers, machine builders, customers
- Introduction of serial production parts
- Suppliers adopting aerospace mentality
  - Manufacturing **plus** materials & process knowledge
  - Quality system , AS9100 certification
  - Fixed processes

14 HONEYWELL - CONFIDENTIAL

Thanks for your time.

Any questions?

TenCate Advanced Composites

TenCate's CCS Composites' Net Molded, Complex Shaped, High Performance Composite Parts for Cost and Weight Reduction

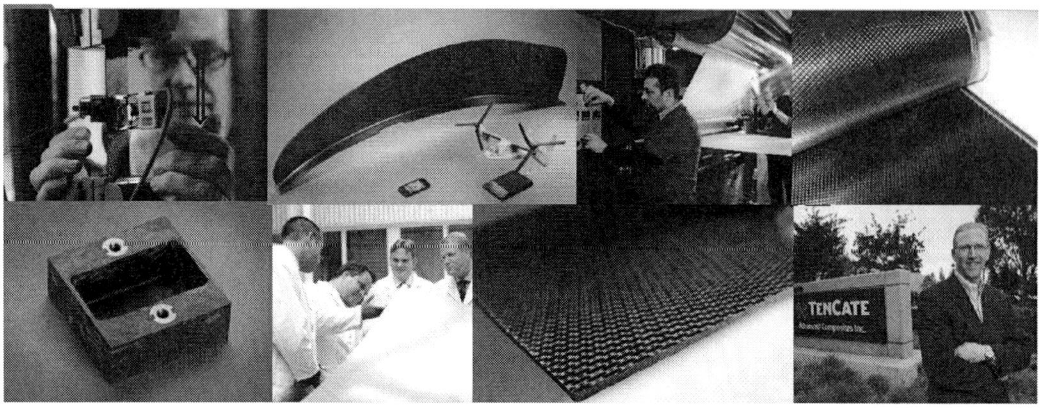

TENCATE
materials that make a difference

TenCate Advanced Composites: Product Offerings

Raw materials to semi-finished products and onto components

PREPREGS

LAMINATES

PARTS

CORE

- Thermoplastic Prepreg Tape, Fabrics and Laminates (Cetex®)
- Thermoset Prepreg Tapes & Fabrics
- Thermoset/Thermoplastic Bulk Molding Compounds
- Compression Molded Parts
- Adhesive Films including Surfacing Films
- Syntactic Foams
- Honeycomb Core
- Tooling Prepregs

Aerospace and Defense Industry Experience

Development to legacy production materials and parts

Key Markets

- Aircraft – Civil and Military
- UAV's
- Space/Satellites
- Radomes & Antennas
- Aircraft Interiors
- Industrial Composites

BOEING
AIRBUS
CIRRUS AIRCRAFT
SPACE SYSTEMS LORAL
LOCKHEED MARTIN
Orbital
GENERAL ATOMICS AERONAUTICAL
EMBRAER
AASC Applied Aerospace Structures Corp
Raytheon
Integrated Composites
NORTHROP GRUMMAN
MDA
Bell Helicopter A Textron Company
NASA
ATK

TenCate Material Data

Material information readily available on web

- Data Sheets for All Materials
  - Typical properties
  - Typical processing cycles
  - Test data
  - Product guides
- Several Locations for Data
  - www.tencate.com
  - Apple Itunes – TenCate Product Selector App
  - Google Play (Android) TenCate Product Selector App

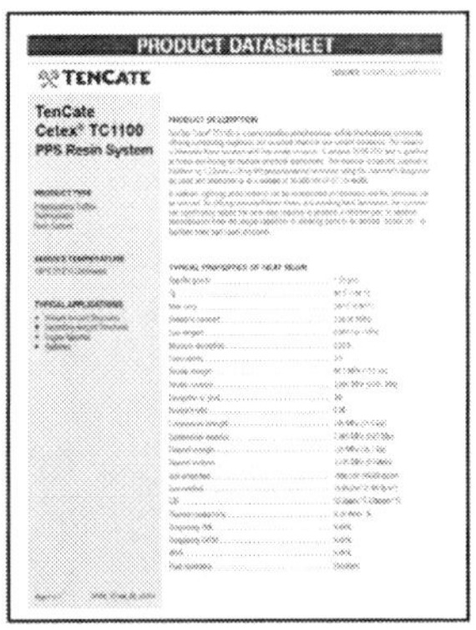

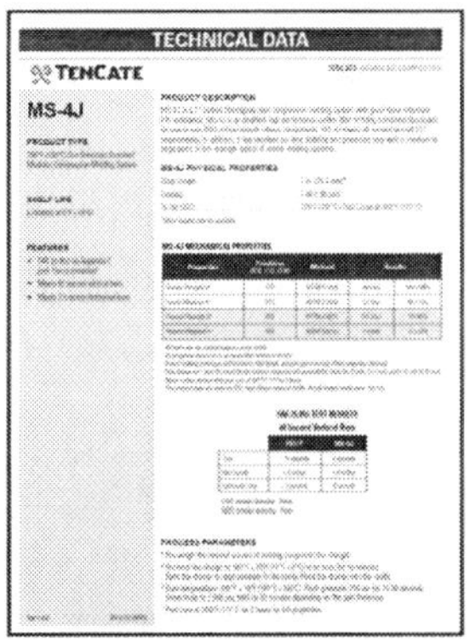

TenCate Advanced Composites North America

Heritage in thermoset and thermoplastic materials

TenCate Advanced Composites

TCAC-Europe

BRYTE Technologies, Inc. Founded in 1990

Phoenixx TPC

YLA Advanced Composite Materials

CCS Composites

Composix Co.

AMBER COMPOSITES

PMC BAYCOMP

1972 TenCate begin first thermoset prepregging.

1982 First thermoplastic prepregs by TenCate Europe.

In 1989, first thermoplastic aerospace usage.

1999 TenCate acquires Bryte Technologies; changes name to TenCate Advanced Composites USA, Inc. (TCAC-USA).

2007 TenCate acquires Phoenixx TPC for thermoplastic unitapes.

2008 TenCate acquires YLA, Inc. & CCS Composites, LLC for space prepregs and compression molded parts.

TenCate acquires Composix, a leader in armor for military vehicles, aircraft and personnel.

2012 2013 TenCate acquires PMC/Baycomp for thermoplastics in industrial markets and AMBER Composites for tooling.

TENCATE CORPORATE www.ccscomposites.com 5

CCS Composites Profile

Well equipped to meet aerospace and military product needs

- Established in 1994
- Located in Fairfield, CA
  - Compression Molding, Machining
  - Prepreg and Testing
- 100% Compression Molding Focus
- TenCate materials; and other suppliers
  - Thermoset and Thermoplastic
  - SMC and TenCate BMC
- Full Capability Tool & Die
- Multiple Compression Molding Presses
- ISO 9001:2008 and AS9100:2009 Rev. C
- Specializing in Carbon & Glass Fiber Molding, High Performance, Tight Tolerance, Difficult Geometries, & Syntactic Foam Molding
- Part Sizes: Under 1" to over 5'; Under 1 pound to over 12 pounds
- Quantities: Ten to tens of thousands per year

TENCATE
CCS Composites

TENCATE CORPORATE www.ccscomposites.com 6

Example Compression Molded Shapes

TENCATE CCS Composites

Complex shapes, molded inserts, paint/coatings

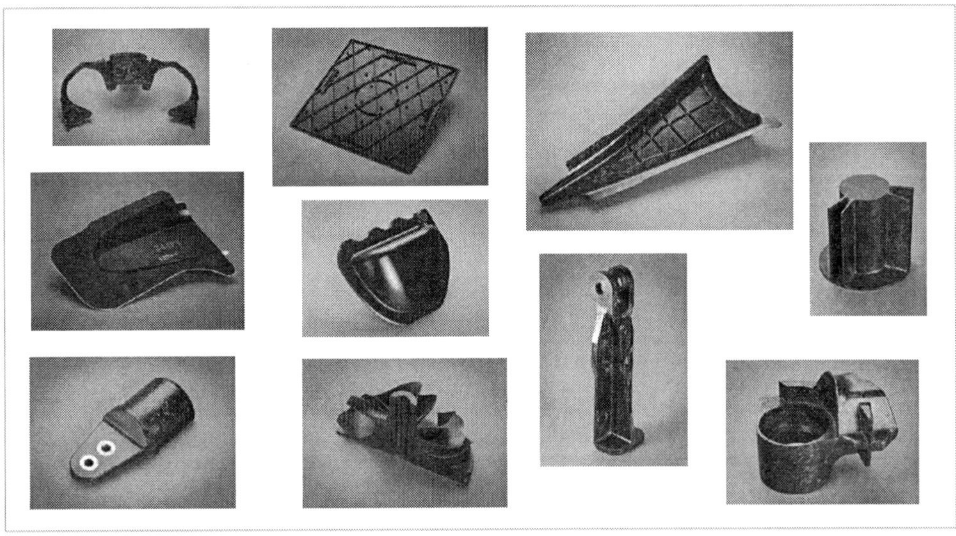

TENCATE CORPORATE www.ccscomposites.com 7

TenCate Thermoset Bulk Molding Compound (BMC)

CCS Composites

Advanced composite material designed specifically for complex compression molding

- Unique Material Form Made From Aerospace Grade Unidirectional Prepreg Tape
  - Custom compression molding resin
  - Completely wet-out fibers; High fiber content (typically 60% wt); straight fibers
- Prepreg Tape That Is Converted To A Specific Chip Size
  - Typical Width = 1/8"; Lengths = ½", 1", or 2"
  - Long chips enhance load carrying ability
- Highly Customized Preform Construction
  - Similar to continuous fiber prepreg layups
  - Increases structural performance
  - Little flow required vs sheet molding compound
  - Allows for more complicated geometries
  - 100% compatible with TenCate's press moldable prepregs for added performance

TENCATE CORPORATE www.ccscomposites.com 8

TenCate Thermoset BMC Molding Process

Fast, simple, repeatable manufacturing

TenCate CCS Composites

1. **Weigh material**
2. **Preheat then preform**
3. **Position in mold**
4. **Press molding**
5. **Remove and deflash part**
6. **(If desired) Post cure**

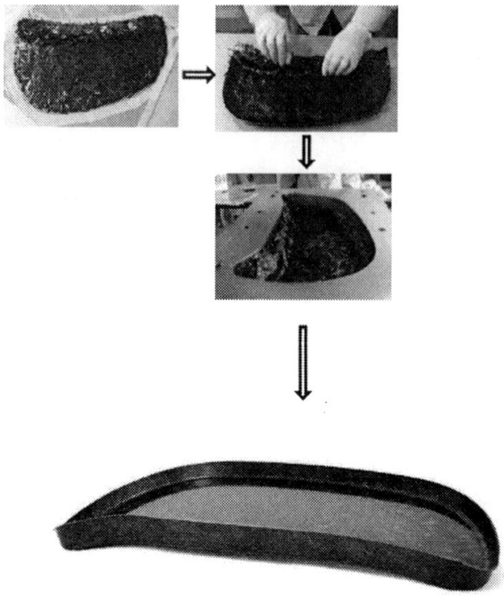

- Typical press molding cycles: 15-30 minutes
- Multicavity tools
- Can mold directly:
  - Metal inserts
  - Lightning strike mesh
  - Fabrics

TENCATE CORPORATE www.ccscomposites.com 9

TenCate's Thermoplastic Tapes and BMC

Highest performance thermoplastic composites

TenCate CCS Composites

Cetex TC series of Unidirectional Tapes

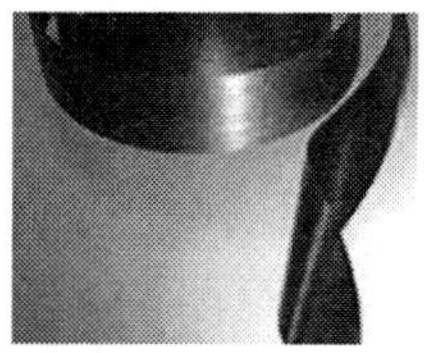

- PEEK, PEI, PPS, others; carbon and glass reinforcement
- High volume fraction ~ 60%
- Extremely high mechanical properties
- TC1220 PEEK: Extremely high compression after impact
- Most pass FST/OSU

Cetex MC series of Molding Compound

- PEEK, PPS, others; carbon reinforcement
- Chopped version of TC series UD tapes - growing
- Made for fast compression molding of complex shaped parts
- Stronger/stiffer than injection molding

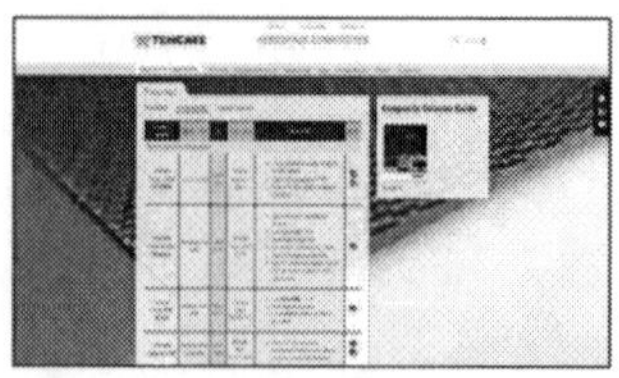

- Multi-lot mechanical property data

TENCATE CORPORATE www.ccscomposites.com 10

Xpress Thermoplastic Molding System

Speed, control and quality in one process

TENCATE CCS Composites

- Speed
  - Proprietary mold design and material
  - World class heating/cooling speed
  - Faster molding = lower cost
- Control
  - Over 30 controlled zones
  - Optimized and unique molding cycle
  - Reduces warping, internal and surface defects
  - Increases speed
- Quality
  - Continuous monitoring of all zones
  - Slight process changes noted early
  - Provides higher consistency and repeatability versus other compression molding processes

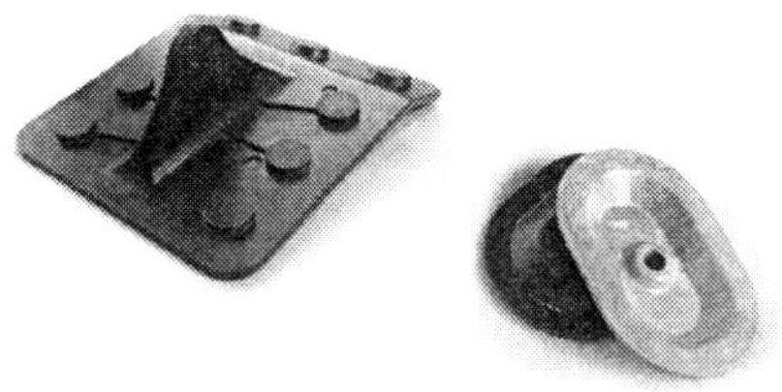

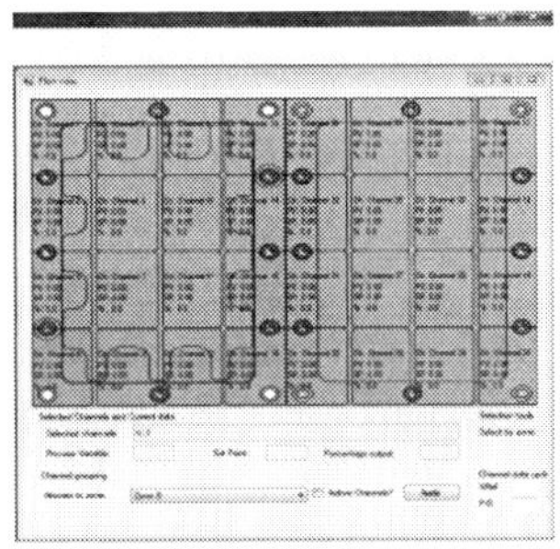

TENCATE CORPORATE www.ccscomposites.com 11

Why use Compression Molded BMC Parts?

Cost savings and increased performance

TENCATE CCS Composites

Reduce Final Cost of Composite Manufacturing

- Lower part labor content
  - reduce/eliminate kitting, layup, final trim, post machining, and inspection steps
- Consolidated part count
  - molded in attachment features, reduced assembly
- High yields
  - Reduced Scrap and MRB with High Pressure Molding
- Mold controlled dimensions
  - Reduced Inspection Time with SPC
- Quick Press Molding Times
  - Efficient Capital Utilization

Enable Composites to Displace Complex Metal Parts

- Competes with structural aluminum final part costs
- Reduces redesign requirements to transition from metal to composites
- Advantages
  - 45% volumetric density reduction over Aluminum
  - Radio Translucency (no x-ray artifacts)
  - Non-corrosive benefits (galvanic, chemical resistance)
  - Higher complex shape performance than other plastic/composite alternatives

TENCATE CORPORATE www.ccscomposites.com 12

Compression Molding Enhances Design Options

High performance materials AND high performance shapes

TENCATE
CCS Composites

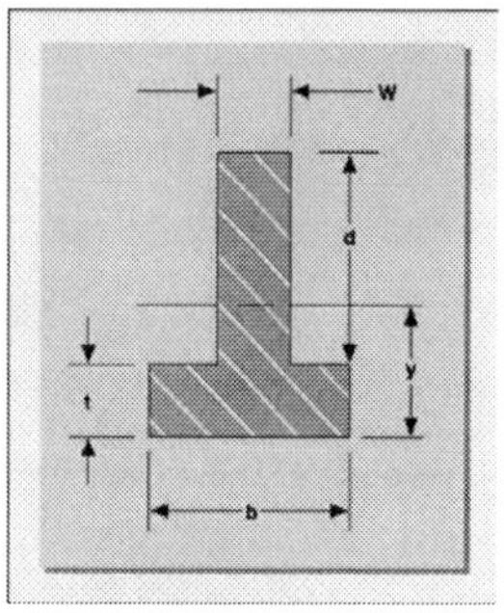

Dimension	Case 1	Case 2	Case 3	Case 4	Case 5	Case 6
b (mm)	40	40	40	40	40	40
t (mm)	4	8	4	4	4	4
d (mm)	0	0	3	6	9	12
w (mm)	0	0	2	2	2	2
Area mm²	160	320	168	172	178	184
y (mm)	2	4	2.12	2.35	2.65	3.04
Momentum of Inertia	213.3	1706.7	288.7	528.4	1018.4	1837.0
% increase stiffness	NA	700	35.3	147.7	377.4	781.1
% increase weight	NA	100	3.75	7.5	11.25	15

Higher performance or lighter weight; more shape complexity at similar cost

Replace Honeycomb Access Panel

Cost savings and fewer parts to inventory/manage

TENCATE
CCS Composites

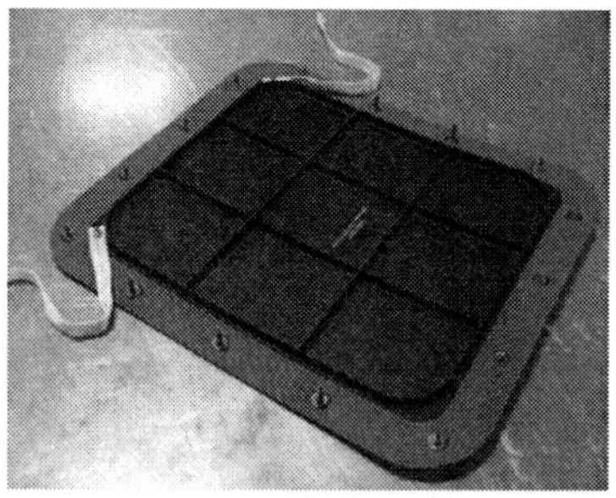

Working with CCS Composites

Years of experience in design to full low to high quantity aerospace production

TENCATE CCS Composites

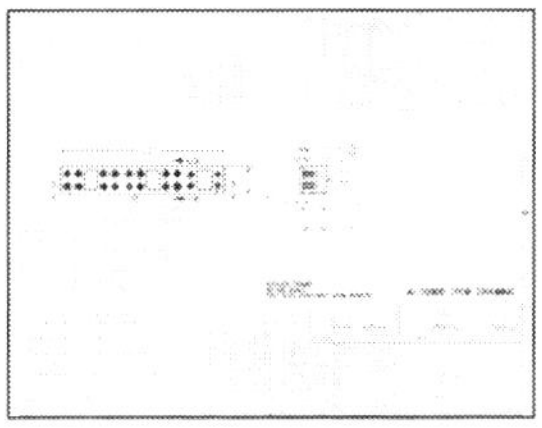

Design Engineering

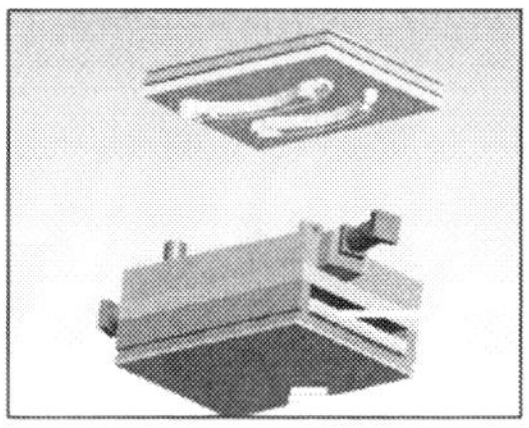

Tooling Design

Tooling

CCS Composites - Core Competencies

Complex, High Tolerance Results

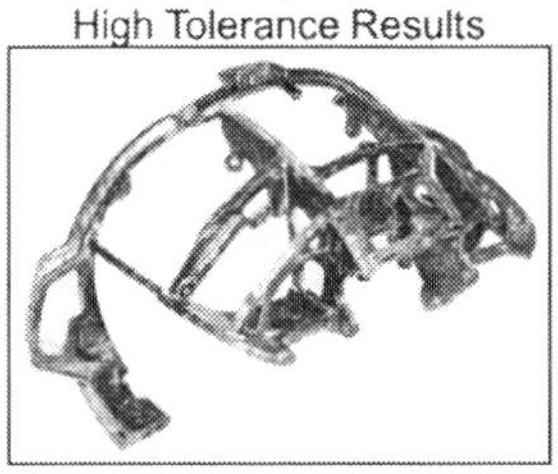

Part Machining

Part Molding

How to learn more?

Typical property, processing, case studies and general information in one place

TENCATE CCS Composites

- All TenCate BMC's listed
  - Click name to see typical properties and processing
- Guides and Case Study PDF's
- Interiors:
  - Thermoplastic: MC1100 PPS, MC1200 PEEK
    - Additional BMCs can be made from tape
  - Thermoset: MS-4J Epoxy/CF
    - Passes FST; not OSU

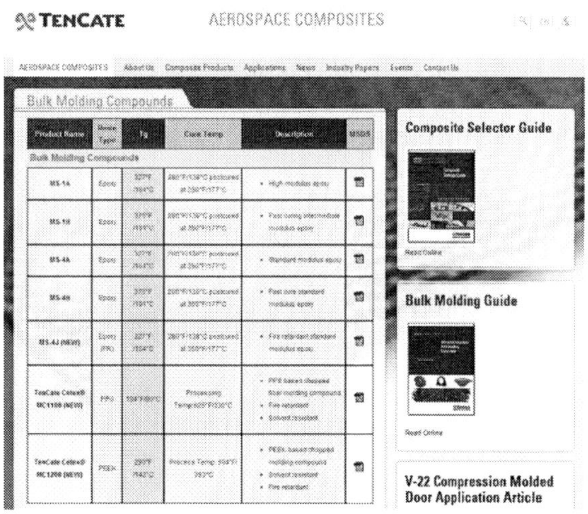

http://tencate.com/amer/aerospace-composites/applications/compression-molded-parts-design/BMC_Selector_Guide_12.aspx

What applications make sense?

Or, contact CCS to discuss further

TENCATE CCS Composites

Where is Compression Molding Best Suited?

- Complex parts that have a changing cross-sectional thickness
- Applications where co-curing of inserts, bushings, or other special material tailoring is desired
- Applications where mechanical properties must outperform injection molding and low fill composites but does not compete with high Vf laminated composites
- Medium to high quantity parts (Hardened Steel Tooling)
- Low quantity, prototype parts (Anodized Aluminum Tooling)

Design Rules of Thumb

- Must perform a part design review and update before manufacturing
- General Tolerance:
  - +/-0.007-0.010 in. (0.18-0.25mm)
  - Select Features: +/- 0.005 in. (0.13mm)
  - Iterative Tool Mods: +/- 0.003 in. (0.08mm)
- Routine Minimum Wall Thickness : ~ 0.050 in. (~1.3mm)
- Draft Wall & Ribs When Possible (1-3°)
- Feature Transition Radii Preferred: 0.025-0.050 in. (0.6-1.3mm)
- CTE Values Range From 0-6 ppm/°F (0-11 ppm/°C)

TENCATE CORPORATE www.ccscomposites.com 17

Contact Information

TENCATE CCS Composites

For Information on TenCate Molding Compounds or CCS Molded Parts

Jason Gabriel
Account Manager – Eastern Region
Philadelphia, PA Area
+1 610-705-2177 j.gabriel@tencate.com

Customer Service
Reyna Landers: Customer Service Representative
Fairfield, CA Facility
+1 707-359-3433 r.landers@tencate.com

www.tencateadvancedcomposites.com www.ccscomposites.com

*****Visit us at Booth 212*****

Electroplated Additive Manufactured Parts
A Synergy of Technologies to Make Robust Composites Without Tooling

Dr. Sean Wise
President
RePliForm Inc.
www.repliforminc.com

Outline

- Why electroplate organic resin AM parts?
  - (qualifier – very difficult to incorporate significant fiber content due to build processes)
- Components
  - Additive Manufactured organic resins & processes
  - Structural electroplated metals
- Rule of mixtures analysis to estimate tensile strength
- Estimate of skin stiffening in bending
- Review of select applications
  - Structural
  - Fire resistance
  - EMI

Additive Manufactured Plastic Resins

- Photopolymer Systems
  - Vat builds (SLA – 3D Systems, Envisiontec, Form 1)
    - Particulate filled resins available
  - Ink Jet Systems (Objet, ProJet)
- Themoplastic
  - Powder (SLS – 3D systems, EOS)
    - Chopped fiber reinforced & particulate filled systems available
  - Filament (FDM – Stratasys, Consumer printers)
    - Resins largely unfilled except ESD materials

Tensile Strength & Sp. Tens. Strength

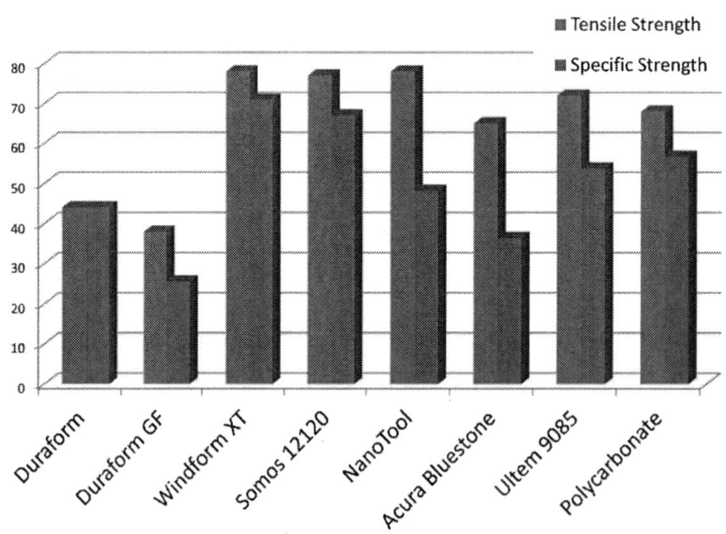

Stiffness, Sp. Stiffness & Elongation

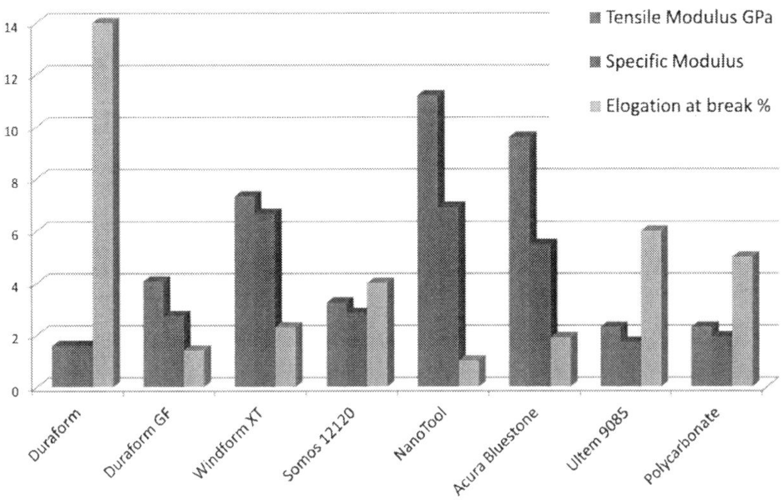

Characteristics of Resin AM Materials

- Narrow Range of strengths – 35 to 80 MPa UTS
- Particulate filled SLA & SLS
  - Stiffness gained at expense of elongation
- Limited fiber volume in SLS
- FDM runs best without filler
- Maximum Sp. Modulus less than 7 Gpa/(g/cc)
- Elongation SLS > FDM > SLA
- Properties in build direction Z < X-Y plane
  - Fibers & fillers worsen this issue

Structural Electroplated Coatings

- Copper base layer (acid copper)
  - Show strength and ductility, ease of throwing into deep areas on parts
  - Low stress layer
- Nickel
  - Structural nickel is ductile, 600 Mpa tensile yield strength, low stress
  - Hard nickel, less ductile, 1.2 Gpa tensile yield strength, compressive stress in coating
- Proprietary Ni-Co and Ni-Fe alloys, strong & ductile

UMBC
AN HONORS UNIVERSITY IN MARYLAND

Paper in Preparation for Submission: **Investigation into Nickel and Copper Electrodeposition using Micro Tension Samples and Microstructure Characterization**

Microsample Testing

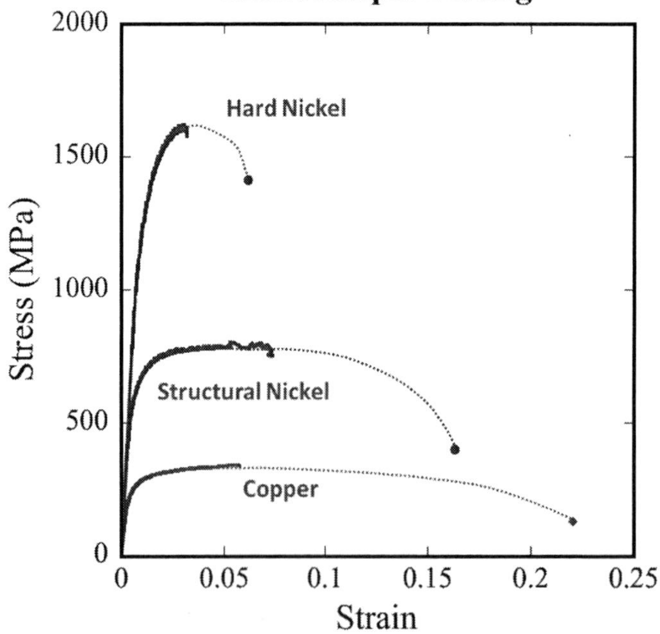

Author Contact Information
Steven Storck
Storck1@umbc.edu
Marc Zupan
Mzupan@umbc.edu

High Strength Ni, Ni-Fe & Co coatings

- Proprietary metal coatings developed with 20 nm grain size
- Combine with molded fiber reinforced engineering TP
- Produce low density composites with properties similar to Mg and Al alloys

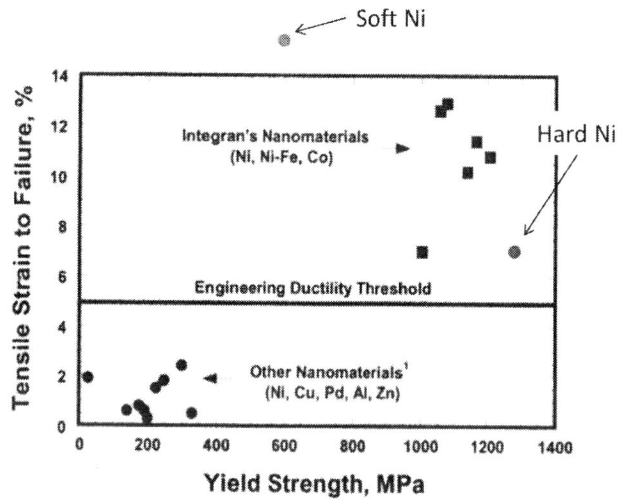

Plating process relative to AM plastics

- Processing in water near room temperature
- Generic method for surface etching, compatible with most resins
  - Chemical methods should be developed that do not degrade resin properties
- Ni coatings as "strong as steel"
  - but which alloy?
    - Low carbon steel ~250 MPa, ductile
    - High hard steel 1-2 GPa, brittle
    - Advance alloys 1-2 Gpa, toughened
  - Electroformed coatings cover steel property ranges

Metal coated AM plastic composite testing

- Testing on coated 3D printed parts has been done over several years with different vendors
- Earliest systematic testing 2004
  - electroplated Cu + Ni (ductile) on filled SLA
  - Vary part wall thickness and coating volume fraction independently to establish rule of mixtures relationship in tension
- Thermal testing of coated SLS parts showed stability of coated parts at elevated temperature
- Varied coating strength & ductility over FDM built specimens

Composite Analysis

- Rule of Mixtures for Tensile Stiffness/Strength

$$E_c = E_p V_p + E_m V_m$$

$$\sigma_c = \sigma_p V_p + \sigma_m V_m$$

- Moment of Inertia – Bending*

$$E_c I_c = E_p I_p + E_m I_m = E_p \left\{ \frac{bh^3}{12} + 2 \frac{E_m}{E_p} \frac{bt(h+t)^2}{4} \right\}$$

* Bending equation only appropriate for very thin coatings rel. to x-section

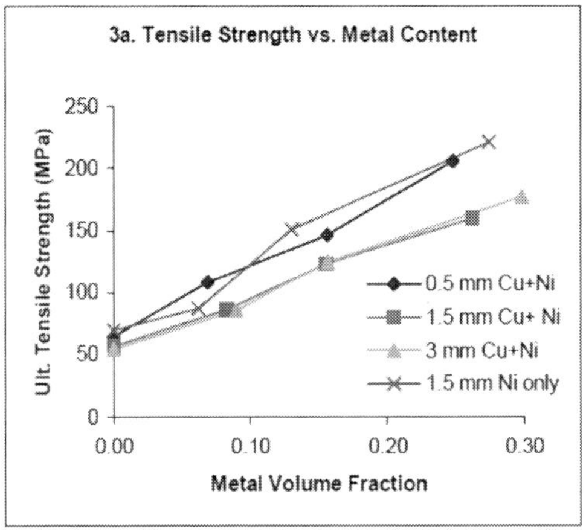

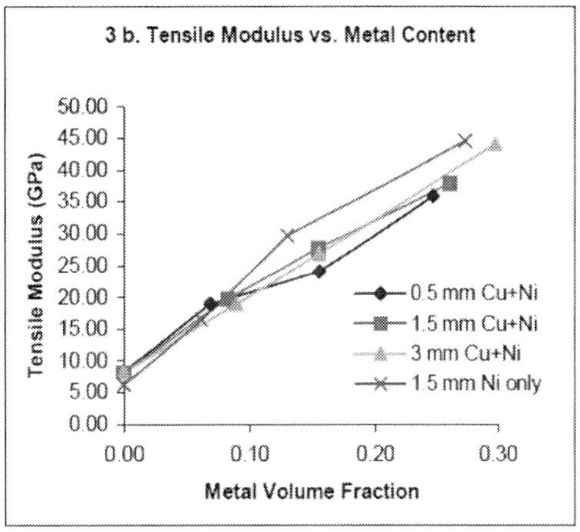

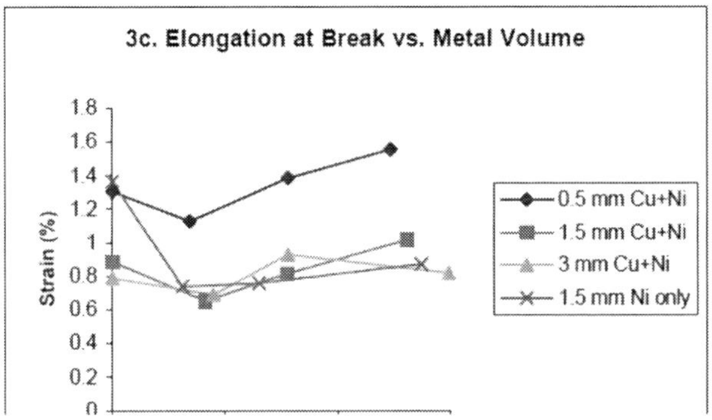

Figure 3a-c. Tensile Strength (a), Modulus (b) and Elongation at Break (c) vs. metal volume for copper and nickel electroform coated ProtoTool 20L.

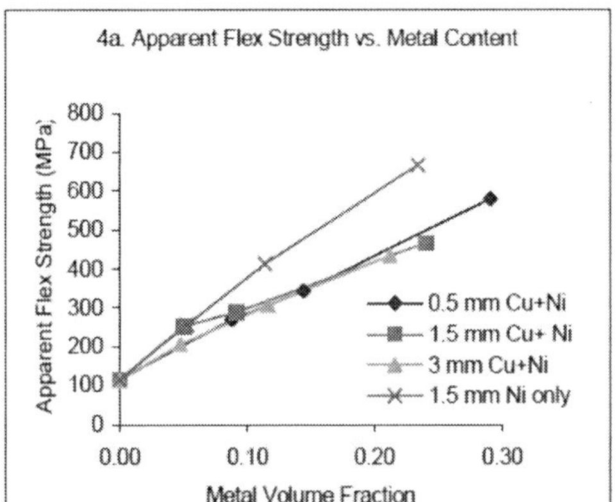

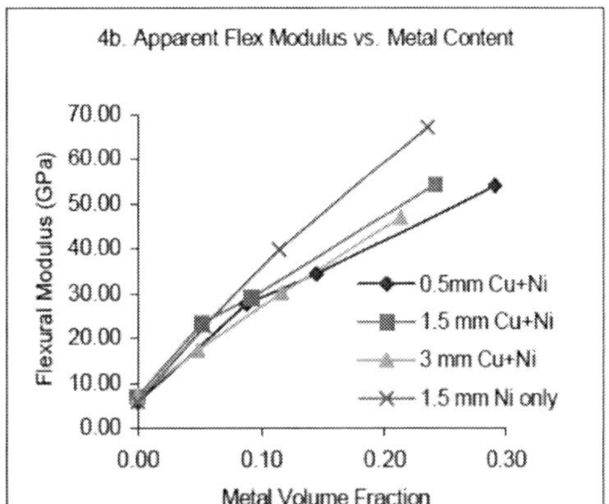

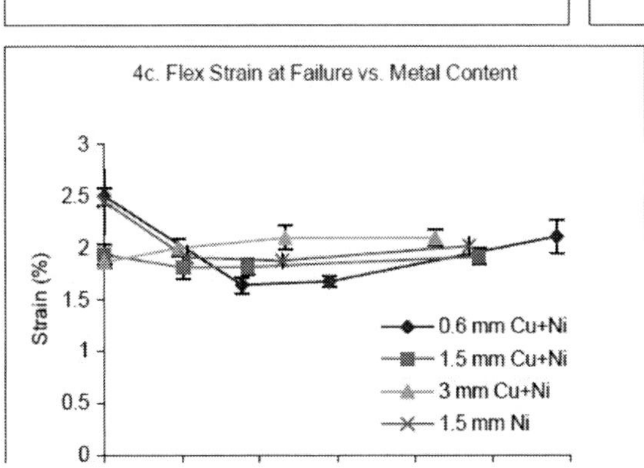

Figure 4a-c. Apparent Flexural Strength (a) Modulus (b) and Strain at Failure (c) vs. metal volume for copper and nickel electroform coated ProtoTool 20L.

Fatigue Cycling of Cu-Ni coated NanoTool

- Boeing Internal Fatigue Cycle Testing (bending)
  - increased 250 fold with ProtoTool resin coated with metal volume fraction of 8%.
- DSM – RePliForm Prelim. Tens. Fatigue Results

MC^2 Type III
24% Cu+ Ni over NanoTool

Stress	Cycles to Failure
21000	9311
21000	12917
21000	10134
28000	481
28000	85
28000	88
17500	17194
10500	197055
10500	424020
10500	140859
8750	653006
8750	1000000
8750	662630

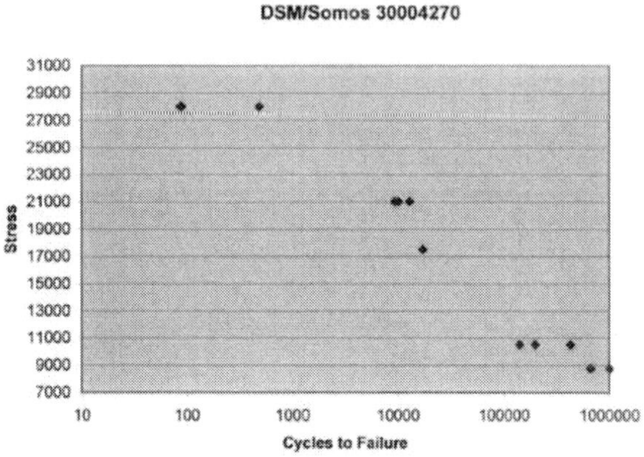

Impact Strength of coated SLA Resins

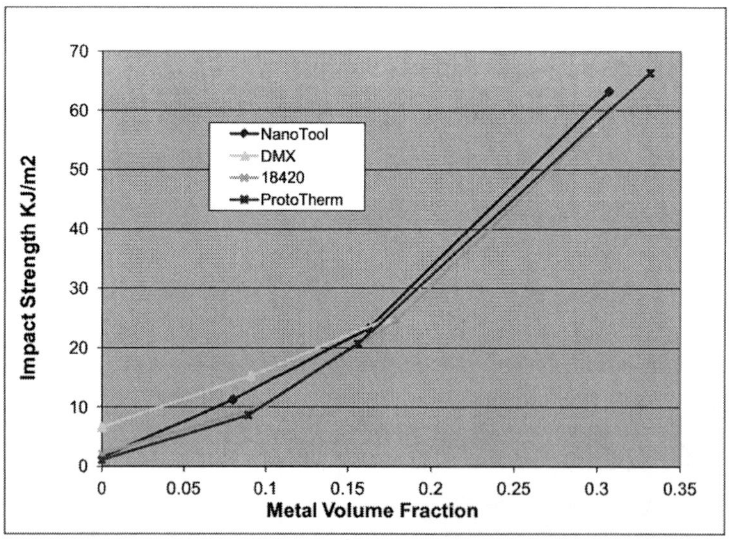

Hot Wet Testing – SLS Duraform

- Most common AM materials have limited utility at elevated temperature
  - Could structural electoplating increase the temperature capability of the system?
- Condition:
  - Hot/Wet testing at 132˚C (270˚F)
- Materials:
  - Duraform
  - Duraform +10% (Vol.) Cu-Ni coating

Hot Wet Testing of Coated and Uncoated Duraform

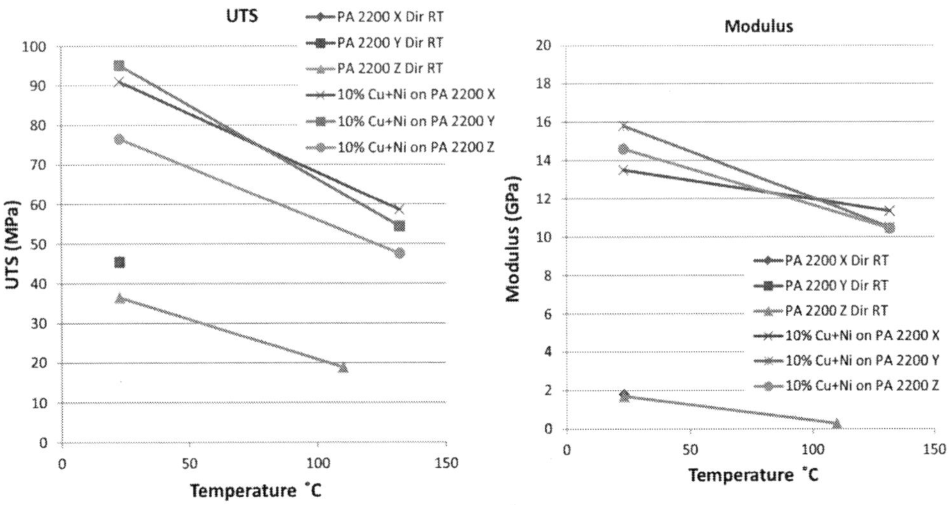

Hard Vs. Ductile Coating on Ultem 9085

- Test two coatings
  - Standard Cu + Ni, 50:50 (high strain coating)
  - High strength Bright Nickel (low strain coating)
- Test 3 different coating thicknesses
  - Control, ~6%, ~12%, ~22% metal
- Flexure & Tensile testing

Coated Ultem variable thickness, hard vs. ductile coating

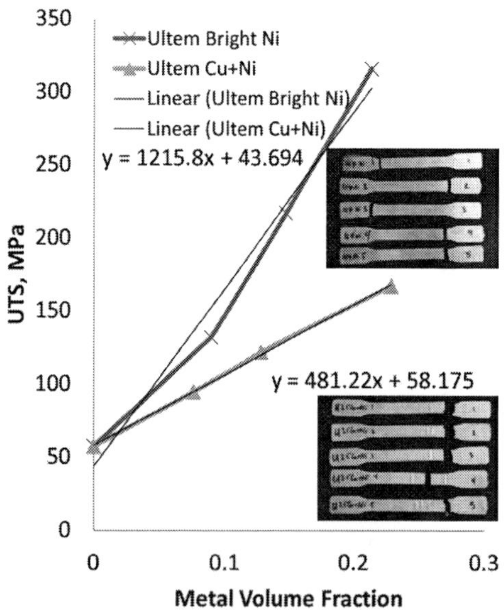

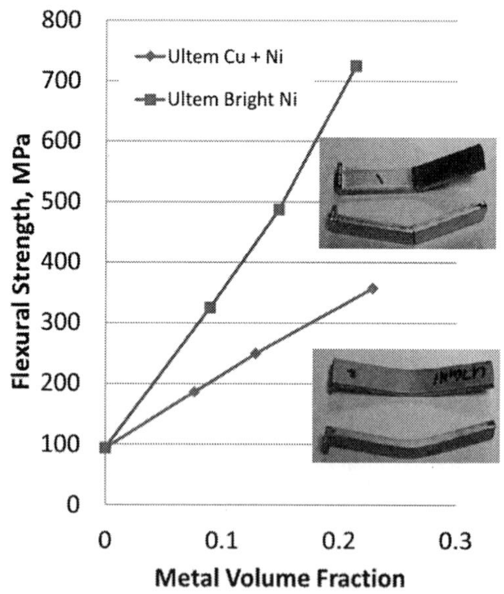

Specific Strength

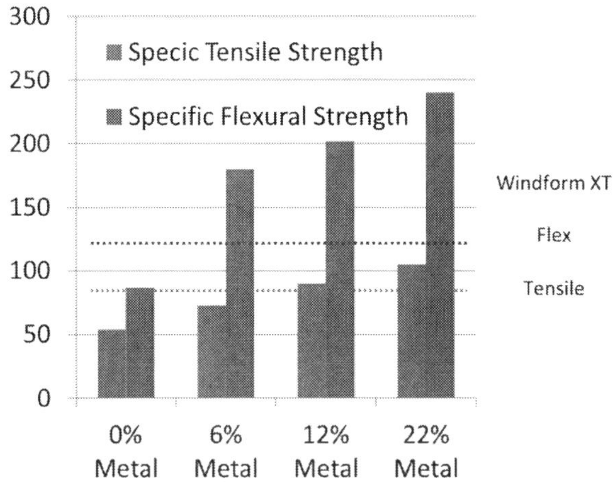

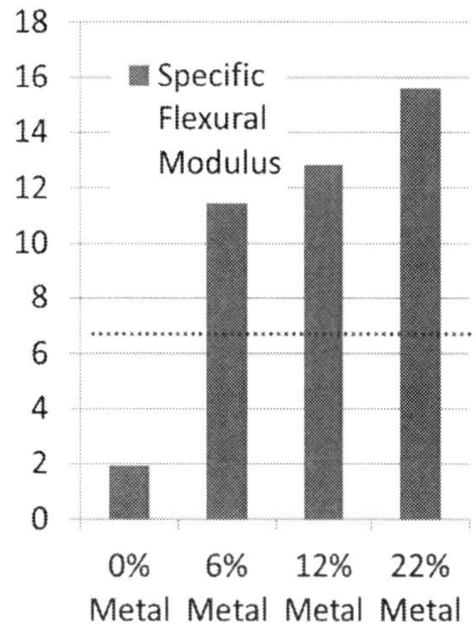

- Hard Nickel coated Ultem 9085

Lessons learned testing coated AM resins

- Rule of Mixtures analysis can estimate tensile strength of coated parts based on component properties
- Temperature capability can be improved with a rigid metal outer layer
- Hard nickel layers is twice as effective stiffenning & strengthening as softer coating but is brittle
- Structural Cu+Ni coating gives 2x strain, deforms plastically

Aerospace Application Examples

- Wind Tunnel models
- Micro gravity space experiment
- Aircraft parts

Wind Tunnel Models

Wind Tunnel Testing

~10% Scale Thrust Reverser Cascades

- Very thin air foils ⇨ max flow
- Cascade elements differ depending on location
- Scaled airfoil shapes, 0.014-0.016” (0.35-0.4 mm) thick
- strong and durable enough for handling and sustain 4-5 psi (0.3 bar) differential across cascade in testing
- Good news - 0.002” (50 μm) coating yields 25% metal
- Bad News – Airfoil wall thickness only 0.010-0.012” (0.25-0.3 mm) prior to coating

FINELINE PROTOTYPING

Results

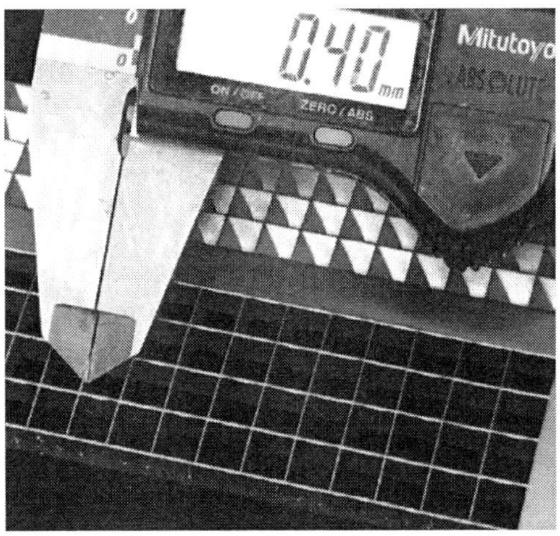

- Overall, parts held to tolerance ±0.005”
- Vane Thickness held to 0 to +0.003”
- Tolerance and Quality suitable for reverser performance and loads testing
  - “huge cost/cycle time reduction” over alternative fabrication method – EDM metal
  - Nearly 90% savings on wind tunnel cascade models allowed testing of more designs

Metal Coated AM Materials for Spaceflight Applications

- NASA GRC and Zin Tech looked at AM as a solution to hardware development challenges
  - must meet stringent flammability, electrical bonding and environmental requirements
- Material qualification
  - 5530 HT, coated with 0.004" +0.003/-0.001" Cu + Ni (made by TTH)
    - Flammability – "A" rating at 10 psi, 30% O_2
    - Toxic Off-gassing – "A" rating
    - Adhesion
    - Durability
    - Protective Capability

Test piece intentionally damaged and subjected to upward flammability testing

Microgravity Science Glovebox (MSG) Mockup with Smoke Aerosol Measurement Experiment (SAME)

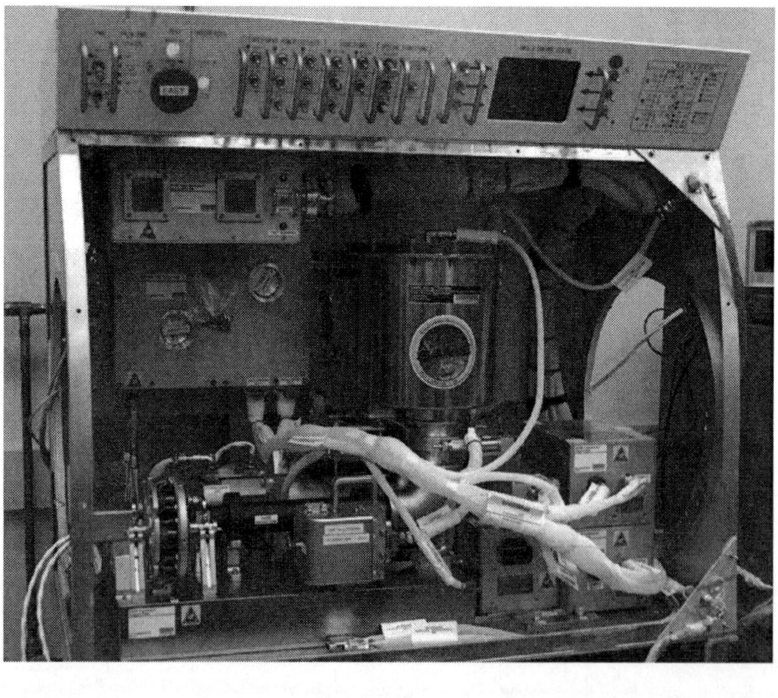

Preliminary System Checkout

Astronaut Clay Anderson checks out SAME set-up in microgravity science glove box , Fall 2007

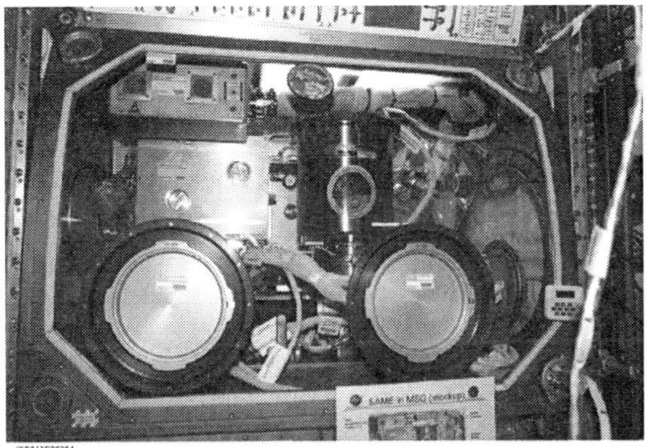

SAME Experiment Status

- SAME Experiment qualified 30 SLA parts
  - More than half Cu+Ni coated
- ~15% of the cost of custom made, low quantity parts made from conv. materials
  - CAD to coated parts < 3 weeks
  - 25% weight savings
- Launched on STS 118, August '07
- SAME experiments conducted on ISS ~30 days after delivery

Aircraft Parts ??

- Yes but non-structural, small components
  - EMI Shielding
  - Wear surface improvement
  - Stiffening/Strengthening
  - Potential for cosmetic interiors

Ultem 9085

Sealed & Sanded (top) plated & polished (bottom)

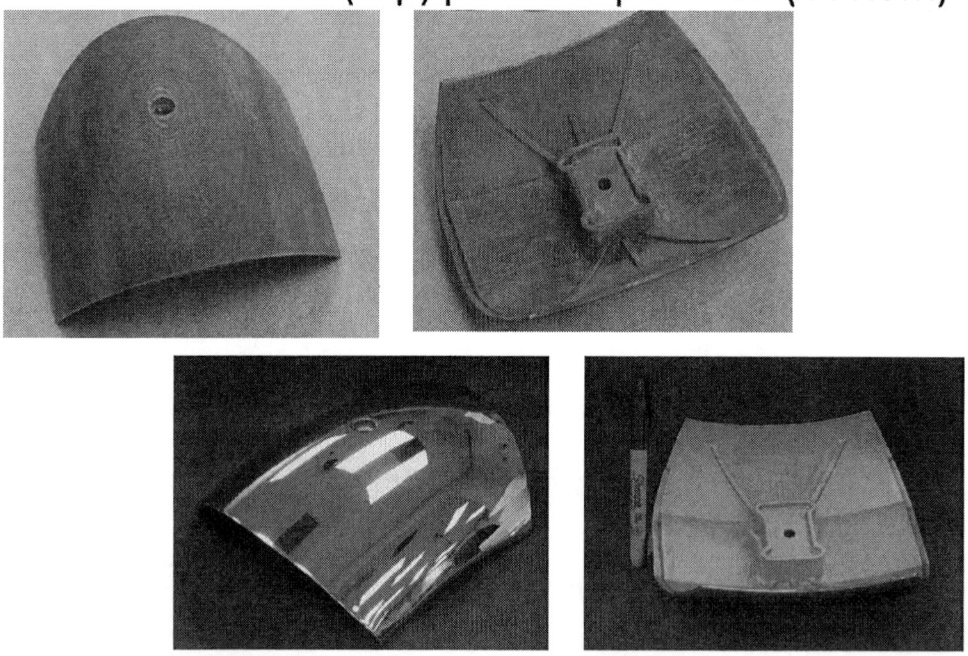

Summary

- Electroplated coatings are a good compliment to 3D printed resin parts
  - Improve finish
  - EMI shield
  - Abrasion resistant surface
  - Environmental barrier
  - Envelops AM laminate with high tensile strength reinforcing coating
  - Improves specific strength and stiffness

Can we step up our game going forward?

- Create a better composite system
  - Look at improved AM materials
    - Parts will likely remain 75-95% plastic
  - Look at improved coatings
    - Commercial / proprietary metal plating systems that deliver higher strength AND ductility
- Better understand/control metal distribution with modeling and field control application
- Improve the metal to plastic interface
  - Purposely formulated resins to improve bond and process efficiency

Acknowledgements

- Steve Stork, UMBC
- Frank Medina, UTEP
- DSM Somos
- Stratasys
- PPI/Harvest Technologies
- Fineline Prototyping
- Zin Technologies / NASA GRC
- Boeing

Production Rate Up, Price Down

Improved process management, from the freezer to the autoclave, reduces costs and enables better quality control in high-rate production environments

Plataine®
Production Optimization Solutions

Avner Ben-Bassat
President & CEO , Plataine Technologies Ltd

Prepared for AeroDef Manufacturing Summit, Long Beach, CA
February 2014

Agenda

- Introduction: Rate Up, Price Down
- Benchmark Analysis and 3 Key Challenges:
  1. Material Tracking
  2. Production Tracking
  3. Bridging the Gaps between Engineering and Manufacturing ("From Product to Production")
- Integrated Production Optimization and Balancing Tradeoffs

Plataine®
Production Optimization Solutions

2

With the increase in production rates and cost pressures manufacturers must now develop <u>manufacturing excellence</u>

1970s – 1980s → **1990** → **2000** → **Today**

A NEW TECHNOLOGY

"Can we build it?"

- Building Aerostructures from composites is an **engineering challenge**

TRANSITION TO MANUFACTURING

"Can we build it consistently?"

- Volumes are low
- Processes are engineering and manual intensive
- Focus on quality, less on cost

MASS PRODUCTION

"Can we build it consistently, at a high rate and at a reduced cost?"

- composite part fabrication is a common reality.
- engineering capabilities to make a single part are no longer enough
- Fabricators must maintain OPERATIONAL AND MANUFACTURING EXCELLENCE

Plataine®
Production Optimization Solutions

3

Loss of control with volume growth eliminates economies of scale and results in accelerated increase in waste

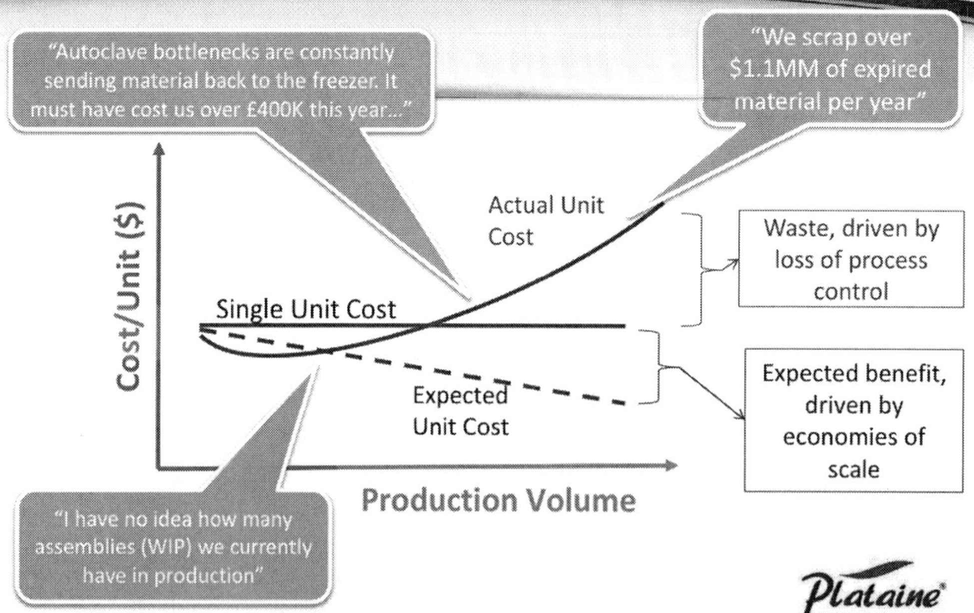

Plataine®
Production Optimization Solutions

4

Description of Benchmark Pilot

- Conducted at a UK-based manufacturer
- **60 individual work orders** were selected, containing numerous parts with each part manufactured from numerous kits.
  - The typical period for completing this work was **one week**.
  - Total of approximately **300 Meters Squared** of material*
  - **Materials used:** Carbon, Engineered Carbon, Glass and RC 660
  - Materials cost range from **$35 to $85** per meter squared.

- In this paper, the production volumes were slightly modified to conform to the manufacturer's confidentiality requirements. However, the savings ratios were kept consistent.

Current Process – Serial decisions made by separate teams and using loosely integrated systems

Planning
(SAP ERP)

Engineering
(Nesting Software)

Material Inventory
(MS Excel)

Production Floor

Schedule Order for Production

No visibility into inventory or kitting optimization opportunities

Retrieve Nests from Library

All Nests are for **single** kits and **single** rolls

No visibility into material inventory or schedule

Select Suitable Rolls from Inventory

Retrieves rolls 'Long Enough' to cover all nests

No visibility to full schedule or future orders

Cut, Kit and Fabricate…

Remaining material returns to the freezer.

Kits flow downstream or are returned to the freezer when having 'obvious' bottlenecks.

Production is reported on paper travelers, then typed into the system

Challenge #1: Current Inventory Management Software/ERP-Modules lack roll or kit-level visibility in real time

1. ERPs typically manage composite inventory at the overall stock levels (length or weight):
   - Lack of roll or kit specific data such as specific length, width or expiration dates
   - ERP system often supplemented by ad-hoc systems like Excel, Access to manage 'shop-floor level freezer'
2. Critical shortage of real-time information
   - Actual usage of raw material is reported retroactively
   - Shop-floor or executive management receive information too late for any meaningful action

The Result

- Reduced levels of visibility, control and quality
- Wasted material as rolls and kits expire before reaching the autoclave
- Endless paper-trail, virtually impossible to audit

Plataine® Production Optimization Solutions

7

Challenge #2 - Tracking Production in Real Time

1. Production Tracking is typically reliant on human input
   - Typical entry methods include: [ERP/MES] terminals, Barcode, Paper
   - Highly dependent on employee discipline
   - Naturally Error Prone
2. Critical shortage of real-time information
   - Reports are generated retroactively
   - Shop-floor or executive management receive information too late for any meaningful action

The Result

- Reduced levels of visibility, control and quality
- Wasted resources – time, material and labor
- Endless paper-trail, virtually impossible to audit or identify opportunities for improvement

Plataine® Production Optimization Solutions

8

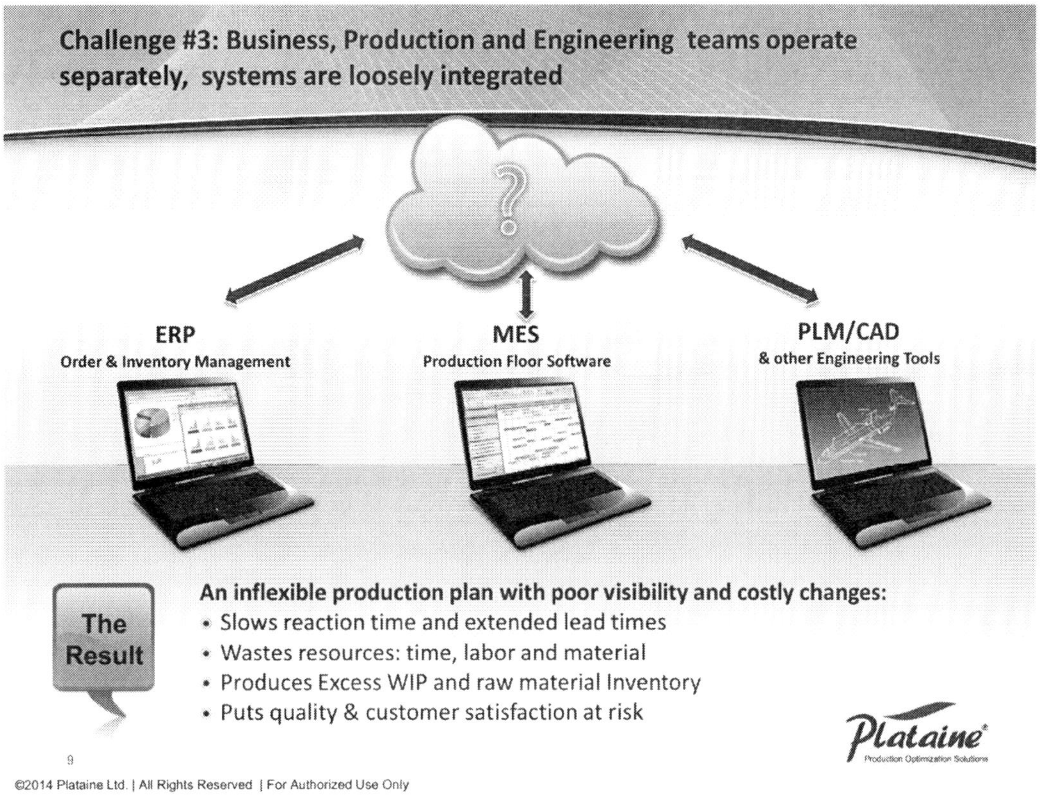
Challenge #3: Business, Production and Engineering teams operate separately, systems are loosely integrated
ERP
Order & Inventory Management
MES
Production Floor Software
PLM/CAD
& other Engineering Tools
The Result
An inflexible production plan with poor visibility and costly changes:
• Slows reaction time and extended lead times
• Wastes resources: time, labor and material
• Produces Excess WIP and raw material Inventory
• Puts quality & customer satisfaction at risk
Plataine
9
©2014 Plataine Ltd. | All Rights Reserved | For Authorized Use Only

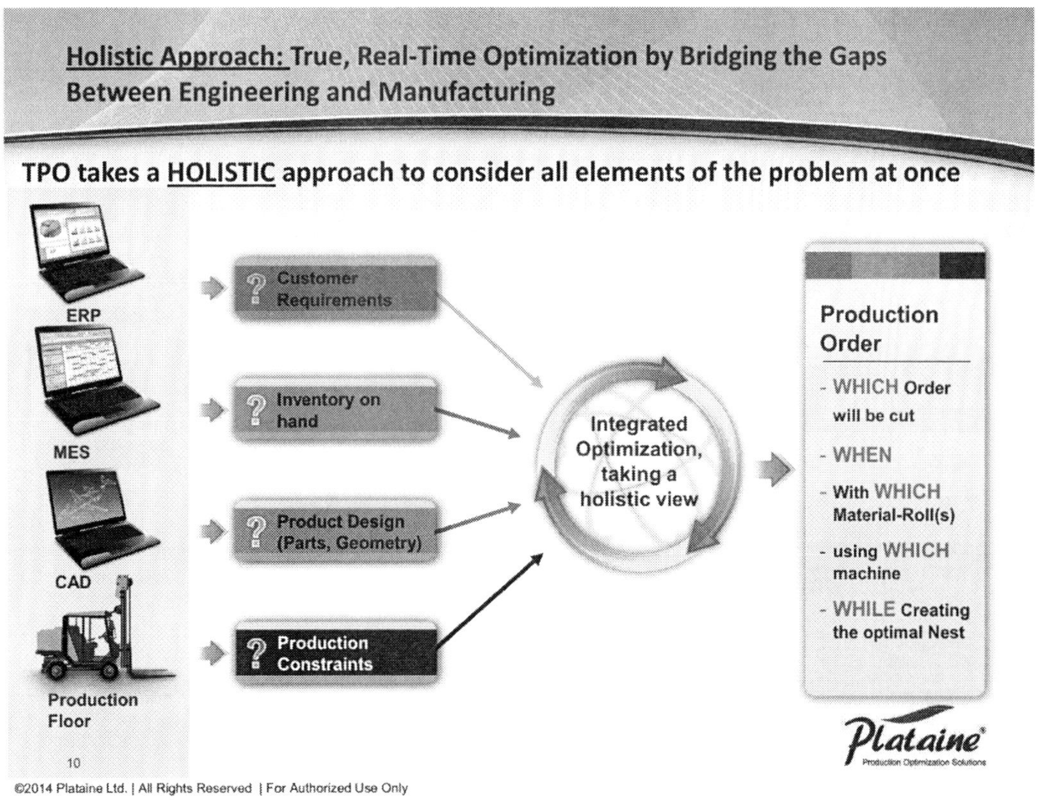
Holistic Approach: True, Real-Time Optimization by Bridging the Gaps Between Engineering and Manufacturing
TPO takes a HOLISTIC approach to consider all elements of the problem at once
ERP
MES
CAD
Production Floor
Customer Requirements
Inventory on hand
Product Design (Parts, Geometry)
Production Constraints
Integrated Optimization, taking a holistic view
Production Order
- WHICH Order will be cut
- WHEN
- With WHICH Material-Roll(s)
- using WHICH machine
- WHILE Creating the optimal Nest
Plataine
10
©2014 Plataine Ltd. | All Rights Reserved | For Authorized Use Only

TPO leverages RF-ID technology to provide real-time, accurate information of kits and rolls in the freezer

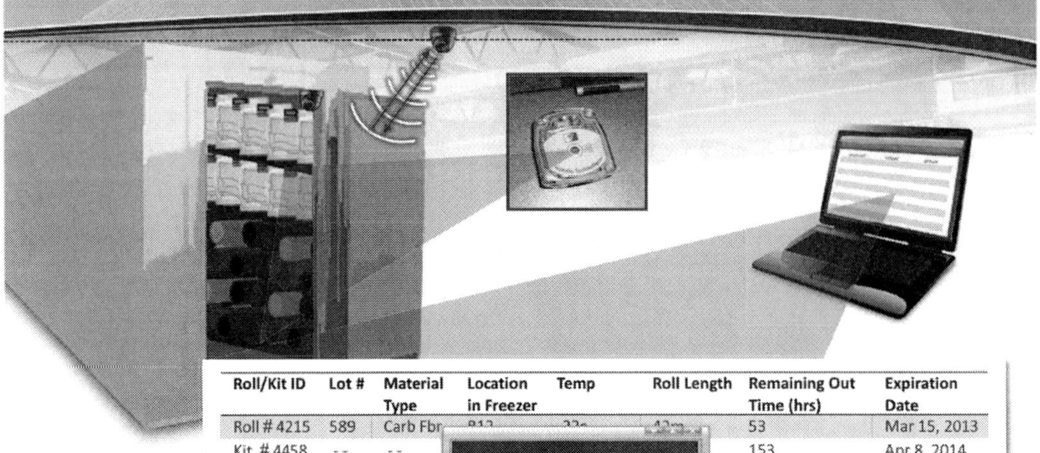

Roll/Kit ID	Lot #	Material Type	Location in Freezer	Temp	Roll Length	Remaining Out Time (hrs)	Expiration Date
Roll # 4215	589	Carb Fbr	[illegible]	[illegible]	[illegible]	53	Mar 15, 2013
Kit # 4458	- -	- -				153	Apr 8, 2014
Roll # 4698	458	Gls Epox				33	Apr 22, 2013
Roll # 5621	147	Gls Fbr				72	May 8, 2013
Kit # 6198	- -	- -				114	May 16, 2014
Roll # 6892	567	Carb Fbr				187	May 25, 2013
Kit # 7522	- -	- -				29	Jun 6, 2013
Roll # 7893	964	Gls Fbr				63	Jun 9, 2014
Roll # 8931	178	Carb Fbr	B3	-4c	22.1m	158	Jun 18, 2013
Roll # 9526	311	Gls Epoxy	C14	-24c	4.3m	22	June 29, 2013

TPO ALERT

Roll #4698
Remaining Shelf Life: 20 hrs

TPO automatically tracks inventory and assemblies on the production floor to further increase productivity and reduce waste

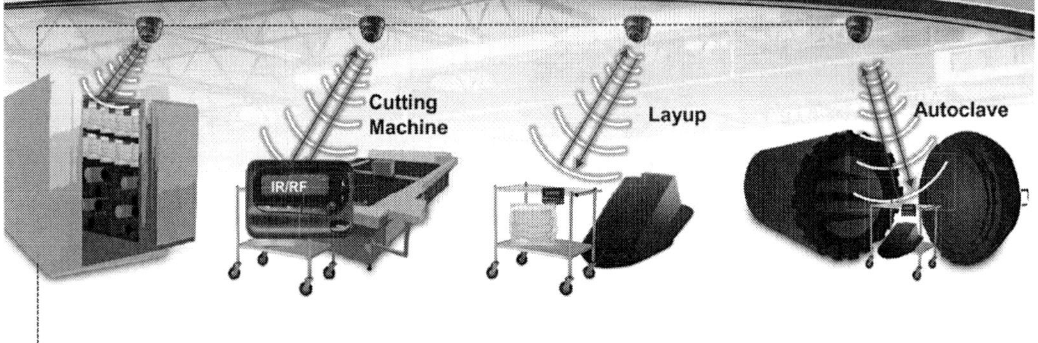

Part ID	Station	Time
Kit # 4215	Cutting	14:17
Kit # 4458	Freezer	15:01
Kit # 4698	Layup	15:22
Kit # 5621	Autoclave	16:18
Kit # 6198	Freezer	17:43
Kit # 6892	Cutting	18:20
Kit # 7522	Layup	18:36
Kit # 7893	Freezer	18:59
Kit # 8931	Autoclave	19:17
Kit # 9526	Cutting	19:42

Sample RFID Coverage map (Aerospace facility in Los Angeles)

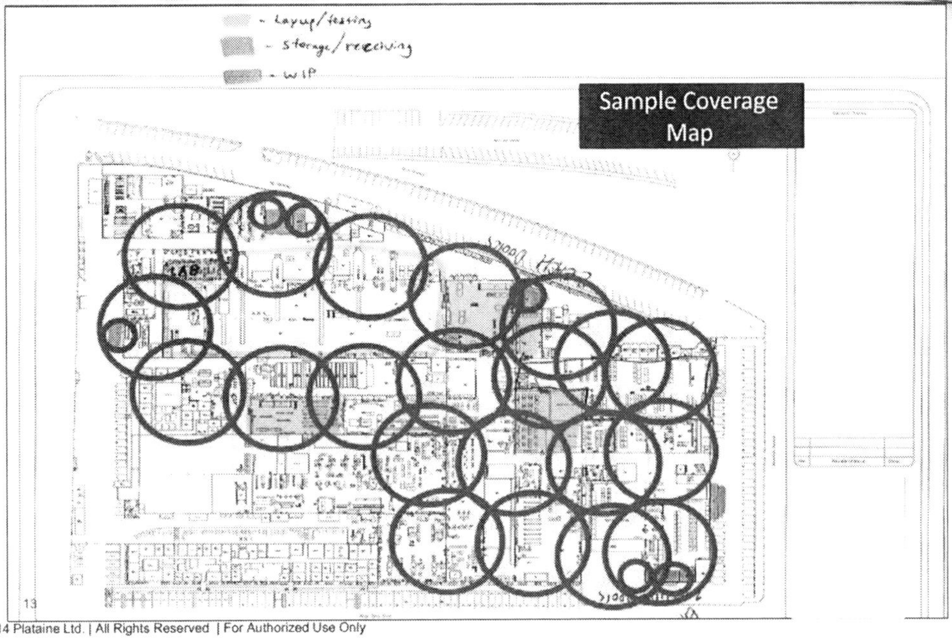

13

Savings Example #1: Multi-Kit Nesting (Real life experiment - UK Composite Parts Fabricator)

Multi-Kit Mixing

- Material Savings driven by multi-kit nesting

Scenario	Length (Meter)	Cost (USD)	%-Material Savings (Length)	%-Cost Savings
Basis (Single Kit Nesting)	322.51	$ 16,932	--	--
Mixing Scenario 1 - Mixing simple pairs of sub-kits	305.09	$ 16,001	5.40%	5.50%
Mixing Scenario 2 - Mix all Kits by Material	295.45	$ 15,459	8.39%	8.70%

Basis (single kit nest)	
Mixing Scenario 1 – mixing simple pairs of sub kits	
Mixing Scenario 2 - mix all kits by material	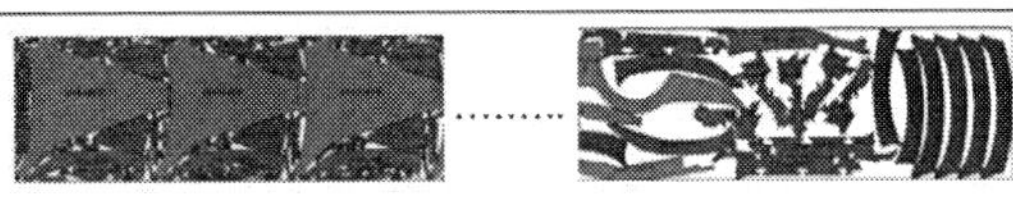

Plataine Production Optimization Solutions

14

TPO's *Kitting Solution* enables order mixing and increases productivity

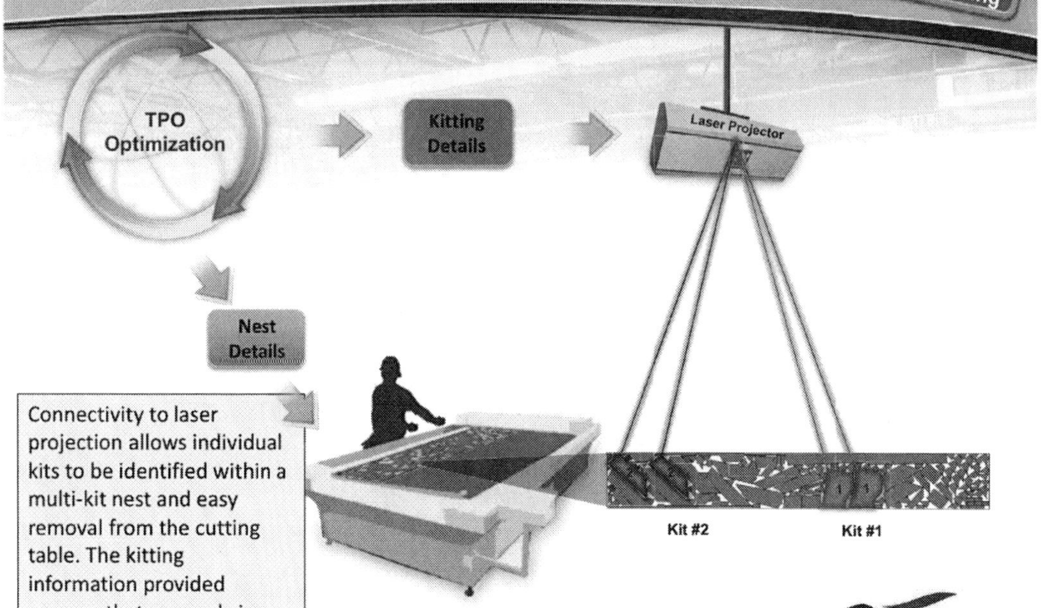

Connectivity to laser projection allows individual kits to be identified within a multi-kit nest and easy removal from the cutting table. The kitting information provided ensures that every ply is placed in the correct kit.

Plataine® Production Optimization Solutions

Roll Management for Optimized Material Usage:

Configurable, optimized remnant (short-end) selection options

- A given kit requires about 15 meters of material to be nested
- Traditionally, each kit is nested to a single roll, leaving behind a growing stock of short rolls
- With TPO, **nesting is dynamic & utilizes all rolls in stock**, keeping short-roll inventory to a minimum

	Starting inventory	Shelf Life	Lot #	Used rolls	Final inventory
Single-roll nesting	Roll A: 5.9 meters Roll B: 9.9 meters Roll C: 44.1 meters	22 Hours 18 Hours 73 Hours	A A B	--- --- C: 14.6 meters	A: 5.9 meters B: 9.9 meters C: 29.5 meters
TPO's Multi-roll nesting	Roll A: 5.9 meters Roll B: 9.9 meters Roll C: 44.1 meters	22 Hours 18 Hours 73 Hours	A A B	A: 5.9 meters B: 9.3 meters ---	--- --- C: 44.1 meters

Plataine® Production Optimization Solutions

16

Roll Management for Optimized Material Usage:
(Example - Continued)

Single-Roll Nest utilizes 14.6 meters of a single roll:

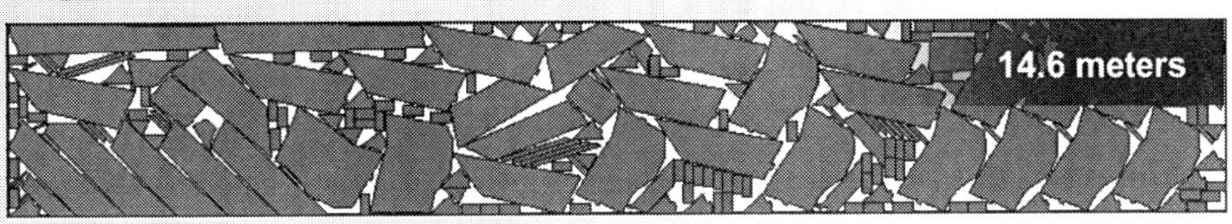

TPO dynamic nesting splits the nest to two separate short-rolls, fully utilizing them:

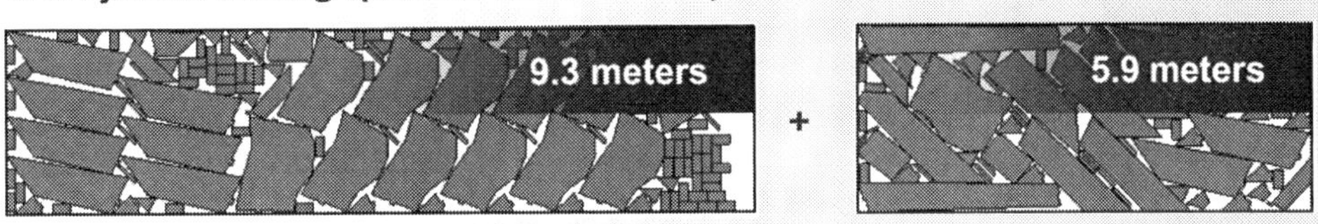

Plataine
Production Optimization Solutions

17

Savings Example #2: Short Roll Utilization

Short-Roll Utilization

- Material Savings driven by Short-Rolls (Remnants) Utilization
  - Around 4% savings without mixing multiple kits in a single next, and over 9% when combining both capabilities.

Scenario	Length (Meter)	Cost (USD)	%-Material Savings (Length)	%-Cost Savings
Basis (Single Kit Nesting)	322.51	$ 16,932	--	--
Scenario 3 - Utilizing Short Rolls w/o kit mixing	309.45	$ 16,271	4.05%	3.90%
Scenario 4 - Utilizing Short Rolls with kit mixing	292.19	$ 15,366	9.40%	9.25%

Basis (single kit nest)	
Utilizing Short-rolls without kit mixing	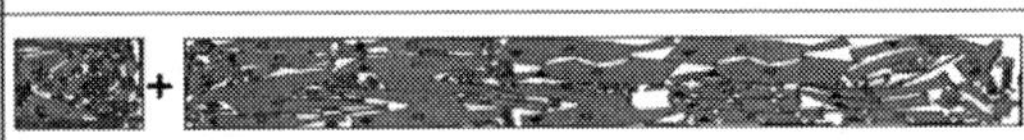
Short-rolls with kit mixing	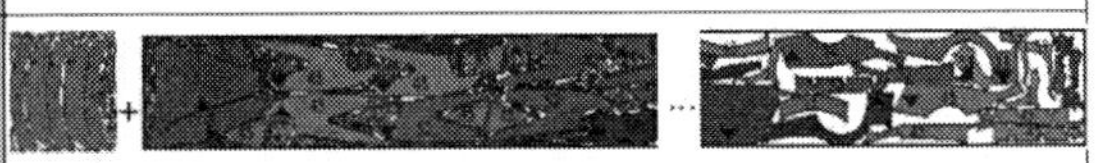

Plataine
Production Optimization Solutions

18

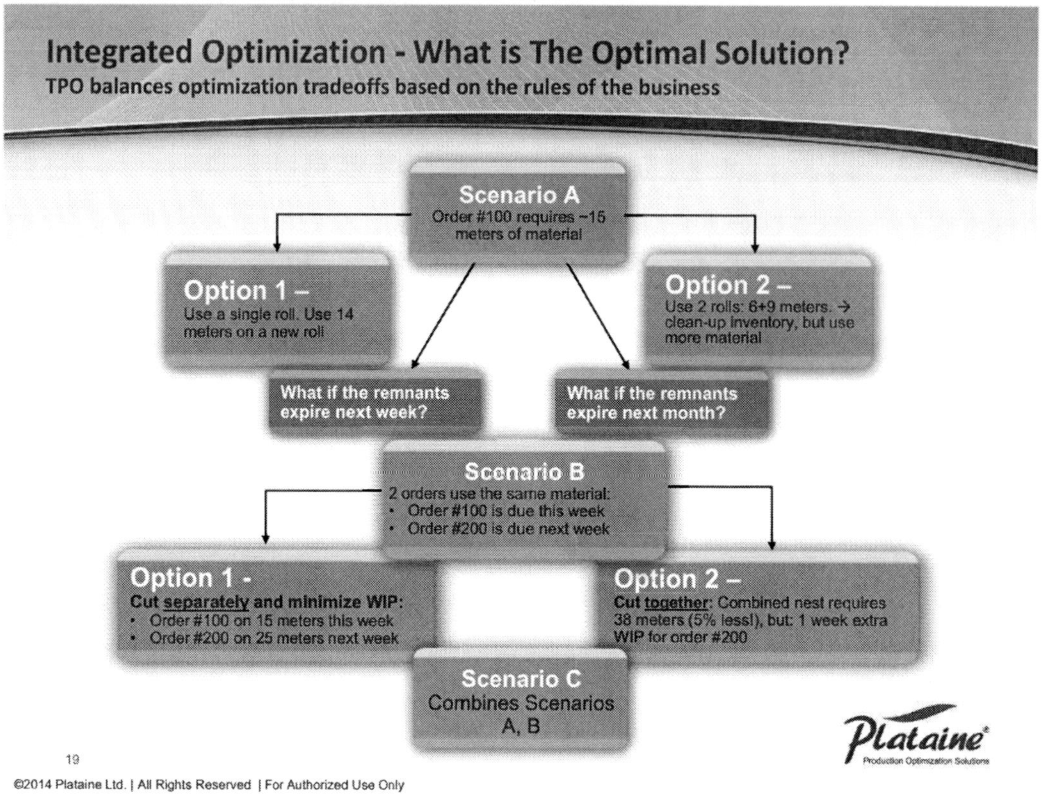
Integrated Optimization - What is The Optimal Solution?
TPO balances optimization tradeoffs based on the rules of the business
Scenario A
Order #100 requires ~15 meters of material
Option 1 –
Use a single roll. Use 14 meters on a new roll
Option 2 –
Use 2 rolls: 6+9 meters. → clean-up inventory, but use more material
What if the remnants expire next week?
What if the remnants expire next month?
Scenario B
2 orders use the same material:
• Order #100 is due this week
• Order #200 is due next week
Option 1 -
Cut separately and minimize WIP:
• Order #100 on 15 meters this week
• Order #200 on 25 meters next week
Option 2 –
Cut together: Combined nest requires 38 meters (5% less!), but: 1 week extra WIP for order #200
Scenario C
Combines Scenarios A, B
Plataine
Production Optimization Solutions
19
©2014 Plataine Ltd. | All Rights Reserved | For Authorized Use Only

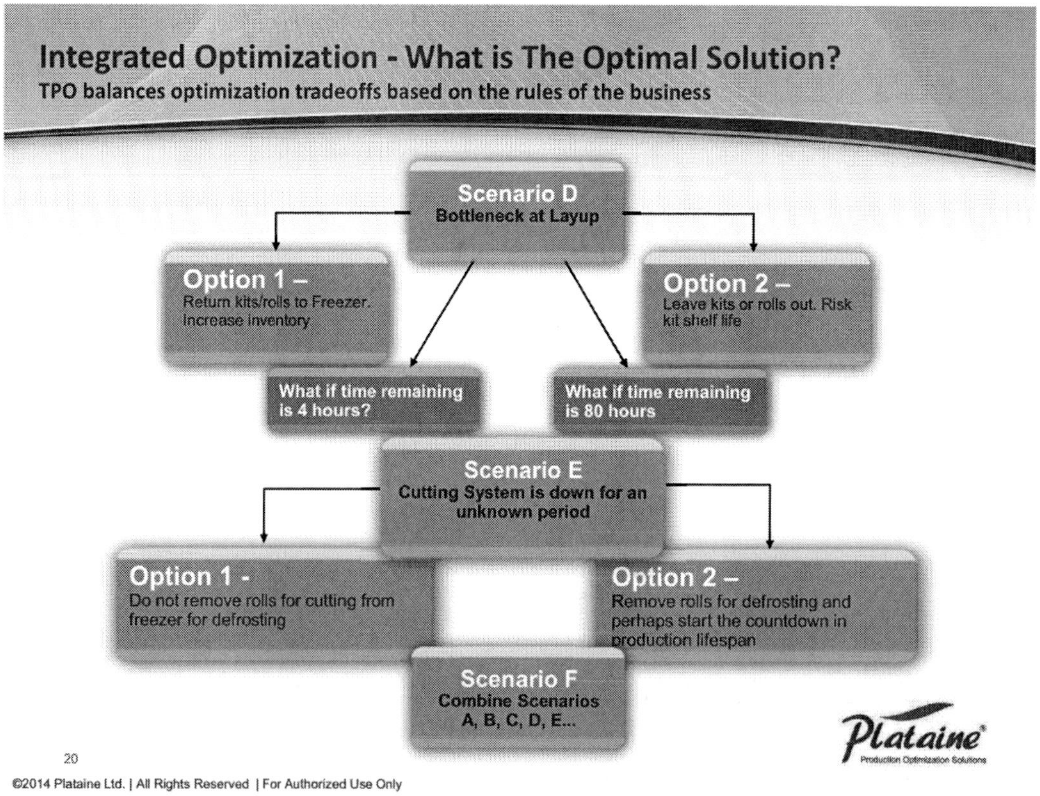
Integrated Optimization - What is The Optimal Solution?
TPO balances optimization tradeoffs based on the rules of the business
Scenario D
Bottleneck at Layup
Option 1 –
Return kits/rolls to Freezer. Increase inventory
Option 2 –
Leave kits or rolls out. Risk kit shelf life
What if time remaining is 4 hours?
What if time remaining is 80 hours
Scenario E
Cutting System is down for an unknown period
Option 1 -
Do not remove rolls for cutting from freezer for defrosting
Option 2 –
Remove rolls for defrosting and perhaps start the countdown in production lifespan
Scenario F
Combine Scenarios A, B, C, D, E...
Plataine
Production Optimization Solutions
20
©2014 Plataine Ltd. | All Rights Reserved | For Authorized Use Only

Loss of control with volume growth eliminates economies of scale and results in accelerated increase in waste

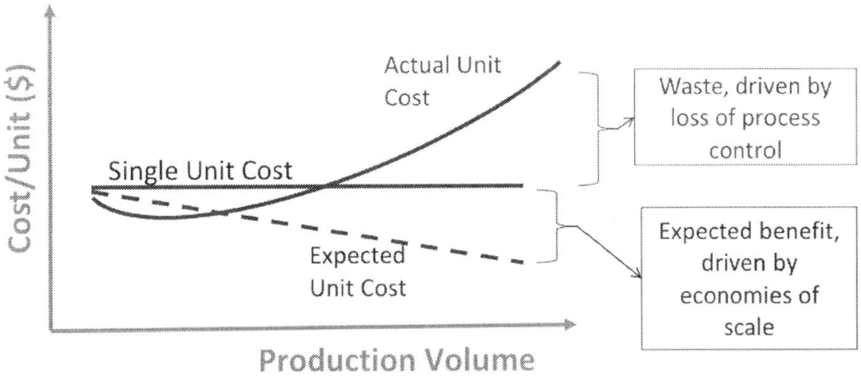

Plataine
Production Optimization Solutions

Regaining process control and establishing consistent optimization practices captures tremendous savings

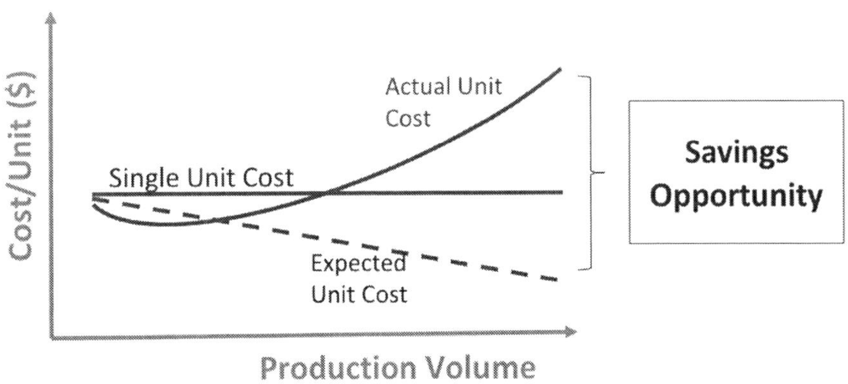

Plataine
Production Optimization Solutions

For each $10MM in annual material costs, manufacturers can save $500K - $1.5MM per year

Annual Cost Savings by Annual Material Costs and % Material Savings

% Material Savings	Total Cost of Raw Material per year			
	$10M	$50M	$100M	...
5%	$500K	$2.50M	$5.00M	
7.5%	$750K	$3.75M	$7.50M	
10%	$1.00M	$5.00M	$10.00M	
12.5%	$1.25M	$6.25M	$12.50M	
15%	$1.50M	$7.50M	$15.00M	

Not including:

- ✓ **Optimized raw-material purchases**
- ✓ **Reduced inventory-holding costs**
- ✓ **Increased productivity**
- ✓ **Improved Quality**

Plataine® Production Optimization Solutions

What our Customers Say...

“ *We chose Plataine's TPO solution as we were looking to* ***improve broader manufacturing metrics, beyond the focus on material of nesting-only software****. TPO has been tested* ***to save us over 5% in raw material consumption, while systematically utilizing our remnant materials and increasing our overall production throughput****. On top of these, we've found Plataine's novel approach to end-end automation as a key enabler of improved process visibility and quality control.*

-- **Beni Manes**
Deputy General Manager, Director of Operations
Aero-Assemblies Division
Israel Aerospace Industries, Commercial Group ”

Plataine® Production Optimization Solutions

Plataine Company Overview

- Specializing in production optimization solutions for over 20 years
- **Today:**
  - Over 1000 customers , including: **Composites Horizons, Hexcel, Hitco, Israeli Aircraft Industries (IAI), Steelcase, Swift Engineering and Vermont Composites (Kaman Aerospace)**
  - Offices, partners & distributors worldwide
  - Specialist team includes industrial and process engineers

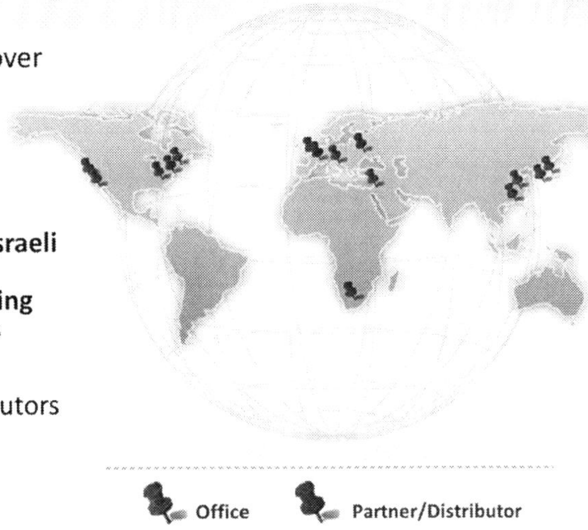

Plataine®
Production Optimization Solutions

25

Thank You

www.plataine.com

aerospace@plataine.com

avner.benbassat@plataine.com

Plataine®
Production Optimization Solutions

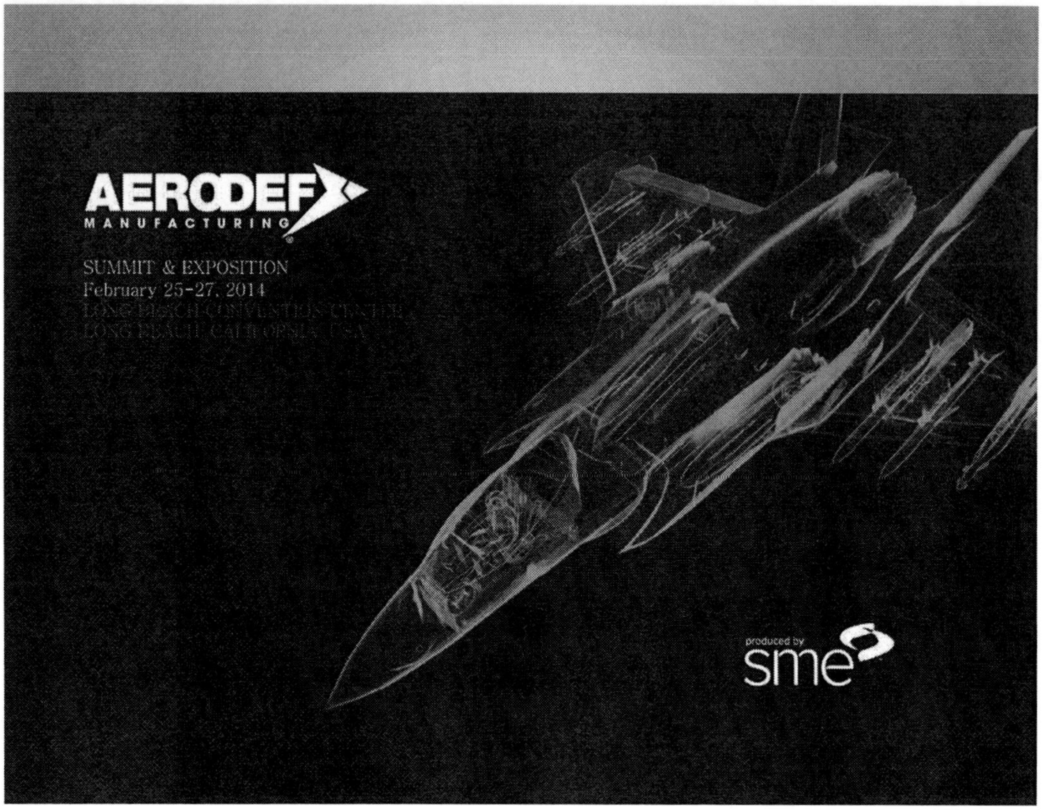

Reducing Autoclave Cycle Times Through the Use of Pressurized Water Mold Temperature Controllers

Kip Petrykowski
980-307-2981
kpetrykowski@regloplasusa.com
Director of National Sales - Variotherm & Composites

www.regloplasusa.com

Rev A 2

AERODEF MANUFACTURING

Current

- Strong incentive to replace Autoclaves with a faster less expensive method for curing parts
- For many companies with large investments in Autoclaves and long process certification times, Autoclave replacement seems unlikely
- What drives the long cycles and costs - Electrical heating of Nitrogen/Air and replenishment cost for lost Nitrogen/Air

Rev A 3

AERODEF MANUFACTURING

New Old Approach

Pressurized water

- Utilized in injection molding for over 50 years
- Proven to be the most consistent, safest, and most energy efficient method for heating tools as compare to air, steam, cartridge, or oil heating
- Has been successfully qualified and used in production for commercial/corporate aircraft primary structure parts

Rev A 4

Experimentation

Goal:

1. Determine energy consumed for both heating methods – air, water
2. Determine temperature profiles across tool at standard industry ramp rates (1.7 C/min.)
3. Determine maximum ramp rates with both heating methods
4. Determine as molded material properties

Rev A 5

Experimentation

Equipment:

- Weber nickel shell integrally plumbed plaque tool
- Used Regloplas P160M - 8 kW Heating / 60 kW Cooling
- Autoclave
- Pre-preg carbon
- Bagging material
- K type TC's
- NI Data logger
- Fluke 1735 Power Meter

Rev A 6

Experimentation

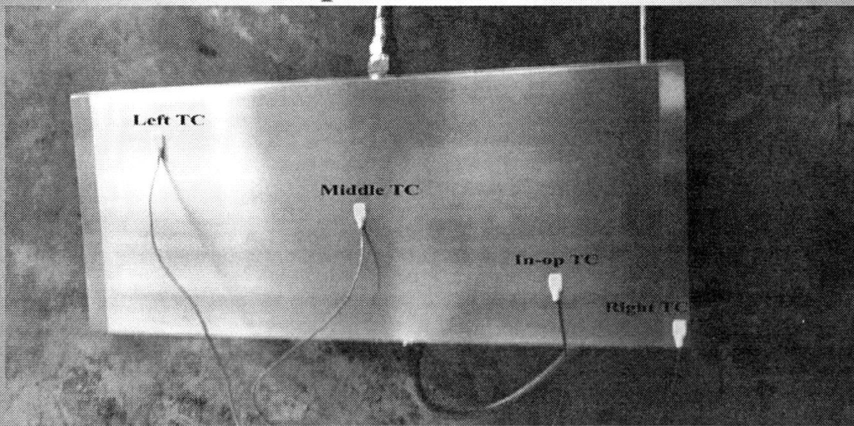

Figure 1. Weber nickel shell test tool

Rev A 7

Experimentation

Figure 2. Weber nickel shell test tool

Rev A 8

AERODEF MANUFACTURING

sme

Experimentation

Test Conditions:

- Measured flow through mold, 22 l/minute
- Ambient air temperature for test, 23 C
- Chilled water supply temperature, 20 C
- Chilled water flow rate, 12 l/minute
- Vacuum, - .76 bar

AERODEF MANUFACTURING

sme

Experimentation

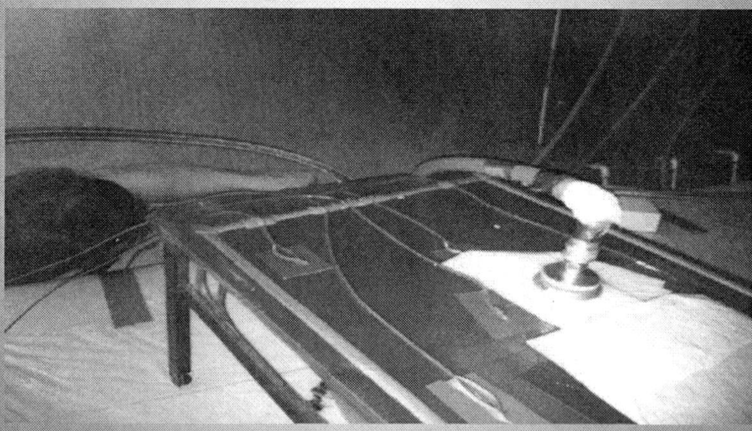

Figure 3. Weber nickel shell test tool with bagged pre-preg carbon sample

AERODEF MANUFACTURING

sme

Experimentation

Figure 4. Vacuum reading

Rev A

11

AERODEF MANUFACTURING

sme

Experimentation

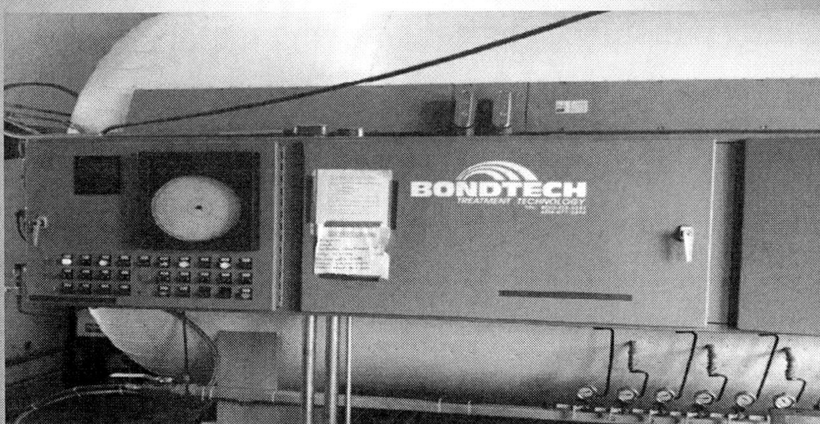

Figure 5. Autoclave

Rev A

12

Experimentation

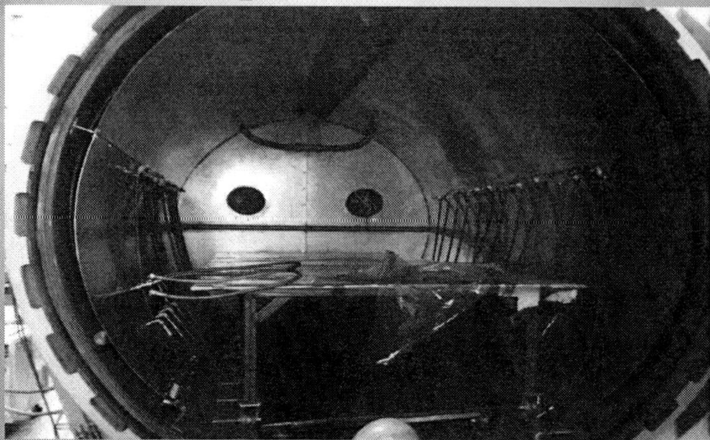

Figure 6. Autoclave

Results

Energy consumed for both heating methods – air, water

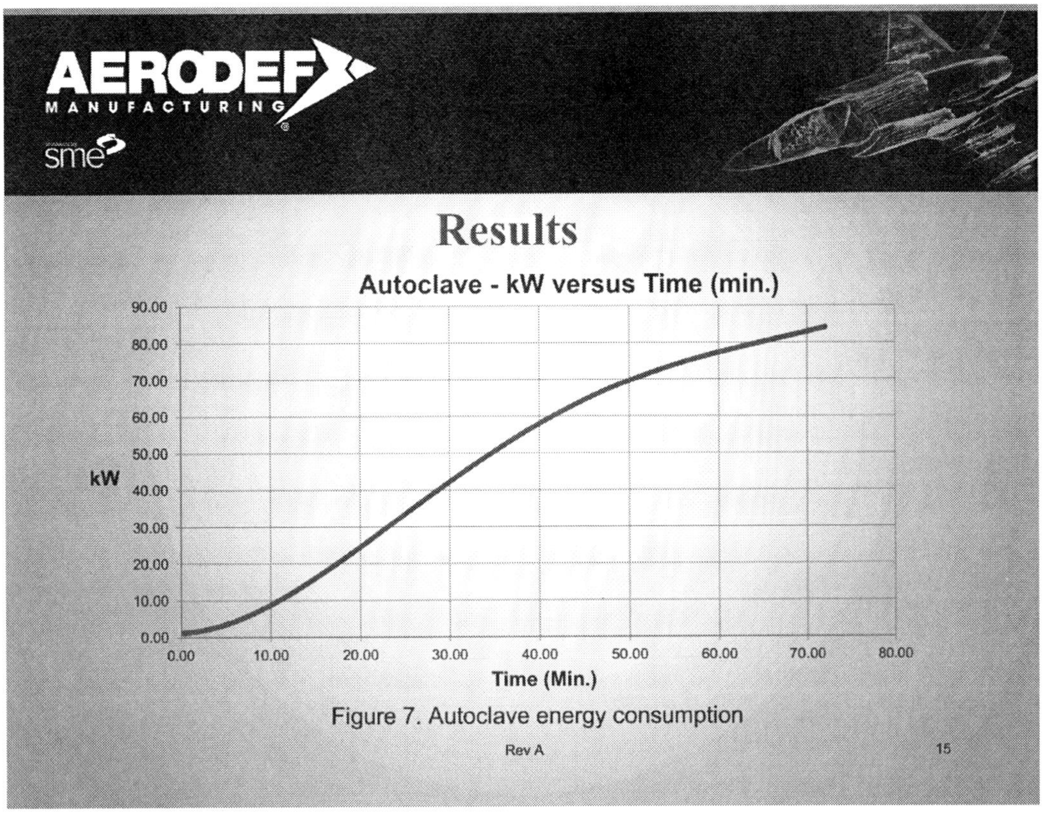

Figure 7. Autoclave energy consumption

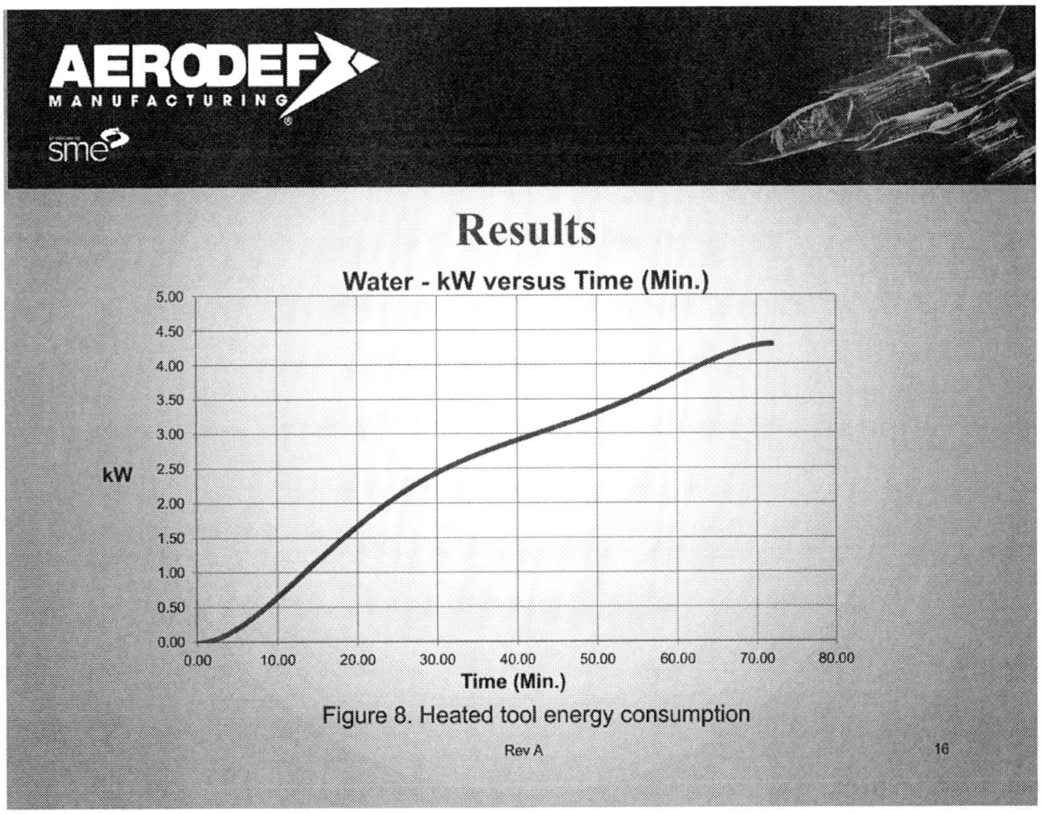

Figure 8. Heated tool energy consumption

AERODEF MANUFACTURING
produced by sme

Results

- Temperature profiles across tool
- Maximum ramp rates
- As molded material properties

In-process????

AERODEF MANUFACTURING
produced by sme

Heat Transfer Refresher

Pressure drop/flow rates			
fv_{fluid} = flow volume	0.07 liters/sec		
V_{fluid} = (fv fluid / passage cross sectional area	0.50 m/sec		
V_{fluid} through ext line= (fv fluid / passage cross sectional area	0.14 m/sec		
L/D	160 >10	40.0	Pump Liters/min.
Reynolds No. thru tool = (ρ_{fluid})(V_{fluid})(D)/v_{v}	15,000 10,000<Reynolds no.<120,000		
Reynolds No. thru lines = (ρ_{fluid})(V_{fluid})(D)/v_{v}	0 10,000<Reynolds no.<120,000		
Prandtl No. = (v_{fluid})/[k_{fluid}/(Cp_{fluid})(ρ_{fluid})]	0.91 .6<Prandtl No.<160	200° C	
Nu = Nusselt number = .023(Reynolds no.)$^{.8}$(Prandtl No.)$^{.35}$	49		
Nu = Nusselt number = (friction factor/8)(Reynolds No)(Prandtl No)/1.07+12.7(friction factor/8)$^{.5}$(Prandtl No)$^{2/3}$-1)	67		
h = heat transfer coeff = Nu(k_{fluid}/D)	12 W/m²°C		
f = friction factor Tool = (.790 ln Reynolds no.-1.64)$^{-2}$	0.0423 3,000<Reynolds no.<5,000,000		
f = friction factor Lines = (.790 ln Reynolds no.-1.64)$^{-2}$	0.0000 3,000<Reynolds no.<5,000,000		
Tool Data			
Cp_{tool} = specific heat at temperature	0.900 kJ/kg-C	200° C	
k_{tool} = thermal conductivity at temperature	0.500 W/m°C	200° C	
T_1 = Tool starting temperature	25° C		
T_2 = Tool ending temperature	180° C		
Room Temperature	25° C		
$\alpha = k_{tool} / (\rho_{tool})(Cp_{tool})$	0.00000035 m²/sec		
T_{Fluid} = Fluid temperature	200° C		
x = distance in tool from heat source	0.013 m		
Heating			
t = Time to heat	300 sec		
Base t = Time to heat	3100 sec		
Required $Kw_{heating}$ = [($Mass_{tool}$)($C_{p\ tool}$)($T_2 - T_1$)/t]*FS	10 Kw		
T(@f,t) = (erfc[(x)/(2)((α)(t))$^{1/2}$] - exp[(h)(x)/(k) + (h)²(α)(t)/k²][erfc[(x)/(2)((α)(t))$^{1/2}$ + (h)((α)(t))$^{1/2}$/k])*(T$_{Fluid}$-Ti))+T$_i$	36.29° C		

Figure 9. Carbon tooling calculated ramp rate

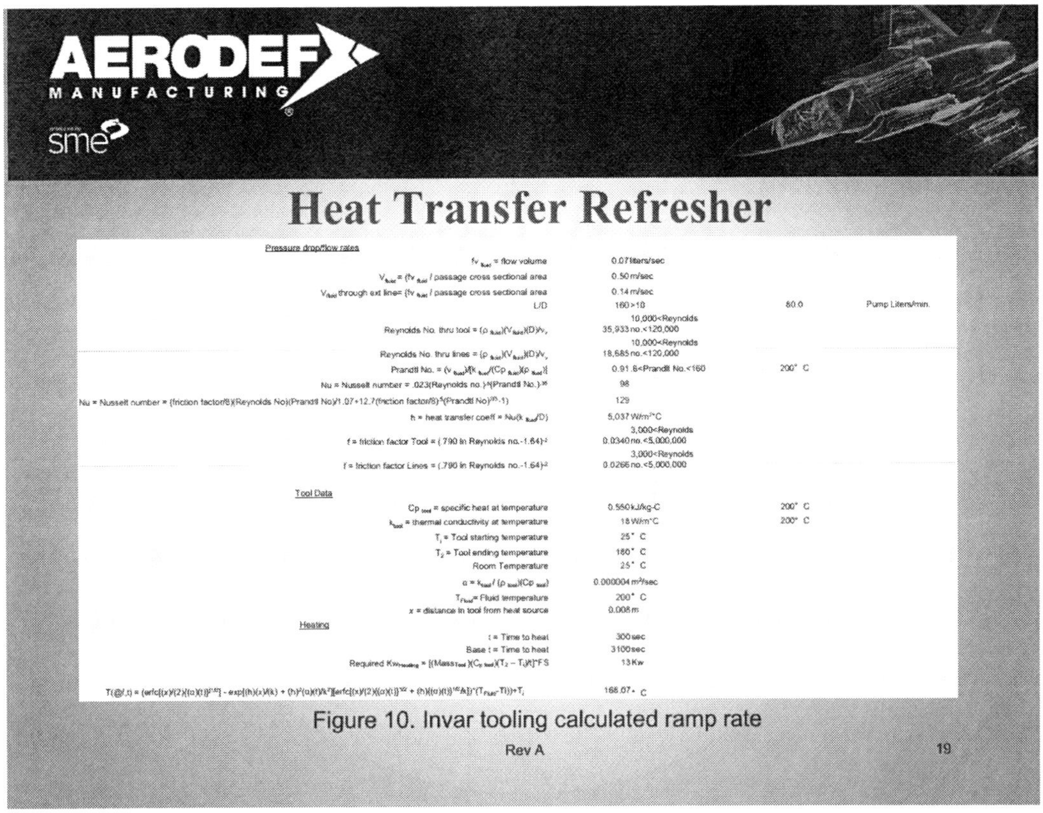

Figure 10. Invar tooling calculated ramp rate

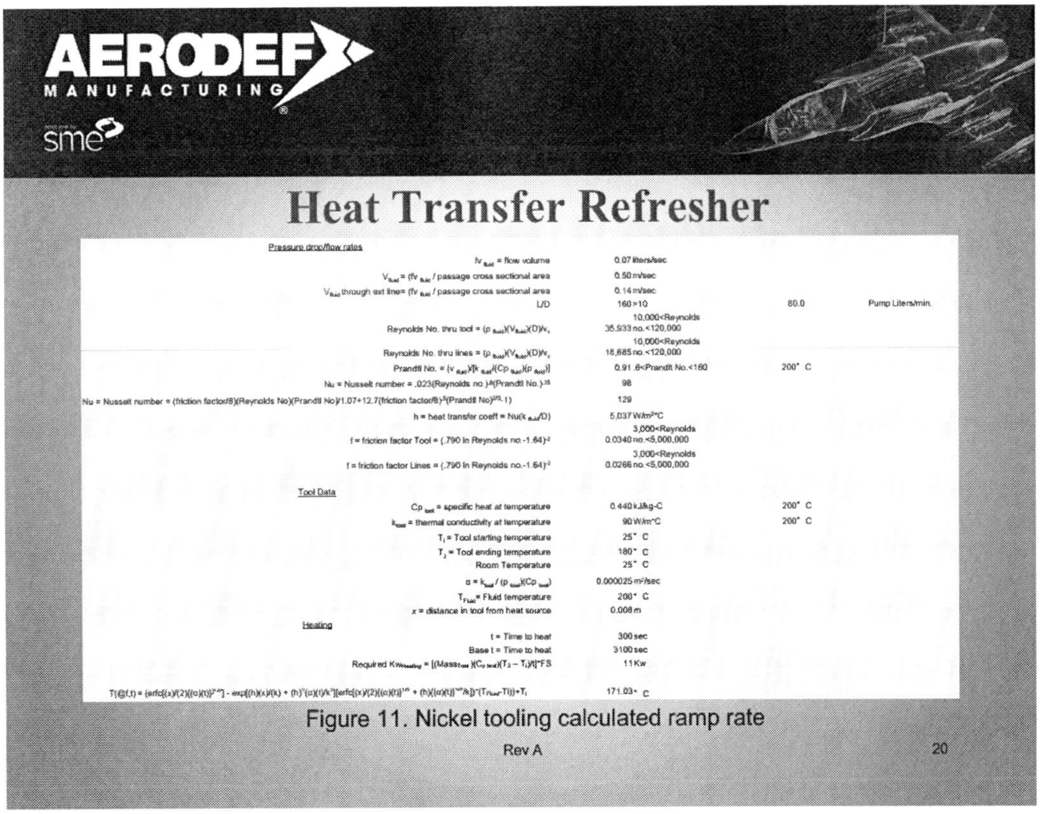

Figure 11. Nickel tooling calculated ramp rate

AERODEF MANUFACTURING

Benefits of New Method

- Better part consistency
- More flexible production
- Reduced energy consumption
- Shorter cycle times

Rev A 21

AERODEF MANUFACTURING

Hurdles to Implementation

- Belief that integrally plumbed tools are too expensive for low volume processes
- Energy costs not high enough to motivate change
- Process changes can be expensive and timely
- Belief that differing expansion rates between metal tool and part cannot be managed

Rev A 22

Technological Breakthroughs Required for Widespread Adoption

- The ability to retro-fit existing invar tooling with plumbing (Several tool makers are currently working on this)
- Integrally plumbed composite tooling (currently in production with a Regloplas customer)

Rev A 23

Conclusions

- Autoclaves burn a lot of energy
- Heat slowly
- Are scary to be inside

Rev A 24

AERODEF MANUFACTURING

sme

Questions?

Rev A 25

AERODEF MANUFACTURING

sme

Pressurized Water

Safety

- Much like an expansion valve in a refrigeration system, water, although not as effective as Freon, when decompressed, shows dramatic reductions in temperature
- Further cooling occurs through evaporation in the air
- The amount of water which can pass through the opening is ½ of the amount if it was cold water
- The combination of these effects reduces the amount energy at depressurization as compared to saturated steam to $^1/_{10}$

Rev A 26

AERODEF MANUFACTURING

sme

Project Partners

- **Abaris Training Resources, Inc.**
  5401 Longley Lane, Ste. 49
  Reno, NV 89511
  Phone: 971-261-2053
  Email: lou@abaris.com
  Web: www.abaris.com

- **TNT Motorsports**
  1095 6th Street Ct SE
  Hickory, NC 28602
  Phone: 828-323-1277
  Email: todd@tntmotorsportsnc.com
  Web: www.tntmotorsportsnc.com

- **Weber Manufacturing Technologies Inc.**
  16566 Hwy 12, P.O. Box 399
  Midland, Ontario, CA, L4R 4L1
  Phone: 705-527-2154
  Email: Tom.Schmitz@webermfg.ca
  Web: www.webermfg.ca

Rev A 27

AERODEF MANUFACTURING

sme

References

- http://www.iccm-central.org/Proceedings/ICCM17proceedings/Themes/Manufacturing/RESIDUAL%20STRESS%20&%20PROCE/INT%20-%20RESIDUAL%20STRESSES/IC4.4%20AnoushPoursartip.pdf

- http://controlhouse.com/sites/default/files/sampe_-_a_study_comparing_electric_cartridge_oil_and_pressurized_water_heating_for_composite_molding.pdf

Rev A 28

AERODEF MANUFACTURING

sme

SUMMIT & EXPOSITION
February 25-27, 2014

Composite Processing Utilizing Real-Time Visibility of Viscoelastic Properties During Cure

Presented by: Louis C Dorworth
Abaris Training Resources, Inc
Reno, Nevada

ABARIS TRAINING

Acknowledgements:

- **Avpro, Inc., Norman OK**
  - Thomas Rose
  - Hans Rose
  - Bob Kelpine
- **Alpha Technologies, Akron, OH**
  - James Caslow
  - Henry Pawlowski

AvPro

ALPHATECHNOLOGIES
A Dynisco Company

sme

ABARIS TRAINING

Definition of "Cure"

- For a thermoset resin it is the process of changing from a liquid state (viscous) to a solid state (elastic).
- "Full Cure" is realized when the resin will stay solid (vitrified) up to a prescribed service temperature (Ts).
  - Ts $\leq$ Onset of T_g

sme

ABARIS TRAINING

Development of Legacy "Recipes"

- For over 30 years, legacy cure recipes have been developed in the lab based on rheology of neat resin
- These recipes are then specified using a very limited process window:
  - Example: Process recipe calls for heat-up at 2°-5°F/min rate to 350°F ±10°F, Hold x 120 mins +15/-0 mins, then cool at < 5°F/min to < 140°F
    - This results in a very limited set of recipe boundaries, when in fact, the actual cure range of a thermoset resin may be accomplished within a much broader boundary

sme

ABARIS TRAINING

Rheometric Profile of Neat Resin

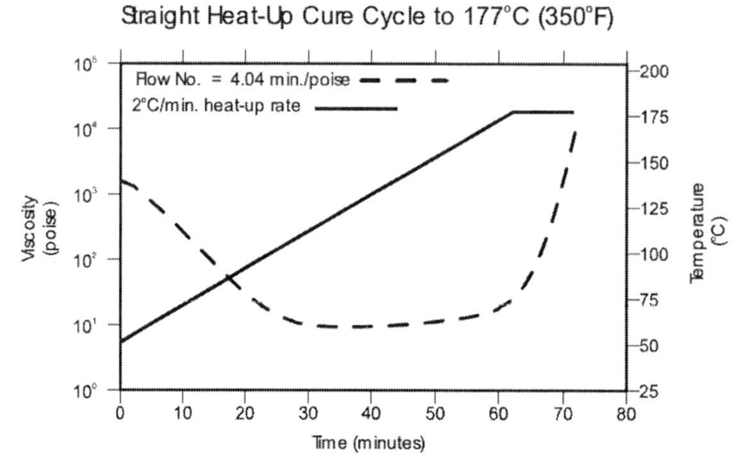

Note: neat-resin data does not take into account the effect of the fiber on thermal properties

Development of Legacy Cure Recipes

Lab Analysis Converted to Limited Time/Temperature Recipe:

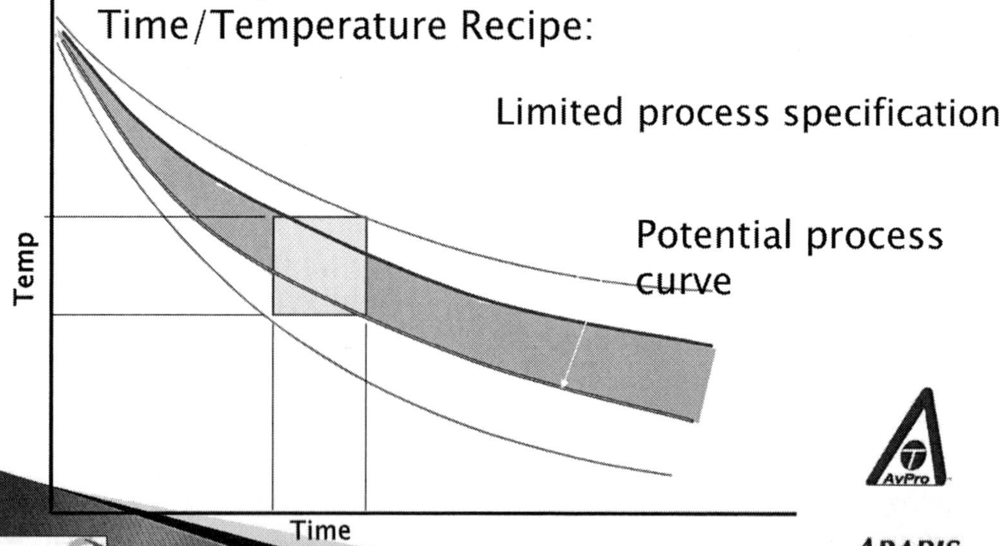

Resulting Process Specification

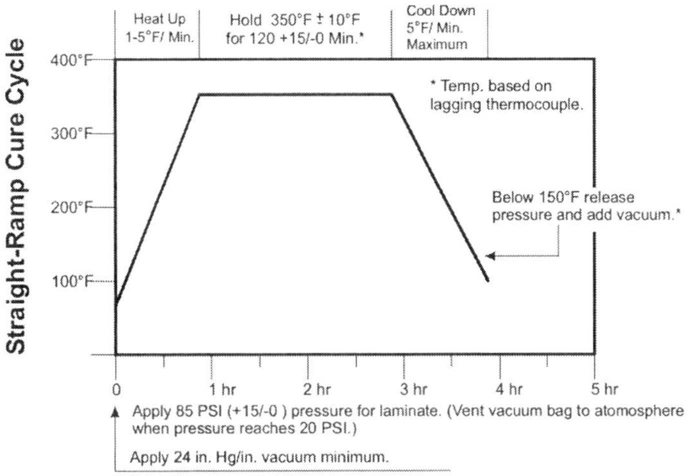

Legacy Recipes Vs. Material State Management (MSM) Tools

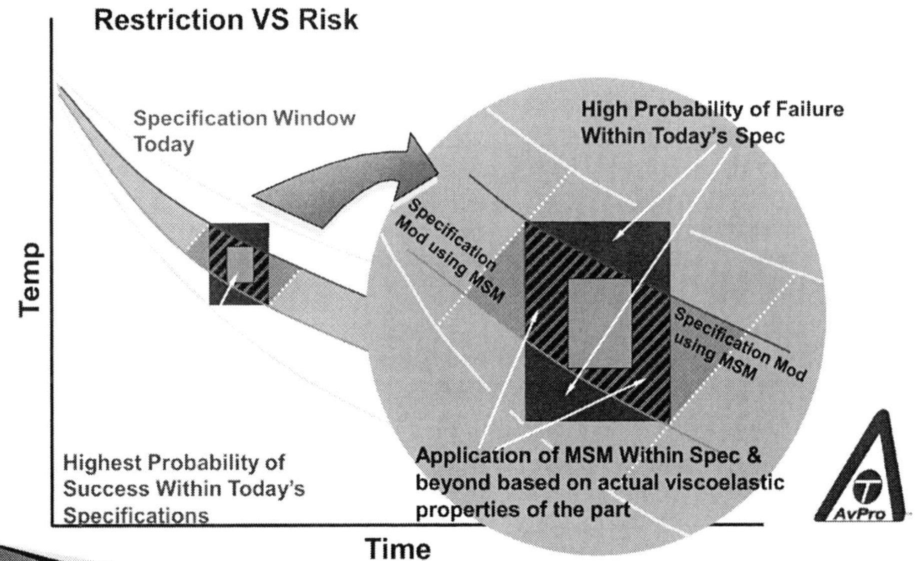

Legacy Vs. Goal

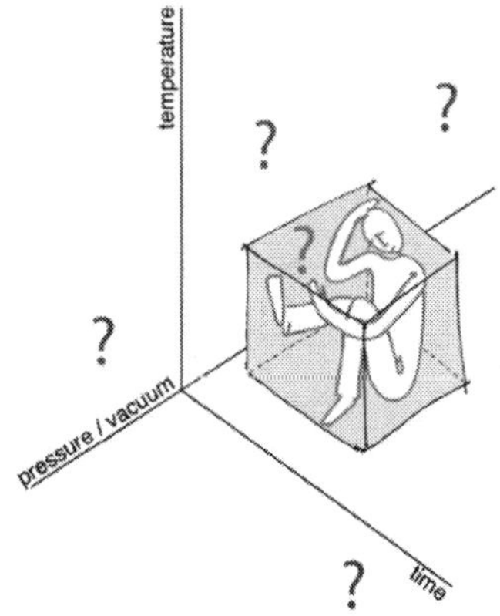

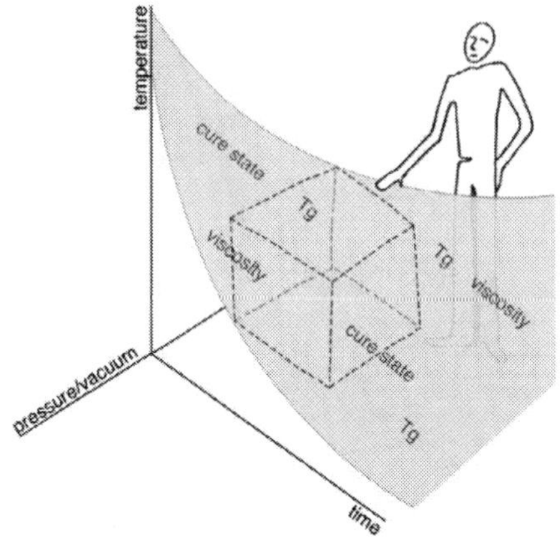

Viscoelastic Properties

- Viscosity of a liquid resin is simply the measure of the resistance to flow
  - Measured in cgs units of Poise (P) or Centipoise (cP)
- Elastic properties of a resin are realized when the resin no longer flows and it regains its shape after being stressed
  - Measured as shear modulus (G), or the ratio of shear stress to shear strain.
- Viscoelastic properties are a measurement of both of these regions on the same material
  - On an uncured, partially cured, or fully cured resin or prepreg materials

More on Viscoelastic Properties

- Why do we want to know when a material exhibits liquid, viscous (sticky liquid), and elastic (solid) properties during the cure?
  - The knowledge of these properties are important in determining how and when certain process events should occur:
    - Vacuum or venting during processing
    - Pressurization at an optimized point in the process
    - Change in viscosity and resin gelation
    - Vitrification or solidification of the resin
    - Cure/Tg

Thermal Analytical Instruments Used for Real Time and Post-Process Analysis

Differential Scanning Calorimeter (DSC)

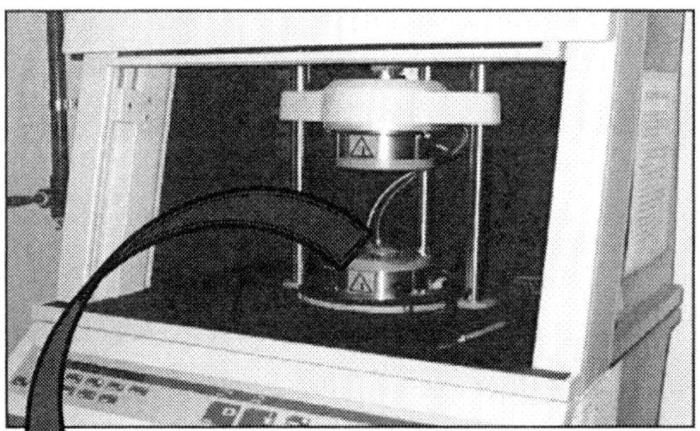

ATD ESR Torsion Plate Rheometer

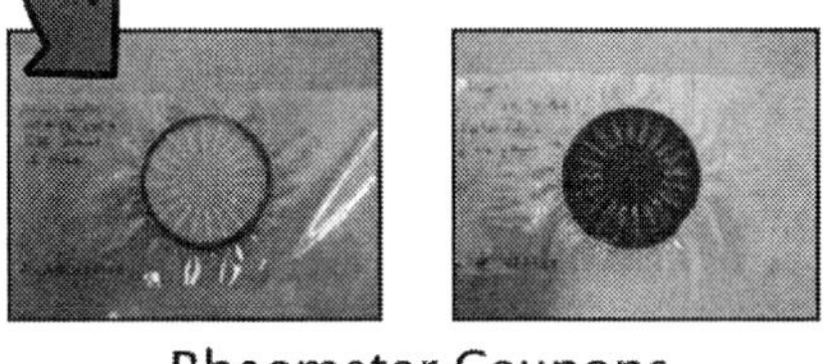

Rheometer Coupons

ATD ESR with Dies Open

Place Prepreg sample onto Lower Die

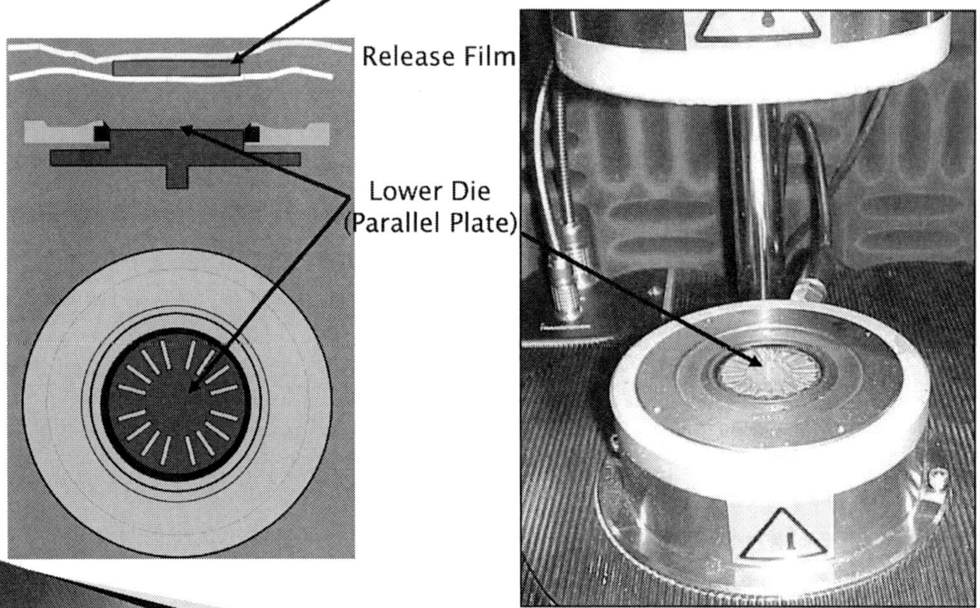

ATD ESR with Dies Closed

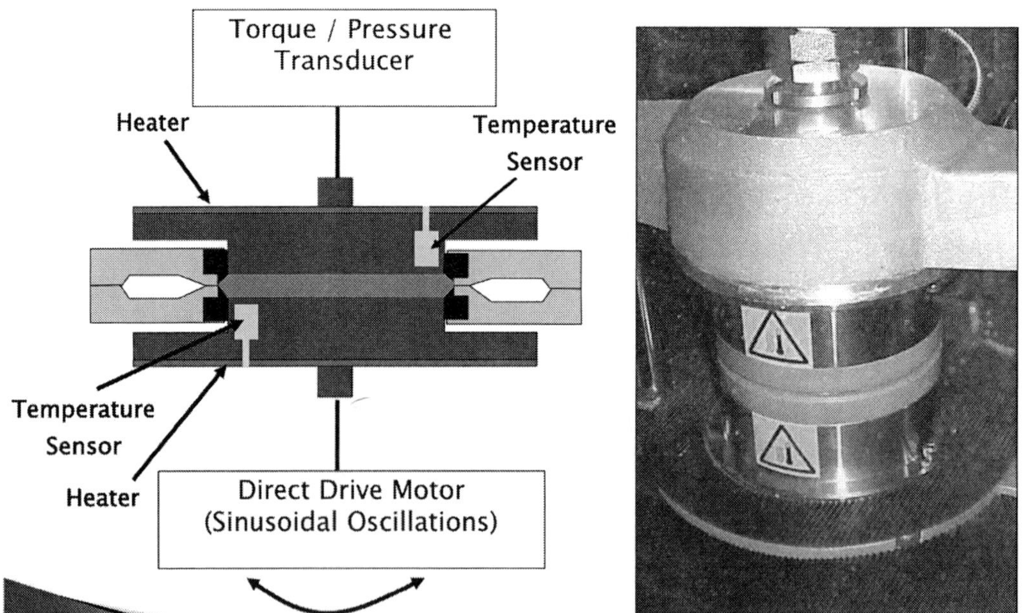

Key Rheometric Measurements

- Dissipative loss moduli (G_{pp})
  - Referenced as viscous modulus
    - Used to calculate viscosity ($Visc_p$)
  - Peak G_{pp} is indicative of vitrification & partial maturation of cross-linking molecules.
    - Glassy state ~ not necessarily fully matured.
- Elastic storage moduli (G_p)
  - Achieves full cure at the prescribed temperature.
    - Max value in asymptotic trace on graph.

sme

ABARIS TRAINING

Example: Difficulty Maintaining Spec Requirements in Real Life Cycles

- What happens if process specifications are not met?
  - Time: Heat rate or soak time is non-compliant:
    - Thermal mass of tooling in vessel causes lag in rate or large delta temperatures between leading and lagging sensors
  - Temperature: Sensors are outside of boundaries:
    - Leading or lagging TC's are just outside of limits causing discrepancy requiring disposition
  - Vacuum: Vacuum too high or low or not vented on time:
    - A time-triggered vacuum event may affect part quality
  - Pressure: Too high or low or not applied on time:
    - A time-only triggered pressure event may affect part quality

sme

ABARIS TRAINING

Result: Rejected Part or Cycle

- When any of the process sensors measure outside the specified boundaries, the part or load is rejected and usually tagged to wait for an engineering disposition.
  - This typically spawns a series of post-process lab tests and a review by a Material Review Board (MRB) before a disposition can be made.
  - This process can hold up production for days before a disposition is reached and the part is either accepted or scrapped.

Remedy:

- Bring real-time visibility of material state condition (viscoelastic properties) to equipment operator, M&P engineering, and quality assurance personnel to streamline part/process disposition.
- Develop a database on all specified materials that will eventually allow engineering and QA to perform a disposition based on statistical analysis of past process data.

With Material State Visibility:

- We have actual viscoelastic data from the discrepant cycle for immediate analysis and comparison to the specified recipe
- Or, we can compare the discrepant condition to the ideal condition by running a post-process analysis using process simulation software to model the "ideal" viscoelastic change
  - Using "Run Compare" software we can overlay the two graphs to Compare viscoelastic, time, temperature data

sme

ABARIS TRAINING

Case Study # 1: Discrepancy in High-Temperature Cure Cycle

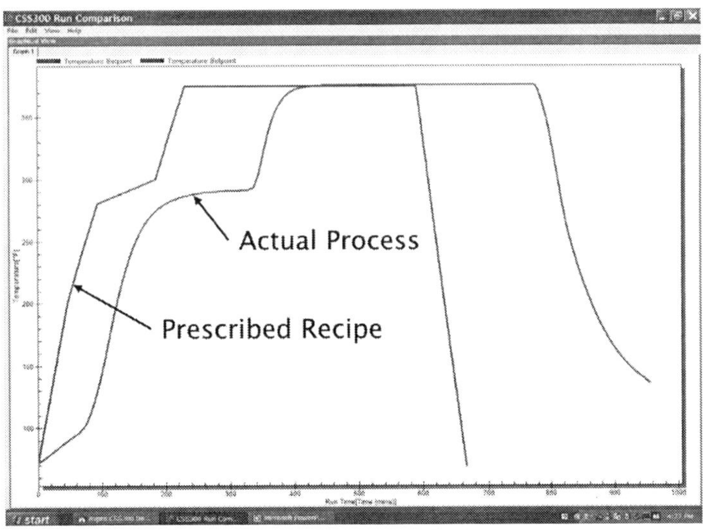

sme

ABARIS TRAINING

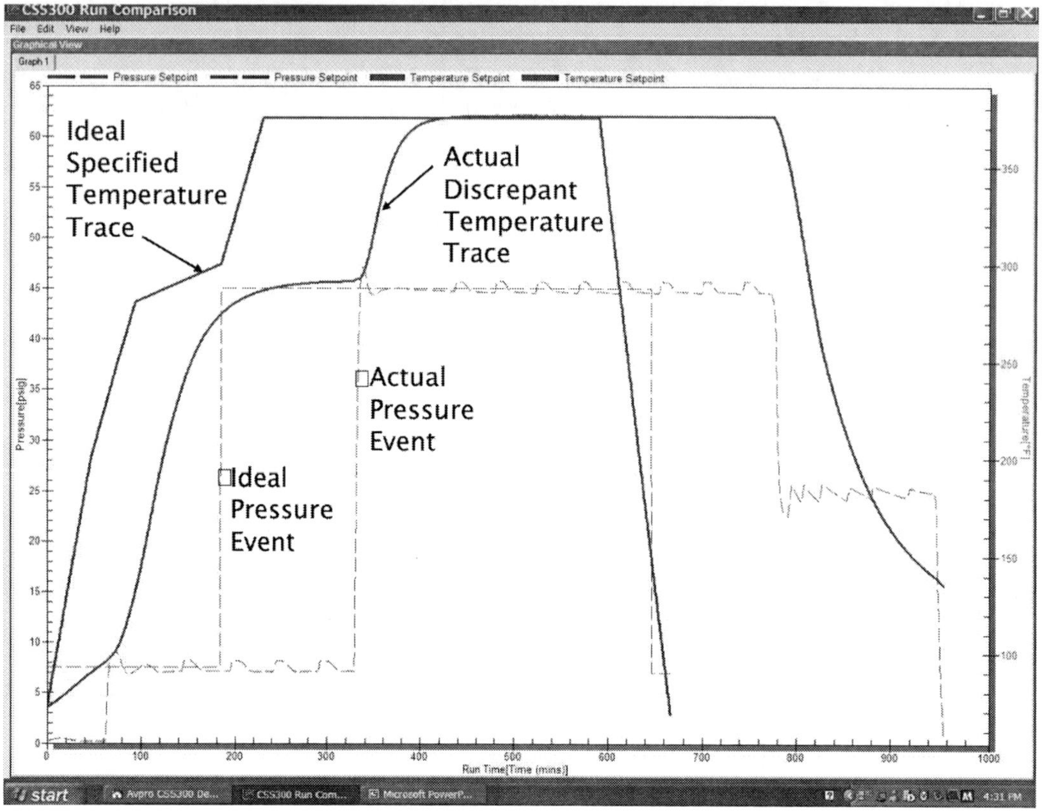

Viscoelastic Properties of Ideal Run

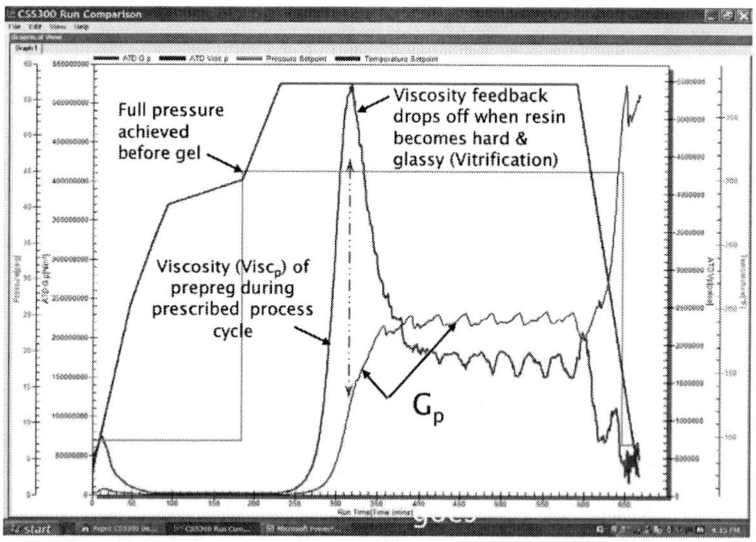

Viscoelastic Traces of Discrepant Run

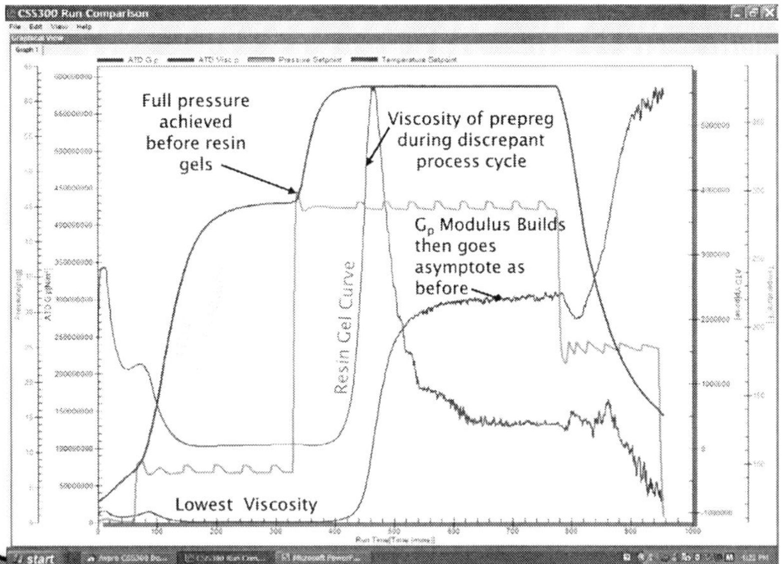

Case Study #2: Loss of Pressure

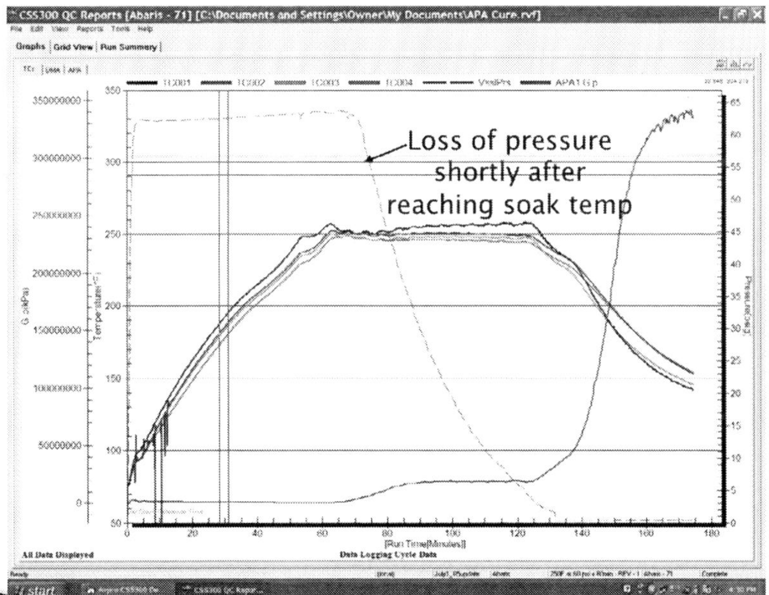

Post Process Evaluation of Event

- Within minutes of completion of the cycle, using the lagging TC sensor from the discrepant run we are able to run a Differential Scanning Calorimeter (DSC) sample using run data to evaluate the time at which the chemical reaction took place during the cycle.
- By using "Run Compare" software we were able to overlay the original senor inputs on a single graph and analyze the affect of the time at which the pressure was lost.

sme

ABARIS TRAINING

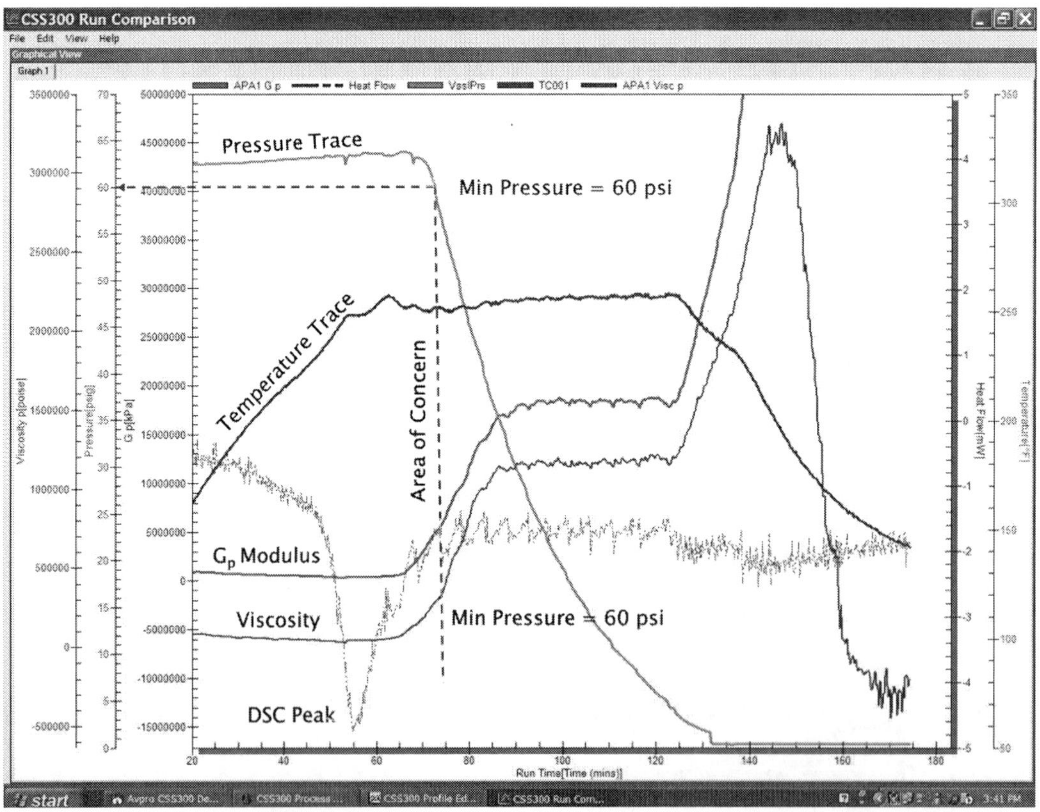

Summary

- Brings visibility to cure-properties during actual production or repair cure cycles.
- Equipment and software is capable of defining state of cure:
  - Vitrification and T_g of composite material.
- Can be used to streamline cure cycle based on actual material state:
  - Max G_p at cure temperature rather than time in recipe.
- Provides a reliable tool for post-process quality assurance and MRB activity.

sme

ABARIS TRAINING

Other Opportunities Using MSM

- Quality control for incoming material thru final process of part:
  - Fast, cost effective receiving inspection
  - Shelf life & out time requalification
    - Flow/gel testing of new or time-x materials
- Load optimization-tool thermal profiling
- Collection of cure data for later use in creating statistical disposition model.

sme

ABARIS TRAINING

Questions?

sme

ABARIS TRAINING

Vision Guided Robotic Part Marking System

David Siedal

Technical Fellow

Lockheed Martin Aeronautics

AERODEF MANUFACTURING
February 25-27, 2014 - SME

Vision Guided Robotic Part Marking System

AERODEF MANUFACTURING
February 25-27, 2014 - SME

Overall Project Objectives

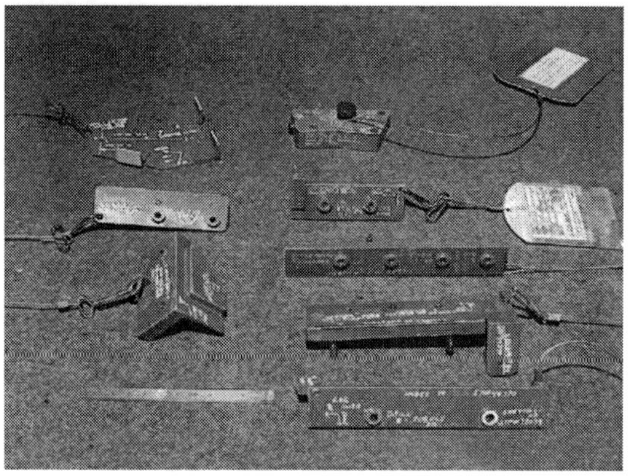

Conventional Bracket Locating Tools
(ASTP - Assembly Templates)

- Eliminate Conventional Hard Tooling To Locate Brackets (2000 Tools)
- Develop A "Digital" Manufacturing Process
- Increase Accuracy Of Bracket Placement
- Reduce Rework
- Reduce Assembly Span Time

AERODEF MANUFACTURING
February 25-27, 2014 - SME

Detailed Project Objectives

- Develop A Robotic System To Mark The "Footprint" Or Attachment Location On Bulkheads Where Brackets And Other Items Are Secured (2000 Total)
- Robotically Mark All Locations Without Touching The Part
- Mark Over 90 Percent Of The Bracket Locations (Largest Parts Have Over 300 Brackets Each)
- Dramatically Reduce The Time (Cost) To Locate Each Bracket
- Take Advantage Of Digital Thread (CATIA Designed Parts)

AERODEF MANUFACTURING
February 25-27, 2014 - SME

How Well Were The Project Objectives Met?

- All Brackets Robotically Marked Without Touching The Part
- One-Hundred Percent Of All Brackets Are Marked
- Time (Cost) To Mark Brackets Dramatically Reduced
  - Manual Method = 1 Hour Per Bracket
  - Robotic Method = Less Than 1 Minute Per Bracket
- Improperly Located Brackets (Rework) Cut By Over Ninety-Five Percent
- Significant Schedule Span Time Reduction
- Digital Thread Utilized For All Robot Processes

AERODEF MANUFACTURING
February 25-27, 2014 - SME

Project Benefits

- Money - Immediate Multi-Million Dollar Cost Avoidance By Not Building Hard Tools (Bracket Locating Templates)
- Money - Thousands Of Labor Hour Savings Per Aircraft
- Money - Brackets Precisely Located Per Engineering Data Model
- Money - Bracket Rework Dramatically Reduced
- Money - Design Changes Integrated Without Tooling Rework (Brackets Locations Change Frequently Early In The Program)
- Money - Significant Schedule Span Time Reduction
- Money - Aircraft Delivery Schedules Met

AERODEF MANUFACTURING
February 25-27, 2014 - SME

Typical Aircraft Bulkhead

AERODEF MANUFACTURING
February 25-27, 2014 - SME

Additional Aircraft Parts

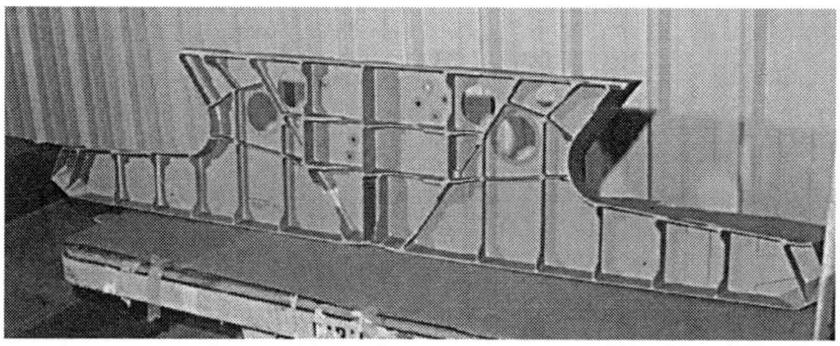

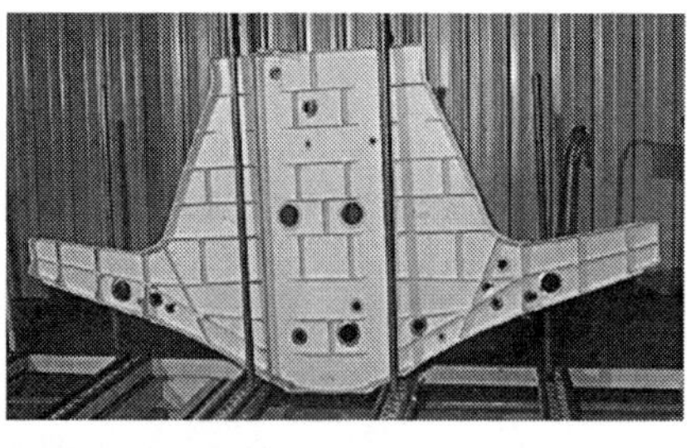

AERODEF MANUFACTURING
February 25-27, 2014 - SME

Project Milestones

F-22

- 161 Consecutive Ship Sets Of F-22 Parts Marked
- 12,200 Individual Parts Marked
- 400,000 Bracket Locations Marked
- 2,000 Hours Per Aircraft Labor Savings

F-35

- 100 Consecutive Ship Sets Of F-35 Parts Marked
- 1,600 Individual Parts Marked
- 32,000 Bracket Locations Marked
- 300 Plus Hours Per Aircraft Labor Savings

AERODEF MANUFACTURING
February 25-27, 2014 - SME

Key Vendors

- PaR Systems
  - Gantry Robot Manufacturer And Systems Integrator
    - Robot, End Effectors And Peripheral Equipment
    - Vision System Hardware & Software
    - Inkjet Marking System Integration
- MicroFab
  - Piezoelectric Printing Device Manufacturer
  - Commercial Printing Systems
  - Medical And Electronics Fabrication

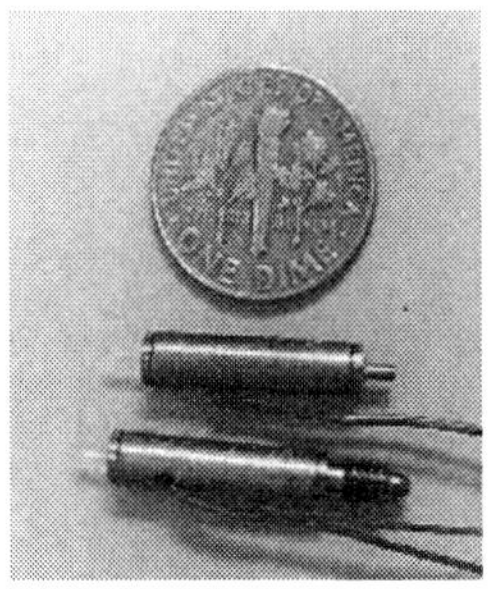

AERODEF MANUFACTURING
February 25-27, 2014 - SME

Major System Components

- PaR Systems Custom Gantry Robot - 16'x 9'x 3' Envelope
- Tooling Table Shuttle Positioner - 50' Travel
- Two Vacuum Tooling Tables - 16'x 6' Each
- PaR Systems Developed Vision System
- Tooling Pins And Targets
- End Effectors - Side And Down Firing
- End Effector Calibration Station
- Safety Fencing And Entrance Gate Interlocks

Typical Part And Bracket Locations

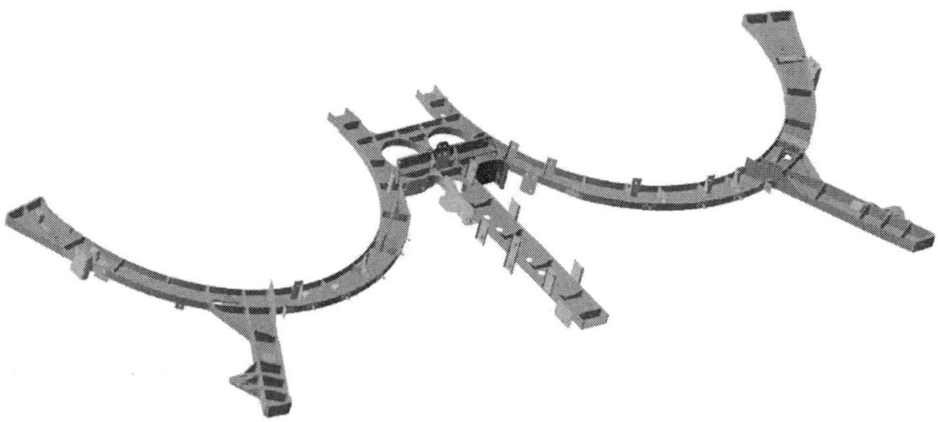

AERODEF MANUFACTURING
February 25-27, 2014 - SME

Robot Program Process Flow

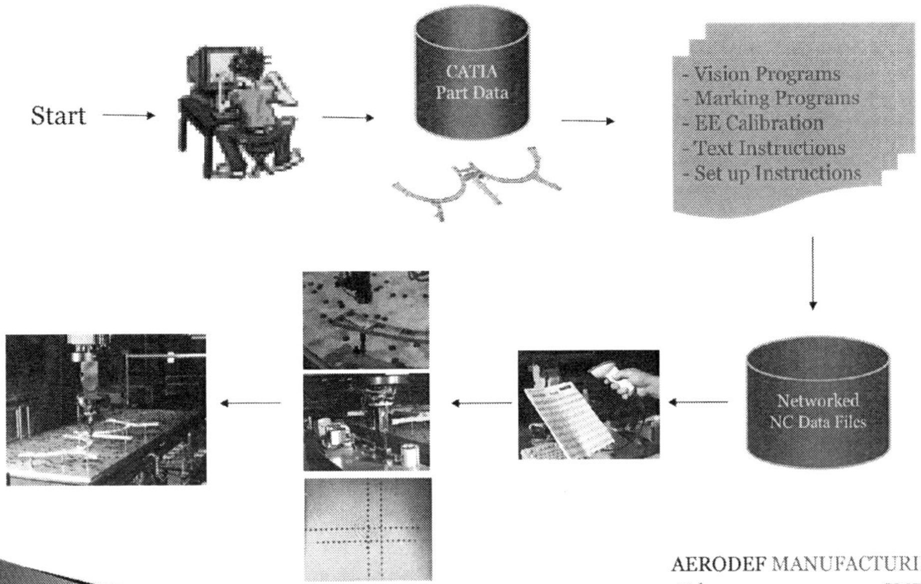

AERODEF MANUFACTURING
February 25-27, 2014 - SME

Operator Interface

Operator Interface Station

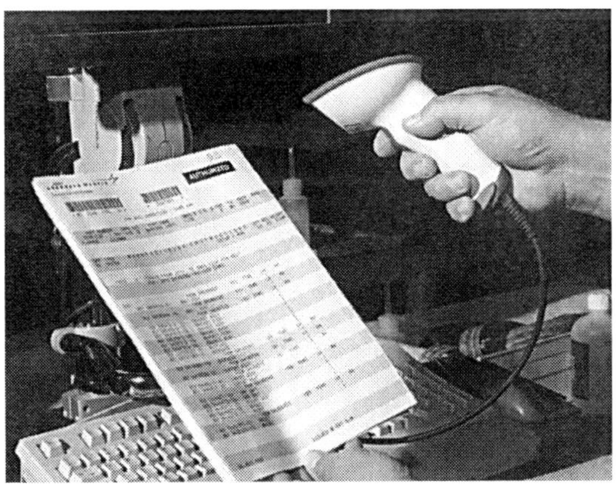

Retrieve Robot Programs From Database

AERODEF MANUFACTURING
February 25-27, 2014 - SME

Tooling Pin Set-up Instructions

End-Effectors	
Seq #	T-Num
10	T 31
20	T 1
30	T 0

Seq #	Hole #	Loc.
1	433 T1	2F
2	434 T2	17F
3	435	3D
4	436	3B
5	437	6C
6	438	8E
7	439	11E
8	440	9B
9	441	13C
10	442	15B
11	443	16D
12		
13		
14		
15		
16		
17		
18		

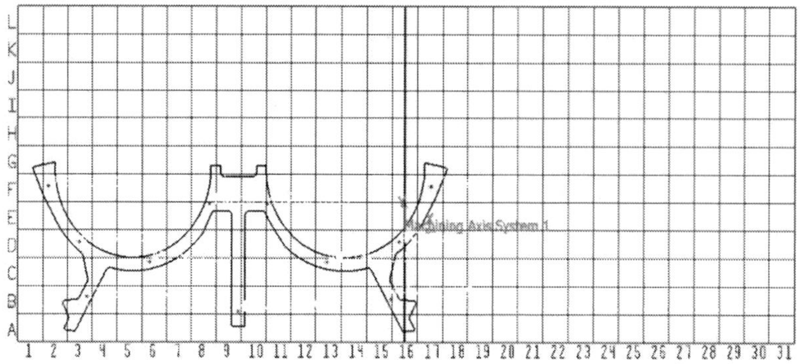

VIEW LKG FWD

AERODEF MANUFACTURING
February 25-27, 2014 - SME

Tooling Pins And Vision Targets

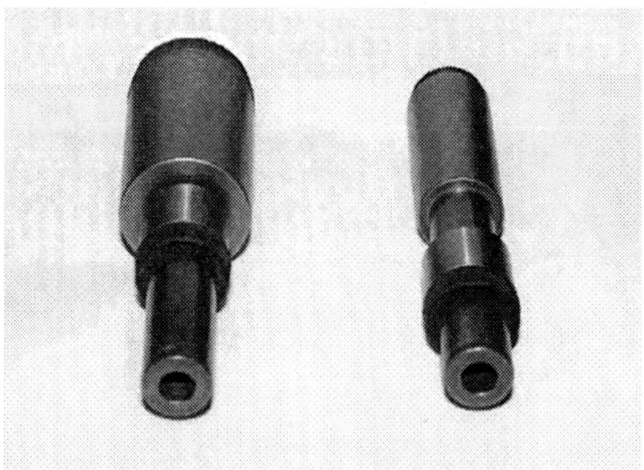

AERODEF MANUFACTURING
February 25-27, 2014 - SME

Tooling Pins And Vision Target Kits

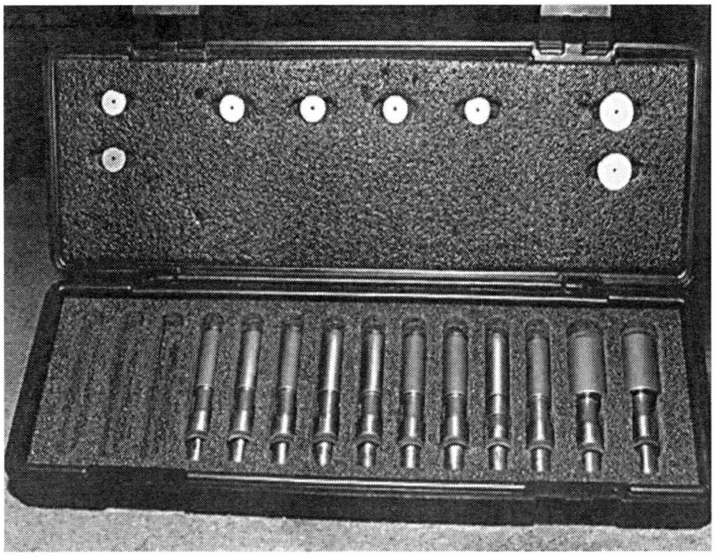

AERODEF MANUFACTURING
February 25-27, 2014 - SME

Tooling Pin Installation

AERODEF MANUFACTURING
February 25-27, 2014 - SME

Existing Tooling Holes In Part

AERODEF MANUFACTURING
February 25-27, 2014 - SME

Vision Target Installation In Part

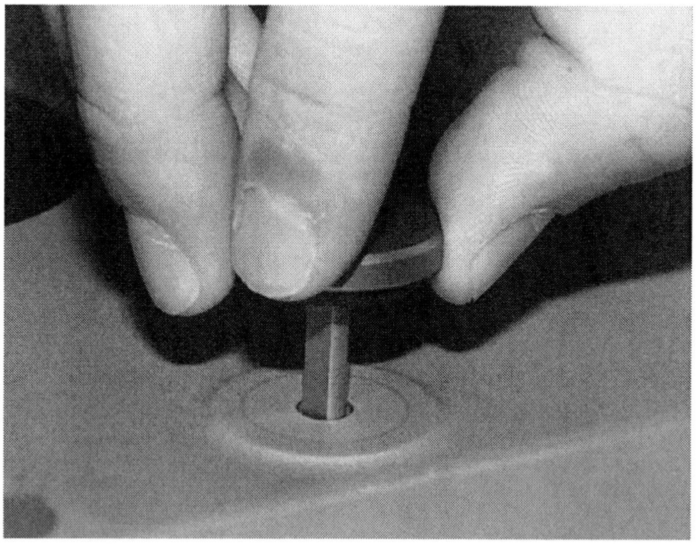

AERODEF MANUFACTURING
February 25-27, 2014 - SME

Part Ready For Processing

AERODEF MANUFACTURING
February 25-27, 2014 - SME

X & Y Axis Coordinate Transformation

One Of Two X & Y Axis Data Points Obtained Per Part

AERODEF MANUFACTURING
February 25-27, 2014 - SME

Z Axis Coordinate Transformation

One Of Three Z Axis Data Points Obtained Per Part

AERODEF MANUFACTURING
February 25-27, 2014 - SME

Robot Picking Up End Effectors

AERODEF MANUFACTURING
February 25-27, 2014 - SME

End Effector Jetting And Calibration Station

AERODEF MANUFACTURING
February 25-27, 2014 - SME

Ink Jet Drop Formation At 1K Hz
(Viewed With Strobe Light)

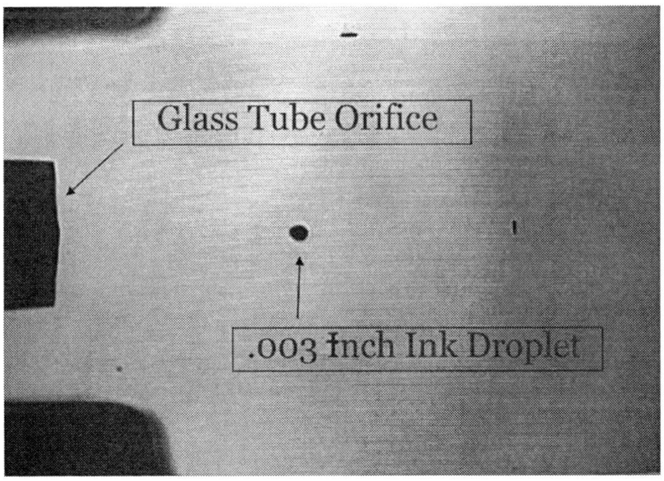

AERODEF MANUFACTURING
February 25-27, 2014 - SME

Side Firing End Effector In Calibration Station

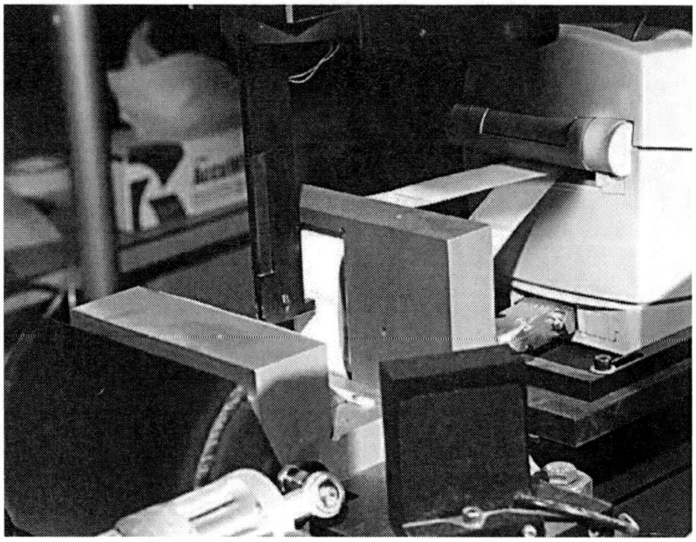

AERODEF MANUFACTURING
February 25-27, 2014 - SME

Jetting Verification And Calibration Marks

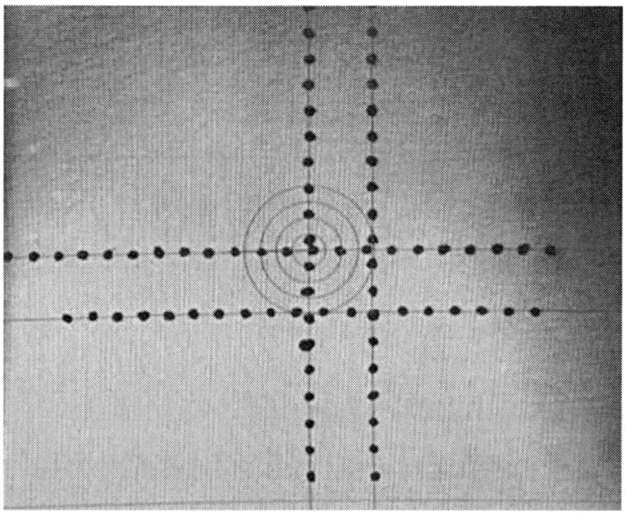

AERODEF MANUFACTURING
February 25-27, 2014 - SME

Application Of Part And Ink Dot Offsets

X, Y, Z Coordinate Transformation

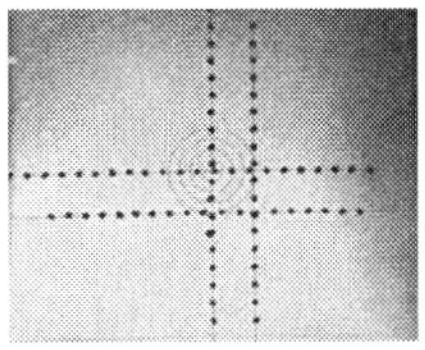

Calibration Station Offsets Established

Nominal NC Program

Part Marking With Part And Ink Dot Location Offsets Applied

AERODEF MANUFACTURING
February 25-27, 2014 - SME

Part Marking In Progress

AERODEF MANUFACTURING
February 25-27, 2014 - SME

Side Firing End Effector Marking Part

AERODEF MANUFACTURING
February 25-27, 2014 - SME

Part Completion Verification Process

AERODEF MANUFACTURING
February 25-27, 2014 - SME

Close Up View Of Ink Dots

Each Dot Equals....

- 4 Droplets Per Dot
- .025" Dot Marking Size
- Vision Calibrated And Marked 4 Times (Twice On Each Table)
- INK DOT on INK DOT! (16 droplets per mark)

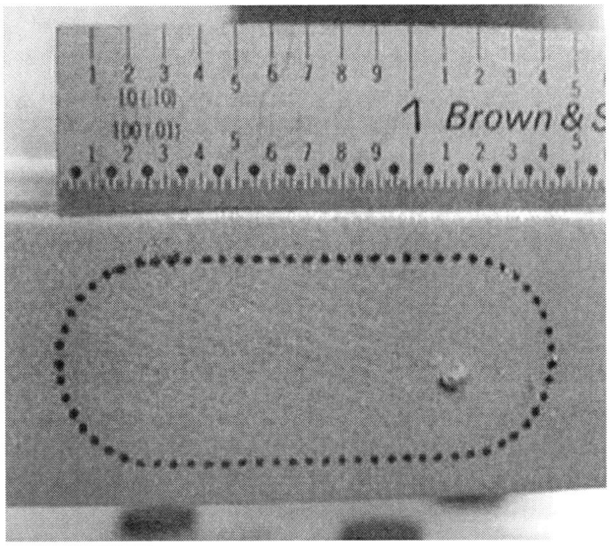

AERODEF MANUFACTURING
February 25-27, 2014 - SME

Key Technology - Drop On Demand Printing

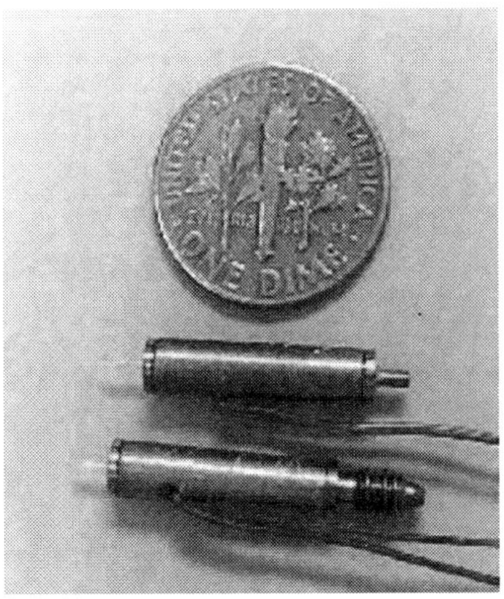

Drop-on-Demand Piezoelectric Ink Dispensing Device

- Manufactured By MicroFab
- Established Technology
- Used In The Biomedical And Electronics Industry
- Extremely Small And Precise
- Pico Liter Drop Volumes
- Excellent Control Characteristics

AERODEF MANUFACTURING
February 25-27, 2014 - SME

F-35 Robotic Part Marking System - Marietta Ga.

AERODEF MANUFACTURING
February 25-27, 2014 - SME

Lockheed Martin Robot Laboratory
(Fort Worth Division)

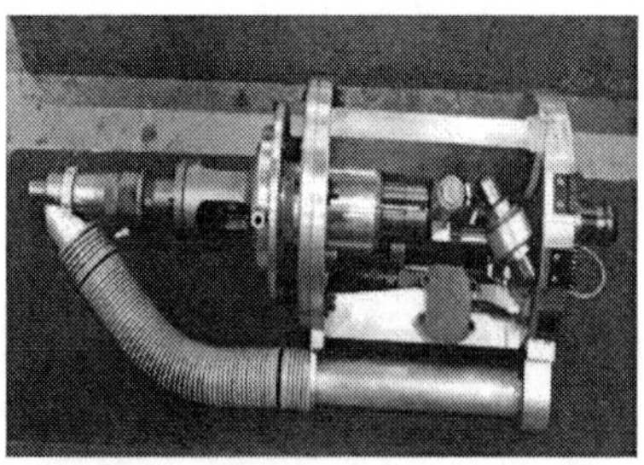

AERODEF MANUFACTURING
February 25-27, 2014 - SME

Robotic Milestones At Lockheed Martin
(Fort Worth Division)

- First "Production" Robot In the Aerospace Industry (1979)
- Robot Development Test Site For Cincinnati Milacron Robotics Division
- In-House Developed Quick Change Adaptors, Drilling End Effectors And Control Systems (1979)
- Product Development Test Site For IBM Robots (1980)
- Robotic Vision System Developed For Machining Center Load/Unload (1982)
- Developed First Production Robot For Spraying LO Coatings - F-16 (1988)
- Drilled Over 50 Million Holes In Several Thousand F-16 Parts (1979- Present)
- Marked Over 400,000 Brackets Locations On The F-22 & F-35 Programs
- Painted Hundreds Of F-35 Detail Parts And 95 Complete Aircraft With LO Coatings
- Over Thirty-Four Years Of Robots In Production Without An Injury

AERODEF MANUFACTURING
February 25-27, 2014 - SME

History Of Robots At Lockheed Martin
(Fort Worth Division)

- 1979 - Vertical Fin - Drill Graphite Composite Skins (F-16)
- 1979 - I/R Panels (2) - Drill/Rout Aluminum Skins (F-16)
- 1980 - I/R Panels (2) - Drill Rout Aluminum Skins (F-16)
- 1980 - IBM Gantry - Drilling Of Sheet Aluminum Panels (F-16)
- 1981 - Horizontal Stabilizer - Drill/Countersink Composite/Aluminum (F-16)
- 1981 - Wing - Drill/Countersink Aluminum Skin/Understructure (F-16)
- 1982 - Wing (3) - Drill/Countersink Aluminum Skin/Understructure (F-16)
- 1982 - Tube Preparation - IBM Robot For Tube End Deburring (F-16)
- 1984 - Vertical Fin Upgrade - Drilling Composite Skins And Edge Assemblies (F-16)
- 1984 - Canopy - Drill Aluminum And Steel Canopy Side Frames (F-16)
- 1984 - Brush Deburr - Abrasive Brush Deburr Of Machined Parts (F-16)
- 1985 - Contact/Filler Insertion - Harness Connector Assembly (F-16)
- 1986 - Vertical Fin Assembly - Drill/Csk, Grip Gauge, Nutplate Drill, Faying Surface Sealant, Install Blind Bolts (F-16)

AERODEF MANUFACTURING
February 25-27, 2014 - SME

© 2014 Lockheed Martin Corporation

History Of Robots At Lockheed Martin
(Fort Worth Division)

- 1987 - Bonded Panels - Bonded Assemblies Added (F-16)
- 1988 - Tube Unloading - Unload Tubes From NC Tube Bender (F-16)
- 1988 - Tube Flaring - Install Fittings On Tubes With Flare Machine (F-16)
- 1988 - Inlet Spray #1 - Spraying Of Specialized Coatings - Inlet Ducts (F-16)
- 1988 - FMS Load/Unload #1 - Load Billets/Unload Finished NC Part (F-16)
- 1989 - FMS Load/Unload #2 - Load Billets/Unload Finished NC Part (F-16)
- 1989 - FMS Tooling Tab - Remove Tooling Tabs From NC Machined Parts (F-16)
- 1989 - I/R Panels (3) - GMF(2) & Westinghouse (1) Electric Robots (F-16)
- 1990 - Horizontal Stabilizer - Relocate FMS Tooling Tab Robot (F-16)
- 1992 - Wing Drilling - Relocate FMS Load/Unload Robots (2) (F-16)
- 1994 - Wing Drilling - Relocate I/R Panel Drilling Robots (2) (F-16)
- 1997 - Bonded Panels - Replace Existing Robot With Refurbished Robot (F-16)
- 1999 - Canopy Drilling - Replace Cincinnati Robot With ABB (F-16)
- 2002 - Inlet Spray #2 - Spraying Of Specialized Coatings - Inlet Ducts (F-16)

AERODEF MANUFACTURING
February 25-27, 2014 - SME

History Of Robots At Lockheed Martin
(Fort Worth Division)

- 2002 - Bulkhead Marking System - Mark Bracket/Clip/Nut Plate Locations (F-22/F-35)
- 2007 - Component Paint Shop Robots (4) - LO Painting Of Detail Parts (F-35)
- 2008 - Aircraft Final Finishes Robots (3) - LO Painting Of Full A/C (F-35)
- 2009 - Bonded Panels - Replaced Cincinnati Robot With Fanuc S420 Lab Robot (F-16)
- 2011 - Component Paint Robot Expansion - Retrofit F-16 Inlet, Booth #1 (F-35)
- 2012/13 - Additional Aircraft Final Finishes Robots (3) - LO Painting Of Full A/C (F-35)
- Future - Several Additional F-35 Painting And Sealant Application Robots

AERODEF MANUFACTURING
February 25-27, 2014 - SME

Lessons Learned Implementing Robots

1. Suggestions For Preparing For A Project ***Prior*** To Purchase Order Placement

2. Suggestions For Managing The Project ***After*** The Purchase Order Is Placed

Lessons Learned *Prior* To Starting Any Robot Project

- Pick Your Automation Applications Carefully
- Never Oversell A Systems Capabilities In Order To Gain Acceptance
- You Cannot Over Prepare For Any Project
- Never Underestimate The Requirements To Automate Even The Simplest Of Tasks
- Know And Understand The Capabilities And Limitations Of Your Team
- Thoroughly Research All Potential Vendors
- Obtain Several Unbiased Third Party Assessments Of The Project Related To Risk, Scope, Cost And Ongoing Support Requirements
- Have A Real Backup Plan In Place Should Issues Delay Implementation

Lessons Learned *Prior* To Starting Any Robot Project

- The End User Must Be The Projects #1 Advocate
- Clearly Communicate To Everyone Up Front That All Projects Carry Some Risk
- First Of A Kind Systems Are Extremely Risky No Matter How Experienced The Vendor And Project Team
- New Robots Will Be Much More Closely Scrutinized Than The Equipment Or Process They Replace
- The System Must Be Simple To Operate And Work Reliably Day After Day In Production
- Meet Early And Frequently With The Personnel Who Will Be Operating The Equipment – Their Input And Support Is Essential
- Know The Skill Level Of The Robot Operators And Design The System Accordingly

Lessons Learned *Prior* To Starting Any Robot Project

- Obtain Vendor Referrals And Visit Their Customers To See Their Systems
- Know If They Are Going To Subcontract Portions Of The Project To Other Vendors
- Be Involved In The Subcontractor Selection Process
- Meet Face-To-Face With The Individuals From Other Companies Who Will Be Working On Your Project
- Go To Their Facilities - See How They Run Their Business
- Laboratory Demonstrations Are Almost Always Misleading And Rarely Replicate Exactly How The System Will Function In Production

© 2014 Lockheed Martin Corporation

AERODEF MANUFACTURING
February 25-27, 2014 - SME

Lessons Learned *Prior* To Starting Any Robot Project

- The Buyers Project Team Should Be An "Expert" In All Phases Of The Existing Process
- The Bid Specification Should Be Extremely Detailed And Include All Process And Performance Requirements
- All Deliverables Should Be Clearly Detailed In The Final Procurement Specification - Both Hardware And Software
- Specify Who Will Be Responsible For Part Programming And What Hardware/Software Programming Packages Are Required
- Do Not Expect Your Vendor To Do Any Significant Work For Free
- All Change Orders Need To Be Priced And Approved By Procurement Personnel Prior To Incorporation To Avoid Funding Issues

Lessons Learned *Prior* To Starting Any Robot Project

- The End User Must Be Able To Pay For Post Installation Support To Keep The System Functioning Properly
- A "System Manager" Should Be Part Of Almost All New Robotic Systems To Maintain Operations And Continuity (Post Installation)
- Reserve Procurement Funding Must Be Included In Planned Cost Estimates To Cover Unforeseen Expenses (Approximately 10% Of Total)
- Make Sure Any Site Preparation And Foundation Work Is Included In Cost Estimates
- Spare Parts And Maintenance Contracts Should Be Purchased As Part Of The System
- Travel Costs Should Also Be Accounted For In The Budget

© 2014 Lockheed Martin Corporation

AERODEF MANUFACTURING
February 25-27, 2014 - SME

Lessons Learned *After* The Purchase Order Is Placed

- A Program Manager With Overall Responsibility For All Issues Must Be Named At Each Company
- It Is Both Program Managers Responsibility To Maintain Open Communication And Good Relationships Between Companies
- A Dedicated Project Team Is A Must - At Both The Primary Vendor And Buyers Facility
- A Single Point Of Contact (Usually The Program Managers) Is Used For All Significant Communications Between Companies
- Agree To A Date Immediately After The PO Is Placed To Finalize All Schedules And Deliverables
- Schedule New Or Risky Hardware / Software Technology Demonstrations To Help Document Development Progress

Lessons Learned *After* The Purchase Order Is Placed

- Remember The Vendor You Selected Is Your Teammate, Not An Adversary
- An Open And Trusting Relationship With Your Vendor Is Critical
- Frequent, Direct, And Honest Communication Will Build That Trust
- A Good Relationship Between Program Managers At Both Companies Is A Must
- The Program Managers Must Not Allow Hidden Agendas At Either Company To Creep Into The Project
- Maintaining Professionalism Throughout The Project Is A Must
- Try To Limit Distractions And Make Sure Goals And Objectives Are Clear To All Project Members - No Scope Creep
- Weekly Teleconferences Documenting Progress Should Be Held With Minutes And Individual Action Items Being Assigned

© 2014 Lockheed Martin Corporation

AERODEF MANUFACTURING
February 25-27, 2014 - SME

Lessons Learned *After* The Purchase Order Is Placed

- Acceptance Testing Of The Complete System At The Vendor's Facility Prior To Shipment Should Be Mandatory
- Every Detail Function And Feature Should Be Tested
- Fully "Automated" Testing Should Also Be Done As If The System Is Operating In A Production Environment
- Acceptance Testing At The Buyer's Facility Should Duplicate The Tests Run At The Vendor's Facility For Comparison Purposes
- Error Recovery Issues Are Very Complex And Should Be Discussed In Detail Early In The Project

Lessons Learned *After* The Purchase Order Is Placed

- Developing & Implementing Automation Requires Thick Skin
- "Skeptics" May Also Appear With Unwarranted Criticism Or Comments
- The Skeptics Will Revel In Any Small Setback And Are Easily Identified By The Words "I Told You So" And "I Knew That Wouldn't Work"
- The Skeptics Rarely Offer Solutions To Problems..... Politely Acknowledge What They Say And Stick To Your Plan
- "Experts" Will Also Appear From Many Places
- Unless The Expert Is A Trusted Technical Resource, Carefully Consider Their Recommendations Before Changing Course
- Project Managers Must Quickly And Tactfully Deal With Project Saboteurs

Lessons Learned *After* The Purchase Order Is Placed

- Include Some Extra Time In The Schedule For Start Up Issues And Debugging Before Making Your First Production Part
- Never Hide Problems - Real Or Perceived
- Immediately Solicit Help To Fix Real Problems If Needed
- Be Prepared To Work Long And Unusual Hours During System Installation And Production Testing
- Avoid Shortcuts - The Problems They Mask Will Surface At The Most Inopportune time

AERODEF MANUFACTURING
February 25-27, 2014 - SME

Thank You!

David Siedal
Technical Fellow

Lockheed Martin

One Lockheed Blvd, MZ 6427
Fort Worth, Texas 76108

817-777-8574
david.s.siedal@lmco.com

AERODEF MANUFACTURING
February 25-27, 2014 - SME

THE VALUE OF PERFORMANCE.

NORTHROP GRUMMAN

Automated Sanding – Robotic Sanding System

Commercialization Pilot Program (CPP)

26 February, 2013

Todd Szallay

Manufacturing Technology

Robotic Sanding System

NORTHROP GRUMMAN

Issue:

- Current confined space coatings are sanded manually resulting in ergonomic and cost impacts

Impact:

- The shape and geometry impact the ergonomics of the technicians who perform the work
- Hours for manual sanding add significant impact to the total hours needed for coatings application
- Automated sanding could reduce the hours needed by as much as 50%

Approach:

- Develop automated sense and sand capability for confined space leveraging commercial technologies
- Mature system business case analysis for two automated sanding systems for installation into existing sanding booths

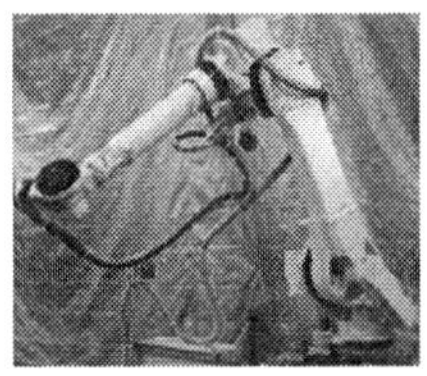

FANUC M710

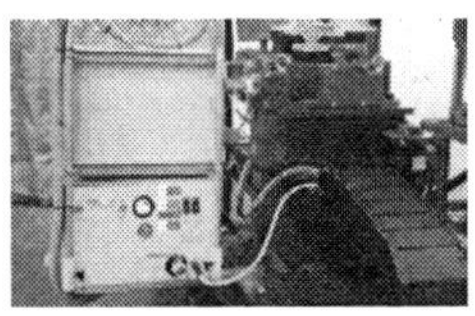

Integrated Vacuum System

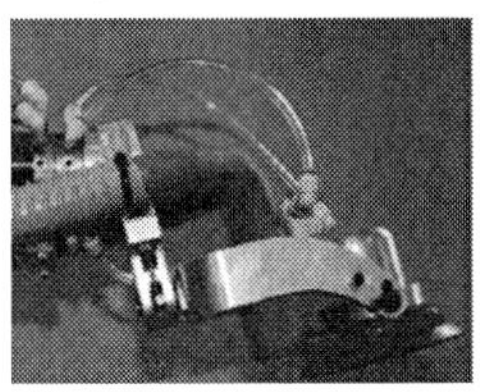

Integrated Vacuum Head

2

Robotic Sanding System

NORTHROP GRUMMAN

Objective:

- Integrate Robotic Sanding System (RSS) into confined space coatings processes where automation is feasible.
- Advance technology to MRL 7 production capability
    - Automated Sanding SBIR Phase III: Commercialization Pilot Program Funded

Problem:

- Coatings are manually sanded amounting to a significant portion of the manufacturing span time for coatings application in the inlet ducts.
- Confined space ergonomics associated with the manual sanding process are less than ideal, resulting in awkward operator positions, reach issues, tool control difficulties, and even operator injury.

Importance:

- Implementation of the robotic sanding system will also minimize potential worker injuries due to the confined work environment and poor ergonomics associated with the current manual sanding processes.
- Labor hours saving opportunities

3

Approved for Public Release; NGAS 14-0334, 2/21/14

Robotic Sanding System

NORTHROP GRUMMAN

Strategy:

- Build upon accomplishments from the Phase I & II SBIR programs
- Transition from a laboratory level system to a production representative system.
- Demonstrate both the field and radius sanders, preliminary HMI and data acquisition systems and vacuum/blow-off capability.
- HMI and QA/QC systems will be fully matured and integrated, and the necessary safety features for the factory floor will be incorporated.
- Prototype production system will be demonstrated in areas of interest on production materials.
- In parallel, a prototype robotic sanding booth will be designed to meet all factory requirements, including footprint, environmental, and safety requirements.
- The program will conclude with a final demonstration of the Robotic Sanding System

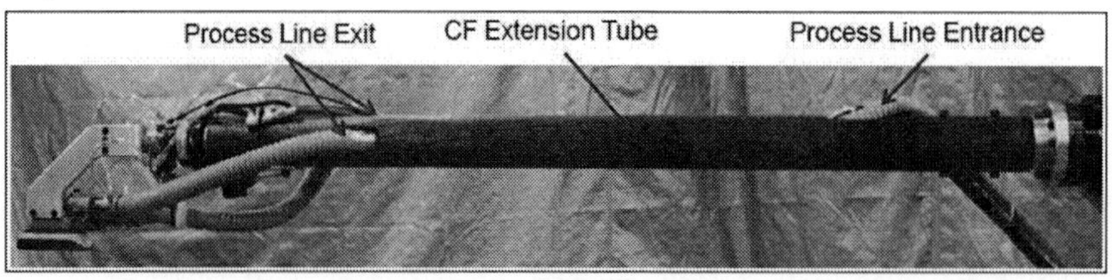

4

Approved for Public Release; NGAS 14-0334, 2/21/14

Robotic Sanding System Components

NORTHROP GRUMMAN

- Sanding System Design Complete (AFRL Funded)
  - Confined space surface geometry (0 deg., -/+ 15 deg. – radius, detail, and field area)
- Robot modifications for CL. 2 Div 1 requirements complete
  - Allows installation into existing CL 2 Div 1 sanding booth
- Control cabinet and vacuum system integrated onto robot
- Incorporates tool change out stand
  - Designed to allow accurate tool changeout
  - Precise location for each sander
- Production turn table concept comparable to existing spray systems

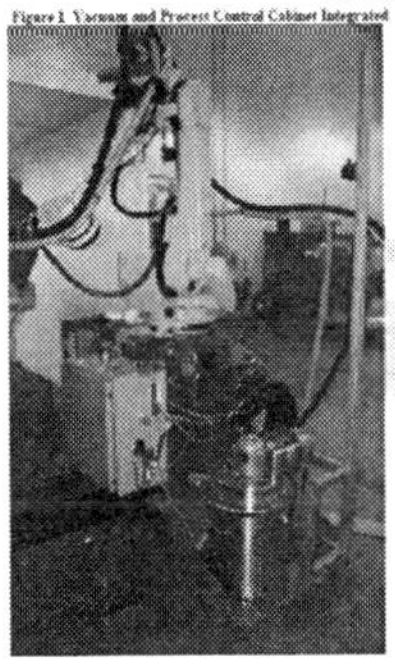

Figure 1 Vacuum and Process Control Cabinet Integrated

Figure 2 Tool Changing Stand

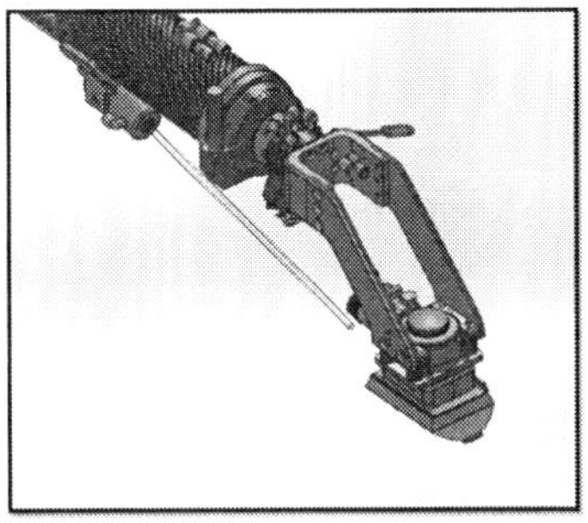

5

Approved for Public Release: NGAS 14-0334, 2/21/14

Robotic Sanding System MRL 5

NORTHROP GRUMMAN

- Completed MRL 5 Demonstration on 02-May-2013. Video documentation of sanding operations was captured
  - Production capability demonstrated at MRL 5 demonstration
  - MRL 6 and MRL 7 will productionize the sanding system
  - Majority of sanding paths executed
  - < Link > ***to video***
- Verified HMI functionality and data reporting

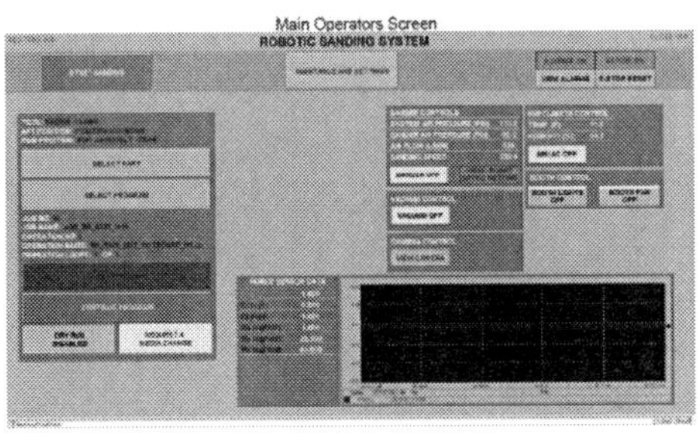

6

Approved for Public Release: NGAS 14-0334, 2/21/14

Robotic Sanding System

NORTHROP GRUMMAN

- On track to meet production and ergonomic requirements
- Process was validated in MRL 5
- MRL 6 and MRL 7 will productionize system and process
  - MRL 6 Tasks:
    - Complete Manufacturing Processes Defined, Demo, ROI Analysis, Long Lead Item ID
  - MRL 7 Tasks:
    - Critical Design Incorporation, Risk Assessment, Cost Model, Procurement Plans, Final Design Initiation

7

Approved for Public Release; NGAS 14-0334, 2/21/14

THE VALUE OF PERFORMANCE.

NORTHROP GRUMMAN

AERODEF 2014 – Long Beach, CA
February 26, 2014

Advancements in Additive Manufacturing Materials

Stratasys

Presented by:

Luke Rodgers
Stratasys, Inc.

Stratasys
FOR A 3D WORLD™

Outline

Stratasys
FOR A 3D WORLD™

- FDM Process
- Historical FDM Limitations
- Key Material Relationships
- SSYS Patented Solutions

Fused Deposition Modeling

- Material Delivery
- Print Engine
- Movement
- Oven

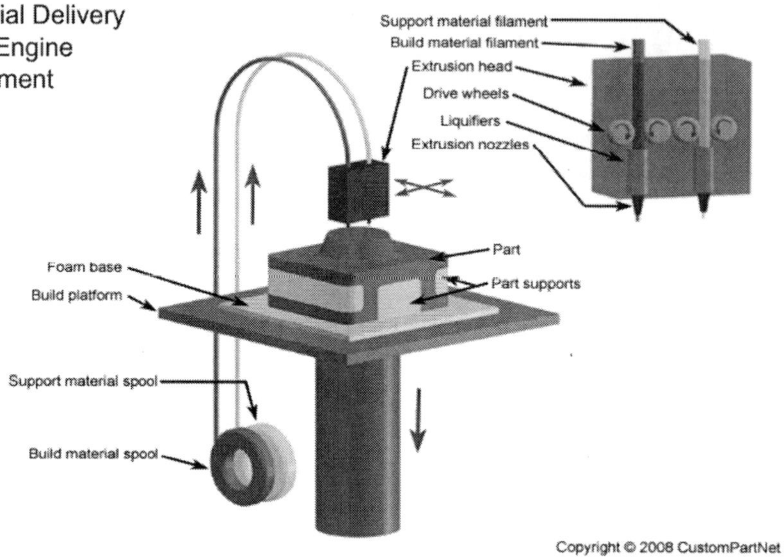

Historical Limitations

- Generally, the FDM process produces parts that are:
  - Non-dense
    - ~92-94%
  - Anisotropic
    - ~60% of XY axis strength in the Z axis

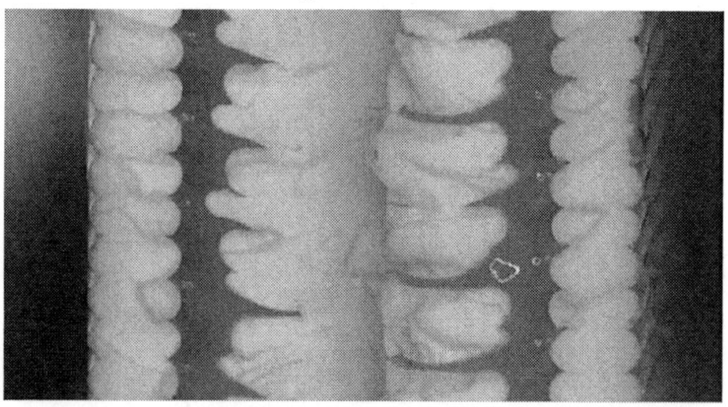

ABS fracture surface

Layer to Layer Bonding

Stratasys
FOR A 3D WORLD™

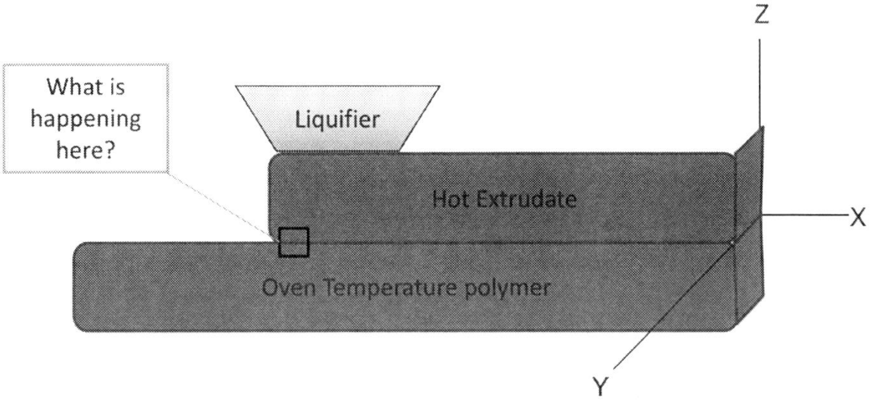

Stages of Interfacial Width

Stratasys
FOR A 3D WORLD™

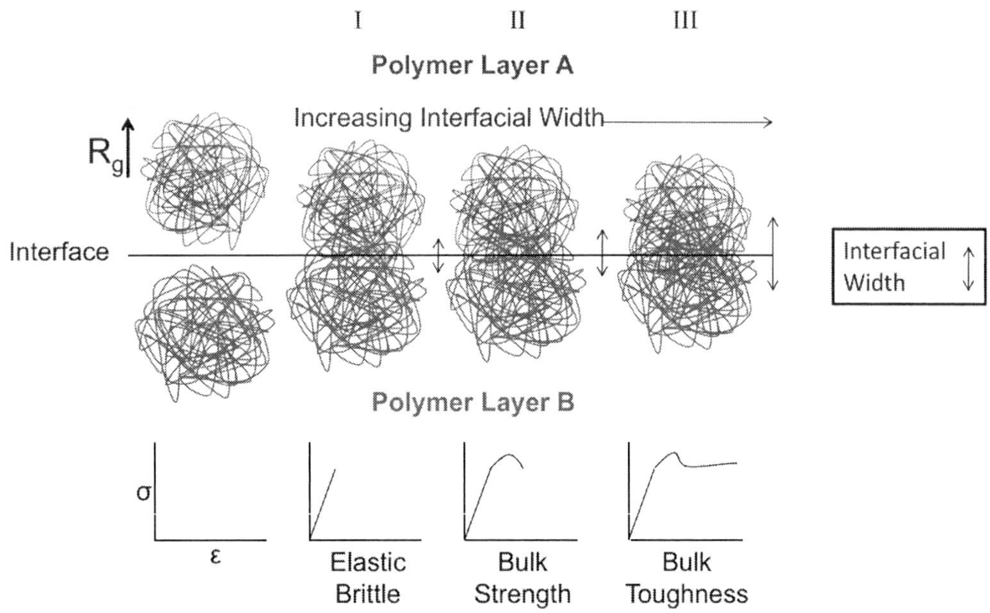

Key Relationships

Stratasys FOR A 3D WORLD

$$Interfacial\,Width = \sum_{T_{int}}^{T_{oven}} \frac{t_{T_{int}}}{\tau_{rep@T_{int}}}$$

$$\tau_{\,rep@T_{int}} = \frac{12}{\pi^2} * \frac{\eta(T_{int})}{G_n(T_{int})}$$

Time it takes for diffusing one R_g has a direct relationship to viscosity and an indirect relationship to modulus

Can find this relationship using dynamic, parallel plate, rheology

$$\eta(T_{int}) = \underline{A} * exp\left(\frac{\underline{E}_a}{R\underline{T}_{int}}\right)$$

Temperature can be measured empirically

These relationships are known, we can measure the variables empirically, and use them to predict interlayer bonding

Predicted Interlayer Bonding for Ultem 9085

Stratasys FOR A 3D WORLD

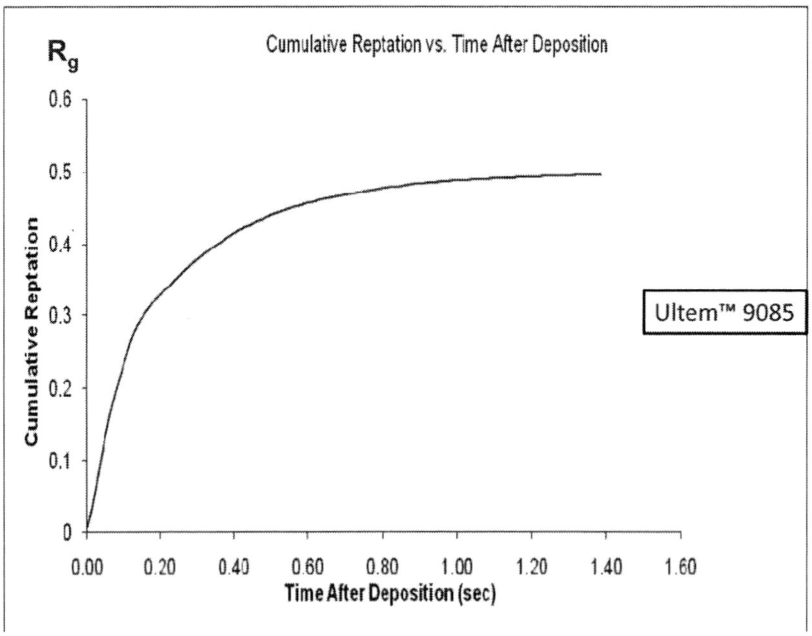

Increasing Z Bonding

Stratasys FOR A 3D WORLD

Z axis bonding improves when viscosities are lower. Therefore, increases in the time that the polymer is at an elevated temperature will increase bonding.

Choose polymers with:

↑ Heat Capacity

↑ Thermal Conductivity

↑ **Heat of Fusion**

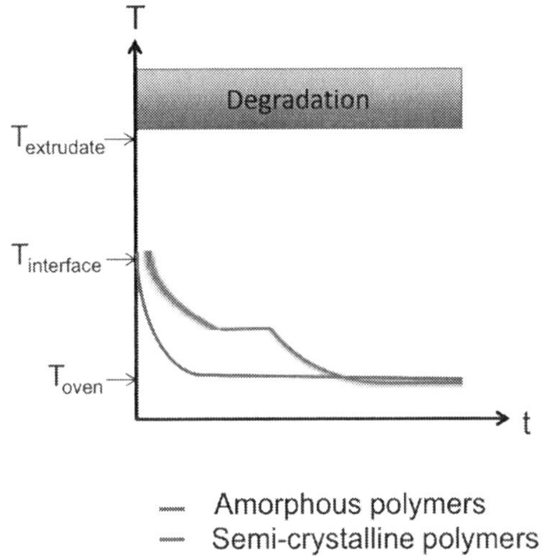

Raster to Raster Bonding

Stratasys FOR A 3D WORLD

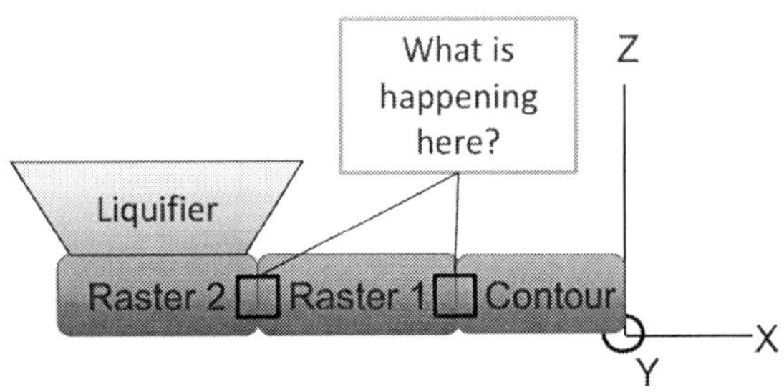

Increasing Intra-Layer Bonding

Intra-layer bonding and bead spreading follow the same underlying principles of interlayer bonding.

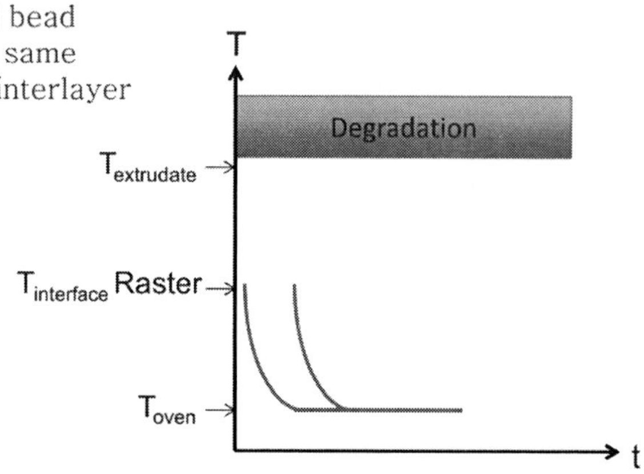

Want:

↑ Heat Capacity

↑ Thermal Conductivity

Increasing Intra-Layer Bonding

Intra-layer bonding and bead spreading follow the same underlying principles of interlayer bonding.

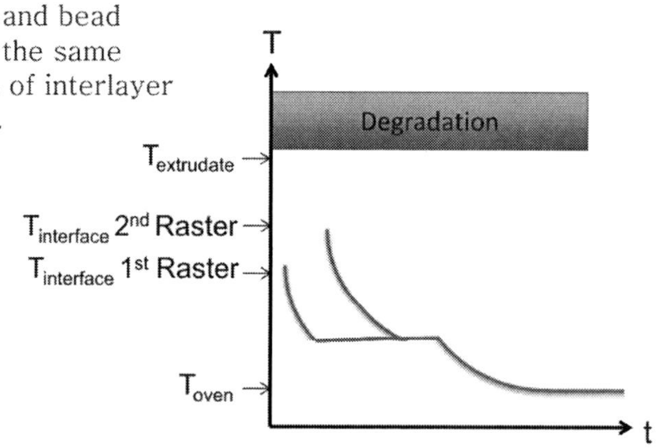

Want:

↑ Heat Capacity

↑ Thermal Conductivity

↑ Heat of Fusion

Semi-crystalline polymers have significantly better bonding between layers and intra-layer as well as better road spreading in the FDM system.

Historically, the change in volume that occurs during crystallization leads to residual stresses that induce part accuracy issues.

Differential Scanning Calorimetry

Stratasys
FOR A 3D WORLD™

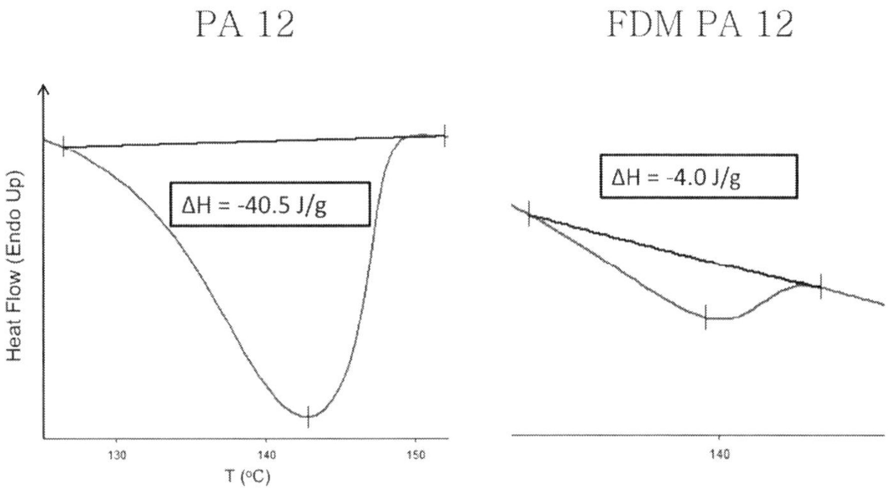

The Heat of Fusion (ΔH) is retarded by 90%, yet still yields exciting results

Thermal Imaging

Stratasys
FOR A 3D WORLD™

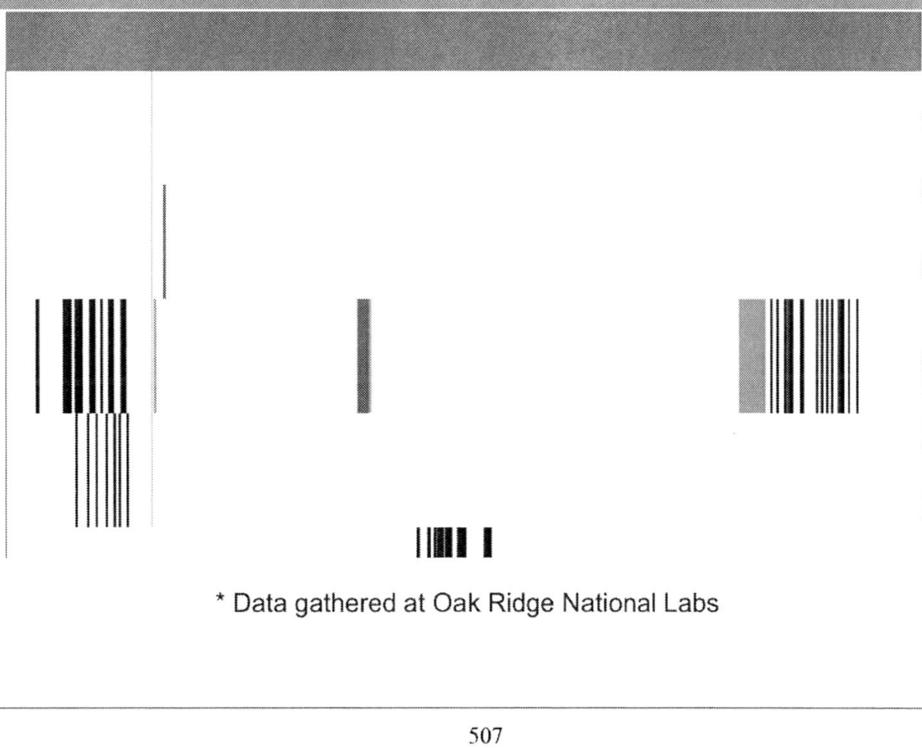

* Data gathered at Oak Ridge National Labs

FDM™ Nylon 12

Stratasys
FOR A 3D WORLD™

Predicted Interlayer Bonding for FDM Nylon 12

Stratasys
FOR A 3D WORLD™

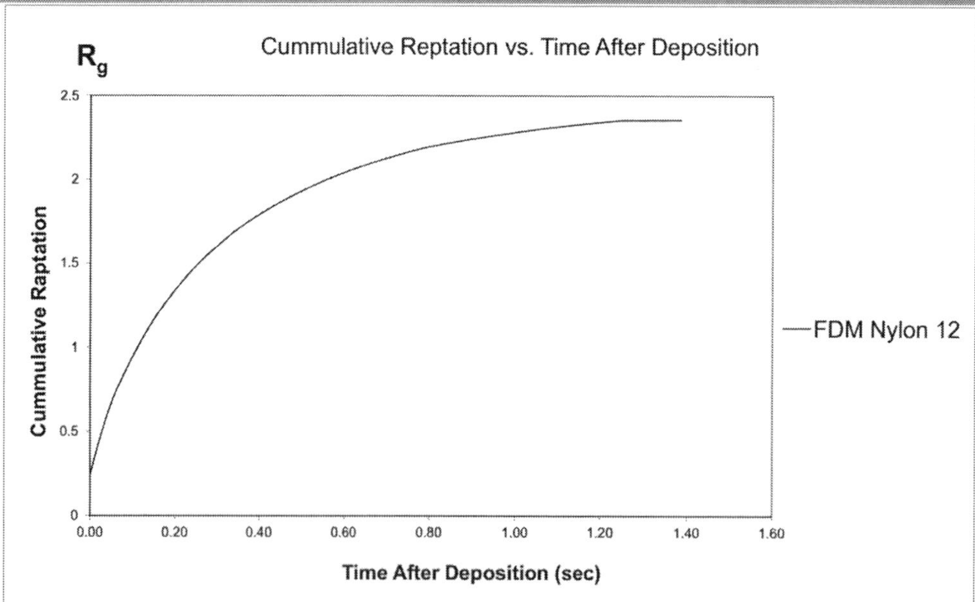

Property Data

Stratasys
FOR A 3D WORLD™

	FDM™ Nylon 12		SLS Nylon 12 PA	Injection Molding Rilsan® PA 12
Orientation	XY	Z	XY	NA
Strength (psi)	7000	6400	6800	6000
elongation at break (%)	30	5.5	15	50
Modulus (ksi)	190	190	246	210
IZOD Unnotched (ft-lb/in)	37	7	8.24	NB

Historically, Z tensile strengths for amorphous materials in the FDM system are ~60% of XY.

Raster to raster bonding is near bulk toughness.

Layer to layer bonding is near bulk strength, but not bulk toughness as predicted.

Benefits

Stratasys
FOR A 3D WORLD™

Increased density
- ↑ Toughness
- ↑ Elongation at break
- ↑ Design allowables

Reduced anisotropy

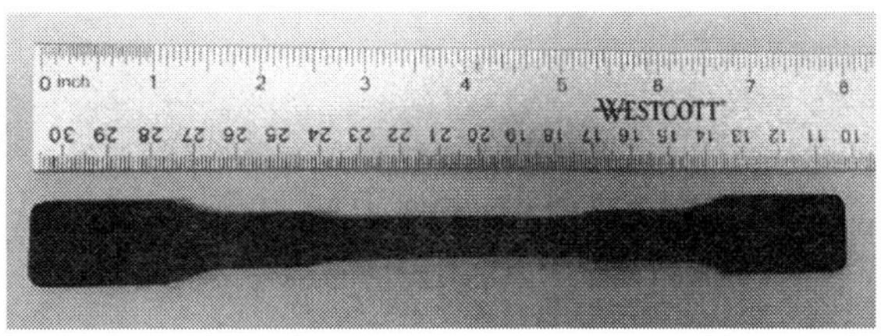

>90% XY tensile strength in Z axis

Enables filled materials, without the detrimental losses in Z axis properties that are seen in amorphous composites

Enabled Filled Materials

Stratasys
FOR A 3D WORLD™

Structural Materials
- Short fiber or milled carbon and glass filled composites

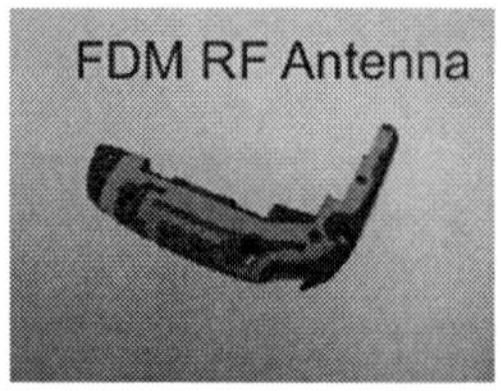

Thermally Conductive Materials
- Thermal conductivity up to 4 W/m-K in-plane as FDM output

Laser Direct Structuring (LDS) Materials
- Controlled activation of plating surfaces on parts of arbitrary geometry

Electrically Conductive Materials
- Resistivity <10^2 Ohm/sq

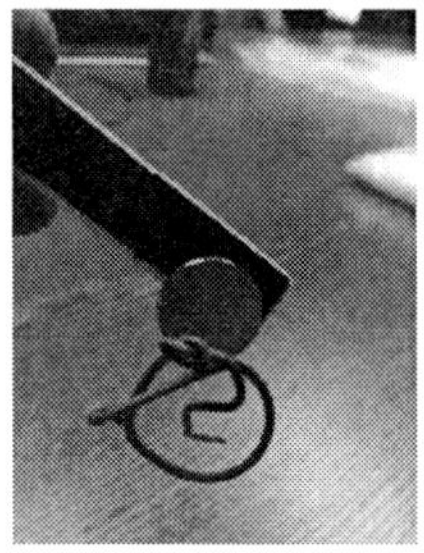

Medical Materials
- Materials with drug-eluting and/or antimicrobial properties

Particle-filled materials
- Metals, minerals, ferrites, glass beads/bubbles

Semi-crystalline Enabled Applications

Stratasys
FOR A 3D WORLD™

Structural-lightly loaded
- FDM carbon fiber composites yield a higher modulus and strength due to the use of short fibers vs. milled in other AM technologies.

	FDM™ Nylon 12 CF		SLS Nytek 1200 CF	Injection Molding RTP® 283 F
Orientation	XY (45/-45)	Z	XY	NA
Strength (psi)	12000	8500	6800	16000
elongation at break (%)	2.9	2.5	3.6	2-3
Modulus (Msi)	1.2	0.5	0.42	1.2

Applicable SC Polymers

Stratasys FOR A 3D WORLD

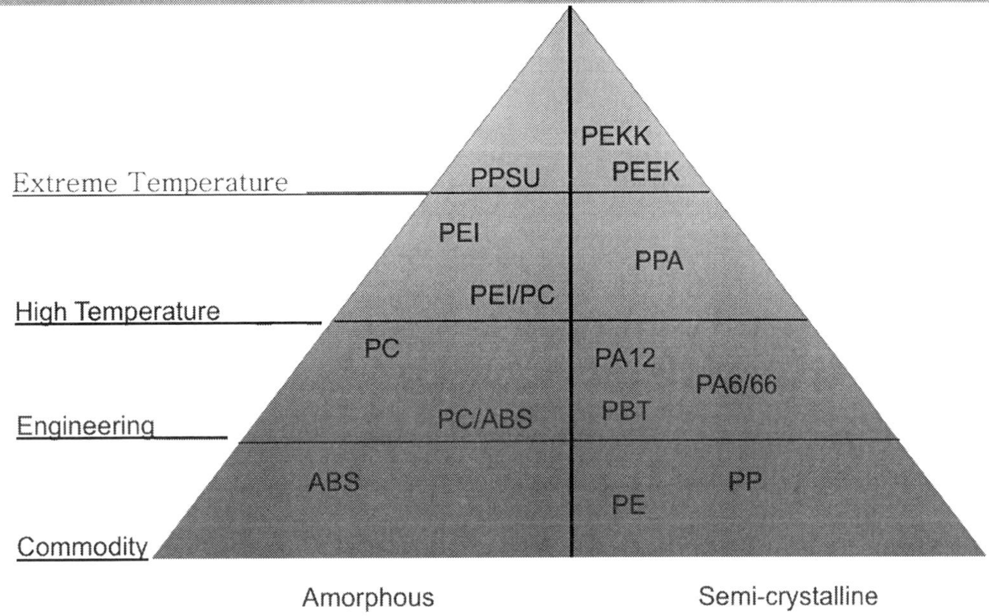

Co-Extrusion Technology

Stratasys FOR A 3D WORLD

Core-shell Consumable Materials for Use in Extrusion-Based Additive Manufacturing Systems

- Patent US20120231225 A1
  - Enables combinations of multi-material and filled systems utilizing either amorphous or semi-crystalline polymers.

Example: PEI shell, PEI carbon fiber filled core

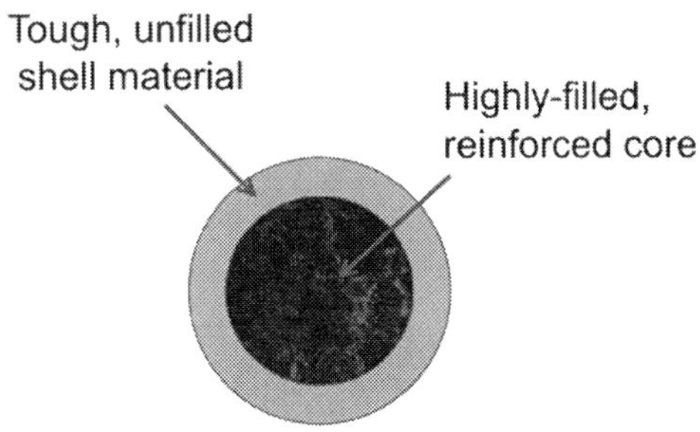

Co-Extrusion Technology

Stratasys FOR A 3D WORLD

PEI core with a PEEK shell.

The PEI core provides
1. Low residual stresses
2. HDT boost for unfilled system
3. Reduced cost

The PEEK shell provides
1. Superior Z strength over PEI
2. Chemical resistance
3. Better bead spreading

Coextruded structure is retained in composite part!

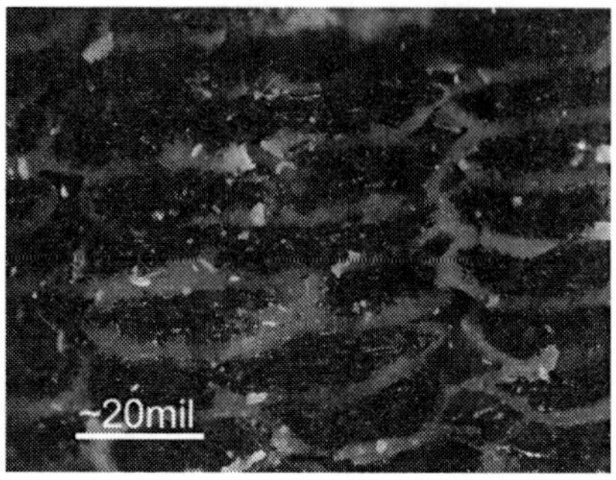

Core/Shell Enabled Applications

Stratasys FOR A 3D WORLD

Structural
- A higher loading of fillers can be delivered to the liquifier utilizing standard FDM systems.

Soluble Composite Cores
- Unique combinations can yield higher HDT soluble cores for use in higher temperature cure cycles.

Tooling
- Higher loadings of fillers possible with this technology, leading to desired CTE values while maintaining Z strength of the virgin resin.

Summary

Stratasys
FOR A 3D WORLD™

- We have recently developed techniques for controlling the crystallization kinetics of semi-crystalline polymers.
- These approaches enable virgin semi-crystalline polymers and semi-crystalline composites to build reliably and accurately on FDM systems.
- We expect this work will lead to a large number of new FDM semi-crystalline materials and new applications served by those materials.
- Further, utilization of unique filament structures can lead to interesting material combinations and structures that can be used on existing FDM equipment.

Thank You

Stratasys
FOR A 3D WORLD™

For More Information Contact:

Luke Rodgers
Luke.rodgers@stratasys.com
+1.952.906.2720

Or see our website at
www.stratasys.com

Stratasys
FOR A 3D WORLD™

AERODEF MANUFACTURING

SUMMIT & EXPOSITION
February 25-27, 2014
LONG BEACH CONVENTION CENTER
LONG BEACH, CALIFORNIA, USA

CSI: More effective manufacturing collaboration for the Model Based Environment

produced by sme

John Gray
ITI TranscenData
John.Gray@transcendata.com

AERODEF MANUFACTURING

produced by sme

SUMMIT & EXPOSITION
February 25-27, 2014
LONG BEACH CONVENTION CENTER
LONG BEACH, CALIFORNIA, USA

- CSI project
- MBE
- Challenges for MBE
- Migrating to MBD/MBE
- Manufacturing Producibility
- ECO Documentation
- Translation / Validation
- TDP Creation / Validation /Delivery / Security
- Additive Manufacturing
- Benefits
- Availability

2

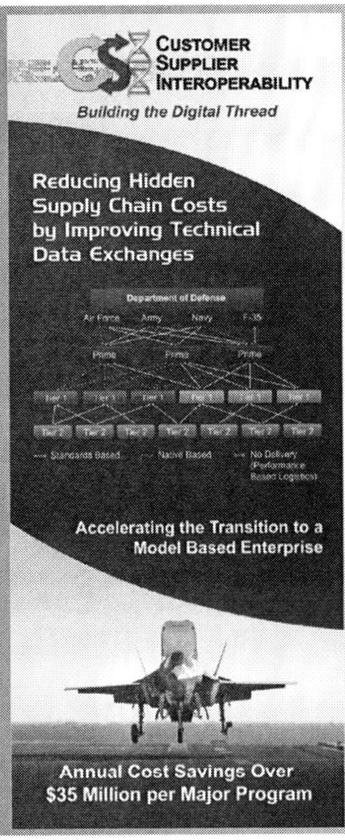

CSI (Customer Supplier Interoperability)

- Inefficient technical data exchange in DoD supply chain
  - Increases large program costs by as much as $ 1 billion
  - Delays delivery of weapons systems
  - Manual tasks repeated thousands of times per month over 30+ year lifespan of major programs
  - Automating these tasks can save 50 to 75% of labor by eliminating these non-value added tasks
- Technology gaps delay transition to more efficient model based enterprise (MBE technology)
  - Migration of existing 3D / 2D models to model based definition (MBD) models
  - Specifying engineering change orders (ECO's) on MBD models
  - Producibility issues with MBD models

Sponsor: AFRL and DoD
Contract/Project Manager: PDES, Inc./ATI
Technical Lead: International TechneGroup Incorporated (ITI)
Industry Prime: Lockheed Martin
Industry Suppliers: Rockwell Collins & Honeywell

"DISTRIBUTION STATEMENT A. Approved for public release; distribution is unlimited. Case Number 88ABW-2013-1251."

3

2012 Defense Manufacturing Achievement Award

The CSI program developed software solutions to address the massive engineering inefficiencies associated with data exchanges in defense supply chains. Through the automation of manual tasks and by creating a toolset to advance current technical data exchanges from 2D to 3D, the CSI program has paved the path toward Model-Based Definition (MBD) for the defense industrial base. Through side-by-side comparisons with current best practices, the CSI technologies generated projected cost savings of ~$22M per year for the three CSI manufacturers, and ~$50M per year when fully adopted by all supply chain members in a large, defense acquisition program.

4

What Is MBE?
… aka the "Digital Thread"

- MBE = Model Based Enterprise
  - Utilizing MBD models / processes throughout lifecycle
  - 3D MBD master instead of 2D Drawing master
- MBD = Model Based Design
  - Full product definition in 3D model with PMI
- DoD Thrust
  - Revised MILSTD 31000a
  - Enhanced Standards (NIST/PDES/ISO)
  - New DoD contract requirements
- Benefits
  - Reuse throughout lifecycle
  - Less ambiguous
  - More effective collaboration

5

Requirements for MBE

- MBD models
- Processes utilizing MBD models
- MBD skills
- MBD compatible partners / suppliers
- MILSTD 31000a compatible TDP

6

MBE Challenges For Primes / Supply Chain

- Getting to MBE Readiness
  - Tools
  - Skills
- Getting to MBD Models
- Manufacturing with MBD models
- ECO process and documentation
- Supplier / partner MBD readiness
- MBD TDP exchange with suppliers / partners
- Certification

7

MBE Scenario

- Prime receives new project award
  - Will reuse many existing models
- Prime wants to establish Model Based Environment
  - Migrate 3D model plus Drawing to MBD model
  - Perform producibility analysis on model
  - Make ECO change
  - Translate MBD models between CAD systems
  - Deliver MBD TDP

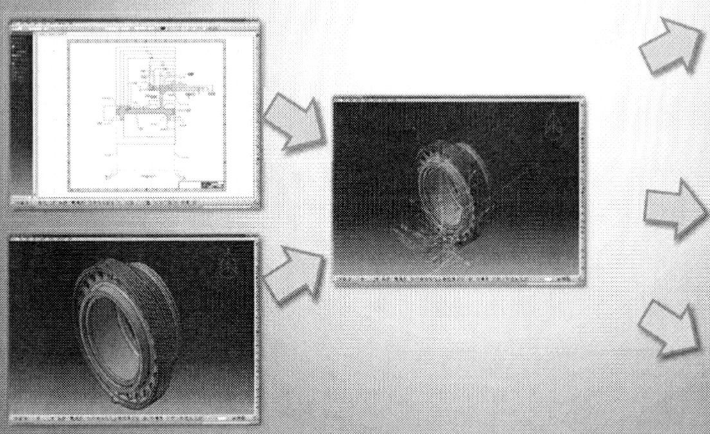

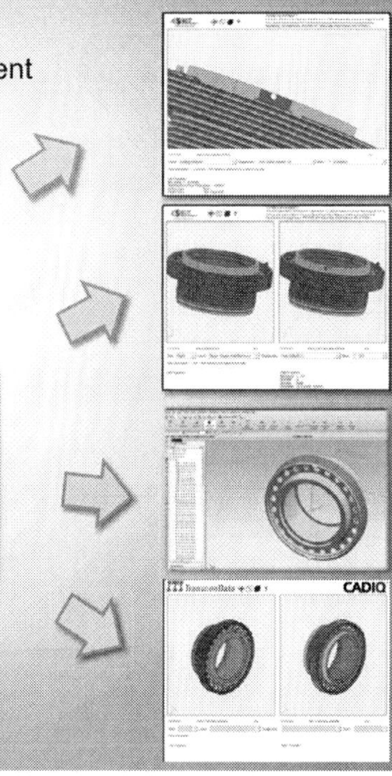

8

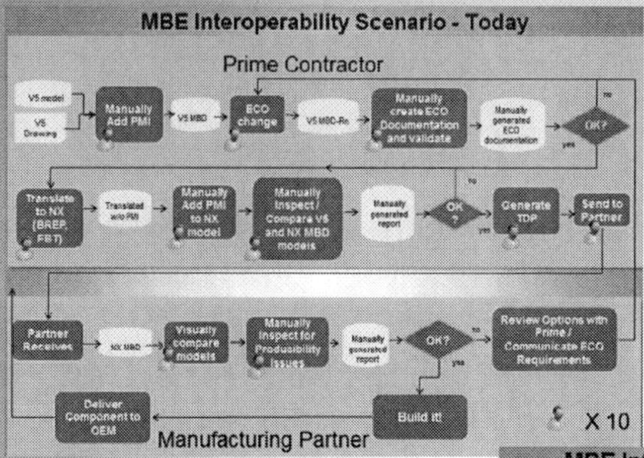

Key MBE Interoperability Processes

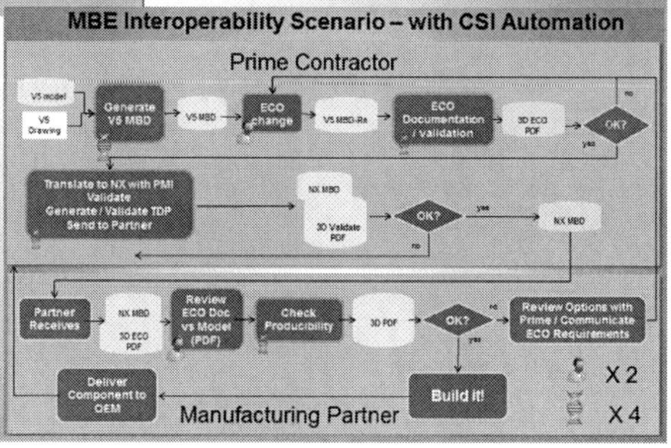

CSI Can Reduce Manual Actions From 10 to 2

- Reduced effort
- Shorter schedules
- More reliability

Migrating to MBE

- Acquire / upgrade tools (PLM/CAD/Collaboration)
- Personnel skills development
- New processes built around MBD
- Selecting / migrating projects to MBD
  - Legacy 2D based projects
  - New DoD projects require MBD
- Existing Project Migration Options
  - Remodel with MBD tools / techniques
    - Time consuming / costly
    - Error prone (manual interpretation/creation)
  - Convert existing models to MBD
    - Faster / less expensive
    - Programmatic method more complete / reliable

10

Generate MBD Capability

- 3D models with associative 2D drawings
- Annotation moved to 3D model and attached to corresponding entities
- CSI Process: Submit 2D/3D -> 3D MBD

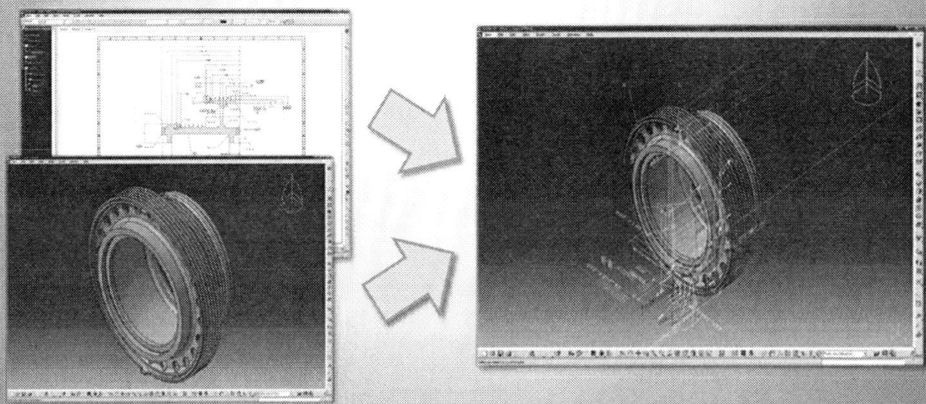

11

MBD Scenario With and Without CSI (Honeywell)

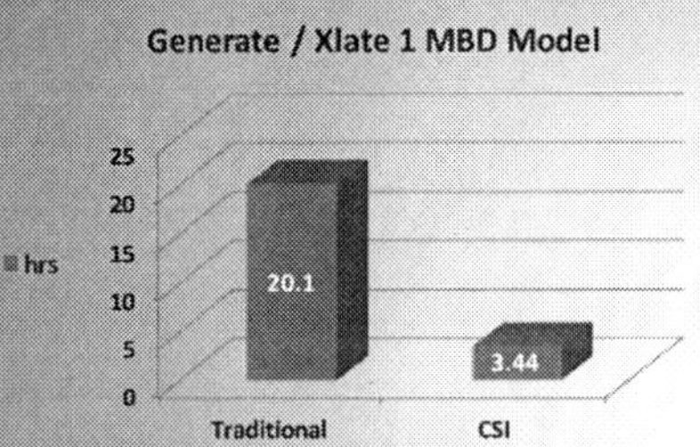

Traditional Process
- Add PMI to source
- FBT translate
- Add PMI to target

CSI Process
- Generate MBD (A)
- Touchup source
- Translate MBD (A)
- Touchup target

80% cost reduction

(A) = automated process

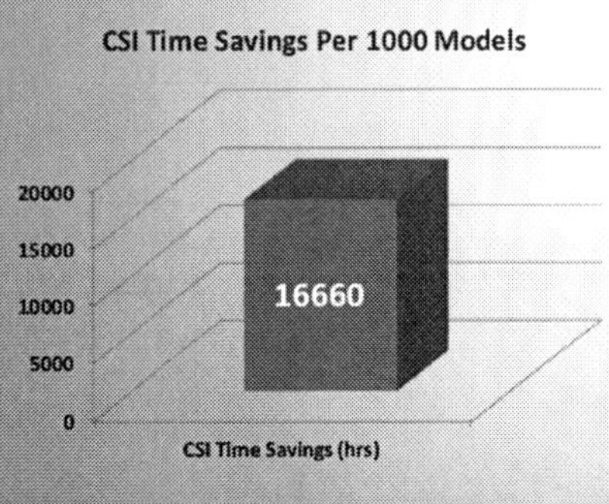

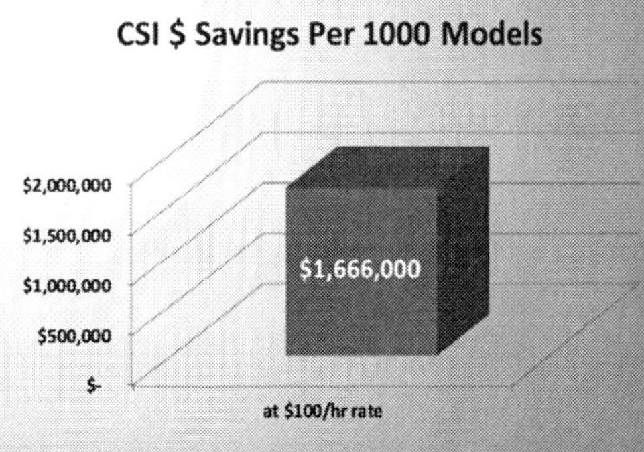

12

ECO Documentation Process

- Traditional 2D ECO Documentation Process
  - Mark up drawing
  - Provide to manufacturing group / supplier
- Corresponding MBD Process
  - Capture screen images
  - Add notes
  - Manufacturing Group / Supplier reviews side by side or overlaid

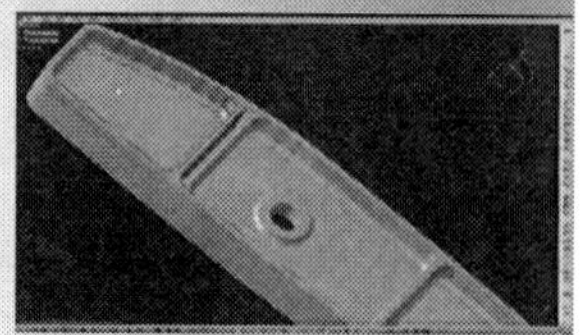

- Improved CSI MBD ECO Documentation Process
  - Make desired ECO changes in model
  - Submit original and modified models to CSI
  - Receive 3D PDF report documenting ECO changes
    - Dynamic viewing with Adobe Acrobat Reader
    - Each change identified separately
    - Improved completeness / reliability / communication

13

CSI ECO Documentation Process

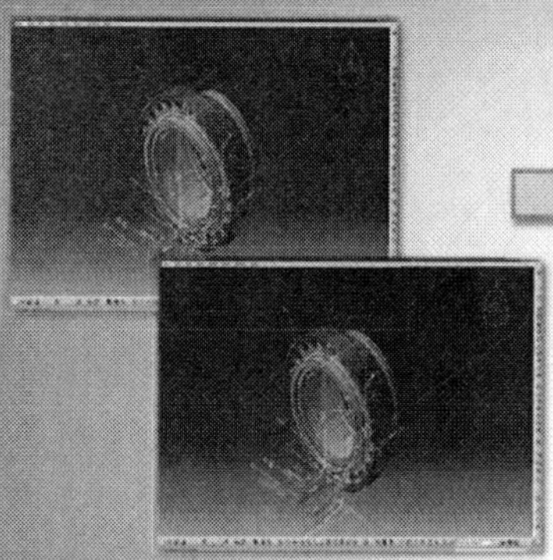

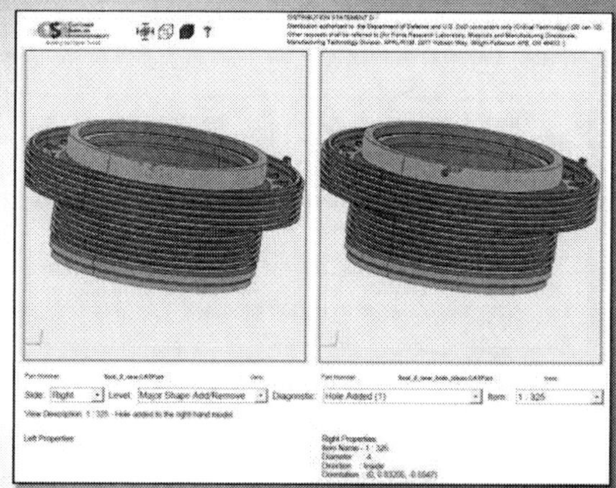

- CSI Process
  - Submit original and modified models
  - Analyzed programmatically
  - Return 3D PDF with side by side comparison
  - Each issue shown separately
  - Unintentional changes identified

14

ECO Scenario With and Without CSI (Lockheed)

ECO Change Hours Per Model

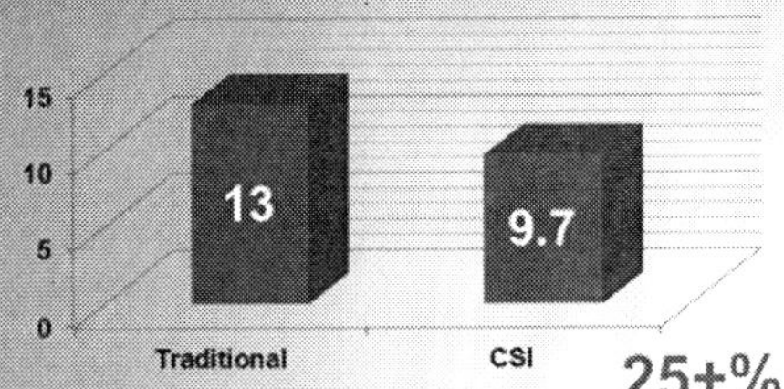

Traditional Process
- Document via screen shots
- Overlay models to understand / review
- Make changes

CSI Process
- ECO Documentation (A)
- Review 3D ECO PDF (E)
- Make Changes

25+% cost reduction

(A) = automated process
(E) = significant efficiency gains

CSI Time Savings Per 1000 Models

CSI Savings Per 1000 Models

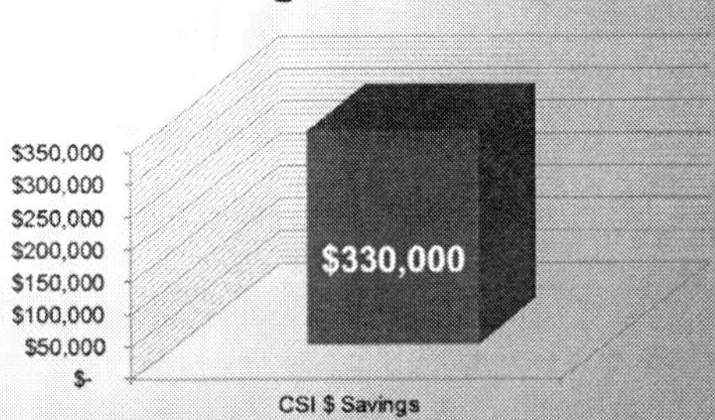

Models are an average of 95% simple/medium complexity and 5% complex.

15

ECO Documentation Video

- http://www.transcendata.com/solutions/csi

(This video has been approved for public release: 88ABW-2012-4974.)

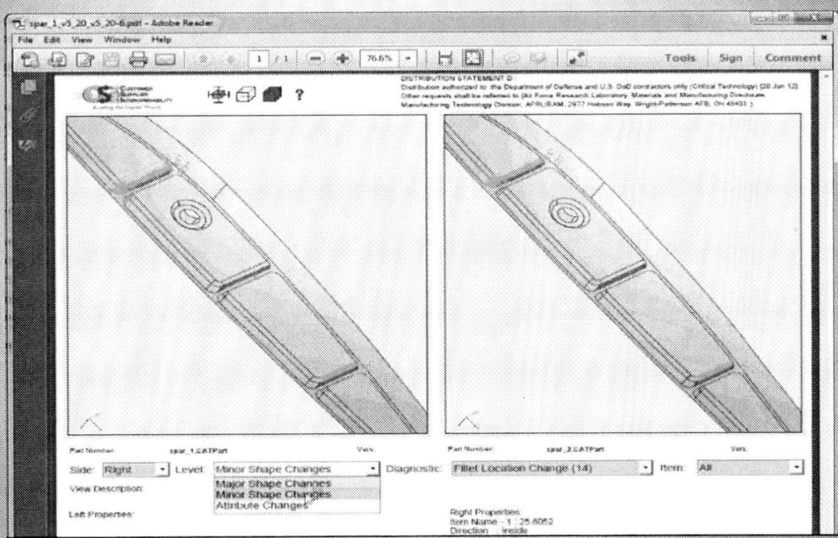

16

Manufacturing Producibility

- Traditional 2D Based Producibility Process
  - Inspect drawings -> un-manufacturable / expensive mfg conditions
  - Interpret ambiguous conditions
  - Develop models / tooling to support manufacturing
- Corresponding MBD Producibility Process
  - Inspect 3D models -> un-manufacturable / expensive mfg conditions
    - More complex / difficult / error prone
  - Models used directly for tooling / manufacturing
    - MUST NOT contain ambiguous conditions
- Improved CSI MBD Producibility Process
  - Programmatic inspection / identification
    - Rules based
    - Un-manufacturable
    - Expensive
    - Ambiguous
  - 3D PDF completely / clearly identifying problematic area

17

CSI MBD Producibility

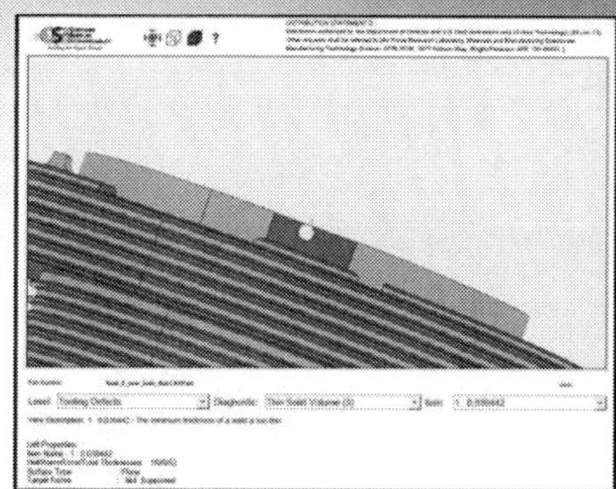

- CSI Automated Process
  - Submit MBD model
  - Analyzed versus identified producibility rules
    - Rules can be enabled / disabled / adjusted for different scenarios
  - Results Reported via 3D PDF
    - Dynamic viewing via Adobe Acrobat Reader
    - Each issue identified individually

18

Producibility Scenario With and w/o CSI (Rockwell Collins)

Producibility Hrs Per Model

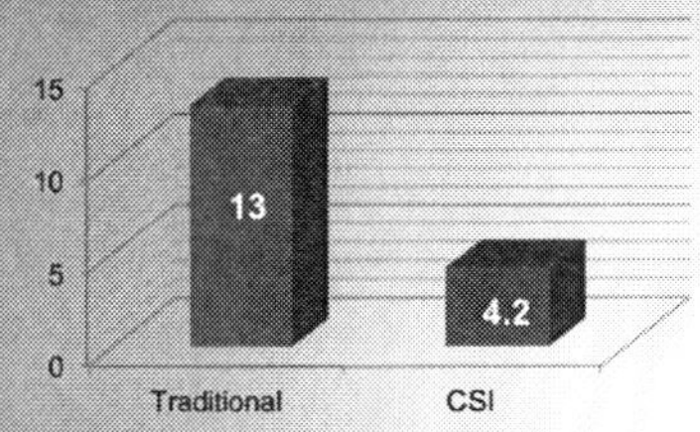

Traditional Process
- Inspect
- Document
- Make changes
- Re-inspect

CSI Process
- Producibility (A)
- Make Changes
- Producibility (A)

68% cost reduction plus fewer escapes

(A) = automated process

CSI Time Savings Per 1000 Models

CSI $ Savings Per 1000 Models

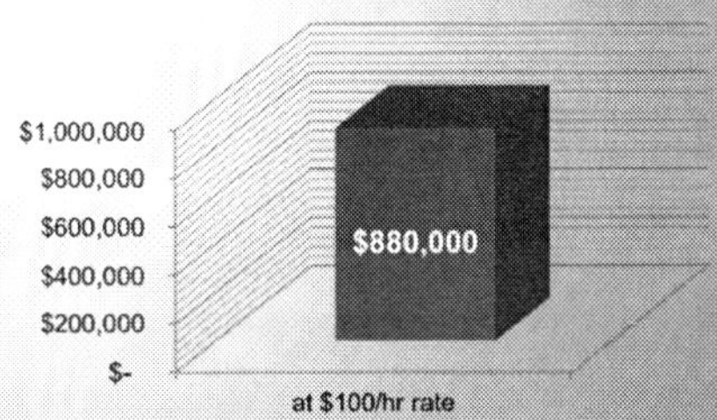

Models are an average of 95% simple/medium complexity and 5% complex.

19

Relative Defect Resolution Cost vs Lifecycle

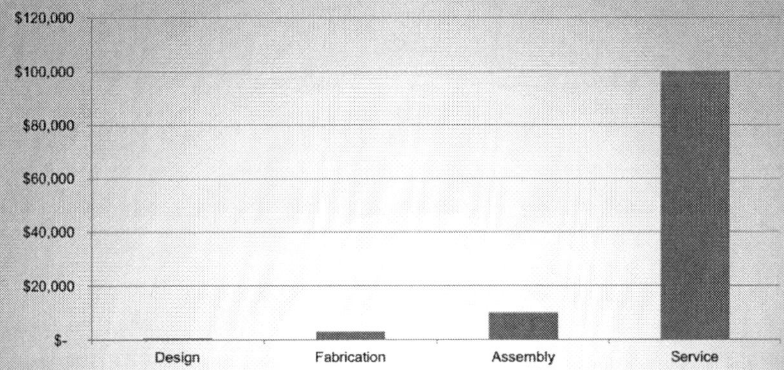

Finding and Resolving Problems Early Has Great Benefits

	Identification	Time to Resolve	Process	Scrap	Warranty
Design	Internal	Hours	Producibility Check	None	None
Fabrication	Supplier	Days	ECO	None	None
Assembly	Assembly Test	Weeks	ECO	Prototypes	None
Service	Customer	Months	ECO	Significant	Significant

20

TDP Exchange Processes

- Traditional 2D Based TDP Exchange
  - Native or neutral drawing provided / maybe 3D reference
  - Variety of file exchange methods (email, ftp, web, PLM)
- MBD TDP Process
  - MILSTD 31000A minimum content:
    - MBD native and / or Neutral (STEP AP203/AP214/AP242)
    - Visualization (3D PDF, JT, other)
    - Manifest (contents/IP/Export Authorization)
  - Validation
    - Derivative files must be validated against originals
  - TDP container (zip, gzip, PDF (DoD favorite))
  - Security importance greater due to TDP completeness
    - IP Protection
    - ITAR / EAR regulatory requirements
    - etc

21

ITI's Automated TDP Process

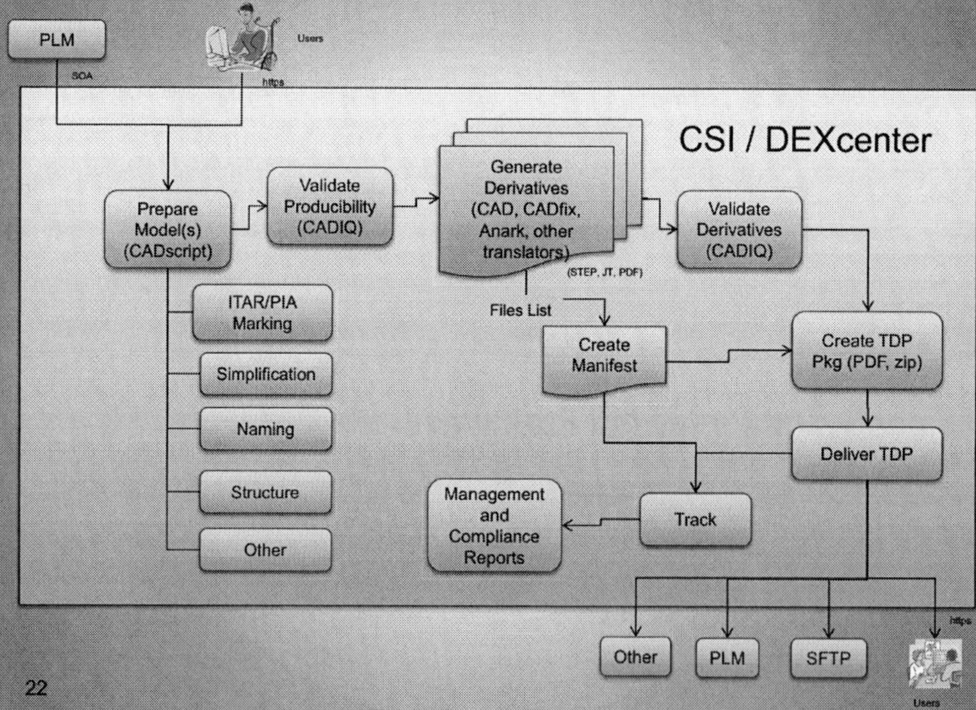

22

CSI Supporting Additive Manufacturing

- Secure Managed File Transfer (Automated TDP)
  - MBE models are larger
    - Often exceed email limits
    - CSI has unlimited size capability
  - Secure
    - More complete models warrant greater security
  - Automated
  - Tracked
  - Easy
    - Only a browser and username required

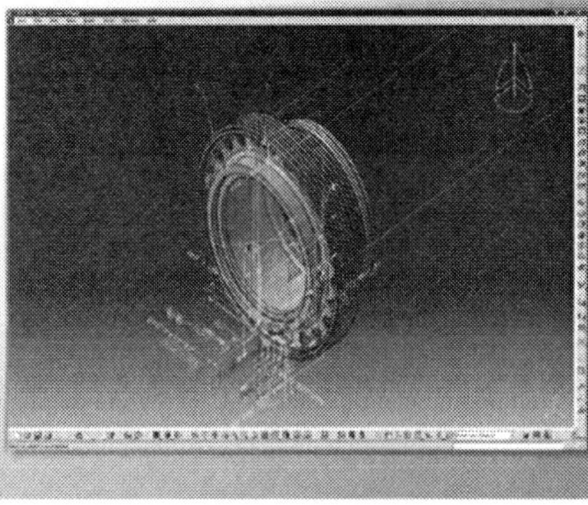

CSI Supporting Additive Manufacturing

- 3D Printing: Slices
  - Slice generation
  - Slice analysis
  - Slice compensation
  - Slice export
- Build box
  - Optimal layout and alignment
- Point cloud comparison
  - Slices vs CAD – will we print what we want?
  - Scan vs CAD – did we print what we want?
- Post-processing/Finishing
  - "Rest volume" extraction for laser milling
- Near Net Shape
  - Construct near net shape geometry
  - Complex compensation

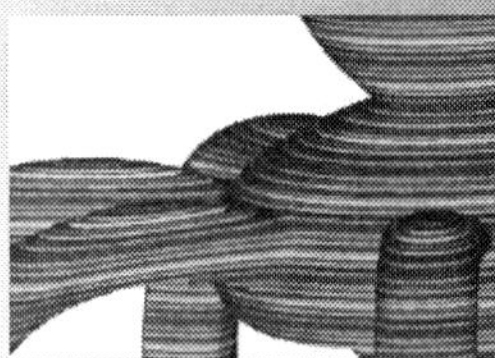

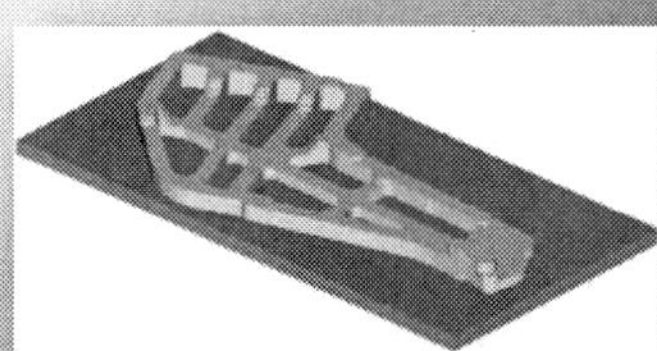

24

Cumulative estimated savings of large project over lifetime

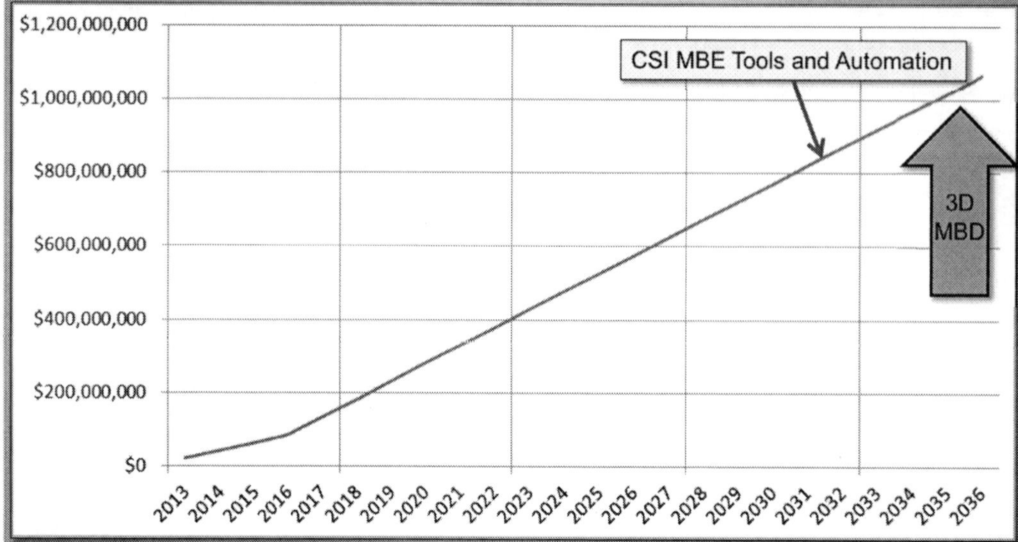

25

CSI Framework

- CSI was built upon existing commercially available technologies / standards
  - ITI:
    - DEXcenter: secure portal and automation framework
    - CADIQ: validation / comparison, quality
    - CADfix: BREP translations and healing
    - Proficiency: Feature Based Translations
    - PDElib: toolkit for neutral standard access / modification
    - CADscript: toolkit for native model access / modification
  - CAD vendors:
    - Neutral standard translators available from Dassault, Siemens, PTC, etc
  - Standards
    - STEP, JT, PDF, IGES, DXF, etc
- The CSI project identified gaps in available technology and piloted prototype solutions to fill these gaps.
- **ITI committed to commercialize these capabilities.**

DEXcenter
CADIQ
CADfix
Proficiency
PDElib
NIST
PDES, Inc.

26

CSI Availability

- CSI capabilities currently available in ITI solutions include:
  - General automation capability for most CAD systems
    - Translation / validation between native CAD and neutral formats
  - TDP preparation / creation / secure delivery with IP / ITAR / EAR compliance
  - Generate MBD [Catia V5]
  - ECO Documentation
  - Producibility [machining]
  - Simplification
  - Generalized CAD native and neutral format utilities for model preparation and parameter extraction
  - SOA and batch capability for PLM connectivity

27

CSI Extensions

- ITI is committed to continue extending these capabilities
- CSI capabilities for which interested partners are needed to further refine requirements and drive commercialization include:
  - Feature Based Translate MBD
  - Generate MBD for other CAD systems
  - Further 3D PDF utilization / customization (Anark, others)
  - Additive Manufacturing support programs
  - PLM / PLCS extensions
  - Producibility for other manufacturing methods (molding, cast, additive, …)
  - Pilots to refine productivity estimates

28

Questions?

- For more information:
  - http://www.transcendata.com/solutions/csi
  - John.Gray@transcendata.com

CAD and PLM Interoperability Solution Partner

29

AeroDef 2014

"Technical Direction at America Makes"

Rob Gorham
Deputy Director – Technology Development
National Center for Defense Manufacturing and Machining (NCDMM)
rob.gorham@ncdmm.org

- *A Defense-wide Manufacturing S&T team-led, Multi-agency collaboration between industry, government and universities*
- *Public-private partnership*

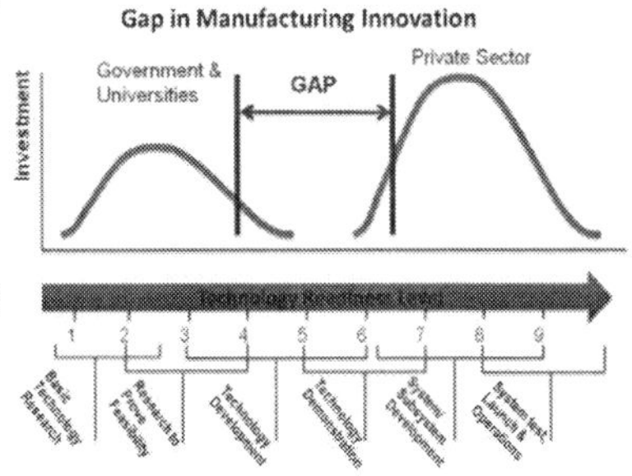

- Shared facilities open to industry
  - Especially attractive to small businesses
- Enabling technology transition and commercialization
- Addressing Technology Readiness Level (TRL) / Manufacturi Readiness Level (MRL) 4-7
  - Bridge the gap in Manufacturing Innovation
- Educational outreach and workforce development
- $ 45M gov't investment pledged for pilot, $30M CA awarded

A Model for Subsequent Manufacturing Innovation Institutes

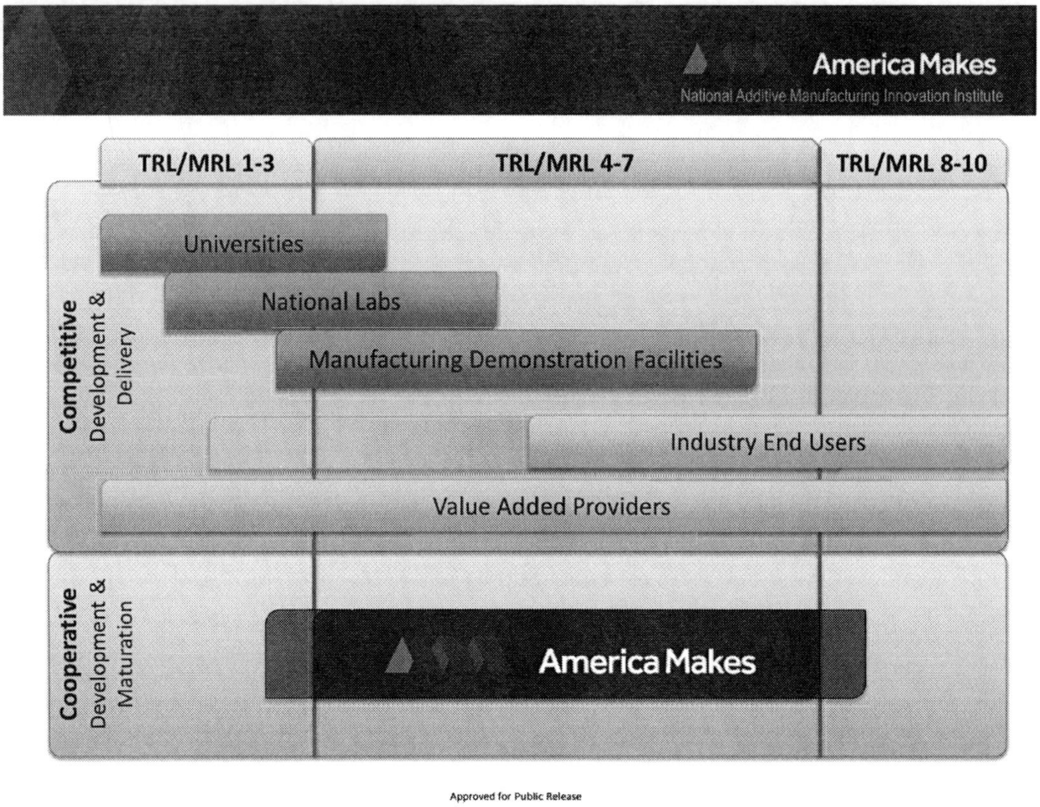
America Makes
National Additive Manufacturing Innovation Institute
TRL/MRL 1-3
TRL/MRL 4-7
TRL/MRL 8-10
Competitive
Development & Delivery
Universities
National Labs
Manufacturing Demonstration Facilities
Industry End Users
Value Added Providers
Cooperative
Development & Maturation
America Makes
Approved for Public Release

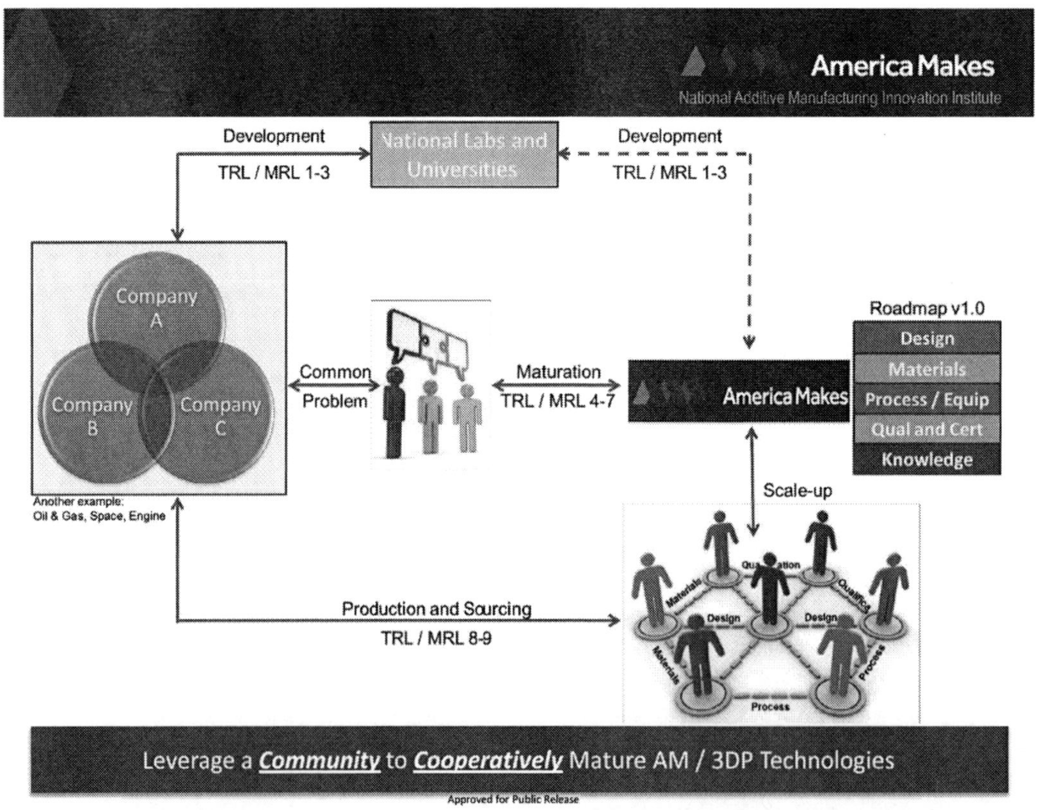
America Makes
National Additive Manufacturing Innovation Institute
Development
TRL / MRL 1-3
National Labs and Universities
Development
TRL / MRL 1-3
Company A
Company B
Company C
Another example:
Oil & Gas, Space, Engine
Common Problem
Maturation
TRL / MRL 4-7
America Makes
Roadmap v1.0
Design
Materials
Process / Equip
Qual and Cert
Knowledge
Scale-up
Production and Sourcing
TRL / MRL 8-9
Leverage a Community to Cooperatively Mature AM / 3DP Technologies
Approved for Public Release

America Makes
National Additive Manufacturing Innovation Institute

Why America Makes

America Makes creates mechanisms for collaboration...

Cooperative Development of
- Training
- Assessments
- Case Studies

Pooling Resources / Pooling Risks

Cooperative Development of
- Material Specs
- Process Specs
- Material Databases
- Design Rules
- Application Guides

Solving Problems Collaboratively

Work Shops, Work Groups, Projects

Public/Private Funded Projects
Crowd Funded Projects

Leveraging Community Knowledge

Knowledge Base, Online Collaboration Tools, Databases, Specifications, Application Guides, Curriculum

Approved for Public Release

America Makes
National Additive Manufacturing Innovation Institute

America Makes Project Models

- Multiple Investment Options Attracting Different Audiences
- Project Calls Well Underway
- Challenge Options Being Developed
- Crowd-Sourced Project Model Developed

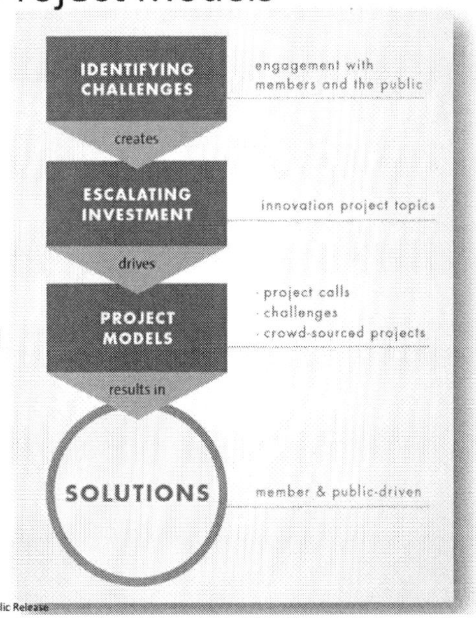

Approved for Public Release

America Makes
National Additive Manufacturing Innovation Institute

Technology Roadmap: Overall Philosophy

- "Voice of the Customer"
- Long-term Investment Strategy
- Strategic Planning, Decision Making, and Collaboration Framework
- Big Impact Outcomes

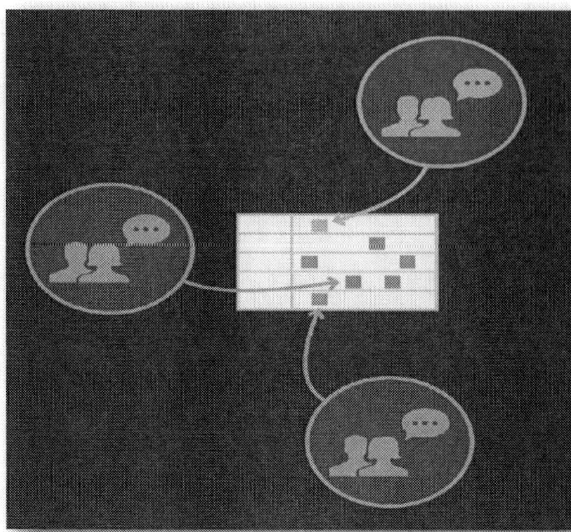

Approved for Public Release

America Makes
National Additive Manufacturing Innovation Institute

Technology Roadmap: Workshop Topics

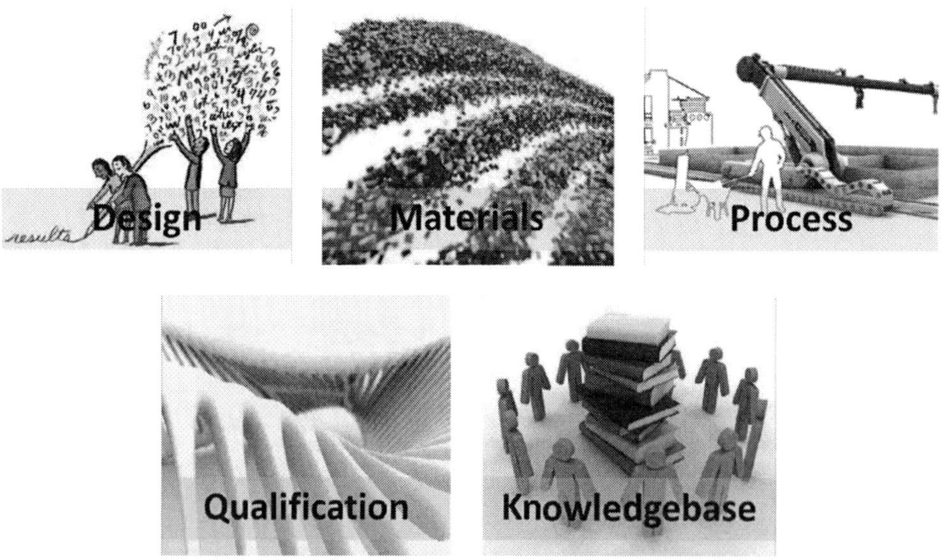

Approved for Public Release

America Makes
National Additive Manufacturing Innovation Institute

Technology Roadmap: Process

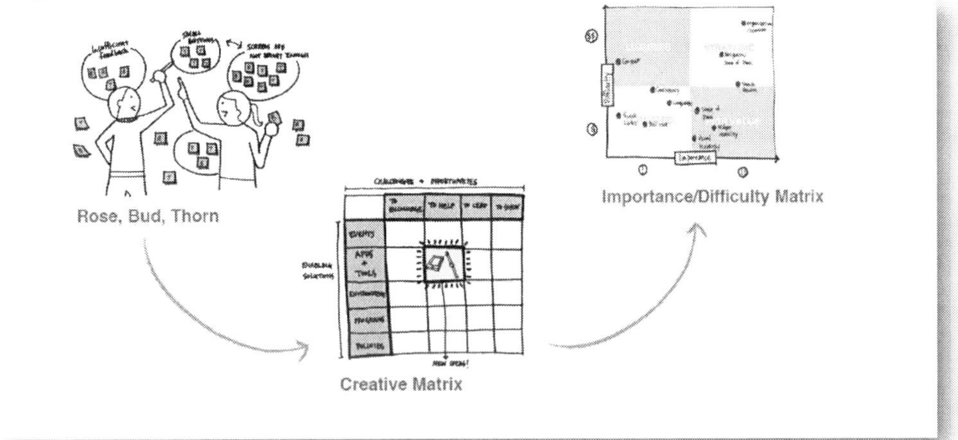

Structured Workshops Leveraging Human-Centered Design

Approved for Public Release

America Makes
National Additive Manufacturing Innovation Institute

Technology Roadmap: Workshop Participation

Approved for Public Release

America Makes
National Additive Manufacturing Innovation Institute

Technology Roadmap: Prioritization

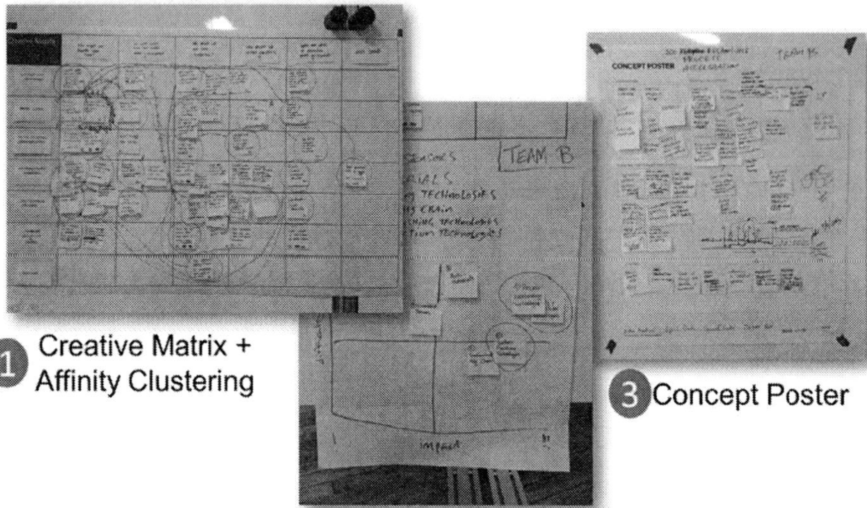

1 Creative Matrix + Affinity Clustering

2 Impact / Difficulty Matrix

3 Concept Poster

Approved for Public Release

America Makes
National Additive Manufacturing Innovation Institute

Technology Roadmap: v1.0

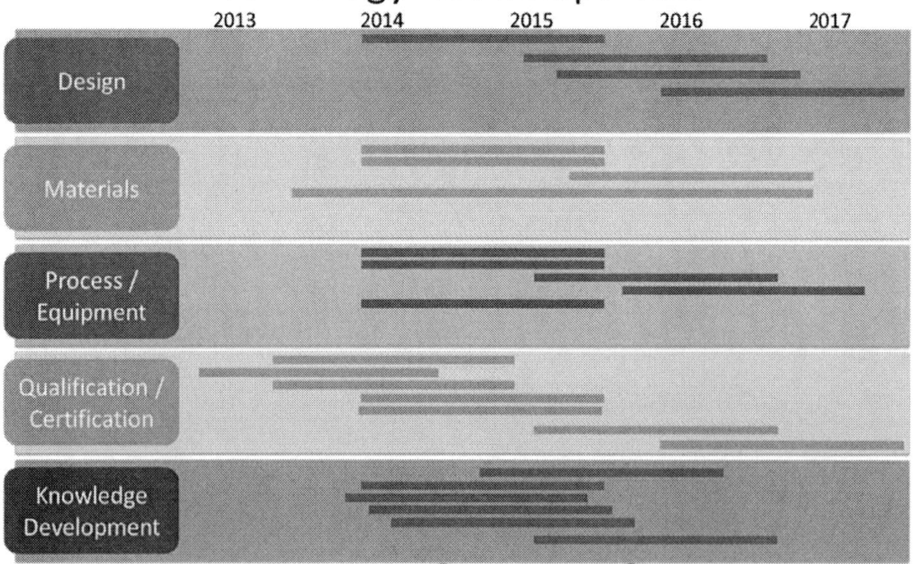

Approved for Public Release

America Makes
National Additive Manufacturing Innovation Institute

Technology Roadmap: Continuous Improvement

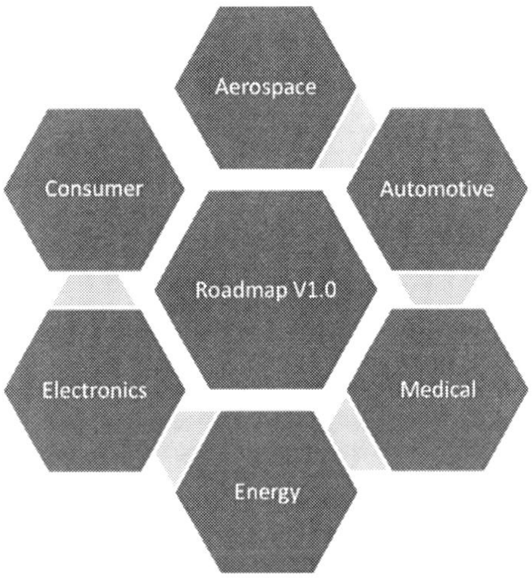

Technology Roadmap Advisory Group Being Formed

Approved for Public Release

America Makes
National Additive Manufacturing Innovation Institute

Status of Project Call #1

- Kickoff April 2013
- 6 Projects with 35 Partners
- $4.75M Government Share, $5.0 Cost Share
- Material Characterization, Process Qualification, Quality Control
- Excellent Teaming

rpm
PENNSTATE
MISSOURI S&T University of Science & Technology
CASE WESTERN RESERVE UNIVERSITY EST. 1826
NORTHROP GRUMMAN

Approved for Public Release

America Makes
National Additive Manufacturing Innovation Institute

Rapid Qualification Methods for Powder Bed Direct Metal Additive Manufacturing Processes

Problem:
- Qualification of a process takes an exorbitant amount of time, holding back adoption of promising AM processes
- Industry struggles to control microstructure and mechanical properties across direct metal laser sintering (DMLS) and electron beam melting (EBM)

Approach:
- Thermal property modeling, Power/Velocity (P/V) process mapping to melt pool geometry
- Determine effect of powder size distribution on P/V process map and resulting microstructure
- Construct process maps for DMLS and EBM
- Develop process-based cost models
- Demonstrate on "Industry Application"

Lead Organizations: Case Western Reserve University
Principal Investigators: John Lewandowski / Jack Beuth (CMU)
Team: CMU, Pratt & Whitney, GE Aviation, Lockheed Martin, Kennametal, Bayer, U. Louisville, Wright State U, NC State U.

Future State:
- Verified methods for tight control of microstructure and mechanical properties
- Reduction in amount of testing/time to qualify
- Ability to switch between high P/V (high rate, low res) and low P/V (low rate, high res) will provide overall faster build times and ability to serve new markets
- Industry uses cost models to make decisions on adopting AM into their factory or supply chain

Approved for Public Release

15

America Makes
National Additive Manufacturing Innovation Institute

Thermal Imaging for Process Monitoring and Control of Additive Manufacturing

Problem:
- Lack of control of microstructure due to lack of control of thermal gradients causing variability in mechanical properties
- Qualification is slow and expensive

Approach:
- Integrate thermal sensors to adjust process real time for increased quality/consistency of microstructure and mechanical properties
- Harden sensors for EBDM process
- 3D visualization and correlation of data
- Expand process maps to real-time microstructure control

Lead Organization: Penn State U
Principal Investigator: Ted Reutzel
Team: Stratonics, Sciaky, Optomec, Carnegie Mellon U, Wright State U, Siemens, Lockheed Martin, Pratt&Whitney, GE Aviation

Future State:
- Multi-sensor thermal imaging and control installed within Optomec LENS and Sciaky EBDM machines
- Increased adoption due to greater confidence in part properties and quality
- Rapid qualification due to consistency in microstructure and mechanical properties

Approved for Public Release

16

America Makes
National Additive Manufacturing Innovation Institute

Maturation of Fused Depositing Modeling (FDM) Component Manufacturing

Problem:
- Lack of materials data for AM high temp thermoplastics because no industry group has yet agreed to a build/design standard

Approach:
- Generate B-basis allowables for ULTEM™ 9085
- FEA analysis and manufacturing of select industry relevant sub-element component parts – common part geometry, thick/thin wall
- Create a design guide parts using ULTEM™ 9085, design out structural defects due to build path
- Develop certification procedures and validate

Lead Organization: rp+m
Principal Investigator: Ed Herderick
Team: University of Dayton Research Institute, Stratasys, Boeing, GE Aviation, Lockheed Martin, Northrop Grumman

Future State:
- AM community gains deeper understanding of properties and market opportunities of high temp FDM polymer parts through access to materials knowledgebase
- Certification procedures widespread
- Market expands for ULTEM ™ 9085 parts including aerospace, automotive, medical and customized consumer products

Approved for Public Release

17

America Makes
National Additive Manufacturing Innovation Institute

Maturation of High Temperature Selective Laser Sintering (SLS) Technologies and Infrastructure

Problem:
- PEEK is a promising high temp polymer used for SLS but foreign, sole source
- PEEK cost has risen 2.5X
- PEEK powder is non-recyclable
- PEKK (similar chemistry) has been FDA approved, but little qualified matl data

Approach:
- Generate B-basis allowables for PEKK
- Process parameter determination such as lowest processing temperature for PEKK for faster cool down
- Determine recyclability
- Demonstrate on "Industry Application"
- Topology optimization of relevant designs for requirements

Lead Organization: Northrop Grumman
Principal Investigator: Pedro Gonzalez
Team: Oxford Performance Materials, Arkema, AST2

Future State:
- AM community gains deeper understanding of properties and market opportunities of high temp SLS polymer parts through access to materials knowledgebase
- Two certified, domestic sources of PEKK, recyclability determined, costs driven down, opening up new markets including aerospace and customized medical implants

Approved for Public Release

18

America Makes
National Additive Manufacturing Innovation Institute

Complex, Rapid Tooling Using Fused Deposition Modeling

Problem:
- Tooling has long lead times and high cost - barrier for low volume manufacturing
- Near term opportunity may be missed for applying AM for composite and sheet metal tooling
- Polymeric AM tooling has been demo'd, but not matured

Approach:
- Manufacturing of AM tools and testing to validate Finite Element models
- Demonstration of sparse-filled tools for VaRTM and autoclave composite manufacturing
- Create a design guide for sparse-built tools, disseminated to composite manufacturers, metal forming, and AM community

Lead Organizations: Northrop Grumman and Missouri S&T
Principal Investigators: Pedro Gonzalez and Ming Leu
Team: Boeing, Stratasys, rp+m, RCBI

Future State:
- Expand AM into new market for low rate production tooling
- Composites industry realizes 15x cost savings and 5x lead time reduction vs. Invar tooling
- Common practice throughout industry and services

Approved for Public Release

19

America Makes
National Additive Manufacturing Innovation Institute

Qualification of AM Processes and Procedures for Repurposing and Rejuvenation of Tooling

Problem:
- Emerging near term opportunity with repurposing and rejuvenation (RR) of tooling that is beyond the risk of one organization
- Die casting tools are very expensive (> $1M, automotive industry) and have long lead times

Approach:
- Compare mechanical properties and microstructure of various AM repair type methods (LENS, POM, etc) on H13 and maraging steel dies, to baseline welding repair
- Evaluate bond strength and die life of each
- Comp model residual stress/distortion for heat treatment
- NADCA and Amer Weld Society to provide guidelines and training

Lead Organization: Case Western Reserve University
Principal Investigator: David Schwam
Team: Sciaky, Optomec, Keystone, Lincoln Electric, 6 die casting companies, Deform. Control Tech, Magma, North American Die Casting Assn (NADCA)

Future State:
- NADCA published guide, workshops, and webinars on RR of tooling using AM
- Industry-wide adoption of RR of tooling results in $500M/year cost savings for die casting manufacturers
- Use of AM within DoD facilities for RR of tooling
- Life cycle energy cost reductions through many manufacturing sectors including automotive from extending die cast tool life

Approved for Public Release

20

Status of Project Call #2

- Award Announcement January 2014
- 15 Projects with 75 Partners
- $9.0M Government Share, $10.3 Cost Share
- Even Better Teaming!

Academic vs Industry Split	
Total Indsutry	8
Total Academic	7
Total	**15**

Roadmap Alignment	
Process and Equipment	6
Design for AM	3
AM Materials	1
Qual and Cert	2
Knowledgebase	3
Total	**15**

CASE WESTERN RESERVE UNIVERSITY EST. 1826

NC STATE UNIVERSITY

Carnegie Mellon University

stony creek labs

GE Global Research

OPTOMEC

EWI. We Manufacture Innovation

NORTHROP GRUMMAN

GE Aviation

MichiganTech

UTEP

PITT

McGowan Institute for Regenerative Medicine

YOUNGSTOWN BUSINESS INCUBATOR ybi cultivate innovate accelerate

Stony Creek Labs

- Title – Optimization of Parallel Consolidation Methods for Industrial Additive Manufacturing
- Problem / Need – Limitations in industrial advanced manufacturing printing (build speed and volume)
- Objective(s) – Development of a next generation 3D printing method by consolidating powder at many points on a part simultaneously
- Partners - Grid Logic; Michigan Economic Development Corporation; MSC; Oakland University; and Raytheon Missile Systems

America Makes
National Additive Manufacturing Innovation Institute

Optomec

- Title – Development of Knowledgebase of Deposition Parameters for Ti-6Al-4V and IN718
- Problem / Need – Minimizing defects in AM still require "trial-and-error" or "cut-and-try" approaches. Process parameters need to be developed to attain defect-free deposition in metallic AM LENS
- Objective(s) – Generate the knowledgebase for Ti-6Al-4V and IN718 that consists of a matrix of permissible process parameter combinations for defect-free deposits using LENS
- Partners – Applied Optimization Inc.

Approved for Public Release

America Makes
National Additive Manufacturing Innovation Institute

EWI

- Title – Refining Microstructure of AM Materials to Improve Non-Destructive Inspection

- Problem / Need – Complex microstructure developed during Electron Beam Directed Energy Deposition (EB-DED) reduces ultrasonic inspection limits by 66% resulting in radiography as the principle inspection method. This method increases inspection burden by 2X degrading the business case for implementation.

- Objective(s) – Address the need to improve the ability to ultrasonic inspect titanium alloy components for high-performance aerospace applications, which feature a complex microstructure created during the electron beam directed energy deposition and subsequent heat treatment processes, through the modification of deposition process parameters and advancement of ultrasonic inspection techniques

- Partners – Lockheed Martin and Sciaky, Inc.

Approved for Public Release

America Makes
National Additive Manufacturing Innovation Institute

Case Western Reserve University

- Title – High Throughput Functional Material Deposition Using a Laser Hot Wire Process
- Problem / Need – Laser Hot Wire (LHW) lacks the robustness and reliability as a viable Additive Manufacturing Process that can be used in Heavy Fabrication, Aerospace, and Energy Sectors

- Objective(s) – Deliver LHW feasibility assessment and a guideline and test methods protocol will be developed for assessing new processes

- Partners – AZZ|WSI; Lincoln Electric Company; rp+m, Inc.; and RTI International Metals

Approved for Public Release

America Makes
National Additive Manufacturing Innovation Institute

Optomec

- Title – Wide-Spread Utilization of Metal Additive Manufacturing
- Problem / Need – Integration of Metal Laser Deposition into existing production methods is currently limiting adoption
- Objective(s) – Facilitate expansion of AM through development of a LENS Engine that will be a self-contained, machine-tool-agnostic laser deposition system, integrate the LENS Engine with the installed machine tool base, and shorten training cycle of new AM machine operators
- Partners – Lockheed Martin Missiles & Fire Control; MachMotion; TechSolve, Inc.; and U.S. Army Benet Laboratories

Approved for Public Release

America Makes
National Additive Manufacturing Innovation Institute

North Carolina State University

- Title – Automatic Finishing of Metal AM Parts to Achieve Required Tolerances and Surface Finishes
- Problem / Need – The need exists to produce a mechanical product to final geometries and removing costly post processing requirements
- Objective(s) – Develop a "Make Button" process with the use of a hybrid manufacturing system using both AM and subtractive processing to "digitally manufacture" final product
- Partners – Advanced Machining; CalRAM Inc.; FineLine Prototyping, Inc.; Iowa State University; John Deere; Kennametal Inc.; and Productivity Inc

Approved for Public Release

America Makes
National Additive Manufacturing Innovation Institute

GE Global Research

- Title – Development of Distortion Prediction and Compensation Methods for Metal Powder-Bed Additive Manufacturing
- Problem / Need – A fundamental challenge in metal powder-bed AM is distortion mitigation. An absence exists with suitable distortion prediction tools resulting in ad-hoc rules of thumb to compensate and mitigate the problem.
- Objective(s) – Benchmark and validate physics-based thermal distortion prediction and mitigation tools for commercialization. Results will lead to a 3X to 4X reduction in development time. A wiki with distortion strategies will be made available to the Digital Estate.
- Partners – 3DSim, Inc.; CDI Corporation; Honeywell Aerospace; Pan Computing LLC; Penn State University; United Technologies Research Center; and the University of Louisville

Approved for Public Release

America Makes
National Additive Manufacturing Innovation Institute

Carnegie Mellon University

- Title – A Database Relating Powder Properties to Process Outcomes for Direct Metal AM
- Problem / Need – The most severe supply chain restriction is requirements to procure quality powders from machine manufacturers. Therefore, a large variety of powder systems must be usable in direct metal machine for broad adoption.
- Objective(s) – Develop a first-of-its-kind database relating powder properties to process outcomes and identify process variable changes need to make a current incompatible powder, comparable to standard powders
- Partners – AMETEK Specialty Metal Products; ATI Powder Metals; CalRAM Inc.; Carpenter Powder Products Inc.; FineLine Prototyping, Inc.; Medical Modeling Corporation; North Carolina State University; Oxford Performance Materials; Pratt & Whitney; Robert C. Byrd Institute; TE Connectivity Ltd.; United Technologies Research Center; and Walter Reed National Military Medical Center

Approved for Public Release

America Makes
National Additive Manufacturing Innovation Institute

Michigan Technological University

- Title – Metal Alloys and Novel Ultra-Low-Cost 3-D Weld Printing Platform for SME Rapid Prototyping and Production
- Problem / Need – The US does not have a low cost metal 3D printer offering. While access to polymer printing is affordable and accessible, nothing exists for metal.
- Objective(s) – Reduce the cost of metal 3D printing by a factor of 100 by maturing current proven concept to commercialization
- Partners – Aleph Objects, Inc.; ASM International; Miller/ITW; ThermoAnalytics, Inc.; and The Timken Company

Approved for Public Release

America Makes
National Additive Manufacturing Innovation Institute

GE Aviation

- Title – In-Process Quality Assurance (IPQA) for Laser Powder Bed Production of Aerospace Components
- Problem / Need – Lack of commercially available, platform independent, Quality Assurance technology for high volume AM production of mission critical aerospace components
- Objective(s) – maturation of an in-process quality assurance technology solution leveraging a development approach incorporating multiple AM OEM machines, multiple super alloys, and multiple product intent aerospace companies
- Partners – Aerojet Rocketdyne; B6 Sigma, Inc.; Burke E. Porter Machinery Company; Honeywell Aerospace; Montana Tech of The University of Montana; and TechSolve, Inc

Approved for Public Release

America Makes
National Additive Manufacturing Innovation Institute

University of Texas El Paso

- Title – 3D Printing Multi-Functionality: Additive Manufacturing for Aerospace Applications
- Problem / Need – Long lead manufacturing of UAV's require the integration of multiple manufacturing processes, building on 3D printing flexibility
- Objective(s) – Comprehensive manufacturing suite will be integrated into AM that include (1) extrusion of wide variety of materials, (2) micromachining, (3) laser ablation, (4) embedding for wires and fine-pitch meshes, and (5) robotic component placement
- Partners - Lockheed Martin; Northrop Grumman Corporation; rp+m, Inc.; Stratasys; The University of New Mexico; and Youngstown State University

Approved for Public Release

America Makes
National Additive Manufacturing Innovation Institute

Northrop Grumman

- Title – Electron Beam Melted Ti-6Al-4V Additive Manufacturing Demonstration and Allowables Development Project
- Problem / Need – Data is lacking for E-Beam AM required for use in airframe, space, and propulsion applications limiting current implementations to non-structural, non durability rated components for static strength at operating temperatures.
- Objective(s) – Leverage progress gained in the IBIF program (Surface Finish Enhancement for the E-BEAM Digital Direct Manufacturing of Ti-6Al-4V Alloy Structural Components) to develop a B-Basis set of Ti-6Al-4V materials design allowables, full scale demonstration of the surface finish attainable via E-Beam AM and post AM surface finishing processes, and validate NDE methods on full scale E-Beam AM demonstration components.
- Partners – CalRAM Inc.; Concurrent Technologies Corporation; General Electric; and Robert C. Byrd Institute

Approved for Public Release

America Makes
National Additive Manufacturing Innovation Institute

University of Pittsburgh

- Title – Additive Manufacturing of Biomedical Devices from Bioresorbable Alloys for Medical Applications

- Problem / Need – Biocompatible, bioresorbable magnesium allows with controlled degredation rates and strength characteristics suitable for biomedical devices exist, but lack manufacturing methods of converting materials into required shapes, such as bone plates, tracheal stents, and scaffolds

- Objective(s) – A powder bed AM technology (ExOne) will be used to demonstrate the feasibility of manufacturing required shapes

- Partners - McGowan Institute for Regenerative Medicine at the University of Pittsburgh, ExOne and Magnesium Elektron Powders

Approved for Public Release

America Makes
National Additive Manufacturing Innovation Institute

University of Pittsburgh

- Title – Developing Topology Optimization Tools that Enable Efficient Design of Additive Manufactured Cellular Structures

- Problem / Need – Existing M&S tools lack efficiency in designing complex geometries, do not account for residual mechanical effects due to specific AM processes, and have not integrated cost modeling and design requirements for AM

- Objective(s) – Develop robust software for design and optimization of AM structural designs based on cellular structures utilizing micromechanics models for capturing the effective behavior of cellular structures in FEA

- Partners - Acutec Precision Machining Inc.; Alcoa Inc.; ANSYS, Inc.; and ExOne

Approved for Public Release

America Makes
National Additive Manufacturing Innovation Institute

Youngstown Business Incubator

- Title – Accelerated Adoption of Additive Manufacturing Technology in the American Foundry Industry
- Problem / Need – US metalcasting industry facing severe challenges resulting in over 1000 facilities closing.
- Objective(s) – Recapitalize the metalcasting industry by developing a framework (education, management, marketing, business) for the establishment of a AM regional network for U.S. foundry industry grounded in AM
- Partners - American Foundry Society; ExOne; Humtown Products; Janney Capital Markets; the University of Northern Iowa; and youngstown State University

Approved for Public Release

America Makes
National Additive Manufacturing Innovation Institute

NIST MSAM CAP Overview

"Holistic Approach to Solving Measurement Science Challenges in Additive Manufacturing"

- **Informative In-Situ Process Monitoring**, led by EWI, will develop, integrate, and validate a robust, informative in-situ process monitoring capability standard for AM
- **Non-Destructive Evaluation**, led by Concurrent Technologies Corporation (CTC), will develop and validate a non-destructive evaluation standard for AM
- **Layerwise Quality Certification for Additive Manufacturing**, led by the University of Louisville (UofL), will develop a layerwise quality certification standard for AM

Quality Parts Produced and Certified for Use in End Product(s)

Approved for Public Release

America Makes
National Additive Manufacturing Innovation Institute

Project Team

In Situ Process Monitoring
Lead: EWI

- Georgia Tech
- UNC Charlotte
- Stratonics
- B6Sigma
- Paramount Industries
- GE Aviation

Non Destructive Evaluation
Lead: CTC

- North Carolina State University
- GE Aviation
- GE Inspection Technologies
- Lockheed Martin
- Pratt & Whitney

Certification Standards
Lead: University of Louisville

- Carnegie Mellon University
- University of Texas El Paso
- Penn State University
- GE Aviation
- Northrop Grumman
- Boeing
- Stryker
- Harvest Technologies
- Solid Concepts

Other Collaborators

- Oak Ridge National Laboratory
- Lawrence Livermore National Laboratory
- Imaginistics
- M7 Technologies
- Ingersoll Machine Tools

Approved for Public Release

America Makes
National Additive Manufacturing Innovation Institute

Digital Estate

- Replaces the NAMII Website
- New Look – New Capabilities to Serve the Community
- Launched Last Week

www.AmericaMakes.us

Approved for Public Release

Summary

- *Huge Opportunity with High Stakes*
- *"None of us can do it alone. For this to succeed, it needs to scale, and for it to scale, we need a passionate and motivated community"*

  \- Avi Reichental, President and CEO, 3D Systems
- *"Rather than companies having to bear all of the cost and burden of qualifying these processes and materials, much can be done through collaborative efforts"*

  \- Terry Wohlers, Principle Consultant and President, Wohlers Associates

When America Makes America Works

America Makes

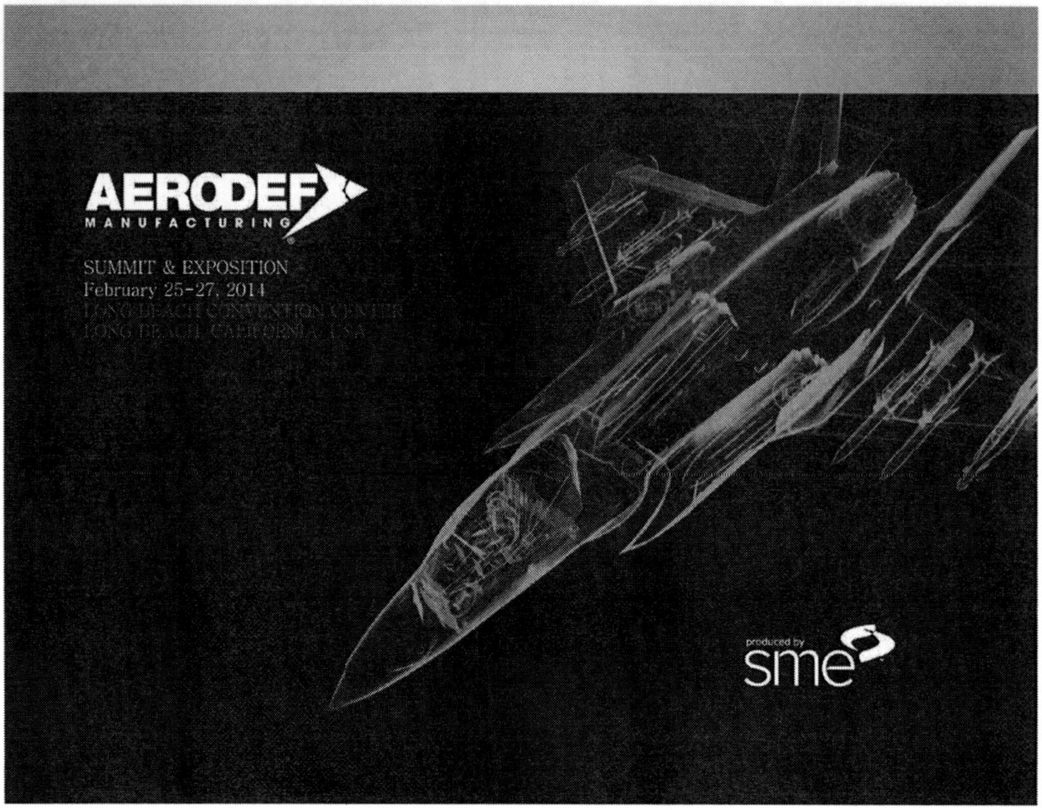

AERODEF MANUFACTURING
produced by sme

Real-Time Adaptive Control for Tool Optimization

Presented by:
Rob Caron, PE
President
Caron Engineering Inc.

CARON ENGINEERING

Caron Engineering Inc. was started in 1986 by founder and President Rob Caron, PE, a licensed professional engineer

Our company has a strong electrical engineering background with experience in both hardware and software design

Our focus has always been to enhance the ability of any CNC machine by providing advanced technology that can be added to new or existing machines

CARON ENGINEERING

- **Our products include:**
  - Tool Monitoring and Adaptive Control (TMAC)
  - Automatic Tool Compensation (AutoComp)
  - Laser Tool Setting sales & integration
  - Touch Probe sales & integration
  - Tool Identification Systems
  - Custom hardware and software design

About This Presentation

1. Discuss common machining problems related to tooling
2. What are the biggest challenges?
3. Introduce adaptive control
4. Discuss how adaptive control can improve the process
5. How can it be implemented at your shop
6. Real world examples

Tooling Problems

Tooling problems!

- How long does a tool last?
  - Many factors affect tool life
    - Material variations
    - Coolant flow/concentration
    - New vs. re-ground
    - Tool run-out
    - Many other factors

Tooling Feedrate Challenges

- What is the best feedrate?
  - o Feed and speed calculations are for a new tool
  - o The tool wears over time so a compromise on feedrate is required
  - o If the CAM program adjusts feedrates in the program they may be too abrupt causing excessive tool wear

```
G01 X12.5 Y40.3 F500.0
X20.2 Y45.0 F1000.0
X35.0 F1800.0
```

- What is the best feedrate for castings and forgings?

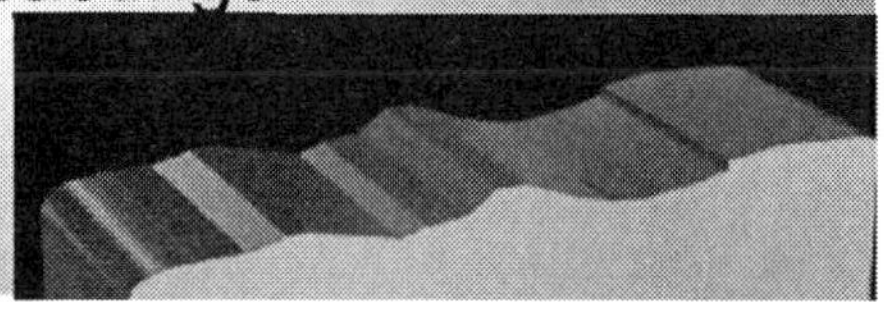

What is Adaptive Control?

- The tools cutting power is measured in real-time
- A target power level is selected
- The control system takes over the feedrate
- If the cutting power is **below** the target the feedrate is **increased**
- If the cutting power is **above** the target the feedrate is **reduced**

Adaptive Control

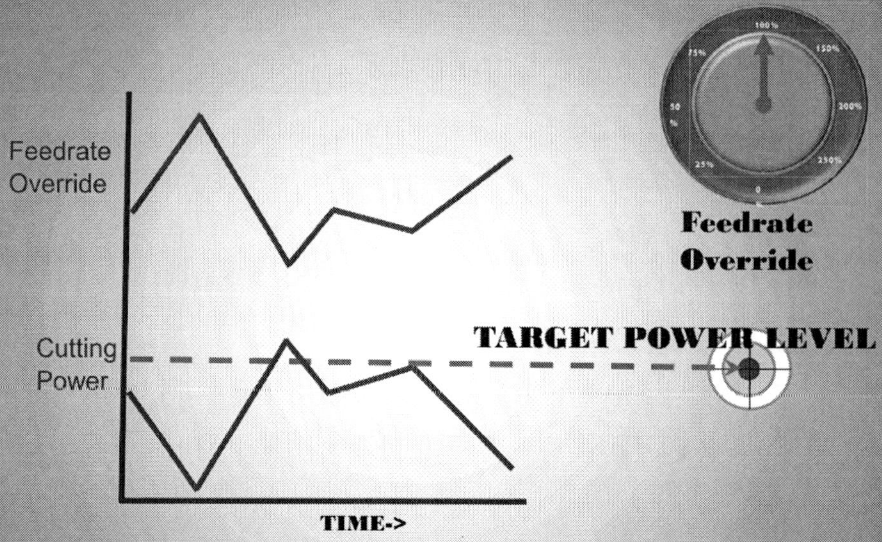

Basic Principles

- The tool cutting power is measured accurately and with a high resolution (0.001 HP)
- The system makes feedrate adjustments at a high rate (10msec)
- The response is different when below normal cutting power versus above cutting power
- Any change in cutting conditions will affect the power that the tool requires

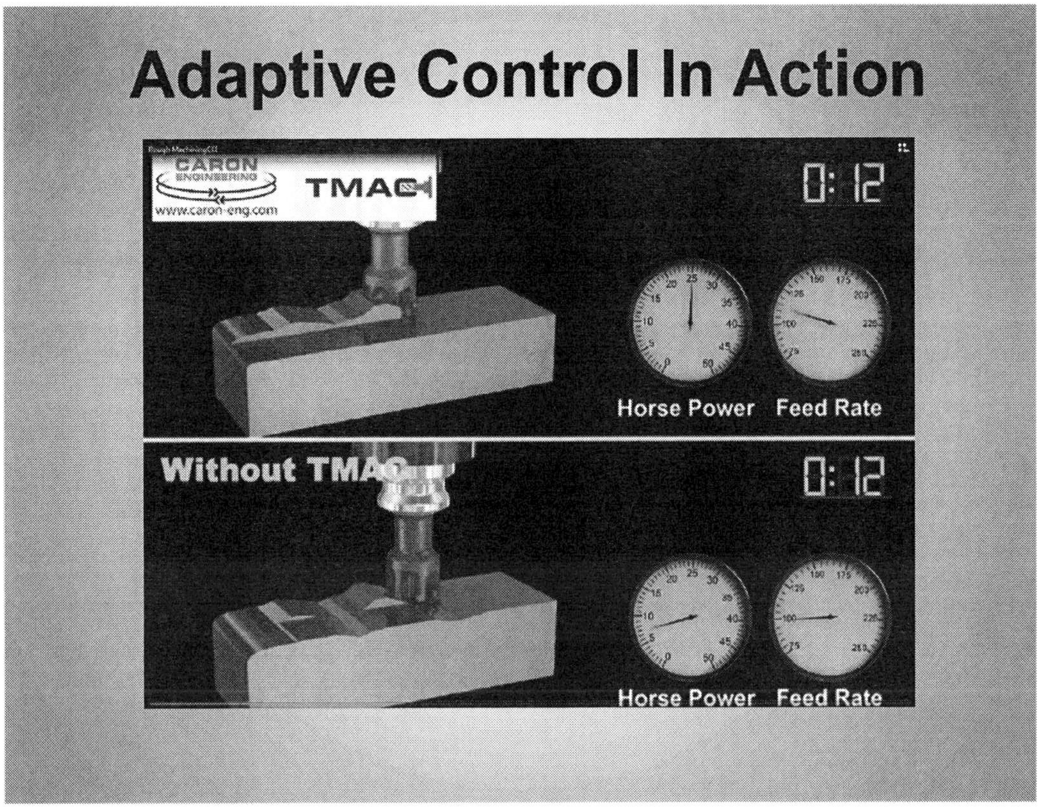
Adaptive Control In Action
CARON ENGINEERING
TMAC
www.caron-eng.com
0:12
Horse Power
Feed Rate
Without TMA
0:12
Horse Power
Feed Rate

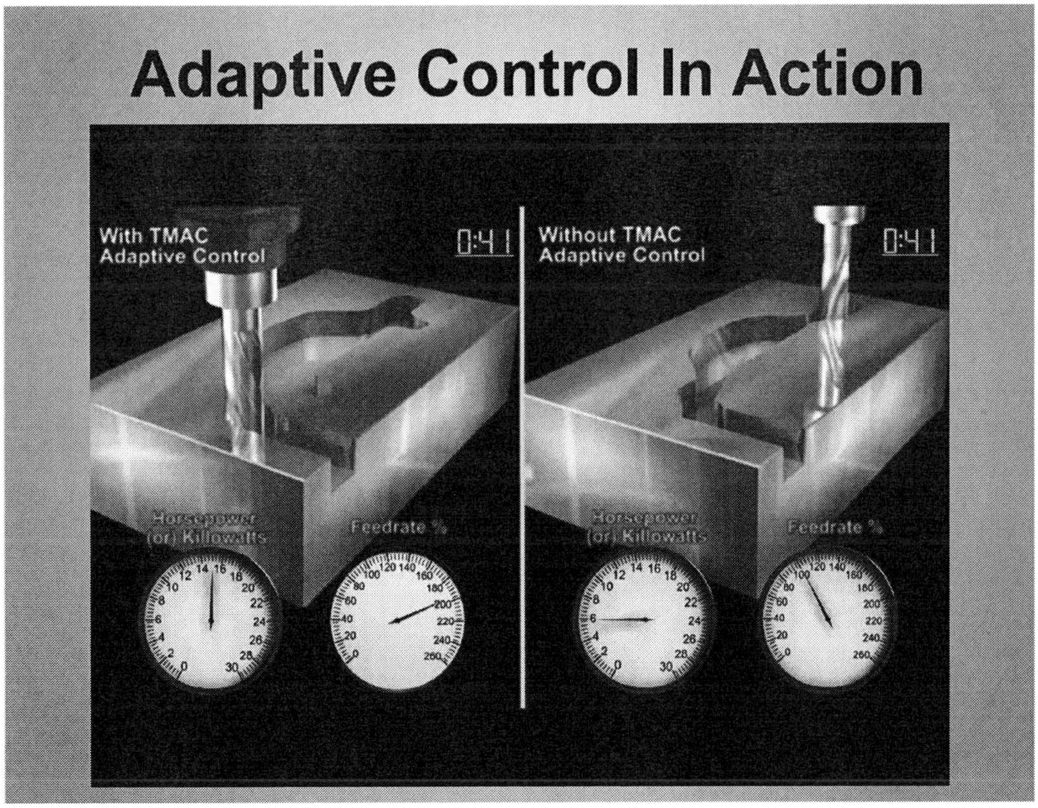
Adaptive Control In Action
With TMAC
Adaptive Control
0:41
Without TMAC
Adaptive Control
0:41
Horsepower
(or) Killowatts
Feedrate %
Horsepower
(or) Killowatts
Feedrate %

What are the Benefits?

- Cycle time is reduced to the best level for each cut
- The system compensates for any changes in the cutting environment
- The tool life will be extended because the tool is controlled through varying conditions

Real World Results

- Typical cycle time savings 15%-40%
- Allows tools to run at optimum feedrates
- Adjusts for variations in material and tooling and cutting conditions
- Gives operators the confidence to let machines run unattended
- Adjusts feedrates smoothly versus CAD/CAM programs that can only adjust per line in the program

Customer Experience # 1

38% Savings

Industry - Weapons

Material - Steel

Machine Tool - Horizontal CNC machining center

Tooling – 3/8"high feed end mill machining 8 parts on a fixture

MAX FEEDRATE – 175%

Before Adaptive

Total cycle time for this tool cutting **8** parts was **11 minutes and 45** seconds.

After applying adaptive control

Minimum feedrate settings to detect worn tooling & catastrophic conditions were set as well.

The result was a decrease in cycle time to **7 minutes and 18 seconds**, a reduction of over **38%.**

Additionally the customer stated that the cutting now sounds better and that the tool life increased as a result of the more consistent load being placed on the tool.

Customer Experience # 2

84% Savings

Industry - Aerospace

Material - Titanium

Machine Tool - Horizontal CNC machining center

Tooling – 3.5mm dia. x 133mm length drill peck drilling six "through" holes in two parts

Before Adaptive

The drill was being fed very slowly to reduce the possibility of deflection. The time required to process the two parts was originally

18 minutes and 15 seconds.

After applying adaptive control

The result was an increase in both the cutting feedrate and the aircut feedrates.

The processing time was decreased to **2 minutes and 52 seconds**, a savings in cycle time of **84%.**

Not all of the cycle time reduction came directly from adaptive control.

By viewing the chart data from TMAC the programmer was able to re-evaluate the machining methods and optimize some feedrates & moves. In his estimation, this only occurred as a result of analyzing the process with TMAC.

Customer Experience # 3

23%
Savings

Industry - Aerospace
Material - Titanium
Machine Tool - Horizontal CNC machining center
Tooling – 10mm & 32 mm carbide end mills

Before Adaptive
Cycle time for 32mm end mill = 77 min., 52 sec.
Cycle time for 10mm end mill = 12 min., 65 sec.

After applying adaptive control
Cycle time for 32mm end mill = 60 min., 08 sec. - time savings **22.5%**
Cycle time for 10mm end mill = 10 min., 53 sec. - time savings **16.7%**

Customer Experience # 4

28%
Savings

Industry - Heavy Vehicle Transmission Housings
Material - Cast Iron
Machine Tool - Horizontal CNC machining center
Tooling – 8mm Thread Mill

Before Adaptive
Cycle time for 8mm thread mill 71.6 seconds

After applying adaptive control
Cycle time for 8mm thread mill 49.3 seconds
A 28.36% cycle time savings

What did we learn?

- Many factors make selecting the correct feedrate a huge challenge
- Constant machining variations have caused customers to reduce feedrates in many part programs over time
- By measuring the cutting power of a tool in real-time, the optimal cutting force can be maintained
- Adaptive Control allows your tools and machine to perform at their best level all of the time with no operator intervention

AERODEF MANUFACTURING

sme

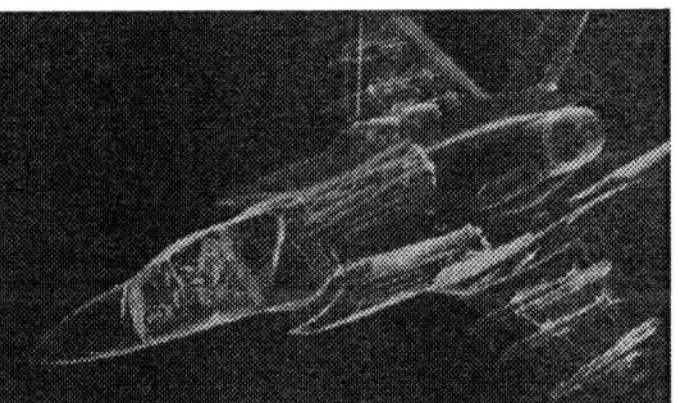

Conclusion

- You can apply this technology to most machine tools on your shop floor
- Once some basic calls are added to the part program the tools are learned automatically
- Adaptive control can be selected for any tool

AERODEF MANUFACTURING

sme

Questions

???????

AERODEF MANUFACTURING

sme

For more information

Caron Engineering Inc.

www.caroneng.com

(207) 646-6071

marketing@caroneng.com

SANDVIK
Coromant

Innovative Processes & Product for Machining Titanium Aero Components

Brian Hoefler & Mike Standridge SANDVIK Coromant

Leveraging Partnerships to Expand the Limits of Technology

Aerospace manufacturers fundamental challenges

- *Complicated parts with no margin for error*
- *Tough materials that limit machining efficiencies*
- *Demanding workloads to meet growing delivery schedules*
- *Resource and time constraints to process improvements*

Leveraging Partnerships to Expand the Limits of Technology

Keys to successful partnerships

- *Capabilities & Resources*
- *Focus on Total Cost Drivers*
- *Teamwork & Dialog*
- *Open Mind to Change*

Application Success Stories

Innovative Processes & Products = Positive Outcomes

- *Large Dovetail "Tulip" Slots – Engine Fandisk*
- *Profile Roughing – Frame Structure*
- *Deep Grooving Slots – Engine Fandisk*
- *Extended Length Finishing – Frame Structure*

Large Dovetail "Tulip" Slots – Engine Fandisk

Brian Hoefler & Mike Standridge

SANDVIK Coromant

Large Dovetail "Tulip" Slots – Engine Fandisk

Challenge

- *Outdated method & equipment*
- *Complicated form with close tolerances*
- *Cycle time constrained*
- *Flexible process required*

Brian Hoefler & Mike Standridge

SANDVIK Coromant

Large Dovetail "Tulip" Slots – Engine Fandisk

Technical Obstacles

- Significant metal removal in a cycle time target of 20 min.
- Cutting Forces and Power
- Large chip volume
- Tight Form tolerances generated by milling

Brian Hoefler & Mike Standridge

SANDVIK Coromant

Large Dovetail "Tulip" Slots – Engine Fandisk

Technical Achievements

- Multi tool rough strategy
- Climb cut "Up & Down"

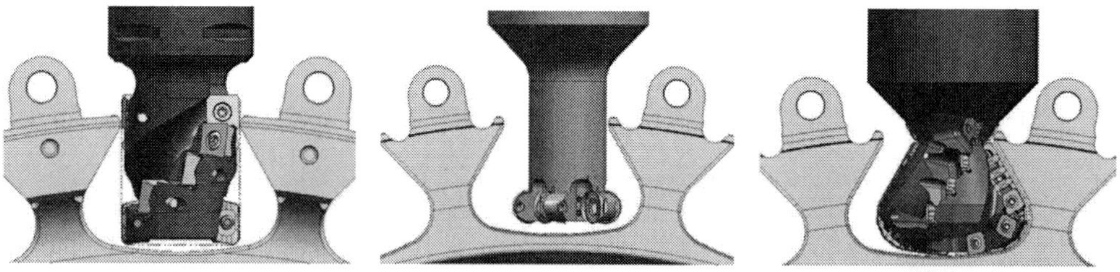

Brian Hoefler & Mike Standridge

SANDVIK Coromant

Large Dovetail "Tulip" Slots – Engine Fandisk

Technical Achievements

- *Cycle time met @ 20 minutes with room to improve*
- *Flexible & Robust*

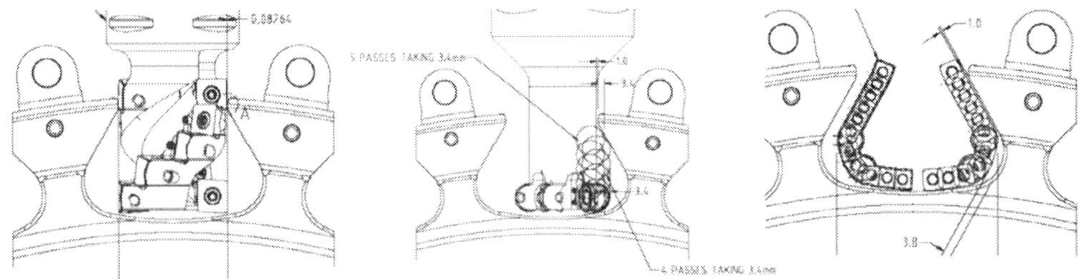

Large Dovetail "Tulip" Slots – Engine Fandisk

Technical Achievements

- *Precise solid carbide form finisher*

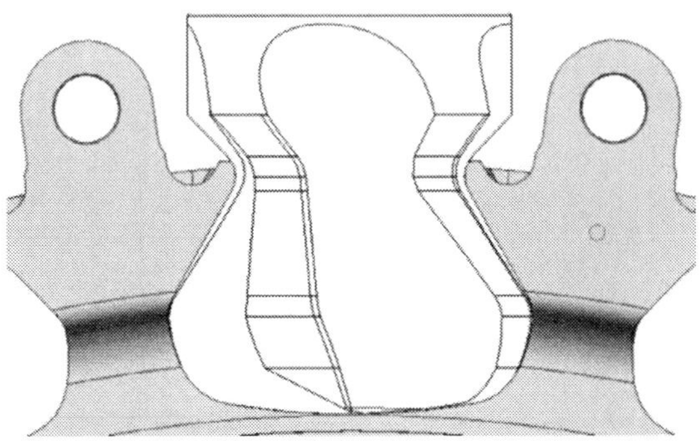

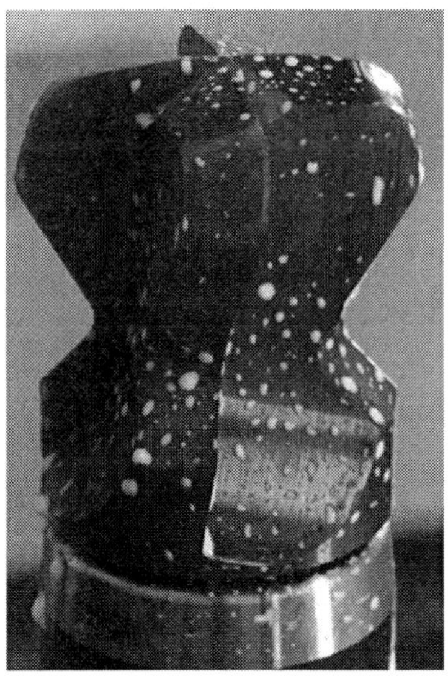

Large Dovetail "Tulip" Slots – Engine Fandisk

Summary Results

- *Proven process that achieves cycle time target even though the customer initially expected to not measure up*
- *Combination of standard & special tools with standard inserts*
- *Complex finishing solution that challenged manufacturability*
- *Simple method that is easily expanded to other parts*
- *Offline evaluation in our application center*
- *Team effort, many things tested, performance drove final solution*

Profile Roughing – Frame Structure

Brian Hoefler & Mike Standridge

SANDVIK Coromant

Profile Roughing – Frame Structure

Challenge

- *Robust end insert*
- *Simple, low cost tool repair*
- *120+ minutes tool life*

Profile Roughing – Frame Structure

Brian Hoefler & Mike Standridge

Profile Roughing – Frame Structure

Technical Obstacles

- *Insert translation stop*
- *Tough grade with high strength*
- *Hardened pocket hardware, easily exchangeable*
- *High pressure coolant through to each insert*

Profile Roughing – Frame Structure

Technical Achievements

- *Insert constrained*

Patent pending

Profile Roughing – Frame Structure

Technical Achievements

- *Grade 1040 proven to be superior!*
  - *Extreme toughness*
  - *Sub-micron substrate for strength*
  - *Tool life > 180 minutes*

Brian Hoefler & Mike Standridge

SANDVIK Coromant

Profile Roughing – Frame Structure

Technical Achievements

- *Hardened seat*

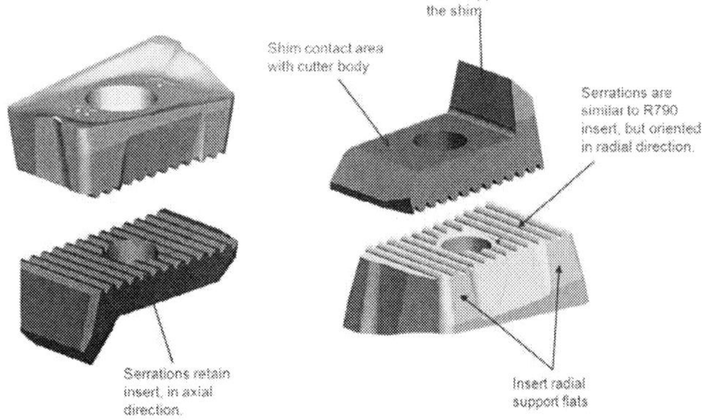

Patent pending

Brian Hoefler & Mike Standridge

SANDVIK Coromant

Profile Roughing – Frame Structure

Technical Achievements

- High Pressure coolant to every pocket
  - Adjustable with exchangeable nozzles

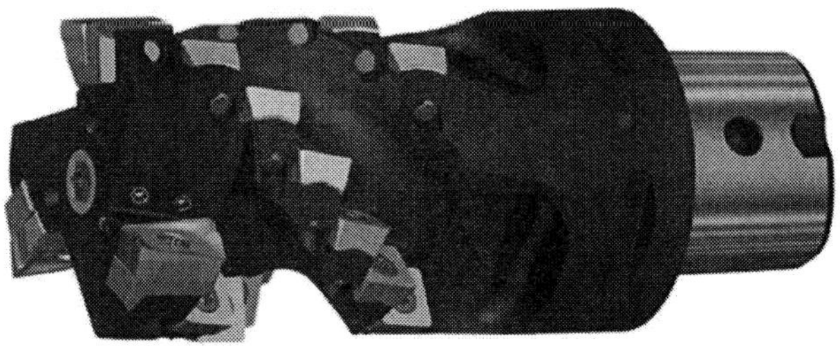

Brian Hoefler & Mike Standridge

SANDVIK Coromant

Profile Roughing – Frame Structure

Summary Results

- Extensively validated robust end insert invention
- Cycle time reduction due to no "slow down areas"
- Development of grade technology beyond current boundaries
- Leveraged competence of HP coolant technologies
- Open dialog on process adjustments yielded significant contributions
- Experienced total solution mindset

Brian Hoefler & Mike Standridge

SANDVIK Coromant

Deep Grooving Slots – Engine Fandisk

Deep Grooving Slots – Engine Fandisk

Challenge

- *New generation engine not producible with existing methods due to depth of groove and thickness of slot*
- *Minimum blend areas*
- *High quality surface finish*
- *Predictable tool deflection for compensation*

Deep Grooving Slots – Engine Fandisk

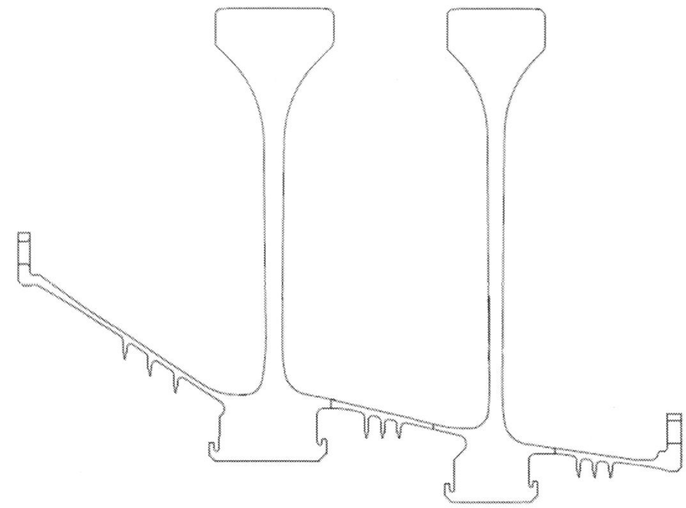

Brian Hoefler & Mike Standridge

SANDVIK Coromant

Deep Grooving Slots – Engine Fandisk

Technical Obstacles

- Extended length with minimum blade thickness
  - Excessive deflection
  - Uncontrolled vibration
- Coolant facilitation
  - Aim
  - Pressure
- Predictable & consisted performance

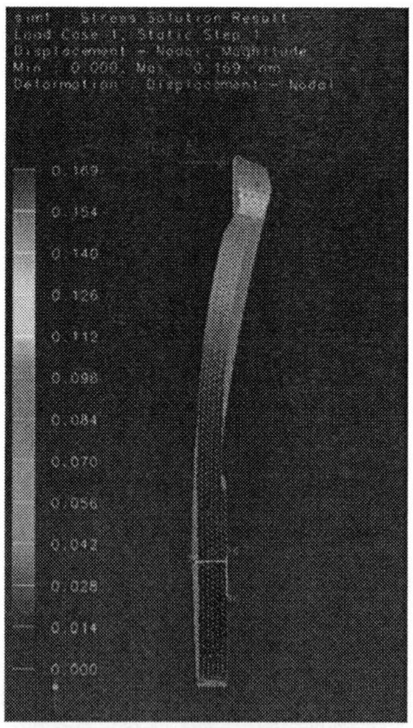

Brian Hoefler & Mike Standridge

SANDVIK Coromant

Deep Grooving Slots – Engine Fandisk

Technical Achievements

- *Laminated construction with high modulus material*
  - *Extreme increase in stiffness*
  - *Critically damped*
  - *Simple construction method*
  - *Coolant routing simplified*
- *Increased height for additional stiffness*

Deep Grooving Slots – Engine Fandisk

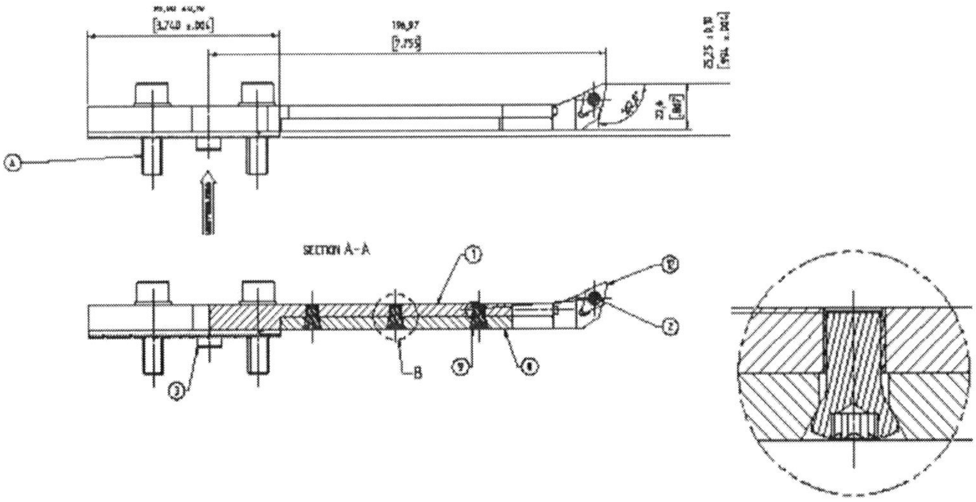

Deep Grooving Slots – Engine Fandisk

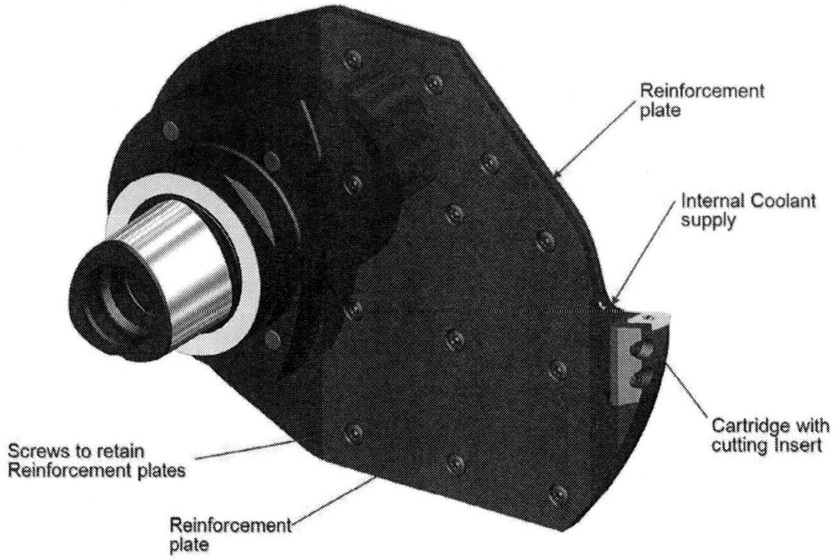

Deep Grooving Slots – Engine Fandisk

Summary Results

- *Development of a robust process to successfully product parts (2 years of service)*
- *Minimal and predictable deflection*
- *No uncontrolled vibration*
- *Adequate coolant distribution*

- *Global best practice technology created*

Extended Length Finishing – Frame Structure

Challenges

- *Outdated tool design driving indexable option*
- *Significant tool path length*
- *Minimal component taper allowed*
- *Tight finish required without mismatch lines*
- *Length of finished walls about 1.5X current worst case*

Brian Hoefler & Mike Standridge

SANDVIK Coromant

Extended Length Finishing – Frame Structure

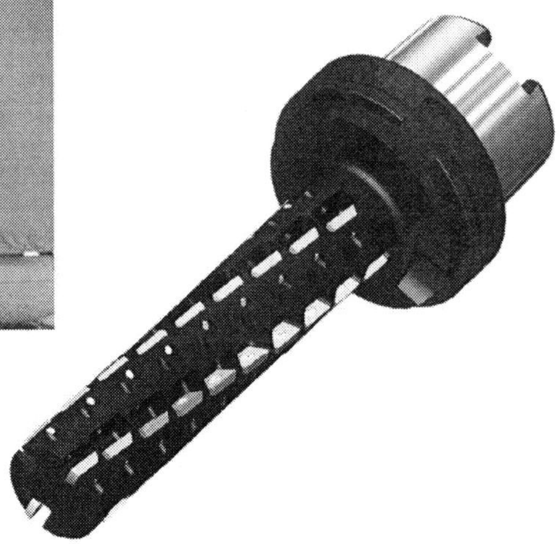

Brian Hoefler & Mike Standridge

SANDVIK Coromant

Extended Length Finishing – Frame Structure

Technical Obstacles

- *Wall waviness <.001"*
- *Limited taper*
- *L:D > 3:1*
- *Indexable*
- *Vibration "chatter" resistant*

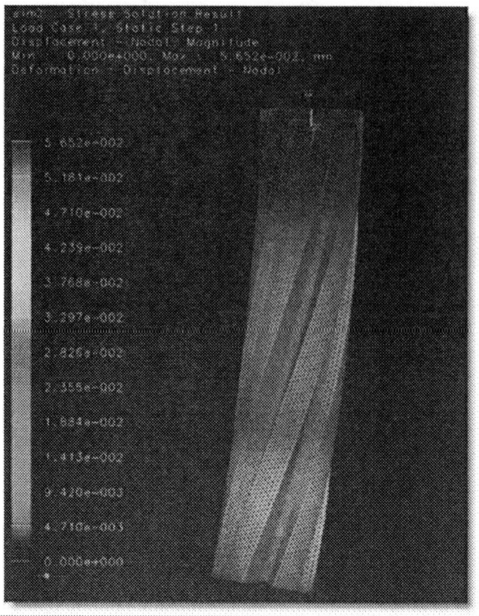

Extended Length Finishing – Frame Structure

Technical Achievements

- *Optimized Stiffness via FEA*
  - *Helix*
  - *Increased core*
- *Internally damped*
  - *"lost motion" frictional absorption technique*

Extended Length Finishing – Frame Structure

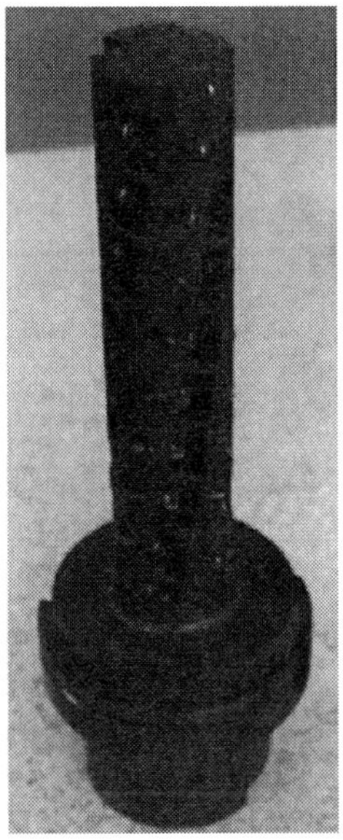

Summary Results

- *Implemented Technology*
- *Wall taper acceptable*
- *Waviness < .001"*
- *Vibration adequately damped*

Brian Hoefler & Mike Standridge

SANDVIK Coromant

Leveraging Partnerships to Expand the Limits of Technology

Aerospace manufacturers face common, fundamental challenges, the demands are not getting any easier

Teamwork with competent & capable partners can have a powerful effect with enormous payoffs

Attention on Total Cost Drivers allow for clear opportunities for enhancement

Open communication with strategic partners can expose simple methods to achieve goals

Brian Hoefler & Mike Standridge

SANDVIK Coromant

Leveraging Partnerships to Expand the Limits of Technology

The best solutions, the difference making developments of the future can be realized with the right partners, teamwork and open mind to change.

Brian Hoefler & Mike Standridge

SANDVIK Coromant

Appendix

Brian Hoefler & Mike Standridge

SANDVIK Coromant

Large Dovetail "Tulip" Slots – Engine Fandisk

Application Details

- *DMU 125 FD duoBLOCK*
- *C10 spindle, horizontal*
- *52Kw, 12,000RPM, 5 axis*

- *Ti6Al4V titanium*
- *Eco cool S761 w/ 12% concentration*

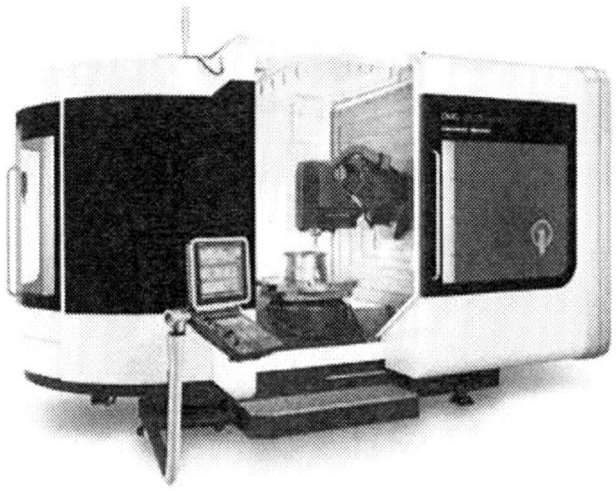

Large Dovetail "Tulip" Slots – Engine Fandisk

Application Details

- *Open slot with facemill R690-051C6-1449H 2eff.*
  - *110 SFM, 1.25 IPM (.003 IPT), 50mm DOC*
- *Undercut form with special R300-056 16H 5eff*
  - *Light radial engagement = 3.3mm*
  - *289 SFM, 20 IPM (.006 IPT) 6 passes each side*
- *Tulip form tool rougher*
  - *75 mm diameter, 4 flute, 2 eff.*
  - *130 SFM, 2 IPM (feed multiplied)*
- *Tulip finisher440 2090R514*
  - *217 RPM (168 SFM),15 IPM (.012 IPT), .5 mm form stock*

Large Dovetail "Tulip" Slots – Engine Fandisk

Application Details

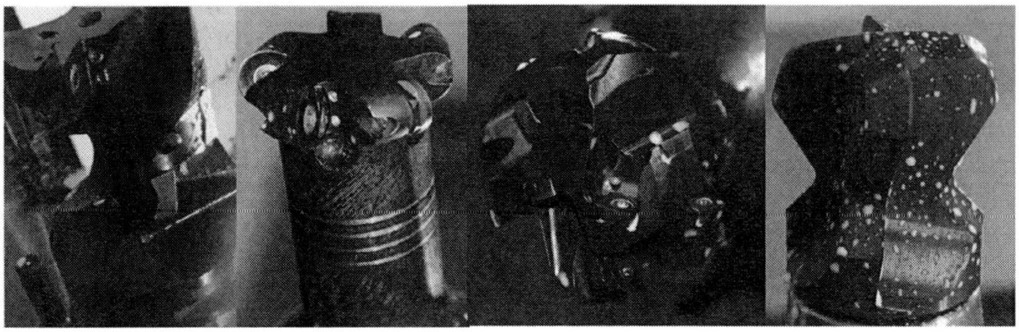

Brian Hoefler & Mike Standridge

SANDVIK Coromant

Large Dovetail "Tulip" Slots – Engine Fandisk

Application Details

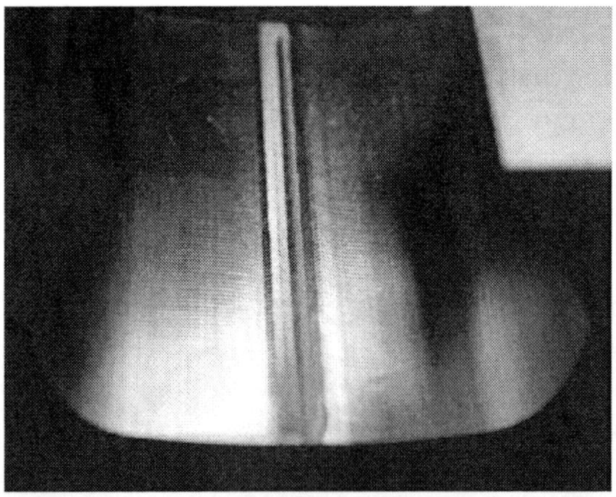

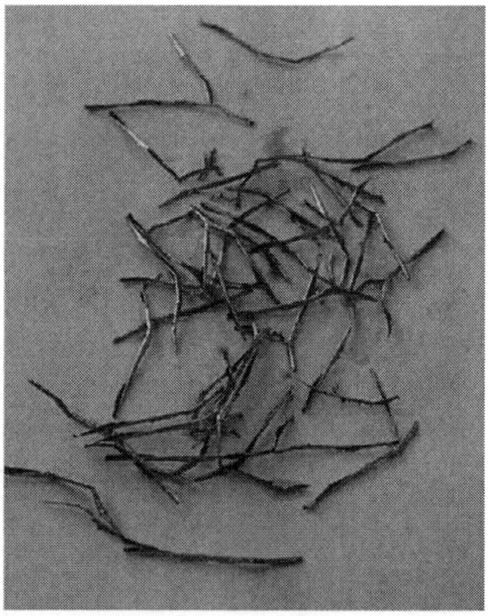

Brian Hoefler & Mike Standridge

SANDVIK Coromant

Profile Roughing – Frame Structure

Application Details

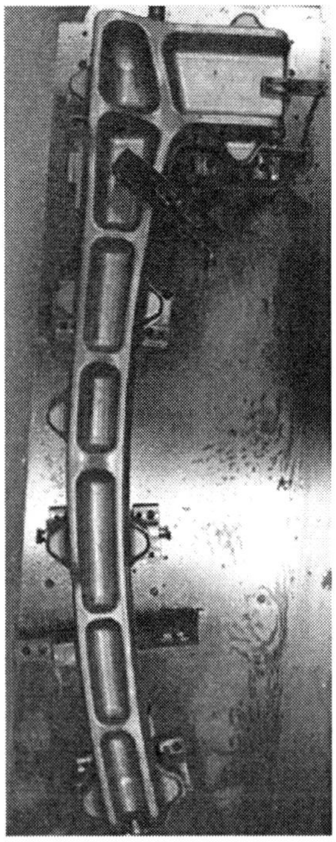

- *Bulkhead Frame*
- *Ti6Al4V forging 155-149 KSI*
- *Vc =165 (315RPM)*
- *Feed 5.2 = IPM (.0055IPT)*
- *198 minutes tool life*
- *Ae = .35-.60"; Ap = 3"*

Brian Hoefler & Mike Standridge

SANDVIK Coromant

Profile Roughing – Frame Structure

Application Details

Brian Hoefler & Mike Standridge

SANDVIK Coromant

Profile Roughing – Frame Structure

Application Details

Brian Hoefler & Mike Standridge

SANDVIK Coromant

Profile Roughing – Frame Structure

Application Details

Brian Hoefler & Mike Standridge

SANDVIK Coromant

Profile Roughing – Frame Structure

Application Details

Brian Hoefler & Mike Standridge

SANDVIK Coromant

Deep Grooving Slots – Engine Fandisk

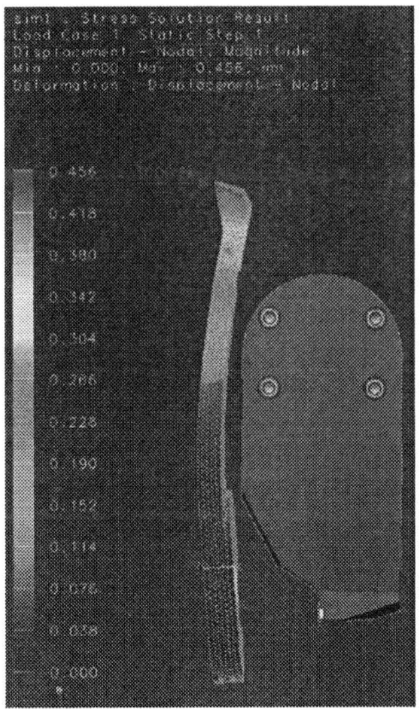

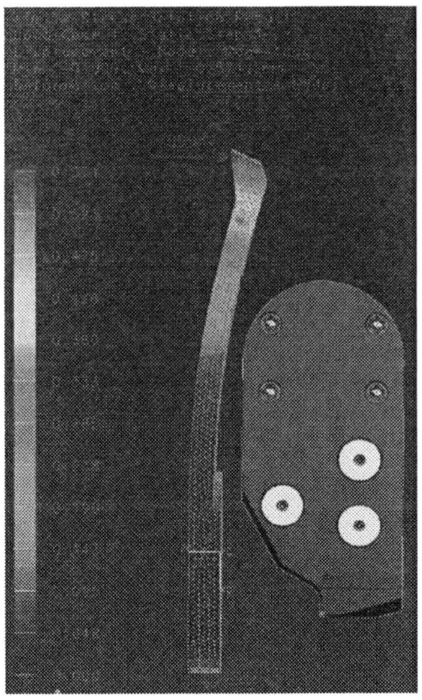

25% worse

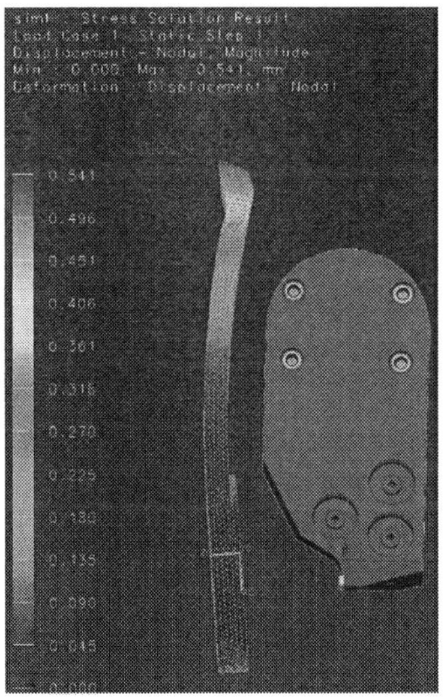

18% worse

Brian Hoefler & Mike Standridge

SANDVIK Coromant

Deep Grooving Slots – Engine Fandisk

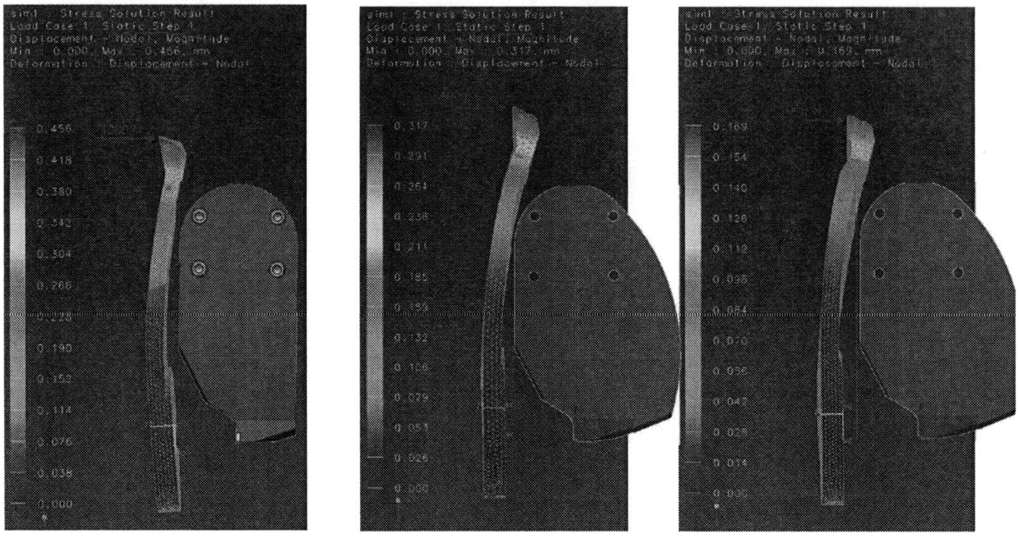

Base line | Taller - 30% stiffer | Material - 63% stiffer

Brian Hoefler & Mike Standridge

SANDVIK Coromant

Deep Grooving Slots – Engine Fandisk

Application Details

- *Tungsten alloy vs. steel*
  - *Material modulus close to 2x steel*
  - *Considerably less strength (40% less yield)*
  - *Considerably reduced fracture toughness (cracking prone)*
  - *Damping characteristics*

- *Significant gains can be made by using stiffer material*
- *Significant gains can be made by adding height*
- *Damping disks contribute to blade deflection regardless of placement*

Brian Hoefler & Mike Standridge

SANDVIK Coromant

Extended Length Finishing – Frame Structure

Application Details

- *416 RPM*
- *12 IPM*
- *.020" radial cut*
- *Flood coolant*
- *Ti6Al4V forging*

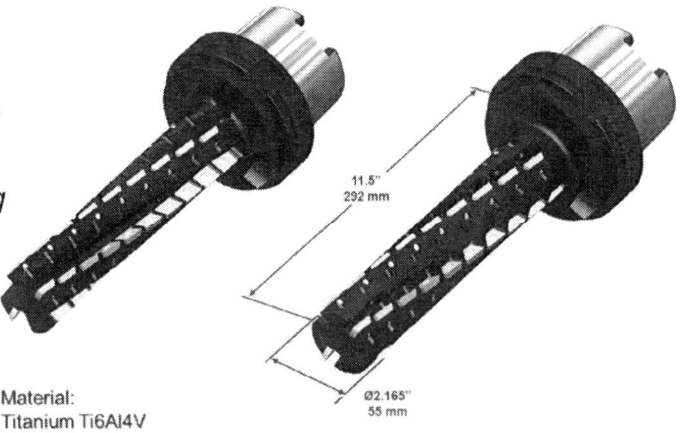

Material:
Titanium Ti6Al4V

Extended Length Finishing – Frame Structure

Application Details

- Eliminating of flutes results 29% in reduction of deflection or increase rigidity

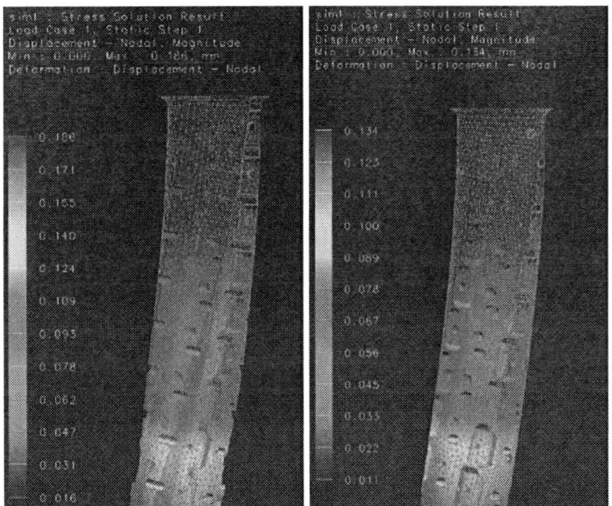

Practical Adaptive Machining
2014 Aerodef Conference

AERODEF MANUFACTURING produced by sme

THE VALUE OF PERFORMANCE.
NORTHROP GRUMMAN

Northrop Grumman Aerospace Systems

February 26, 2014

Presenter: Dan Braley, ManTech Development Engineer

Contributors: Dale Perkins (Manufacturing Engineering), Bill Louder (JWST Test Engineering), Jason Heron (ManTech Engineering), Ben Megel (Mgr Site 3 ManTech Engineering), Todd Szallay (Mgr NGAS ManTech Engineering)

Outline

NORTHROP GRUMMAN

- What is Adaptive Machining & Why is it Needed?
- Background on Vector Measurement and Kellering Approach
- Overview of Vector-Based Models
  - CATIA Cutter Contact Point Model
  - Seek and Destroy Targets Model
- NGAS Adaptive Machining Process Overview
- Effect of the Net Profile Machine Path Vector Shift (Adaptive Machining)
- NGAS Machining Data Flow (With and Without Adaptive Machining)
- Digital Thread & the Primary Goal of Adaptive Machining
- Digital Thread Benefits
- Application of Adaptive Machining Technology to Other Automated Processes
- Questions

1

What is Adaptive Machining & Why is it Needed?

NORTHROP GRUMMAN

- **Practical Adaptive Machining Definition:**

  A machining technology using current, off-the-shelf metrology equipment, that can sense the key characteristics of as-built detailed parts and assemblies, and make real-time machine compensations/adjustments to produce products correctly the first time

- **History:** Yoram Koren from the University of Michigan published a paper on adaptive controls for machining processes in 1989 through ASME

- **Why Are We Still Interested in Adaptive Machining?:**

  Per the Defense-Wide Manufacturing Science & Technology (DMS&T) Organization

  - As complexity of composites increases to enable performance, machining has become increasingly more difficult, resulting in the following challenges:
    - Bulk residual stresses from the composite production process cause component distortion as material is removed
    - As composites tend to deflect under the clamping forces induced during fixturing for machining and inspection processes, thin-walled parts are produced
    - Increased part-to-part differences vs. metallics
  - Machining issues result in longer cycle times, higher cost and scrap rates, problems meeting production rates, large amounts of Work in Process, and increased touch labor
  - Current deficiencies are in-process sensing, monitoring, and availability of correcting methodologies for errors

2

Background – Vector Measurement and Kellering

NORTHROP GRUMMAN

- A vector based model is created when a complex, multiple intersecting curved surface is measured after being machined using the milling technique of Kellering
- The vector based measurement model duplicates the cutting tool's normal vector as opposed to the cutting tool's centerline axis vector
- This vector based modeling technique eliminates an induced cosine measurement error and produces a more accurate dimensional feature measurement value result

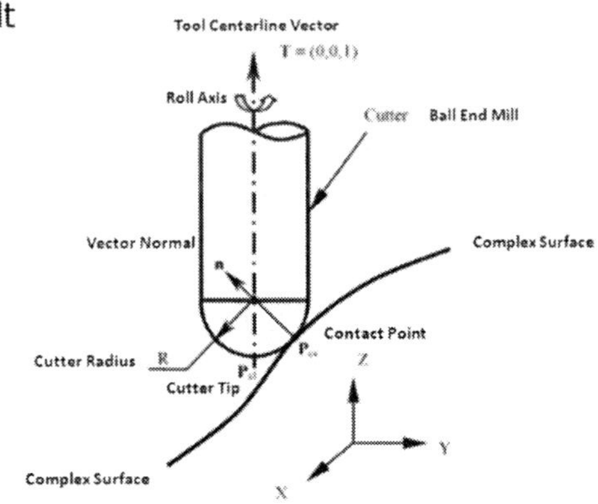

3

Overview of Vector-Based Models

NORTHROP GRUMMAN

2 Types of Vector-Based Models:

CATIA Cutter Contact Point Model (.CATPart File Output)

***"Seek & Destroy"* Targets Model (.mcx File Output)**

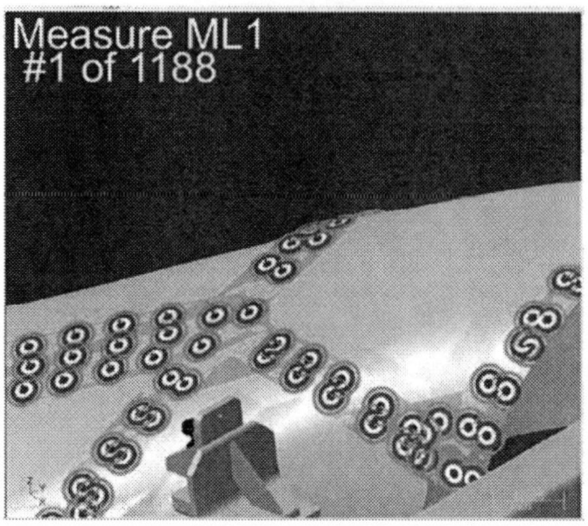

4

NGAS Adaptive Machining (AM) Process

NORTHROP GRUMMAN

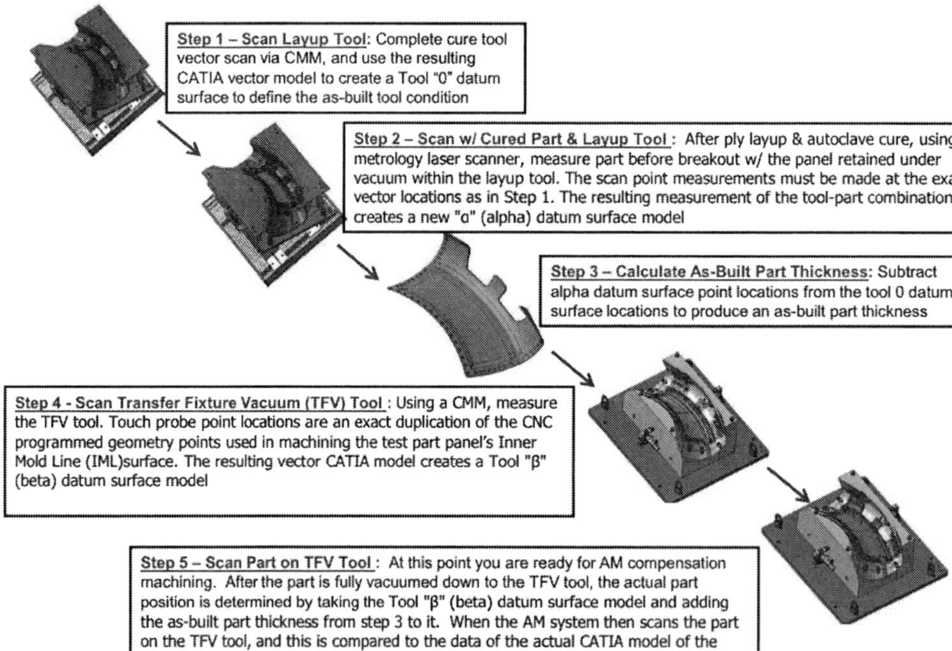

Step 1 – Scan Layup Tool: Complete cure tool vector scan via CMM, and use the resulting CATIA vector model to create a Tool "0" datum surface to define the as-built tool condition

Step 2 – Scan w/ Cured Part & Layup Tool: After ply layup & autoclave cure, using a metrology laser scanner, measure part before breakout w/ the panel retained under vacuum within the layup tool. The scan point measurements must be made at the exact vector locations as in Step 1. The resulting measurement of the tool-part combination creates a new "α" (alpha) datum surface model

Step 3 – Calculate As-Built Part Thickness: Subtract alpha datum surface point locations from the tool 0 datum surface locations to produce an as-built part thickness

Step 4 - Scan Transfer Fixture Vacuum (TFV) Tool: Using a CMM, measure the TFV tool. Touch probe point locations are an exact duplication of the CNC programmed geometry points used in machining the test part panel's Inner Mold Line (IML)surface. The resulting vector CATIA model creates a Tool "β" (beta) datum surface model

Step 5 – Scan Part on TFV Tool: At this point you are ready for AM compensation machining. After the part is fully vacuumed down to the TFV tool, the actual part position is determined by taking the Tool "β" (beta) datum surface model and adding the as-built part thickness from step 3 to it. When the AM system then scans the part on the TFV tool, and this is compared to the data of the actual CATIA model of the part, a compensation or shift can be made to the machine path

5

Vector Shift of Net Profile Machine Path Using Adaptive Machining

NORTHROP GRUMMAN

Prior to Adaptive Machining Compensation (Ideal Scenario But Not Reality)

Effect of Adaptive Machining Compensation to Tool Cutting Path (Vector Shift for First Time Part Quality)

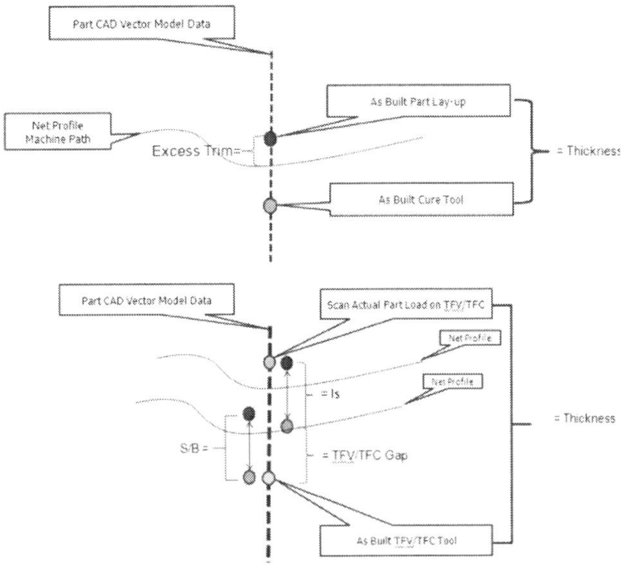

6

NGAS Adaptive Machining Data Flow

NORTHROP GRUMMAN

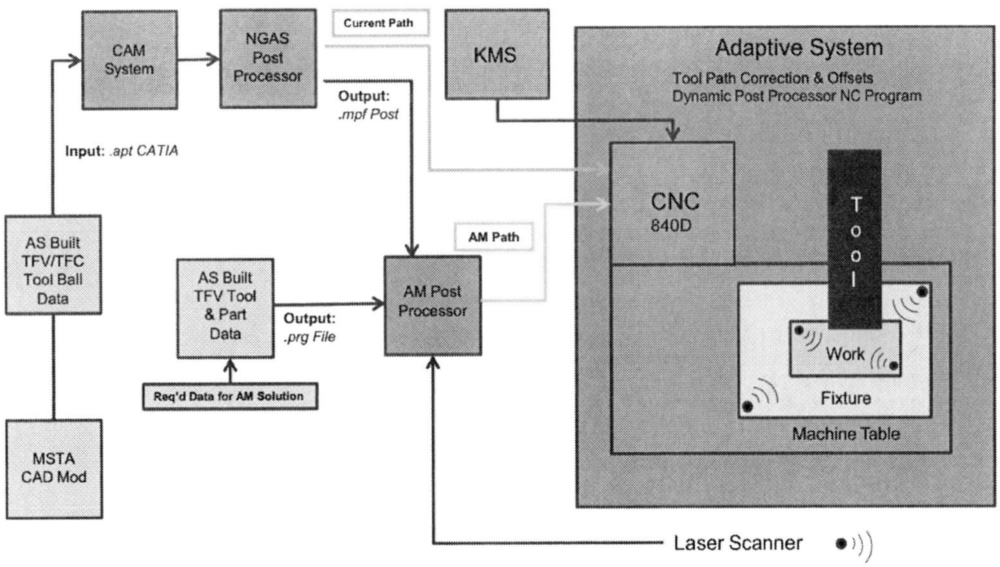

7

Digital Thread & the Primary Goal of Adaptive Machining (AM)

NORTHROP GRUMMAN

- **Government Definition of Digital Thread:**

  The Digital Thread is the creation and use of a digital surrogate of a material system to allow dynamic, real-time assessment of the system's current and future capabilities to inform decisions across all acquisition phases

- **Goal of Adaptive Machining at NGAS:**

  Develop machining technologies using current, off-the-shelf metrology equipment, that can sense the key characteristics of as-built detailed parts and assemblies, and make real-time machine compensations/adjustments to produce products correctly the first time

- **How does Adaptive Machining Help Achieve Digital Thread Goals?:**
  - The same inspection points are used from step 1 to step 5 from defining the inspection points in the part CATIA model, to scanning the tools, to scanning the part on the tools
  - If a vector point's location is off, the AM system will adjust the machining profile to the actual part position
  - Knowing the actual thickness of the part prior to machining (from previous retained scan data), and the ultimate thickness goal (located in the retained CATIA model data), an accurate assessment of how much material needs to be machined is determined
  - AM increases the chances of falling in tolerance the first time, and reduces rework

8 The use of Digital Thread methodology in the Adaptive Machining process enables first-time part quality, reducing tags, reducing product costs, and improving schedule

NGAS Digital Thread Process & Benefits

NORTHROP GRUMMAN

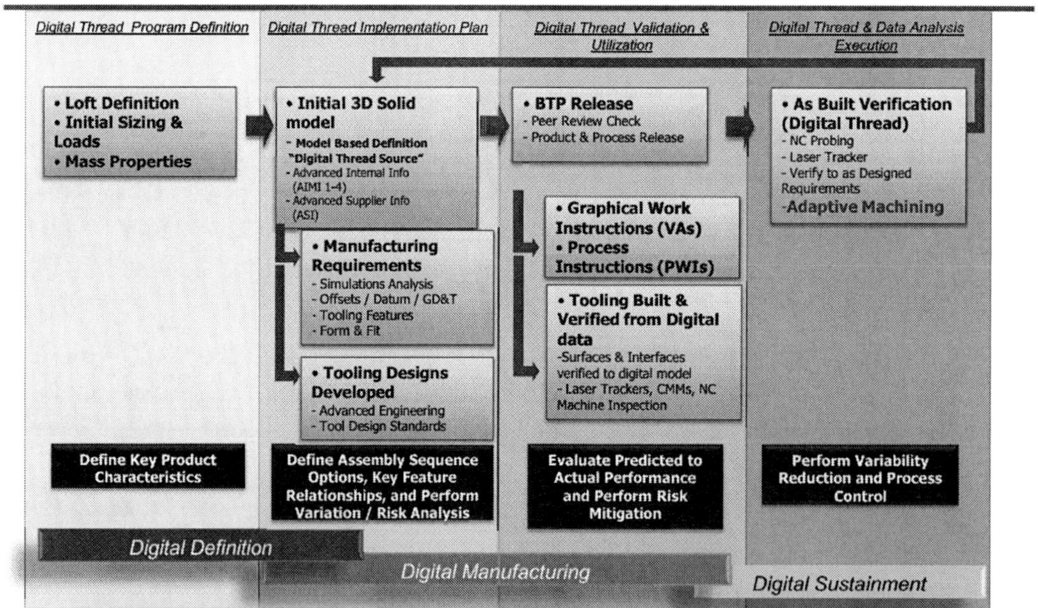

The push towards Digital Thread can significantly reduce part manufacturing costs, and ultimately help to reduce military systems total life cycle costs

Application of Adaptive Machining Methodology to Other Automated Processes

NORTHROP GRUMMAN

- **Automated Sanding:**

  Manual sanding is labor intensive and prone to ergonomic strain. NGAS, in conjunction with Aerobotix and the AFRL, is in the process of developing an Automated Sanding solution that will theoretically reduce cycle time to sand performance coatings as well as the manufacturing costs associated with sanding operations

- **Robotically Applied Coatings:**

  Thickness of coatings is highly effected by the application process, including the angle of spray, the distance of the spray nozzle from the part surface, etc. Small changes in part x,y, or z location can significantly effect the effectiveness and correct thickness of the applied coatings. An approach like that of the Adaptive Machining vector shift compensation could have a positive effect on robotically applied coating processes

10 The Adaptive Machining approach can be tied to many automated processes, and can significantly improve first time quality and associated manufacturing costs

NORTHROP GRUMMAN

Questions?

THE VALUE OF PERFORMANCE.

NORTHROP GRUMMAN

Automated drilling and countersinking of composite titanium material stacks one shot.

"All things are difficult before they are easy." –Thomas Fuller

Abstract

This paper will describe in detail the machine end effector physical components, algorithm including subroutines, drill set lengths, speeds and feeds necessary to drill through complex material stacks of varying thickness "one shot". The presentation will provide the necessary data and information including pecking subroutines, chip removal, and drill types to successfully drill close tolerance holes with diameters from .187 inch (4.75 mm) to .500 inch (12.7 mm) with tolerance of -.000 to +.001 inch (-.000 mm to .025 mm) through stacks of dissimilar material of up to 1.25 inch thick including complex material stacks of composite (GFE)-titanium-composite (GFE).

Introduction

On-assembly automated drilling and countersinking is the application of specialized Numeric Control (NC) machinery into the manufacturing process to drill and countersink a finished hole "one shot" for insertion of the fasteners that holds the airframe together. Applying automation replaces the traditional hand methods for drilling and countersinking.

There can be well over 100,000 holes that need to be drilled and countersunk in a modern military fighter/attack airframe and more than 2,000,000 in a large commercial airframe. Large commercial and military airframes can exceed 3 million holes. The number of holes and the hand method for drilling and countersinking drives cost, quality, schedule, complexity, and safety. Sixty-five percent of the cost of airframe assembly is derived from drilling and countersinking holes by hand. Eighty-five percent of the quality issues and eighty percent of the lost time injuries are the result of the hand process for drilling and countersinking. (Bullen, 2013)

This paper will focus on the end effector and algorithms specifically designed for the automated drilling and countersinking of airframes. The other components of on-assembly drilling and countersinking such as controllers, software, and machine hardware and mode of operation derive from mature technologies that have been well documented, presented and published. Rarely have the operational components, their relationship to the work-piece, and the sequence of commands necessary to acquire and successfully perform the drill and countersink operation been presented or published. This is because of proprietary and intellectual property concerns of the machine originators and operators to maintain a perceived or real competitive position.

End Effector

Once the machine motion is established an action must be performed to accomplish the drill and countersinking operations. The actions necessary to carry out all the operations associated with drilling and countersinking a hole are performed by an end effector.

Parts are combined to form assemblies. They are held in position by tools while holes are drilled and countersunk. Fasteners of differing types are installed into the holes. The assembly then becomes a stand-alone component of the airplane. Skin thickness in metallic parts is very uniform as a product of its manufacturing method. Skins made with composite materials vary in thickness based on the number of plies that are used in their fabrication. A .375 inch (9.53 Millimeter) thick part can vary from .345 inch (8.76 millimeter) to .405 inch (10.29 millimeter), with thickness variability increasing in proportion to the part thickness. Therefore the end effector must be designed to closely replicate the actions of the hand operation by sensing the variability of the part thickness and compressing (compaction) the pieces of the assembly together. Sensing the precise skin surface location is an essential part of the end effector design and operation.

The end effector performs a complex set of actions that are coordinated to successfully drill and countersink a hole. An end effector's functions must be packaged small enough to access holes in and around structure and to minimize weight that adds to deflection at the tool tip. Actions that the end effector performs include holding the tool that drills and countersinks, vacuum to collect dust, sensors to determine over-limits, scales to provide feedback, a pressure-foot to compress the work-piece, and a coolant distribution system. Figure 1 shows an end effector used for drilling and countersinking holes on airframe assemblies with the pressure foot closed and figure 2 shows an end effector with the pressure foot open to facilitate a tool change.

Figure 1 End effector with pressure foot closed

Figure 2 End effector with pressure foot open

Size

The access required to acquire as many holes as possible places design pressures on end effector engineers. In many cases, the end effector must be designed small enough to fit into tight spaces. And minimizing size also provides more flexibility and freedom when moving in and around parts, tooling, and obstructions. The design requires size minimization while including all the attributes necessary to carry out the requirements of drilling and countersinking. Included in the design requirements are the hole evaluation requirements to maintain continuous speeds and controlled feeds through the material designated for the end effector.

The speeds and feeds regime the end effector must perform are high speed to low speed and high torque. While minimizing size, the design engineer must also provide a strong enough motor to supply continuous power and torque to the spindle for the stacked materials it drills and countersinks. This includes high speeds and feeds for drilling through composites and low speeds with high torque when used for drilling through titanium and steel. The positional accuracy and diameter requirements for holes drive the requirement for a motor that can sustain operations without degrading performance or accuracy due to heat.

Capabilty

The capability of the end effector includes a quill that feeds out and is integrated with a pressure foot attached to a measurement device that provides feedback to the controller. As the quill is fed out, the pressure foot advances and touches the structure. Upon touching the structure the scale reader/linear encoder (figure 3) is compressed alerting the controller to spin the motor and begin the drill countersink routine for that position. The mechanical detection of linear movement is done by a front guided push rod with inside spring return device. To reach high accuracy the push rod is guided with precise sleeve bearings. The precise sleeve bearings are placed inside the shaft. On the push rod an accurate glass scale is mounted. Inside the sensor housing the glass scale passes an electronic board with optical detector which generates counting pulses. Due to the glass material the measurement is very temperature stable. The scale continues to send signals to the controller as it progresses through the material stacks changing speed and feed rates as different materials are encountered. In addition, the scale feedback mechanism alerts the controller to activate a pecking routine to clear chips. A pecking routine drills through material to a specified depth and rapidly retracts so that the drill is clear of the hole to clear chips as the result of drilling. Once the chip is cleared the drill reenters the hole and begins drilling at a point just short of where it left off before retraction. This is to compensate for drill pressure deflection of the combined work-pieces and prevents impact when the drill returns to re-engage the remaining material.

While not as glamorous as all the other capabilities of the end effector, it also includes a lighting component for illuminating the work-piece during operation.

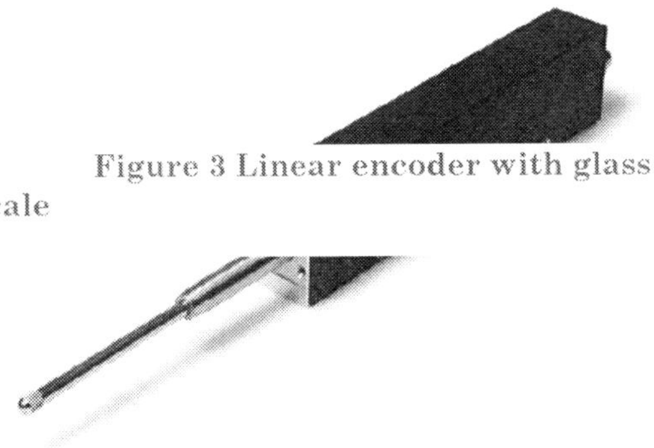

Figure 3 Linear encoder with glass scale

Pressure Foot

The pressure foot is the finger that touches the surface of the part. The pressure foot performs several functions. It is a vibration-absorbing mechanism of an automated drill and countersink machine having a shock to mitigate communications of vibrations to the drill tool, while simultaneously dampening the vibrations at their source, in order to increase control of the drill tool during drilling and countersinking operations. (Bullen, 2001). The pressure foot of the end effector is upwardly or outwardly rotatable to access the drill tool to the chuck.

The pressure foot also has a pneumatically or hydraulically pressurized cylinder to facilitate pressure against the part and substructure to minimize separation.

The end effector must be designed as to allow airflow in and around the spindle for chip evacuation. The end effector when installed is a precision removable (for maintenance) part of the machine where its length is coordinated with the scale feedback system.

Tool holders come in different configurations and sizes. Precision tool holders are used to clamp the tool bit and countersink. They are inserted into the machine and their size is dependent on the size of the end effector and the size of the drill they will hold. The tool holders are preset in the tool setup cribs to a precise depth to insure the correct activation of speeds and feeds as the machine progresses through the material stacks. The drill tip is set in relationship to the surface of the pressure foot. The countersink is also set in relationship to the pressure foot surface. This is done to ensure the machine "knows" when it has touched the surface of the product, precisely how far it is before the drill impacts the product.

Algorithm

Gottfried Wilhelm von Leibniz (July 1, 1646 – November 14, 1716) was a German mathematician and philosopher. He occupies a prominent place in the history of mathematics and the history of philosophy. Leibniz developed the infinitesimal calculus independently of Isaac Newton, and Leibniz's mathematical notation has been widely used ever since it was published. He became one of the most prolific inventors in the field of mechanical calculators. He also refined the binary number system, which is at the foundation of virtually all digital computers. By most accounts he is the father of the modern algorithm.

The use of the algorithm in automated drilling and countersinking is best described by David Berlinski as "A finite procedure, written in a fixed symbolic vocabulary, governed by precise instructions, moving in discrete steps, 1, 2, 3, . . ., whose execution requires no insight, cleverness, intuition, intelligence, or perspicuity, and that sooner or later comes to an end." (Berlinski, 2001)

In this paper I have reduced the drill and countersinking algorithms to narrative for clarity. I have also added a model of one algorithm (Figure 4) to provide a visual representation to the narrative. There are variables associated with drill and countersink algorithm application such as drill bit types, coolant application, and hole diameter. Also, algorithms as defined by Berlinski are absolutes that are aggravated by the variability of the assembled airframe and dissimilar materials. Therefore before activating the drilling algorithm a sensitivity analysis is integrated into the process to "acquire" the surface of the work-piece to be drilled as a precursor to the drill bit engaging the material of the airframe. Finding and quantifying the location of the work-piece surface is the first step for every hole to be drilled by automated systems.

Drill (Same Material)

Position Motion (X,Y,Z,I,J,K)

- Position vector – normal to surface.

DRILL ROUTINE (ALGORITHM) – SAME MATERIAL

- Signal start ((Locks-out four axis motion (X,Y,A,B) mechanical and software))
- Approach skin (Machine 5th Axis)
- Pressure foot compression window achieved 25/10000 = .003" (.076 millimeter)
  - ❖ Internal position set at 0 = -.003" (-.076 mm) from pressure foot front surface to drill tip.
  - ❖ Drill tip offset .250" (6.35 mm) to front surface of pressure foot.
- Lock fifth-axis (mechanical and software)
- Activate spindle RPM (4000), mister, vacuum, and approach skin (spindle) at 1.5 IPM (38.1 MMPM).
- Drill hole @ constant speed one shot.
  - ❖ Hole should exit backside of material .100 inch (.2.54 mm) past the end of the drill tip.
  - ❖ Calculation for exit criteria should be engineering material thickness plus six percent to account for variability.
- Dwell .5 second.

- Return spindle to zero position. Turn off spindle and mister.
- Unlock fifth axis
- Return fifth axis to tracking position.
- Dwell .5 seconds and turn off vacuum.
- Signal operation complete (Unlocks machine four-axis of motion).

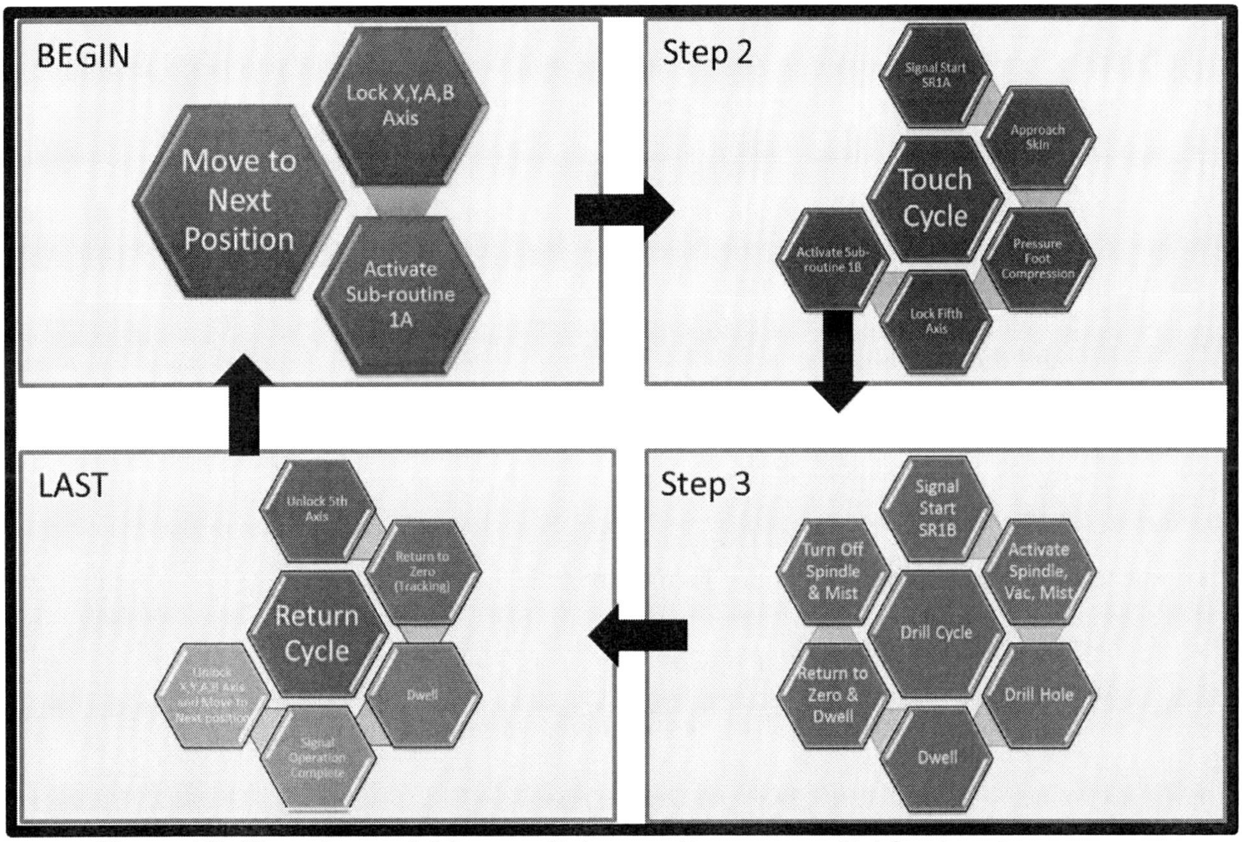

Figure 4 Drill routine model (composite/composite or composite/aluminum)

Drill & Countersink (Composite/Titanium Stack)

Position Motion (X,Y,Z,I,J,K)

- Position vector – normal to surface.

DRILL/COUNTERSINK ROUTINE (ALGORITHM) – COMPOSITE/TITANIUM WITH PECK

- Signal start (Locks-out machine four-axis of motion)
- Approach skin (Machine 5th Axis)
- Pressure foot compression window achieved 25/10000 = .003" (.076 millimeter)
  - ❖ Internal position set at 0 = -.003" (-.076 mm) from pressure foot front surface to drill tip.
  - ❖ Drill tip offset .250" (6.35 mm) to front surface of pressure foot.
- Lock fifth-axis (mechanical and software)
- Activate spindle RPM (4000), mister, vacuum and approach skin at 1.5 IPM (38.1 MMPM).
- Drill hole one-shot in composite material.
  - ❖ Reduce speed and feed at low end of composite material stack tolerance.

- ❖ Example: thickness -6% (1 inch reduce speed at .940 inch drill penetration). (25.4 mm reduce speed at 23.876 mm drill penetration)
- ❖ Continue drilling at reduced speed (450 RPM – titanium) and feed IPM through remaining composite material.
- ❖ Continue drilling into titanium at lower IPM and RPM to end of drill tip
- Rapid retract spindle to zero set point
- Rapid return (into hole) to last end point minus .050 inch (1.27 mm).
  - ❖ This action compensates for drill pressure caused spring-back that reduces "bumping")
- Continue drilling at titanium IPM and RPM for .100 to .150 inch (2.54 to 3.81 mm)
- Rapid retract spindle to zero set point
- Rapid return (into hole) to last end point minus .050 inch (1.27 mm).
- Continue peck drilling at titanium IPM and RPM for .100 to .150 inch (2.54 to 3.81 mm)
- Continue process until drill exits backside of titanium material .100 inch (.2.54 mm) past the end of the drill tip.
  - ❖ Calculation for exit criteria should be engineering material thickness plus six percent to account for variability.
- Rapid to preset countersink transition "tip"
- Slow spindle to countersink specified RPM and IPM.
- Complete countersink to depth.
- Dwell .5 second.
  - ❖ Dwell cleans up residue and polishes countersink.
- Return spindle to zero position. Turn off spindle and mister.
- Unlock fifth axis
- Return fifth axis to tracking position.
- Dwell .5 seconds and turn off vacuum.
- Signal operation complete (Unlocks machine four-axis of motion).

DRILL/COUNTERSINK ROUTINE (ALGORITHM) - COMPOSITE/TITANIUM/COMPOSITE WITH PECK

- Signal start (Locks-out machine four-axis of motion)
- Approach skin (Machine 5th Axis)
- Pressure foot compression window achieved 25/10000 = .003" (.076 millimeter)
  - ❖ Internal position set at 0 = -.003" (-.076 mm) from pressure foot front surface to drill tip.
  - ❖ Drill tip offset .250" (6.35 mm) to front surface of pressure foot.
- Lock fifth-axis (mechanical and software)
- Activate spindle RPM (4000), mister, vacuum and approach skin at 1.5 IPM (38.1 MMPM).
- Drill hole one-shot in composite material.
  - ❖ Reduce speed and feed at low end of composite material stack tolerance.
  - ❖ Example: thickness -6% (1 inch reduce speed at .940 inch drill penetration). (25.4 mm reduce speed at 23.876 mm drill penetration)
  - ❖ Continue drilling at reduced speed (450 RPM – titanium) and feed IPM through remaining composite material.
  - ❖ Continue drilling into titanium at lower IPM and RPM to end of drill tip
- Rapid retract spindle to zero set point
- Rapid return (into hole) to last end point minus .050 inch (1.27 mm).

- ❖ This action compensates for drill pressure caused spring-back that reduces "bumping")
- Continue peck drilling at titanium IPM and RPM for .100 to .150 inch (2.54 to 3.81 mm)
- Continue process until drill exits backside of material .100 inch (.2.54 mm) until engineering defined thickness for first composite/titanium stack.
  - ❖ Calculation for stack criteria should be engineering material thickness plus six percent to account for variability.
- Continue drilling through titanium into composite material to end of drill tip.
- Increase spindle speed to RPM (4000) at 1.5 IPM (38.1 MMPM).
- Drill hole @ constant speed one shot.
  - ❖ Hole should exit backside of material .100 inch (.2.54 mm) past the end of the drill tip.
- Rapid to preset countersink transition "tip"
- Slow spindle to countersink specified RPM and IPM.
- Complete countersink to depth.
- Dwell .5 second.
  - ❖ Dwell cleans up residue and polishes countersink.
- Return spindle to zero position. Turn off spindle and mister.
- Unlock fifth axis
- Return fifth axis to tracking position.
- Dwell .5 seconds and turn off vacuum.
- Signal operation complete (Unlocks machine four-axis of motion).

Summary

While every other aspect of airplane manufacturing was taking advantage of the advances in computer controlled machine technology, until recently airframe assembly resisted the move away from hand processes such as drilling and countersinking. There were many reasons for the resistance to remove the mechanic from the airframe assembly process. The mechanic had the touch, sight, and ability to adjust and adapt to the variability of the complex array of very expensive work-pieces placed together and held during the drilling, countersinking, and fastening operations. The ability of the mechanic to place their tools directly against the skin of the assembly, feel the resistance change when the drill bit went through one material into the other, and continually check and adjust the tools of their trade gave them the advantage. The use of a countersink cage that could be normalized to the surface and reapplied if additional depth was needed added to the challenge to automate the process.

The encoder sensitizes the pressure foot and enables the machine to identify the location of the skin and dampen vibration to provide data for precision drilling of one shot holes.

Replicating the ability to identify the location of the skin in relationship to the drill tip was an essential first step to automating the process. Drilling holes through stacks of material where the substrate material is harder than the surface material such as in a composite skin over a titanium bulkhead was a bigger challenge. The harder sharpened chip is drawn through the softer composite material and can tear and open the hole in the skin to an unacceptable dimension. The precise pecking routine described in this paper solves the problem with very little or no difference in the diameter of the outer composite skin and the composite under-structure.

Countersinking became the final challenge to automating the drilling and countersinking process and added a level of complexity that still frustrates a level of success that drilling by machine or robot has realized. The challenges are derived from many sources including detritus residue left from the previous hole that underlies the pressure foot and defeats its ability to contact the skin. The pressure foot radius is often greater than the span between holes leaving overlaps of debris reminiscent of a Venn diagram. The residue thickness sends a false signal to the controller for depth control resulting in progressively shallow countersinks as the residue accumulates. In the hand operation the mechanic simply wiped away the residue from the previous countersink and cleaned the surface of the countersink cage before each hole was countersunk.

Many solutions have been designed, tested, and transitioned to production with varying success. The ability of the machine to provide a near depth countersink still provides economic value as shallow countersinks can be touched-up by the mechanic to account for fastener variability and shallow machine made countersunk holes.

The end effector attributes, functionality, and operation as directed by the appropriate algorithm described in this paper are valid operational elements necessary to complete the on-assembly drilling and countersinking of airframes.

References

Bullen G.N. *Automated/Mechanized Drilling and Countersinking of Airframes*. SAE International, Warrendale, PA, 2013. ISBN: 976-0-7680-7646-2.

Bullen, George N., *Vibration-absorbing end effector of an automated drill and countersink machine*, US Patent number: 6231280, Filing date: Aug 10, 199I. Issue date: May 15, 2001.

Berlinski D. *The Advent of the Algorithm: The 300-Year Journey from an Idea to the Computer.* Harvest Books, New York, 2001. ISBN: 0-15-601391-6.

Damer, T. D. *Attacking Faulty Reasoning: A Practical Guide to Fallacy-Free Arguments* (3rd. ed.). Wadsworth Publishing, Belmont, CA. 1995, pp. 131. ISBN 978-0-534-21750-1. OCLC 30319422.

Figure Captions

Figure 1 End Effector (example)-pressure foot closed (Courtesy of M Torres).
Figure 2 End Effector (example)-pressure foot open (Courtesy of M Torres).
Figure 3 Linear encoder (example)
Figure 4 Drill routine algorithm model (composite/similar material) (Bullen, 2013)

Key Terms & Definitions

Algorithm: A finite procedure, written in a fixed symbolic vocabulary, governed by precise instructions, moving in discrete steps, 1, 2, 3, . . ., whose execution requires no insight, cleverness, intuition, intelligence, or perspicuity, and that sooner or later comes to an end. (Berlinski, 2001)

End Effector: An end effector is a device or tool that's connected to the end of a robot arm where the hand would be. The end effector is the part of the robot or machine that interacts with the environment. The structure of an end effector and the nature of the programming and hardware that drives it depend on the task the robot will be performing.

One Shot: One shot means drilling a finished hole in a single operation without pilot holes or tool changes. The hand method of drilling and countersinking requires multiple steps and tool changes. (Bullen, 2013)

Peck: Peck drilling involves plunging the drill part way through the work-piece, no more than five times the diameter of the drill, and then retracting it to or beyond the surface to remove the chips.

Author

George N. Bullen is the author of the recently released book titled, *Automated/Mechanized Drilling and Countersinking of Airframes*. He is a Fellow in the Society of Manufacturing Engineers (FSME) and Certified in Production and Inventory Management (CPIM). He has written over 100 articles that have been published in magazines and peer review journals on airframe assembly, automation, robotics, and composites fabrication. George retired from Northrop Grumman Corporation in 2011 as Principal Engineer, Technical fellow, Technical Expert for Advanced Programs and Manufacturing Technology Development. He is now President and CEO of Smart Blades Inc. located in Oxnard, California.

AERODEF MANUFACTURING

SUMMIT & EXPOSITION
February 25-27, 2014
LONG BEACH CONVENTION CENTER
LONG BEACH, CALIFORNIA USA

Composite Tooling Assembly Strategies

James Diedesch
Sr. Manufacturing Engineer

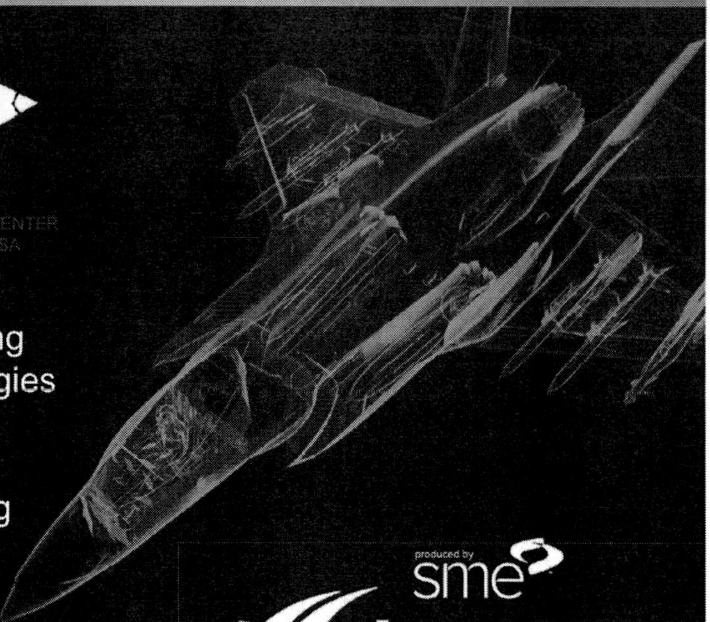

produced by sme

JANICKI INDUSTRIES

AERODEF MANUFACTURING

produced by sme

SUMMIT & EXPOSITION
February 25-27, 2014
LONG BEACH CONVENTION CENTER
LONG BEACH, CALIFORNIA USA

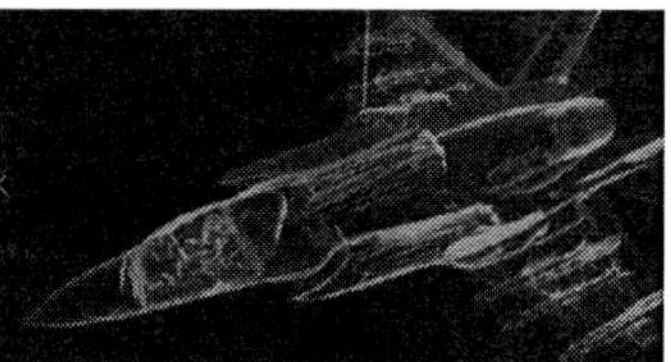

Composite Tooling Assembly

1. Room Temperature Patterns and Molds.
2. Intermediate Temperature / Light Production Tools.
3. High Temperature Interchangeable Tooling.

AERODEF MANUFACTURING
produced by sme

SUMMIT & EXPOSITION
February 25-27, 2014
LONG BEACH CONVENTION CENTER
LONG BEACH, CALIFORNIA USA

Room Temperature Patterns and Molds

AERODEF MANUFACTURING
produced by sme

SUMMIT & EXPOSITION
February 25-27, 2014
LONG BEACH CONVENTION CENTER
LONG BEACH, CALIFORNIA USA

Tooling For Mega Yachts

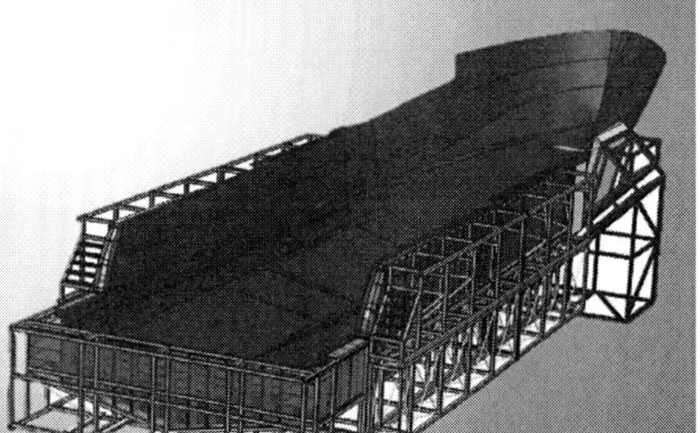

AERODEF MANUFACTURING
produced by sme

SUMMIT & EXPOSITION
February 25-27, 2014
LONG BEACH CONVENTION CENTER
LONG BEACH, CALIFORNIA USA

Tooling For Mega Yachts

AERODEF MANUFACTURING
produced by sme

SUMMIT & EXPOSITION
February 25-27, 2014
LONG BEACH CONVENTION CENTER
LONG BEACH, CALIFORNIA USA

Patterns for Large Molds

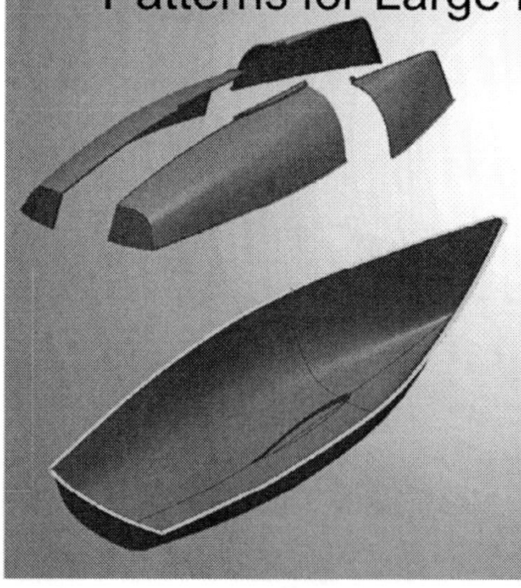

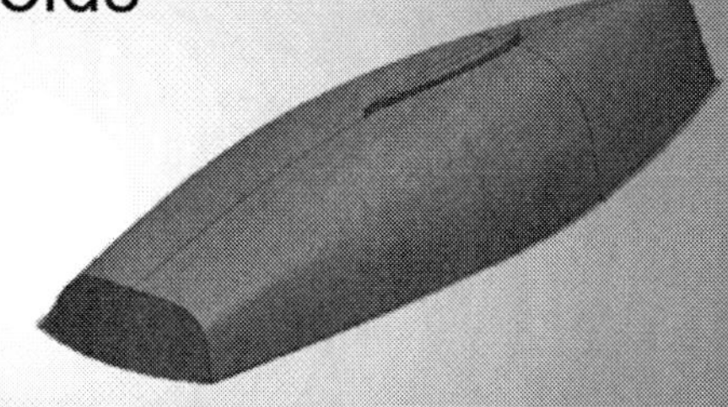

AERODEF
MANUFACTURING
sme
SUMMIT & EXPOSITION
February 25-27, 2014
LONG BEACH CONVENTION CENTER
LONG BEACH, CALIFORNIA USA
Simple Rebate Taping Joints

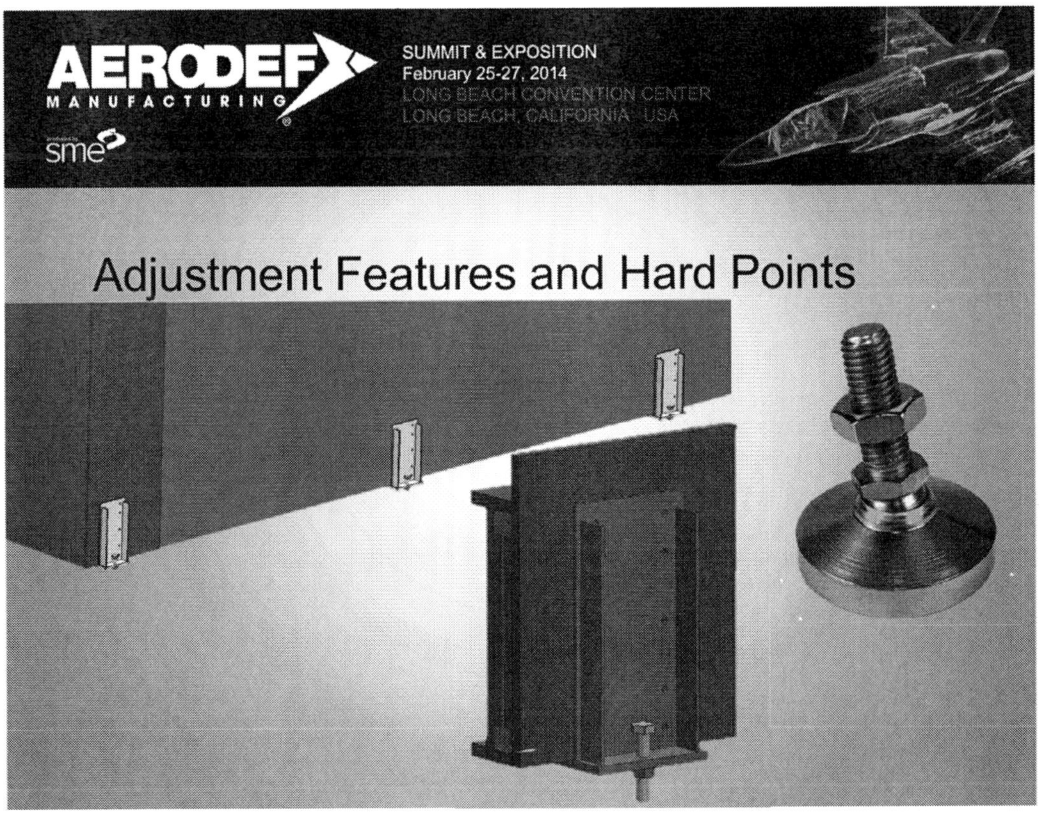
AERODEF
MANUFACTURING
sme
SUMMIT & EXPOSITION
February 25-27, 2014
LONG BEACH CONVENTION CENTER
LONG BEACH, CALIFORNIA USA
Adjustment Features and Hard Points

AERODEF MANUFACTURING
produced by sme

SUMMIT & EXPOSITION
February 25-27, 2014
LONG BEACH CONVENTION CENTER
LONG BEACH, CALIFORNIA USA

Water Line Leveling

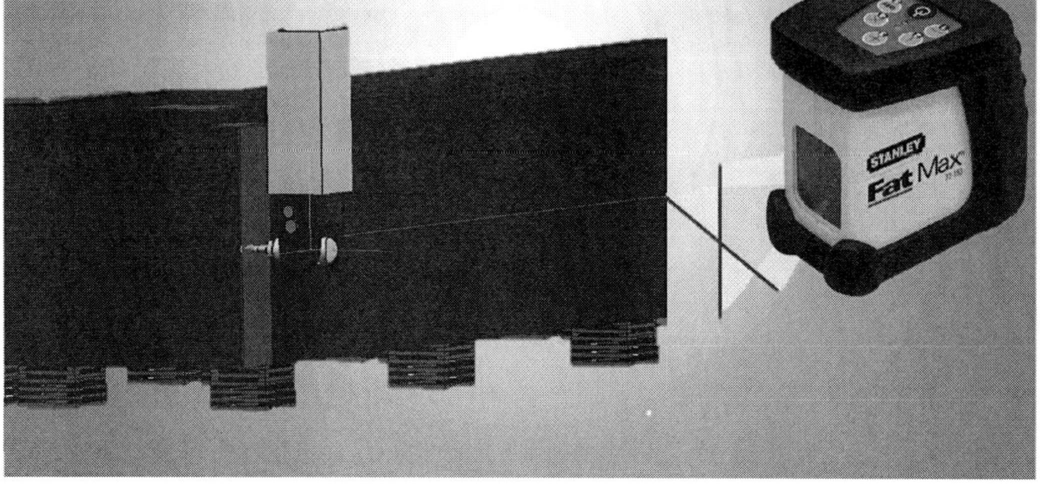

AERODEF MANUFACTURING
produced by sme

SUMMIT & EXPOSITION
February 25-27, 2014
LONG BEACH CONVENTION CENTER
LONG BEACH, CALIFORNIA USA

Intermediate Temperature and Light Production Tooling

AERODEF MANUFACTURING

SUMMIT & EXPOSITION
February 25-27, 2014
LONG BEACH CONVENTION CENTER
LONG BEACH, CALIFORNIA USA

Light Production Tools

AERODEF MANUFACTURING

SUMMIT & EXPOSITION
February 25-27, 2014
LONG BEACH CONVENTION CENTER
LONG BEACH, CALIFORNIA USA

Alignment Features

- Trapezoids
- Cones

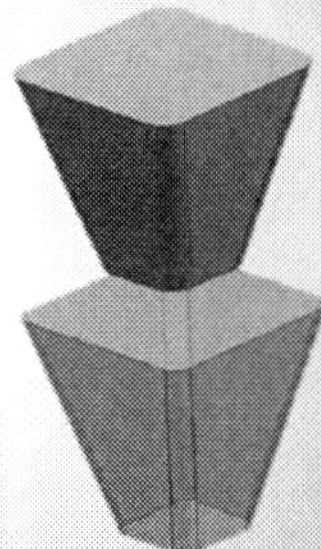

AERODEF MANUFACTURING
produced by sme
SUMMIT & EXPOSITION
February 25-27, 2014
LONG BEACH CONVENTION CENTER
LONG BEACH, CALIFORNIA, USA

Alignment Features

- Ball Bearings/Spheres
- Slotted Bushings

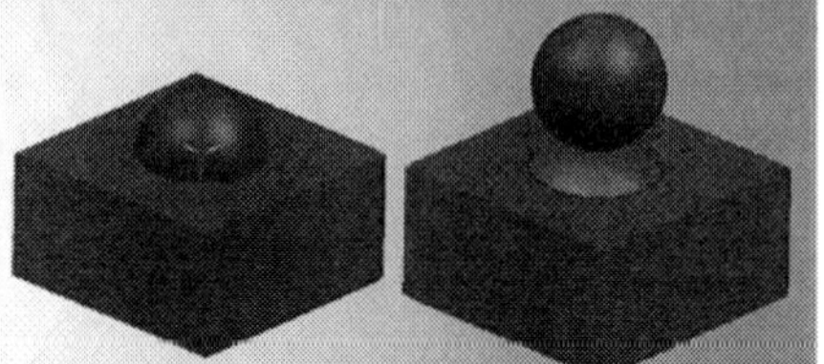

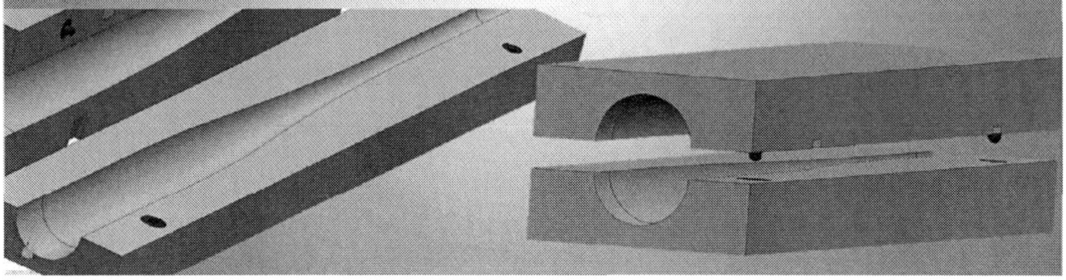

AERODEF MANUFACTURING
produced by sme
SUMMIT & EXPOSITION
February 25-27, 2014
LONG BEACH CONVENTION CENTER
LONG BEACH, CALIFORNIA, USA

High Temperature Composite Tooling with Interchangeable Segments

AERODEF MANUFACTURING
produced by sme

SUMMIT & EXPOSITION
February 25-27, 2014
LONG BEACH CONVENTION CENTER
LONG BEACH, CALIFORNIA USA

3-2-1 Method of Assembly

- 3- Fully Constrained with Load
- 2 Partially Constrained with Load
- 1-Partially Constrained without Load

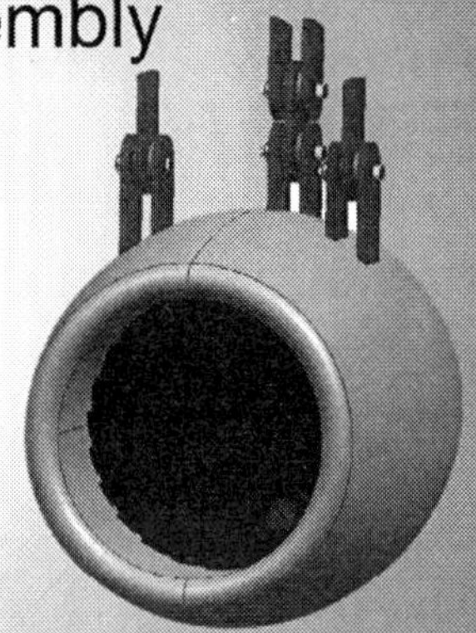

AERODEF MANUFACTURING
produced by sme

SUMMIT & EXPOSITION
February 25-27, 2014
LONG BEACH CONVENTION CENTER
LONG BEACH, CALIFORNIA USA

3- Fully Constrained with Load

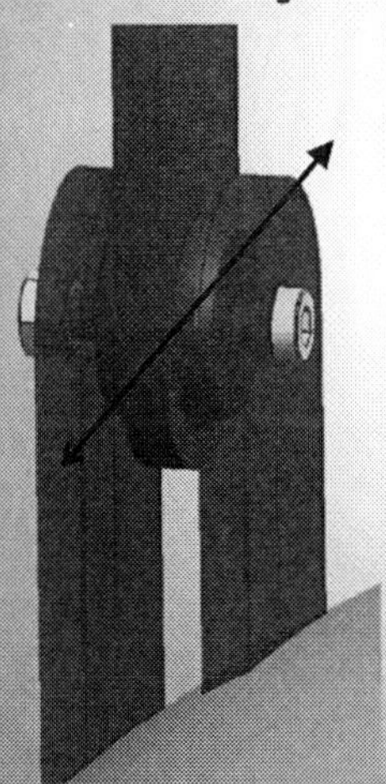

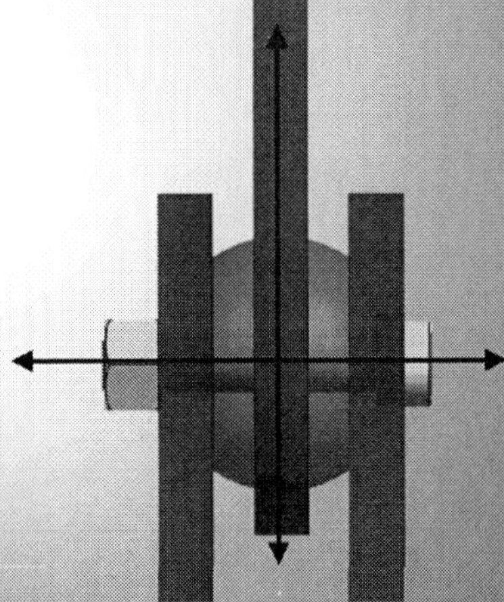

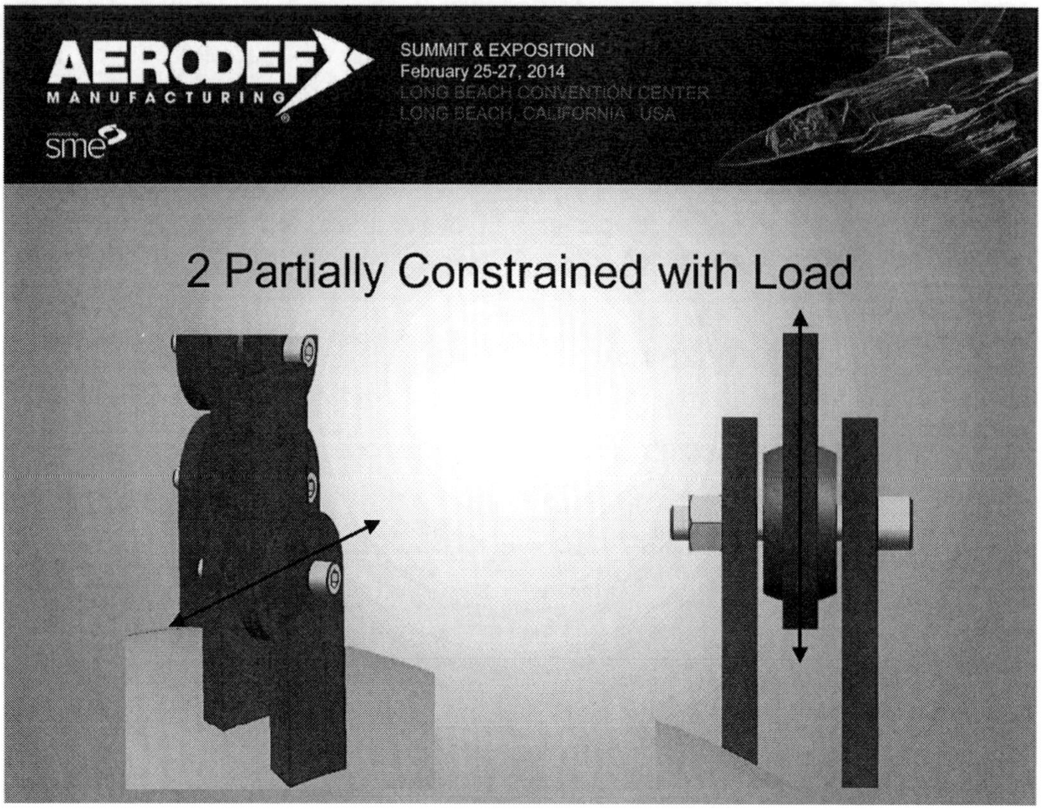
AERODEF
MANUFACTURING
sme
SUMMIT & EXPOSITION
February 25-27, 2014
LONG BEACH CONVENTION CENTER
LONG BEACH, CALIFORNIA USA
2 Partially Constrained with Load

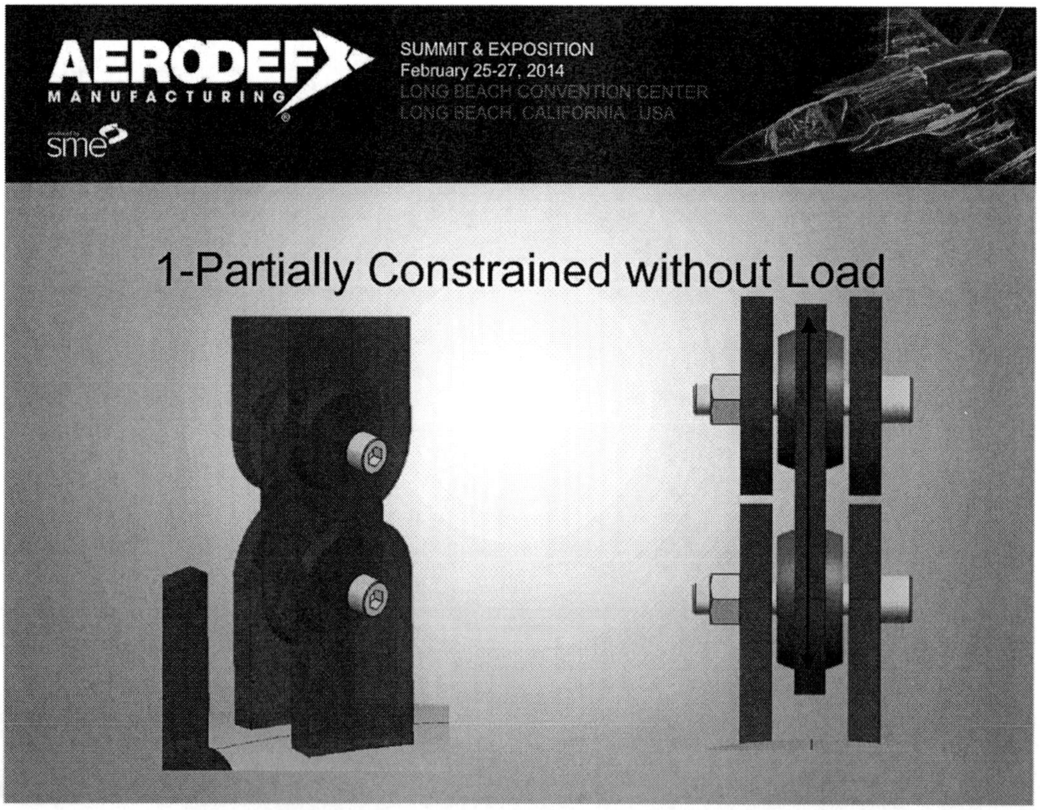
AERODEF
MANUFACTURING
sme
SUMMIT & EXPOSITION
February 25-27, 2014
LONG BEACH CONVENTION CENTER
LONG BEACH, CALIFORNIA USA
1-Partially Constrained without Load

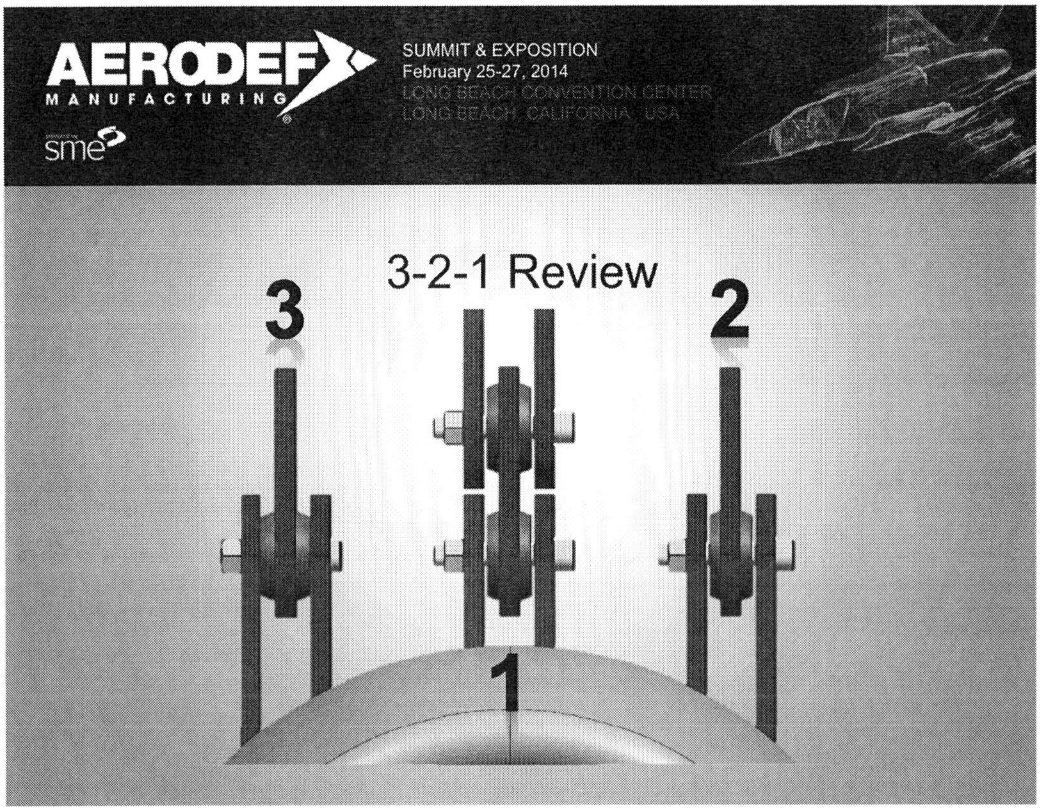
AERODEF
MANUFACTURING
sme
SUMMIT & EXPOSITION
February 25-27, 2014
LONG BEACH CONVENTION CENTER
LONG BEACH, CALIFORNIA USA
3-2-1 Review
3
2
1

AERODEF
MANUFACTURING
sme
SUMMIT & EXPOSITION
February 25-27, 2014
LONG BEACH CONVENTION CENTER
LONG BEACH, CALIFORNIA USA
Level 3 Assembly with
Sphere and Cones

Finding the Center

With Sphere and cone, location needs to control only one location, the center of the Spherically Mounted Retroreflectors (SMR)

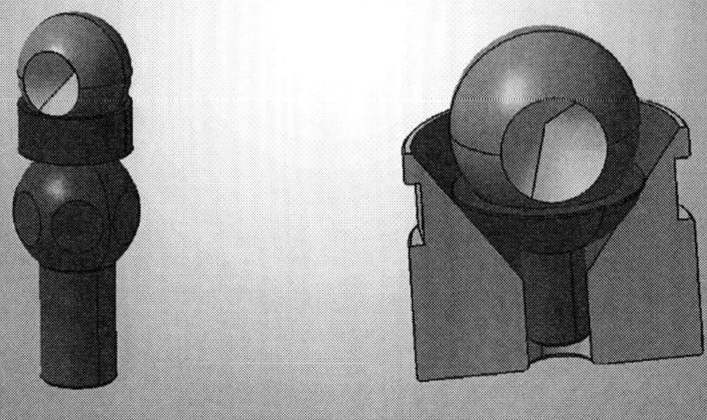

AERODEF MANUFACTURING
sme
SUMMIT & EXPOSITION
February 25-27, 2014
LONG BEACH CONVENTION CENTER
LONG BEACH, CALIFORNIA USA

Examples

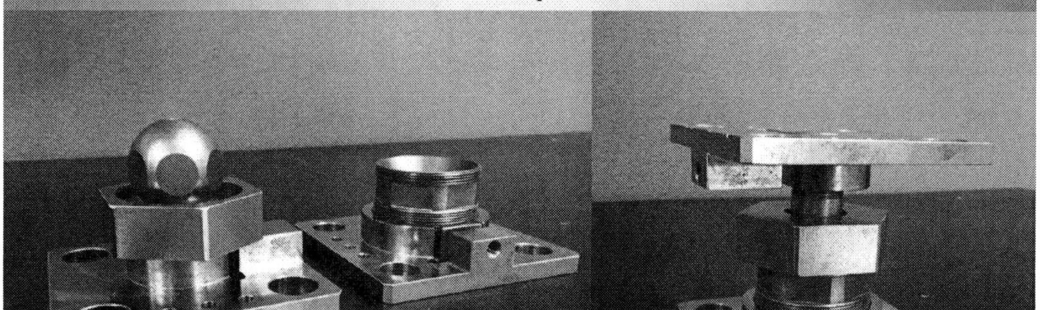

AERODEF MANUFACTURING

produced by sme

SUMMIT & EXPOSITION
February 25-27, 2014
LONG BEACH CONVENTION CENTER
LONG BEACH, CALIFORNIA · USA

Metrology Assisted Assembly

AERODEF MANUFACTURING

produced by sme

SUMMIT & EXPOSITION
February 25-27, 2014
LONG BEACH CONVENTION CENTER
LONG BEACH, CALIFORNIA · USA

Questions?

AERODEF
MANUFACTURING
SUMMIT & EXPOSITION
February 25-27, 2014
LONG BEACH CONVENTION CENTER
LONG BEACH, CALIFORNIA, USA
Thank you,
James Diedesch
Sr. Manufacturing
Engineer
produced by
sme
JANICKI
INDUSTRIES

AERODEF MANUFACTURING

Dimensional Engineering for Determinant Assembly:

Utilizing Dimensional Engineering Tools to Optimize Assembly Processes

produced by sme

Don Jasurda
Dimensional Control Systems

DCS ENGINEERING IN NEW DIMENSIONS

AERODEF MANUFACTURING SUMMIT & EXPOSITION February 25-27, 2014 sme

Determinant Assembly

- What is Determinant Assembly?

AERODEF MANUFACTURING SUMMIT & EXPOSITION February 25-27, 2014 sme

The Cost of Tooling

Tooling can account for up to 10% of an aircraft's total production cost.

EMBRAER
LOCKHEED MARTIN
AIRBUS
NORTHROP GRUMMAN
Gulfstream A GENERAL DYNAMICS COMPANY
BOEING
BOMBARDIER

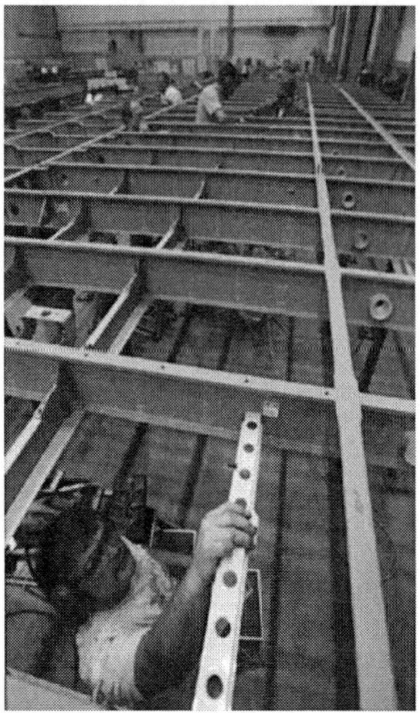

AERODEF MANUFACTURING SUMMIT & EXPOSITION February 25-27, 2014 sme

Benefits of Determinant Assembly

- Remove the Cost of Tooling
- Reduction in Lead Times
- Faster Assembly Processes
- More Efficient Assembly
- Faster Design Changes

AERODEF MANUFACTURING SUMMIT & EXPOSITION February 25-27, 2014 sme

Case Study: Boeing 747

Accuracy

Flexibility

Cost Advantages

	DETERMINANT ASSEMBLY	CONVENTIONAL ASSEMBLY	
Assembly Technique	Snap Together Parts Part-to-Part Indexing	Parts-to-Tool Indexing	
Tooling Characterization	Gaugeless Tooling	Hard Tooling	Soft Tooling
Tooling Requirements	Free Form Manipulation	Indexing	Indexing
Benefits	Precision, Speed, Flexibility (versus non-automated tooling)		Less Constraining Less Routining (vs. hard tooling)
FAIT Program Solution	Precision Automated Motion	⊘	Laser Measurement

AERODEF MANUFACTURING SUMMIT & EXPOSITION February 25-27, 2014 sme

Achieving Broad-Scale Utilization

Traditional

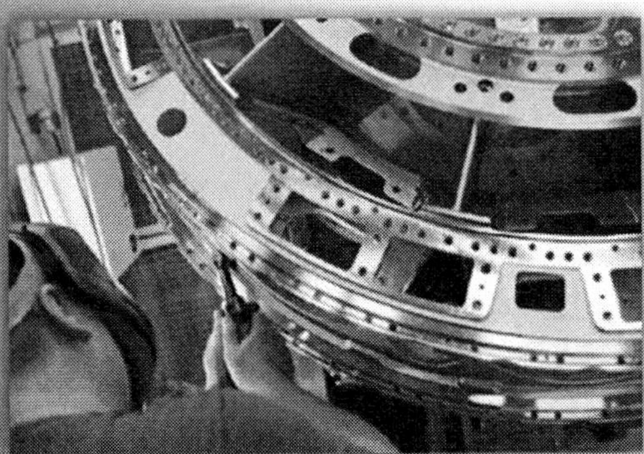

Modern Application

AERODEF MANUFACTURING SUMMIT & EXPOSITION February 25-27, 2014

Dimensional Engineering Process

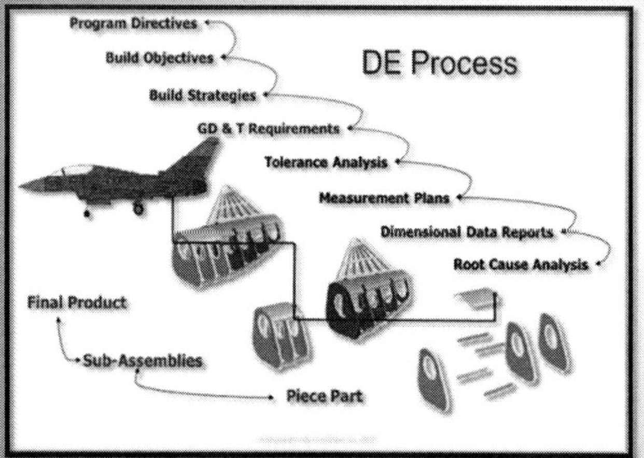

A well defined dimensional engineering process utilizing 3D simulation is a powerful enabler of determinant assembly.

AERODEF MANUFACTURING SUMMIT & EXPOSITION February 25-27, 2014

DE Process Flow

Program Directives

Applied Early and Throughout Product Lifecycle

Build Objectives

Build Strategies

GD&T Requirements

Variation Analysis

Measurement Plans

Dimensional Data Reports

Root Cause Analysis

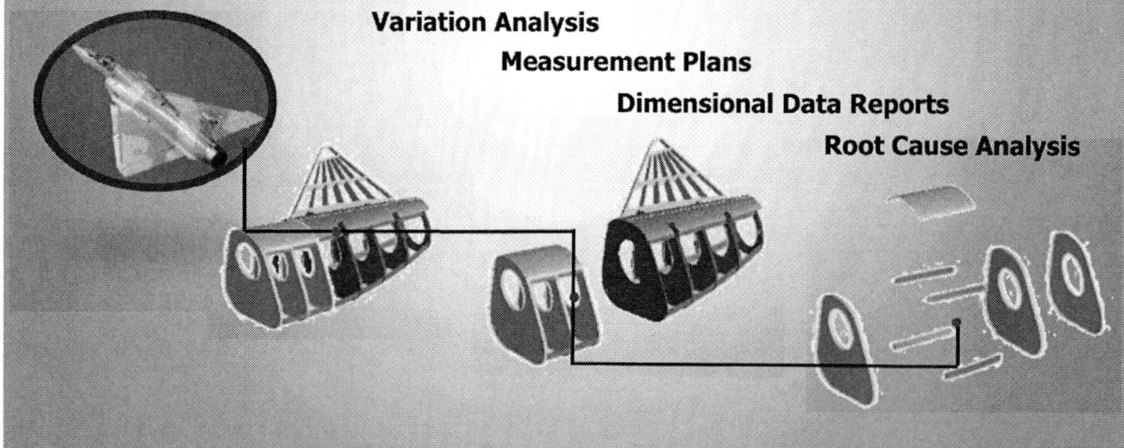

Validating with Dimensional Analysis

Build Objectives
Process Strategies
GD & T Requirement
Dimensional Analysis

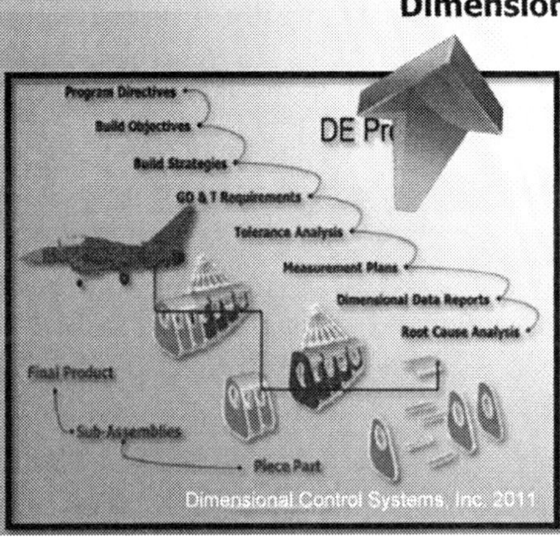

Dimensional Analysis Validates:

- Build Objectives (Technical Specification)
- Process Strategies
- GD&T Requirements

9

AERODEF MANUFACTURING SUMMIT & EXPOSITION February 25-27, 2014 sme

GD&T

Common language of manufacturing quality

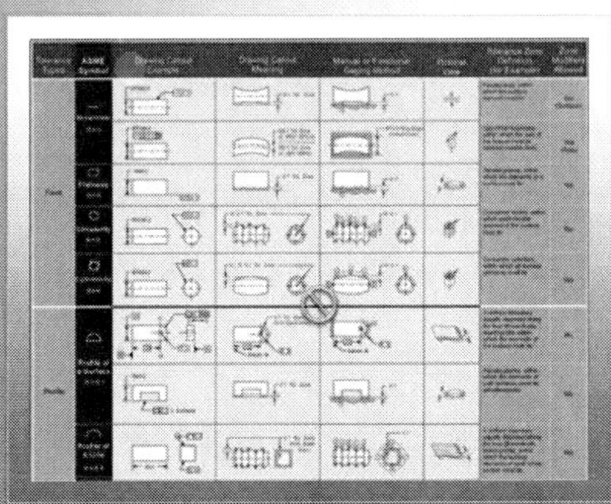

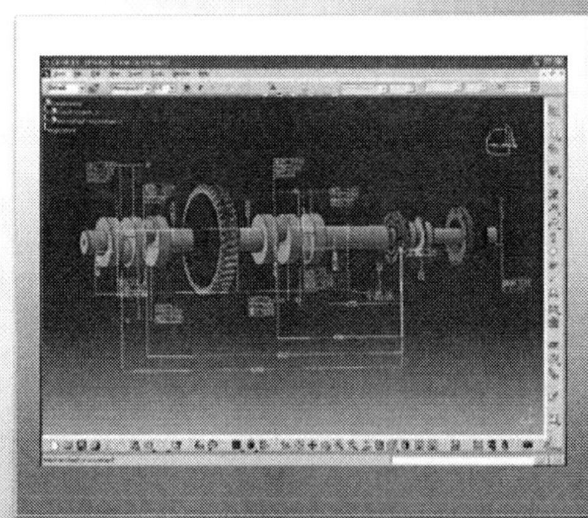

3D Simulation

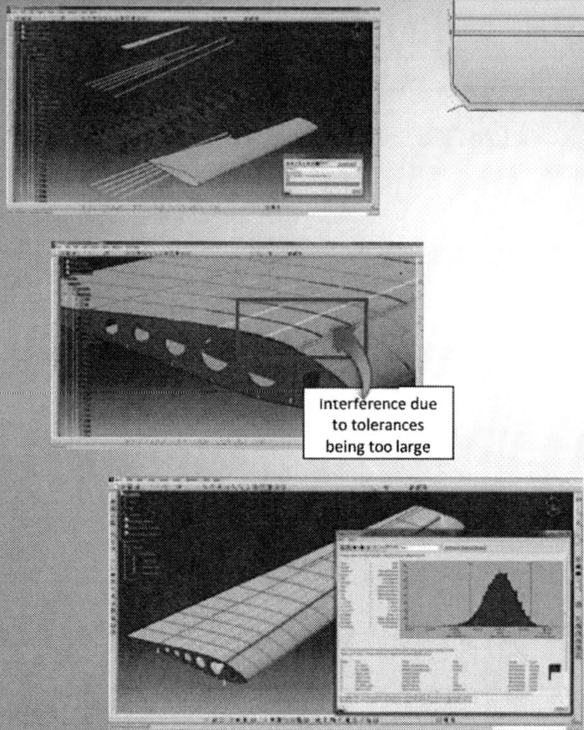

- ✓ CAD Model Based
- ✓ Variation Analysis
- ✓ Consider Process
- ✓ Validate Determinant Assembly
- ✓ Test Multiple Builds Scenarios

Identifying Key Contributors to Variation

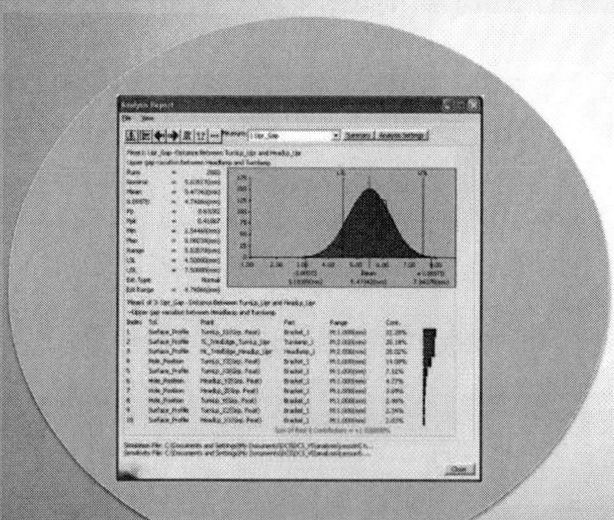

- ➢ Test and optimize various build scenarios.
- ➢ Test determinant assembly application virtually.

Determinate Assembly of Tooling Allows Concurrent Design of Airbus Wings and Major Assembly Fixtures

- The original fixtures used to locate the stringers and wing skins followed a conventional design.
- Individual details for stringer locations were designed and field set using laser trackers.

- Over two thousand individual points had to be aligned in the field.
- This was a difficult, labor intensive and time-consuming process.

AERODEF MANUFACTURING SUMMIT & EXPOSITION February 25-27, 2014 sme

"Pegboard" Formboard

In the field, only the formboard must be located as opposed to the up to twenty individual details per board.

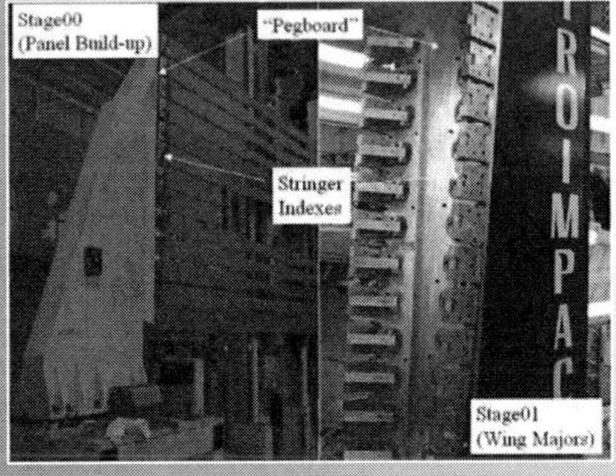

Airbus Final Results

- Majority of tooling designed and manufactured early in the program.
- Only index plates require final design prior to integration with formboards.
- Overall field time reduction.

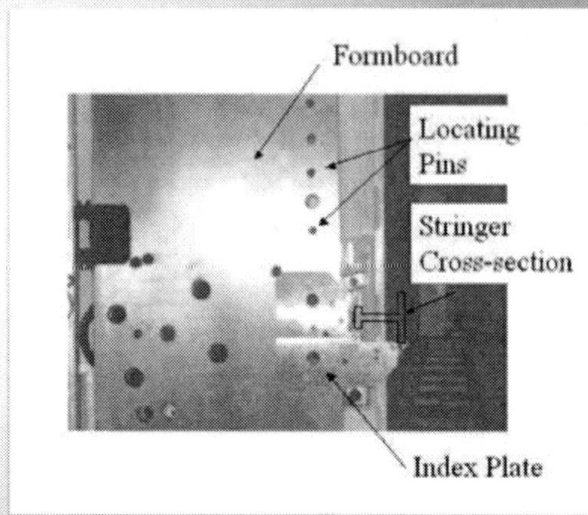

Boeing Poseidon 8a AAS Radar System

New Variant Retrofit

Determinant Assembly in attachment of Raytheon Advanced Airborne Sensor (AAS), an advancement to the previous AGS-149 radar system flown on P-3's.

AERODEF MANUFACTURING

SUMMIT & EXPOSITION
February 25-27, 2014

sme

Conclusions

- Determinant Assembly reduces costs and lead times
- Applicable in multiple sub-systems
- Modern software makes application possible in new ways
- Means of improving efficiency and profitability

Don Jasurda
Dimensional Control Systems Inc.
580 Kirts Blvd.
Troy, Michigan
48315 USA

Main Office: 248-269-9777
Web: www.3dcs.com
Email: sales@3dcs.com

AERODEF MANUFACTURING

Computer Aided Inspection for
Better Precision Parts,
Much Faster,
at Lower Cost &
with Greater Confidence

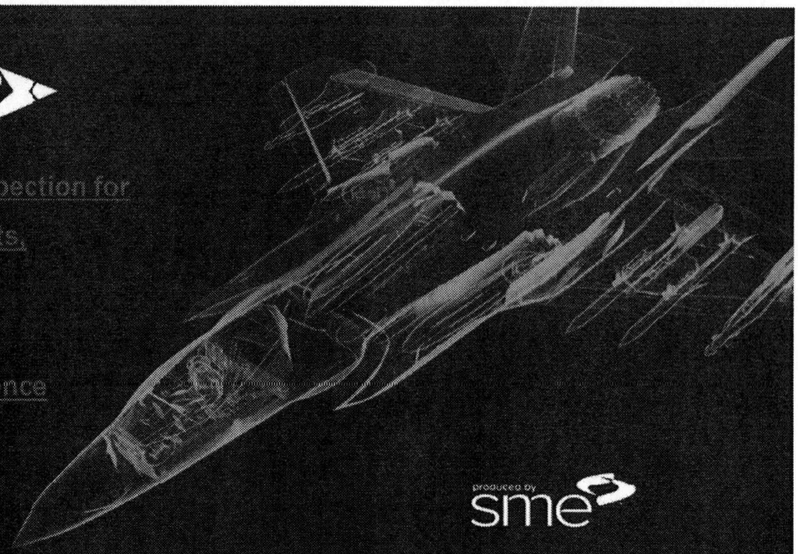

produced by sme

Level 3 Inspection

Bill Greene
Level 3 Inspection LLC
Computer Aided Inspection
Master Practitioners

Accelerating **Process Optimization**
Accelerating **Production Throughput**
Value Added Partnership
Unique Value Promise

Level 3 Inspection

Our Unique **Value Promise**

- Better precision parts
- Much ***Faster***
- at Lower Cost
- with Greater Confidence
- Cutting scrap & iterations
- Saving months (30%) & millions.

When do you want us to start?

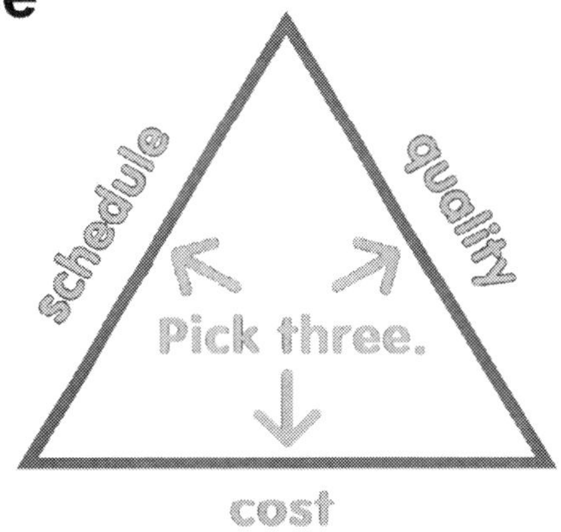

Beat The Triple Constraint Problem

 3

Level 3 Inspection

L3I Introduction

- Mechanical & Metallurgical Engineers + Jet Engine OEM
  - ➢ Design Intent
  - ➢ Precision Mfg. Processes
- GD&T Professionals
  - ➢ Reqs & Specs
- Over 12 years Experience
  - ➢ Thousands of CAI Layouts
- ***Fast*** Problem Solvers
  - ➢ Working with 10,000-times more information (for 12 years)

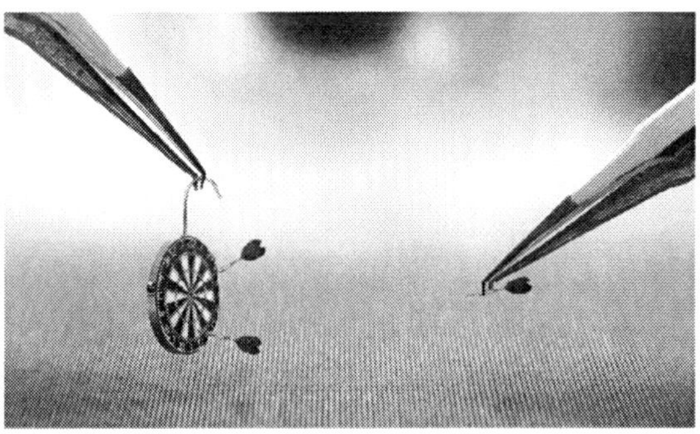

 4

Master CAI Practitioners
Expert Partners: Professional Services

Level 3 Inspection

- Computer Aided Inspection (CAI) from
- Structured Light Metrology (SLM = WLS)
- Dimensional & First Article Inspections – ***Customer Ready***
  - ➢ Cut 95% of inspection report ***review time*** *(10 days*

Structured Light Metrology (SLM)

Level 3 Inspection

- Optical 3D geometric measurement
  - ➢ Based on principle of triangulation
  - ➢ Projected light pattern
  - ➢ Hi-Res digital camera
    - ❖ Millions of data points per 3D scan
    - ❖ Multi-scans merged into single 3D SmartScan ™ file
      - Actual produced part in .stl format

21st Century Computer Aided Processes

Level 3 Inspection

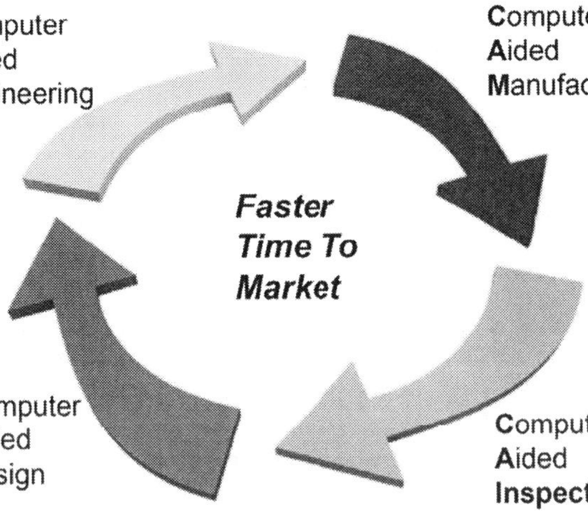

We *accelerate* each step:

- CAD
- CAE
- CAM
- CAI (back to CAD)

to *accelerate* your:

1. Throughput
2. Inspections
3. Deliveries
4. Profits

 7

Computer Aided Inspection (CAI)

Level 3 Inspection

- Comprehensive data processing
- Align SmartScan™ file to CAD
  - Generate SmartCompare™
- Extract Dimensions from SmartPrint™
  - Computer-Itemized BluePrint
    - integrates to SmartReports
- Develop SmartInspect™ or SmartFAI™ Reports
  - .pdf & .xls
  - Delivery in SmartInspection Portfolio (.pdf)
- Trackable, Trendable, Traceable, Auditable, SPC Integration

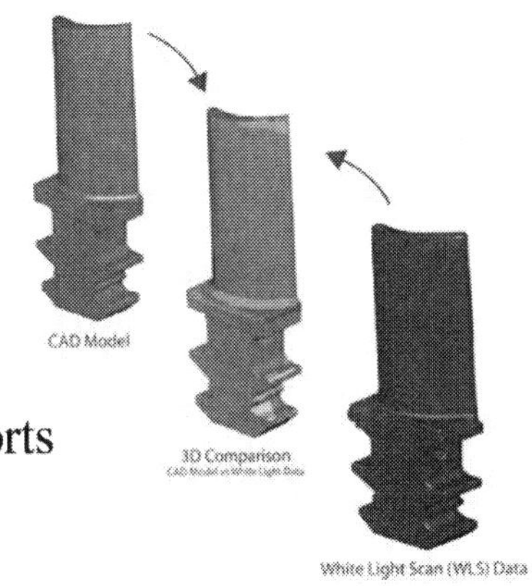

 8

Highly ***Integrated*** CAI Processes

Level 3 Inspection

SmartPrint™ - computer itemized BP, feeds AS9102 reports

SmartCompare™ is SmartScan™ aligned to CAD

SmartFAI™ or SmartInspect™ dimensional extract

AS9102 Form with conditional formatting

Automated data processing to myriad reports

Results ***highlights*** the SmartPrint™ balloons

All done with minimal human input for ***no data entry error***

Cuts 95% of dimensional inspection report review time

OK, What About Accuracy?

Level 3 Inspection

- Vastly Superior to CMM ("industry standard")
- Proven by ANOVA Gauge R&R - (MSA)
- ISO 17025 accredited to 0.000078" (2 microns)
- Routine System Calibration
  - NIST certified artifacts
  - AAAM certified artifacts

- *Distinct Advantage*: Computer Aided Inspection over any other method
  - Comprehensive, Complete part geometry
  - Illustrative, Advanced results visualization
  - ***Confidence*** & improved ***comprehension*** lead to ***SPEED***

Experienced: thousands of layouts

Level 3 Inspection

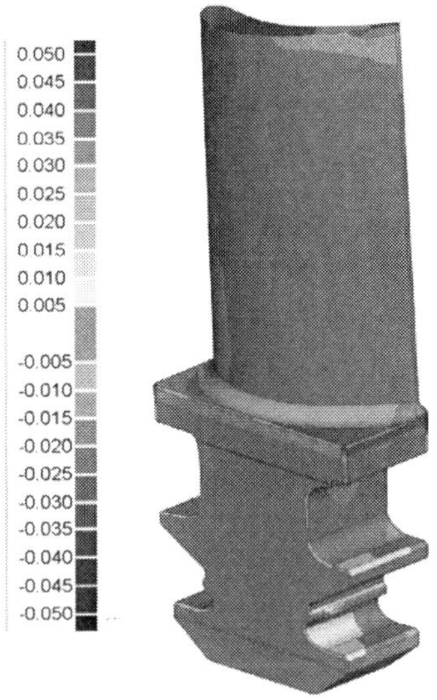

- **SmartInspection™ CAI Reports (portfolio)**
  - Executive Summary (.pdf)
  - SmartCompare™ (.pdf)
  - SmartFAI™ (.pdf)
  - SmartCorrelate™ (.xls)
  - *Customer-Ready* Dimensional Inspection Report AS9102 (.xls)

What Our Clients Are Saying (1)

Level 3 Inspection

"Relative to the results L3I is providing, the current CMM and optical comparator are only a small improvement over guesswork."

K.J., Mfg. Engineer, Orthopaedic OEM

"Studies have proved that the Level 3 Inspections system and method of inspection for implants is as accurate, or better than our CMMs. Since the white light scanning processes have the required accuracy and can cover so much more of the feature surface, their data is a more accurate representation of the feature true size and form."

I.H., Metrology Director, Orthopaedic OEM

What Our Clients Are Saying (2)

Level 3 Inspection

"If you know me then you know that I rarely offer endorsements or recommendations. Well, Level 3 (Bill Greene & Company) will be one of those rare times; an exception. Bill is a man of character and, in kind, surrounds himself with associates of that same caliber. Bill, and Scott McAfee, are consummate professionals and, in my opinion, are the industry leaders, and experts, in their chosen technologies."

– Harold Slaterpryce, QA Manager at PCC Airfoils LLC

Move CAI upstream for max benefit

Level 3 Inspection

- Investment Casting: Inspect wax dies & core dies
- Injection Molding: Inspect the molds
- CNC Machining: confirm Step 1 makes conforming parts
  - Before investing in Step 2, feed SPC database
- Working with 10,000-times more information, it's easier to:
  - Make more confident decisions
  - Cut iterations & scrap, Save months & millions

What Our Clients Are Saying (3)

Level 3 Inspection

Thank you for all of the support. Our view of what our process is yielding has increased dramatically thanks to what you have been able to show us, and to quickly get feedback as we make this (DMLS) system rock solid.

Thanks again,

- Z.F., Engineering Manager, Turbine Engine Producer

What Our Clients Are Saying (4)

Level 3 Inspection

I'm very happy with the reports from Level 3 Inspection.

They are very useful when combined with S500 drawing when dispositioning MRB - one can search the report (PDF file) for a given characteristic for more details - very useful.

- Miroslaw "Mirek" Pajak, Design Engineer, GE Aviation US

What Our Clients Are Saying (5)

Level 3 Inspection

"As a GE Product Quality Engineer, I've been extremely happy with the Level3 Inspection reports I've been receiving from my vendors. The speed to completion, thorough nature of the reporting and data accuracy is second to none. To me, Level 3 Inspection is setting the new standard for outsourced inspection services."

- Mike Yacavone, Product Quality Engineer, GE

17

Aerospace & Energy Industry Clients

Level 3 Inspection

Alcoa Howmet

GE Aviation

LOCKHEED MARTIN

ALCOA

PCC Precision Castparts Corp.

GE Energy

CPP Consolidated Precision Products

DONCASTERS A Doncasters Group Limited company

Pratt & Whitney A United Technologies Company

Sikorsky A United Technologies Company

Pratt & Whitney Canada A United Technologies Company

SELMET, INC.

CHROMALLOY

Rolls-Royce

TurboCare

PSM

VERICOR

Honeywell

SIEMENS

Mikro Systems

18

OEM Assemblers Avoid Non-Conformances: Remove problems ***before*** they become yours

Level 3 Inspection

- Have your buyers <u>accept only conforming parts</u>
  - ➢ By viewing dynamic color-plots (.pdf) of conformance
  - ➢ *Before* parts are shipped to your assembly plant.
- Select by Serial Number which parts you want shipped
- Do not accept shipment of any non-conforming parts
- Eliminate Receiving Inspection – have CAI done at supplier
- Improve your Suppliers' business for everyone's benefit

Level 3 Inspection

Level 3 Inspection LLC
Computer Aided Inspection
Master Practitioners

Accelerating Production Throughput

Value Added Technology Business Partnership

Smart Inspection Station™

- Computer Aided Inspection (CAI) from White Light Scanning (WLS)
- Now *FAST* (~12 minutes)
- Integrated & Fully Automated
- Highly Configurable – most any part
- Pre-Programmable
  - ➢ non-technical operator gets CAI results
- Shop-Floor Ready

Enclosed Configuration for small parts up to 21"

Open Configuration s for any size part are available

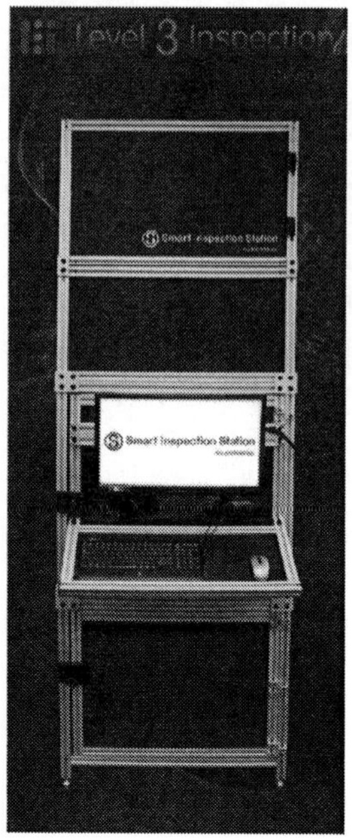

Master CAI Practitioners
Expert Partners: Our lab *In* Your Plant

Level 3 Inspection

- Hybrid Combination of Services & Systems
  - ➢ Our (or your) configured Scanner
  - ➢ Our experienced Scan Operator
  - ➢ Our Engineering Analysis (remote)
  - ➢ ***Fast Answers In Your Facility***
    - ❖ Eliminate logistics challenges
  - ➢ Skip the Learning Curve
  - ➢ Accelerate the Benefits

AERODEF MANUFACTURING

sme

- Questions
- Discussions
- Prioritizing near-term opportunities
- Challenges
- Let us help you strategize

Level 3 Inspection

- Bill@Level3Inspection.com Direct: 772.405.7502

3D Printing of Rocket Engine

Youping Gao
John Malinzak
Jeff Haynes
William Sack

Feb 27, 2014

This document does not contain any ITAR or EAR Technical Information

AEROJET ROCKETDYNE

ABOUT AEROJET ROCKETDYNE

Aerojet Rocketdyne is a world-recognized aerospace and defense leader providing propulsion and energetics to the space, missile defense and strategic systems, tactical systems and armaments areas, in support of domestic and international markets.

DELIVER excellent performance in our core business

DRIVE initiatives across all aspects of the operating model to maximize our competitive posture

CREATE value by enhancing and expanding the core and diversifying into broader markets

AERODEF MANUFACTURING

This document does not contain any ITAR or EAR Technical Information

2

AEROJET ROCKETDYNE

56 YEARS OF POWERING UNITED STATES' SPACE EXPLORATION

AERODEF MANUFACTURING

This document does not contain any ITAR or EAR Technical Information

AEROJET ROCKETDYNE

MORE THAN 5,000 EMPLOYEES ACROSS 14 STATES

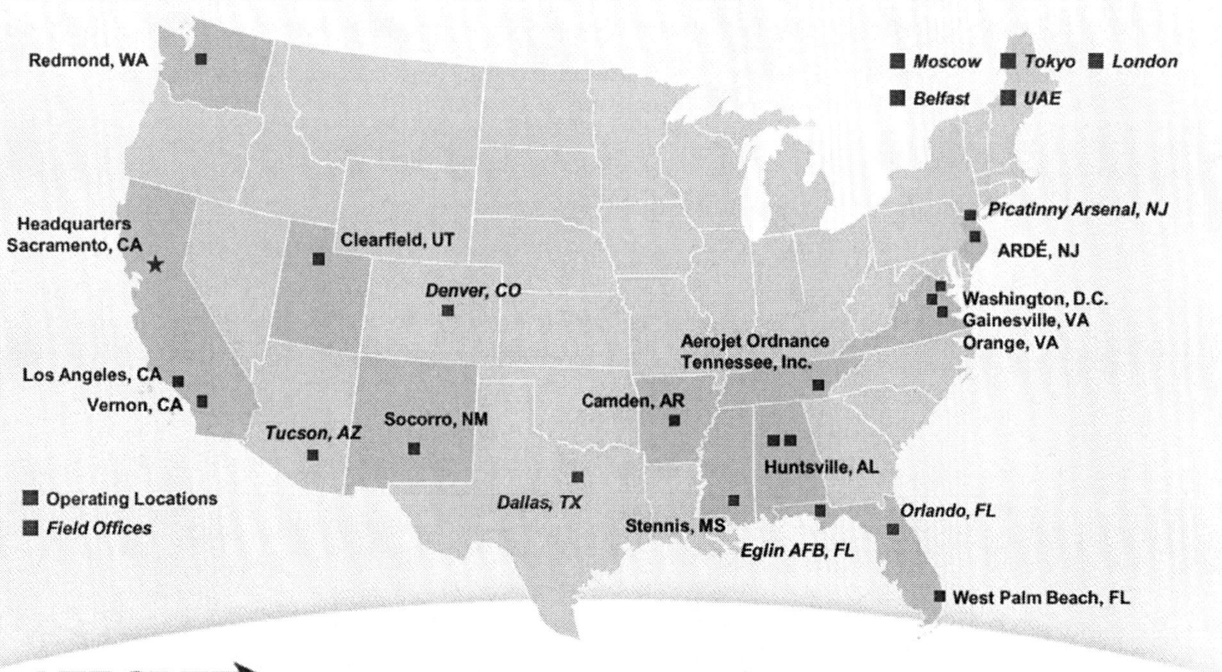

AERODEF MANUFACTURING

This document does not contain any ITAR or EAR Technical Information

4

AEROJET ROCKETDYNE

3D Printing & Rocket Engine

- 3D Printing of Metallic Structures
  - Recent advancement in 3D printing
  - Unprecedented capability in producing geometrically complex monolithic structures
- Liquid Rocket Engines
  - Reaction engines that produce extreme high propulsive jet
    - Injector, combustion chamber, nozzle
  - Typically has complex geometry and uses multiple materials to meet the functional requirements
  - 3D printed components have been launched into space recently

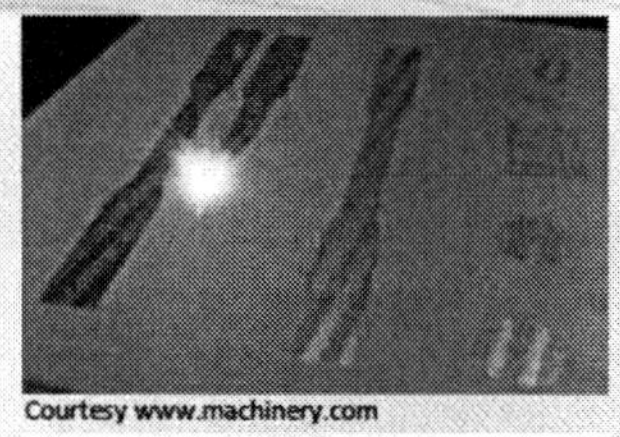
Courtesy www.machinery.com

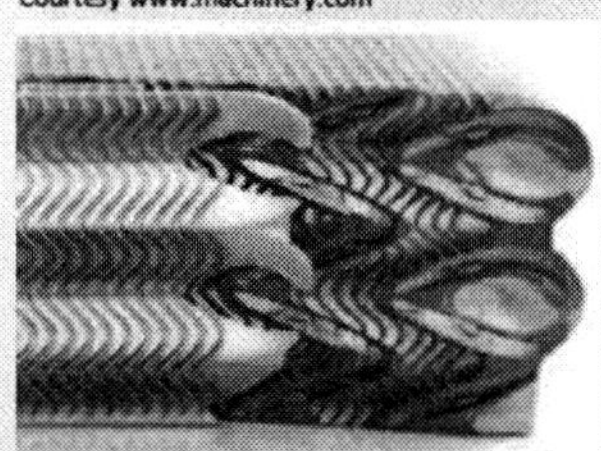
Courtesy Within Lab

AERODEF MANUFACTURING

This document does not contain any ITAR or EAR Technical Information

5

AEROJET ROCKETDYNE

3D Printing Process

- Powder Bed 3D Printing
  - Laser selectively melting metal powder layer by layer
  - Capable of producing geometrically complex monolithic structures that is ideal for challenging, extreme high heat flex with active cooling mechanisms

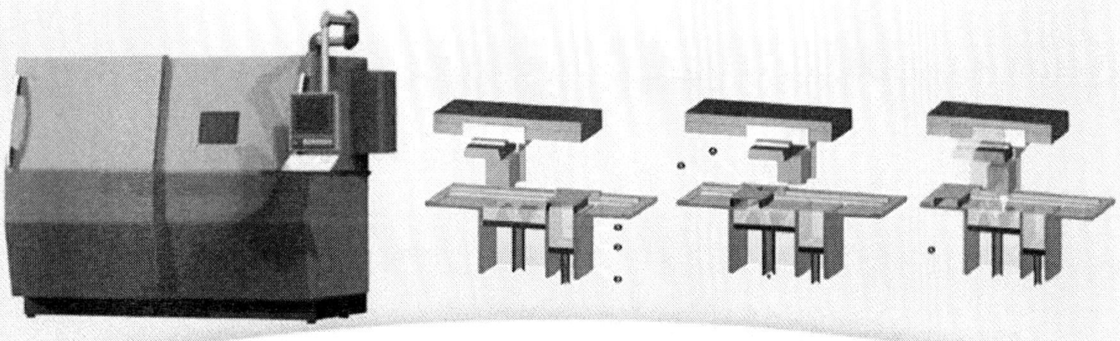

AERODEF MANUFACTURING

AEROJET ROCKETDYNE

3D Printed Metallic Material

- Microstructure of 3D Printing of Metallic Structures in As Built
  - Extremely uniform & consistent

AERODEF MANUFACTURING

This document does not contain any ITAR or EAR Technical Information

7

AEROJET ROCKETDYNE

3D PRINTED METALLIC MATERIAL

- Microstructure of 3D Printing of Metallic Structures
  - Through heat treatment, 3D printed metallic material's microstructures can be homogenized and refined

As Built | Stress Relieved | After HIP

AERODEF MANUFACTURING

This document does not contain any ITAR or EAR Technical Information

8

AEROJET ROCKETDYNE

3D PRINTING OF LIQUID ROCKET ENGINES

- Resurrecting Saturn V Engine Systems and 3D Printing Them
  - F1 Engine Gas Generator Injector
  - Re-designed to enable internal design features without build supports

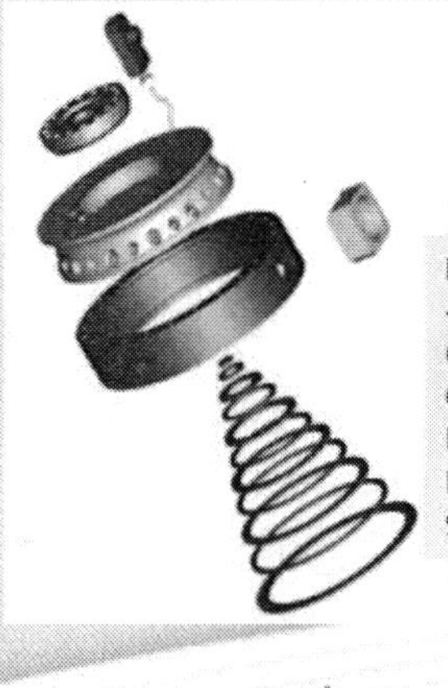

F1 Engine (Saturn V)
- circa 1960's
Gas Generator Injector
Conventional Manufacturing
Forgings + Machined
Brazed structures
9 months Lead Time

↔

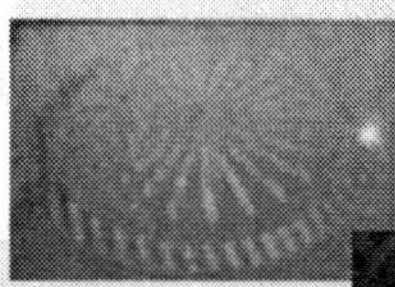

F1 Engine (Saturn V)
- circa 2013
Gas Generator Injector
Additive Manufacturing
Part Reduction
No Braze
60% Lead Time Savings
70% Cost Savings

AERODEF MANUFACTURING

AEROJET ROCKETDYNE

3D PRINTING OF LIQUID ROCKET ENGINES

- New design developed for Liquid Oxygen / Gaseous Hydrogen Injector
  - Functional component tested and validated in a relevant environment
  - Full design properties established for SLM Alloy in high pressure hydrogen

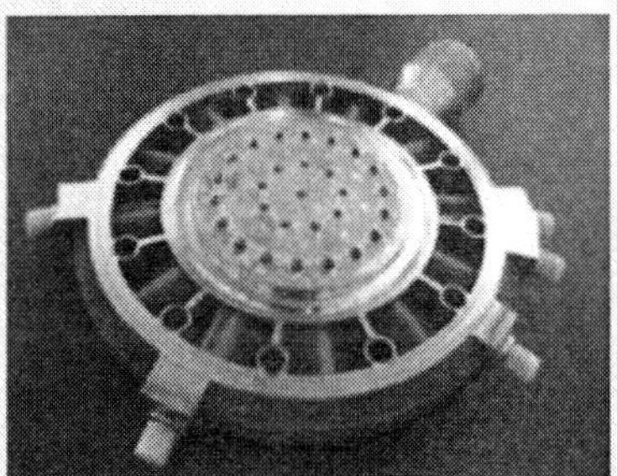

AERODEF MANUFACTURING

This document does not contain any ITAR or EAR Technical Information

AEROJET ROCKETDYNE

3D Printing of Liquid Rocket Engines

- 3D Printing of Entire Liquid Rocket Combustion Systems
  - Clean sheet of design
  - Design to use single alloy to accomplish different functional requirements
  - High dimensional accuracy
  - Minimum post build machine operation
  - Significant development cycle and cost reduction

AERODEF MANUFACTURING

This document does not contain any ITAR or EAR Technical Information

AEROJET ROCKETDYNE

CHALLENGES TO INSERTION

- Barriers to 3D printed liquid rocket engine insertion
  - New product form requires full material design allowable database
  - Transitioning between machine manufacturer platforms
  - Powder metallurgy product and considerations
  - Inspection methods for newly designed complex structures
  - Variability and repeatability demonstrations
  - Customer confidence evolving

AERODEF MANUFACTURING

AEROJET ROCKETDYNE

CONCLUSION

- 3D printing offers significant leap forward in producing geometrically complex monolithic hardware
- 3D printed material is characterized by the confinement of very small printing size that results in fine structures in the as built state which can be further refined through heat treatment
- 3D printed materials' properties is comparable to that of its wrought equivalent
- 3D printing process allows design using monolithic material to achieve traditionally multi materials functional requirements
- 3D printed functional hardware significantly reduces product development cycle and cost

AERODEF MANUFACTURING

Thank you!

AEROJET ROCKETDYNE

AERODEF MANUFACTURING

sme

This document does not contain any ITAR or EAR Technical Information

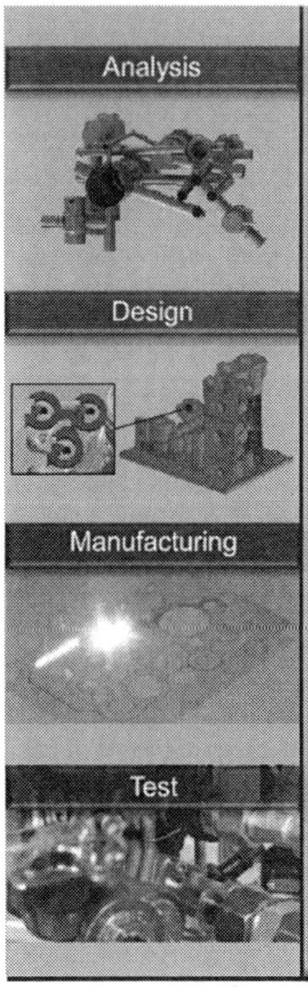

Additive Manufacturing for High Pressure Hydraulic Valve Manifolds

Hummel, G.[1]; Baumbach, V.[1]

Kroll, L.[2]; Schubert, F.[2]

Altmann, A.[3]; Rayczyk, G. F.[3]

[1] Airbus Operations GmbH

[2] Chemnitz University of Technology

[3] Liebherr-Aerospace Lindenberg GmbH

Introduction

Requirements for next generation aircrafts

- Increased energy efficiency
  - → Advanced engines and aerodynamics
  - → Lightweight structures and equipment
- Cost reduction
  - → New manufacturing technologies

Equipment requirements for hydraulic components

- High performance and reliability
- Significant weight reduction
- Reduction of installation space
- High functional density

→ **Additive manufacturing technologies**

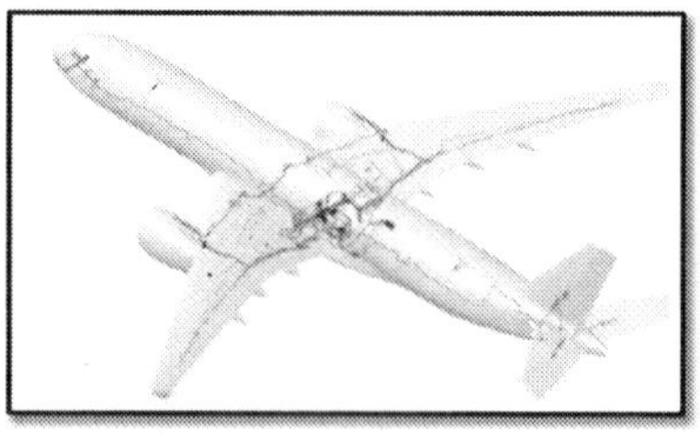

Benchmark Selection

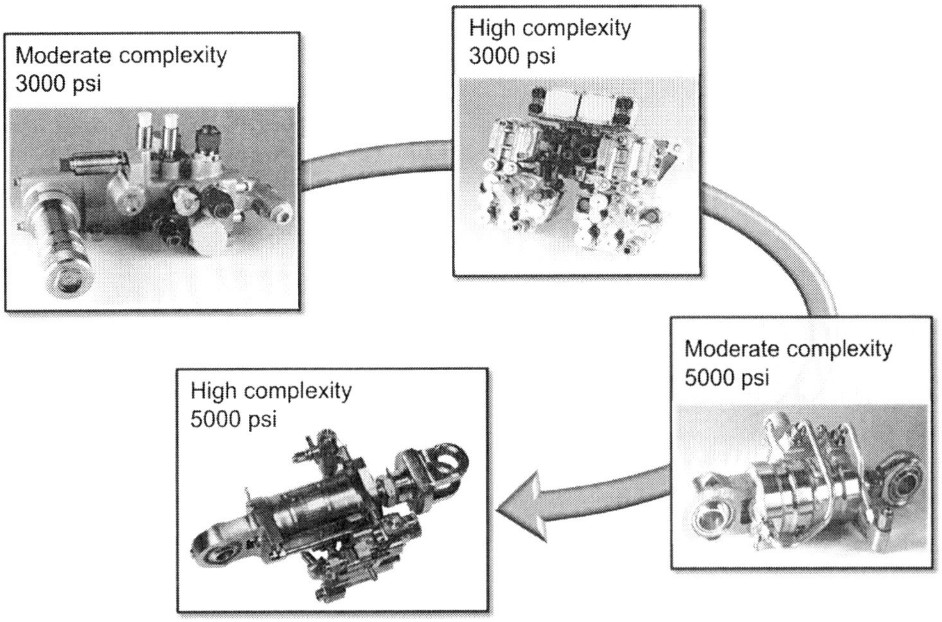

Reference Part

A380 Spoiler Servo Control

Spoiler Functionality

- Primary flight control components
- Reduction of lift

Hydraulic Load Requirements

- Design operation pressure: 350 bar (5,000 psi)
- Proof pressure: 525 bar (7,500 psi)
- Ultimate pressure: 700 bar (10,000 psi)
- Impulse/Fatigue: $12 \cdot 10^6$ pulse cycles

Global Trade Study

	Machining	Selective Laser Melting	Soldered Tube System	Investment Casting	Hybrid Technology
Design Solution					
Weight	-	↓↓↓	↓↓	↓	↓
Costs	-	↓	↓↓	↓↓	↓↓

➡ Analysis showed highest potential for weight and cost saving by use of SLM technology

Selective Laser Melting Process

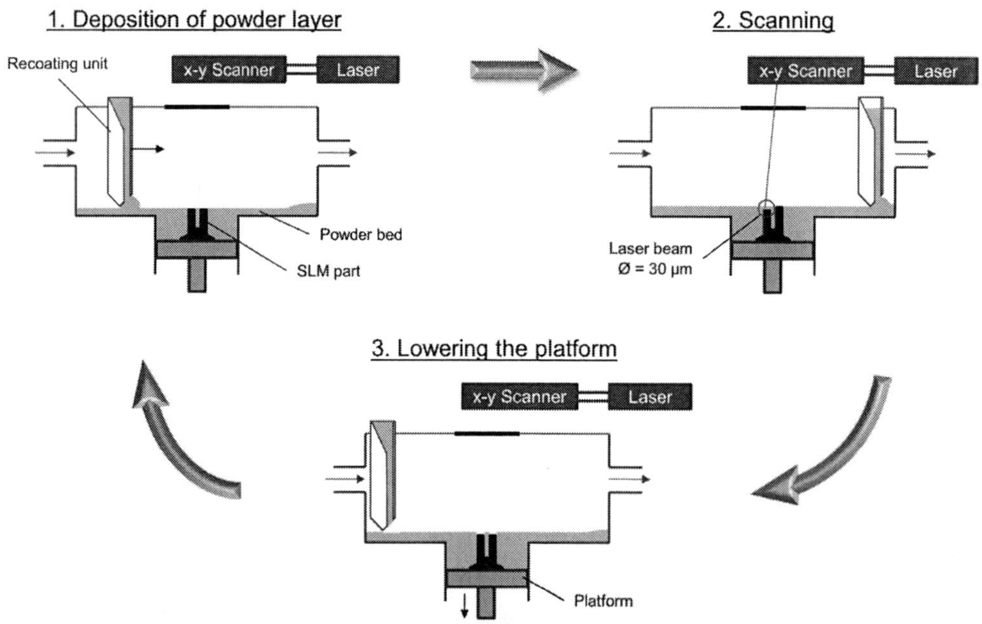

Reference Part Analysis

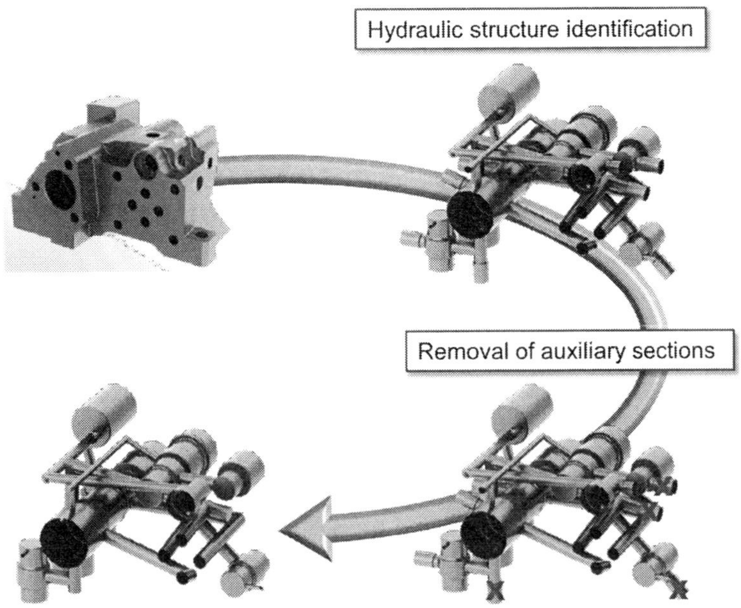

- Reference Part Analysis
- Positioning of Main Components
- Design of Connection Lines
- Design Verification
- Design of Support Structures

 7

Positioning of Main Components

- One to one replacement component
- Consideration of installation space and interface requirements
- Vertical orientation preferred to increase geometric accuracy

Intelligent position for short connection lines & support reduction

- Reference Part Analysis
- Positioning of Main Components
- Design of Connection Lines
- Design Verification
- Design of Support Structures

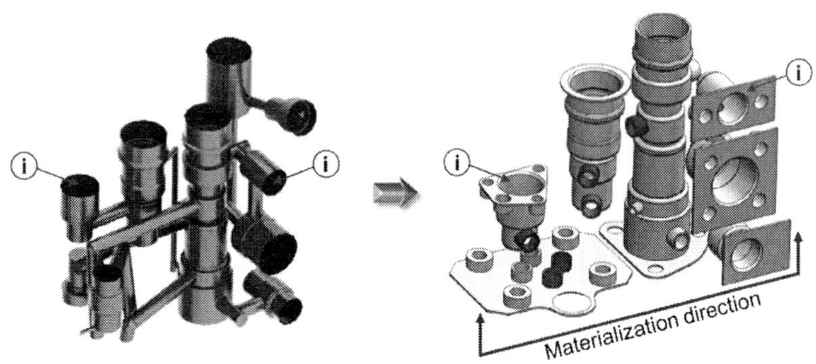

(i) Fixed interfaces

 8

Design of Connection Lines

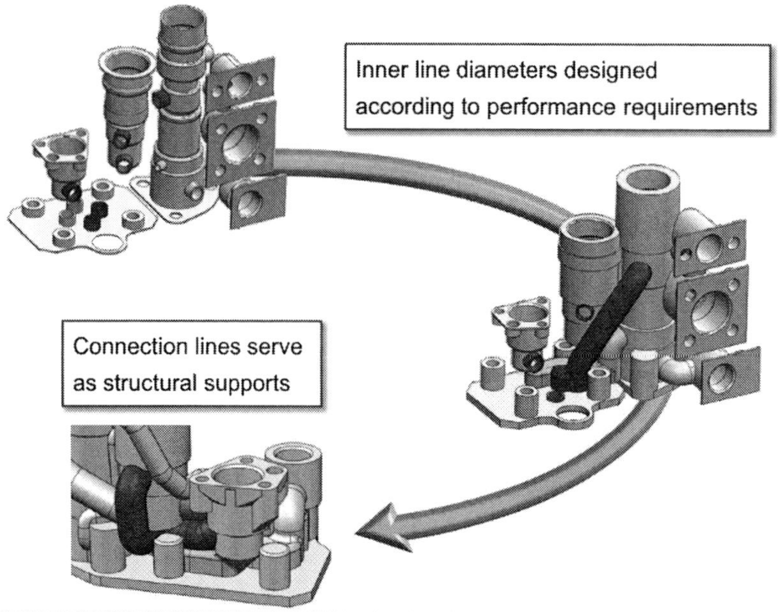

Reference Part Analysis
↓
Positioning of Main Components
↓
Design of Connection Lines
↓
Design Verification
↓
Design of Support Structures

Design Verification

Numerical Calculation of critical Load Cases

- Hydraulic fatigue
- Ultimate pressure
- Vibration (incl. SEI)
- Assembling loads

→ Critical notch stresses at main valve bodies' intersections

→ Enlargement of transition radii required

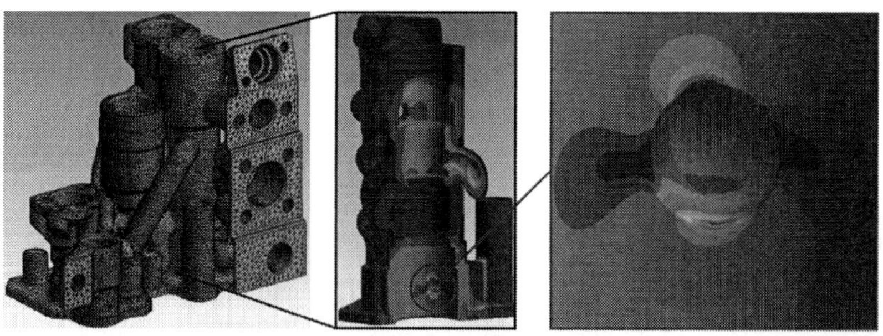

Reference Part Analysis
↓
Positioning of Main Components
↓
Design of Connection Lines
↓
Design Verification
↓
Design of Support Structures

Design of Support Structures

- Development of open support structures
  - → Improvement of metal powder retrieval
  - → Increase of process efficiency

Reference Part Analysis
↓
Positioning of Main Components
↓
Design of Connection Lines
↓
Design Verification
↓
Design of Support Structures

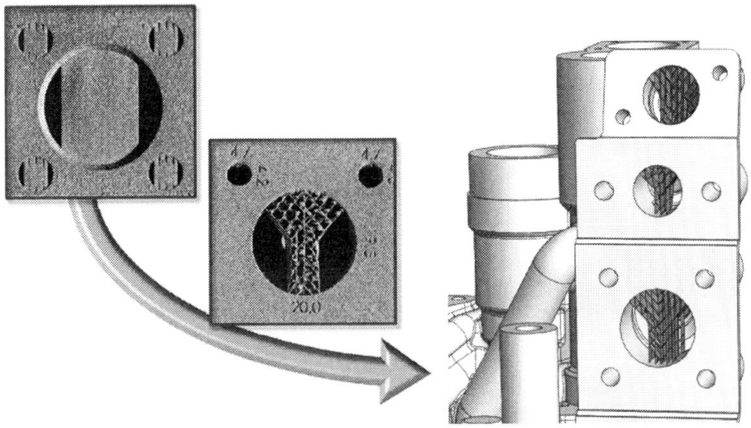

Manufacturing of Prototype Component

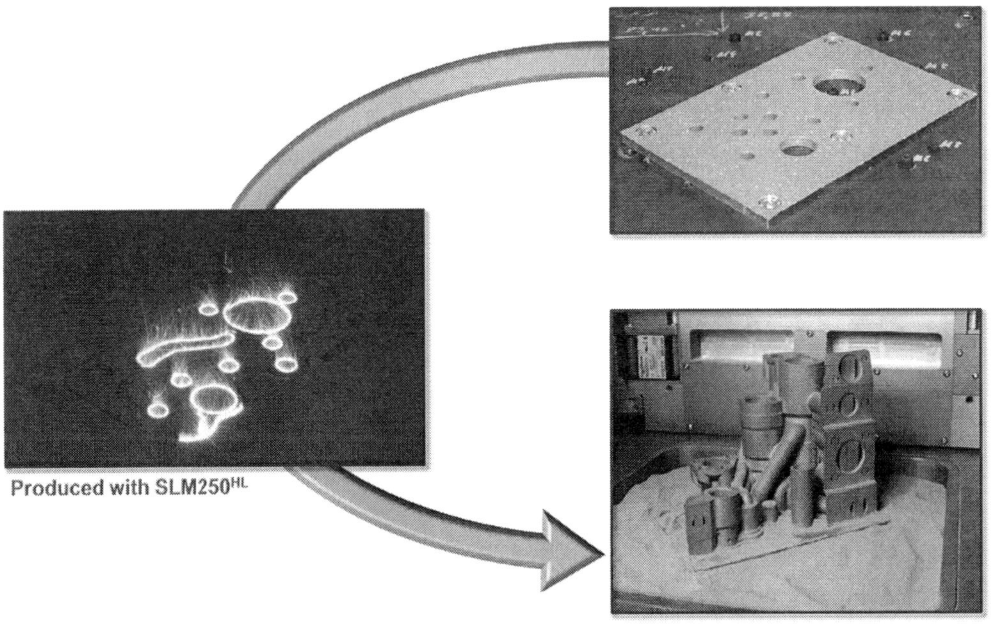

Produced with SLM250^{HL}

Quality Assurance

Microfocus Computed Tomography

- Detection of material defects before and after Hot Isostatic Pressing (HIP)
  - → Identification of small pores requires high resolution (voxel size ≈ 20 μm)
  - → After HIP no pores detectable

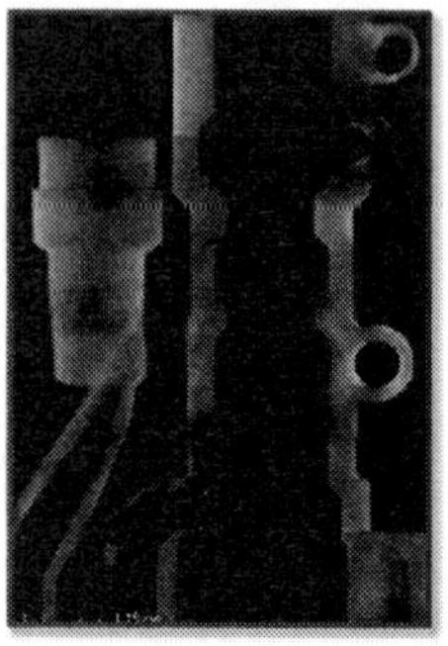

© Copyright by AIRBUS AN EADS COMPANY CUT CHEMNITZ UNIVERSITY OF TECHNOLOGY LIEBHERR 13

Quality Assurance

Optical Layer Evaluation

- Grey scale image of every applied powder layer
  - → Identification of large material defects during SLM process
  - → Detectable defect size depends on powder layer thickness

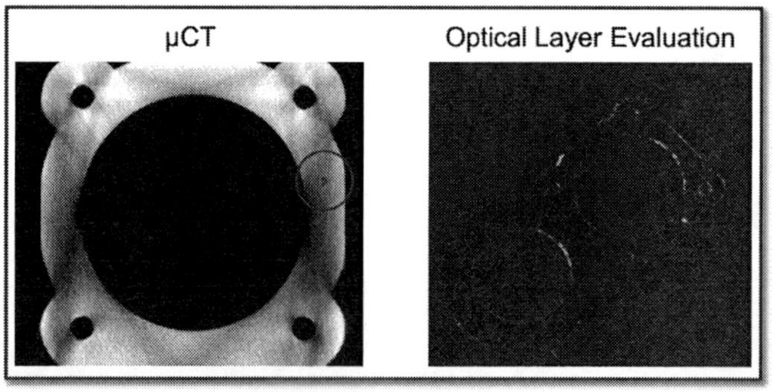

© Copyright by AIRBUS AN EADS COMPANY CUT CHEMNITZ UNIVERSITY OF TECHNOLOGY LIEBHERR 14

Assembly and Testing

<u>Acceptance Test</u>

- Application of original Acceptance Test Procedure
  - → Functionality and full performance was reached

<u>Ultimate Pressure Test</u>

- Static pressure of 700 bar (10,000 psi)
  - → No cracks or deformations

<u>Impulse/Fatigue Test</u>

- Dynamic pressure of 350 bar (5,000 psi)
  - → Cracks at high pressure intersections
    Root cause: Surface quality

Conclusion

<u>Specification Fulfillment</u>

- Weight reduction: 55% (≥ 30%)
- Manufacturing costs: Comparable to benchmark
- Functionality: Identical
- Interface: Equal to reference
- Reliability: Depending on post processes

→ **SLM technology is applicable for hydraulic aircraft manifolds**

→ **High lightweight capability demonstrated**

→ **Surface quality need to be improved to reach impulse/fatigue cycles**

Outlook

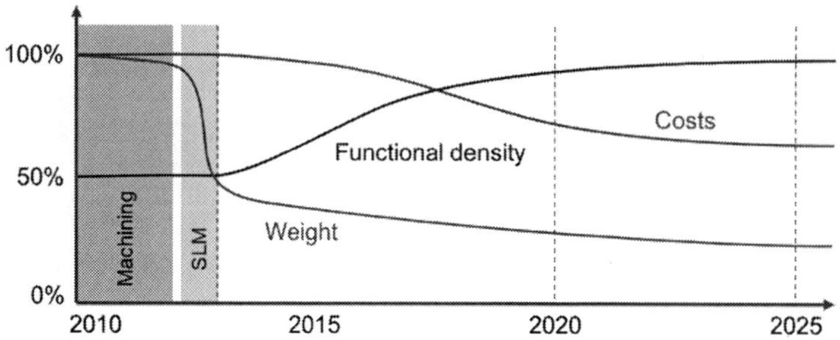

Machining	**SLM 2014**	**SLM 2020**
Benchmark	**Component replacement**	**System integration**

Federal Ministry of Economics and Technology

Funded by the Federal Ministry of Economics and Technology in the frame of LuFo IV

Thanks for your attention

AIRBUS AN EADS COMPANY

LIEBHERR

Additive Metal Applications for Gas Turbine Engine Components

By: Greg Morris

imagination at work

GE Proprietary Information
Export Authorization, ECCN: 9E991, NLR

Who We Were..Who We Are

imagination at work

Pre-GE Acquisition

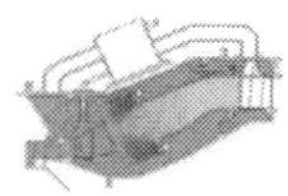

MTI & RQM

- ➢ Founded 1994
- ➢ Introduced DMLM to NA in 2003
- ➢ 135 employees
- ➢ Services included:
  - ➢ Engineering & Design
  - ➢ Additive (Metals/Polymers) (25 additive machines)
  - ➢ Post-Processes (Machining, Thermal Processing, Finishing)
  - ➢ Collaborations/JV's..i.e. MicroTek Finishing w/BinC, etc.
- ➢ Customer Base: Aerospace, Medical, Consumer Goods, Energy

GE Acquisition of MTI & RQM (Nov 2012)

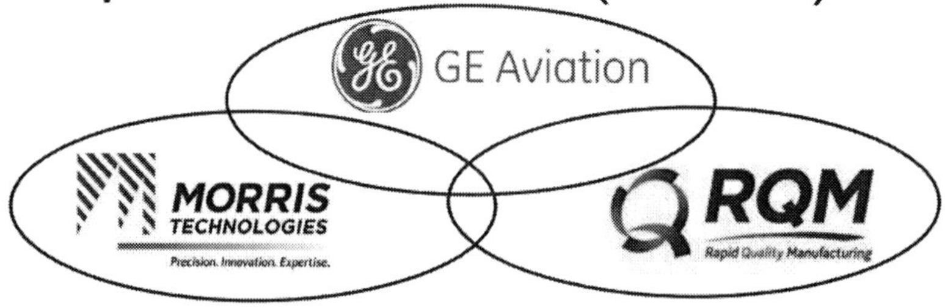

Additive Development Center
- GEA focus
- Other GE
- Select Externals

Additive Lean Lab / Pilot Production
- Low rate initial production

GE Aviation ADC:

SLA

Arcam EBM

FDM

Phenix

PolyJet

EOS DMLS

Concept Laser

SLM

imagination at work

GE Proprietary Information

Photos Courtesy: 3D Systems, Stratasys, EOS, SLM

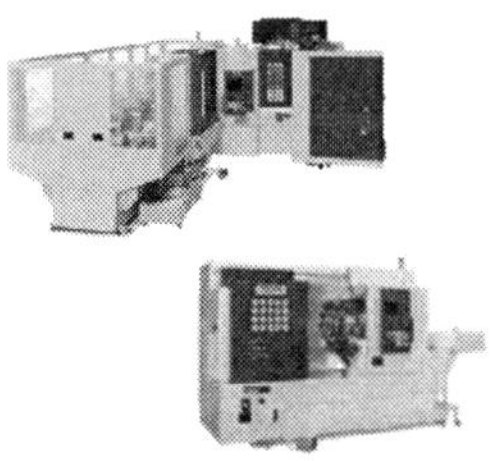

Machining

Thermal Processing

Metallurgical Lab

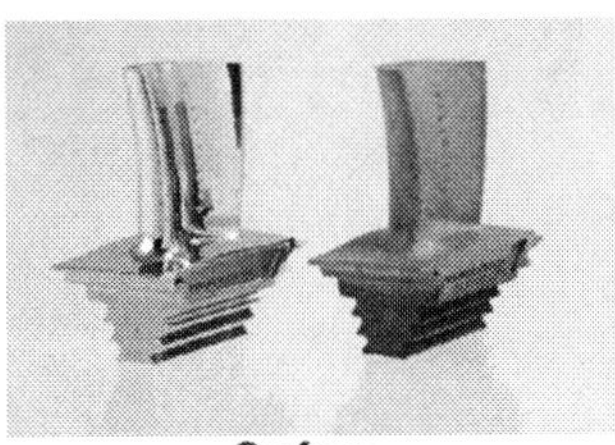

Surface Finishing

Special Processes

imagination at work

GE Proprietary Information

Production Prep:

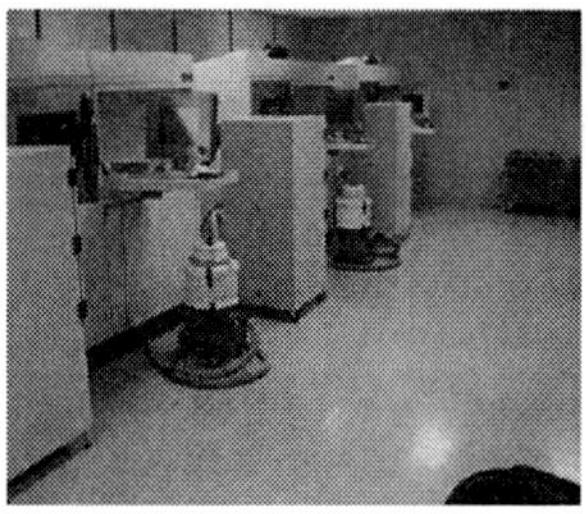

DMLM: (7) M270's

Inspection

Thermal Processing: (1) Vacuum Stress, (1) HIP

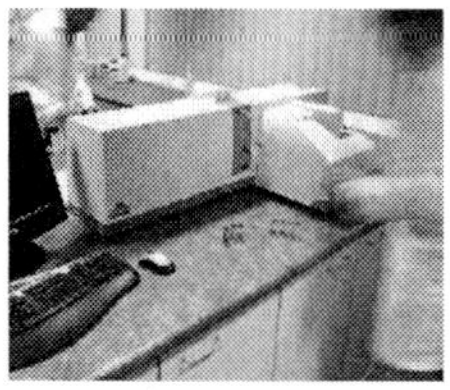

Powder Size Analysis

imagination at work

GE Proprietary Information

Material Properties Enable Applications

imagination at work

GE Proprietary Information

Material Properties Drive Applications

- Super Alloys (2005)
- Many similar to wrought strengths
- Some materials characterized
- Large opportunity for novel alloy development

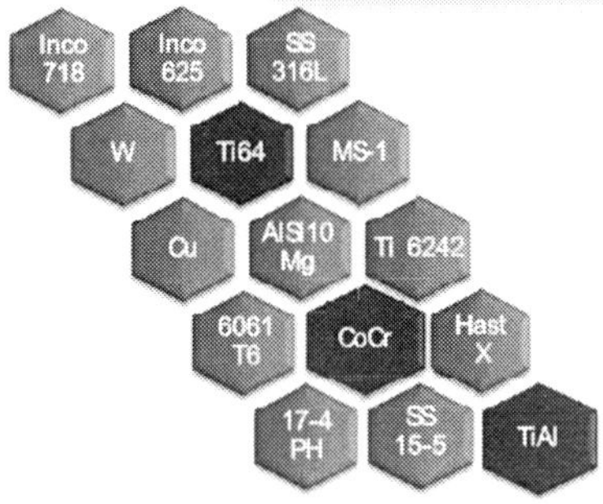

imagination at work

GE Proprietary Information
Export Authorization, ECCN: 9E991, NLR

Material Properties Drive Applications

Co-28Cr-6Mo

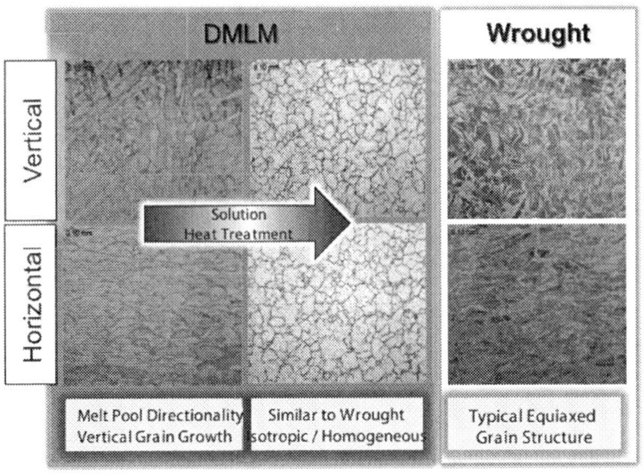

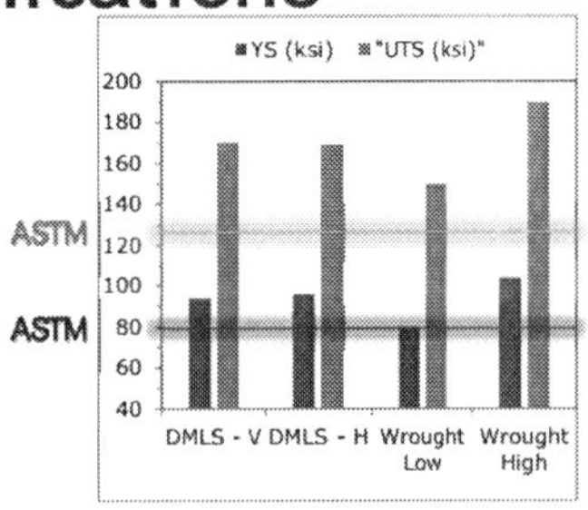

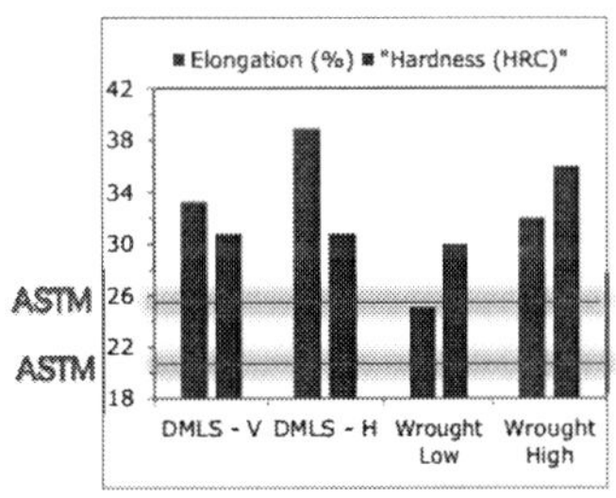

imagination at work

GE Proprietary Information
Export Authorization, ECCN: 9E991, NLR

Material Properties Drive Applications

Ti-6Al-4V

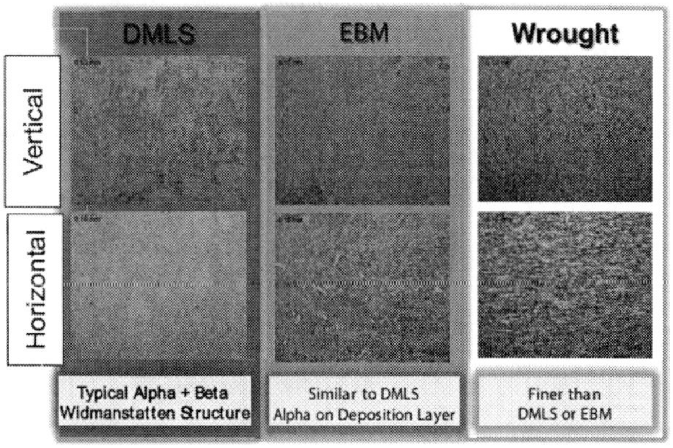

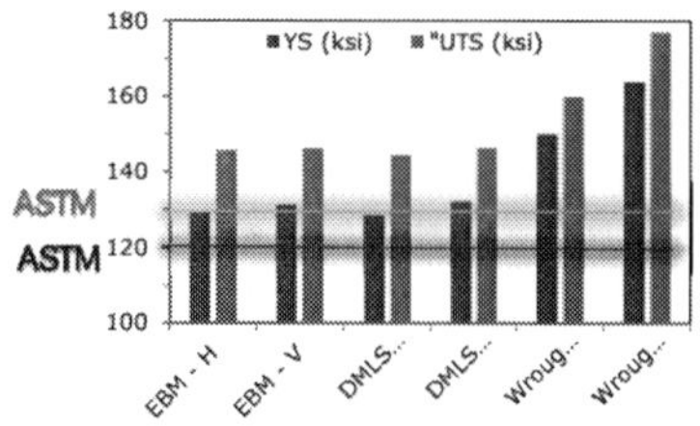

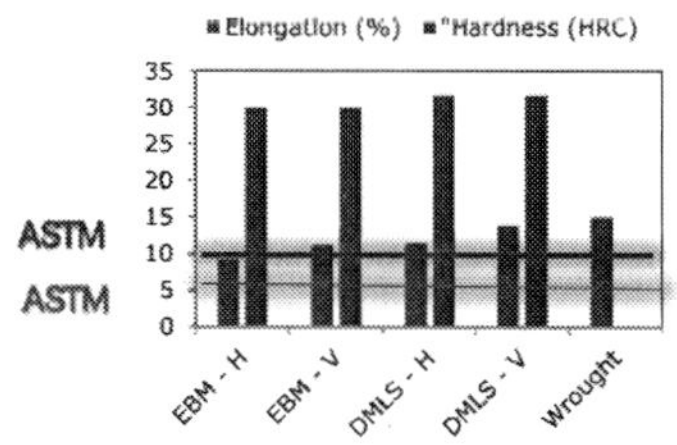

imagination at work

Additive - NPI Hardware

imagination at work

Instrument Probe - 2005

imagination at work

GE Proprietary Information

Instrument Probe - 2007

- Successful builds (no crashes)
- Less cost to produce
- Higher test life
- More accurate probes
- Lower number of probes required
- Better data collection
- Shorter fabrication time

imagination at work

GE Proprietary Information

Leveraging AM – NPI Hardware

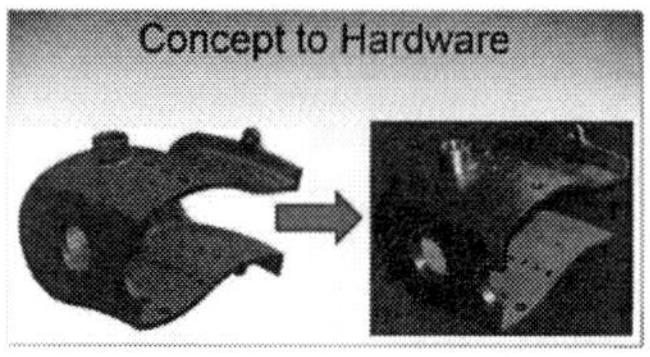

- 7th set of HPT's in 6 years
- Substantial cost and lead time reduction
- Design refinement options

Fuel Swirlers

- Land-based Gas Turbine Engines
- 90 parts in (3) months
- Met testing timetable
- Eventually will be castings

imagination at work

GE Proprietary Information

Engine Component Repair Tool

- Inco 718 DMLM
- 20% cost of traditional
- Optimal design for longer life
- Days vs. weeks acquisition time

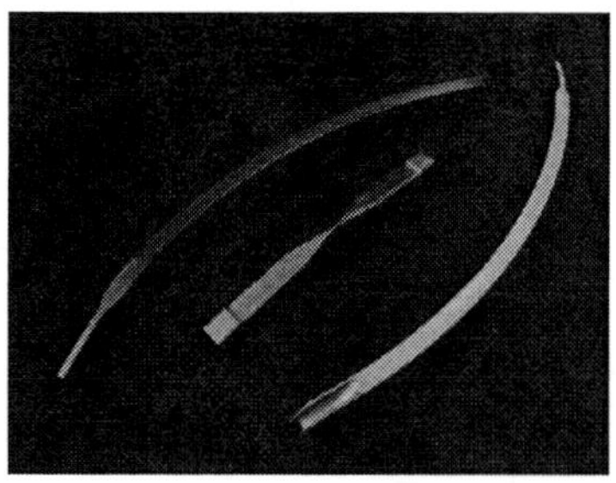

imagination at work

GE Proprietary Information

2/4/2014

Moving Toward Production...

imagination at work

Design For Additive

Topology & Shape Optimization

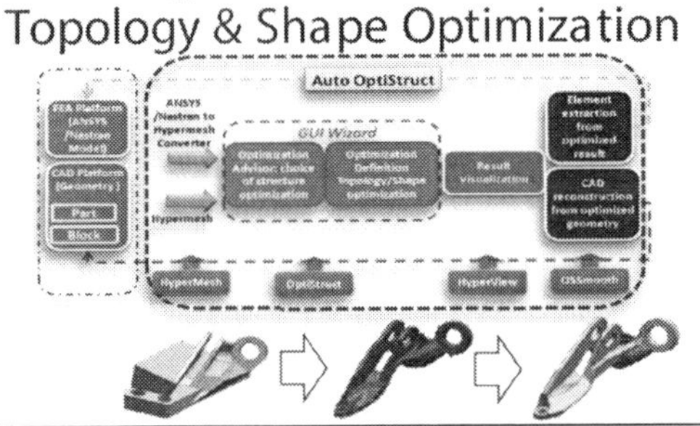

Crowdsourcing Design Challenge

Public challenge to find new design strategies for additive manufacturing

150 entries in less than 30 days

Top designs will be manufactured and tested by GE

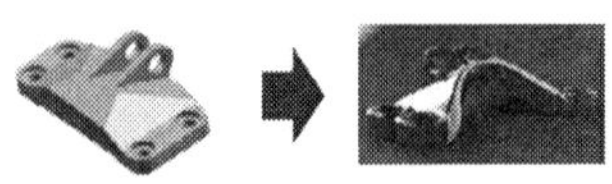

Original design

Example entry

Makers Guild

Corporate initiative to gain exposure and expertise in the world of 3D printing

Learn about:

- Build orientation
- Shape capability
- Support structure

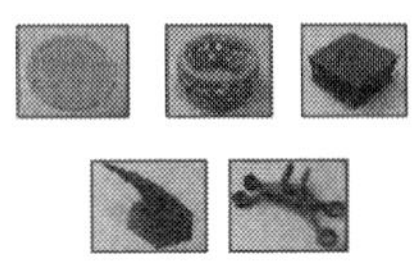

imagination at work

GE Proprietary Information

Design Challenge and the Potential of Additive

- Tapped into 750k+ online community
- Over 700 entries received
- Many submissions over 80% weight reduction

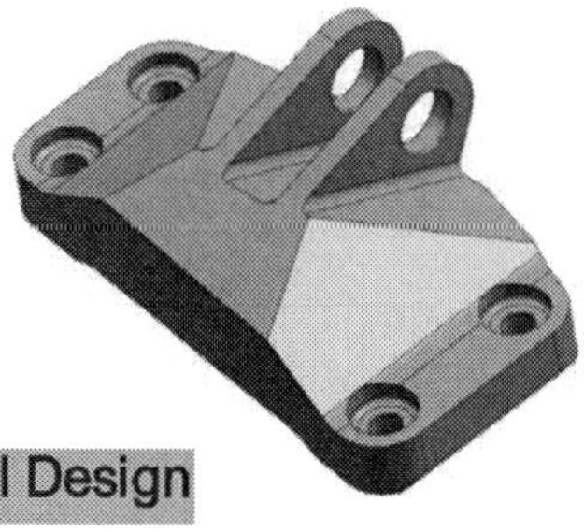

Traditional Design

imagination at work

GE Proprietary Information

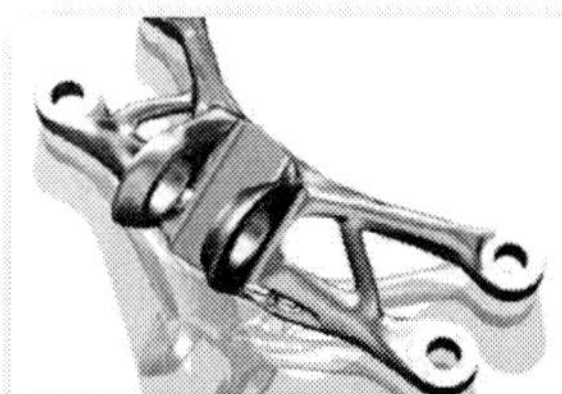

imagination at work

GE Proprietary Information

Ti64 Grown Brackets

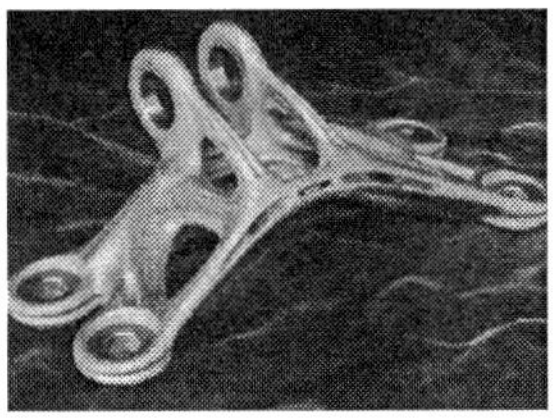

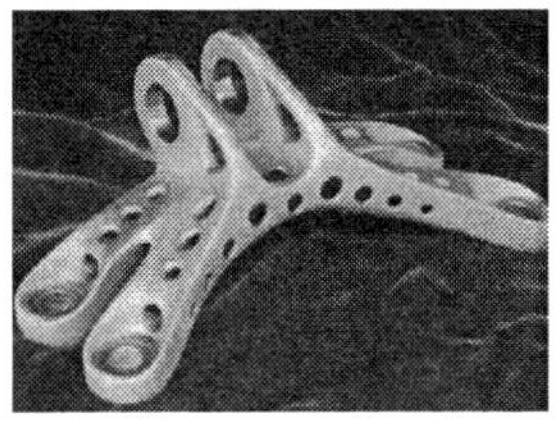
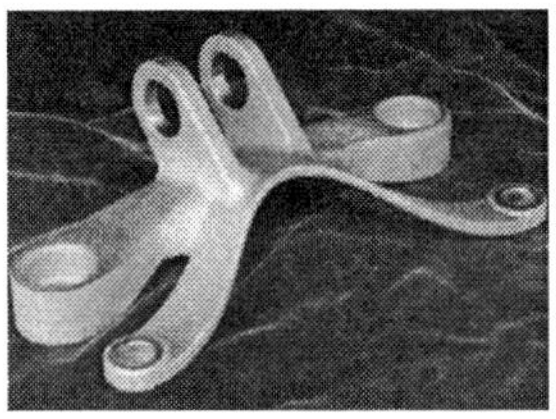
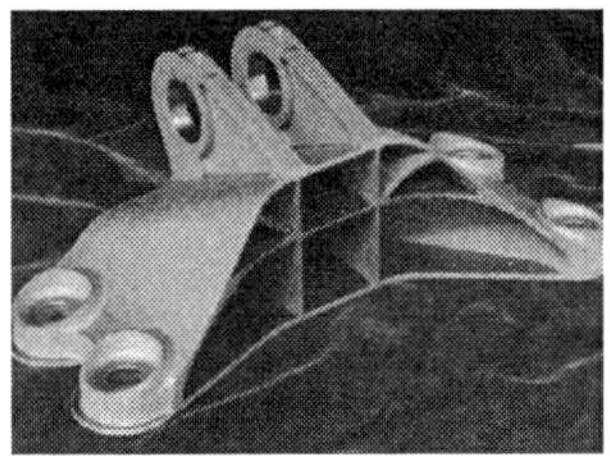

imagination at work

GE Proprietary Information

2/4/2014

DMLM Design Advantage

LEAP Fuel Nozzle

imagination at work

GE Proprietary Information
Export Authorization, ECCN: 9E991, NLR

DMLM Fuel Nozzle Design History

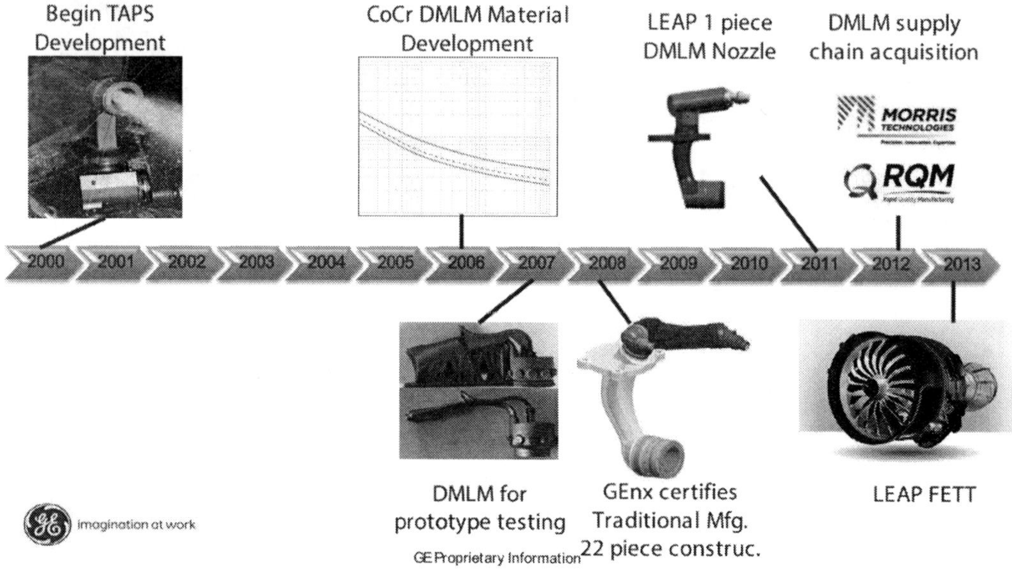

imagination at work

GE Proprietary Information

LEAP 56 Next Generation Fuel Nozzles

- 19 Fuel Nozzles in each engine – 40K full production
- 20 pieces to 1
- 5x durability improvement
- 25% weight reduction (compared to GEnx)
- Fuel Nozzle is a very efficient heat exchanger

imagination at work

GE Proprietary Information

Additive Blade Mfg

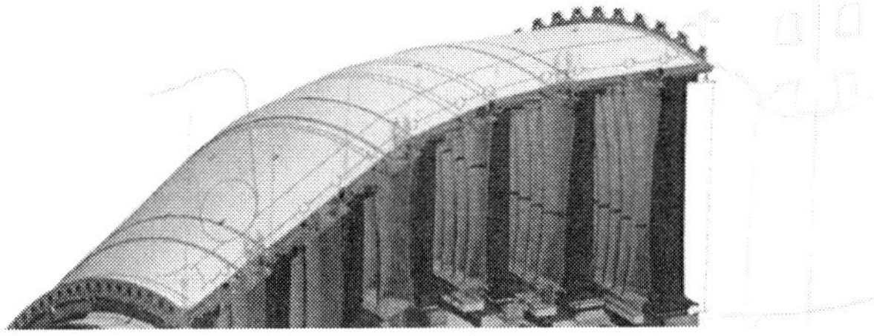

Ti Alloys

imagination at work

GE Proprietary Information

2/4/2014

Improved Fuel Consumption via Weight Reductions

- Combine parts to eliminate weight
- Performance gains
- Maintenance reduction
- Optimal structure designs

Common Materials, New to Additive
1000# Identified in Leap Configuration

imagination at work

GE Proprietary Information
Export Authorization, ECCN: 9E991, NLR

Final Thoughts...

imagination at work

GE Proprietary Information

What Phase Are We In?

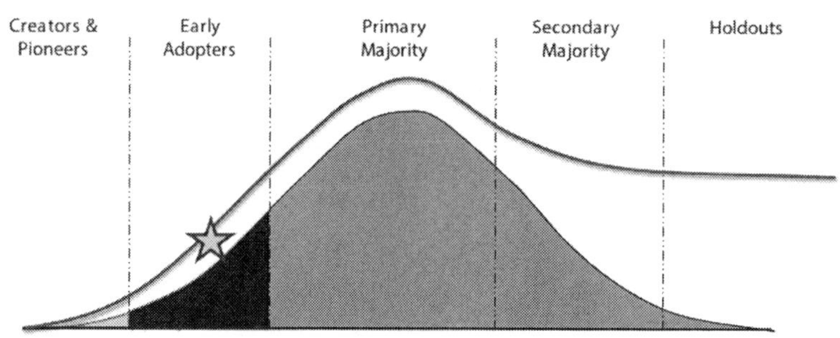

Technology Maturation and Acceptance Curves

imagination at work

GE Proprietary Information
Export Authorization, ECCN: 9E991, NLR

Accelerating Adoption

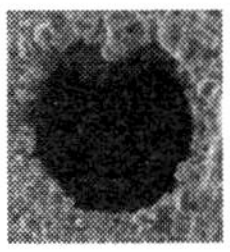

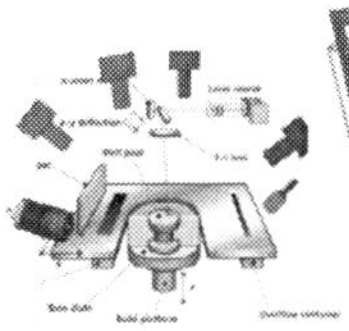

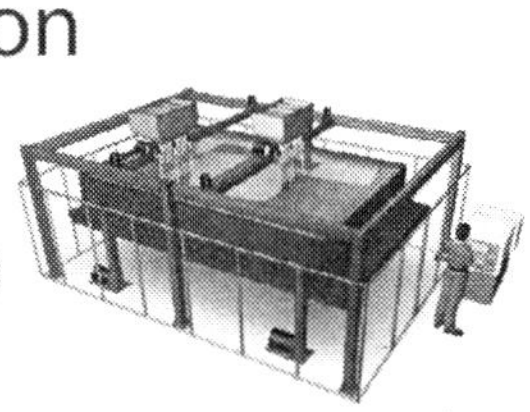

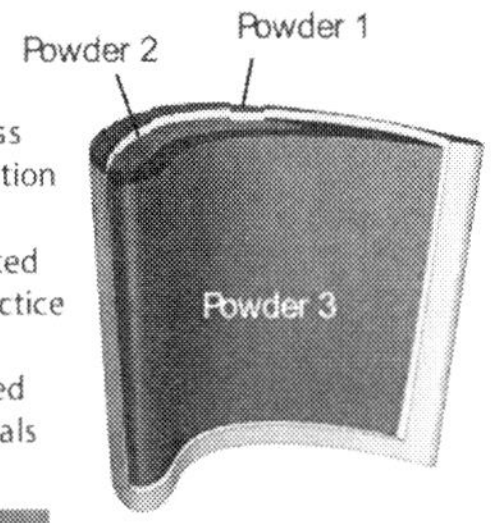

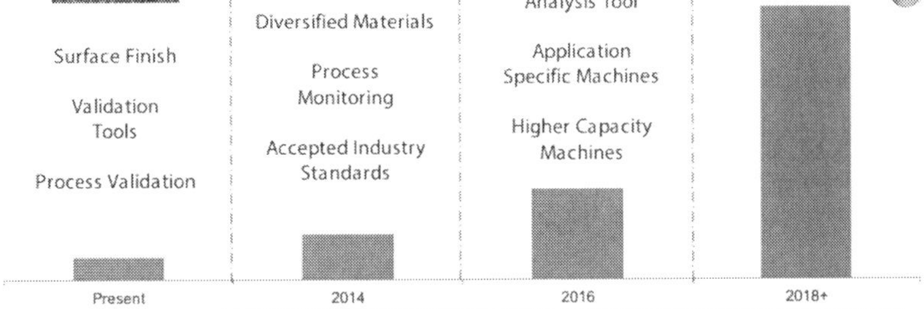

Growing Acceptance with Continuous Innovation

imagination at work

GE Proprietary Information
Export Authorization, ECCN: 9E991, NLR

Supply Chain Shift?

From

To

imagination at work

GE Proprietary Information
Export Authorization, ECCN: 9E991, NLR

Broad Trends

- ➢ Challenging IP landscape
- ➢ Further consolidation (all areas)
- ➢ Increased adoption of technologies
- ➢ Accelerating innovation
- ➢ New business models

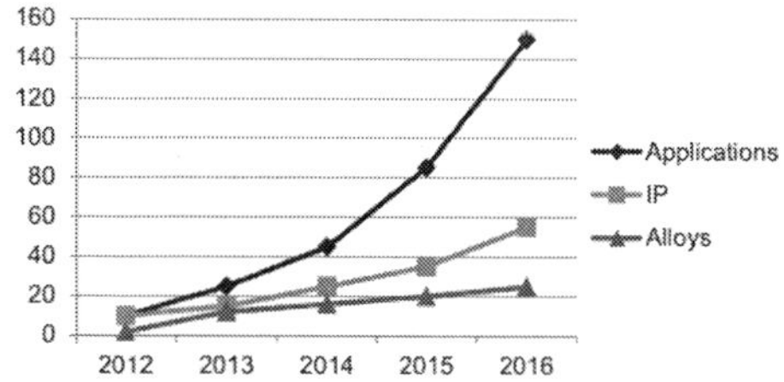

imagination at work

GE Proprietary Information

Thank you!

imagination at work

GE Proprietary Information
Export Authorization, ECCN: 9E991, NLR

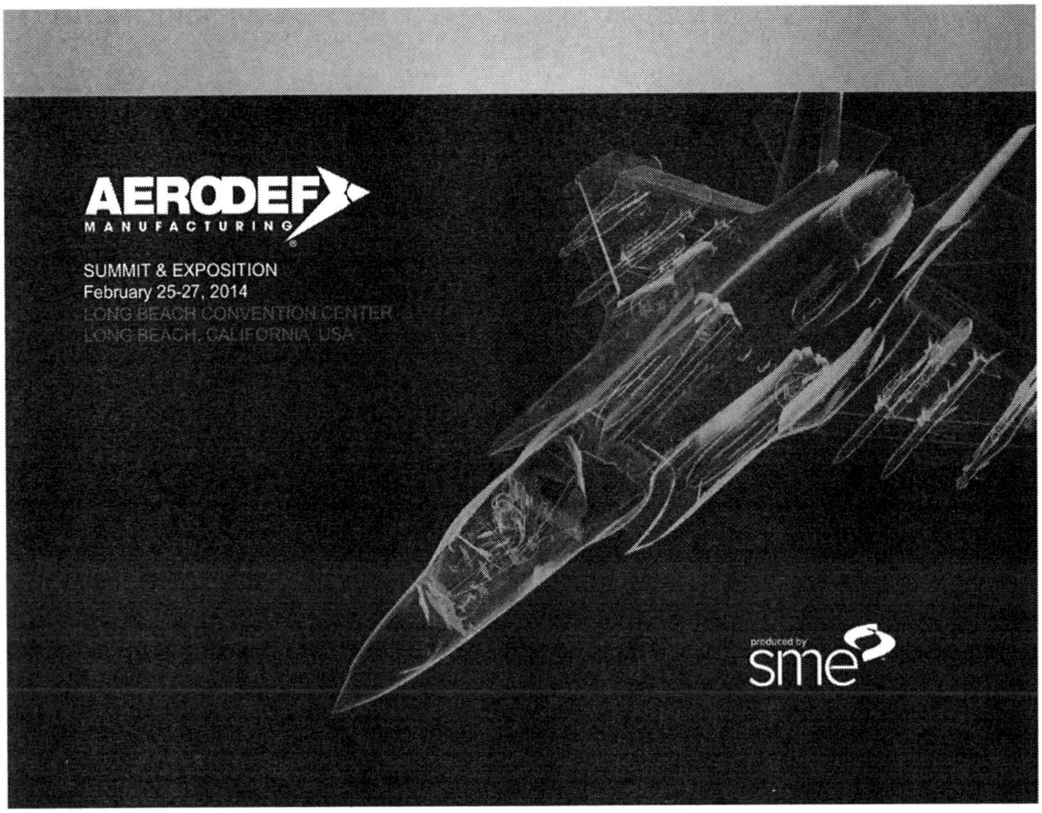
AERODEF
MANUFACTURING
SUMMIT & EXPOSITION
February 25-27, 2014
LONG BEACH CONVENTION CENTER
LONG BEACH, CALIFORNIA, USA
produced by
sme

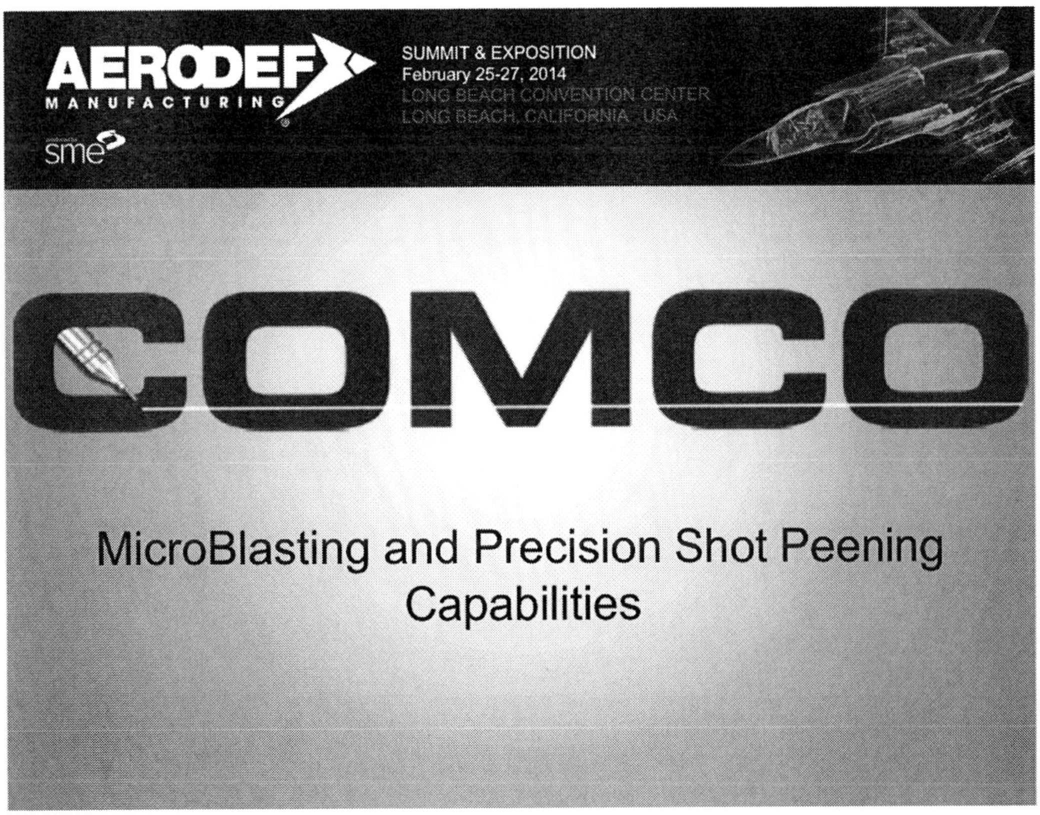
AERODEF
MANUFACTURING
sme
SUMMIT & EXPOSITION
February 25-27, 2014
LONG BEACH CONVENTION CENTER
LONG BEACH, CALIFORNIA, USA
COMCO
MicroBlasting and Precision Shot Peening
Capabilities

Introduction

- Who is Comco
- Blasting vs. MicroBlasting
- Peening vs. Precision Peening
- Key Factors for the processes
- The processes: Surface Texturing, Controlled Erosion, Precision Deburring, Selective Cleaning and Shot Peening
- Automation and Application examples
- Conclusion

Our First Manufacturing Site

Manufacturing MicroBlasting systems since 1968

What do most people think when we say blasting?

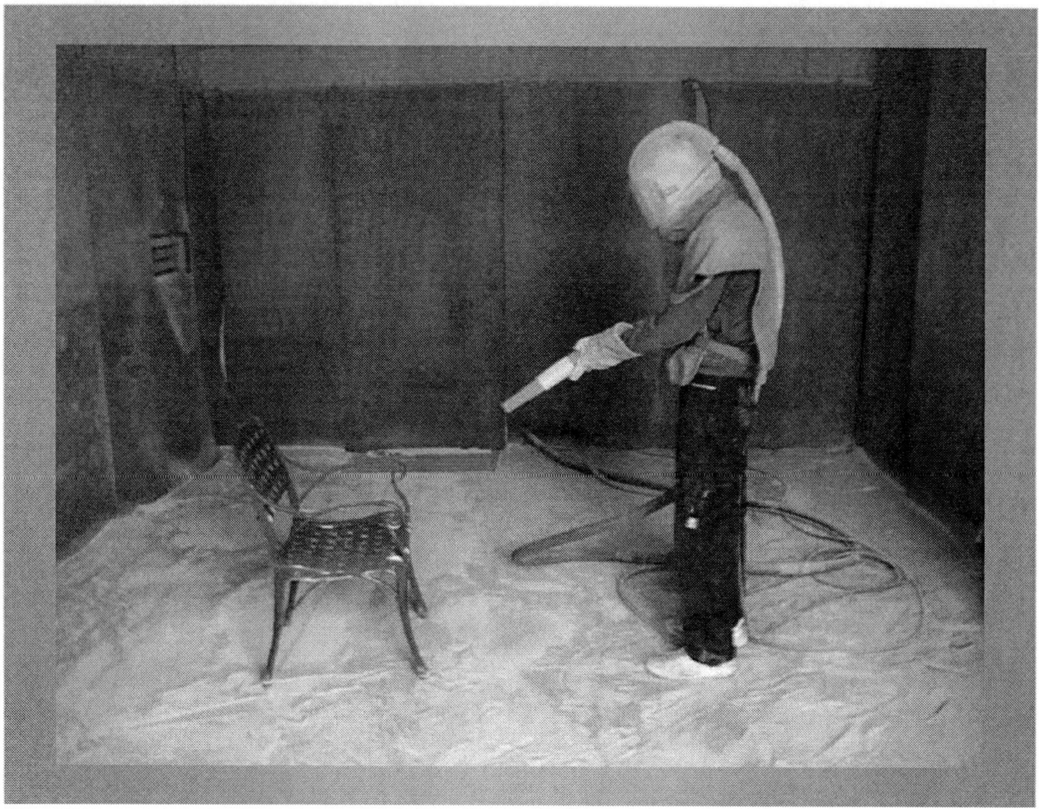

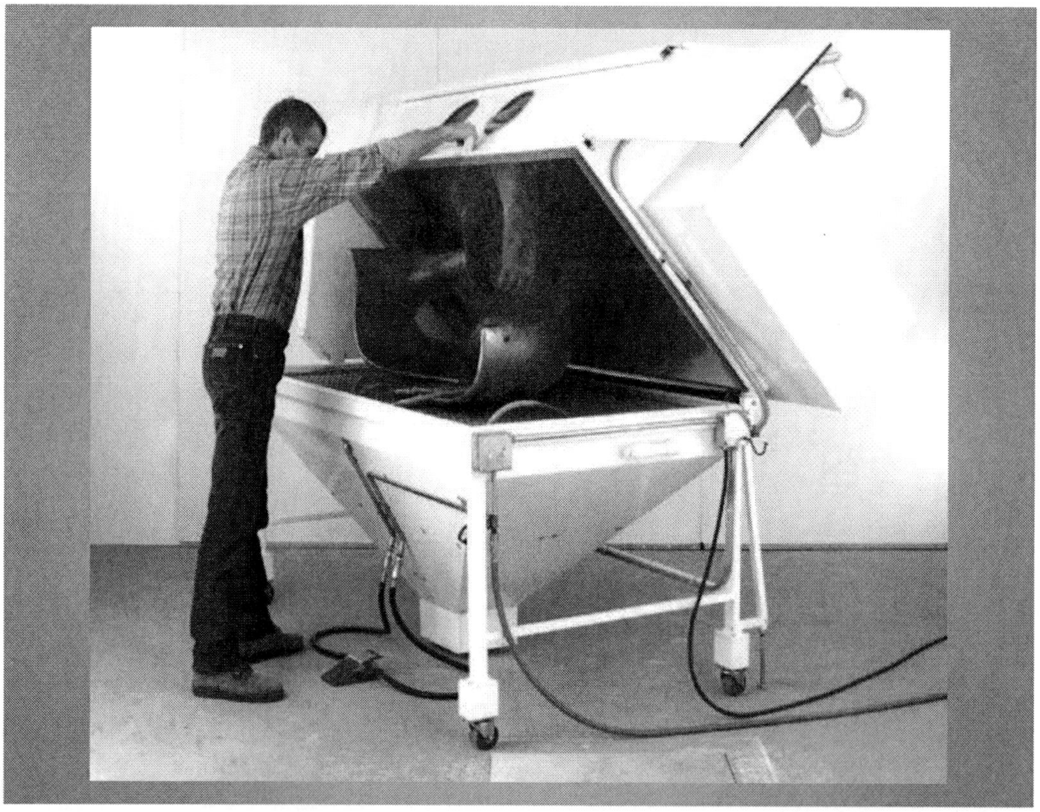

What is MicroBlasting?

MicroBlasting projects a blast of clean, dry air mixed with abrasive media in a focused stream

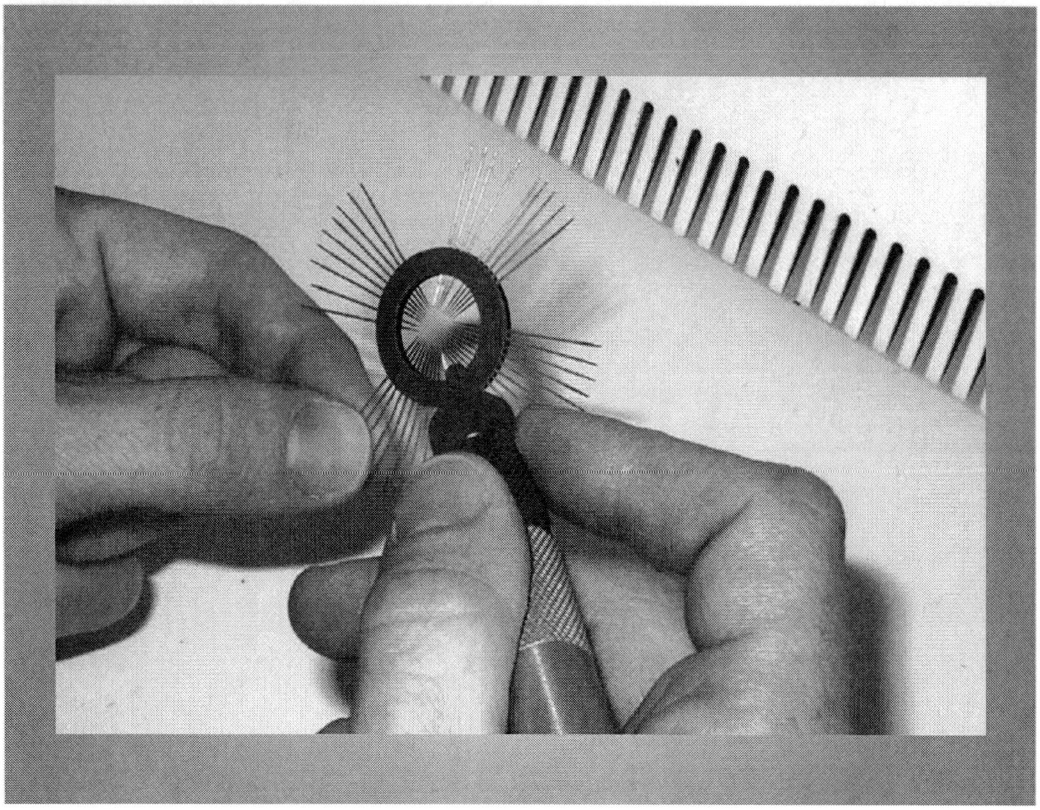

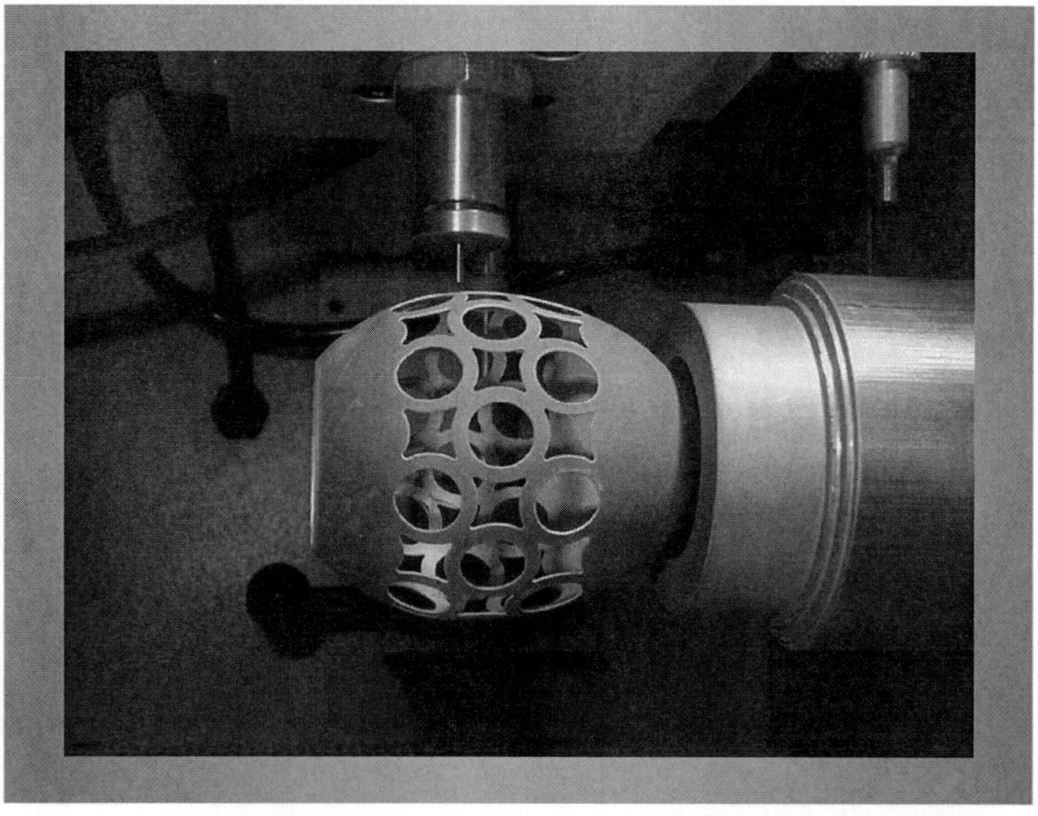

- Focused abrasive stream to minimize the number of particles that miss the part

- Creates a consistent media delivery system to reduce over blasting

- Uses nozzle sizes appropriate for the part/application

- Reducing media flow requirements makes everything else less expensive to operate (Less to convey, less to process, lower air consumption, less media waste)

- Works on a wide range of materials from Nylon to hardened Steel

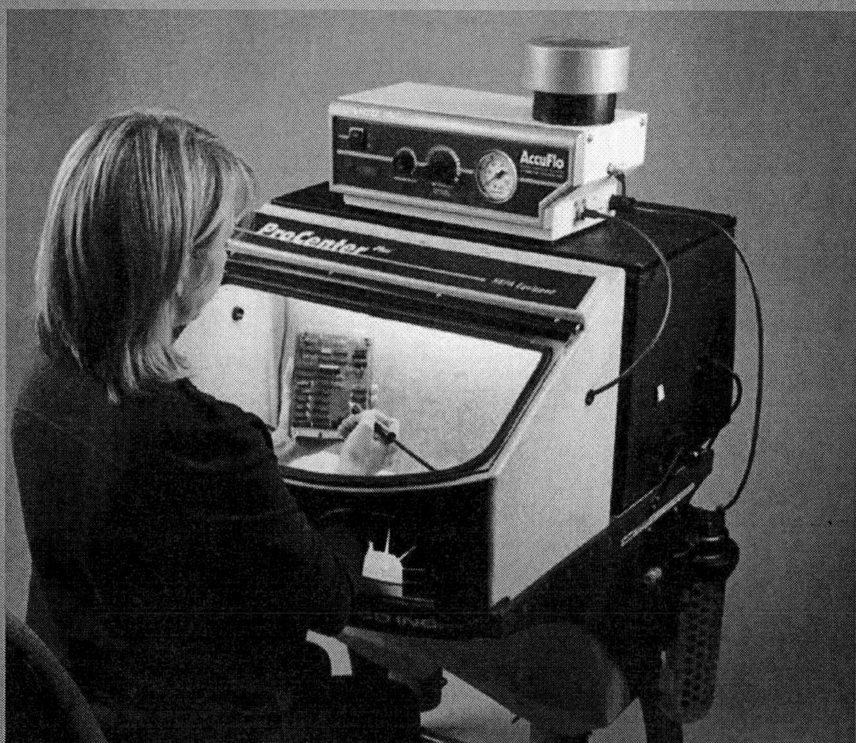

Most people think large when we talk about shot peening (especially for Aerospace)

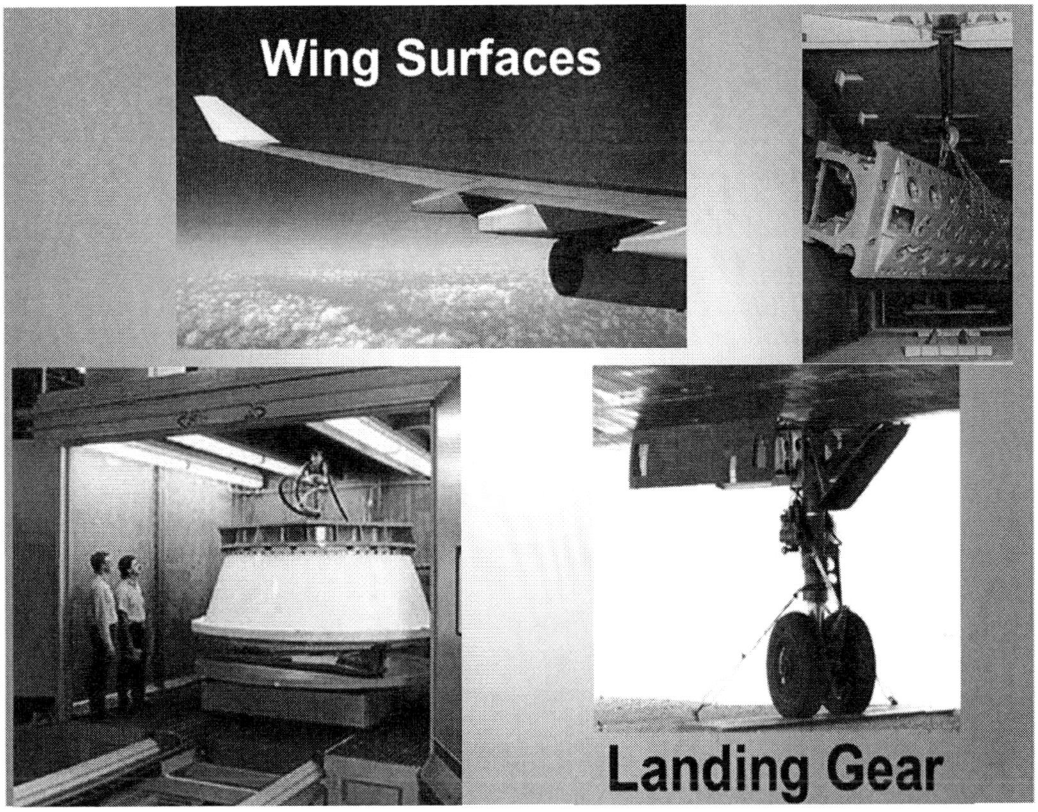

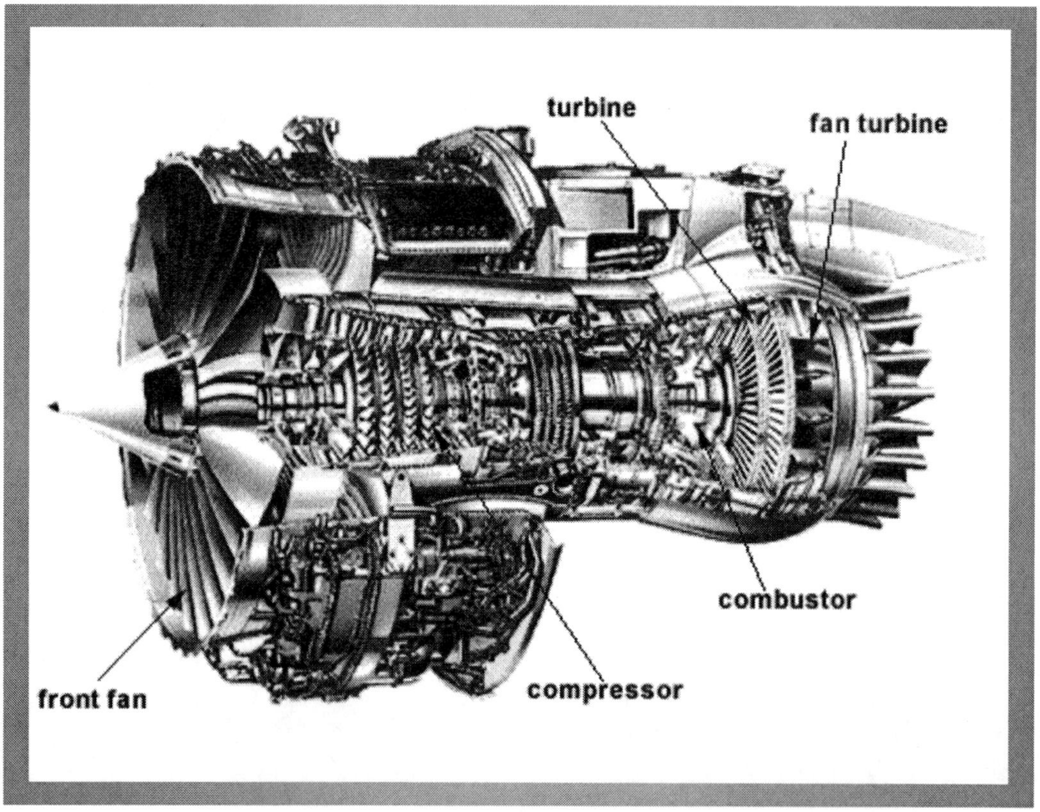
turbine
fan turbine
combustor
front fan
compressor

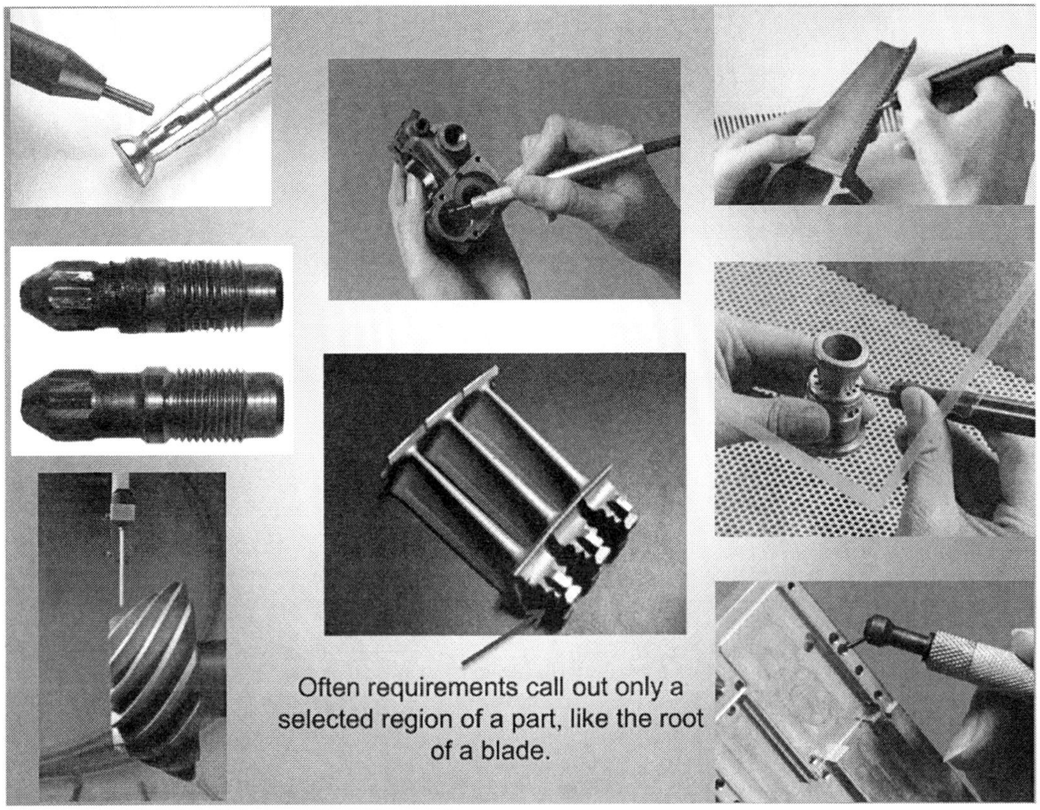
Often requirements call out only a
selected region of a part, like the root
of a blade.

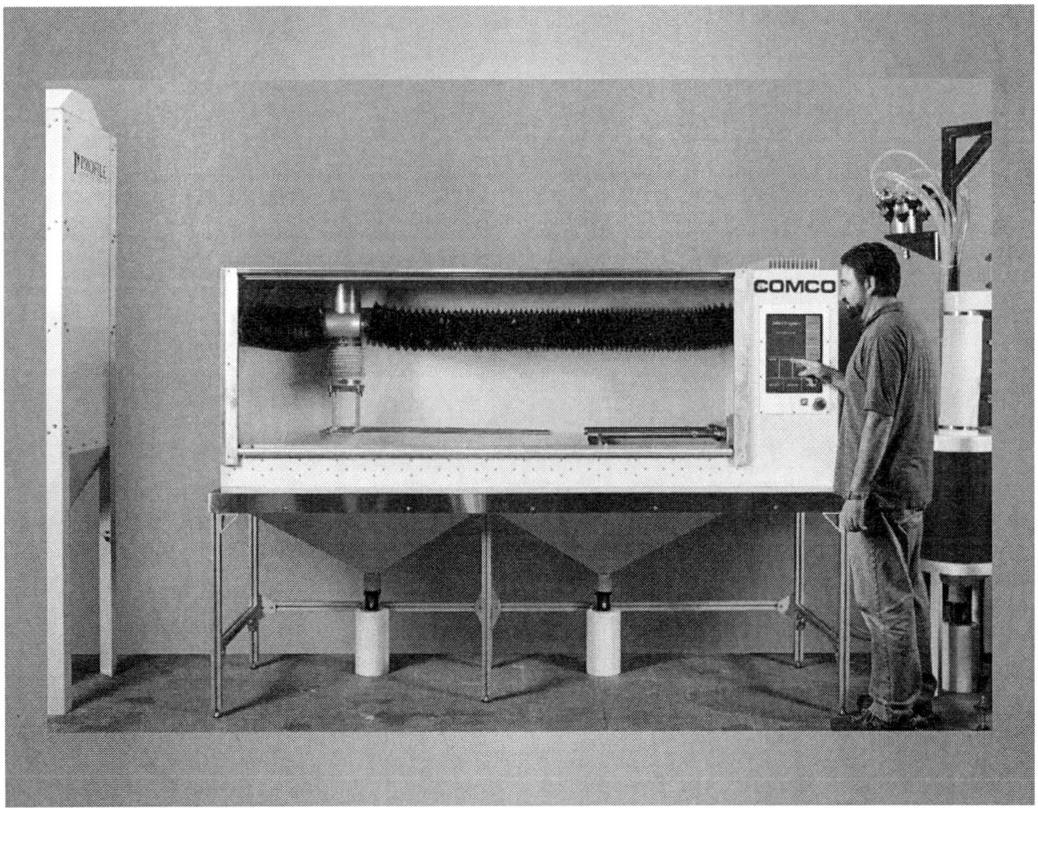
COMCO

What type of parts are ideal candidates for MicroBlasting?

- Parts that are intricate or fragile
- Where excessive or misplaced abrasive can be harmful
- Surfaces that require a highly consistent finish
- High value
- Parts with tight dimensional tolerances
- Precision edges or surfaces
- Parts that have no other good alternative ways to produce

What do we mean by precision?

- Has to be controlled
- Must be repeatable
- Typically difficult to achieve

MicroBlasting and Shot Peening use the same basic fundamentals

The 4 Fundamentals of MicroBlasting and Peening

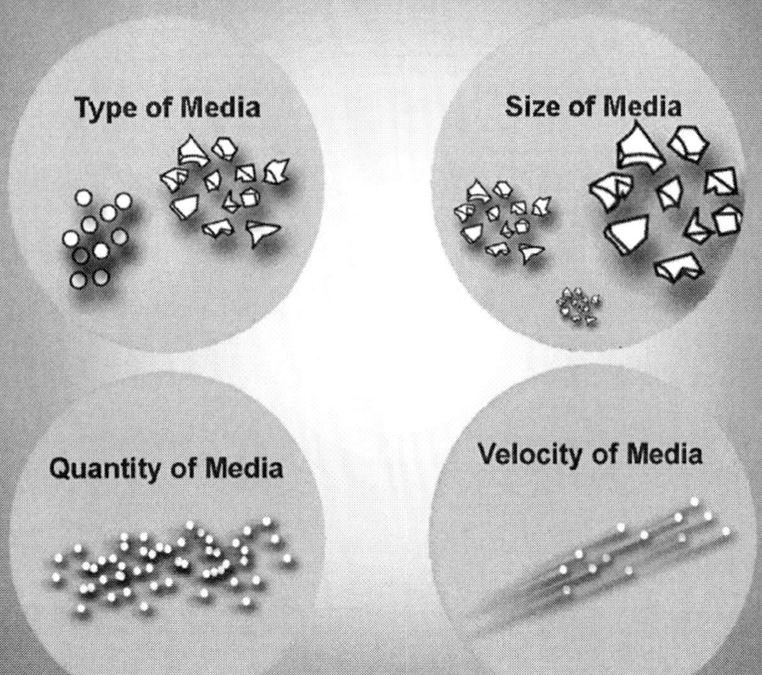

Type of Media
Common MicroBlast Medias
Aluminum Oxide
Plastic
Glass Bead
Sodium Bicarbonate
Silicon Carbide
Walnut Shell

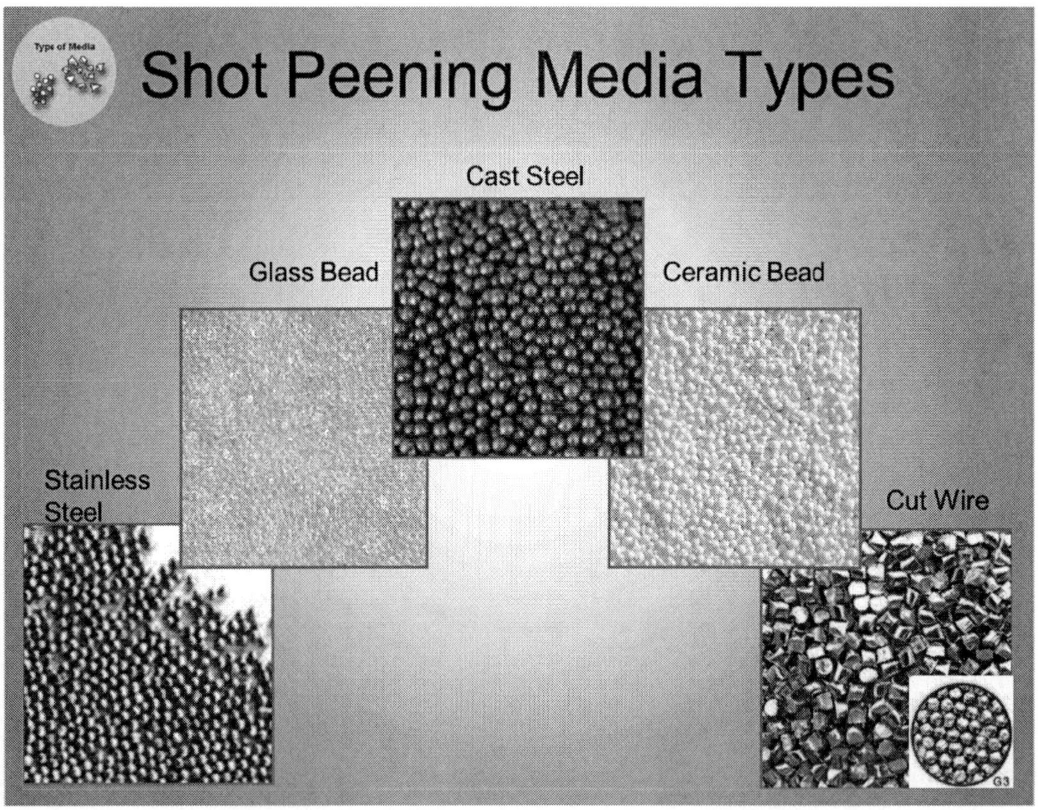
Type of Media
Shot Peening Media Types
Cast Steel
Glass Bead
Ceramic Bead
Stainless Steel
Cut Wire
G3

Selecting Media Type

- Blasting goal (application)
  - Cutting
  - Creating an Ra Finish
  - Inducing Compressive Residual Stress to increase fatigue life
  - Deburring
  - Material Removal
- Process requirements

Range of Media Sizes

- Shot Peening media as fine as 35 microns up to S550 (1,100 microns)
- MicroBlasting abrasives range from 10 – 300 microns

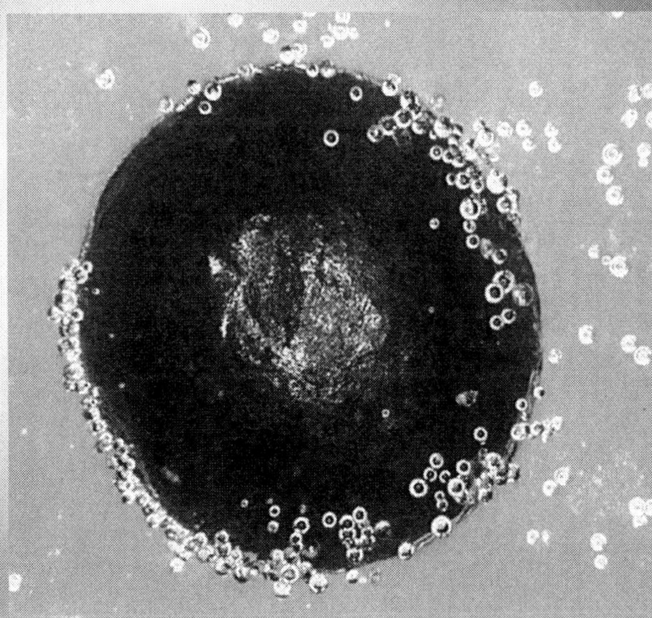

Selecting Size

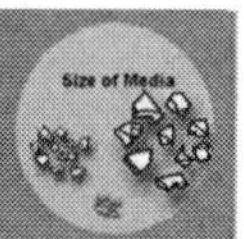

- Intensity (shot peening)
- Roughness
- Removal rate

Selecting Quantity

- Coverage
- Process time

❖Process specific (ineffective at drilling)

❖Does not impact roughness

❖Increased stream density reduces velocity

Selecting Velocity

- Intensity
- Roughness
- Process rate

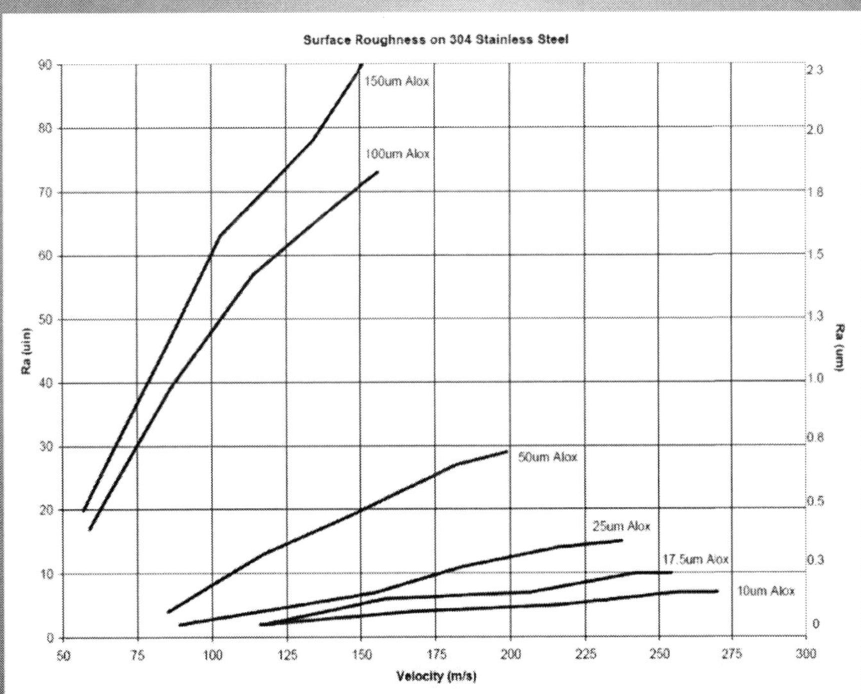

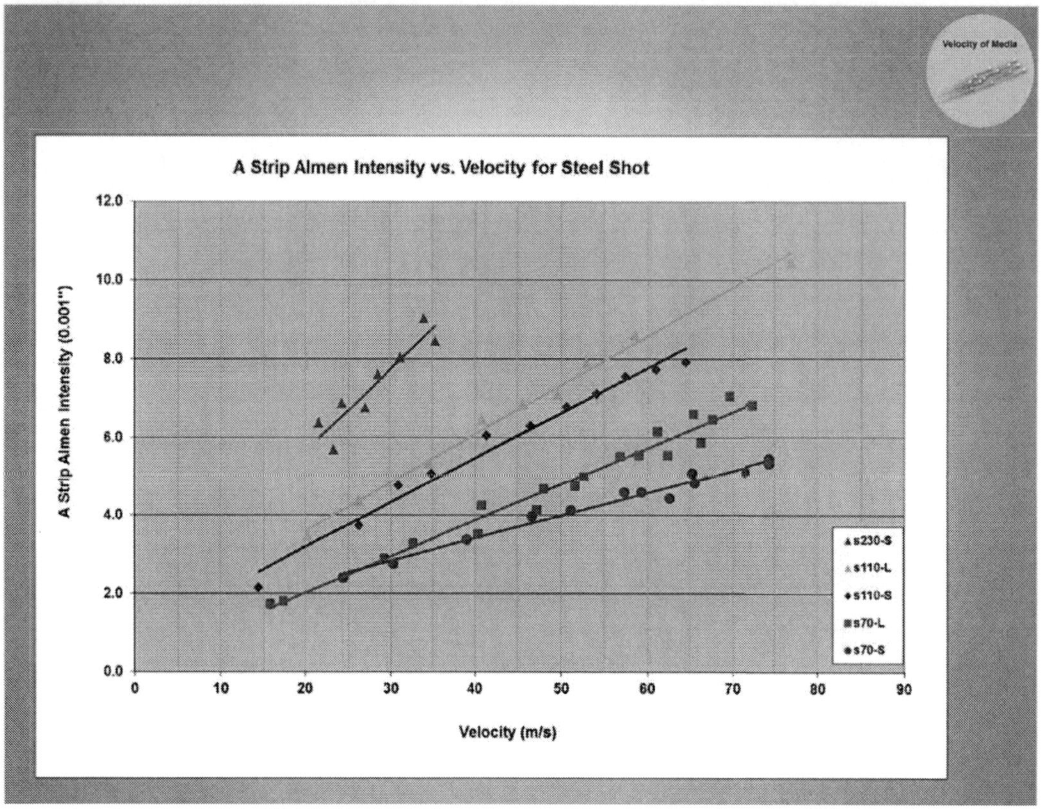
Velocity of Media
A Strip Almen Intensity vs. Velocity for Steel Shot
A Strip Almen Intensity (0.001")
Velocity (m/s)
s230-S
s110-L
s110-S
s70-L
s70-S

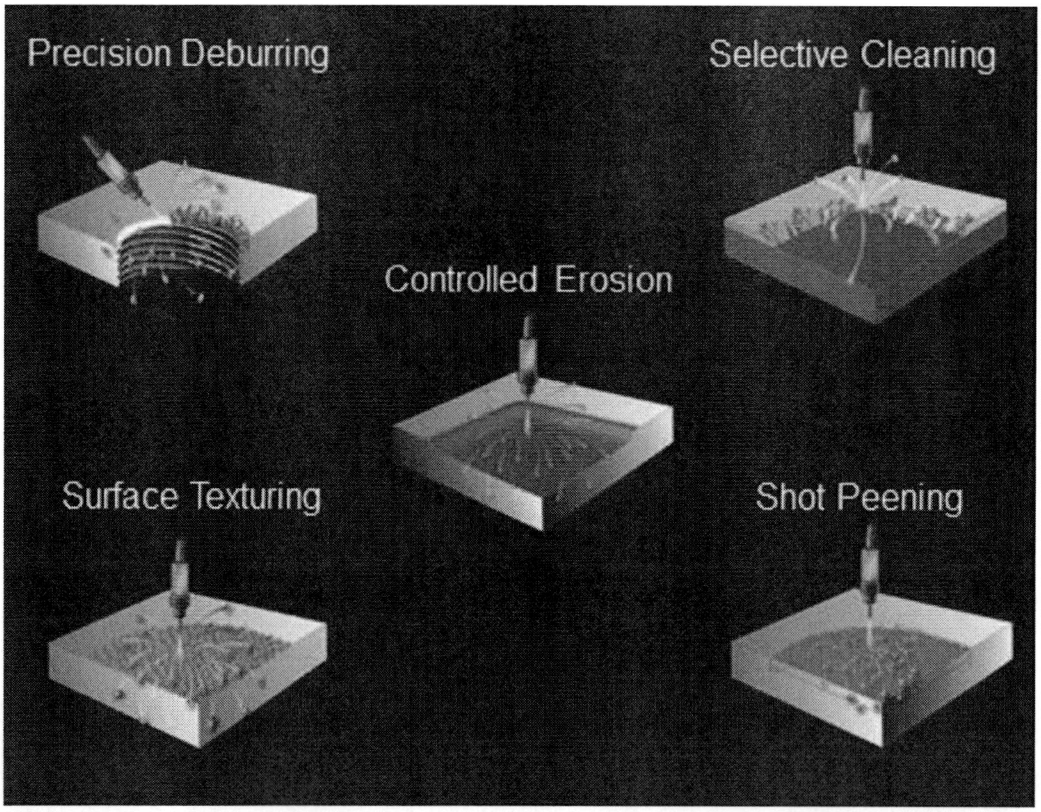
Precision Deburring
Selective Cleaning
Controlled Erosion
Surface Texturing
Shot Peening

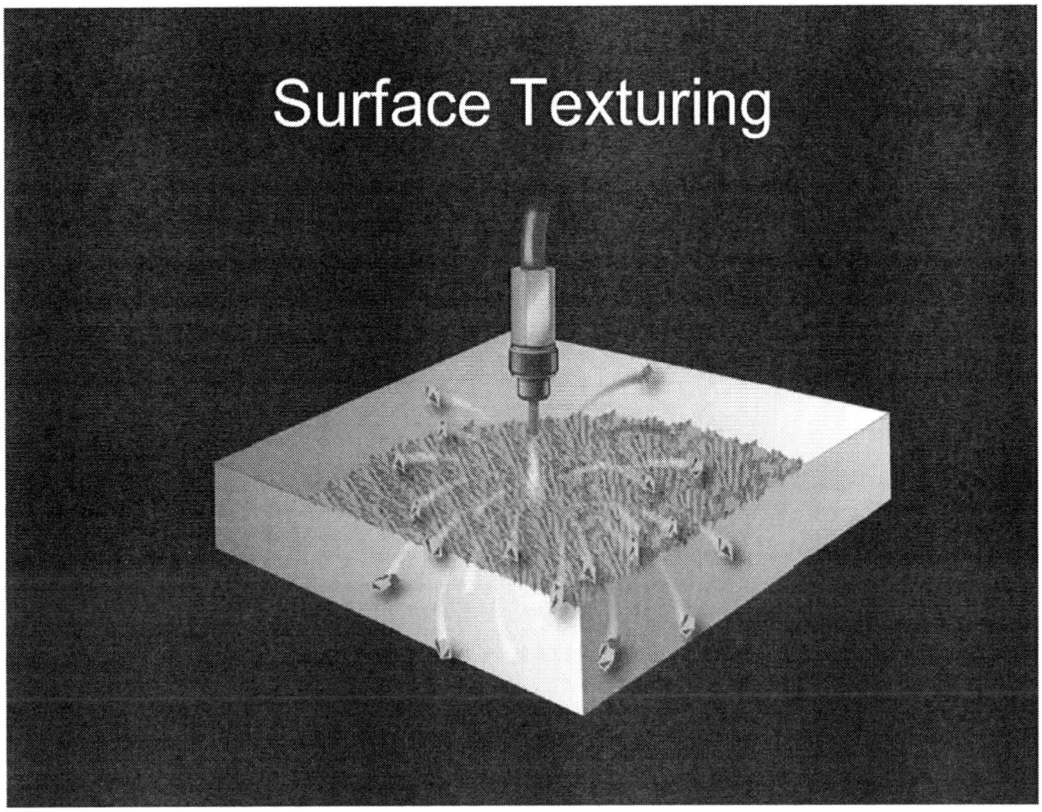
Surface Texturing

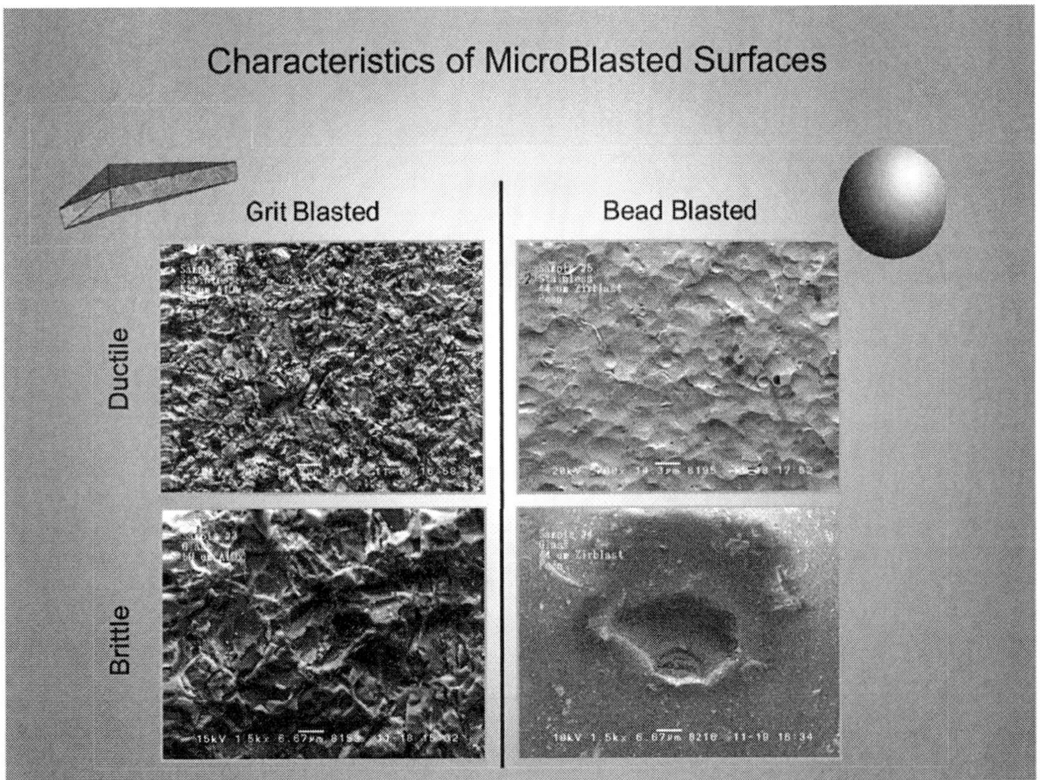
Characteristics of MicroBlasted Surfaces
Grit Blasted
Bead Blasted
Ductile
Brittle

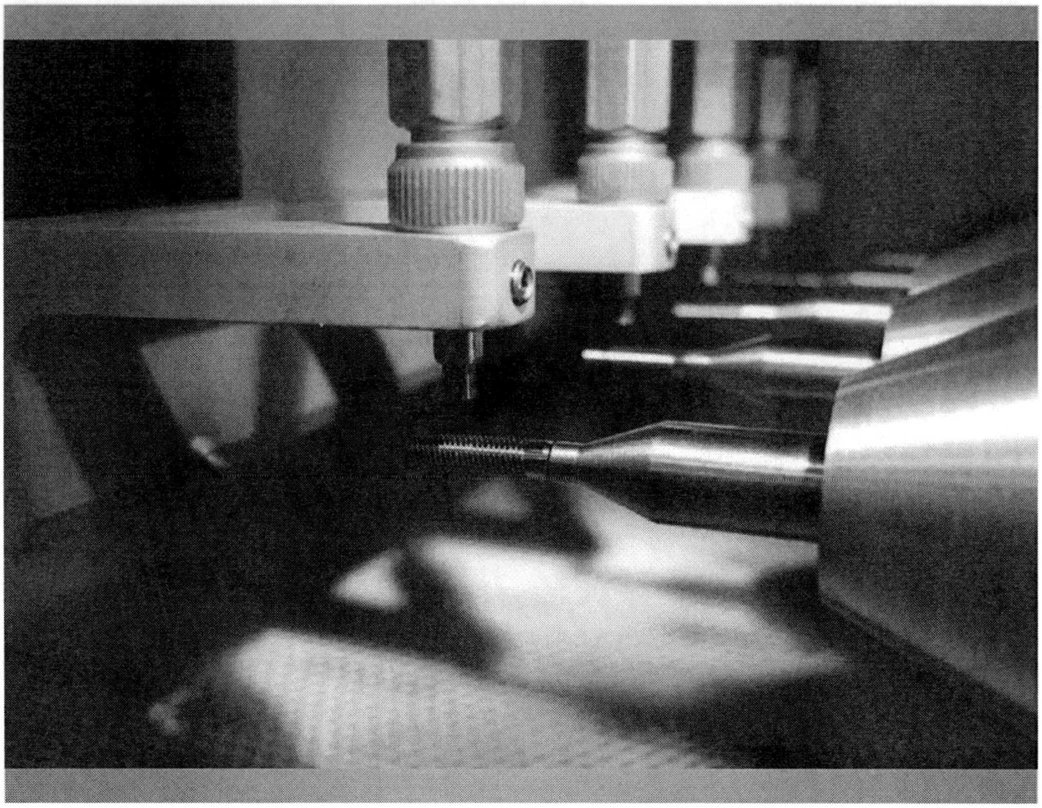

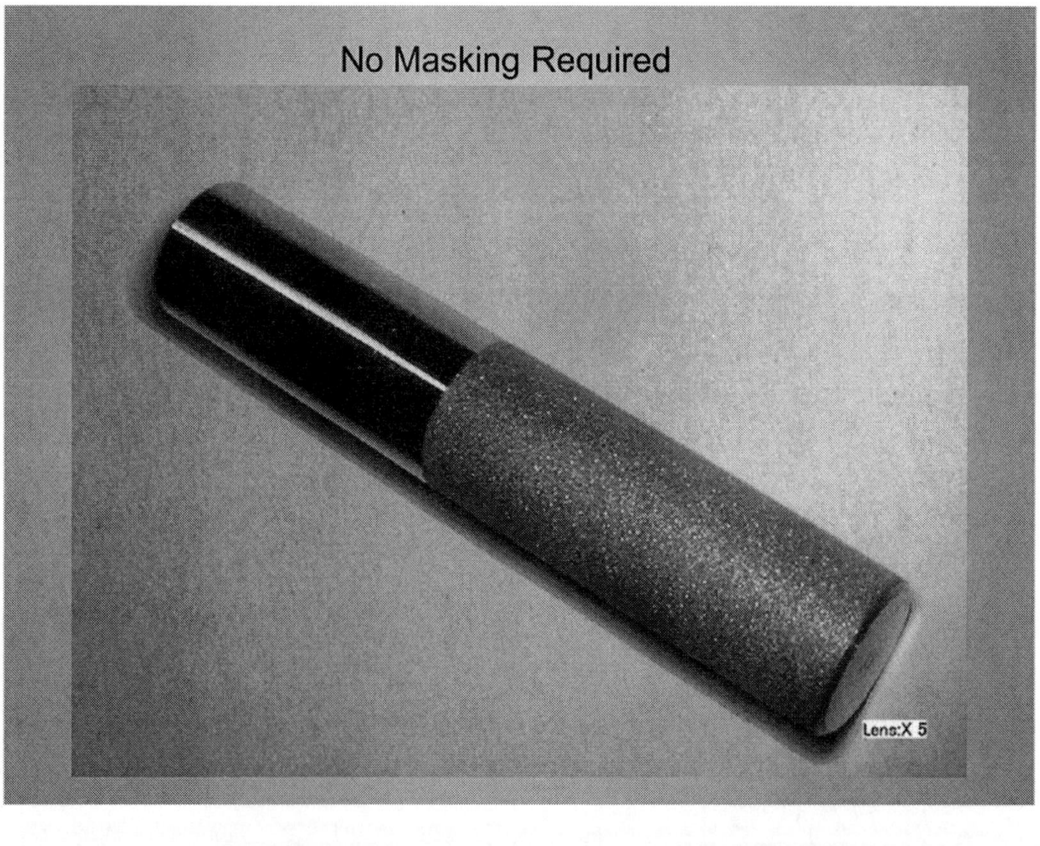
No Masking Required
Lens:X 5

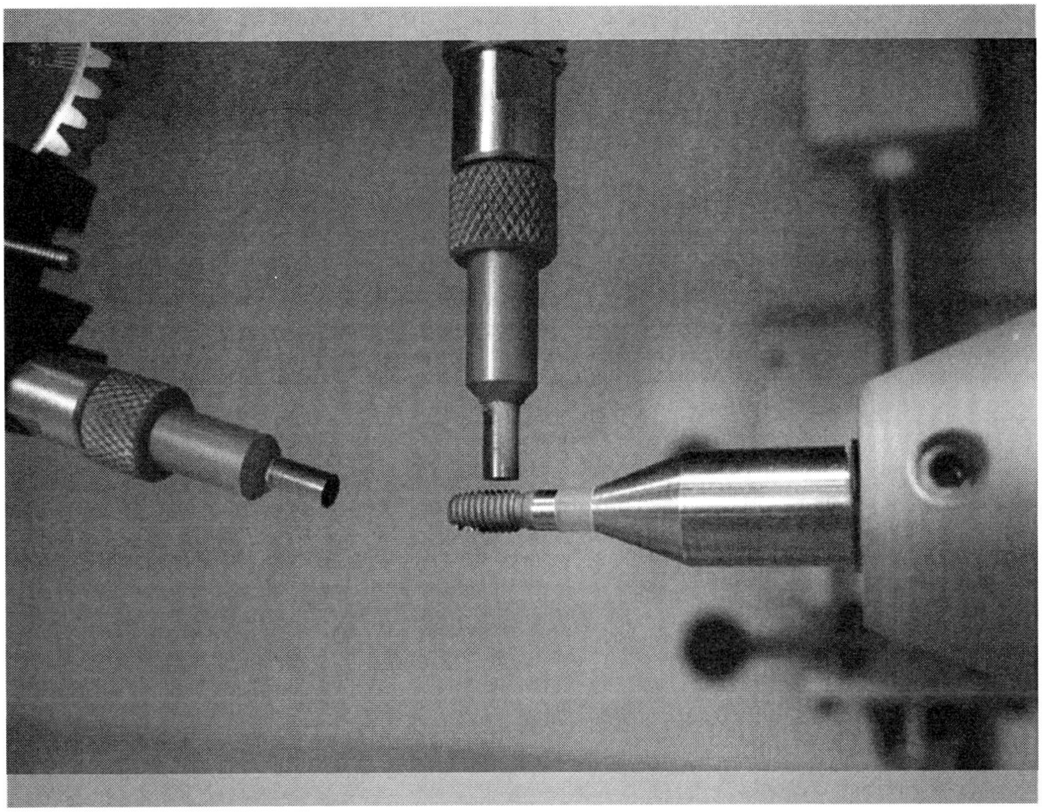

Surface Texturing with MicroBlasting

- Creates a very consistent surface texture
- Optimizes both Ra and Developed surface for specific applications
- Effective results can be achieved without masking
- Avoid dimensional changes, even in threaded areas

Precision Deburring

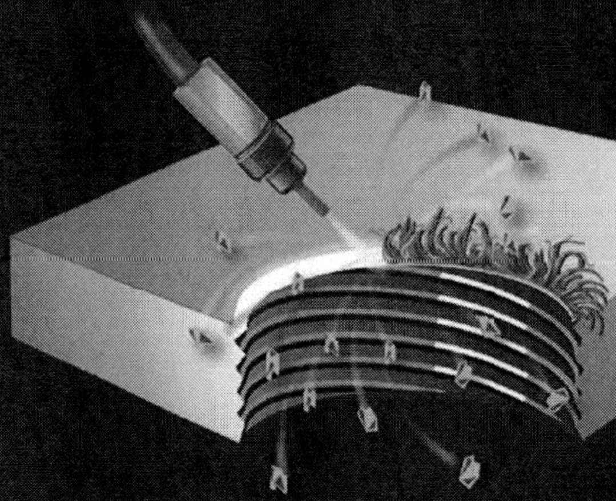

Precision Deburring

- Fine, feather or sliver burrs
- Parts with difficult or intricate geometries
- Hard-to-reach areas
- Cross-drilled holes
- Precision edges or surfaces
- Swiss precision-machined parts
- High value parts and materials
- Tight dimensional tolerances

Heel Burr Removal

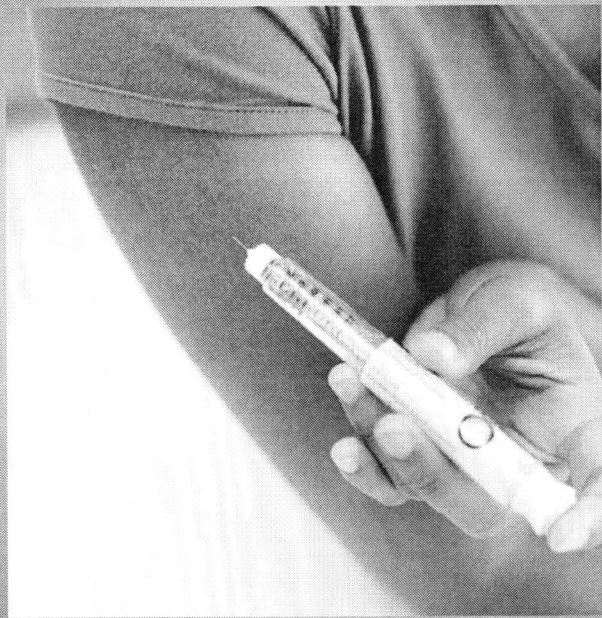

Hypodermic Needle (insulin needle)
Manufacturing challenges:
The needle portion, or cannula, has to be very small. Ideally from 28 - 32 gauge. Roughly .014 - .009". The smaller the tube diameter the easier the injection.

Have to creat a tip that is very sharp minimizing friction. A series of grinds at the tip creates multiple cutting edges.

A perfectly sharp needle is very problematic. It will core the skin rather than creating a trap door. The heel region needs to be slightly blunted without affecting the rest of the needle.

Millions of insulin needles need to be produced every year. The process needs to be able to handle high volume.

The challenges that we had to address to make MicroBlasting work for this application

Has to be very accurately located

Media selection must create the appropriate edge finish without damaging the rest of the cannula.

Requires very consistent intensity to avoid damaging the tip or failing to roll the edge.

Has to be able to handle high volume hundreds of millions per year.

The abrasive stream needs to focus on only the heel of the grind. If too much of the tip is blasted the needle will become dull and cause pain for the patient. In this application a small nozzle is used to very accurately target the glass bead abrasive

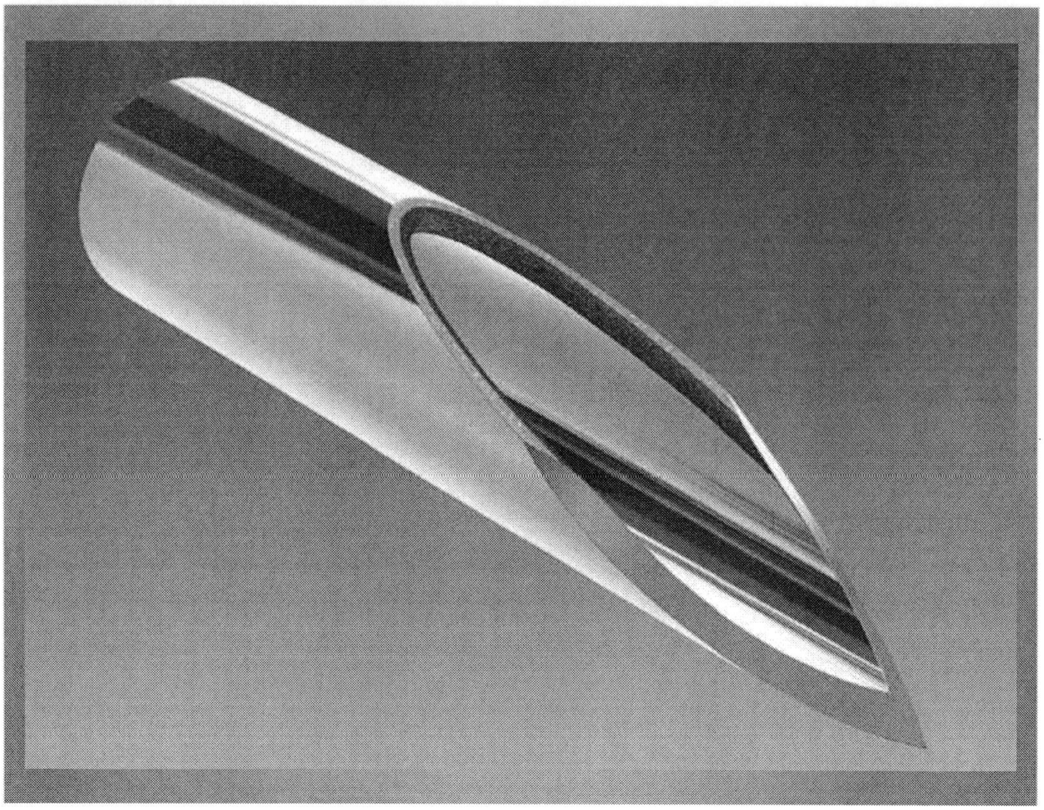

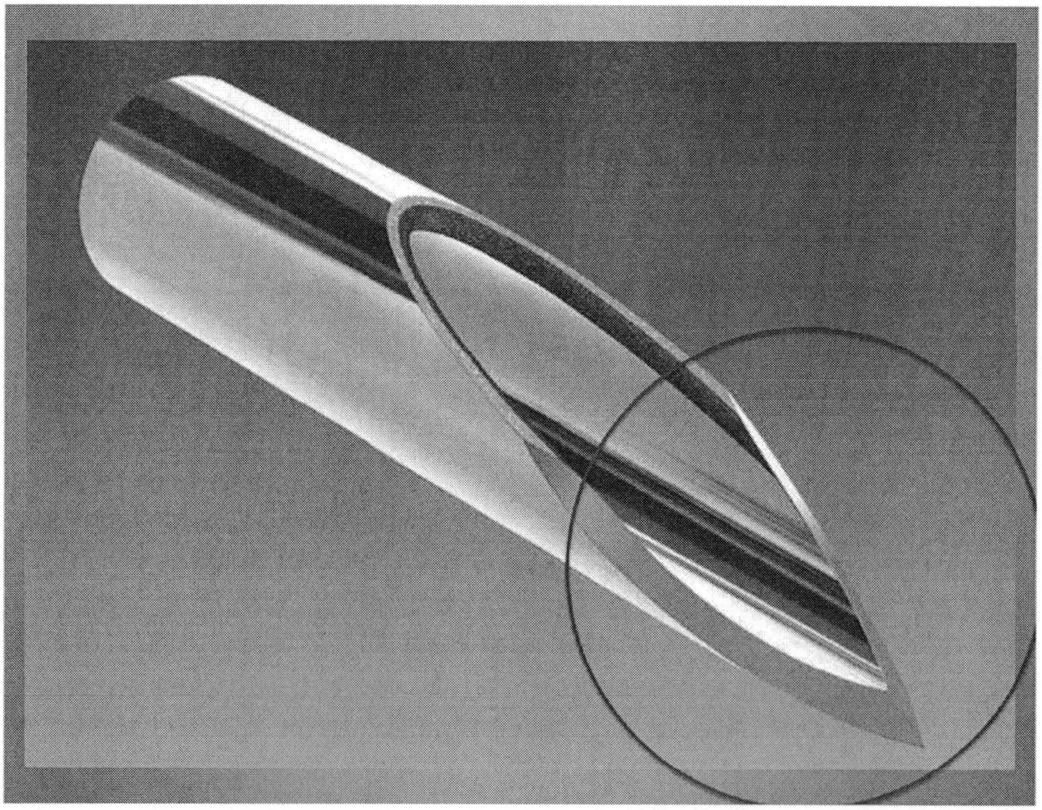

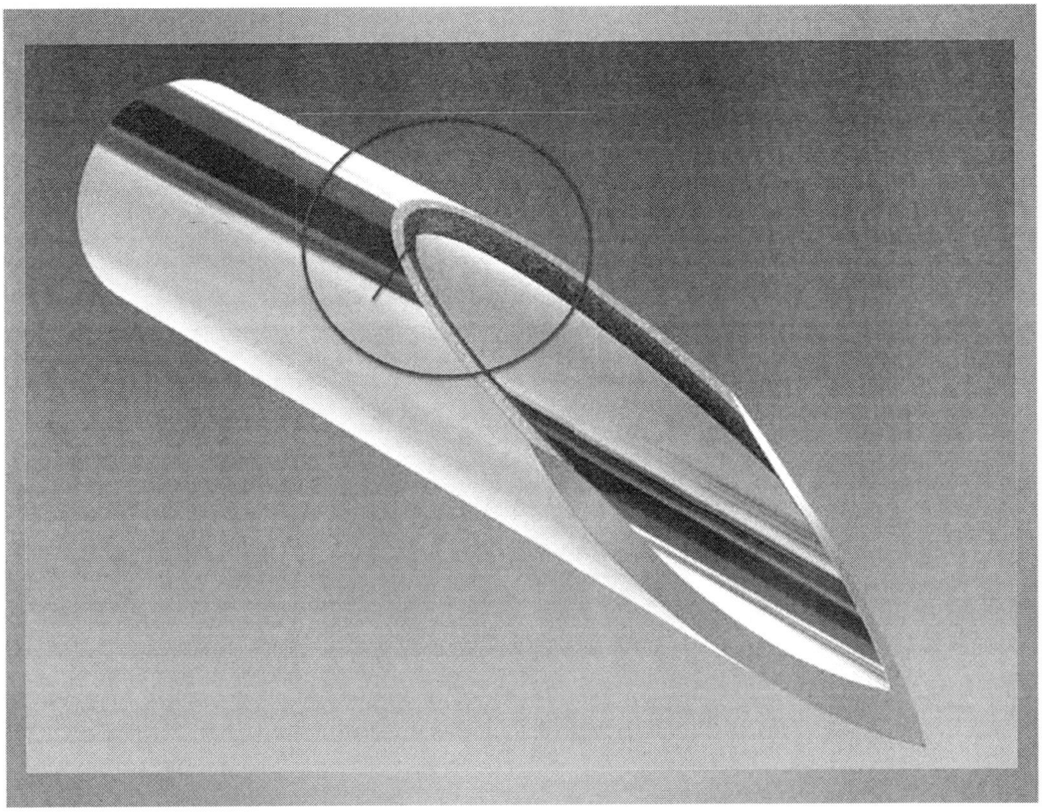

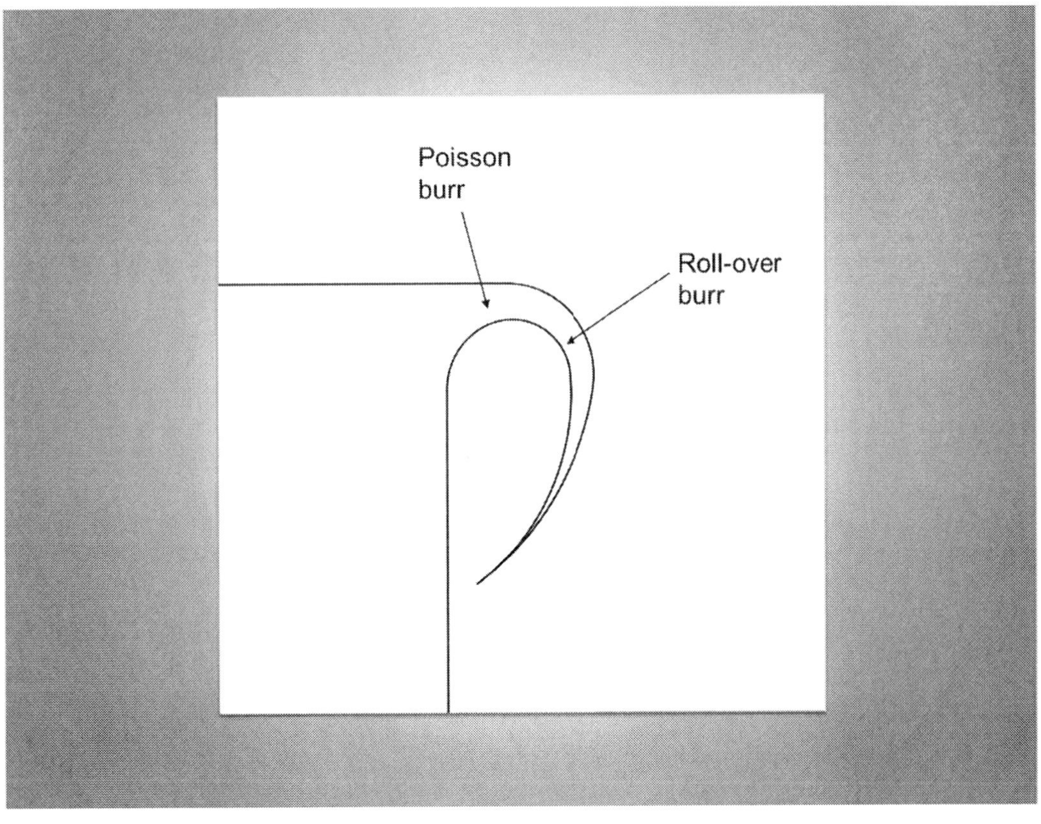
Poisson
burr
Roll-over
burr

Feather burr removal

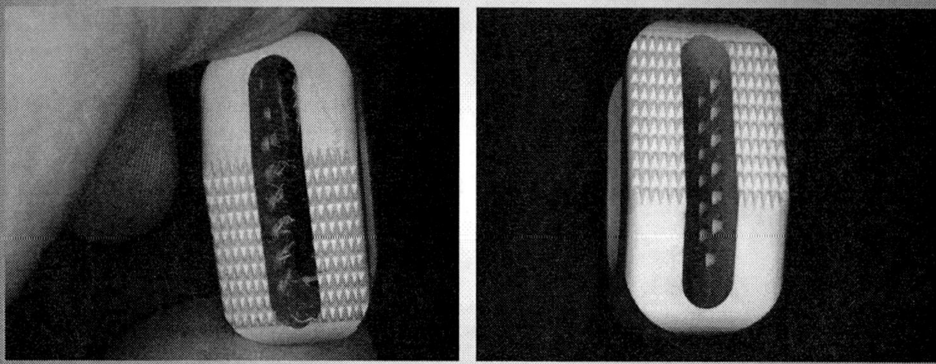

Precision Deburring with MicroBlasting

- Ability to target selected areas
- Less potential for collateral damage than manual deburring processes
- Access difficult internal geometries
- Avoid damage to the underlying material

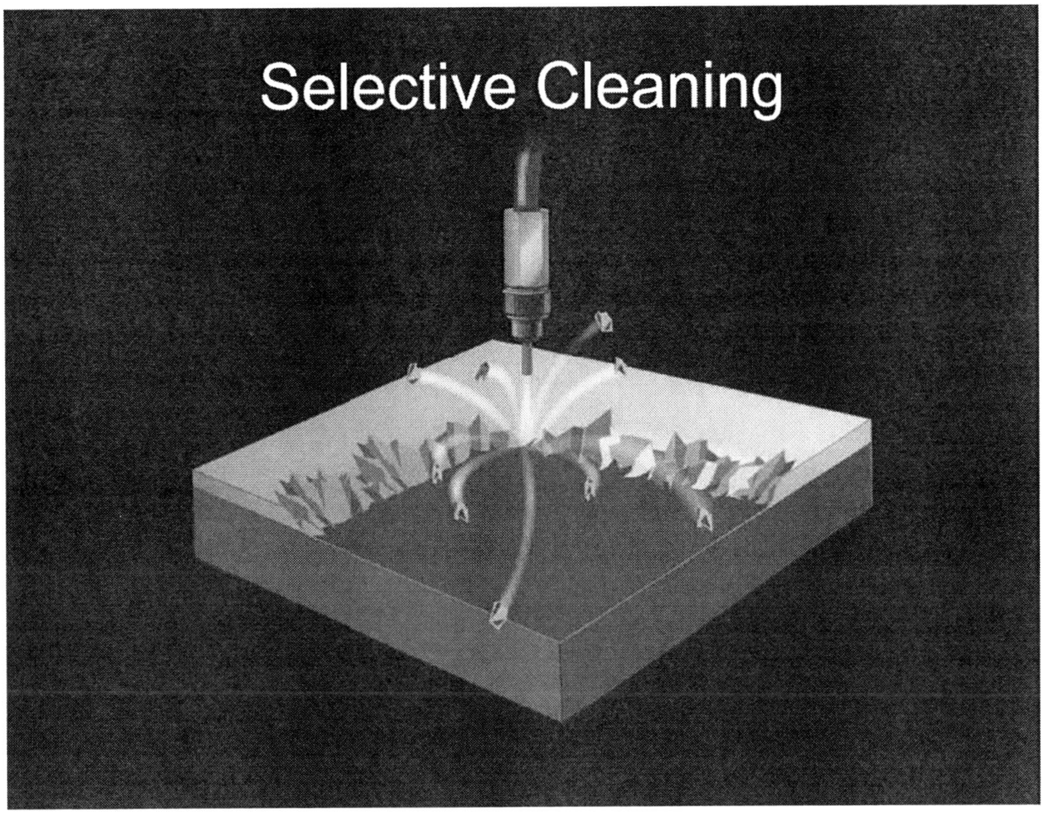
Selective Cleaning

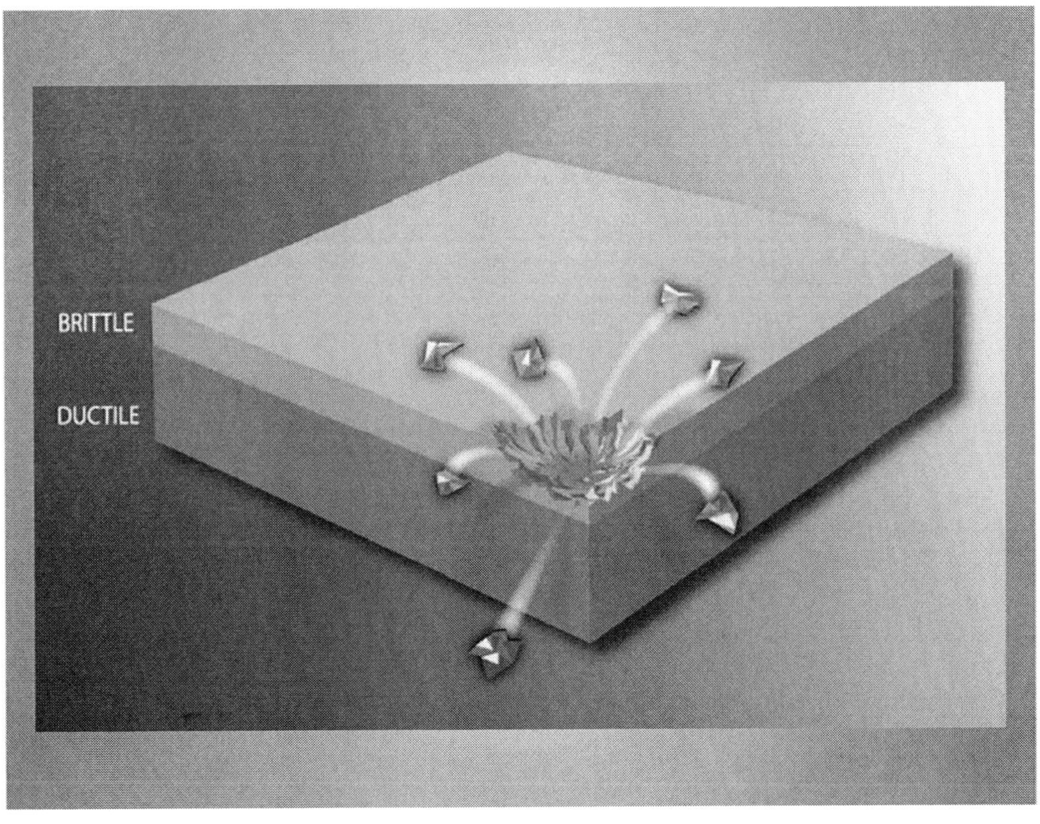
BRITTLE
DUCTILE

Selective removal of conformal coatings

Selective Cleaning with MicroBlasting

- Targets specific properties of the top layer either brittle or ductile
- Avoid damage to the underlying material
- Will not leave a residue

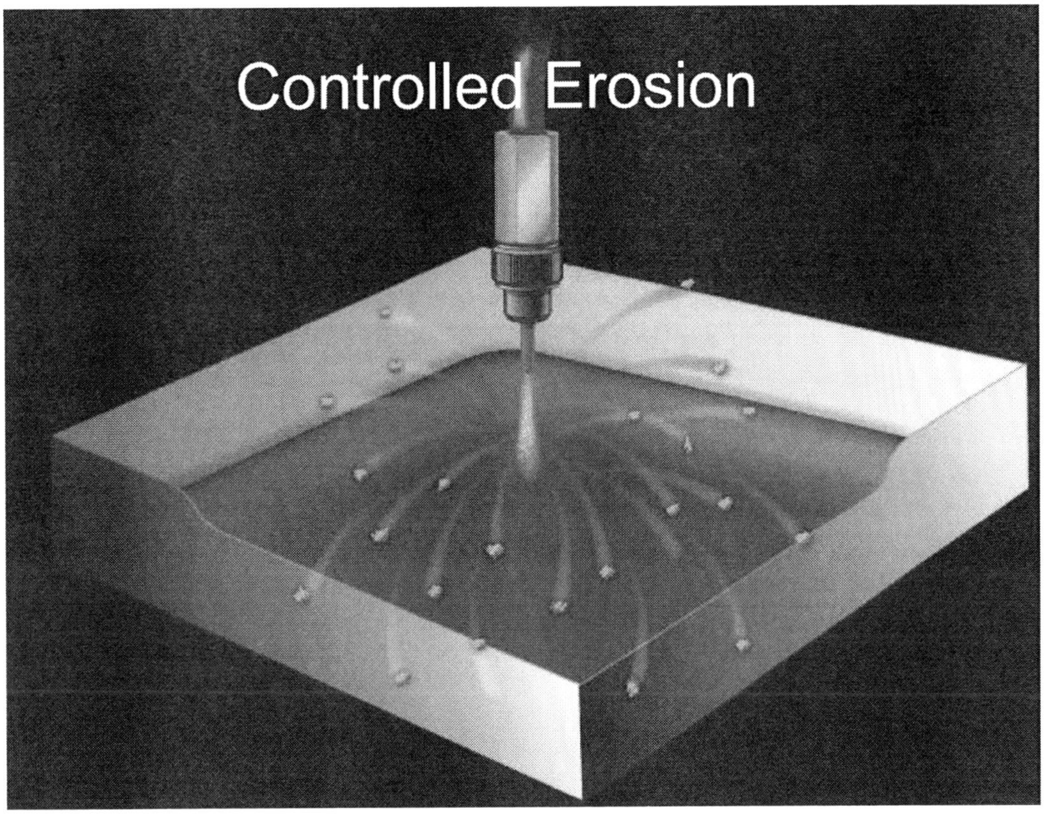
Controlled Erosion

Stent Surface Engineering

Targeted Surface Modification

Laser Cut Stent MicroBlasted Stent

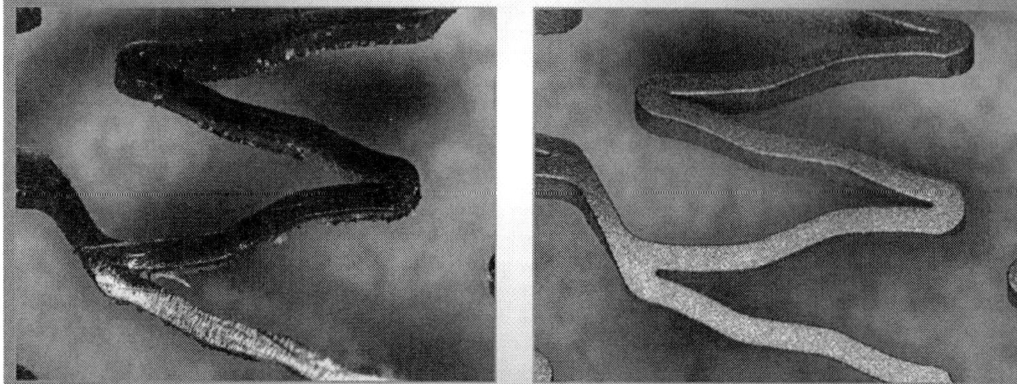

Precision Machining

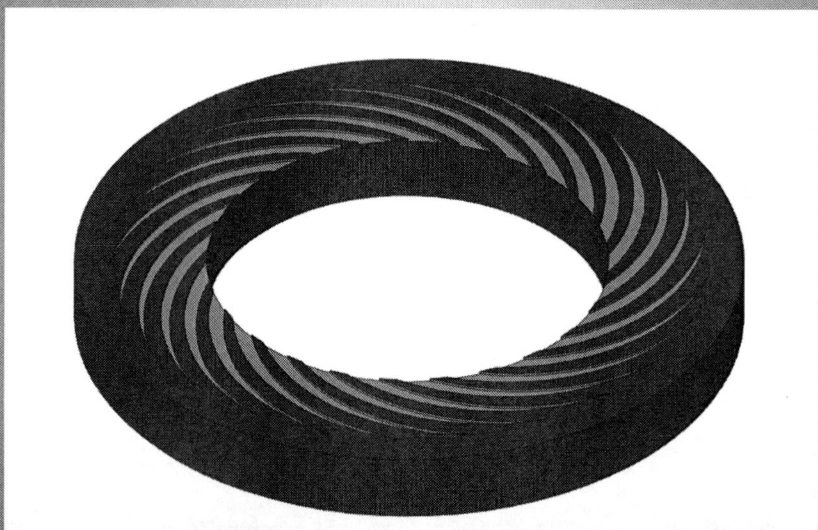

Hydro Dynamic Seal

Hydrodynamic ring seals

The hydrodynamic seal creates an air bearing in turbines. This technology has the ability to create significant improvements in engine efficiency.

An effective hydrodynamic seal requires a very specific engineered design to work properly. The flow dynamics are very complicated and change with altitude. The computational flow models

Once the proper design is developed it has to be applied to the ring surface.

- Etched surface must be very uniform; consistency is measured using a profilometer.
- Hold a tolerance of plus or minus 0.5 microns; etching 10-15 microns deep.
- May require ramped features

How to achieve this with MicroBlasting

- This process requires automation
- Feedback loop monitoring both media consumption as well as consistency
- Tightly controlled abrasive media.

At a fundamental level we are talking about a particulate in an air stream. Variation and application comes from the shape and properties of the media, how fast it is going, and how much is there.

Etch Resolution: 70 nm

Controlled Erosion with MicroBlasting

- Blast resolution as fine as 70 nanometers
- Does not generate heat or vibration
- Fine media size avoids chipping
- Effective on both brittle and ductile material
- Able to erode contoured surfaces uniformly

Automation

- Typical media for blasting in the medical industry is 10-25 microns in diameter and almost as hard as diamond.
- We got into automation because no one else could build equipment that would hold up in production environments. We saw that our future required automation.
- This material is similar to handling lapping powder. Without proper control it gets everywhere.
- Automation provides a consistent, repeatable process. Faster process times and eliminates human error.
- Challenges in shot peening are similar but much less acute

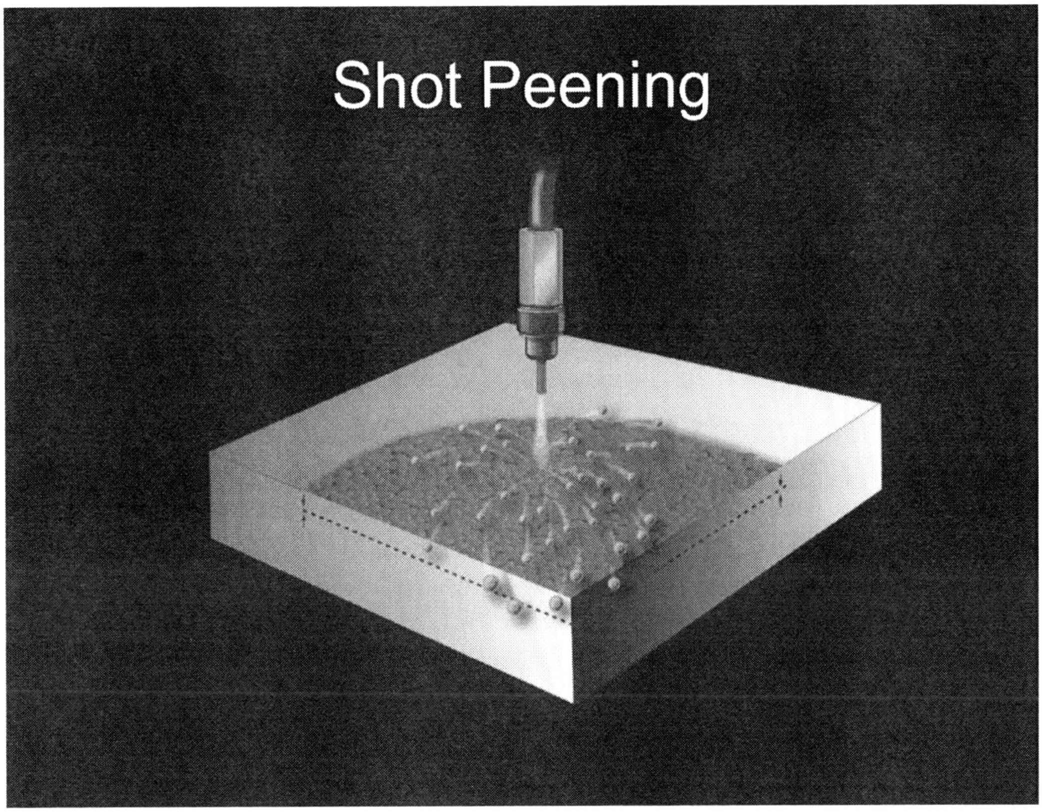
Shot Peening

Localized Peening

Shotgun vs. Machine Gun

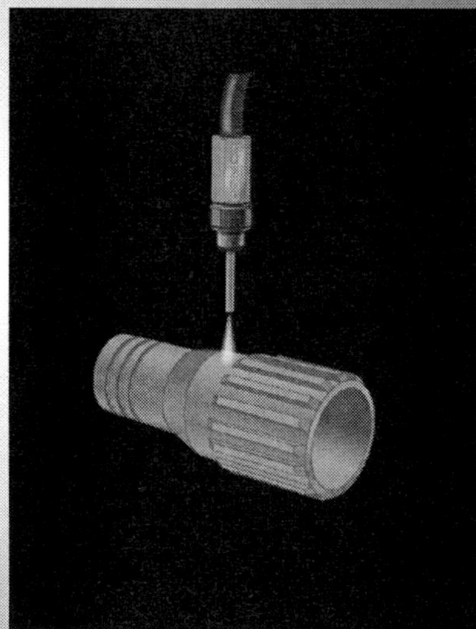

Access Difficult Geometries

Precision Shot Peening

- Reduce or eliminate masking
- Consistent media flow avoids potential for part deformation
- Case study: parts processed 50% faster with less air and media consumption
- Avoid variation due to interplay of different nozzles, creating turbulence

Conclusions

- Effective;
  - Wide range of materials
  - Intricate or fragile parts
  - Difficult part geometries
  - Process without affecting part geometries
  - Wide range of applications; texturing, eroding, deburring, cleaning and shot peening
- Typically focused on projects that require localized treatment
- Process Factors; media type, size, quantity and velocity
- Automation provides consistency and high volume
- Applying the capabilities of MicroBlasting to shot peening will create new uses for the technology

COMCO

Comco Inc, Burbank, CA (800) 796-6626 www.comcoinc.com

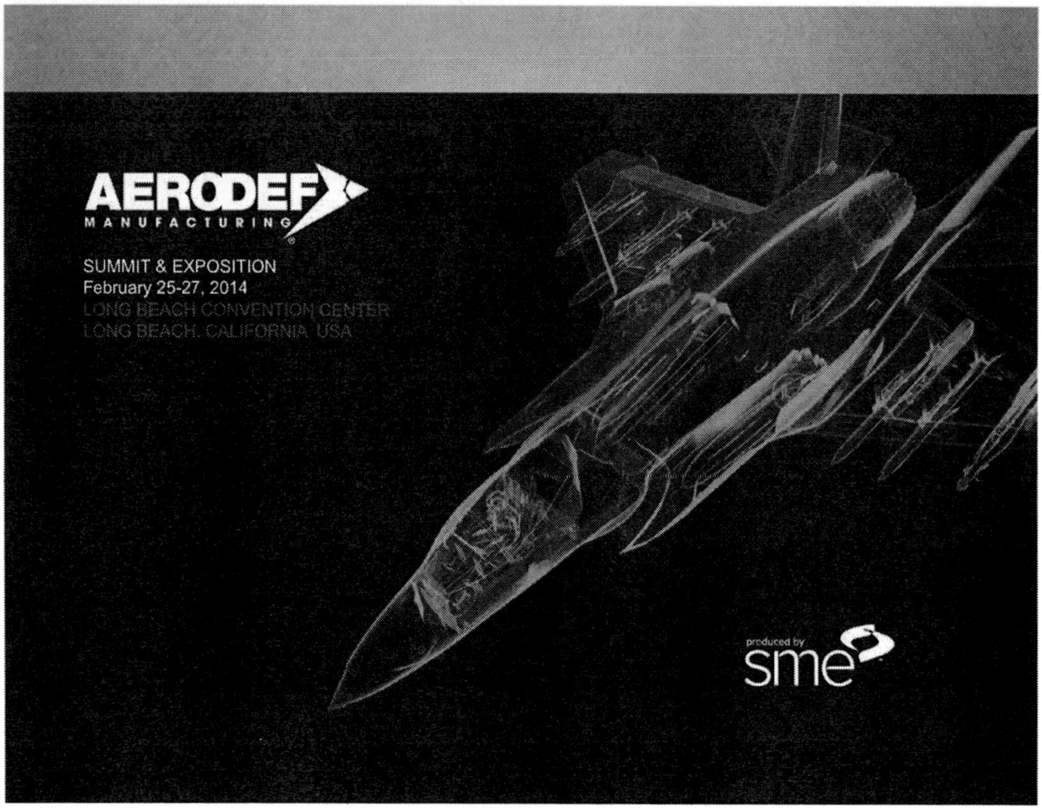
AERODEF
MANUFACTURING
SUMMIT & EXPOSITION
February 25-27, 2014
LONG BEACH CONVENTION CENTER
LONG BEACH, CALIFORNIA USA
produced by
sme

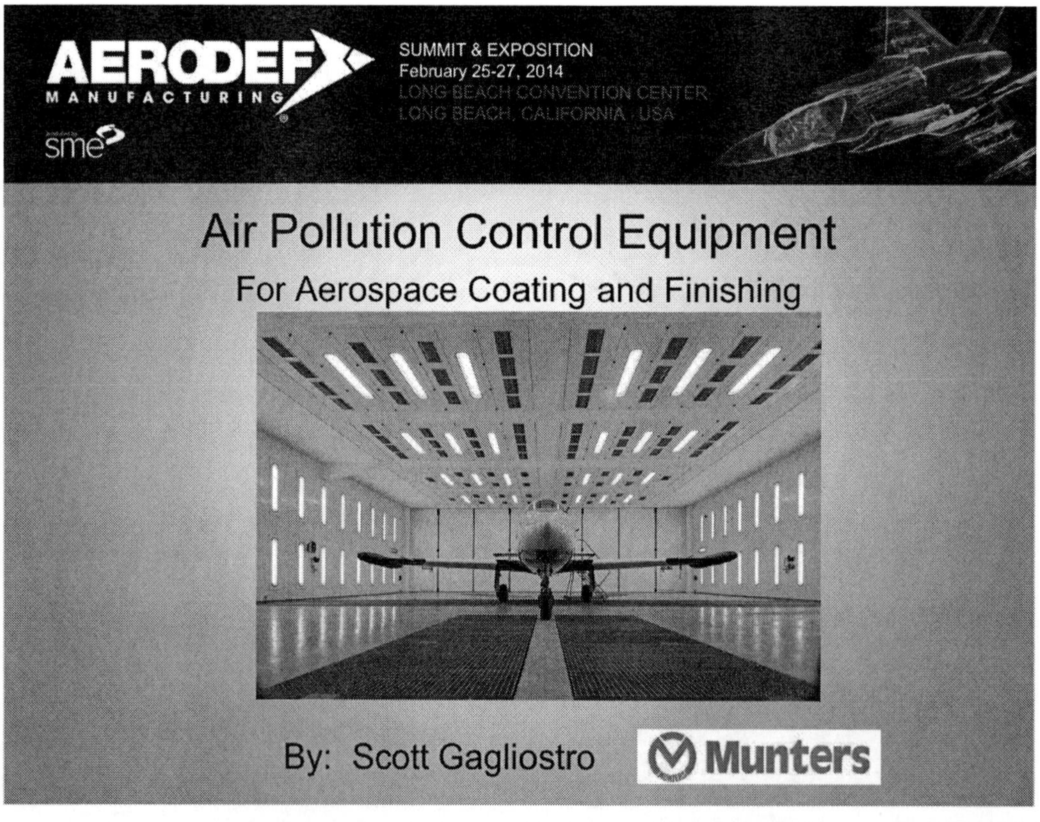
AERODEF
MANUFACTURING
sme
SUMMIT & EXPOSITION
February 25-27, 2014
LONG BEACH CONVENTION CENTER
LONG BEACH, CALIFORNIA, USA
Air Pollution Control Equipment
For Aerospace Coating and Finishing
By: Scott Gagliostro
Munters

Topics of Discussion:

- VOC Abatement –What/Why?
- Methods of abatement
- Benefits of each technology
- Ideal process conditions
- Environmental impacts

AERODEF MANUFACTURING
produced by sme
SUMMIT & EXPOSITION
February 25-27, 2014
LONG BEACH CONVENTION CENTER
LONG BEACH, CALIFORNIA USA

Why do we need to abate VOCs?

- VOC + Nox + Sunlight (UV) = Ozone (smog)
- Health and other respiratory issues

Common VOCs

- Toluene
- Xylene
- Heptane
- Mineral spirits
- Naphta
- MEK
- Ethanol
- Propanol
- Butanol
- Acetone
- Methanol
- Isopropyl acetate
- Hexane
- Benzene

Methods of Abatement

- Incineration
  - Thermal Recuperative
  - Catalytic
  - RTO (Regenerative Thermal Oxidizer)
- Concentrator Technology (Adsorption / Incineration)
- Biofiltration / Alt Technologies

AERODEF MANUFACTURING
sme
SUMMIT & EXPOSITION
February 25-27, 2014
LONG BEACH CONVENTION CENTER
LONG BEACH, CALIFORNIA USA

Incineration Technology:

VOC + Fuel → (1375° F) → $CO_2 + H_20$

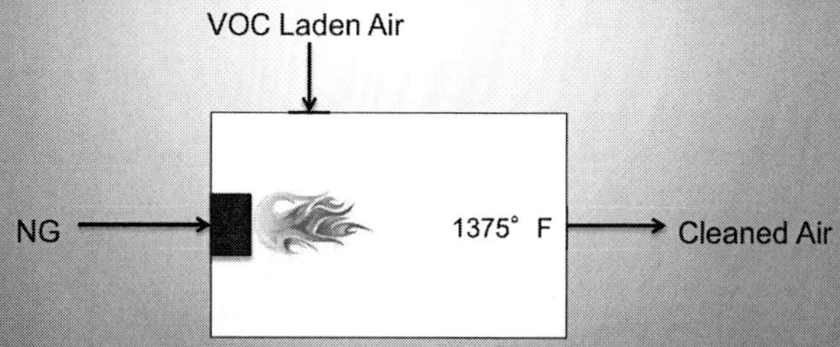

AERODEF MANUFACTURING
sme

SUMMIT & EXPOSITION
February 25-27, 2014
LONG BEACH CONVENTION CENTER
LONG BEACH, CALIFORNIA USA

Understanding LEL:

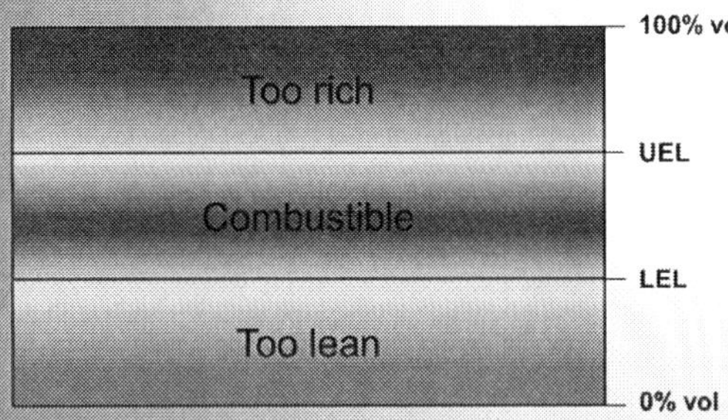

	LEL	UEL
Acetone	2.15%	13.0%
Benzene	1.2%	8.0%
IPA	2.0%	12.7%
Methane	5.0%	15.0%
Toluene	1.2%	7.0%

A typical paint application yields about .4% LEL

AERODEF MANUFACTURING
sme

SUMMIT & EXPOSITION
February 25-27, 2014
LONG BEACH CONVENTION CENTER
LONG BEACH, CALIFORNIA USA

Incineration :

RTO
- 98% + DRE
- 95% thermal eff
- 1500° F chamber

Thermal Recup
- 99% +DRE
- 70% thermal eff
- 1375° chamber

Catalytic
- 98% +DRE
- 70% thermal eff
- 800° chamber

- Best for applications
  - Greater than 3-5% LEL
  - Less than 20,000 SCFM

- At .4% LEL, NG input = 70,500 BTU/hr per 1,000 SCFM

(Assumes a stand-alone RTO and typical Aerospace VOCs)

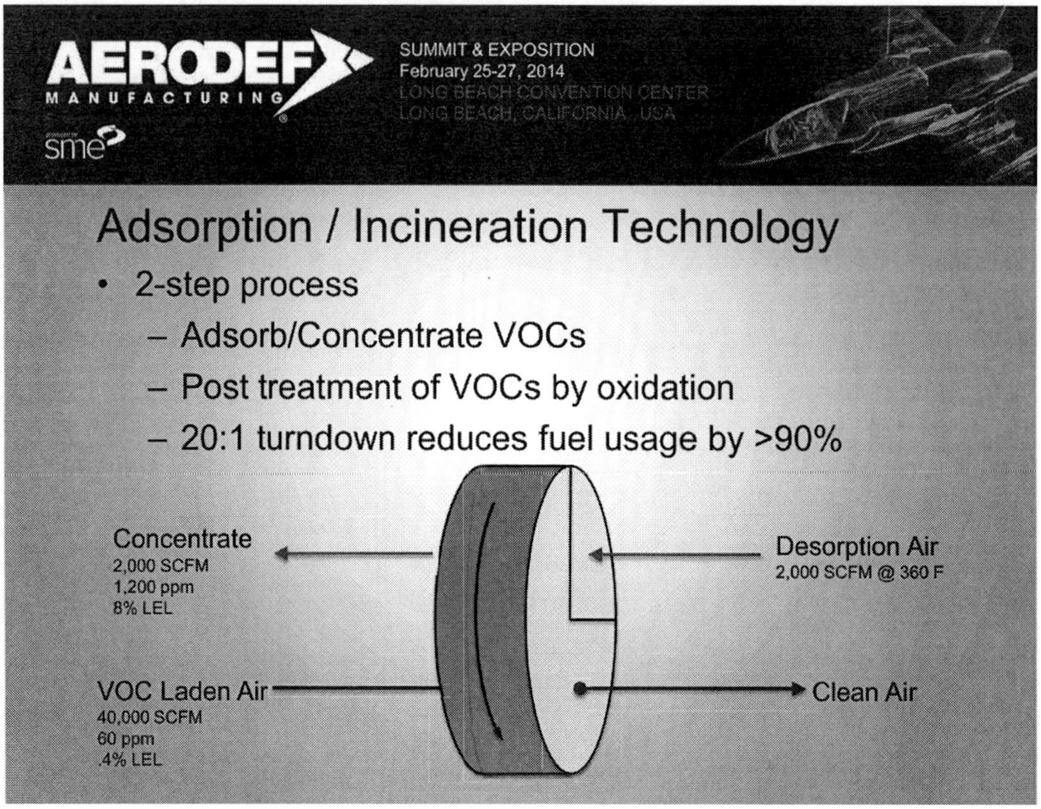
AERODEF MANUFACTURING
sme
SUMMIT & EXPOSITION
February 25-27, 2014
LONG BEACH CONVENTION CENTER
LONG BEACH, CALIFORNIA, USA
Adsorption / Incineration Technology
• 2-step process
– Adsorb/Concentrate VOCs
– Post treatment of VOCs by oxidation
– 20:1 turndown reduces fuel usage by >90%
Concentrate
2,000 SCFM
1,200 ppm
8% LEL
Desorption Air
2,000 SCFM @ 360 F
VOC Laden Air
40,000 SCFM
60 ppm
.4% LEL
Clean Air

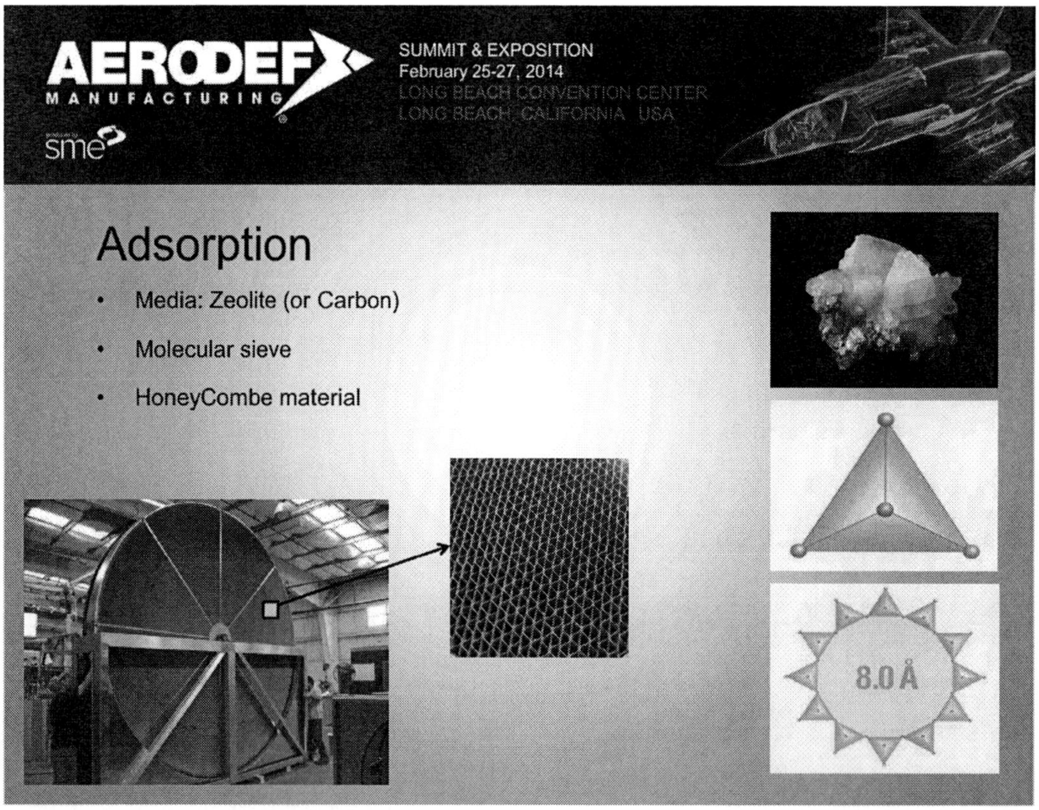
AERODEF MANUFACTURING
sme
SUMMIT & EXPOSITION
February 25-27, 2014
LONG BEACH CONVENTION CENTER
LONG BEACH, CALIFORNIA, USA
Adsorption
• Media: Zeolite (or Carbon)
• Molecular sieve
• HoneyCombe material
8.0 Å

AERODEF MANUFACTURING

SUMMIT & EXPOSITION
February 25-27, 2014
LONG BEACH CONVENTION CENTER
LONG BEACH, CALIFORNIA, USA

Zeolite	Carbon
• Inert	• Potential fire risk with ketones
• Hydrophobic	• Poor performance with high humidity
• Less Zeolite required	• Heel effect
• More expensive than carbon	• Low efficiency with alcohols

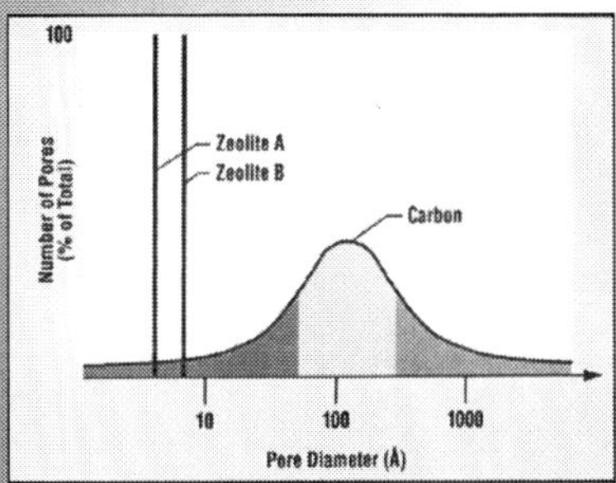

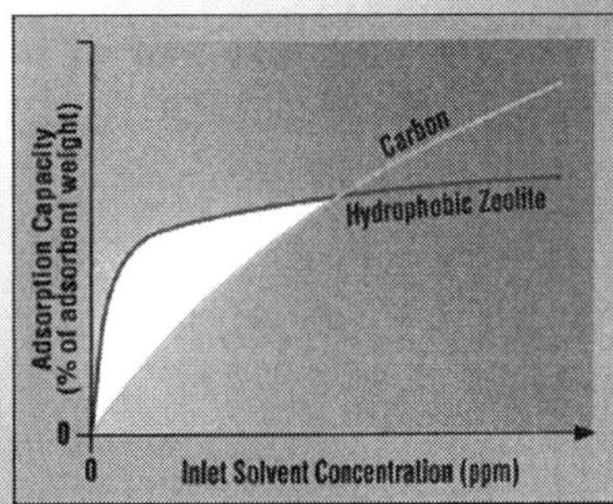

AERODEF MANUFACTURING

SUMMIT & EXPOSITION
February 25-27, 2014
LONG BEACH CONVENTION CENTER
LONG BEACH, CALIFORNIA, USA

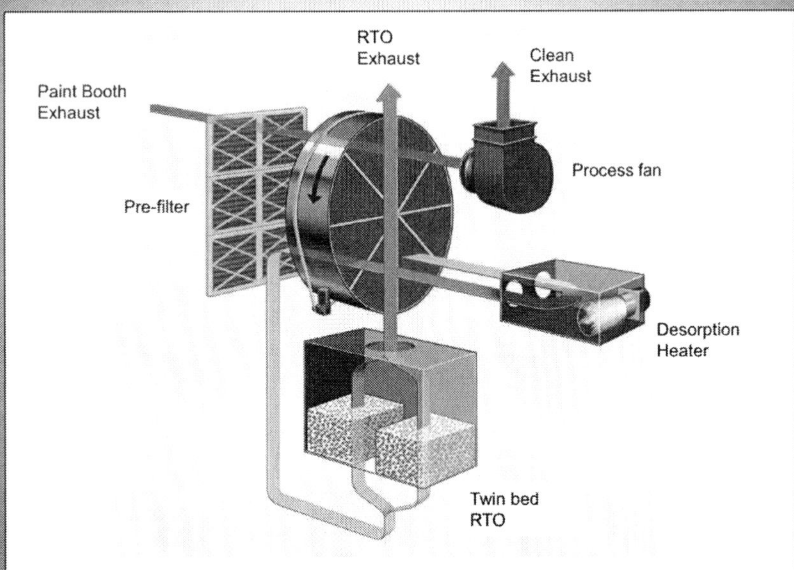

AERODEF MANUFACTURING
sme
SUMMIT & EXPOSITION
February 25-27, 2014
LONG BEACH CONVENTION CENTER
LONG BEACH, CALIFORNIA USA

Ideal Process conditions

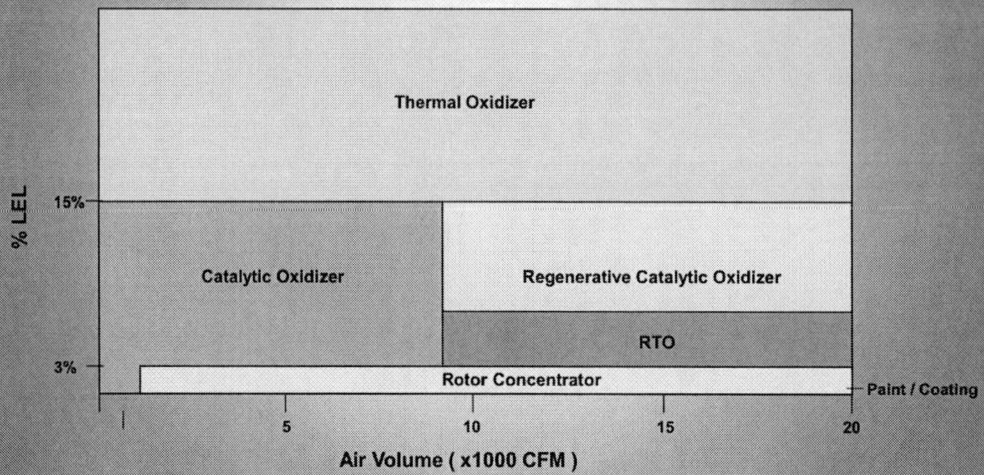

AERODEF MANUFACTURING
sme
SUMMIT & EXPOSITION
February 25-27, 2014
LONG BEACH CONVENTION CENTER
LONG BEACH, CALIFORNIA USA

Operating Cost Example

	RC with RTO	Stand-alone RTO
Process Exhaust Flow	44,390 scfm	44,390 scfm
RTO Size	4,930 scfm	44,390 scfm
System Pressure Drop	8 in. w.c.	22 in. w.c.
BHP for Fans	101 BHP	197 BHP
Electrical Use	75 kW	148 kW
Natural Gas Usage (40 lb/hr VOC loading)	0.9 MM Btu/hr	3.8 MM Btu/hr
Annual Operating Cost • 4800 hrs/year • \$0.08/kW • \$7.00/MM Btu	**\$59,040**	**\$184,512**

AERODEF MANUFACTURING
SUMMIT & EXPOSITION
February 25-27, 2014
LONG BEACH CONVENTION CENTER
LONG BEACH, CALIFORNIA USA

Concentrator Environmental Impact:

- Less NG and electrical consumption
- Smaller Carbon footprint
- Reduction in NOx

AERODEF MANUFACTURING
SUMMIT & EXPOSITION
February 25-27, 2014
LONG BEACH CONVENTION CENTER
LONG BEACH, CALIFORNIA USA

Concentrator Technology Benefits

- > 95% destruction efficiencies
- Low operating cost (~ 50% reduction vs RTO)
- Electricity and gas savings
- Less secondary emissions
- Lighter, smaller footprint

AERODEF MANUFACTURING

sme

SUMMIT & EXPOSITION
February 25-27, 2014
LONG BEACH CONVENTION CENTER
LONG BEACH, CALIFORNIA USA

Recap/Conclusion

- Methods of VOC Abatement
- Technologies available
- Concentrator environmental advantages
- Concentrator advantages
- Questions?

Munters

"redefining the limits of ultrasound"

Application of Non-Contact (Air-coupled) Ultrasound for In-line Inspection and Material Qualification

AeroDef Conference

Long Beach, CA

February 27, 2014

the ultran group

Agenda

We will explore non-contact ultrasound (NCU) and the advantages of non-destructive testing for material qualification

Topic Agenda

1 Non-Contact Ultrasound (NCU)	• Significant advancements in non-contact ultrasound now allow for analysis of composite and other materials in the early stages to final stages of their formation
2 Qualification of Material Properties	• Vital material properties can now be measured and verified using NCU –Porosity, density, bond quality, etc...
3 Applying NCU for Continuous Inspection in Production	• Application of NCU to continuous production allows for a safe, reliable, and relatively inexpensive way to save money, improve manufacturing and performance, and gain competitive advantage

the ultran group

1 NCU Advancements

Significant advancements in non-contact ultrasound allow for high performance and widespread applicability

Elements of Non-Contact Transducers*

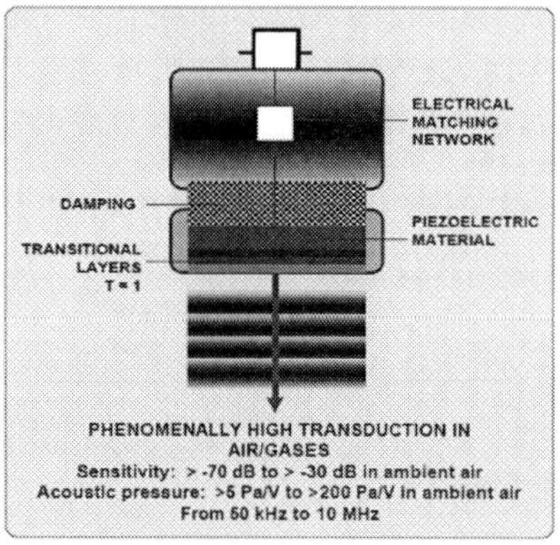

High Performance from 30 kHz to 5 MHz

Features

- Transition layer and matching layers provide efficient transmission through air
- NCU transducers optimized for frequencies between 30 kHz and 5 MHz
- Gas matrix piezoelectric (GMP)* composite allows for enhanced performance at frequencies between 30 kHz and 500 kHz
- High quality results achieved with many composite materials

*US and International Patents

the ultran group

1 NCU Applications

Through transmission is the most applicable and robust method of non-contact analysis

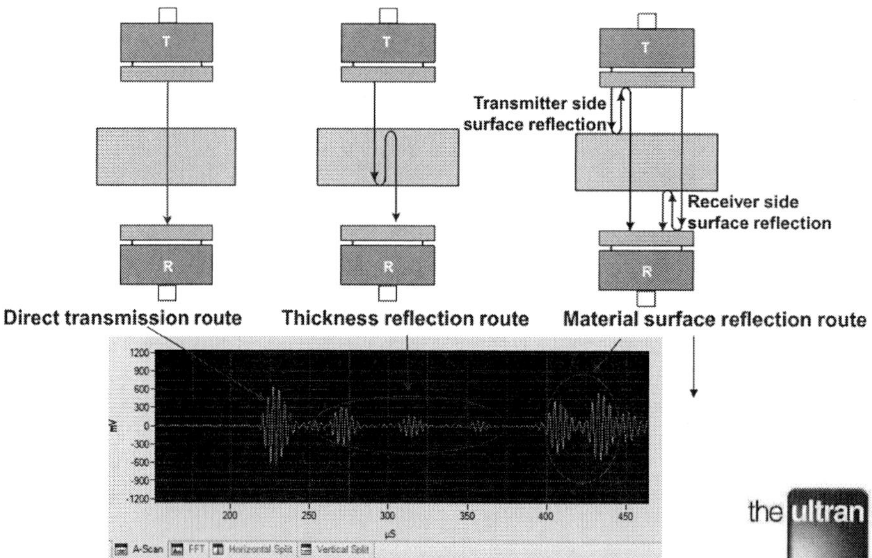

the ultran group

1 Non-Contact Analysis Modes

For most analyses in non-contact ultrasound, it is easiest to use the direct transmission route

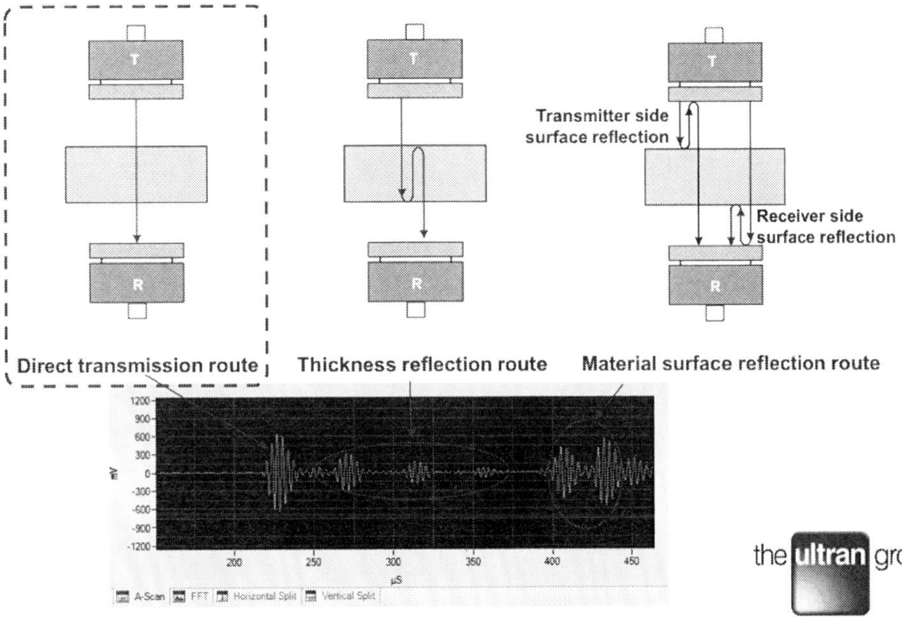

the ultran group

1 Ultrasonic Amplitude

The key ultrasonic measurement through non-contact through transmission is attenuation or transmittance

Transmittance in Material, T_m (dB)*

$$T_m = T_c - T_a$$

T_c (dB) transmission in air + material
T_a (dB) transmission in air column

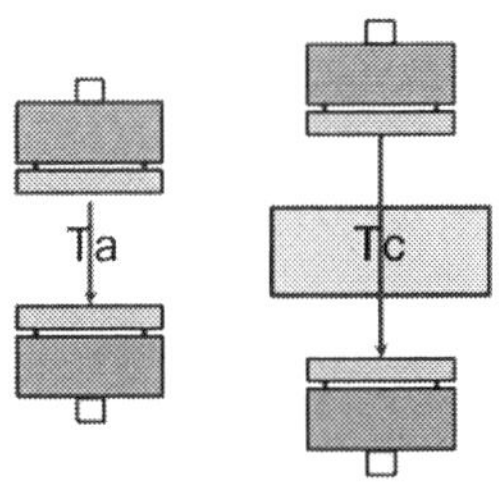

Material Transmittance is related to material texture, Z, homogeneity, and other physical characteristics

the ultran group

① Velocity Measurement

Material Velocity is often directly related to material density

Material Velocity, V_m when thickness is known

$$V_m = \frac{d_m}{t_{am} - (t_a - t_c)}$$

$$t_{am} = \frac{d_m}{V_a}$$

d_m material thickness
t_{am} tof in air corresponding to d_m
t_a tof in air
t_c tof in air + material
V_a air ultrasound velocity

Material Velocity, V_m when thickness is unknown

$$V_m = \frac{d_m}{t_{am} - (t_a - t_c)} \qquad d_m = V_a \times t_{am}$$

$$t_{am} = t_a - \left(\frac{t_1 + t_2}{2}\right)$$

t_1 round trip tof from transducer 1 to materials surface
t_2 round trip tof from transducer 2 to material surface

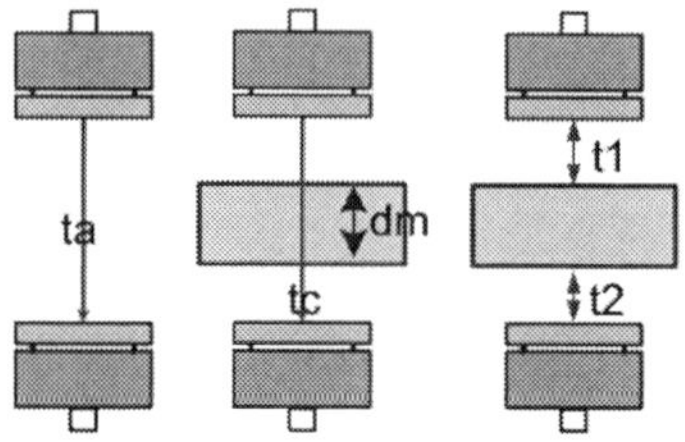

***Material Velocity Equivalent, V_e when thickness is known**

$$V_e = \frac{d_m}{\delta t} \qquad \delta_t = t_a - t_c$$

*Indirectly proportional to V_m
Easy to measure, does not require air/gas velocity

the ultran group

② Material Qualification

Non-Contact Ultrasound can measure key material properties in many composite materials

- **Prepreg:** Carbon Fiber, Glass Fiber, etc...
- **CFRP & GFRP**
- **Sandwich structures**
  - –Nomex and aluminum core honeycomb with composite & Al skins
  - –Ceramic Matrix (CMC) structures
  - –Foam Core with various skins
- **Carbon-Carbon composites**
  - –Autoclave oven fixtures
  - –Disk Brakes (aircraft and automobile)

the ultran group

2 Material Qualification

Using the Ultran M510x bench-top C-Scan system, we can characterize various composite materials

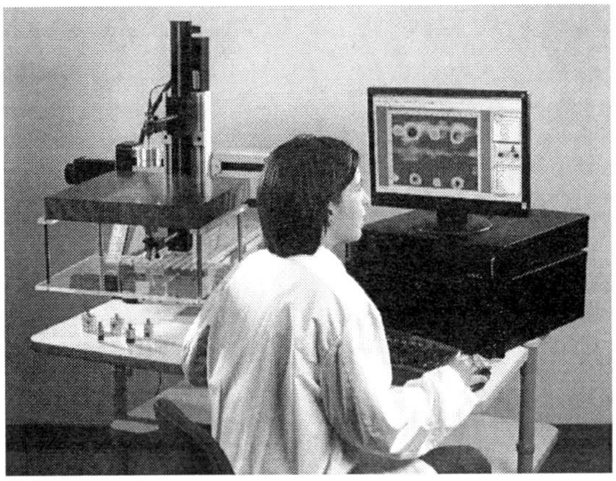

System Features

- Tone-burst pulser up to 390V, with frequency range from 20 kHz to 6.25 MHz
- 2-16 channel receiver up to 84 dB dynamic gain
- Software features:
  - Cross-sectional profiles for quantitative analysis
  - Absolute transmittance and reflectance measurements
  - Palette selection for easy accept-reject limits
  - Parametric correlation of acoustic vs. material characteristics
  - Statistical Quality Control
  - Numerous features for detailed localized region analysis
  - X-Y Scanning capability can be provided at customer request (various sizes available)

2 Composite Qualification

The below composite section demonstrates bonded and dis-bonded regions detected by NCU

C-Scan and Line Scan Images of CFRP-GFRP Cylindrical composite section (19mm thick)

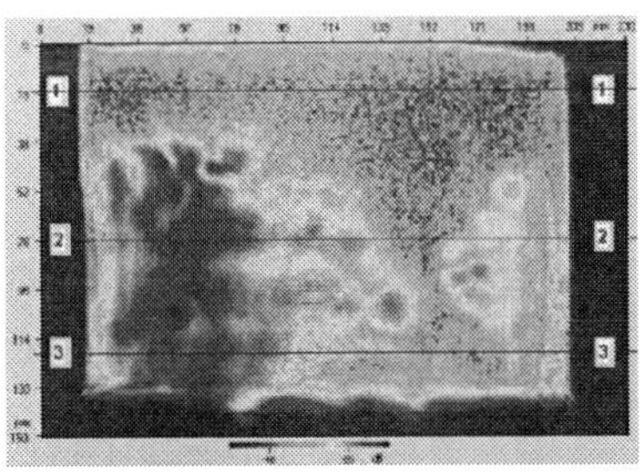

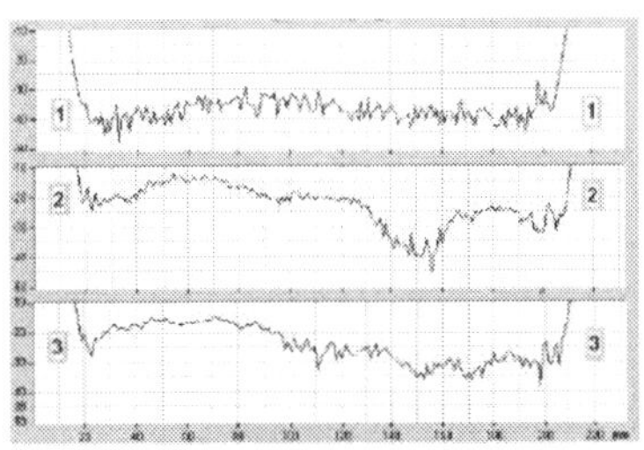

1. Complete disbond across top region of part
2. Well-bonded area on left side with disbonded region on right
3. Well-bonded area on left side with disbonded region on right

2 Composite Qualification

Delamination can be detected within foam core structures

C-Scan and Line Scan Images of GFRP Foam Core Sandwich Composites

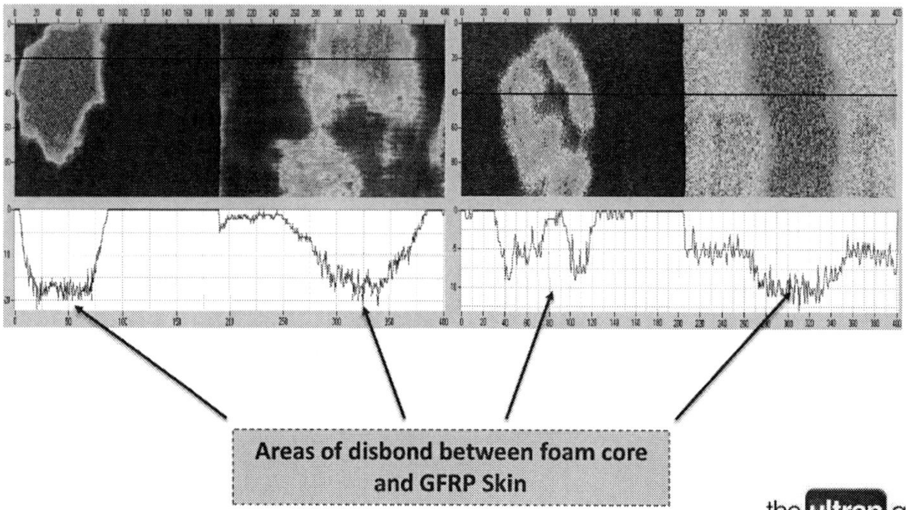

the ultran group

---

2 Composite Qualification

Delamination between layers for carbon-carbon plates can easily be detected using NCU

Carbon-Carbon Plates for Oven Fixtures (~10mm thick)

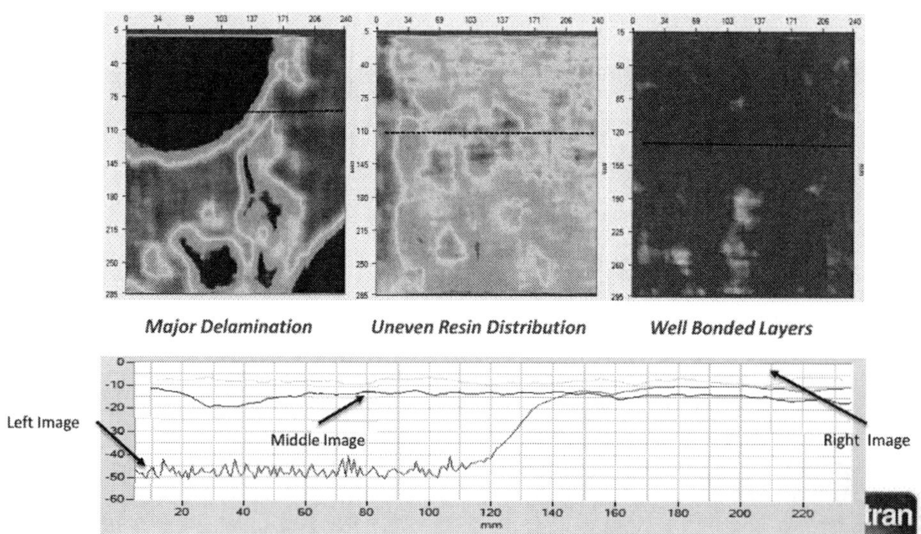

the ultran group

2 Composite Qualification

NCU can depict areas of delamination between layers of carbon-carbon disc brakes

Carbon-Carbon Aircraft Disk Brakes

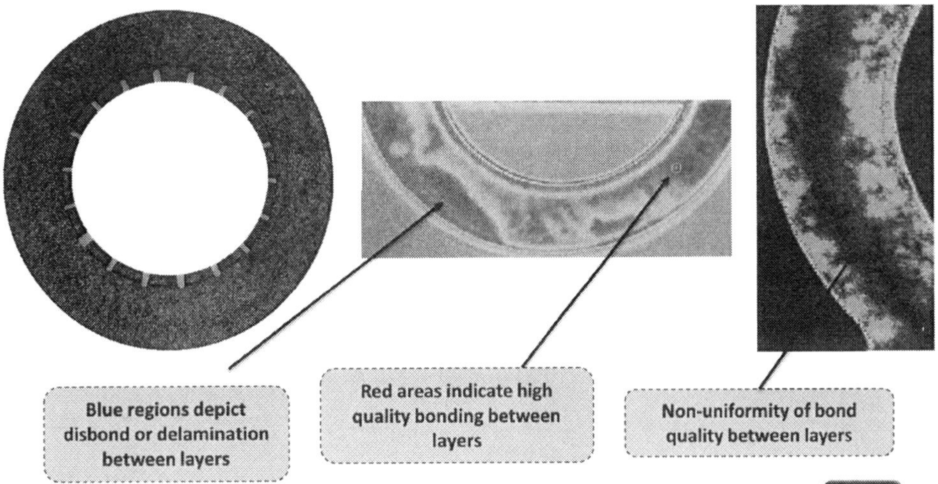

the ultran group

2 Correlation Representation

The relationship between the desired material property and ultrasonic amplitude can be formulated using statistical analysis on experimental results

Graphical Representation of Material Property vs. Ultrasonic Transmittance

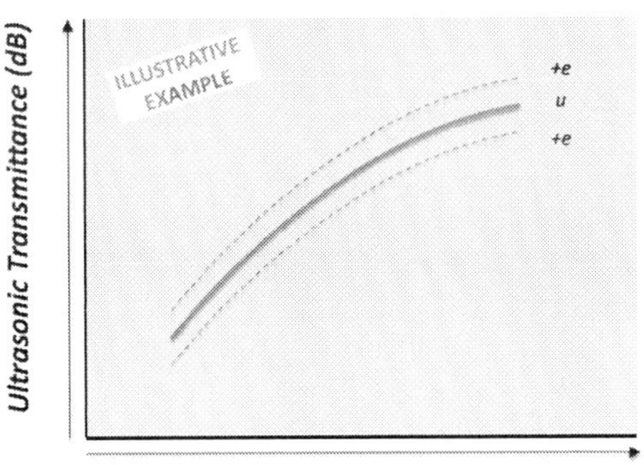

Correlation Function

- Transmissivity is expected to decrease as porosity increases or bond quality decreases
  - Low porosity (dryer material) and disbonded layers will have high attenuation and low transmissivity

the ultran group

2 Actual Correlation

The following data represent actual correlation between green ceramics and their key material properties

NCU Velocity vs. Density

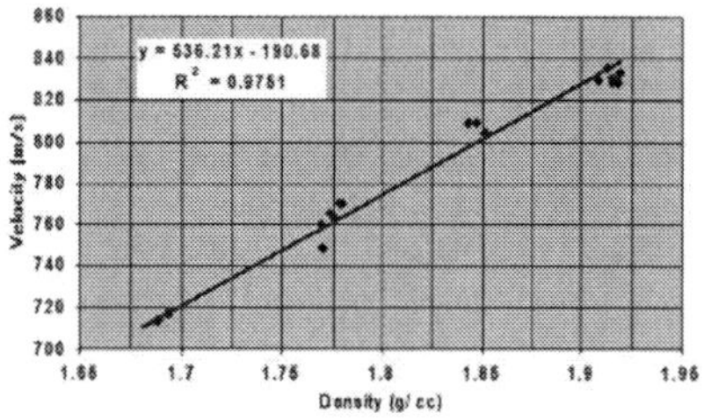

NCU Transmittance vs. Density

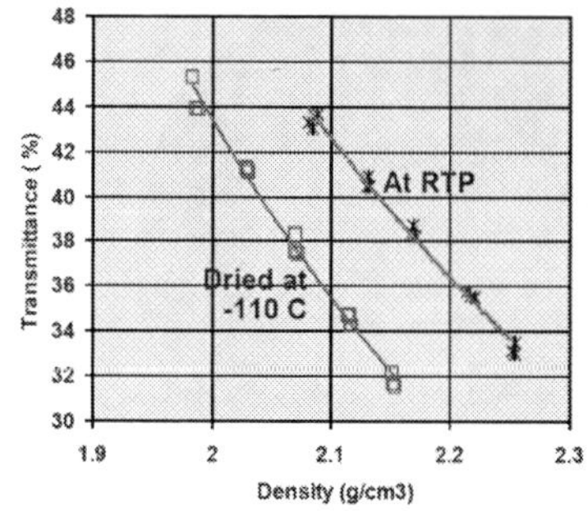

NCU Transmittance vs. Porosity

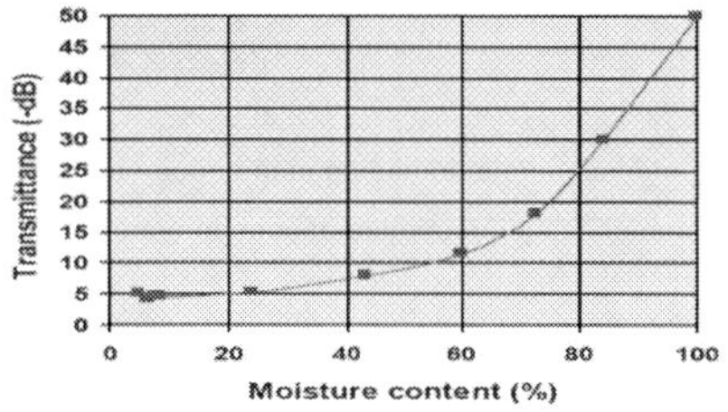

the ultran group

2 Prepreg Qualification

The wetness or porosity of carbon fiber prepreg can be directly correlated to NCU transmittance

C-Scan and Line Scan Images of Carbon Fiber Prepreg Samples of Varying Impregnation level

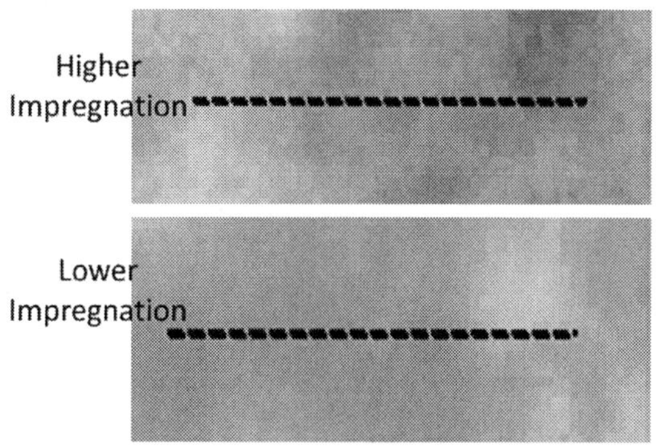

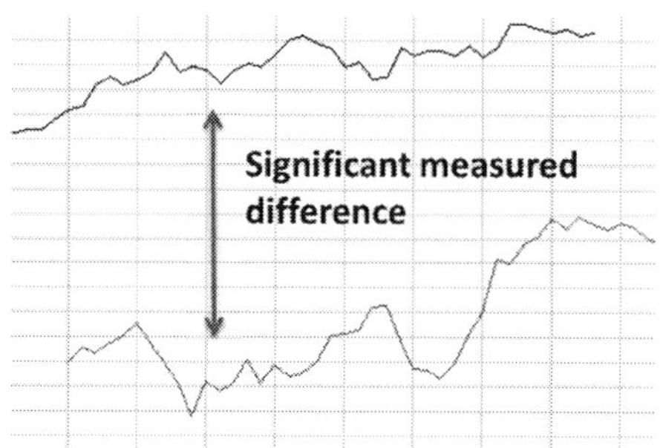

Subtle impregnation level differences demonstrate significant variation NCU Transmittance

an group

Prepreg Sample Set 1 (Roll A)

Left Sample **Middle Sample** **Right Sample**

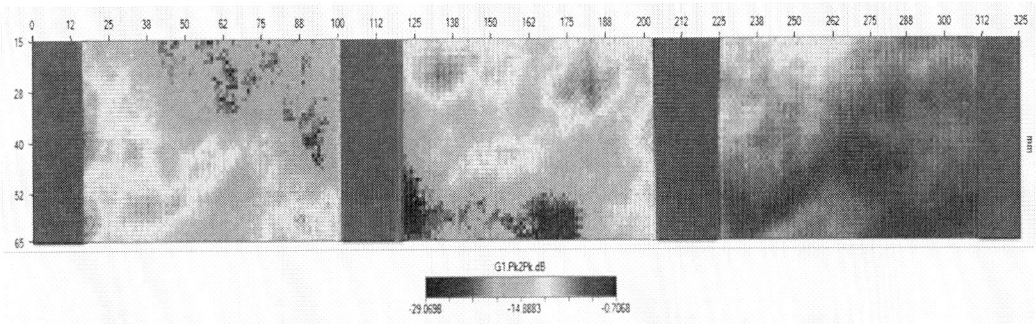

Scan Statistics	Left	Middle	Right
Average Amplitude (dB)	-14.14	-14.05	-4.55
Standard Deviation	4.52	5.29	2.00
Highest Amplitude (dB)	-4.46	-2.97	-1.31
Lowest Amplitude (dB)	-27.73	-28.64	-11.25

the ultran group

Prepreg Sample Set 2 (Roll A)

Left Sample **Middle Sample** **Right Sample**

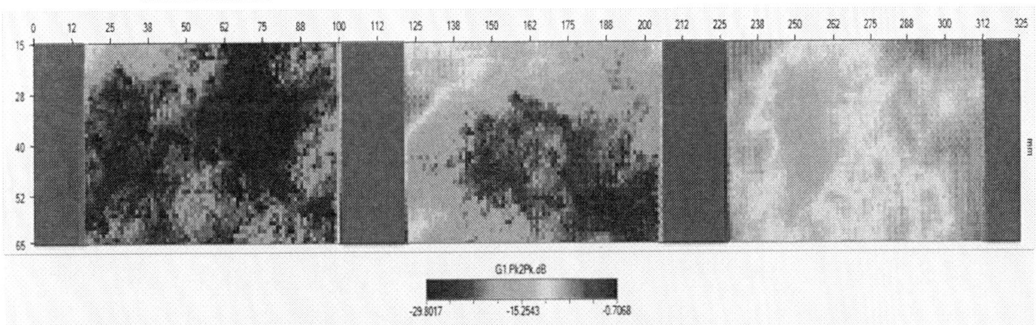

Scan Statistics	Left	Middle	Right
Average Amplitude (dB)	-24.90	-20.29	-10.15
Standard Deviation	2.92	5.25	3.14
Highest Amplitude (dB)	-10.84	-3.79	-3.97
Lowest Amplitude (dB)	-33.61	-29.80	-21.32

the ultran group

2 Prepreg Qualification

Prepreg Sample Set 3 (Roll A)

Left Sample **Middle Sample** **Right Sample**

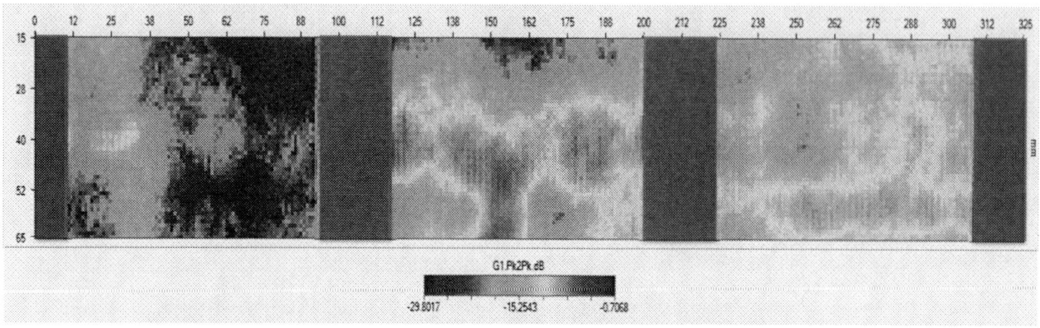

Scan Statistics	Left	Middle	Right
Average Amplitude (dB)	-22.23	-12.27	-12.46
Standard Deviation	4.63	5.11	3.43
Highest Amplitude (dB)	-10.03	-2.99	-5.95
Lowest Amplitude (dB)	-32.97	-27.03	-24.15

the ultran group

2 Prepreg Qualification

Prepreg Sample Set 4 (Roll A)

Left Sample **Middle Sample** **Right Sample**

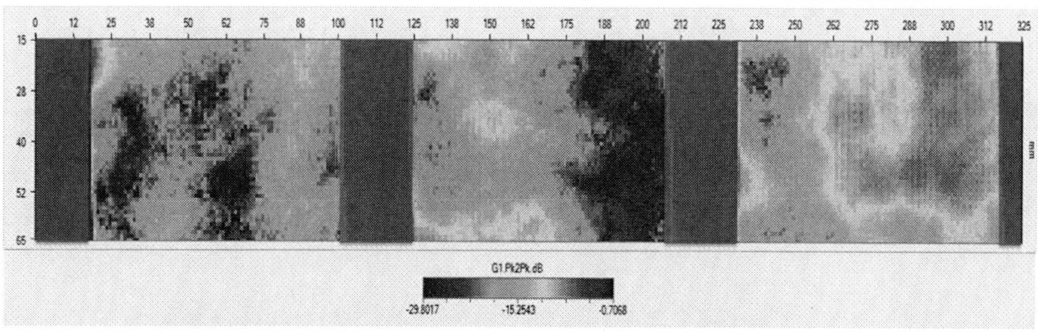

Scan Statistics	Left	Middle	Right
Average Amplitude (dB)	-20.25	-19.24	-11.22
Standard Deviation	4.14	5.66	5.04
Highest Amplitude (dB)	-3.95	-3.96	-2.87
Lowest Amplitude (dB)	-31.52	-31.50	-28.82

the ultran group

Cured Laminates from Left and Right sides of Prepreg Roll A

Laminate from Prepreg on Left Side of Roll

Laminate from Prepreg on Right Side of Roll

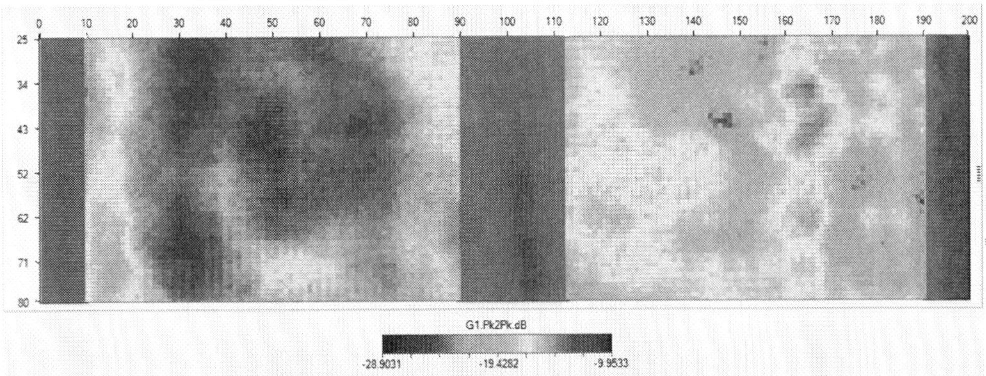

Scan Statistics	Left	Right
Average Amplitude (dB)	-13.57	-18.04
Standard Deviation	1.71	2.27
Highest Amplitude (dB)	-9.95	-12.12
Lowest Amplitude (dB)	-20.59	-26.71

the ultran group

Closing the loop on a manufacturing process allows for instant feedback and process control

- Can make adjustments during process to remain within control limits
- Enables continuous process improvement
- Provides further product information and creates opportunity for product improvement
- Allows for 100% inspection of manufactured product
  - –Identify regions of defective material
  - –Certification of sold product

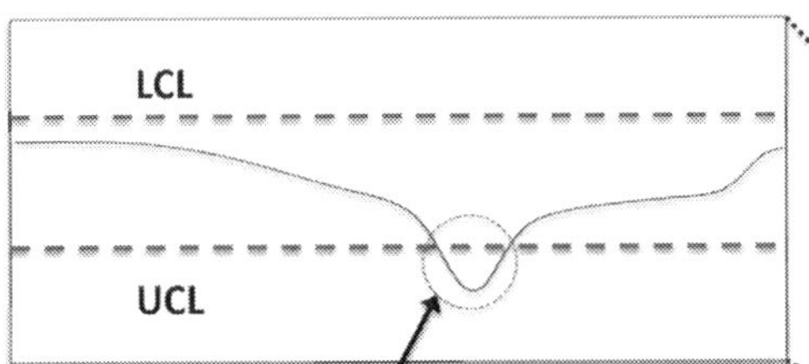

Process out of control

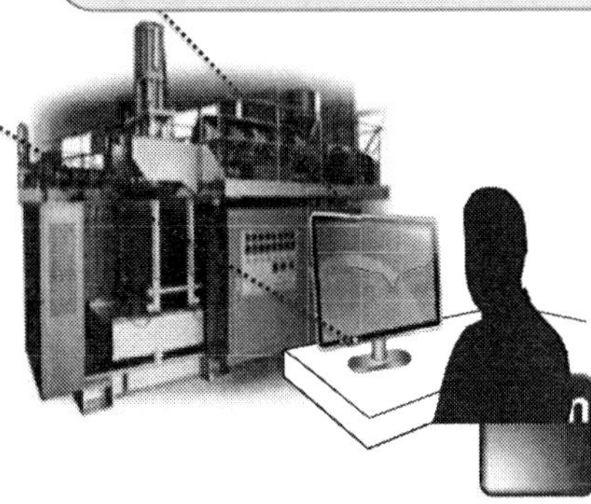

n group

3 NCU for Continuous Inspection

A multi-channel non-contact array can continuously analyze parts or web-lines in the downstream direction

Representation of Multi-Channel Array for Continuous Inspection

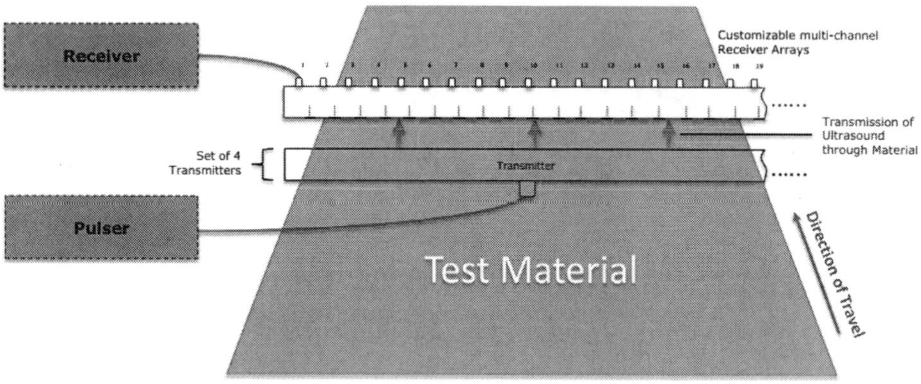

Transducer array can be arranged in a brick pattern for continuous cross-web coverage

the ultran group

3 NCU for Continuous Inspection

Application of non-contact ultrasound provides a safe and reliable method of continuous inspection

Brick Pattern Array for Continuous Inspection Cross-web

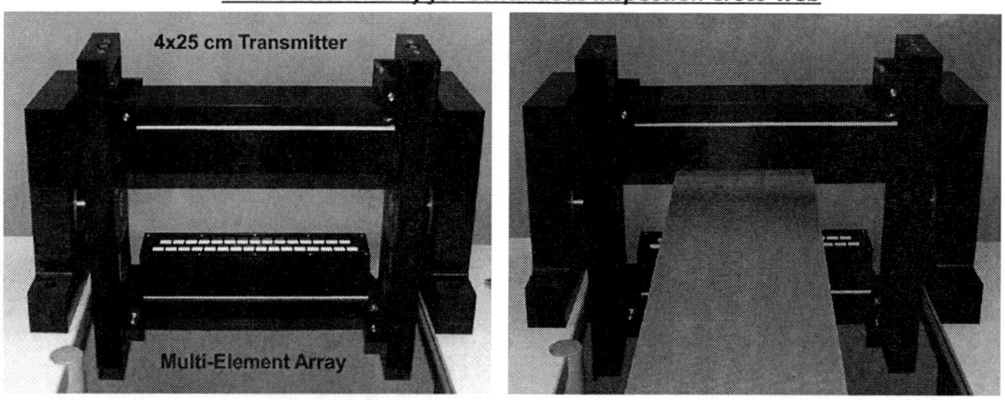

Designed and Produced by The Ultran Group

the ultran group

3 NCU for Continuous Inspection

A multi-channel non-contact array can continuously analyze parts or web-lines in the downstream direction

Representation of Multi-Channel Linear Array for Continuous Inspection

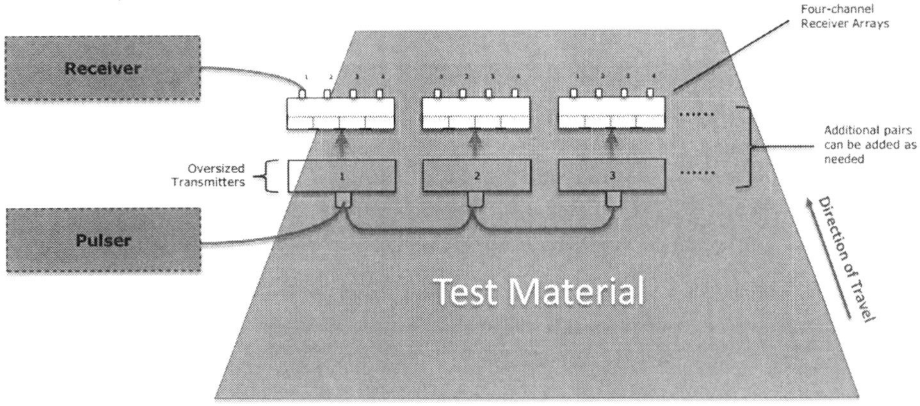

Linear array pattern allows increased modularity across the web-line or test material

the ultran group

3 NCU for Continuous Inspection

Our 4-channel array pair is fully modular and can be used with mechanism for alignment in rotational axes

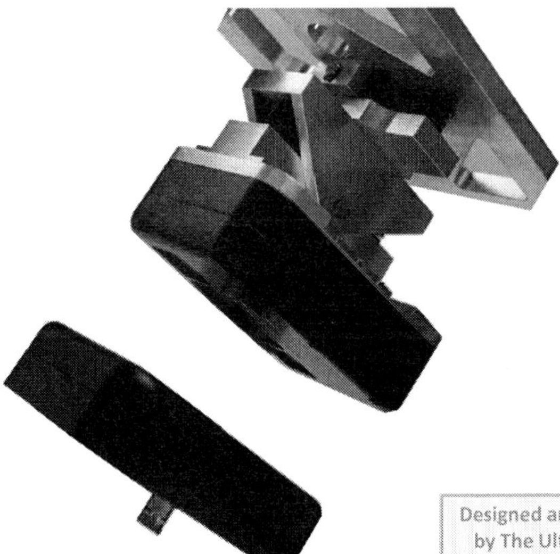

- 4-channel receiver array, can be built at frequencies between 50 kHz and 1 MHz
- Fully modular to allow for addition of increased number of channels
- Receiver alignment mechanism allows for adjustment in two axes of rotation
  - Alignment mechanism can be mounted to fixture across production line

Designed and Produced by The Ultran Group

the ultran group

At each channel we can continuously record the peak-to-peak amplitude across the product

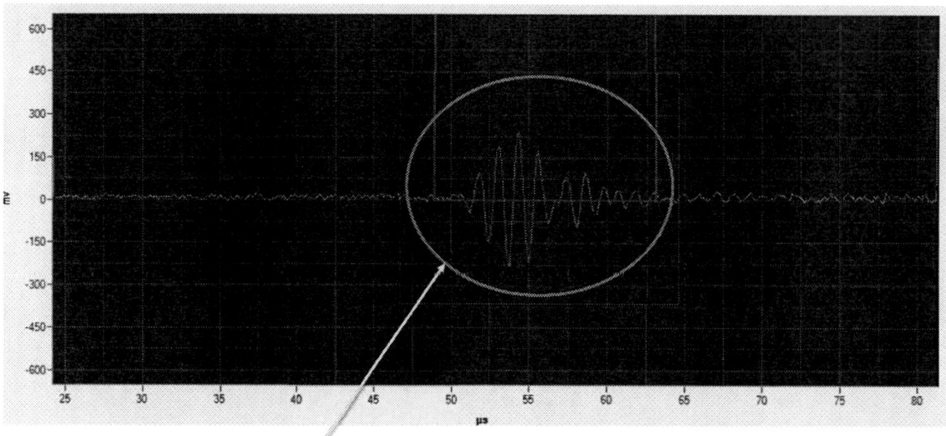

Peak to Peak value is Recorded at channel at specified time intervals

Designed and Produced by The Ultran Group

the ultran group

3 NCU for Continuous Inspection

Plotting the peak-to-peak values over time, we can continuously monitor materials and products via user-friendly software

Continuous Line Scans of Material

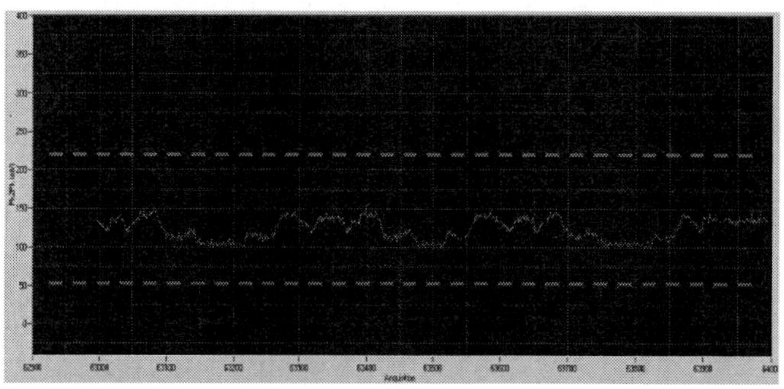

Features

- Continuous line scan for up to 32 or more channels simultaneously
- Adjustable upper and lower control limits
- Alarm output if readings reach limits
- Y-axis units can be converted to distance or other desired units
- Y-axis units can be converted via a correlation function to directly measure desired material property

Designed and Produced by The Ultran Group

the ultran group

Non-Contact Ultrasound provides a safe and reliable method of qualifying and measuring key material properties

Non-Contact Ultrasound

Non-Contact Improvements

- High performance between 30 kHz and 5 MHz
- Capable of measuring properties of many composite materials
- Very high signal to noise ratios obtained

Correlation of NCU to Material of Interest

- It is relatively simple to correlate NCU data with material properties
  - For Example: Change in porosity, delamination, air gap, etc...

Continuous Inspection & Qualification

Improve Process and Product

- Obtain data earlier during manufacturing process
- Improve process with immediate feedback
- Can improve product performance with better knowledge of manufacturing process, gaining competitive advantage

Waste Reduction

- Locate specific areas with defects or poor performance
  - Create product maps and product certifications
- Eliminate destructive tests and need to discard untested product

Close the Loop

- Multi-channel non-contact ultrasonic analysis is readily available
  - Products can be customized with relative ease for numerous applications
- NCU is robust, reliable, and relatively low cost
- Ultrasound is one of the safest technologies for inspection

the ultran group

Thank you!

the ultran group

America Makes
National Additive Manufacturing Innovation Institute

Breaking the Ground in the Execution of the Mission
(i.e., building a plane while flying it)

Ralph Resnick
President & Executive Director
National Center for Defense Manufacturing and Machining (NCDMM)
Founding Director
America Makes

Approved for Public Release

America Makes
National Additive Manufacturing Innovation Institute

- *A Defense-wide Manufacturing S&T team-led, Multi-agency collaboration between industry, government and universities*
- *Public-private partnership*

- Shared facilities open to industry
  - Especially attractive to small businesses
- Enabling technology transition and commercialization
- Addressing Technology Readiness Level (TRL) / Manufacturing Readiness Level (MRL) 4-7
  - Bridge the gap in Manufacturing Innovation
- Educational outreach and workforce development
- $ 45M gov't investment pledged for pilot, $30M CA awarded

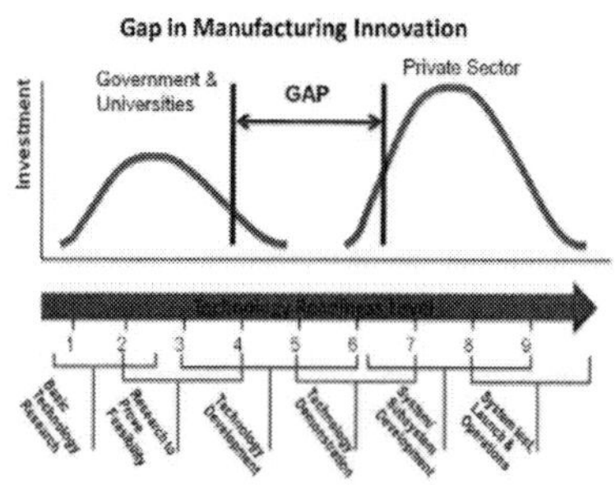

A Model for Subsequent Manufacturing Innovation Institutes

Approved for Public Release

America Makes
National Additive Manufacturing Innovation Institute

Why Additive?
A Manufacturing Revolution!

Projected AM Sales (products and services)

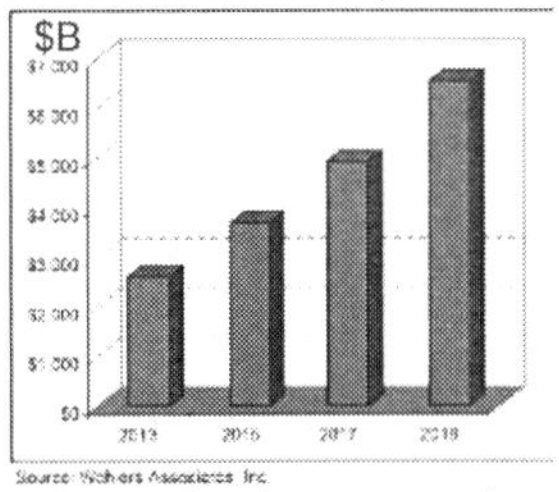

"20% of output of 3D printers is now final products, rather than prototypes. By 2020 it may be 50%." – *The Economist (2011)*

Government agency investments and interest - throughout NASA, DoD, and DOE

Consumer Product Market 3

Approved for Public Release

America Makes
National Additive Manufacturing Innovation Institute

GE Aviation

100k Additive parts will be manufactured by GE Aviation by 2020

1k lbs. Potential reduction in weight of a single aircraft engine through additive production

300+ 3D Printing Machines currently in use across GE

http://www.ge.com/stories/additive-manufacturing

Approved for Public Release

America Makes
National Additive Manufacturing Innovation Institute

Why America Makes?

America Makes creates mechanisms for collaboration...

Pooling Resources / Pooling Risks

Cooperative Development of
- Training
- Assessments
- Case Studies

Cooperative Development of
- Material Specs
- Process Specs
- Material Databases
- Design Rules
- Application Guides

Solving Problems Collaboratively

Work Shops, Working Groups, Projects

Public/Private Funded Projects
Crowd Funded Projects

Leveraging Community Knowledge

Knowledge Base, Online Collaboration Tools, Databases, Specifications, Application Guides, Curriculum

5

Approved for Public Release

America Makes
National Additive Manufacturing Innovation Institute

Strategy for Innovation

- Innovation is..."the ability to create a desired future"
  "Innovation Nation" author John Kao

- Not limited to technology development:
  - –Strategic planning
  - –Business Practices
  - –Collaboration
  - –Membership
  - –STEM
  - –Workforce Training
  - –Sustainability
  - – Technology Transition
  - – Environment / Collaboration spaces
  - – Small business engagement
  - – Intellectual Property
  - – Start-up companies
  - – Revenue streams
  - – Venture Capital

Approved for Public Release

America Makes
National Additive Manufacturing Innovation Institute

Governance Structure – Shared Leadership

Executive Committee*
"Champions" - Vision, Policy and Long-term Strategy and Planning

OSD ManTech (Mfg and Ind. Base Policy)
Gov't Co-op Agreement Program Manager (AFRL/RX)

Technical Advisory Board

Governance Board**
Technical strategy, program operating guidance

Director America Makes

Deputy Director: Technology Development
- Project Management
- IP Management
- Sustainability Thread

Deputy Director: Technology Transition
- Outreach & Engagement
- Technology Dissemination
- Conferences & Events

Deputy Director: Advanced Manufacturing Enterprise
- SME Coordination
- Incubation/Commercialization
- Supply chain and design modeling
- Digital Thread

Deputy Director: Workforce/Educational Outreach
- Education Outreach
- STEM activities
- Workforce Training
- Integrated Education thread

——— = Direct Oversight / Responsibility
- - - - - = Guidance / Direction

***Executive Committee (11)**: Industry, for-profit organization (2), Non-profit association (2), Academic (2), Government (3), At-large (2)
****Governance Board (35):** Lead and Full Members, states ex-officio

7

Approved for Public Release

America Makes
National Additive Manufacturing Innovation Institute

Membership Benefits

Community
- Formalized community looking to leverage AM Technologies
- Leverage learning curves
- Cooperatively solve similar issues
- Direct future equipment, material, and software development
- Technical strategy input

Work Force Development
- Pre-K to grey training resources
- Immersive training/learning at America Makes facility

Projects
- Lead America Makes directed project calls
- Team with America Makes & members to win additional government project calls
- Use America Makes resources for targeted company projects

Enterprise Data
- Processing & Application data
- Material properties
- Supply chain
- Analysis

8

Approved for Public Release

America Makes
National Additive Manufacturing Innovation Institute

87 Member Organizations (as of 2/3/14)

3D Systems Corporation*
3M
Alcoa
Allegheny Technologies Incorporated*
Applied Systems and Technology Transfer (AST2)*
Arkema, Inc.
ASM International
Association of Manufacturing Technology*
Bayer Material Science*
The Boeing Company
Carnegie Mellon University*
Case Western Reserve University*
Catalyst Connection*
Concurrent Technologies Corporation*
Deformation Control Technology, Inc.
DSM Functional Materials
Energy Industries of Ohio*
EWI
The ExOne Company*
General Electric Company (GE)*
General Dynamics Ordnance and Tactical Systems
Hoeganaes Corporation
Illinois Tool Works, Inc.
Johnson Controls, Inc.*
Kennametal*
Kent Display*
Lehigh University*
The Lincoln Electric Company
Lockheed Martin*
Lorain County Community College
M-7 Technologies*
MAGNET*
Materion Corporation
MAYA Design Inc.
Michigan Technological University
Missouri University of S&T
MIT Lincoln Laboratory
Moog, Inc.
NorTech*
North Carolina State University
Northern Illinois Research Foundation
Northrop Grumman*
Ohio Aerospace Institute*
Optomec*
Oxford Performance Materials*
Pennsylvania College of Technology
Pennsylvania State University*
PTC ALLIANCE
Raytheon Company*
Rhinestahl Corporation
Robert C. Byrd Institute (RCBI)*
Robert Morris University*
RP+M
RTI International Metals, Inc. *
Sciaky, Inc.
SME*
Solid Concepts
South Dakota School of Mines & Technology
Stony Creek Labs
Stratasys, Inc.
Strategic Marketing Innovations, Inc.
Stratonics*
TechSolve*
Texas A&M Univeristy
The Timken Company*
Tobyhanna Army Depot
United Technologies Research Center
University of Akron*
University of California, Irvine
University of Connecticut
University of Dayton Research Institute
University of Louisville
University of Maryland – College Park
University of Pittsburgh*
University of Texas – Austin
University of Texas at El Paso
University of Toledo
USA Science and Engineering Festival
Venture Plastics, Inc.
Westmoreland County Community College*
West Virginia University
Wohlers Associates, Inc.*
Wright State University
Youngstown Business Incubator*
Youngstown State University*
Zimmer, Inc.

*__Platinum__ Members ($200K) = RED; __Gold__ Members ($50K) = BLUE; __Silver__ Members ($15K) = BLACK ($15K); * __Original__ Members (39)*

Approved for Public Release

America Makes
National Additive Manufacturing Innovation Institute

Technology Innovation Factory (9/27/12 Ribbon Cutting) Youngstown Business Incubator Annex Building

Approved for Public Release

Member & Community Outreach: Manufacturing Day Open House & Tours

- Over 1000 Visitors since April PMR for Business Meetings & Tours
- Business Meetings & Tours Have Grown from 16 per Month to 59 per Month

Technology Roadmapping to Drive Investments

- Strategic Planning, Decision Making, and Collaboration Framework based on "Voice of the Customer"
- Near-, Mid-, & Long-term Investment Strategy
  - Design
  - Materials
  - Process/Equipment
  - Qualification/Certification
  - Knowledge Development
- Structured Workshops Target Big Impact Outcomes

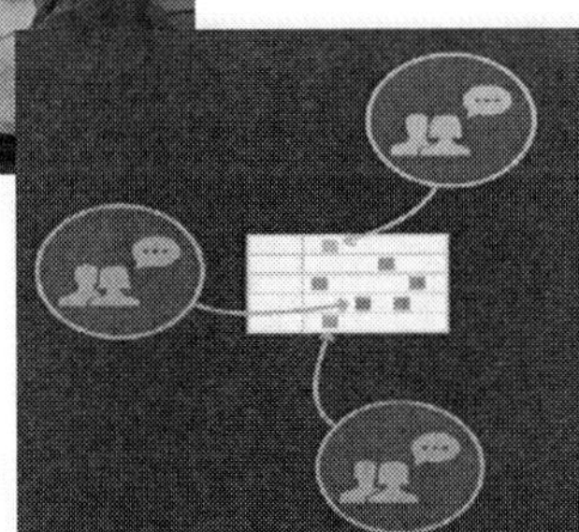

America Makes
National Additive Manufacturing Innovation Institute

America Makes
National Additive Manufacturing Innovation Institute

Technology Roadmap: Workshop Participation

Approved for Public Release

America Makes
National Additive Manufacturing Innovation Institute

Technology Roadmap: Prioritization

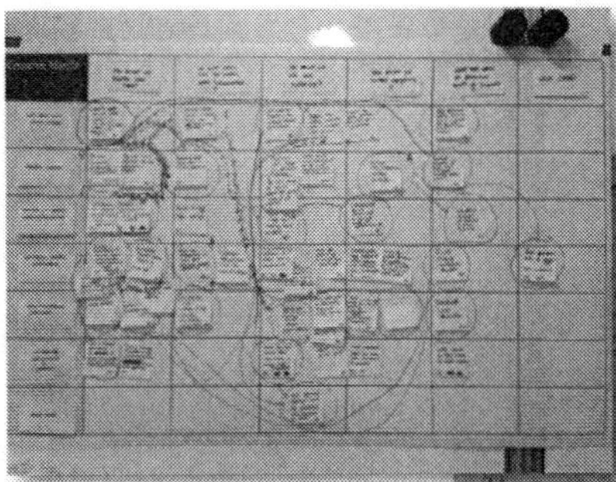

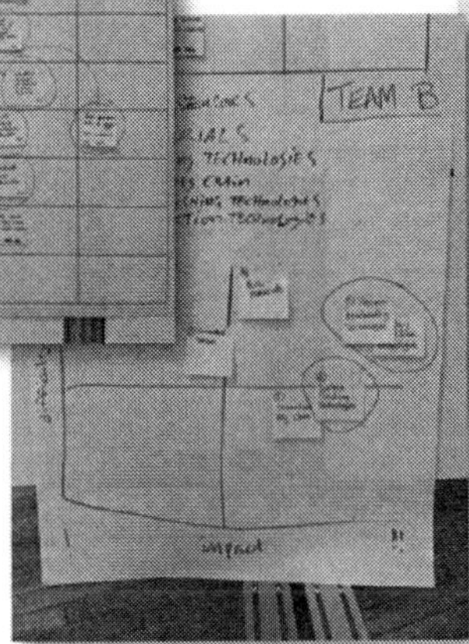

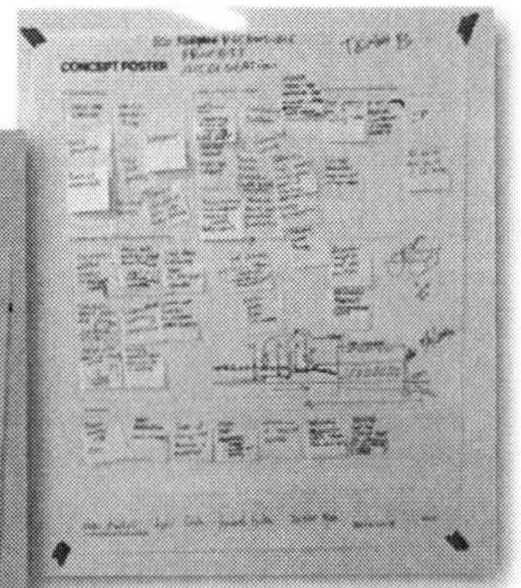

1 Creative Matrix + Affinity Clustering

2 Impact / Difficulty Matrix

3 Concept Poster

Approved for Public Release

America Makes
National Additive Manufacturing Innovation Institute

Technology Roadmap: v1.0

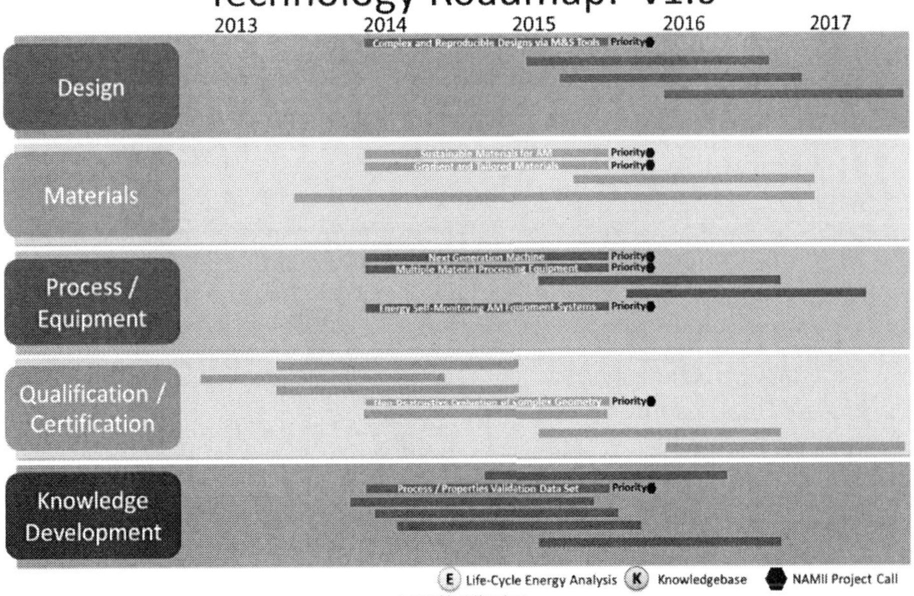

Approved for Public Release

America Makes
National Additive Manufacturing Innovation Institute

America Makes Project Models

- Multiple Investment Options Attracting Different Audiences
- Project Calls Well Underway
- Challenge Options Being Developed
- Crowd-Sourced Project Model Developed

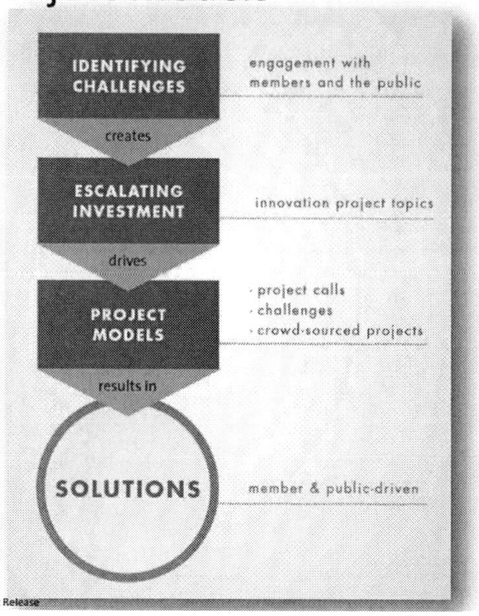

Approved for Public Release

America Makes
National Additive Manufacturing Innovation Institute

Status of Project Call #1

- Kickoff April 2013
- 6 Projects with 35 Partners
- $4.75M Government Share, $5.0 Cost Share
- Material Characterization, Process Qualification, Quality Control
- Excellent Teaming

rpm
the new way to innovate

PENNSTATE
1855

MISSOURI S&T
University of Science & Technology

CASE WESTERN RESERVE UNIVERSITY EST. 1826

NORTHROP GRUMMAN

Approved for Public Release

America Makes
National Additive Manufacturing Innovation Institute

Status of Project Call #2

- Award Announcement January 2014
- 15 Projects with 75 Partners
- $9.0M Government Share, $10.3 Cost Share
- Even Better Teaming!

Academic vs Industry Split	
Total Indsutry	8
Total Academic	7
Total	**15**

Roadmap Alignment	
Process and Equipment	6
Design for AM	3
AM Materials	1
Qual and Cert	2
Knowledgebase	3
Total	**15**

Approved for Public Release

America Makes
National Additive Manufacturing Innovation Institute

Case Western Reserve University EST 1826

NC STATE UNIVERSITY

Carnegie Mellon University

stony creek labs

GE Global Research

OPTOMEC

Additive Manufacturing Systems—from NANO to MACRO

EWI

We Manufacture Innovation

NORTHROP GRUMMAN

GE Aviation

PITT

MichiganTech

UTEP

Youngstown Business Incubator

ybi cultivate innovate accelerate

McGowan Institute for Regenerative Medicine

A program of the University of Pittsburgh and UPMC

Approved for Public Release

America Makes
National Additive Manufacturing Innovation Institute

Technology Roadmap: Continuous Improvement

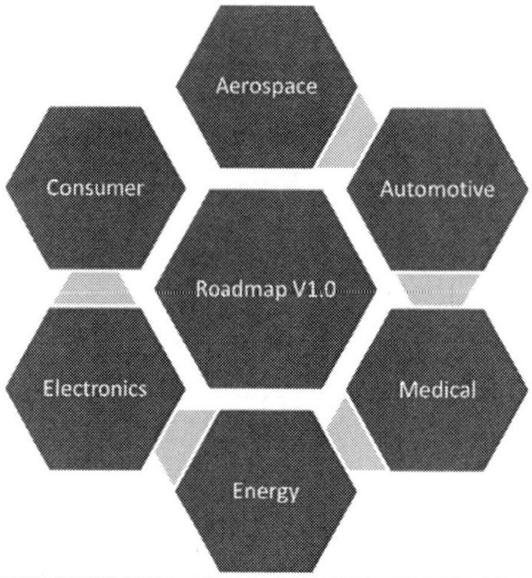

Technology Roadmap Advisory Group Being Formed

Approved for Public Release

America Makes
National Additive Manufacturing Innovation Institute

Digital Estate

- Replaces the NAMII Website
- New Look – New Capabilities to Serve the Community

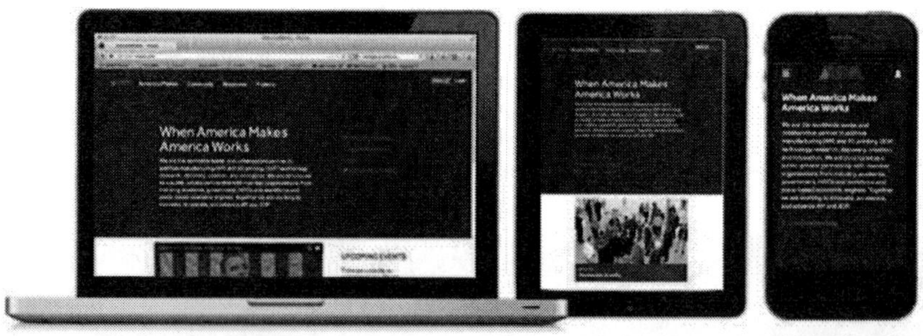

www.AmericaMakes.us

Approved for Public Release

America Makes
National Additive Manufacturing Innovation Institute

The ORNL and America Makes Partnership

Additive Manufacturing Shared Resources Network

- AM partners have open access to ORNL MDF equipment, as they do Youngstown center equipment
- AM shop floor grows by 530 miles

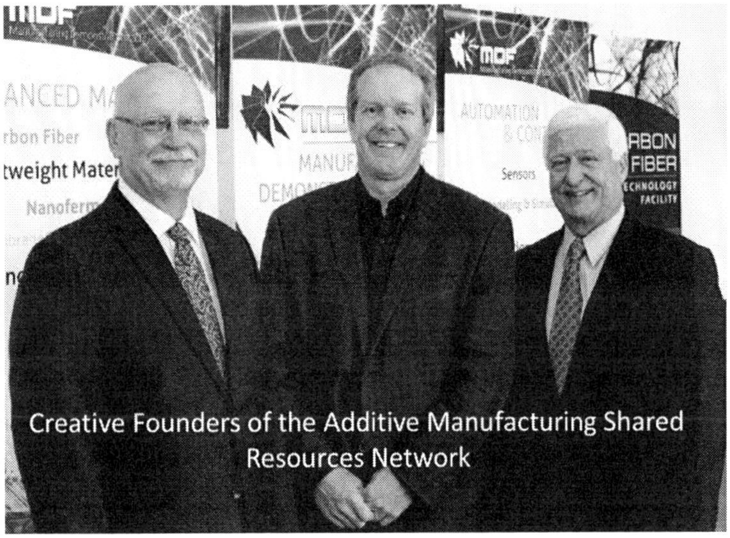

Creative Founders of the Additive Manufacturing Shared Resources Network

America Makes
National Additive Manufacturing Innovation Institute

Introducing Additive Mfg to the FIRST Robotics Championship

America Makes & ORNL Manufacturing Demonstration Facility (MDF) Joint Effort

OAK RIDGE National Laboratory

26

Approved for Public Release

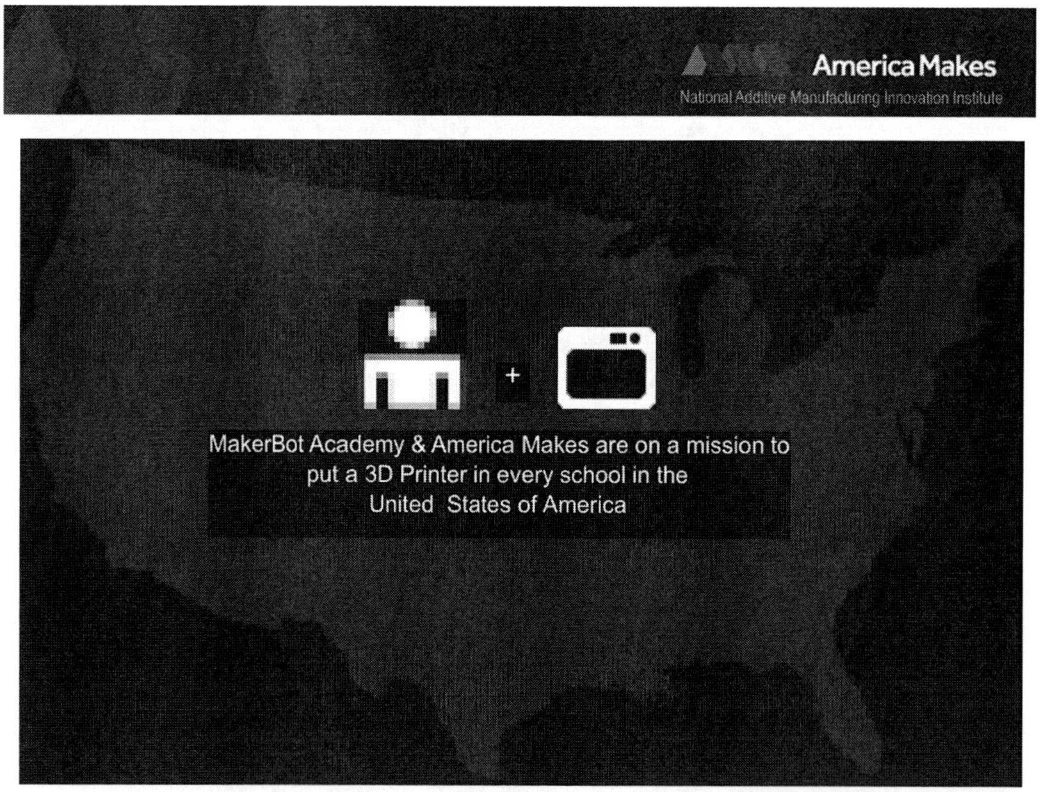

Approved for Public Release

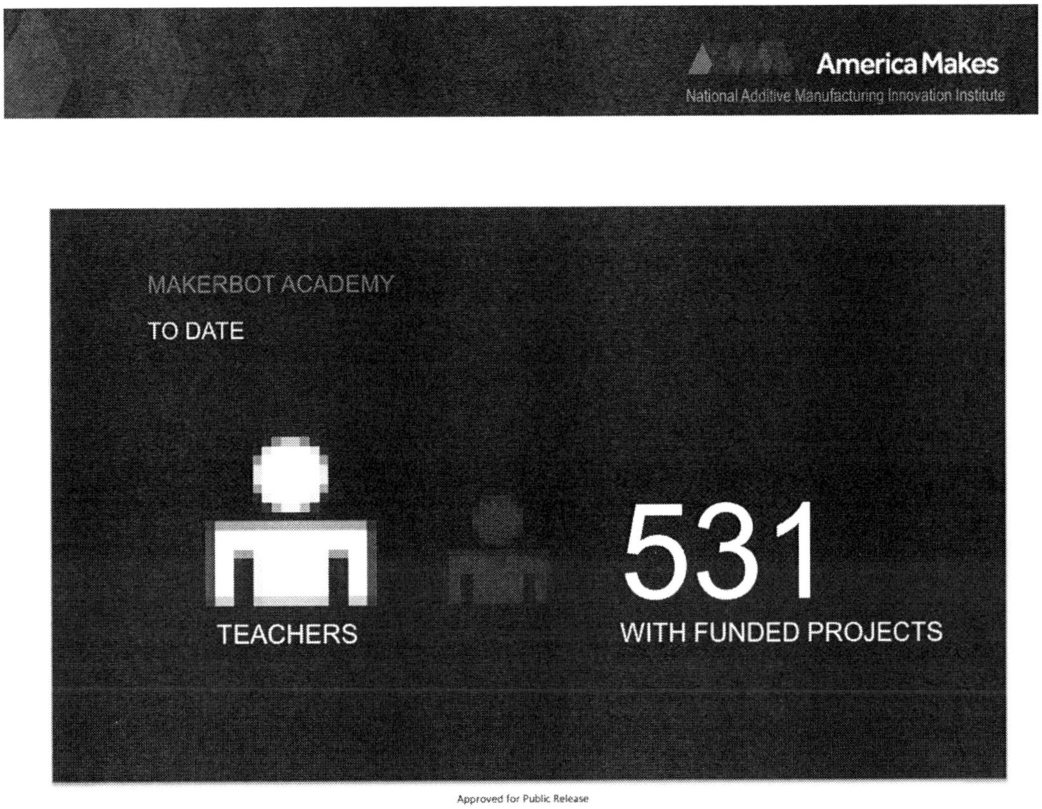
America Makes
National Additive Manufacturing Innovation Institute
MAKERBOT ACADEMY
TO DATE
TEACHERS
531
WITH FUNDED PROJECTS
Approved for Public Release

America Makes
National Additive Manufacturing Innovation Institute
MAKERBOT ACADEMY
TO DATE
TEACHERS
STUDENTS
106,268
IMPACTED
Approved for Public Release

America Makes
National Additive Manufacturing Innovation Institute

MAKERBOT ACADEMY

TO DATE

$1.3 million

MONEY RAISED

Approved for Public Release

America Makes
National Additive Manufacturing Innovation Institute

Strong, Holistic, Integrated Technology Plan

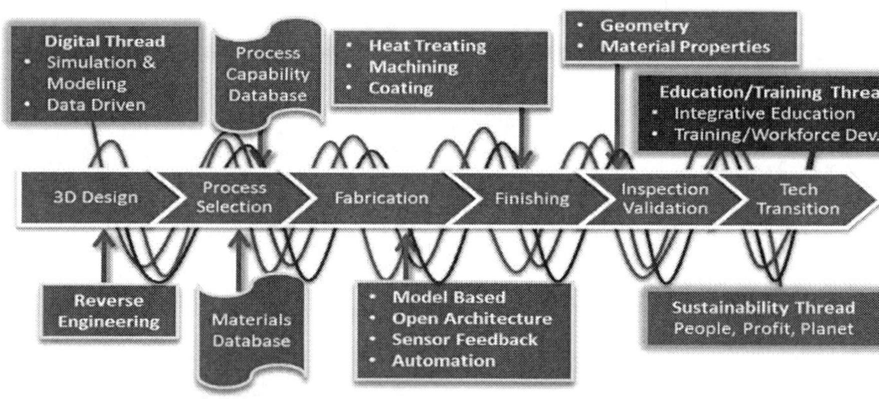

- Process development: metals, polymers, ceramics, electronics, hybrid
- Digital thread / Advanced Mfg Enterprise
- Specialized, portable AM systems
- Open Architecture
- Process planning

- Process Control
- Material Development
- Component Design

Approved for Public Release

America Makes
National Additive Manufacturing Innovation Institute

Summary

- *Huge Opportunity with High Stakes*
- *"None of us can do it alone. For this to succeed, it needs to scale, and for it to scale, we need a passionate and motivated community"*

  \- Avi Reichental, President and CEO, 3D Systems
- *"Rather than companies having to bear all of the cost and burden of qualifying these processes and materials, much can be done through collaborative efforts"*

  \- Terry Wohlers, Principle Consultant and President, Wohlers Associates

Approved for Public Release

When America Makes America Works

America Makes

Air Force Global Strike Command

To Deter and Assure

Making History

Mr. Yancy Mailes
AFGSC/HO
25 February 2014

So You're a Historian???

UNCLASSIFIED

How We Romanticized our Job

UNCLASSIFIED *To Deter and Assure* 3

UNCLASSIFIED

The Reality of the Air War

UNCLASSIFIED *To Deter and Assure* 4

UNCLASSIFIED

The Post 9-11 Evolution

UNCLASSIFIED *To Deter and Assure* 5

UNCLASSIFIED

Enter the Combat Historian

UNCLASSIFIED *To Deter and Assure* 6

UNCLASSIFIED

Home Away From Home

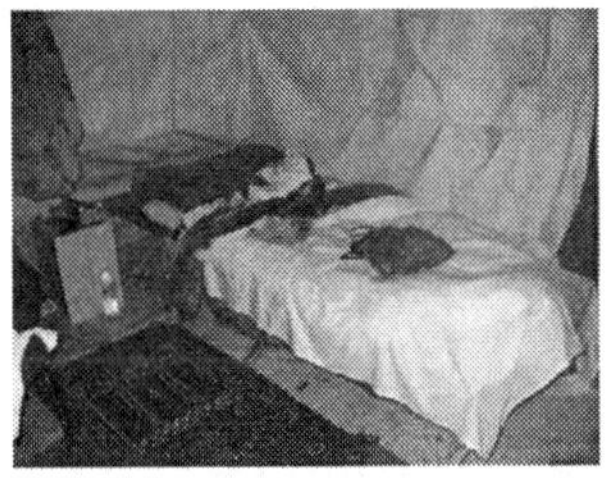

UNCLASSIFIED *To Deter and Assure* 7

UNCLASSIFIED

Armed Tourism

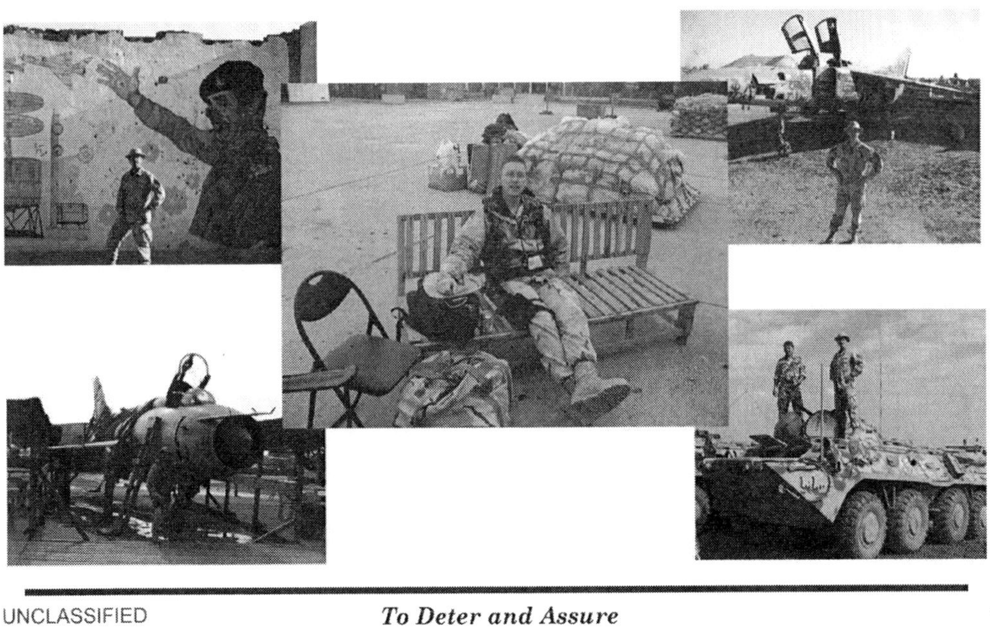

UNCLASSIFIED *To Deter and Assure* 8

UNCLASSIFIED

Witness the Events

UNCLASSIFIED *To Deter and Assure* 9

UNCLASSIFIED

Art As a Document

UNCLASSIFIED *To Deter and Assure* 10

UNCLASSIFIED

Outside the Wire

UNCLASSIFIED *To Deter and Assure* 11

UNCLASSIFIED

What I Thought Was Important

UNCLASSIFIED *To Deter and Assure* 12

UNCLASSIFIED

Look Off The Flightline

UNCLASSIFIED *To Deter and Assure* 13

UNCLASSIFIED

Operationalizing History

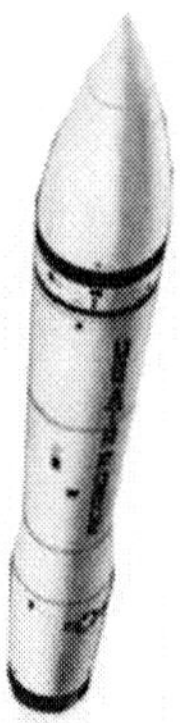

US suspends 34 nuclear missile officers over exam cheating

Washington (AFP) Jan 15, 2014

The US military has suspended 34 officers in charge of launching nuclear missiles for cheating on a proficiency test, Air Force leaders said Wednesday.

The scandal at Malmstrom Air Force Base in Montana marked the latest in a series of damaging revelations dogging the country's nuclear force, including a separate probe into illegal drugs that came to light last week.

"There was cheating that took place with respect to this particular test," Air Force Secretary Deborah Lee James told a news conference.

"Some officers did it. Others apparently knew about it, and it appears that they did nothing, or at least not enough, to stop it or to report it."

She called the cheating "absolutely unacceptable behavior" but insisted that the nuclear weapons had been safe all along.

"And very importantly, I want you to know that this was a failure of some of our airmen. It was not a failure of the nuclear mission," said James.

Officials said it was the largest scale cheating scandal ever in the nuclear force, as it implicated roughly 20 percent of the Malmstrom base's 190-member launch officer corps.

The Air Force also stripped the officers of their security clearances over the allegations.

UNCLASSIFIED *To Deter and Assure* 14

UNCLASSIFIED

Closing Thoughts…

UNCLASSIFIED *To Deter and Assure* 15

UNCLASSIFIED

Questions?

UNCLASSIFIED *To Deter and Assure* 16

Summary

- Background and Problem Statement
- Current Process
- System Specifications
- An Innovative Solution for Legacy Program
- Future Applications and Conclusion

BACKGROUND AND PROBLEM STATEMENT

Over 2,400 C130, since August 23 1954

Limited evolution on the design of the aircraft structure

Different type of challenges compare with new aircraft development

CURRENT PROCESS

Bulkheads with 15,000 pilot holes, not visible from the skin side

Manual drilling from the inside out

Manual drilling to final size from the skin side

SYSTEM SPECIFICATIONS

Available Space Constraint

Maintaining the Pilot Holes and Avoid re-Engineering of Many Bulkheads

Maintaining Capability for Jack Pads Drilling

Safely Maintain Access for Manual Work concurrent with Automated Operation

Minimum Disruption of Current Production

PAR SYSTEMS

HISTORY OF PaR SCANNING TECHNICS

1998 - FAS

Defastening

AN INNOVATIVE SOLUTION FOR LEGACY PROGRAM

A Unique 2-axis Motion System:

- 72-degree Curved Axis to match constant radius of the fuselage
- 50-foot Linear Axis

→ specific coordinate system

AN INNOVATIVE SOLUTION FOR LEGACY PROGRAM

Nexen drive, zero backlash on both axis

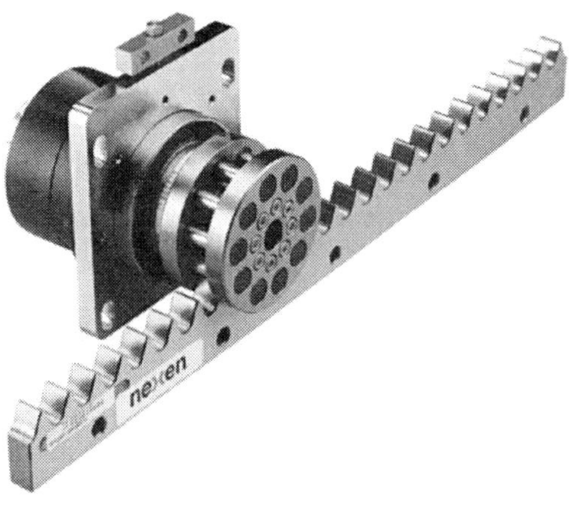

A FULLY INTEGRATED SOLUTION

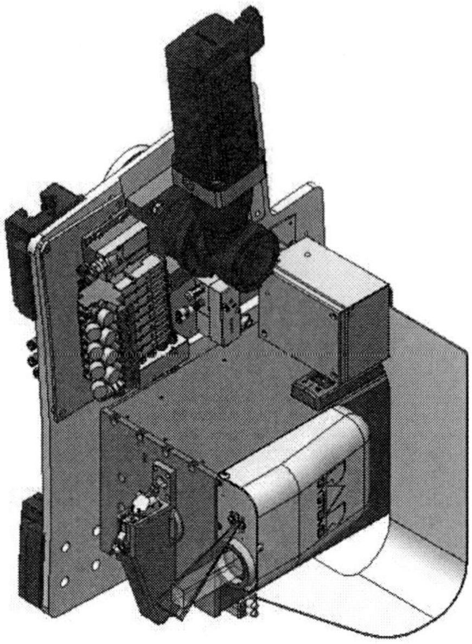

Drilling and Vision end-effector:

- 2D vision system
- Precision Drilling Capabilities
- Pneumatic Clamp Axis
- Servo Feed Axis
- Spherical Compliance Nosepiece

A FULLY INTEGRATED SOLUTION

Operator Station

- PaR Standard Design
- Located Above Deck near tool change location

A FULLY INTEGRATED SOLUTION

Safety Integrated:

- Laser Safety Scanner
- Column Guarding
- Walk on way covers
- Safety guarding with interlocked access points

SCANNING ON THE FLY PILOT HOLES TO MAXIMIZE PRODUCTIVITY

2 ips Scanning Speed,

Synchronize Position and Offset

Validate Compare with Theoretical Hole Position

High Fidelity Between Spindle and Vision System

Repeatability and/or Accuracy

Auto-Calibration

SCANNING ON THE FLY PILOT HOLES TO MAXIMIZE PRODUCTIVITY

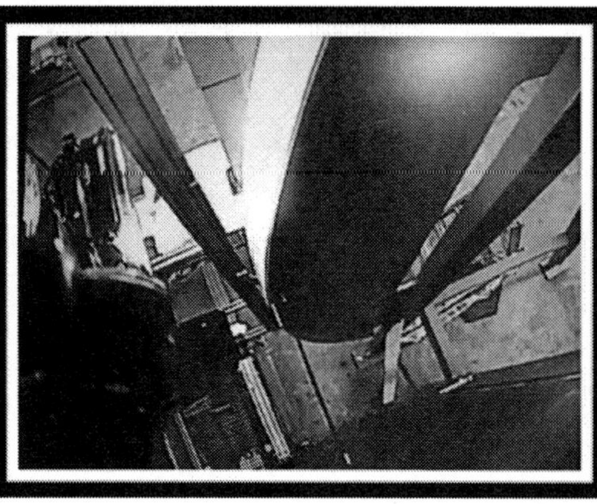

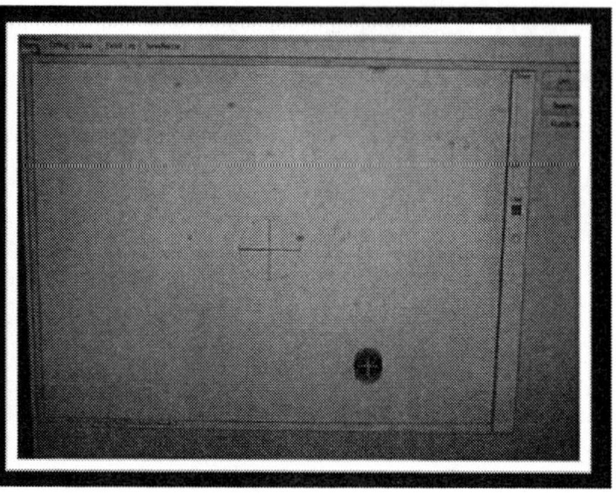

MULTIPLE POSITIONING CORRECTION TYPES

Type 1: No correction

Type 2: Transformation Based on Limited Number of Points

Type 3: Single Hole Offset

Type 4: Transformation and Single Hole Combination

→ Accurately Drill Non-Visible Holes

MULTIPLE HOLES SCANNING

FUTURE APPLICATIONS AND CONCLUSION

- System in Production by Spring
- Faster scanning speed
- Not "just" a 2-Axis Drilling System
- Highly Collaborative Process

Additive Manufacturing at Lockheed Martin Aeronautics

AERODEF Manufacturing Summit

February 25-27, 2014
Long Beach, CA

Brian Rosenberger
Lockheed Martin Aeronautics Company

Outline

- *Background & History*
- *What We've Done*
- *What We're Doing Now*
- *Where We Think This Technology Can Go*

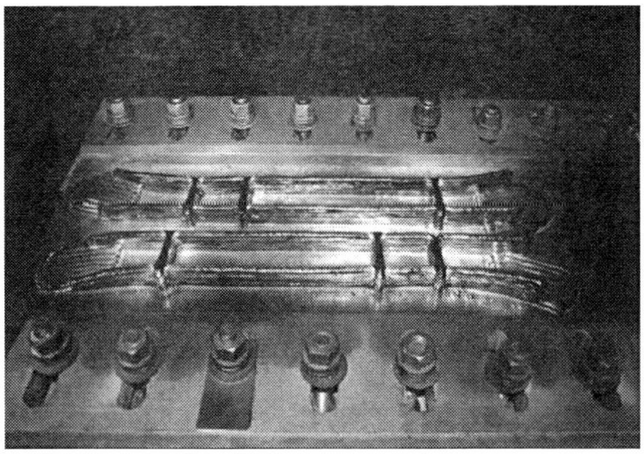

Direct Manufacturing History at LM Aero

3

Original Direct Manufacturing Vision

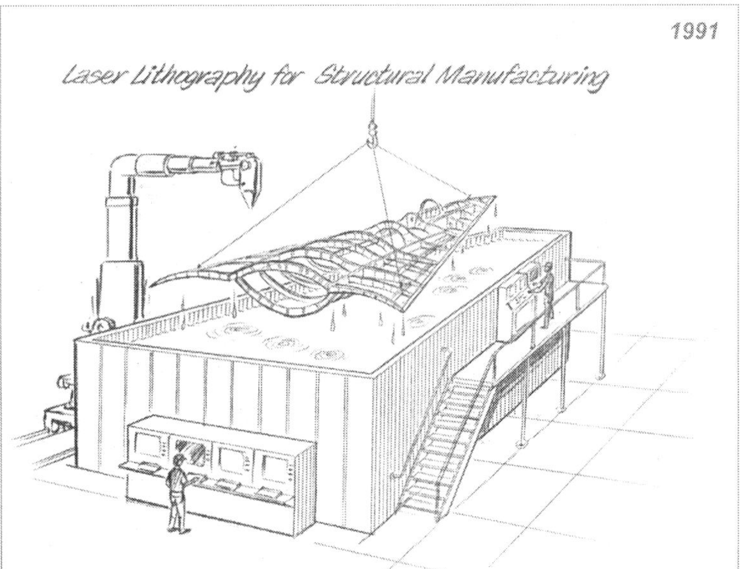

4

Large-Scale Metal Systems

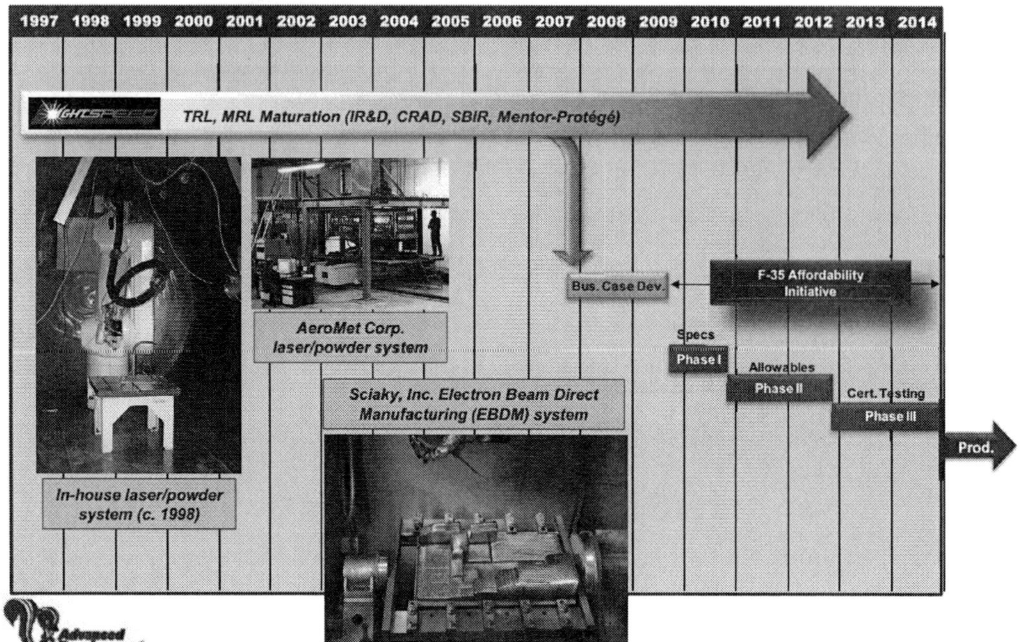

Advanced Development Programs

5

What We've Done - Metals

Advanced Development Programs

6

What We've Done - Polymers

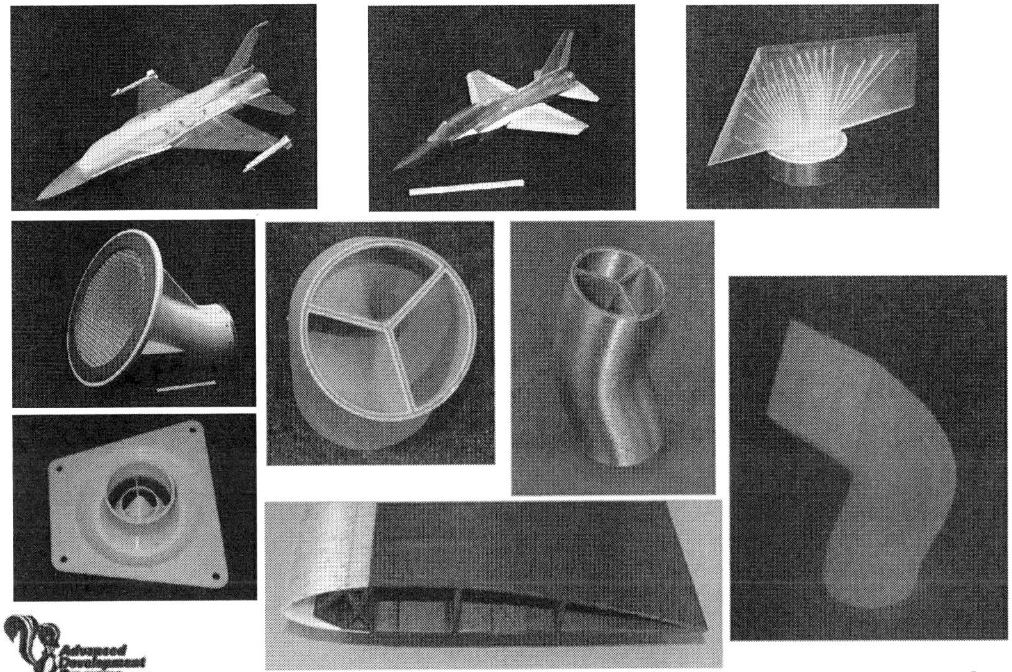

7

What We're Doing Now – Large Metal Parts Transition to Production

8

F-35 EBDM Applications

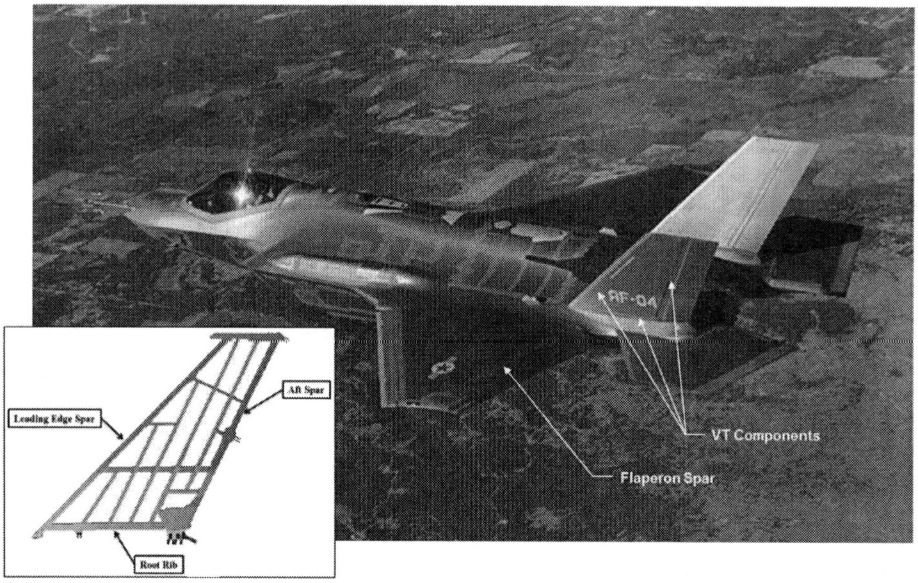

Advanced Development Programs

Direct Manufacturing Process Flow

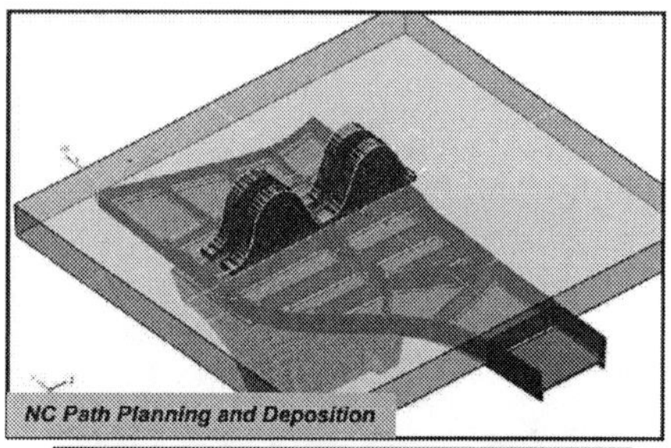
NC Path Planning and Deposition

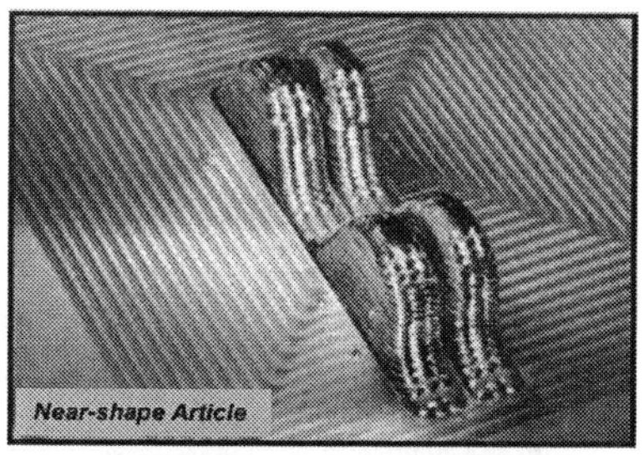
Near-shape Article

Near-shape Article

NC Machining

Advanced Development Programs

Potential F-35 EBDM Applications

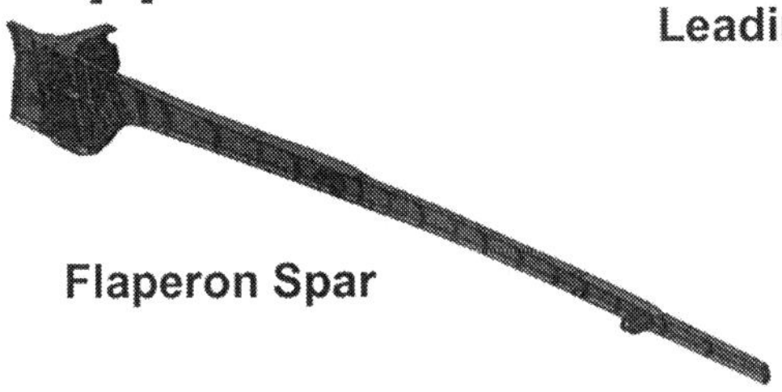

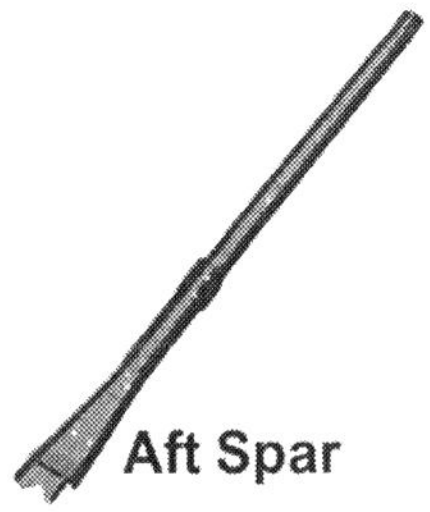

Part Name	Height (in)	Width (in)	Length (in)
Flaperon Spar	12	13	118
Leading Edge Spar	4.2	6	73
Aft Spar	4.2	6	73
Root Rib	4.5	10	46

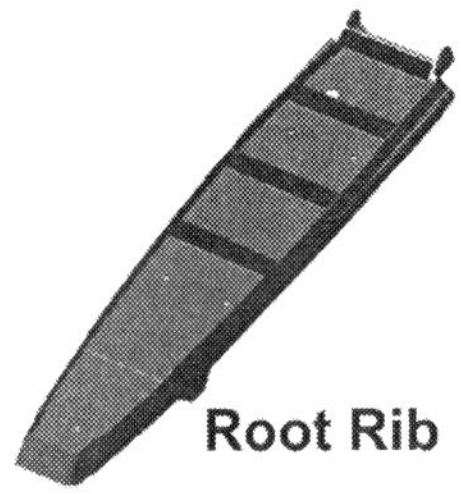

Advanced Development Programs

Benefits of EBDM

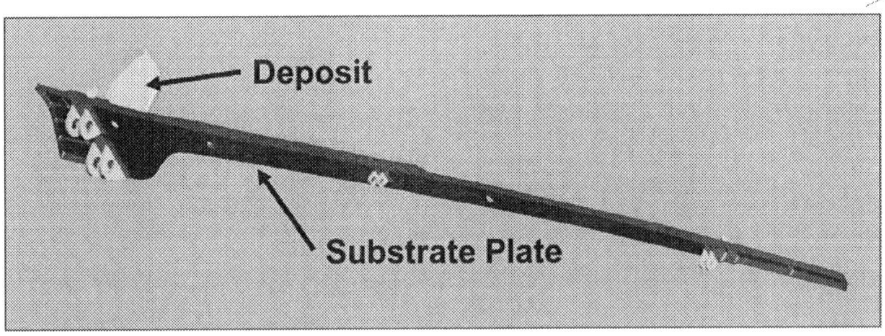

- **Minimal Tooling Required**
- **Cost Savings**
  - ***Up to 60%***
    - ***Reduced BTF***
    - ***Reduced Machining Costs***
- **Lead Time Savings**
  - ***Up to 80% vs Die Forgings***

Advanced Development Programs

DM/Forging Comparison

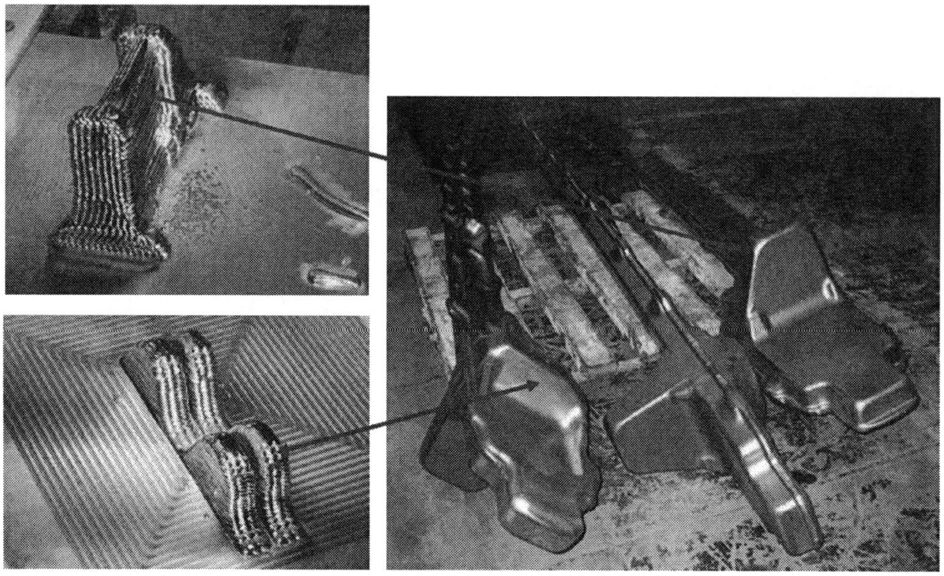

EBDM Fabrication Studies (Sciaky SBIRs)

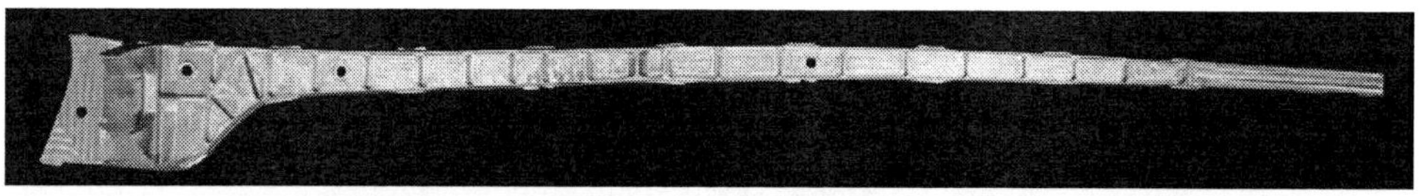

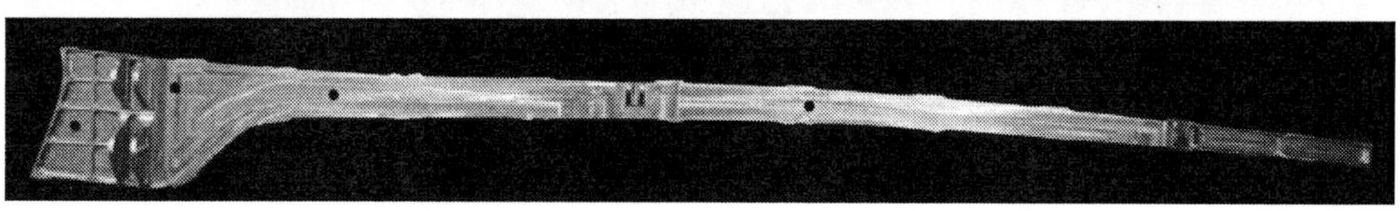

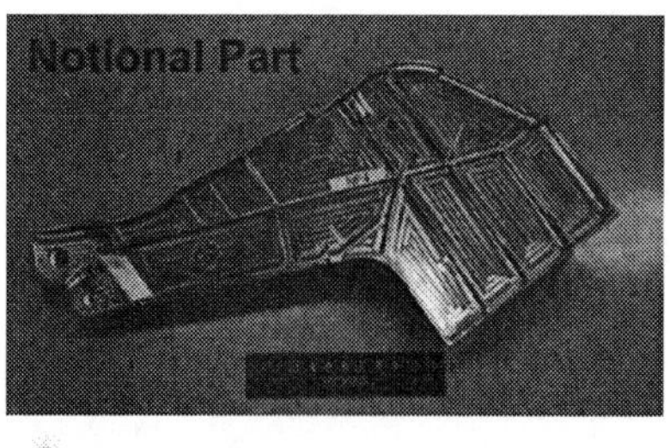

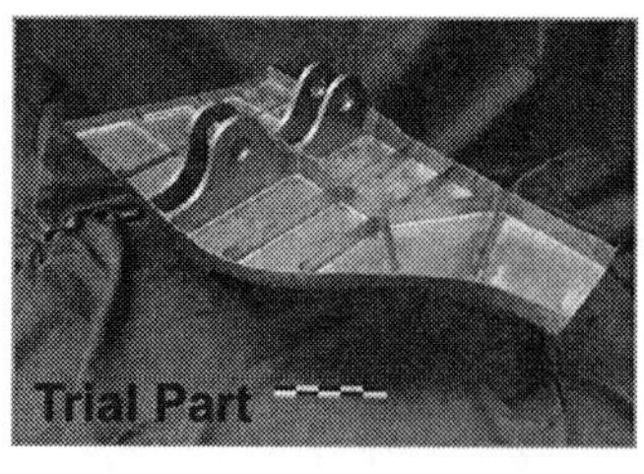

Materials Qualification Effort

- Deposition of Large Preforms
- Preforms
  - *11 Preforms*
  - *Up to 270 lbs per Deposit*
  - *3 Wire Sources*
  - *Beta Annealed*
- Specimen Testing
  - *Up to 139 Test Coupons per Preform*
  - *Static Testing*
  - *DaDT Testing*
  - *Material Testing*

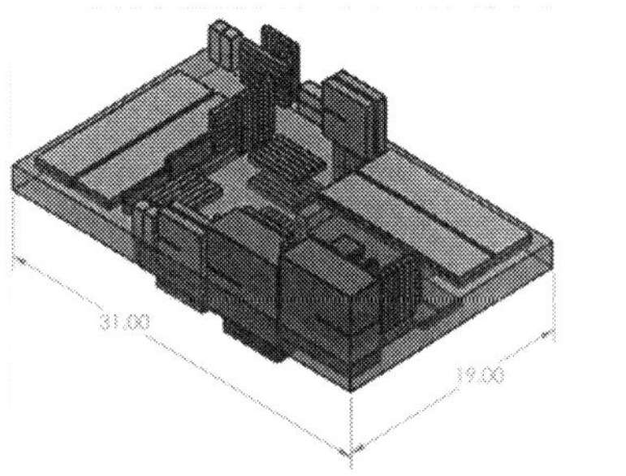

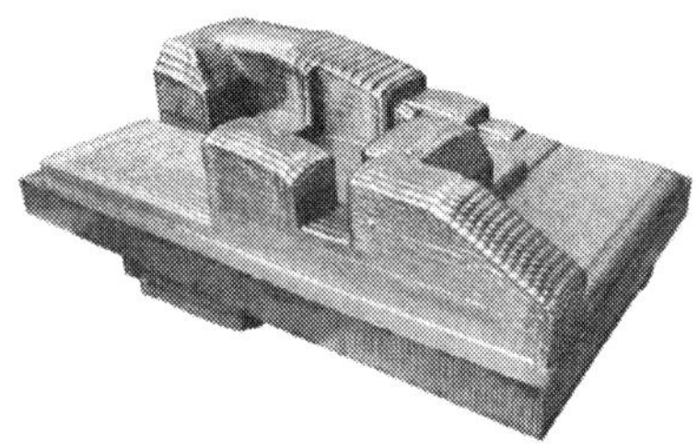

Ref: Needler, Steve, "F-35 Direct Manufacturing: Material Qualification Results" presented at Aeromat 2012 Conference

Where Do We Think This Technology Can Go?

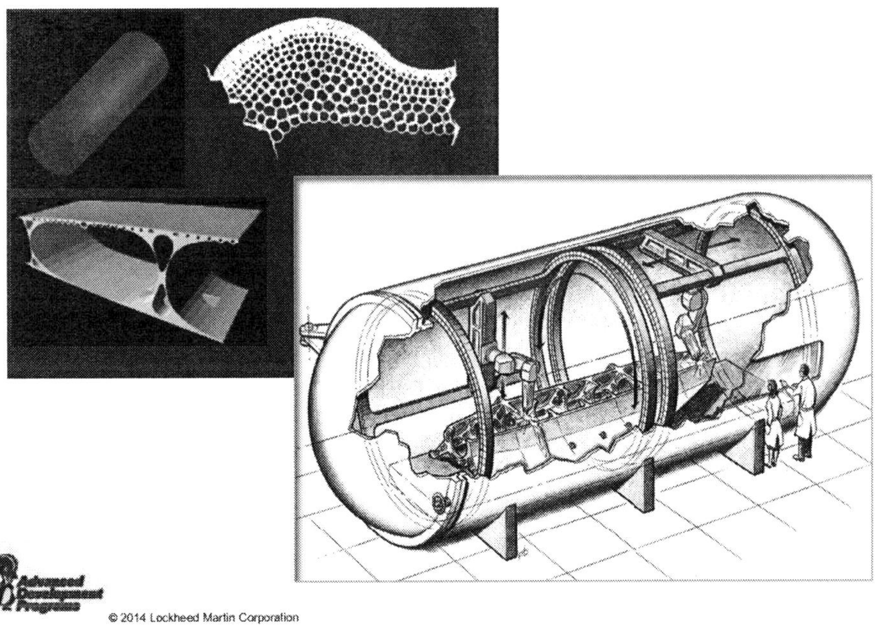

EBDM Space Application

- **Spacecraft Forward Bay Cover Deck**

Advanced Development Programs

17

Society of Manufacturing Engineers (SME)
One SME Drive
Dearborn, Michigan 48128
USA

ISBN 978-1-5108-5900-5

CURRAN ASSOCIATES INC.
proceedings.com